String Theory and Its Applications

From meV to the Planck Scale

String Theory and Its Applications

From meV to the Planck Scale

Editors

Michael Dine & **Thomas Banks**
University of California at Santa Cruz, USA

Subir Sachdev
Harvard University, USA

World Scientific

NEW JERSEY · LONDON · SINGAPORE · BEIJING · SHANGHAI · HONG KONG · TAIPEI · CHENNAI

Published by

World Scientific Publishing Co. Pte. Ltd.
5 Toh Tuck Link, Singapore 596224
USA office: 27 Warren Street, Suite 401-402, Hackensack, NJ 07601
UK office: 57 Shelton Street, Covent Garden, London WC2H 9HE

British Library Cataloguing-in-Publication Data
A catalogue record for this book is available from the British Library.

STRING THEORY AND ITS APPLICATIONS
TASI 2010
From meV to the Planck Scale

ISBN-13 978-981-4350-51-8
ISBN-10 981-4350-51-6

Printed in Singapore by B & Jo Enterprise Pte Ltd

Preface

Theoretical particle physics is a rapidly developing subject. Students planning to work in the field need intense exposure to a broad array of topics. Since 1984, the Theoretical Advanced Study Institute (TASI) has provided a four week program of pedagogical lectures aimed at advanced Graduate Students in elementary particle physics. Since 1989, the school has been located at the University of Colorado at Boulder. Each year, a theme for the school and directors are selected by the Scientific Advisory Board. In 2010, it was decided that the school should have a string theory emphasis.

Two developments guided the structure of the program. First, the school took place just as the Large Hadron collider was beginning to ramp up. So lectures included experimental and theoretical reviews of LHC physics, while many of the string and supersymmetry talks had a phenomenological emphasis. Second, the past few years have seen striking progress in the application of gauge/gravity dualities to questions in condensed matter and high temperature/density physics (heavy ion physics). About 1/3 of the lectures were aimed at providing students with the tools to understand these developments. Given these themes, we selected the title "String Theory and its Applications: from the meV to the Planck scale". The result was an eclectic, intense but coherent program.

Almost all of the lectures are reproduced in this volume. The exceptions are those by Nathan Seiberg and Krishna Rajagopal. The lectures by Seiberg introduced important recent developments in our understanding of four dimensional supersymmetric theories. Interested readers might wish to consult his review with K. Intriligator, "Lectures on Supersymmetry Breaking", in *Classical and Quantum Gravity* **24**, BS741 (2007), [arXiv:hep-ph/0702069]. Rajagopal focused on heavy ion physics and its understanding within QCD, including, but not limited to, connections with duality. The review article, "Gauge/String Duality, Hot QCD and Heavy Ion Collisions" by H. Liu, D. Mateos, K. Rajagoal and U.A. Wiedemann, arXiv:1101.0618 provides a pedagogical introduction to the material in these lectures. We thank Cambridge University Press for permission to publish the lectures by Minwalla et al., which will appear also in the volume *Black Holes in Higher Dimensions* (G. Horowitz, ed.).

We are very appreciative of the efforts of the lecturers, both in their presentations and for their contributions to this volume. We are grateful to all of the work of K.T. Mahanthappa in making this school work. His wisdom ranged from physics issues to the day to day details of the school. Efforts of other Colorado faculty, especially Tom Degrand (who, as always, organized mountain adventures for the students), Shanta DeAlwis and Oliver De Wolfe, were greatly appreciated. Susan Spika provided vital support before, during and after the school. TASI received partial support from the National Science Foundation, the Department of Energy, and the University of Colorado.

Michael Dine
Thomas Banks
Subir Sachdev

1) Andrew Hamilton
2) Pontus Ahlqvist
3) Xi Dong
4) Usama Al-Binni
5) Dan Wohns
6) Idse Heemskerk
7) Yue Zhao
8) Michael Dine
9) Antun Skanata
10) Shanta de Alwis
11) Tarun Grover
12) Jae Hyuk Oh
13) Gang Xu
14) Francisco Rojas
15) Dan Gulotta
16) Sho Yaida
17) Tom Kellex
18) Steve Avery
19) Bart Horn
20) Oscar Chacaltana
21) Subir Sachdev
22) Kewang Jin
23) Phil Szepietowski
24) Ibrahima Bah
25) Zuhair Khanhker
26) Shannon McCurdy
27) Tomas Rube
28) Sohang Gandhi
29) Gokce Basar
30) K.T. Mahanthappa
31) Chi-Ming Chang
32) Eric Perlmutter
33) David Marsh
34) Marko Djuric
35) Mark Mezei
36) Congkao Wen

37) Vladimir Rosenhaus
38) Hajar Ebrahim
39) Anastasios Taliotis
40) Oliver De Wolfe
41) Archisman Ghosh
42) Daniel Harlow
43) Dusan Simic
44) Yu Guo
45) Wenbin Yan
46) Yang Zhang
47) Nabil Iqbal
48) Alberto Faraggi
49) Michael Gary
50) Tom DeGrand
51) Riei Ishizeki
52) Kentaro Hanaki
53) Andreas Stergiou
54) Josh Powell
55) Koushik Balasubramanian
56) Yi Yin
57) Joe Polchinski
58) Richard Eager
59) Christopher Rosen
60) Paul McGuirk
61) George Bruhn
62) Abhijit Gadde
63) Stefan Leichenauer
64) Jim Halverson
65) Nima Lashkari
66) Curtis Asplund
67) Hong Liu
68) Victor Alexandrov
69) Dam T. Son
70) Sung-Sik Lee
71) Thomas Banks
72) Umesh Vijayashanker

Contents

Overview

Chapter 1

Introduction to Gauge/Gravity Duality

Joseph Polchinski

Kavli Institute for Theoretical Physics, University of California, Santa Barbara, CA 93106-4030, USA

joep@kitp.ucsb.edu

These lectures are an introduction to gauge/gravity duality. The first three sections present the basics, focusing on $AdS_5 \times S^5$. The last section surveys a variety of ways to generate duals of reduced symmetry.

1.1. Generalities

1.1.1. *The greatest equation*

A few years back, Physics World magazine had a reader poll to determine the Greatest Equation Ever, and came up with a two-way tie between Maxwell's equations

$$d * F = j\,, \quad dF = 0\,, \tag{1.1}$$

and Euler's equation

$$e^{i\pi} + 1 = 0\,. \tag{1.2}$$

The remarkable appeal of Euler's equation is that it contains in such a compact form the five most important numbers, $0, 1, i, \pi, e$, and the three basic operations, $+, \times,$ ^. But my own choice would have been Maldacena's equation

$$\mathrm{AdS} = \mathrm{CFT}\,, \tag{1.3}$$

because this contains all the central concepts of fundamental physics: Maxwell's equations, to start with, and their non-Abelian extension, plus the Dirac and Klein-Gordon equations, quantum mechanics, quantum field theory and general relativity. Moreover, in addition to these known principles of nature, it contains several more that theorists have found appealing:

supersymmetry, string theory, and extra dimensions, and it ties these all together in an irreducible way. Also, while Euler's equation is a bit of an oddity, the relation AdS = CFT is just the tip of a large iceberg, it can be deformed into a much large set of gauge/gravity dualities.

So I get to teach you about this, trying to focus on things that you will need in the many upcoming lectures which will use the left-hand side of the equation, string theory and gravity, to learn more about the the right-had side, quantum field theory. Of course it's also interesting to use the equation in the other direction, and maybe some of that will sneak in. Roughly speaking, today's lecture will be a conceptual overview, lecture 2 will give some essential details about the two sides of the duality, lectures 3 and 4 will work out the dictionary between the two sides focussing on the familiar $AdS_5 \times S^5$ example, and lecture 5 will discuss generalizations in many directions.

Let me start by noting a few other reviews. The early MAGOO review[1] contains a detailed summary of the early literature, in which many of the basic ideas are worked out. The 2001 TASI lectures by d'Hoker and Freedman[2] are thorough and detailed, particularly with regard to the constraints from supersymmetry and the conformal algebra, and the calculation of correlation functions. McGreevy's course notes[3] are similar in approach to my lectures.

1.1.2. *A hand-waving derivation*

I am going to first motivate the duality in a somewhat unconventional way, but I like it because it connects the two sides, gauge theory and gravity, without going directly through string theory (as do many of the applications), and it allows us to introduce many ideas that will be important later on. So let me start with the question, is it possible to make the spin-2 graviton as a bound state of two spin-1 gauge bosons? With the benefit of generous hindsight, we are going to make this idea work. To start off, there is a powerful no-go theorem that actually forbids it.[4] Theories without gravity have more observables than theories with gravity (local operators, in particular, since there is no invariant local way to specify the position in general relativity), and this leads to a contradiction. Specifically, Weinberg and Witten show that if there is a massless spin-2 particle in the spectrum, then the matrix element

$$\langle \text{massless spin } 2, k | T_{\mu\nu} | \text{massless spin } 2, k' \rangle \tag{1.4}$$

of the energy momentum tensor (which exists as a local observable in the gauge theory) has impossible properties.

Of course, to prove a no-go theorem one must make assumptions about the framework, and it often proves to be the case that there is some assumption that is so natural that one doesn't even think about it, but which turns out to be the weak link. The Coleman-Mandula theorem, classifying all possible symmetries of the S-matrix,[5] is a classic example. This paper played an important role in its time, ruling out a class of ideas in which spin and flavor were unified in $SU(6)$. However, it made the unstated assumption that the symmetry generators had to be bosonic,[a] which was sufficient for the immediate purposes but missed the possibility of supersymmetry. The more powerful a no-go theorem, the deeper its counterexamples.[b]

The reason for going through this is that the no-go theorem is indeed wrong, but to violate it we have to recognize a deep property of quantum gravity, the holographic principle.[6,7] The entropy of a black hole is proportional to its area in Planck units, and this is the largest possible entropy for a system with given surface area. This suggests that quantum gravity in any volume is naturally formulated in terms of degrees of freedom on its surface, one per Planck area. Thus we see the hidden assumption, that the graviton bound state moves in the same spacetime as its gauge boson constituents; rather, it should move in one additional dimension. Of course, there might be other ways to violate the theorem that will turn up in the future.

So how is the two-gauge boson state in four dimensions supposed to correspond to a graviton in five dimensions? With the benefit of hindsight, there are several places in QCD phenomenology where the size of a gluon dipole, the magnitude z of the separation, behaves like a spacetime coordinate. In color transparency,[8] and in the BFKL analysis of Regge scattering,[9] interactions are approximately local in z and the pair wavefunction satisfies a five-dimensional wave equation. So when you look at the gluon pair you picture it as a graviton four of whose coordinates are the center of mass of the pair, and the fifth is the separation.

We need just two more ingredients to make this idea work, but first we will make an excursion and discuss the shape of the five dimensional spacetime. Quantum field theory is nicest when it applies over a wide

[a]Rereading Ref. 5, this assumption seems to have made its entrance at the point where the symmetry generators are diagonalized, which can't be done for a nilpotent operator.
[b]Perhaps this is what Bohr meant when he said "It is the hallmark of any deep truth that its negation is also a deep truth."

range of scales, so it is natural to consider scale-invariant theories first. If we rescale the system, the c.m. coordinates x and the separations z scale together. The most general metric respecting this symmetry and the symmetries of the four-dimensional spacetime is

$$ds^2 = \frac{L^2 dz^2 + L'^2 \eta_{\mu\nu} dx^\mu dx^\nu}{z^2} \to L^2 \frac{dz^2 + \eta_{\mu\nu} dx^\mu dx^\nu}{z^2}, \tag{1.5}$$

where I have rescaled z in the last form so as to emphasize that there is only one scale R.

This is the metric of anti-de Sitter space, in Poincaré coordinates. If we replace the z with t in the denominator we get de Sitter space, the approximate geometry of our own accelerating spacetime. Certainly one of the major frontiers in gauge/gravity duality, though not one that I will focus on, is to figure out how to interchange z and t in the dual field theory: it appears to require great new concepts. Anti-de Sitter spacetime is not expanding but warped: clocks that run at the same rate in inertial coordinates run at different rates in terms of x^0, depending on where they are in z.

To finish off our 'derivation' of the duality, we need two more ingredients. The first is a large number of fields. We want the AdS scale L to be large compared to the Planck length $L_{\rm P}$, so that we can use Einstein gravity. This means that we can fit a large black hole into the space, one with many Planckian pixels and so a large entropy, and so the field theory had better have a correspondingly large number of degreees of freedom. As we will discuss in more detail later, the number of fields is a power of $L/L_{\rm P}$, depending on the example. The other ingredient we need is strong coupling, so that the gauge boson pair behaves like a graviton and not like a pair of gauge bosons.[c] I should say VERY strong coupling, much larger than one, to get a limit in which the gravitational description is quantitative.

Thus we have two necessary conditions for the duality, many fields and very strong coupling. The second condition can actually be made a bit stricter, as we will see: we need that most operators get parametrically large anomalous dimensions. In this form, these necessary conditions are actually likely to be sufficient, as shown in part in Ref. 10. Large anomalous dimensions clearly require very strong coupling, but do not necessarily follow from it as we will see in an example later.

These conditions of many fields and strong coupling will reappear at various points in these lectures. Clearly they play a controlling role in the

[c]Of course, it will then mix with states having more constituents, but one can still retain a bit of the basic idea that the graviton spin comes from two 'valence' gauge bosons.

applications, as to whether we have a quantitive description or merely a 'spherical cuprate,' a solvable model that captures the qualitative physics that we wish to understand. For example, in the application to heavy ion physics, the coupling seems to be of order one, midway between weak and strong. The strong coupling picture is not quantitative in a precise way, but seems to do better than perturbation theory on many qualitative properties. Naively the number of fields is $N^2 - 1 = 8$ for the gluon states, which is a modestly large number, though it has been noted that the parameter N/N_{flavor} is only unity, and there should be corrections of this order. Fortunately the large-N approximation seems to be fairly robust even at small values.

In the condensed matter applications, again the relevant couplings are of order one at a nontrivial fixed point, where a strong coupling expansion has a chance to capture things that a weak coupling expansion cannot. There is no large N, but condensed matter theorists in the past have not been above introducing a large-N vector index in order to get a tractable system. The large-N vector is a mean-field approximation; this is true for the large-N matrix limit as well, though in a more subtle and perhaps more flexible way:[11] expectations of products of color singlets factorize, but there is a very large number of color singlet fields.

Notice that I have not yet mentioned supersymmetry. However, it tends to enter necessarily, through the requirement of very strong coupling. Quantum field theories tend to become unstable at strong coupling, through the production of pairs whose negative potential energy exceeds their kinetic energy. In continuum theories this can happen at all scales, and the theory ceases to exist. Supersymmetry protects against this: schematically the Hamiltonian is the sum of the squares of Hermitian supercharges, $H = \sum_i Q_i^2$, so the energy is bounded below. Thus, we can break the supersymmetry softly and still have a duality, but if we break it at high energy we lose the theory. We may have to be satisfied with metastability, but that is OK, we probably live with that in our own vacuum.

We have also not mentioned strings, we seem to have found a theory of quantum gravity that uses only gauge theory as a starting point, but in a range of parameters for N and coupling that is not so familiar. But of course what happens is that when we get gravity in this way we get everything else as well, the strings, branes, extra dimensions and so on. Again this was anticipated by 't Hooft,[12] who argued that the planar structure of large-N gauge theory made it equivalent to a theory of strings.

1.1.3. *A braney derivation, for $AdS_5 \times S^5$*

Let me now outline, much more briefly, the original argument for the duality.[13,d] This will have the advantage that it gives specific constructions of gauge/gravity duals. Consider a stack of N coincident D3-branes, in the IIB superstring. In string perturbation theory, each additional world-sheet boundary on the branes brings in a factor of the string coupling $g = e^{\Phi}$ from the genus and a factor of N from the Chan-Paton trace. Thus perturbation theory is good when $gN \ll 1$ and breaks down when $gN \gg 1$.[e] Now, there are black branes[15] that source the same Ramond-Ramond fluxes as D-branes.[16] The black 3-brane metric for N units of flux (in string frame where the physical string tension is α') is

$$\begin{aligned} ds^2 &= H^{-1/2}(r)\eta_{\mu\nu}dx^\mu dx^\nu + H^{1/2}(r)dx^m dx^m\,, \\ &\quad \mu,\,\nu \in 0,\ldots,3\,, \quad m,\,n \in 4,\ldots,9\,, \\ H &= 1 + \frac{L^4}{r^4}\,, \quad L^4 = 4\pi g N \alpha'^2\,, \quad r^2 = x^m x^m\,. \end{aligned} \tag{1.6}$$

You will have to take my word for this solution here, but we will make a useful check later.

The harmonic function H is singular as $r \to 0$, but the metric has a nice behavior in the 'near-horizon' small r limit,

$$ds^2 \to \frac{r^2}{L^2}\eta_{\mu\nu}dx^\mu dx^\nu + \frac{L^2}{r^2}dr^2 + L^2 d\Omega_{S^5}^2\,. \tag{1.7}$$

The first two terms are again the AdS_5 metric, with the coordinate $r = R^2/z$, while the last term is the metric of S^5. The curvature radius is L for both factors. When $gN \gg 1$ this length is large in string units, so the low energy supergravity theory is a good effective description, but it breaks down when $gN \ll 1$. This is just complementary to the perturbative description: the system looks like D-branes in one regime and black branes in another. By varying g one can move adiabatically between these. This was the strategy used to count the states of a black hole,[17] though in a different brane system that has a finite zero-temperature entropy, as we will discuss later.

This is not yet a duality; we have a single theory, the IIB string theory, describing both limits, but in different approximations: perturbation theory around flat spacetime with branes at small gN, and perturbation theory

[d] Note also Ref. 14, which reaches a similar picture by very different reasoning.

[e] Of course the same is true for other Dp-branes, but the rest of the story is simplest for $p = 3$. We will consider other cases later.

around the curved black brane spacetime (with no branes and no open strings) at large gN. To find a duality let us look at the low energy limit of both descriptions. In the small-gN description, this consists of the massless open and closed strings. The massless open strings, ending on the D3-branes, are the usual $U(N)$ adjoint gauge field and collective coordinates, as well as their fermionic partners, while the massless closed strings are the supergravity multiplet. The open strings remain interacting at low energy, because the gauge coupling is dimensionless in 3+1 dimensions, but the closed strings have irrelevant interactions and decouple. In the large-gN description, we again have the massless closed strings away from the brane, but there are also states whose energy is small because they are in the $AdS_5 \times S^5$ region at small r, where the warp factor g_{00} is going to zero; these include not just the massless states, but any massive string state will have an arbitrarily small energy as $r \to 0$.

Let us ignore the massless closed strings away from the brane, which are decoupling in both pictures. Concretely, consider the scaling $r \to r/\zeta$, $x^\mu \to \zeta x^\mu$. The only effect on the black brane metric (1.7) is that the 1 in the harmonic function becomes ζ^{-4}, and scales away as $\zeta \to \infty$ leaving $AdS_5 \times S^5$. In the D3-brane picture this is a symmetry of the low energy gauge theory, while the closed string phase space volume scales to zero, and massive open string effects are suppressed by powers of $\alpha'/x^2 \sim \zeta^{-2}$. In Ref. 13 this limit is described in terms of scaling α' to zero while holding x^μ fixed, with appropriate scalings elsewhere; I have always found it harder to think this way, but it is the same in terms of dimensionless ratios.

Now, if we make the innocuous-sounding assumption that taking the low energy limit commutes with the adiabatic continuation in g we get a remarkable result. At weak coupling we have the gauge theory, and at strong coupling we have all the string states in the $AdS_5 \times S^5$ region. The assumption that the limits commute means that the strongly coupled gauge theory is identical to the full string theory in $AdS_5 \times S^5$. This is now a duality, the statement that two seemingly different theories describe the same system, but in different limits.

This 'derivation' of the duality seems rather slick, but it explained why very different calculations in the two pictures were giving identical answers, and has been supported by many further checks over time. How could the argument fail? The most obvious issue is that the gauge theory could have a phase transition as we vary from weak to strong coupling, but the supersymmetry strongly restricts this possibility. In particular, the property $H = \sum_i Q_i^2$ implies that any supersymmetric vacuum is a minimum of the energy, zero, so we can't have some other vacuum cross to lower energy at some intermediate coupling.[f] We can imagine more intricate ways that the duality might fail, and I will say a bit more about this later, but it is very hard to think of an alternative that is consistent with all the evidence, and the simplest conclusion is that the duality, remarkable though it be, is true.

There are many other weak-strong duality conjectures. Some relate one field theory to another, and some relate one string theory to another or to M theory, but here we have a duality between a quantum field theory and a string theory. From the point of view of trying to construct a theory of quantum gravity this duality is particularly striking, because it allows us to reduce it to the problem of constructing a QFT, solved by Wilson; of course there is much more to say on this point, but it is not the focus of these lectures. What most of these dualities have in common is that there is no explicit derivation, but rather there is the kind of plausibility argument we have just made, combined with many tests giving circumstantial evidence. Some QFT-QFT dualities can be derived explicitly, like bosonization and Ising self-duality in 1+1 dimensions, and Abelian dualities in higher dimensions. These seem as though they should be prototypes for all the other dualities, but somehow the direct steps used to derive them are not enough, and some big new idea is needed.

1.1.4. *Statement of the duality*

The precise statement is that

$$D=4,\ \mathcal{N}=4,\ SU(N)\ \text{Yang-Mills} \ = \ \text{IIB string theory on}\ AdS_5\times S^5\,. \tag{1.8}$$

We will give more detail about each side in the next lecture. Again, this is just one example of a very large number of such dualities. We focus on it, as do most other introductions, because it is both the simplest example, and the one whose field theory side is the most familiar and relevant to many applications. The equality means a one-to-one mapping of the spectra, at any given value of the energy and other quantum numbers. It also includes equality of observables, namely the correlation functions of operators with an appropriate dictionary between the two sides.

A minor aside that may occur to the reader: we have replaced the $U(N)$ of the brane stack with $SU(N)$. The missing $U(1)$ represents the

[f]This would become an issue however if we try to extend the argument to nonsupersymmetric configurations, as we might want to in condensed matter systems. I will discuss an examples in Sec. 4.9.

supermultiplet containing the overall collective motion of the stack. In the black brane picture, before we take the near-horizon limit, this is a fluctuation of the spacetime geometry. The mode is peaked in the transition region between flat spacetime, where the 1 in $H(r)$ dominates, and the near-horizon region, where the L^4/r^4 dominates. In the $AdS_5 \times S^5$ limit, this moves away, and the collective mode is trivial there.

Another aside: the holographic argument led to one extra dimension, but now we have five more from the S^5. Still, the Poincaré and radial dimensions will play a central role, and for much of the discussion we can reduce on the S^5. Note that the warping acts only on the Poincaré dimensions, the S^5 having a constant radius.

On the gauge side we have the two parameters $g^2_{\rm YM}$ and N. On the string side these map to the string coupling g and again N, where the latter is now interpreted as the number of units of five-form flux on the S^5, which counts the D3-branes in the stack. The map between the couplings is standard for D3-branes, $g = g^2_{\rm YM}/4\pi$, with action normalized $-\frac{1}{2g^2_{\rm YM}} \operatorname{Tr} F_{\mu\nu}F^{\mu\nu}$.

The parameters on the gravity side are also usefully expressed in terms of ratios of length scales. We have already noted that

$$L/L_{\rm string} \equiv L/\alpha'^{1/2} = (4\pi g N)^{1/4} = (g^2_{\rm YM}N)^{1/4} = \lambda^{1/4}\,, \qquad (1.9)$$

where in the last line we have introduced the 't Hooft parameter $\lambda = g^2_{\rm YM}N$. The other relevant length is Planck's. In general spacetime dimension D it is natural to work with the reduced Planck length $\hat{L}_{{\rm P},D}$, such that the coefficient of the Ricci scalar in the action is $1/2\hat{L}^{D-2}_{{\rm P},D}$; in four dimensions $\hat{L}_{{\rm P},4} = (8\pi)^{1/2}L_{\rm P}$ in terms of the usual Planck length. Then in string theory $\hat{L}^8_{{\rm P},10} = \frac{1}{2}(2\pi)^7 g^2\alpha'^4$, and

$$L/\hat{L}_{{\rm P},10} = 2^{-1/4}\pi^{-5/8}N^{1/4}\,. \qquad (1.10)$$

A classical spacetime description requires the ratios (1.9) and (1.10) to be large, and so as anticipated in the handwaving argument we need the coupling λ and the number of fields N to be large.

The evidence for the duality takes many varied forms, some of which will be discussed as we go along:

(1) The symmetries on the two sides match.
(2) The spectra of supersymmetric states match. This includes, for example, all modes of the graviton in $AdS_5 \times S^5$.
(3) Amplitudes which are protected by supersymmetry and so can be compared between the two sides are equal.

(4) When we perturb the duality in ways that break some of the supersymmetry and/or conformal symmetry, the geometry realizes the behaviors expected in the field theory, such as confinement.
(5) There are higher symmetries on both sides, which allow some quantities to be calculated for *all* g, with apparent consistency.
(6) Matching long string states can be identified on both sides.
(7) The predictions of the duality for strongly coupled gauge theories can be compared with numerical calculations in those theories, using both light-cone diagonalization and lattice simulation (these are computationally challenging, but progressing).
(8) The predictions agree with experiment!

The next-to-last item, and especially the last, are currently at a crude level of accuracy. Probably I have missed some good tests. Essentially, every time one applies the duality one is testing it, because there is always the possibility of some absurd consequence.

It is sometimes asserted that the evidence supports only a weak form of the duality, but it is not clear what a sensible weak form would be. Suggestions include only the supergravity states, but the number of states in the gauge theory is much larger than this, and some can be identified clearly with string states. Another weak form would be to hold only in the extreme large-N limit. However, this is not consistent (unitary) by itself on either side of the duality, and unitarity largely determines the $1/N$ expansion on the QFT side and the gravitational loop expansion on the string side; it follows from (1.10) that these are expansions in equivalent parameters. Yet another weak form would be to hold only perturbatively in $1/N$ and not exactly. However, the most important nonperturbative phenomena is present on both sides: the integer property of N. In the $1/N$ expansion $1/N$ is a continuous parameter, but in reality it is discrete with an accumulation point at zero. The string side knows about this because the 5-form flux satisfies a Dirac quantization condition.

Thus, by far the simplest way to account for all the facts is that the duality is an exact statement. Of course, we only have an explicit construction of the theory on the QFT side,[g] so I mean that the QFT must agree with all of the approximations we have to the string theory, and with

[g]When I say this, I am thinking of the lattice regulator, with supersymmetry broken but restored in the continuum limit. For the 3+1 example that we are considering, a skeptic can still doubt whether this limit is controlled, because the coupling is strong, but we will see other examples where the theory flows from a superrenormalizable gauge theory and so the continuum limit is much simpler.

any future constructions of the theory. Anyway, the QFT is fully quantum mechanical and consistent, and as we have noted it includes all the graviton states (with the right trilinear interactions), so at the very least it is *some* theory of quantum gravity.

One should note that there are local symmetries on both sides of the duality, the $SU(N)$ gauge symmetry in the QFT and coordinate invariance and local supersymmetry on the AdS side. These are different and neither contains the other. It is an important general principle that dualities acts only on the physical quantities and not on the redundant variables that we use to construct them.

We conclude with some homework problems to think about before the next lecture:

Ex. 1: It is obviously absurd to claim that a four-dimensional quantum field theory is the same as a ten-dimensional string theory. Give one or more reasons why it can't be true.

Ex. 2: Figure out why your answer to the previous problem is wrong.

1.2. The two sides

1.2.1. *The gravity side*

The massless sector of the IIB string theory, IIB supergravity, controls the physics on long distance scales on the gravity side. The fields are the metric G_{MN}, the dilaton, and a 3-form Neveu-Schwarz (NS) field strength H_{MNP}, and 1-, 3- and self-dual 5-form Ramond-Ramond (RR) field strengths F_M, F_{MNP}, and F_{MNPQR} (I'll use capital Roman indices for spacetime, saving Greek for the Poincaré directions). You can find the action in your favorite textbook. Of course the theory also includes excited string states, and D(-1), D1, D3, D5, and D7-branes, as well as NS5-branes and the various (p,q) bound states. All these branes are much heavier than the massless fields, but they exist in the $AdS_5 \times S^5$ region and represent various excitations of the gauge theory.

We could write out the field equations here and verify that $AdS_5 \times S^5$ is a solution, but instead I will do something a bit cruder, which gives some useful insight. Leaving out numerical constants, the relevant terms in the action are

$$S \sim \alpha'^{-4} \int d^{10}x \, \sqrt{-G}(e^{-2\Phi} R - F_{MNPQR}F^{MNPQR}) \, . \qquad (1.11)$$

The 5-form is self-dual, which complicates the action principle. One way to deal with this, which is most efficient for the energetic argument that we are going to make, is to treat the fully spatial components as the independent fields.[h] Now, we are going to make a Kaluza-Klein reduction on the S^5, which is taken to have radius r. The Dirac quantization condition gives $F_{MNPQR} \sim N\alpha'^2$, and so

$$S_5 \sim \alpha'^{-4} \int d^5x \sqrt{-G_5} r^5 (e^{-2\Phi} R_5 + e^{-2\Phi} r^{-2} - \alpha'^4 N^2 r^{-10}) . \qquad (1.12)$$

The second term is from the curvature of the sphere, and in the third we include r-dependence from the inverse metric. To interpret this in terms of an effective potential, rescale $G_5 \to r^{-10/3} e^{-4\Phi/3} G'_5$ to get

$$S_5 \sim \alpha'^{-4} \int d^5x \sqrt{-G'_5} (R'_5 + e^{4\Phi/3} r^{-16/3} - \alpha'^4 N^2 e^{10\Phi/3} r^{-40/3}) . \qquad (1.13)$$

Then

$$V(r, \Phi) \sim -\frac{1}{x^{4/3}} + \frac{\alpha'^4 N^2}{x^{10/3}} \qquad (1.14)$$

where $x = r^4 e^{-\Phi}$. The negative term from the curvature of S^5 dominates at large x, while the positive flux term dominates at small x, leaving a minimum at $x \sim \alpha'^2 N$, or

$$r^4 \sim \alpha'^2 N e^{\Phi} . \qquad (1.15)$$

This is in agreement with the exact solution asserted before, both the scaling and the presence of a flat direction in the potential so that the dilaton is undetermined. The minimum is at negative potential, giving rise to an AdS_5 solution.

This energetics argument is similar to the way one studies stabilization of moduli in string compactification.[18] It is crude by comparison to the sophisticated methods that are employed to find anti-de Sitter solutions in supergravity, but it is a useful complement to these. Some examples:

Ex. 3: M theory has just a metric and a four-form flux, and the only length scale is the Planck scale L_{M}. Use the potential method to find $AdS_7 \times S^4$ and $AdS_4 \times S^7$ solutions, and determine how the radii of the spheres scale with the number of flux units (note footnote h).

[h] In fact, even for forms other rank, we have the option of using either a q-form or a Hodge-dual $(D - q)$ form, and for energetics it is always useful to work with the one that is spatial, else there are surface terms that enter. Note that any dilaton or moduli dependence in the field strength action gets inverted for the Hodge dual.

Ex. 4: In the IIB theory find an $AdS_3 \times S^3 \times T^4$ solution with Q_5 units of RR 3-form flux on the S^3 and (again using footnote h) Q_1 units of RR 7-form flux on $S^3 \times T^4$. Identify flat directions, and the scalings of the radii with the number of units of each of the fluxes.

All these solutions have another important property that is more easily seen in 10 dimensions than in the 5-dimensional reduction: the curvature radii of the spherical factors and the AdS factor are of the same order. If one considers the Einstein equations along the AdS space and the sphere, one can conclude that these curvature terms are of the same order unless there are cancellations. In fact, in all explicitly known examples the radii are similar. A framework for constructing AdS/CFT duals with a large hierarchy is given in Ref. 19, but many details remain to be filled in. Again, from study of moduli stabilization[20] we know that there are many string theory solutions where all the compact directions are much smaller than the AdS radius, but the CFT duals are not known. Making progress in this direction is important for the nonperturbative construction of the landscape, and also for the top-down construction of applied dualities.

1.2.2. *More about anti-de Sitter space*

Anti-de Sitter space has unusual properties that play a key role in the duality. Referring to the metric (1.5), the radial coordinate range is $0 < z < \infty$, and the null geodesics are $z = \pm(t - t_0)$. These reach the boundary $z = 0$ in finite time, and so boundary conditions must be imposed there in order to have a well-defined system. These boundary conditions also provide a natural set of observables, which generally are complicated to define in quantum gravity. On the other hand, the geodesics take infinite time to reach or emerge from $z = \infty$.

The gravitational reshift means that invariant time intervals and coordinate time intervals are related as $d\tau = L dx^0/z$. Correspondingly, the energy $\mathcal{E}$ of an excitation as seen by an invariant observer, and its Killing (conserved) energy P_0 conjugate to x^0 are related

$$P_0 = L\mathcal{E}/z\,. \tag{1.16}$$

A given bulk excitation has an energy that increases when it moves nearer the boundary, which is another way of thinking about the origin of z in the CFT. For Kaluza-Klein excitations near rest $\mathcal{E} \sim 1/L$ and so $P_0 \sim 1/z$. It is closely related to the earlier interpretation in terms of gluon separation:

if we work in terms of the internal momenta for the gluon we have $p_z \sim 1/z \sim P_0$.[i]

The Poincaré coordinates (1.5, 1.7) are the natural ones for applications, but this space is not geodesically complete: the $z = \infty$ limit is a horizon, and a bulk excitation can actually fall through it. A nice way to describe the global space AdS_{D+1} is by embedding it in $D+2$ dimensions with signature $(D, 2)$,[j]

$$ds^2 = L^2 \gamma_{ab} dX^a dX^b\,, \quad \gamma_{ii} = +1\,,\ i = 1, \ldots, D\,, \quad \gamma_{00} = \gamma_{D+1,D+1} = -1\,. \tag{1.17}$$

The embedding is

$$\gamma_{ab} X^a X^b = -1\,. \tag{1.18}$$

Defining $U = X^D - X^{D+1}$ and $V = X^D + X^{D+1}$, this is

$$ds^2 = L^2(\eta_{\mu\nu} dX^\mu dX^\nu + dU dV)\,, \quad \eta_{\mu\nu} X^\mu X^\nu + UV = -1\,, \tag{1.19}$$

where μ runs over the Poincaré directions. Solving for V and defining $x^\mu = X^\mu/U$, $z = 1/U$, the metric in the region $U > 0$ takes the Poincaré form (1.5). The extension to negative U doubles this space, and we can go further: the sum $(X^0)^2 + (X^{D+1})^2$ is positive definite, so there is a noncontractible circle and a larger covering space. In terms of

$$X^0 = (1+\rho^2)^{1/2} \cos\tau\,, \quad X^{D+1} = (1+\rho^2)^{1/2} \sin\tau\,, \quad \rho^2 = \sum_{i=1}^{D} (X^i)^2\,, \tag{1.20}$$

the metric is

$$ds^2/L^2 = (1+\rho^2) d\tau^2 + \frac{d\rho^2}{1+\rho^2} + \rho^2 d\Omega^2_{S^{D-1}}\,. \tag{1.21}$$

The range $0 < \tau < 2\pi$ gives the periodic space, and $-\infty < \tau < \infty$ gives global AdS space.

Falling through the horizon would not seem to be relevant to the applications, since these always involve systems that are finite in size and duration: we turn the experiment off before anything reaches the horizon! However, the global picture has several uses, some of which will be developed later: 1) it makes the $SO(D, 2)$ conformal symmetry manifest; 2) it is

[i] Of course there are also bulk excitations associated with other energy scales, like the string scale, which would introduce a possibly large dimensionless constant into the relation $P_0 \sim 1/z$. In this case it seems more general to associate z with the size of the state.[21]

[j] I will always use D for the spacetime dimension of the field theory and $D+1$ for the dimension of the AdS space.

dual to the CFT quantized on the space S^{D-1} rather than $\mathbb{R}^{D-1}$, if this is what interests us; 3) for this reason, it gives a mapping between the local operators of the CFT and the spectrum of states; 4) if we are interested in studying quantum gravity, the global spacetime is the natural place to formulate it. Curiously, although the Poincaré patch is a small subset of the global space, the respective Wick rotations of the Poincaré x^0 and the global τ yield the same Euclidean version of anti-de Sitter spacetime.

Writing the $SO(D,2)$ as $\delta X^a = \epsilon^a{}_b X^b$, we can identify the action of the symmetries on the Poincaré coordinates: the $\epsilon^\mu{}_\nu$ are the Lorentz transformations, the $\epsilon^\mu{}_U$ are the translations, the $\epsilon^U{}_U$ are the scale transformations, and the $\epsilon^\mu{}_V$ are the special conformal transformations.

1.2.3. *Scalar fields in AdS*

Consider a Klein-Gordon field, with Lorentzian action:

$$\begin{aligned} S_0 &= -\frac{\eta}{2L^{D-1}} \int dz\, d^d x \, (G^{MN}\partial_M\phi\partial_N\phi + m^2\phi^2) \\ &= -\frac{\eta}{2}\int \frac{dz\, d^d x}{z^{D+1}} \left(z^2\partial_z\phi\partial_z\phi + z^2\eta^{\mu\nu}\partial_\mu\phi\partial_\nu\phi + m^2L^2\phi^2\right) . \end{aligned} \tag{1.22}$$

I've left in an arbitrary normalization constant η for later use, but you can ignore it for now. To discuss stability it is useful to define $\phi = z^{D/2}\psi$, $z = -\ln y$, so that

$$\begin{aligned} S_0 = -\frac{\eta}{2}\int dy\, d^d x \, &\left(\partial_y\psi\partial_y\psi + e^{-2y}\eta^{\mu\nu}\partial_\mu\psi\partial_\nu\psi + [m^2L^2 + \tfrac{1}{4}D^2]\psi^2\right) \\ &+ \frac{\eta D}{4}\int d^d x\, \psi^2\Big|_{y=-\infty}^{y=\infty} . \end{aligned} \tag{1.23}$$

If we add a boundary term

$$S = S_0 + S_b = S_0 - \frac{\eta\zeta}{2}\int d^D x\, \psi^2|^{y=\infty} \tag{1.24}$$

with $\zeta \geq \frac{D}{2}$, and if m^2 satisfies the Breitenlohner-Freedman (BF) bound

$$m^2 \geq -\frac{D^2}{4} , \tag{1.25}$$

then the Hamiltonian is bounded below term-by-term and the system is stable.[22] In fact, the Hamiltonian can be organized into a sum of squares under the weaker condition $\zeta \geq \Delta_-$, defined in Eq. (1.28) below.

We see that a range of tachyonic masses is allowed. If the mass-squared lies below the BF bound the free-field energy is unbounded below regardless of boundary terms, though it can be stabilized by higher order bulk terms. As you will hear from some of the other speakers, this leads to interesting phase transitions.

The momentum space field equation is

$$z^{1+D}\partial_z(z^{1-D}\partial_z\phi_k) - (m^2 + k^2z^2)\phi_k = 0\,. \tag{1.26}$$

Near the $z = 0$ boundary the k^2 term can be neglected and the solutions behave as

$$\phi_k \sim z^{\Delta}\,, \quad \Delta(\Delta - D) = m^2L^2\,. \tag{1.27}$$

There are two roots,

$$\Delta_{\pm} = \frac{D}{2} \pm \left(\frac{D^2}{4} + m^2L^2\right) \equiv \frac{D}{2} \pm \nu\,, \tag{1.28}$$

so the asymptotic behavior is

$$\phi(z,x) \sim \alpha(x)z^{\Delta_-} + \beta(x)z^{\Delta_+}\,,\ z \to 0\,. \tag{1.29}$$

We will consider the 'standard' boundary condition $\alpha = 0$ and the 'alternate' boundary condition[23] $\beta = 0$. For the first of these the surface terms in the action and the equations of motion vanish due to the falloff of the fields. For the second, the vanishing of the surface term in the equation of motion requires $\zeta = \Delta_-$; as we have noted, this gives a stable system in the tachyonic case.[k] With the alternate boundary condition both the boundary term and the bulk action are divergent, but the divergences cancel (one should put the boundary at $z = \epsilon$ and then take $\epsilon \to 0$) provided that $\Delta_- > \frac{D}{2} - 1$; beyond this point the k^2 terms in the action diverge and only the standard boundary condition can be used. We refer to the z^{Δ_+} solution as normalizable and the z^{Δ_-} solution as non-normalizable, though the later is normalizable in the more restrictive sense just discussed.

Ex. 5: Extend the discussion to the degenerate case $\Delta_+ = \Delta_-$. Show that the alternate quantization is consistent with conformal invariance in the generic case but not in the degenerate one.

The general solution to the field equation is

$$\phi(z,x) = e^{ik\cdot x}z^{D/2}J_{\Delta - D/2}(qz)\,, \tag{1.30}$$

[k]One can think of the boundary condition in either of two ways: one can impose it on the fields from the start, or one can take free boundary conditions and let the equation of motion determine the boundary behavior. These differ only by contact terms in the operator correlations.

where $q^2 = -k^2 < 0$ and Δ is whichever of $\Delta_\pm$ we are using. These are plane-wave normalizable at the horizon $z \to \infty$, and correspond to a particle that comes out of the horizon, reflects off the boundary, and returns to the horizon. If conformal symmetry is broken in such a way that the space is cut off near the horizon at some large value of z, the effective four-dimensional mass spectrum becomes discrete.

1.2.4. *Conformal field theories*

Now we discuss the field theory side of the duality. Again, gauge/gravity duality is much more general than AdS/CFT, but the conformal case gives the cleanest examples to start from, so I will start with few general comments about CFT. The basic observables are the correlation functions of local operators $\mathcal{O}(x)$. Under scale transformations, an operator of dimension Δ transforms as

$$\mathcal{O}(x) \to \zeta^\Delta \mathcal{O}(\zeta x) . \tag{1.31}$$

Thus scale invariance determines the form of the two-point function

$$\langle 0|\mathcal{O}(x)\mathcal{O}(0)|0\rangle = \zeta^{2\Delta}\langle 0|\mathcal{O}(\zeta x)\mathcal{O}(0)|0\rangle \implies \langle 0|\mathcal{O}(x)\mathcal{O}(0)|0\rangle \propto \frac{1}{x^{2\Delta}} . \tag{1.32}$$

Conformal invariance also determines the position-dependence of the three-point function, so that the basic data is a single coefficient c_{ijk} for $\langle 0|\mathcal{O}_i\mathcal{O}_j\mathcal{O}_k|0\rangle$. This is essentially the OPE coefficient

$$\mathcal{O}_i(x)\mathcal{O}_j(0) = \sum_k c_{ij}{}^k O(x^{\Delta_k - \Delta_i - \Delta_j})\mathcal{O}_k(0) , \tag{1.33}$$

the two-point function being used to raise and lower indices (also, the OPE of descendant fields is determined in term of their primaries). The OPE has a nonzero radius of convergence, as it can be regarded under the state-operator mapping as simply the insertion of a complete set of states. Then higher n-point functions can then be reduced to the two-point function using the OPE, and so are determined by the c_{ijk}. The products in different channels have overlapping regions of convergence, so there is an associativity relation

$$\sum_k c_{ij}{}^k c_{kl}{}^m \sim \sum_k c_{il}{}^n c_{nj}{}^m , \tag{1.34}$$

with appropriate functions of x inserted. There is always an energy-momentum tensor $T_{\mu\nu}$ which is traceless and has $\Delta = D$. Its OPE coefficients are determined by the Ward identities.

This gives a purely algebraic description of a CFT, not referring directly to a Lagrangian.

1.2.5. *The $D = 4$, $\mathcal{N} = 4$ theory*

The particular CFT of interest for the $AdS_5 \times S^5$ duality can be compactly defined as the dimensional reduction of the $D = 10$ super-Yang-Mills theory, whose action is simply

$$S = \frac{1}{g_{\rm YM}^2} \int d^{10}x \, {\rm Tr} \left(-\tfrac{1}{2} F_{MN} F^{MN} + i\bar{\chi}\gamma^M D_M \chi \right) . \tag{1.35}$$

To reduce to four dimensions, ignore the coordinates x^M for $M > 3$ and set the corresponding derivatives to zero. The ten-dimensional gauge field separates into a four-dimensional gauge field and six scalars,

$$A_M \to A_\mu, \ \mu \leq 3, \quad A_m, \ m \geq 4 . \tag{1.36}$$

The Majorana-Weyl spinor χ, with 16 real components, separates into four $D = 4$ Weyl spinors. Further, the supersymmetry generators, also a Majorana-Weyl spinor in $D = 10$, separates into four sets of $D = 4$ generators. The scalars and spinors, like the gauge fields, are $SU(N)$ adjoints.

The one-loop β-function vanishes by cancellation between the gauge field and matter contributions; we will give a nonperturbative agrument for this in Sec. 4.1.

It is sometimes useful to single out one supersymmetry and use $\mathcal{N} = 1$ superfields. There is one gauge multiplet and three chiral multiplets $\Phi_{1,2,3}$, and the superpotential is

$$W \propto {\rm Tr}(\Phi_1[\Phi_2, \Phi_3]) . \tag{1.37}$$

The scalar potential

$$V \propto \sum_{m,n} {\rm Tr}|[A_m, A_n]|^2 \tag{1.38}$$

arises from the non-Abelian terms in the field strength in the dimensional ,reduction and as the sum of F-terms and D-terms in the $\mathcal{N} = 1$ form.

Writing the prefactor of the action (1.35) as N/λ, the action is of order N in the 't Hooft limit of large-N fixed λ. Each propagator then contributes a factor N^{-1} and each interaction a factor N. We are interested in expectation values of gauge-invariant operators,

$$\mathcal{O} = N \, {\rm Tr}(\ldots) . \tag{1.39}$$

With this, an operator behaves like N^{-1} times an interaction. There is also a factor of N for each index loop. Filling the index loops in to make the faces of a simplex the total N-dependence is

$$N^{V-O+F-P} = N^{\chi-O} \tag{1.40}$$

where V, F, P, O count the vertices, faces, propagators, and operators, and χ is the Euler number of the simplex. Thus for the $\mathcal{O}^4$ expectation value, the leading part of the fully disconnected term (the product of four one-point functions) has the topology of four two-spheres for N^4, the terms with two connected two-point functions has leading behavior N^0, and the fully connected amplitude has leading behavior N^{-2}For the fully connected contribution for O operators the leading connected amplitude is N^{2-O}.

1.3. The dictionary

In addition to classic Refs. 24, 25, the review Ref. 2 is good for further details.

1.3.1. *Symmetries*

The first check is a comparison of the symmetries of the two systems. On the AdS side there are the $SO(4,2) \times SO(6)$ symmetries of the AdS_5 and S^5 spaces. In the CFT, $SO(4,2)$ is the conformal group: we have already discussed at the end of Sec. 2.2 how the various symmetry transformations are realized. The $SO(6)$ is the symmetry of the scalar field space A_m, with the four fermions transforming as the spinor representation. Equivalently, the fermions are in the fundamental representation of $SU(4) = SO(6)$, and the six scalars in the antisymmetric product of two fundamentals. On both sides of the duality this bosonic symmetry group is extended by supersymmetry to the superconformal $PSU(2,2|4)$. Again, there are gauge/gravity duals with much less symmetry, but it is useful to take account of it in this prototype example.

There is also an $SL(2,\mathbb{R})$ weak-strong duality on both sides, in the $D = 4$, $\mathcal{N} = 4$ gauge theory and in the IIB string theory.

1.3.2. *Matching of states*

The duality implies a 1-1 mapping of the states on the two sides. On the AdS side, since we are expanding in the loop parameter $L_{\rm P}/L$, we can count particle states at weak coupling. On the CFT side, there is the usual isomorphism between states and operators: drawing a $(D-1)$-sphere around an operator, we can go to radial quantization and so we have an isomorphism between local operators and states on S^{D-1}. Combining these, we can express the duality as a 1-1 mapping between the particle species in AdS_{D+1} and the single-trace chiral primary operators in the CFT.

The statement is that a field whose boundary behavior is z^Δ maps to an operator of dimension Δ. The mapping is simply that the scaled boundary limit of the bulk operator is the CFT operator,

$$\mathcal{O}(x) = C_\mathcal{O} \lim_{z\to 0} z^{-\Delta}\phi(x,z)\,. \tag{1.41}$$

This fits with our original picture of the z coordinate emerging from the size of the composite state: a local operator creates a state with zero size. The constant $C_\mathcal{O}$ depends on convention, which we will set later. The scale transformation in the bulk takes $\phi(z,x) \to \phi(\zeta z, \zeta x)$ and so

$$\mathcal{O}(x) \to C_\mathcal{O} \lim_{z\to 0} z^{-\Delta}\phi(\zeta x, \zeta z) = C_\mathcal{O} \lim_{z\to 0} (z/\zeta)^{-\Delta}\phi(\zeta x, z) = \zeta^\Delta \mathcal{O}(\zeta x)\,, \tag{1.42}$$

which is the scale transformation of an operator of dimension Δ. Recall that the boundary behavior is related to the mass of the field by $m^2L^2 = \Delta(\Delta - D)$, so that a given mass corresponds to two possible dimensions depending on the boundary condition taken. The condition $\Delta > \frac{D}{2} - 1$ is precisely the lower limit on the dimension of an interacting scalar field in a unitary CFT.

We have focused on scalar fields, but the scalings work the same way for tensor fields if we use tangent indices rather than coordinate indices. For example, every CFT contains the energy-momentum tensor, of dimension D, which maps to the graviton on the AdS side. In terms of tangent space indices we then have for the metric perturbation $h_{\hat\mu\hat\nu} \sim az^D + b$, or $h_{\mu\nu} = (z/L)^{-2}h_{\hat\mu\hat\nu} \sim az^{D-2} + bz^{-2}$. Note that the larger behavior is the same as that of the AdS metric itself.

When the CFT has conserved currents, there is a corresponding gauge field in the bulk. The dimension of a conserved current is $D-1$, so $A_{\hat\mu} \sim az^{D-1} + bz$ and $A_\mu \sim az^{D-2} + b$. The constant term can be identified directly with a background gauge field in the CFT. The $AdS_5 \times S^5$ theory has the $SO(6)$ global symmetry noted earlier, and correspondingly there are $SO(6)$ Kaluza-Klein gauge fields from the S^5.

There is an important distinction between states that would be massless in $D = 10$ and get effective $D = 5$ masses only due to the spacetime curvature, and those that are already massive in $D = 10$. The former have masses of order $1/L$ and so dimensions of order one. In the $D = 4$,

$\mathcal{N} = 4$ case, there is enough supersymmetry in fact to guarantee that these dimensions are actually independent of the coupling. This is because the maximum spin of such an operator is 2, which requires it to live in a BPS multiplet of the supersymmetry. For duals with less supersymmetry the dimensions of such operators will be $O(1)$ at weak and strong coupling, but can depend on the coupling.

Excited string states, on the other hand, have masses of order $\alpha'^{1/2}$ and so dimensions at strong coupling of order $\lambda^{1/4}$. This is a striking prediction of the duality. It is plausible that anomalous dimensions would become large at strong coupling, but there is no simple analytic argument for this particular behavior. Results based on integrability, which are supposed to interpolate over all couplings, are consistent with it,[26] but there still seems to be some guesswork involved. Eventually we might hope that the numerical approaches will reproduce it.

These large dimensions are essential to have a supergravity limit: in supergravity the maximum spin is two, so any operator of higher dimension must get a large anomalous dimension.[1] This requires very strong coupling, but it is really a stronger statement, as there are theories with couplings that are much larger than one but still without parametrically large anomalous dimensions for most operators. Thus we can now elaborate on how we define strong coupling as a necessary condition for the duality: we need all stringy states, in particular those with spins greater than two, to get large anomalous dimensions.

Focusing now on the specifics of the $D = 4$, $\mathcal{N} = 4$, theory, the gauge kinetic term $F_{\mu\nu}F^{\mu\nu}$ maps to the dilaton; it has dimension 4 at weak coupling, and (with appropriate additional pieces) it is BPS and so its dimension is exactly four. The topological term $F_{\mu\nu}\tilde{F}^{\mu\nu}$ maps to the RR scalar and has dimension 4.

The traces of products of scalars, $\mathrm{Tr}(A_m A_n \dots A_p)$, give a large set of operators. The linear combinations that are traceless on the $SO(6)$ indices map to shape fluctuations of the S^5, mixed with other modes. The operator with l scalars has exact dimension l. The modes with traces, such as the Konishi operator $\mathrm{Tr}(A_m A_m)$ must map to excited string states.

The superpotential perturbations $\delta W(\Phi_{1,2,3})$ map to the perturbations of the harmonics of the NS and RR 3-form fluxes F_{MNP}, H_{MNP} (which mix), and an operator with m fields has dimension $m + 1$.

---

[1]The higher spin theories constructed by Vasiliev may allow one to go beyond this in some cases.

There is a one-to-one correspondence between BPS operators of the CFT and supergravity modes in AdS_5, as required by the duality. It might seem impossible for a four-dimensional field theory to have as many states as a ten-dimension supergravity theory, much less a string theory, and this would be many peoples' answer to Ex. 1, e.g. Ref. 27. The key is that there are a lot of states due to the large matrices. We have seen that the Kaluza-Klein excitations come from traces of many operators, and these traces are all independent as $N \to \infty$. At one level it seems trivial that one can encode anything in a large-N matrix; what is remarkable is that the codebook is just the $\mathcal{N} = 4$ path integral.

This also provides another nice illustration of the duality working non-perturbatively in N. At finite N the the traces with more than N operators are not independent, and so there must be a cutoff on the momentum parallel to the S^5. One can make harmonics up to $l = N$, corresponding to momenta of order $N/L \sim N^{3/4}/L_{\rm P}$. It is might seem wrong that the cutoff is much larger than the Planck scale, but there is no bar to boosting a particle up to highly super-Planckian momenta. The cutoff arises rather from a curious phenomenon:[28] the interaction of the particle with the flux on the S^5 causes it to blow up to the point that it can no longer fit in the space.

Rather than counting supergravity states, we could consider the total partition function of the theory. In the gravity picture this is dominated by the black 3-brane, whose entropy per unit volume is

$$s_{\rm black\,D3} = \frac{\pi^2}{2} N^2 T^3 \,, \tag{1.43}$$

to be compared with the free-field entropy of the CFT,

$$s_{\rm free\,CFT} = \frac{2\pi^2}{3} N^2 T^3 \,. \tag{1.44}$$

These agree up to the famous factor of 3/4:[29] the duality is saying that as the coupling is increased from zero to infinity in the gauge theory, there is this small shift in the density of states. The agreement of the temperature dependences is confirming that the AdS theory is behaving like a four-dimensional CFT, and the agreement of the N-scaling is another satisfying check. However, there could have been a nontrivial dependence on the 't Hooft coupling λ, so that the weak and strong coupling entropies could have differed for example by a power of λ. There is no independent check of this result, so for now it is a prediction of the duality rather than a check.

1.3.3. *Correlators I*

I am going to take the operator relation (1.41) as the starting point for the dictionary. First, I would like to dispel a myth, that the AdS/CFT dictionary is somehow naturally Euclidean, and that there are difficulties with extending it to Lorentzian spaces.[m] In the first place, when we compare the spectra of the Hamiltonian on the two sides, this makes no reference to any signature, since the difference is just whether we evolve forward with e^{-iHt} or $e^{-H\tau}$. Secondly, the basic relation (1.41) holds equally in any signature. It is true that in Lorentzian signature there are several correlators of interest (time-ordered, advanced, retarded), and the prescriptions for their calculation, at finite temperature, are intricate. This will be discussed in the lectures by Son; for a recent overview see Ref. 30.

But these questions are not issues for the duality, which via (1.41) just relates any given correlator in the CFT with the same correlator in the AdS space.

I want to calculate the time-ordered 2-point correlator in several ways. First, the propagator of ϕ in the bulk of AdS_{D+1} is

$$\langle 0|\mathrm{T}\,\phi(z,x)\phi(z',x')|0\rangle = \frac{1}{\eta}G_\Delta(\xi)\,, \tag{1.45}$$

where $\Delta = \Delta_+$ or Δ_- according to the choice of boundary condition. Here

$$\xi = \frac{2zz'}{z^2+z'^2+(x-x')^2} \tag{1.46}$$

is the unique conformal invariant that can be constructed from the two positions. It is a measure of the distance between them, approaching 1 when they become coincident and 0 when they are far apart. Emphasizing again the signature of spacetime, this is equally valid in the Euclidean and Lorentzian theories, with

$$(x-x')^2 \equiv_{\mathrm{E}} \sum_{i=1}^{D}(x^i-x'^i)^2 \equiv_{\mathrm{M}} (i+\epsilon)^2(x^0-x'^0)^2 + \sum_{i=1}^{D-1}(x^i-x'^i)^2\,, \tag{1.47}$$

and we are working in the Poincaré patch.

The propagator $G_\Delta(\xi)$ is hypergeometric, and Ref. 2 is a good source for more details. When the points are far apart, and this includes the case that z and/or z' approach the boundary with $x-x'$ fixed and nonzero, then

$$G_\Delta(\xi) \to \frac{C_\Delta}{2\Delta - D}(\xi/2)^\Delta\,, \quad C_\Delta = \frac{\Gamma(\Delta)}{\pi^{D/2}\Gamma(\Delta - D/2)}\,. \tag{1.48}$$

[m] If anything the reverse is true, since there is still no comprehensive treatment of the contour for the conformal factor in the path integral for Euclidean gravity.

In particular, it behaves as $z^\Delta z'^\Delta$ as required by the boundary conditions. Thus we can form the limit (1.41) and conclude that

$$\langle 0|\mathrm{T}\,\mathcal{O}(x)\mathcal{O}(x')|0\rangle = \frac{C_\Delta C_\mathcal{O}^2}{\eta(2\Delta - D)}(x-x')^{-2\Delta}\,. \tag{1.49}$$

Of course the 2-point correlator is determined up to normalization by scale invariance, but we are just warming up for the later lecturers, who will look at the 2-point function after turning on temperature, densities, or background fields, where it captures much interesting physics.

In some cases the normalization of the 2-point function is interesting, but here it is just the product of a bunch of conventional factors. We now introduce a standard convention. If we take just the primed field to the boundary we get

$$\langle 0|\mathrm{T}\,\phi(z,x)\mathcal{O}(x')|0\rangle = \frac{C_\Delta C_\mathcal{O}}{\eta(2\Delta - D)}\left(\frac{z}{z^2+(x-x')^2}\right)^\Delta\,. \tag{1.50}$$

If we now take $z \to 0$, the correlator goes to zero pointwise except at $x = x'$, and its value integrated d^Dx is $z^{D-\Delta}C_\mathcal{O}/\eta(2\Delta - D)$. It is therefore conventional to set $C_\mathcal{O} = \eta(2\Delta - D)[N^{-1}]$ (ignore for now factors in square braces!), so that

$$\langle 0|\mathrm{T}\,\phi(z,x)\mathcal{O}(x')|0\rangle \stackrel{z\to 0}{\to} z^{D-\Delta}\delta^D(x-x')[N^{-1}]\,. \tag{1.51}$$

At this point I am going to switch to a Euclidean metric, not because I have to but because it is conventional in this subject (also, the Lorentzian form would have a factor of $-i$ in Eq. (1.51) and compensating factors elsewhere); anyway, I hope I have already made my point about the duality being just fine in Lorentizian signature. Note that this also requires that I flip the sign of the action due to standard conventions. Noting that $D - \Delta_+ = \Delta_-$ (and $D - \Delta_- = \Delta_+$), the operator $\mathcal{O}$ can be regarded as a delta-function source for the mode that was previously set to zero.

With this normalization we also have

$$\begin{aligned}\langle 0|\mathrm{T}\,\mathcal{O}(x)\mathcal{O}(x')|0\rangle &= \frac{\eta(2\Delta - D)C_\Delta}{(x-x')^{2\Delta}}[N^{-2}]\,,\\ \langle 0|\mathrm{T}\,\phi(z,x)\mathcal{O}(x')|0\rangle &= C_\Delta\left(\frac{z}{z^2+(x-x')^2}\right)^\Delta[N^{-1}]\,.\end{aligned} \tag{1.52}$$

The constant η arises from the normalization of the Klein-Gordon action.[n]

The relation (1.51) allows us to connect with the more standard way to write the dictionary. Noting that $D - \Delta_+ = \Delta_-$ (and $D - \Delta_- = \Delta_+$), the operator $\mathcal{O}$ can be regarded as a delta-function source for the mode that was previously set to zero. In particular, if we introduce into the CFT path integral a factor

$$e^{\int d^d x\, j(x)\mathcal{O}(x)}\,, \tag{1.53}$$

then we now have

$$\begin{aligned} \alpha(x) &\to j(x) \quad \text{(standard quantization)}\,; \\ \beta(x) &\to j(x) \quad \text{(alternate quantization)}\,. \end{aligned} \tag{1.54}$$

In other words, the coefficient of the fixed ($z^{D-\Delta}$) mode is that with which the operator $\mathcal{O}$ is added to the CFT, while the coefficient (1.41) of the z^Δ quantized mode is the expectation value of $C_{\mathcal{O}}^{-1}\mathcal{O}$.

Evaluating the bulk path integral then gives a generating functional

$$\langle 0|\mathrm{T}\, e^{\int d^D x\, j(x)\mathcal{O}(x)}|0\rangle = Z_j \to e^{-S_{\rm cl}}\,. \tag{1.55}$$

Here Z_j is the bulk path integral with boundary condition (1.54), and in the last form we have evaluated it in the semiclassical approximation, extremizing with respect to the field with given boundary conditions. Using Eqs. (1.52, 1.54), the extremum is

$$\phi_{\rm cl}(z,x) = C_\Delta \int d^D x'\, j(x') \left(\frac{z}{z^2 + (x-x')^2}\right)^\Delta \tag{1.56}$$

This corresponds to the planar approximation in this CFT, but is also exact for the quadratic theory being studied. This is the form in which the dictionary is often given.

Integrating by parts, the classical bulk action can be written as a surface term (again, all signs flipped from the previous discussion)

$$S_{0,\rm cl} = -\frac{\eta}{2L^{D-1}} \int dz\, d^D x\, \phi_{\rm cl}(\Box - m^2)\phi_{\rm cl} + \frac{\eta}{2}\epsilon^{1-D} \int d^d x\, \phi_{\rm cl}(\epsilon, x)\partial_\epsilon \phi_{\rm cl}(\epsilon, x)\,, \tag{1.57}$$

[n]Noting the discussion of operator normalization in Sec. 2.5, we see that η must be of order N^0. On the other hand, it is often convenient to normalize all closed string fields like the graviton, with a $1/G$ in the action, so $\eta \sim L^{D-1}/G_{D+1} \sim L^8/G_{10} \sim N^2$. In this case one must include the factors in braces. In the remainder of these notes we use only the $\eta \sim N^0$ convention.

the first term vanishing by the equation of motion. This is not the whole story because of the need for a boundary term in the action. To regulate potential divergences, we temporarily move the boundary in to a small value $z = \epsilon$. The boundary term must be such that the boundary terms in the variation of the action vanish for variations that respect the boundary conditions; this is in order to have a good variational principle.[o] One good choice of boundary action for the standard quantization is

$$S_{\rm b} = \frac{\eta\Delta_-}{2}\epsilon^{-D}\int d^Dx\,\phi^2(\epsilon,x)\,, \tag{1.58}$$

which also guarantees stability as discussed earlier. Then

$$\delta(S_0 + S_{\rm b}) = -\eta\epsilon^{-D}\int d^Dx\,\delta\phi(\epsilon,x)(\epsilon\partial_\epsilon - \Delta_-)\phi(\epsilon,x)\,. \tag{1.59}$$

The boundary condition fixes $\alpha(x)$, so $\delta\phi(\epsilon,x)$ has only a $\delta\beta$ term, and then only the cross term with α in the second field survives at the boundary, and this is annihilated by $\epsilon\partial_z - \Delta_-$. One can also check that the potentially divergent α^2 terms cancel for the same choice of boundary term. In all

$$S_{\rm cl} = -\frac{\eta}{2}\epsilon^{-D}\int d^Dx\,\phi_{\rm cl}(\epsilon,x)(\epsilon\partial_\epsilon - \Delta_-)\phi_{\rm cl}(\epsilon,x)\,. \tag{1.60}$$

Inserting $\phi_{\rm cl}$ leaves a convolution to do,[3,25] but it is not hard to deduce the answer. The $\epsilon\partial_\epsilon - \Delta_-$ kills the α part of the second $\phi_{\rm cl}$, so we must get precisely the α part of the first $\phi_{\rm cl}$ in order that the term survive at the boundary, and this is just $j(x)$. So

$$\begin{aligned} S_{\rm cl} &= -\frac{\eta}{2}\epsilon^{-\Delta}\int d^Dx\,j(x)(\epsilon\partial_\epsilon - \Delta_-)\phi_{\rm cl}(\epsilon,x) \\ &\to -\frac{\eta}{2}(2\Delta - D)C_\Delta\int d^Dx\int d^Dx'\,j(x)j(x')\frac{1}{(x-x')^{2\Delta}}\,, \end{aligned} \tag{1.61}$$

correctly generating the earlier result. The extensions to the alternate quantization, and to the more general nonconformal boundary condition $\alpha = f\beta + j$, are left as exercises. The solution for the latter is given below Eq. (3.22) of Ref. 31.

---

[o]This may still leave some freedom in the choice of the boundary action, which would just correspond to changes of convention such as operator redefinition. Also, finiteness of the total bulk plus boundary action as $\epsilon \to 0$ might be desirable, but the divergences are local contact terms and so easily subtracted by hand.

1.3.4. *Correlators II*

For gauge fields and metric perturbations it is much the same, except that the symmetries give a natural normalization. We have noted in Sec. 3.2 that the non-normalizable mode of a gauge field scales as z^0, so we can interpret its limit directly as a gauge field in the CFT,

$$\lim_{z\to 0} A_\mu(z,x) = A_{\mathrm{b}\mu}(x)\,. \tag{1.62}$$

This identification is necessary in order that gauge transformations act consistently on charged operators. This mode couples to the corresponding conserved current in the CFT as $\int d^Dx\, A_{\mathrm{b}\mu}(x) j^\mu(x)$. Similarly the nonnormalizable mode of the metric has the same scaling as the AdS metric, and so we interpret

$$\lim_{z\to 0} z^2 h_{\mu\nu}(z,x) = h_{\mathrm{b}\mu\nu}(x)\,, \tag{1.63}$$

which couples to the energy-momentum tensor in the CFT as $\frac{1}{2}\int d^Dx\, h_{\mathrm{b}\mu\nu}(x) T^{\mu\nu}(x)$.

Since later lecturers will be looking the the TT correlator, let us look at one component of this, at zero temperature. We excite $h \equiv h_{12}$ with momentum k in the 3-direction. So far we have worked in position space, but with less symmetry there will not be a simple coordinate form and so we go to momentum space. The relevant terms in the Euclidean action in AdS_5 are

$$\frac{L^3}{4\hat{L}^3_{\mathrm{P},5}} \int \frac{d^4x\, dz}{z^{d-1}} (\partial_z h \partial_z h + k^2 h^2)\,. \tag{1.64}$$

This is exactly like the massless Klein-Gordon field. The usual gravitational extrinsic curvature term has been included implicitly to make the action first order,[32] but no additional surface term is needed because $\Delta_- = 0$.

The extremum can again be written as a surface term,

$$S_{\mathrm{cl}} = -\frac{L^3}{8\hat{L}^3_{\mathrm{P}}} \epsilon^{-3} \int d^dx\, \partial_\epsilon (h^2(\epsilon))\,. \tag{1.65}$$

The classical solution that approaches h_{b} at the boundary and remains bounded at $z = \infty$ is

$$h_{\mathrm{cl}}(z) = h_{\mathrm{b}} \frac{k^2 z^2}{2} K_2(kz)\,. \tag{1.66}$$

Expanding the Bessel function at small z,

$$h_{\mathrm{cl}}(z) = h_{\mathrm{b}} \left\{ 1 + O(k^2 z^2) - \frac{k^4 z^4}{16} \ln k + O(k^4 z^4 \ln z) + \ldots \right\} \tag{1.67}$$

There is a quadratically divergent $\epsilon^{-2}k^2$ piece; this is analytic in k^2 and so is a contact term. There are also $\ln\epsilon$ and finite k^4 contact terms. The term of interest ir $k^4 \ln k$, leading to

$$S_{\rm cl} = \frac{h_{\rm b}^2 L^3 k^4 \ln k}{16 L_{\rm P}^3}\,, \tag{1.68}$$

and

$$\langle 0|T_{12}(k)T_{12}(0)|0\rangle = \frac{L^3}{8\hat{L}_{\rm P}^3} k^4 \ln k\,. \tag{1.69}$$

Using $\hat{L}_{{\rm P},5}^3 = \hat{L}_{{\rm P},10}^8 / V_{S^5}$ where $V_{S^5} = \pi^3 L^5$, this becomes[24]

$$\langle 0|T_{12}(k)T_{12}(0)|0\rangle = -\frac{\pi^3 L^8}{8\hat{L}_{{\rm P},10}^8} k^4 \ln k = -\frac{N^2}{32\pi^2} k^4 \ln k\,. \tag{1.70}$$

In position space this goes as $1/x^8$ as it must for an operator of dimension 4. The central charge c is defined as $-8\pi^2$ times the coefficient of $k^4 \ln k$, i.e. $N^2/4$. It must be independent of the coupling because it is related by supersymmetry to the $SO(6)$ R-anomaly, and indeed agrees with the weak coupling result.[33]

1.4. Breaking symmetries

In this final section I want to generalize the example we have been studying in various ways. Two challenging goals are to construct duals to landscape AdS states, and to construct non-Fermi liquids. We will not reach these, but I will discuss some issues. The landscape states, as mentioned before, have the property that the compact dimensions can be much smaller than the AdS space, and also that one can uplift them to higher AdS states, and also dS, by breaking supersymmetry (though as Tom Banks emphasizes, the uplifted states may not be visible in the AdS construction). The non-Fermi liquids represent a set of observed strongly coupled fixed points that have defied a clear description.

In the applications to condensed matter physics there are two approaches being taken. One is to postulate a gravitational theory with some set of fields, implying that the CFT has the corresponding operators. There will always be the gravitational field, corresponding to $T_{\mu\nu}$, and there may be some gauge fields in the bulk, mapping to symmetry currents in the CFT, and also some charged matter fields, mapping to charged operators of various dimensions. In this case there is no specific Lagrangian in the

CFT, the results apply to any theory with the given set of operators: once one assumes some specific operators (and their dimensions and maybe some couplings) the dual allows one to calculate how the system responds to temperature, electric and magnetic fields, and other perturbations. This has been instructive, and led to interesting work on non-Fermi liquids which Hong Liu will discuss, but it would seem more satisfying to start from some brane system and deduce the Lagrangian and its gravity dual. This top-down approach has been challenging, as I will describe.

1.4.1. *The Coulomb branch*

Starting from the $D = 4$, $\mathcal{N} = 4$ example, there is a way to break the conformal symmetry without changing the theory, but just going to a different vacuum. The scalar potential $\sum_{m,n} \mathrm{Tr}|[A_m, A_n]|^2$ vanishes whenever the six matrices A_m commute, but they can otherwise take any values. This corresponds to separating some or all of the D3-branes in the six transverse directions of the Poincaré picture. There is no force between them due to supersymmetry; the potential gets no quantum corrections. Separating the D3-branes breaks the gauge invariance to a block for each set of coincident D3-branes, a product of $U(1)$ and SU factors, and it breaks the scale and conformal invariance because some of the branes are sitting at nonzero r. A fairly recent discussion of this subject is Ref. 34.

From the point of view of the applications, this is more of a nuisance perhaps: most of them don't have fundamental scalars but in the duality these come along due to supersymmetry. And, when we break the supersymmetry, it may be that the vacuum that we want (where the scalar vacuum expectation values are zero) is no longer stable. The top-down construction of a Fermi sea has been a goal of mine for a long time because of the possibility of constructing a non-Fermi liquid, but these constructions often seem to have this problem.

We can use these vacua to give a modern nonperturbative proof of the conformal invariance of the $\mathcal{N} = 4$ theory. If we make all of the vevs distinct, the gauge group is $U(1)^{N-1}$ and the low energy effective fields are just the gauge fields and their scalar A_m and fermionic partners. The low energy effective action for the gauge fields is

$$-\sum_i \frac{1}{4g_i^2} F_{i\mu\nu} F_i^{\mu\nu} . \tag{1.71}$$

If the couplings run, then the g_i should be evaluated at a scale of order the masses of the higged gauge bosons, which are proportional to the A_m.

But the action will then depend on the scalars, and this is forbidden by the $\mathcal{N} = 4$ supersymmetry.

1.4.2. *Renormalization group flows*

If we start with the $D = 4$, $\mathcal{N} = 4$ theory and perturb the Hamiltonian by a relevant operator, $\Delta < D$, the effect of the perturbation becomes large at low energies. This is readily studied in the dual theory: adding $g \int d^D x\, \mathcal{O}(x)$ to the action for some operator $\mathcal{O}$ shifts the corresponding fixed mode. For an operator of dimension Δ this mode behaves as $z^{D-\Delta}$ which is just right: for relevant operators, $\Delta < D$, their effect grows as we move away from the boundary.[p]

The interesting question is, what happens at low energies where the perturbation gets large? The two broad classes of possibility are that we might flow to a new conformal theory, or that everything just gets massive. We can find examples of each within the $\mathcal{N} = 4$ theory. A perturbation that preserves $\mathcal{N} = 1$ supersymmetry is to add a mass term for one or more of the chiral superfields,

$$\delta W = \sum_{\alpha=1}^{3} m_\alpha \Phi_\alpha^2 \,. \tag{1.72}$$

This is a nice example because the low energy physics depends in an interesting way on how many of the fermions are massive.

First on the field theory side, for one nonzero mass there are field theory arguments that the system flows to a new conformal fixed point.[35] For two equal nonzero masses, there is a $\mathcal{N} = 2$ supersymmetry and the massless sector is just the pure $\mathcal{N} = 2$ super- Yang Mills, whose low energy physics is the $SU(N)$ version of Seiberg-Witten,[36] where the massless fields are described by a $U(1)^{N-1}$ effective theory. (For two unequal masses I am not sure what happens, but I'm guessing that it flows to the same fixed point). For three nonzero masses,[37] there are actually many supersymmetric vacua: the superpotential equations

$$m\Phi_\alpha = \epsilon_{\alpha\beta\gamma}[\Phi_\beta, \Phi_\gamma] \tag{1.73}$$

give a vacuum for each N-dimensional representation of $SU(2)$. The vacuum corresponding to the trivial solution $\Phi_\alpha = 0$ is expected to be confin-

[p]If we try to add an irrelevant operators $\Delta > D$, the effect grows as we approach the boundary, the nonlinearities become large, and generically we will lose the AdS geometry and the duality.

ing, while all the others have the gauge group broken to some subgroup containing non-Abelian factors that confine and always some unbroken $U(1)$'s.

On the gravity side, the mass perturbation corresponds to a perturbation of the 3-form flux on the S^5. One can analyze this in terms of an effective Lagrangian for just this mode coupled to gravity. This seems like a cheat, because this is not a real effective field theory, there is no symmetry that allows us to truncate to just this mode, but there is some supergravity magic that allows one to extend any solution of this truncated system to the full theory. The effective potential looks like the curve $x^3 - x$, with one maximum and one minimum. The $AdS_5 \times S^5$ theory actually sits at the maximum, which looks unnerving but is fine because of the BF bound: the field can't roll down in time, but it can roll in the radial direction, ending up at large z at the minimum, which is a new CFT. The overall geometry in the infrared is $AdS_5 \times X$ where X is some deformed S^5 with three-form flux.[38]

One can repeat this with two masses, but now the field flows off in a different direction, to infinity. The supergravity magic allows this to be lifted to a ten-dimensional solution which is now singular, but the singularities have a nice interpretation: they are just N D3-branes.[39,40] The z-coordinate (defined so that the metric is $L^2\eta_{\mu\nu}/z^2 + \ldots$) is cut off at a finite upper limit, meaning that there is a lower bound on the energies $P_0 = L\mathcal{E}/z$. However, there are massless states on the D3-branes, giving just the expected $U(1)^{N-1}$ theory.

When masses are turned on for all three superfields, one can try to repeat the above effective Lagrangian strategy. It again shows the geometry ending before $z = \infty$, but now the five-dimensional solution[41] lifts to a ten-dimensional solution with an unphysical singularity, it is not quite right. One must work in fully ten-dimensional terms. With some guesswork,[37] guided by the field theory, one can find a sensible solutions where the geometry ends at some finite z, and there are explicit NS5- and/or D5-branes, with D3-branes dissolved in them. The D5's correspond to non-commuting coordinate matrices A_m for the original D3's, as suggested by (1.73): this is the Myers effect,[42] where in a background field it can become energetically favorable for the D3 coordinates not to commute. For the NS5's, there is some strong-coupling dual of this effect, which has no such classical description. The confining vacuum is the one with just a single NS5-brane.

To talk about confinement we need another observable, which measures the force between quark-like sources.[45] This is simply the energy of a string that starts and ends on the the sources (one can give a longer justification, but this is just part of 't Hooft's large-N story, where the string corresponds to a tube of color flux). Now, the effective tension of such a string, as seen in the gauge theory, is $\mu(z) = 2\pi L^2/\alpha' z^2$, since the tension seen by a local inertial observer is $2\pi/\alpha'$. In the conformal theory, as one move the sources apart, the string can drop down closer and closer to the horizon and reduce its tension indefinitely. One then gets $V(r) \propto 1/r$ (as one must in a conformal theory, though it is interesting that at strong coupling the potential goes as $\lambda^{1/2}$ rather than as λ).

In the confining vacua, there is a maximum z and so a minimum tension $2\pi L^2/\alpha' z_{\rm max}^2$. There are many other realizations of confinement in AdS/CFT;[25,43,44] others are simpler on the gravity side, but generally more complicated on the CFT side. In the Coulombic vacua, as with two masses, strings hang down from each source and end on a D3-brane, where they source an explicit flux that can spread out coulombically.

1.4.3. *Multitrace RG flows*

If the CFT has a scalar operator of dimension $\frac{D}{2} - 1 < \Delta < \frac{D}{2}$, this is described in the bulk by the alternate quantization in which the larger mode α fluctuates and the smaller mode β is fixed. There are actually no such operators in the $\mathcal{N} = 4$ theory, but there are in many other examples.

In this case the double-trace operator $\mathcal{O}^2$ is relevant:[46,47]

$$\Delta_{\mathcal{O}^2} = 2\Delta + O(1/N^2)\,, \tag{1.74}$$

because the large-N factorization means that any graph that connects the two single-trace operators is down by $1/N^2$. Thus if we add this to the action there will be some RG flow. To see where it goes, write[47]

$$\begin{aligned} Z_j(g) &= \left\langle e^{\int d^Dx\,\{-g\mathcal{O}^2(x)+j(x)\mathcal{O}(x)\}}\right\rangle_{\beta=0} \\ &= \int \mathcal{D}\sigma \left\langle e^{\int d^Dx\,\{-\sigma^2(x)/4g+(j(x)+i\sigma(x))\mathcal{O}(x)\}}\right\rangle_{\beta=0} \\ &= \int \mathcal{D}\sigma \left\langle e^{-\int d^Dx\,\sigma^2(x)/4g}\right\rangle_{\beta=j+i\sigma} \\ &= \int \mathcal{D}\sigma\, e^{-\int d^Dx\,\sigma^2(x)/4g}\,\langle 1\rangle_{\beta=j+i\sigma} \end{aligned} \tag{1.75}$$

In the first line we have written the generating functional for expectation values of $\mathcal{O}$ with the perturbation $g\mathcal{O}^2(x)$ in the Hamiltonian,[q] and in

[q]It was asked whether g must be positive. Naively, negative g leads to instability, but recently it has been shown[48] that nonlinear effects of backreaction can stabilize the system.

the second line we have introduced an auxiliary field to relate this to the theory with just the single-trace perturbation. Now let use integrate out the auxiliary field. In the planar approximation we can just find the saddle point,[r]

$$\begin{aligned}\sigma(x) &= -2g\partial_\sigma Z_{\beta=j+i\sigma} \\ &= 2gi\,\langle\mathcal{O}\rangle_{\beta=j+i\sigma} \\ &= 2giC_{\mathcal{O}}\alpha_{\beta=j+i\sigma}\,.\end{aligned} \tag{1.76}$$

In the second line we have used the fact that differentiating with respect to j is the same as with respect to σ, and pulls down a $\mathcal{O}$. In the last line we have used the basic dictionary (1.41). Eliminating σ

$$\beta(x) = j(x) - 2gC_{\mathcal{O}}\alpha(x)\,. \tag{1.77}$$

In other words, the earlier pure boundary conditions are now replaced by a mixed condition with a linear relation between α and β.[46,47] In terms of the field we have

$$\phi(z,x) = z^{D-\Delta}\{j(x) - 2gC_{\mathcal{O}}\alpha(x)\} + z^{\Delta}\alpha(x)\,. \tag{1.78}$$

At high energy $z \to 0$ the $z^{D-\Delta}$ term vanishes (we are in the regime $\Delta < D/2$) and we have the alternate quantization. At low energy the $z^{D-\Delta}$ term dominates and we have the usual quantization. Thus we flow from one to the other.

By the way, in the middle two lines of Eq. (1.75) we see an interesting construction: a field σ coupled to the boundary of the supergravity. In this case the field is auxiliary, but we could consider more general actions for σ. This corresponds to coupling some strongly coupled field theory, with a gravity dual, to another field theory (generally weakly coupled) with an explicit Lagrangian description.

1.4.4. *Orbifolds*

Tom Banks discussed some features of these, let me add a little. If we identify the $\mathbb{R}^6$ space transverse to the D3-branes under some discrete group $\Gamma \subset SO(6)$ then the origin becomes a singularity. If we then take the near horizon limit, then we get a new duality.[49] On the supergravity side S^5 is replaced by S^5/Γ. On the gauge side, the orbifolding retains only the Γ-invariant fields from the original gauge theory, leaving a quiver gauge

[r] Loop corrections for auxiliary fields are generally uninteresting, being dominated by the cutoff scale for dimensional reasons, and so absorbable into redefinitions.

theory,[50] e.g. for $\Gamma = \mathbb{Z}_k$ we get $SU(N)^k$, with bifundamental matter that depends on how Γ acts on the different planes in $\mathbb{R}^6$. The surviving supersymmetry can be $\mathcal{N} = 2, 1$, or 0.

In the nonsupersymmetric cases, Tom mentioned that at weak coupling conformal invariance is destroyed by the flow of double-trace interactions. The pathology takes a different form at strong coupling. If a nonsupersymmetric element of Γ has a fixed point there will be a localized tachyon, with the result that a hole develops and consumes the space.[51] If there is no fixed point the same process occurs via tunneling.[52] These are examples of the general phenomenon that nonsupersymmetric states in string theory are almost always unstable on long enough time scales.

There is an interesting generalization where $SU(N)^k$ is replaced by $SU(N_1) \times \ldots \times SU(N_k)$. On the supergravity side this corresponds to wrapping higher dimensional branes (D5 in particular) on cycles hidden in the singularity. The different gauge factors now have different nonzero β-functions and the couplings flow. Figuring out where the flow ends up has been a rich subject.[43]

1.4.5. *Non-spherical horizons. II*

If we combine orbifolding and RG flow we get something new. Some supersymmetric orbifolds have relevant operators in the twisted sector, corresponding to smoothing the fixed point. By turning these on one gets to new manifolds, such at the famous conifold.[53] Here the S^5 is replaced by a manifold $T^{1,1}$. In fact one gets a solution to the supergravity field equations with just metric and five-form if S^5 is replaced by any Einstein space ($R_{mn} = cg_{mn}$ with c a positive constant). Supersymmetry requires a more restrictive Einstein-Sasaki condition.[54] The dual gauge theory can be determined in many cases by a combination of orbifolding, RG flow, moving to the Coulomb branch, and so on. The toric case, where X has a $U(1)^3$ symmetry, is understood best .[55] There has been recent progress on the more general case.[56]

1.4.6. *Nonconformal branes*[57]

Suppose we try to repeat the near-horizon limit for D2-branes. Here the gauge theory is superrenormalizable: at high energy it is weak, and the dimensionless coupling $g^2_{\rm YM}N/E$ becomes small. In the IIA supergravity theory, the curvature is small in some range, but becomes large in string units near the boundary. Thus we have the same complementarity as for D3-branes, but at different scales within a given theory rather than by varying the coupling. The superrenormalizable gauge theory is the Lagrangian description, and the gravitational theory describes the low energy physics.

There is more to the story, though. The growth of the dimensionless gauge coupling is also reflected in the blowup of the dilaton at large z. In this regime one must lift to M theory, and the extreme low energy limit is the M2-brane geometry. Thus the extreme IR of $D = 3$, $\mathcal{N} = 8$ gauge theory provides one Lagrangian description of this geometry. I like this $D = 3$ theory as an 'existence proof' of AdS/CFT as a theory of quantum gravity, because we can regulate it on the lattice without exact supersymmetry, and the superrenormalizability makes it easy to see the restoration of the symmetry in the continuum limit, since only a finite number of perturbative graphs need be canceled.

If we periodically identify one of the Poincaré directions of the $D = 4$ theory, we would expect its low energy physics to be given by the $D = 3$ theory. Correspondingly on the gravity side, the periodic dimension pinches off near the horizon due to the L^2/z^2 in the metric, and a T-duality brings us to the D2-brane geometry. One can repeat this to get down to D0-branes, and I think that it is accurate to say that the BFSS Matrix Theory conjecture[58] is implied by AdS/CFT duality. It is less clear that the reverse is true, because the BFSS duality focuses on certain extreme low energy observables.

If we move up in dimension to D4-branes, the order of things is reversed. The gauge theory is classical at low energy, while at higher energies the effective description is the D4-brane metric, and at higher energies still the dilaton becomes large and we go over to the M5-brane geometry. We do not need the duality to solve the low energy physics because it is free, but it provides a UV completion. Higher dimensional branes become more intricate.

Question: How can the N^2 D2-brane entropy go over to the $N^{3/2}$ M2-brane entropy? Reference 57 finds that the black 2-brane entropy is

$$s_{\mathrm{black\,D2}} \sim N^2 T^2 (\lambda/T)^{-1/3} . \tag{1.79}$$

Unlike the D3-brane, this depends on the 't Hooft coupling, and as λ increases more degrees of freedom freeze out. Why the D2 should be different from the D3 in this regard is a puzzle! But because λ is dimensional, this implies an extra $T^{1/3}$ dependence. Exercise: by following through the calculations in Ref. 57, show that this suppression at low temperatures is just enough to make the D2 result match on to $N^{3/2}$ at the crossover temperature between the D2 and M2 pictures.

1.4.7. *D2-D6*

(The discussion in this and the following section is based in part on unpublished work with E. Silverstein.)

For a $D = 3$ gauge theory with matter, the one-loop running is

$$\mu \partial_\mu \gamma^2 = -\gamma^2 + b_0 \gamma^4 \,, \tag{1.80}$$

where $\gamma^2 = g_{\rm YM}^2/\mu$ is the dimensionless gauge coupling. The constant b_0 is proportional to the one-loop β-function from $D = 4$. If $b_0 > 0$, meaning the that $D = 4$ theory is free in the IR, then the one-loop equation (1.80) exhibits an IR fixed point at nonzero coupling $\gamma^2 = 1/b_0$. This is not reliable in general, but if we take an N-vector of matter fields the one-loop term dominates higher orders and we can believe the result. Thus we generate a large class of conformal theories in $D = 3$,[59] gauge theory analogs of the Wilson-Fisher fixed point.

We can build such theories with branes by taking D2-branes lying within D6-branes, i.e. they are extended in the directions

$$\begin{array}{lcccccccccc} & 0 & 1 & 2 & 3 & 4 & 5 & 6 & 7 & 8 & 9 \\ \text{D2:} & \times & \times & \times & & & & & & & \\ \text{D6:} & \times & \times & \times & \times & \times & \times & \times & & & \end{array} \tag{1.81}$$

This preserves half the supersymmetry of the D2's, namely $D = 3$, $\mathcal{N} = 4$. The 2-2 strings are the $SU(N_2)$ reduction of the $D = 4$ theory and give $b_0 = 0$ for this group. The 2-6 strings provide N_6 fundamentals and give a positive contribution to b_0. The 6-6 strings are free at low energy, because of the high dimension of the D6-brane, and do not of course contribute to the $SU(N_2)$ running. Thus, for $N_6 \gg N_2$ there is a weakly coupled fixed point. In general one cannot guarantee that this fixed point survives to strong coupling, but here there is enough supersymmetry to assure that it does. Moreover there is a gravity dual.[60] Normally it is difficult to describe such localized branes explicitly, but here the M-theory lift of the D6-brane allows to construct the solution as the $\mathbb{Z}_{N_6}$ orbifold of the M2-geometry of charge $N_2 N_6$.[61] For $N_6 \ll N_2 \ll N_6^5$ there is a weakly coupled IIA description, and for $N_6^5 \ll N_2$ the M theory picture is the valid one. These same scalings are found in a different orbifold of this geometry studied in Ref. 62.

1.4.8. *D2-D8*

Now consider D2-D8,

$$\begin{array}{lcccccccccc} & 0 & 1 & 2 & 3 & 4 & 5 & 6 & 7 & 8 & 9 \\ \text{D2:} & \times & \times & \times & & & & & & & \\ \text{D8:} & \times & \times & \times & \times & \times & \times & \times & \times & \times & \end{array} . \tag{1.82}$$

This case is similar to the preceding, except that there is no supersymmetry, and the 2-8 strings are all fermionic. This sounds a good starting point for a Fermi liquid: turn on a chemical potential for these strings and the geometry will tell us what happens when they are in interaction with a strong gauge coupling (in the D2-D6 case, there are also 2-6 bosons, and energetically it is likely that the charge will be carried predominantly be these).

An effective potential analysis along the lines of that in Sec. 2.1 indicates that there is the possibility of an AdS_4 solution before we turn on the chemical potential. The analysis is crude, though, because there is less symmetry (the D8's lie on a particular S^5 within the S^6 surrounding the D2's).

Further, there are several instabilities that may keep us from getting to the desired strong coupling phase. The system may be unstable along the Coulomb branch direction, the D2-branes separating (E. Silverstein has christened this seasickness because the Fermi sea make the AdS throat eject the branes). The D8's may be repelled from the D2's, which is chiral symmetry breaking, a long-standing issue in carrying these $D = 3$ gauge fixed points to strong coupling. Shape modes of the S^6 surrounding the D2's may become BF-forbidden tachyons; this is a common occurence in nonsupersymmetric AdS spaces.[63] These instabilities may be present even before the chemical potential is turned on, and the fermion density seems to make some worse (e.g. the seasickness along the Coulomb branch).

It seems likely that one can eventually, by adding additional ingredients, stabilize any of these and get to a Fermi liquid dual, at least one that is metastable (one additional knob is to add D6-branes in various orientations). Therefore let us ignore them as far as we can and turn on a chemical potential. Something nice then happens,[64] in that the system does exhibit at least one feature of a Fermi surface: there are particle-hole states of finite momentum and arbitrarily small energy, as one gets by taking a fermion from just below the surface and moving it to a point just above the surface somewhere else. The calculation is this: turning on a chemical potential introduces a radial electric field on the D6-brane. One measures the correlator of the fermion current (dual to a 6-6 string) by expanding the D-brane Dirac-Born-Infeld action, and because of the special form of this action (which in particular has a maximum electric field) the correlator has the distinctive behavior of an imaginary part that goes down to zero frequency at finite momentum.

There are still several puzzles, however: 1) this behavior is also seen in the D2-D6 system, where a Fermi liquid might not have been expected; 2) in Fermi liquids there is a maximum momentum of $2k_{\rm F}$, the maximum distance an electron can be moved, while here there is no maximum — perhaps this is a strong-coupling effect; 3) the large field near the origin backreacts on the metric in a divergent way, and the resolution of this singularity is not understood in general, it could gap the surface; 4) even when the backreaction is parametrically suppressed by taking N_2 large, the DBI action breaks down near the horizon due to the field gradients.

Question: What about T-dual configurations, like D3-D7 in the non-supersymmetric orientation

$$\begin{array}{lcccccccccc} & 0 & 1 & 2 & 3 & 4 & 5 & 6 & 7 & 8 & 9 \\ \text{D3:} & \times & \times & \times & \times & & & & & & ? \\ \text{D7:} & \times & \times & \times & & \times & \times & \times & \times & \times & \end{array} \tag{1.83}$$

Good question: the intersection is still $D = 3$, and the fermions live there, but the gauge fields now move in $D = 4$. Things are less IR singular but still interesting: the perturbative analysis seems more tractable. So maybe this is another interesting case. For D4-D6,

$$\begin{array}{lcccccccccc} & 0 & 1 & 2 & 3 & 4 & 5 & 6 & 7 & 8 & 9 \\ \text{D4:} & \times & \times & \times & \times & \times & & & & & , \\ \text{D8:} & \times & \times & \times & & & \times & \times & \times & \times & \end{array} \tag{1.84}$$

the gauge fields move in $D = 5$ and are classical at low energy, they do not give interesting dynamics.

1.4.9. *Fractional D3-D7*

Finally I mention a case that addresses the generality of AdS/CFT. If we replace the $\mathcal{N} = 4$ gauge theory with $\mathcal{N} = 2$, and add $N_f = 2N_c$ fundamentals, we get another theory which is conformal for all values of the coupling. We can get this from N_c fractional D3-branes at an orbifold point plus N_f D7-branes. However, the conformal limit is singular and suggests a T-duality to a IIA configuration.[65] Even then, an effective potential analysis suggests that the scales of both the compact space and the AdS space are stringy, meaning that the anomalous dimensions remain of order one even at large coupling, they do not grow as a power of the coupling. Thus, it appears that the necessary condition of large operator dimensions is rather special. Further this seems to go against the principle that whenever a coupling becomes very large we can find a small expansion parameter in a different picture. Reference 66 discusses this system in much more detail, and proposes a more intricate dual picture.

Acknowledgments

I would like to thank Allan Adams, Alex Buchel, Tom Faulkner, Sean Hartnoll, Idse Heemskerk, Gary Horowitz, Jacopo Orgera, Amanda Peet, Joao Penedones, Eva Silverstein, Matt Strassler and James Sully for collaborations on various aspects of this subject, and Tom Banks, Gary Horowitz, Hong Liu, Don Marolf, Dam Son, and Sho Yaida for useful discussions. I would also like to thank the students at TASI for their incessant questions and all-around enthusiasm, which made this lecture series a rewarding experience. This work was supported in part by NSF grants PHY05-51164 and PHY07-57035.

References

1. O. Aharony, S. S. Gubser, J. M. Maldacena, H. Ooguri and Y. Oz, "Large N field theories, string theory and gravity," Phys. Rept. **323**, 183 (2000) [arXiv:hep-th/9905111].
2. E. D'Hoker and D. Z. Freedman, "Supersymmetric gauge theories and the AdS/CFT correspondence," arXiv:hep-th/0201253.
3. John McGreevy, 8.821, *String Theory,* Fall 2008, http://ocw.mit.edu/courses/physics/8-821-string-theory-fall-2008/.
4. S. Weinberg and E. Witten, "Limits On Massless Particles," Phys. Lett. B **96**, 59 (1980).
5. S. R. Coleman and J. Mandula, "All Possible Symmetries of the S Matrix," Phys. Rev. **159**, 1251 (1967).
6. G. 't Hooft, "Dimensional reduction in quantum gravity," arXiv:gr-qc/9310026.
7. L. Susskind, "The World As A Hologram," J. Math. Phys. **36**, 6377 (1995) [arXiv:hep-th/9409089].
8. G. Bertsch, S. J. Brodsky, A. S. Goldhaber and J. F. Gunion, "Diffractive Excitation In QCD," Phys. Rev. Lett. **47**, 297 (1981);
   A. H. Mueller, "Color Transparency and Nuclear Shadowing," CU-TP-412.

9. L. N. Lipatov, "The Bare Pomeron In Quantum Chromodynamics," Sov. Phys. JETP **63**, 904 (1986) [Zh. Eksp. Teor. Fiz. **90**, 1536 (1986)];
R. C. Brower, J. Polchinski, M. J. Strassler and C. I. Tan, "The Pomeron and Gauge/String Duality," JHEP **0712**, 005 (2007) [arXiv:hep-th/0603115].

10. I. Heemskerk, J. Penedones, J. Polchinski and J. Sully, "Holography from Conformal Field Theory," JHEP **0910**, 079 (2009) [arXiv:0907.0151 [hep-th]].

11. E. Witten, "The 1 / N Expansion In Atomic And Particle Physics," S. R. Coleman, "1/N,"

12. G. 't Hooft, "A planar diagram theory for strong interactions," Nucl. Phys. B **72**, 461 (1974).

13. J. M. Maldacena, "The large N limit of superconformal field theories and supergravity," Adv. Theor. Math. Phys. **2**, 231 (1998) [Int. J. Theor. Phys. **38**, 1113 (1999)] [arXiv:hep-th/9711200].

14. A. M. Polyakov, "String theory and quark confinement," Nucl. Phys. Proc. Suppl. **68**, 1 (1998) [arXiv:hep-th/9711002].

15. G. T. Horowitz and A. Strominger, "Black strings and P-branes," Nucl. Phys. B **360**, 197 (1991).

16. J. Polchinski, "Dirichlet-Branes and Ramond-Ramond Charges," Phys. Rev. Lett. **75**, 4724 (1995) [arXiv:hep-th/9510017].

17. A. Strominger and C. Vafa, "Microscopic Origin of the Bekenstein-Hawking Entropy," Phys. Lett. B **379**, 99 (1996) [arXiv:hep-th/9601029].

18. E. Silverstein, "TASI / PiTP / ISS lectures on moduli and microphysics," arXiv:hep-th/0405068.

19. J. Polchinski and E. Silverstein, "Dual Purpose Landscaping Tools: Small Extra Dimensions in AdS/CFT," arXiv:0908.0756 [hep-th].

20. S. Kachru, R. Kallosh, A. D. Linde *et al.*, "De Sitter vacua in string theory," Phys. Rev. **D68**, 046005 (2003). [hep-th/0301240].

21. L. Susskind, E. Witten, "The Holographic bound in anti-de Sitter space," [hep-th/9805114];
A. W. Peet, J. Polchinski, "UV / IR relations in AdS dynamics," Phys. Rev. **D59**, 065011 (1999). [hep-th/9809022].

22. P. Breitenlohner and D. Z. Freedman, "Positive Energy In Anti-De Sitter Backgrounds And Gauged Extended Supergravity," Phys. Lett. B **115**, 197 (1982);
P. Breitenlohner and D. Z. Freedman, "Stability In Gauged Extended Supergravity," Annals Phys. **144**, 249 (1982).

23. I. R. Klebanov and E. Witten, "AdS/CFT correspondence and symmetry breaking," Nucl. Phys. B **556**, 89 (1999) [arXiv:hep-th/9905104].

24. S. S. Gubser, I. R. Klebanov and A. M. Polyakov, "Gauge theory correlators

from non-critical string theory," Phys. Lett. B **428**, 105 (1998) [arXiv:hep-th/9802109].

25. E. Witten, "Anti-de Sitter space and holography," Adv. Theor. Math. Phys. **2**, 253 (1998) [arXiv:hep-th/9802150].
26. N. Gromov, V. Kazakov and P. Vieira, "Exact AdS/CFT spectrum: Konishi dimension at any coupling," arXiv:0906.4240 [hep-th].
27. R. Penrose, *The Road to Reality,* Knopf, 2004.
28. J. McGreevy, L. Susskind and N. Toumbas, "Invasion of the giant gravitons from anti-de Sitter space," JHEP **0006**, 008 (2000) [arXiv:hep-th/0003075].
29. S. S. Gubser, I. R. Klebanov and A. W. Peet, "Entropy and Temperature of Black 3-Branes," Phys. Rev. D **54**, 3915 (1996) [arXiv:hep-th/9602135].
30. K. Skenderis and B. C. van Rees, "Real-time gauge/gravity duality: Prescription, Renormalization and Examples," JHEP **0905**, 085 (2009) [arXiv:0812.2909 [hep-th]].
31. I. Heemskerk, J. Polchinski, "Holographic and Wilsonian Renormalization Groups," [arXiv:1010.1264 [hep-th]].
32. G. W. Gibbons and S. W. Hawking, "Action Integrals And Partition Functions In Quantum Gravity," Phys. Rev. D **15**, 2752 (1977).
33. S. S. Gubser and I. R. Klebanov, "Absorption by branes and Schwinger terms in the world volume theory," Phys. Lett. B **413**, 41 (1997) [arXiv:hep-th/9708005].
34. K. Skenderis and M. Taylor, "Holographic Coulomb branch vevs," JHEP **0608**, 001 (2006) [arXiv:hep-th/0604169].
35. R. G. Leigh and M. J. Strassler, "Exactly Marginal Operators And Duality In Four-Dimensional N=1 Supersymmetric Gauge Theory," Nucl. Phys. B **447**, 95 (1995) [arXiv:hep-th/9503121].
36. M. R. Douglas and S. H. Shenker, "Dynamics of SU(N) supersymmetric gauge theory," Nucl. Phys. B **447**, 271 (1995) [arXiv:hep-th/9503163].
37. J. Polchinski and M. J. Strassler, "The string dual of a confining four-dimensional gauge theory," arXiv:hep-th/0003136.
38. A. Khavaev, K. Pilch and N. P. Warner, "New vacua of gauged N = 8 supergravity in five dimensions," Phys. Lett. B **487**, 14 (2000) [arXiv:hep-th/9812035];
A. Karch, D. Lust and A. Miemiec, "New N = 1 superconformal field theories and their supergravity description," Phys. Lett. B **454**, 265 (1999) [arXiv:hep-th/9901041];
D. Z. Freedman, S. S. Gubser, K. Pilch and N. P. Warner, "Renormalization group flows from holography supersymmetry and a c-theorem," Adv. Theor. Math. Phys. **3**, 363 (1999) [arXiv:hep-th/9904017].
39. K. Pilch and N. P. Warner, "N = 2 supersymmetric RG flows and the IIB dilaton," Nucl. Phys. B **594**, 209 (2001) [arXiv:hep-th/0004063].

40. A. Buchel, A. W. Peet and J. Polchinski, "Gauge dual and noncommutative extension of an N = 2 supergravity solution," Phys. Rev. D **63**, 044009 (2001) [arXiv:hep-th/0008076].

41. L. Girardello, M. Petrini, M. Porrati and A. Zaffaroni, "The supergravity dual of N = 1 super Yang-Mills theory," Nucl. Phys. B **569**, 451 (2000) [arXiv:hep-th/9909047].

42. R. C. Myers, "Dielectric-branes," JHEP **9912**, 022 (1999) [arXiv:hep-th/9910053].

43. I. R. Klebanov and M. J. Strassler, "Supergravity and a confining gauge theory: Duality cascades and chiSB-resolution of naked singularities," JHEP **0008**, 052 (2000) [arXiv:hep-th/0007191].

44. J. M. Maldacena and C. Nunez, "Towards the large N limit of pure N = 1 super Yang Mills," Phys. Rev. Lett. **86**, 588 (2001) [arXiv:hep-th/0008001].

45. S. J. Rey and J. T. Yee, "Macroscopic strings as heavy quarks in large N gauge theory and anti-de Sitter supergravity," Eur. Phys. J. C **22**, 379 (2001) [arXiv:hep-th/9803001];
"Wilson loops in large N field theories," Phys. Rev. Lett. **80**, 4859 (1998) [arXiv:hep-th/9803002].

46. E. Witten, "Multi-trace operators, boundary conditions, and AdS/CFT correspondence," arXiv:hep-th/0112258.

47. M. Berkooz, A. Sever and A. Shomer, "Double-trace deformations, boundary conditions and spacetime singularities," JHEP **0205**, 034 (2002) [arXiv:hep-th/0112264];
S. S. Gubser, I. R. Klebanov, "A Universal result on central charges in the presence of double trace deformations," Nucl. Phys. **B656**, 23-36 (2003). [hep-th/0212138].

48. T. Faulkner, G. T. Horowitz and M. M. Roberts, "Holographic quantum criticality from multi-trace deformations," arXiv:1008.1581 [hep-th].

49. S. Kachru and E. Silverstein, "4d conformal theories and strings on orbifolds," Phys. Rev. Lett. **80**, 4855 (1998) [arXiv:hep-th/9802183].

50. M. R. Douglas and G. W. Moore, "D-branes, Quivers, and ALE Instantons," arXiv:hep-th/9603167.

51. A. Adams, J. Polchinski and E. Silverstein, "Don't panic! Closed string tachyons in ALE space-times," JHEP **0110**, 029 (2001) [arXiv:hep-th/0108075].

52. G. T. Horowitz, J. Orgera and J. Polchinski, "Nonperturbative Instability of AdS_5xS^5/Z_k," Phys. Rev. D **77**, 024004 (2008) [arXiv:0709.4262 [hep-th]].

53. I. R. Klebanov and E. Witten, "Superconformal field theory on threebranes at a Calabi-Yau singularity," Nucl. Phys. B **536**, 199 (1998) [arXiv:hep-th/9807080].

54. A. Kehagias, "New type IIB vacua and their F-theory interpretation," Phys. Lett. B **435**, 337 (1998) [arXiv:hep-th/9805131];

D. R. Morrison and M. R. Plesser, "Non-spherical horizons. I," Adv. Theor. Math. Phys. **3**, 1 (1999) [arXiv:hep-th/9810201].

55. S. Franco, A. Hanany, D. Martelli, J. Sparks, D. Vegh and B. Wecht, "Gauge theories from toric geometry and brane tilings," JHEP **0601**, 128 (2006) [arXiv:hep-th/0505211].
56. P. S. Aspinwall and D. R. Morrison, "Quivers from Matrix Factorizations," arXiv:1005.1042 [hep-th].
57. N. Itzhaki, J. M. Maldacena, J. Sonnenschein and S. Yankielowicz, "Supergravity and the large N limit of theories with sixteen Phys. Rev. D **58**, 046004 (1998) [arXiv:hep-th/9802042].
58. T. Banks, W. Fischler, S. H. Shenker and L. Susskind, "M theory as a matrix model: A conjecture," Phys. Rev. D **55**, 5112 (1997) [arXiv:hep-th/9610043].
59. T. Appelquist and D. Nash, "Critical Behavior in (2+1)-dimensional QCD," Phys. Rev. Lett. **64**, 721 (1990).
60. O. Pelc and R. Siebelink, "The D2-D6 system and a fibered AdS geometry," Nucl. Phys. B **558**, 127 (1999) [arXiv:hep-th/9902045].
61. N. Itzhaki, A. A. Tseytlin and S. Yankielowicz, "Supergravity solutions for branes localized within branes," Phys. Lett. B **432**, 298 (1998) [arXiv:hep-th/9803103].
62. O. Aharony, O. Bergman, D. L. Jafferis and J. Maldacena, "N=6 superconformal Chern-Simons-matter theories, M2-branes and their gravity duals," JHEP **0810**, 091 (2008) [arXiv:0806.1218 [hep-th]].
63. O. DeWolfe, D. Z. Freedman, S. S. Gubser, G. T. Horowitz and I. Mitra, "Stability of AdS(p) x M(q) compactifications without supersymmetry," Phys. Rev. D **65**, 064033 (2002) [arXiv:hep-th/0105047].
64. M. Kulaxizi and A. Parnachev, "Holographic Responses of Fermion Matter," Nucl. Phys. B **815**, 125 (2009) [arXiv:0811.2262 [hep-th]].
65. M. Grana and J. Polchinski, "Gauge / gravity duals with holomorphic dilaton," Phys. Rev. D **65**, 126005 (2002) [arXiv:hep-th/0106014].
66. A. Gadde, E. Pomoni and L. Rastelli, "The Veneziano Limit of N=2 Superconformal QCD: Towards the String Dual of N=2 $SU(N_c)$ SYM with $N_f = 2N_c$," arXiv:0912.4918 [hep-th].

Chapter 2

TASI Lectures on Holographic Space-Time, SUSY, and Gravitational Effective Field Theory

Tom Banks

Santa Cruz Institute for Particle Physics and Department of Physics, University of California, Santa Cruz, CA 95064, USA
banks@scipp.ucsc.edu

NHETC and Department of Physics and Astronomy, Rutgers University, Piscataway, NJ 08854-8019, USA

I argue that the conventional field theoretic notion of vacuum state is not valid in quantum gravity. The arguments use gravitational effective field theory, as well as results from string theory, particularly the AdS/CFT correspondence. Different solutions of the same low energy gravitational field equations correspond to different quantum systems, rather than different states in the same system. I then introduce *holographic space-time* a quasi-local quantum mechanical construction based on the holographic principle. I argue that models of quantum gravity in asymptotically flat space-time will be exactly super-Poincare invariant, because the natural variables of holographic space-time for such a system, are the degrees of freedom of massless superparticles. The formalism leads to a non-singular quantum Big Bang cosmology, in which the asymptotic future is required to be a de Sitter space, with cosmological constant (c.c.) determined by cosmological initial conditions. It is also approximately SUSic in the future, with the gravitino mass $K\Lambda^{1/4}$.

2.1. Vacuum States in Non-gravitational Quantum Field Theory

QFTs in fixed space-time backgrounds, like Minkowski space, often exhibit the phenomena of degenerate and/or meta-stable vacuum states. In the semi-classical approximation these are solutions of the field equations that preserve all the isometries of the background, and for which there are no exponentially growing small fluctuations. Typically, this requires the model to contain fundamental scalar fields. The potential energy density is a function of these scalars, and multiple solutions occur when this function has multiple minima.

In the semi-classical approximation, this is evidence for multiple *superselection sectors* of the QFT: the Hilbert space breaks up into a direct sum of spaces, each associated with a different minimum. In the infinite volume limit, transitions between sectors vanish because the Hamiltonian is an integral of a local energy density. Actually, this is only true in perturbation theory around the true minima. When non-perturbative physics is taken into account, there are generally bubble nucleation processes, which signal an instability of all but the lowest energy minima. Superselection sectors only exist for minima which are exactly degenerate, including all quantum corrections to the energy (the energy differences between semi-classical vacua do not suffer from renormalization ambiguities).

A more non-perturbative view of these phenomena is afforded by the Wilsonian definition of quantum field theory. A general QFT is defined by a relevant perturbation of a CFT. CFT's in turn are defined by their spectrum of conformal primary operators and their operator product expansions (OPEs). In particular, this includes a list of all the relevant operators, which might be added as perturbations of the CFT, using the GellMann-Low formula to compute the perturbed Green's functions. The OPE allows us to perform these computations. Although there is no general proof, it is believed that these conformal perturbation expansions are convergent in finite volume.

The CFT has a unique conformally invariant vacuum state, which is the lowest energy state if the theory is unitary. However, in the infinite volume limit the Hilbert space of the perturbed theory might again separate into superselection sectors. It might also/instead have meta-stable states, but meta-stability always depends on the existence of a small dimensionless parameter, the life-time of the meta-stable state in units of the typical time scale in the model. In most explicit examples, this parameter is a semi-classical expansion parameter for at least some of the fields in the theory.

The following general properties of degenerate and meta-stable vacua in QFT, follow from these principles:

- The short distance behavior of Green's functions, and the high temperature behavior of the partition function of the theory are independent of the superselection sector. Both are controlled by the CFT. The partition function in finite volume V has the asymptotic form

$$Z = e^{-cV^{\frac{2d-1}{d}} E^{\frac{d-1}{d}}},$$

where d is the space-time dimension and E the total energy. This follows from scale invariance and extensivity of the energy. Extensivity follows from locality. The constant c, roughly speaking, measures the number of independent fields in the theory, at the UV fixed point.

- Tunneling from a meta-stable state produces a bubble, which grows asymptotically at the speed of light, engulfing any time-like observer[a] propagating in the false vacuum. Inside the bubble, the state rapidly approaches the true vacuum. If one excites a local region of the false vacuum to sufficiently high energy, the tunneling rate goes to infinity and meta-stability is lost. This is because the energy density cost to produce a stable expanding bubble of true vacuum is finite.
- If there are two exactly degenerate quantum vacua, separated by a barrier in field space, then, with finite cost in energy, one can produce an arbitrarily large region of vacuum 1, in the Hilbert space of the model which consists of local operators acting on vacuum 2. If the region is very large, it is meta-stable and survives at least as long as the time it takes light to cross that region.

2.2. Are there Vacuum States in Models of Quantum Gravity?

One of the main contentions of this lecture series is that the answer to the above question is NO. In fact, in the end, we will contend that each possible large distance asymptotic behavior of space-time corresponds to a different Hamiltonian, with different sets of underlying degrees of freedom. This is true even if we are talking about two different solutions of the *same* set of low energy gravitational field equations. In the case of Anti-de Sitter

[a]We will often use the word observer in these lectures. We use it to mean a large quantum system with many semi-classical observables. Quantum field theories give us models for a host of such systems, whenever the volume is large in cutoff units. They are collective coordinates of large composites and have quantum fluctuations that fall off like a power of the volume. Quantum phase interference between different states of the collective coordinate falls off like the exponential of the volume, except for motions of the collective coordinates that excite only a small number of low lying states of the system. With this definition of the word, an observer has neither gender nor consciousness.

asymptotics we will see that the models are literally as different from each other as two different QFTs, defined by different fixed points. The most conclusive evidence for this point of view comes from the Matrix Theory [1] and AdS/CFT [2] formulations of non-perturbative string theory, and ITAHO[b] it is overwhelming. However, we can see the underlying reasons for these differences from simple semi-classical arguments, to which this section is devoted.

The essential point is that general relativity is not a quantum field theory, and that the reasons for this can already be seen in the classical dynamics of the system. Again, it is worthwhile making a formal list of the ways in which this is evident[c].

- The classical theory has no conserved stress energy tensor. The covariant conservation law for the "matter" stress energy is not a conservation law, but a statement of local gauge invariance. There is no local energy density associated with the gravitational field. In particular, this implies that there is no gauge invariant definition of an analog of the effective potential of non-gravitational QFT.
- Correspondingly, when we try to define an energy in GR, which could play the role of the Hamiltonian in the quantum theory, we find that we have to specify the behavior of the space-time geometry on an infinite *conformal boundary*. Geometries restricted to such time-like or null boundaries often have asymptotic isometry groups, and the Hamiltonian is defined to be the generator of such an asymptotic isometry, whose associated Killing vector is time-like or null near the boundary. This feature of GR is the first inkling of the *holographic principle*, of which much will be said below. It is already at this level that one begins to see that different solutions of the same low energy effective equations will correspond to different Hamiltonians and degrees of freedom in the quantum theory. I note in passing that asymptotic symmetry groups do not seem to be an absolute necessity in this context. For example, many of the Hamiltonians used in the AdS/CFT correspondence have perfectly well behaved time dependent deformations and one would suspect

[b]ITAHO - In this author's humble opinion.
[c]This list will use language compatible with the idea that the quantum theory of GR is somehow the quantization of the variables that appear in the classical Einstein equations. This idea lies behind all attempts to define quantum gravity outside the realm of string theory, from loop quantum gravity to dynamical triangulations. We will argue below that this idea is wrong.

that these correspond to space-time geometries with no time-like asymptotic isometries.

- More generally, the principle of general covariance shows us that no model of quantum gravity can have local gauge invariant observables. This fact was discovered in string theory, and considered an annoyance by some, long before it was shown to be a model of quantum gravity. All known versions of string theory incorporate this fact. The observables are always defined on an infinite conformal boundary. ITAHO, the fact that other attempts to formulate a quantum theory of gravity do not have this property, is evidence that they are incorrect. Note that this property is in direct contradiction with claims that a proper theory of gravity should be *background independent.* We will argue below that the holographic principle does allow for a more local, background independent formulation of models of quantum gravity, but that this formulation is inherently tied to particular gauge choices.
- More important than all of these formal properties is the nature of the space of solutions of gravitational field theories. It is well known that the mathematical theory of quantization begins by identifying a symplectic structure on the space of solutions, choosing a polarization of that symplectic structure, and identifying a family of Hilbert spaces and Hamiltonians whose quantum dynamics can be approximated by classical dynamics on that phase space. The general structure of ordinary QFT is that the space of solutions is parametrized, according to the Cauchy-Kovalevskaya theorem, in terms of fields and canonical momenta on a fixed space-like slice. The corresponding formulation of GR was worked out by Arnowitt, Deser and Misner (ADM), but it runs into a serious obstacle. Almost all solutions of GR are singular, and in order to define the phase space one must decide which singular solutions are acceptable. There are no global theorems defining this class, but there is a, somewhat imprecise, conjecture, called *Cosmic Censorship.* Here is what I think of as a precise formulation of this conjecture for particular cases:
  Start with a Lagrangian which has a Minkowski or AdS solution with a positive energy theorem. Consider a space-time with a boundary in the infinite past on which it approaches Minkowski or Anti-deSitter space, with a finite number of incoming wave packets corresponding to freely propagating waves of any of the linearized

fluctuations around the symmetric solution[d]*. The amplitudes of these incoming waves are restricted to be small enough so that the following conjecture is true*[e]*. The conjecture is that to each such asymptotic past boundary condition there corresponds a solution which obeys Cosmic Censorship: the future evolution is non-singular, except for a finite number of finite area black holes. The asymptotic future solution corresponds to a finite number of outgoing wave packets plus a finite number of finite area black holes.*

The last item focusses attention on the starring actor in the drama that will unfold in these lectures, the black hole. Our basic contention is that it is the answer to the age old question: How many angels can fit on the head of a pin? In modern language this is phrased: *How many bits (log_2 of the number of quantum states) can fit into a given space-time region?* This is the content of what I will call the Strong Holographic Principle, and we will eventually view it as a crucial part of the *definition* of space-time in terms of quantum concepts.

For the moment, we stick to semi-classical arguments, and revisit our itemized list of the properties of the QFT concept of multiple vacua, but now with a view towards understanding whether this concept makes sense in a theory of quantum gravity.

- As a consequence of general covariance, no quantum theory of gravity can have gauge invariant correlation functions which are localized at a point in space-time. The physical reason for this is the existence of black holes. Quantum mechanics tells us that localized measurements require us to concentrate a large amount of energy and momentum in a small region. General relativity tells us that when the Schwarzschild radius corresponding to the amount of mass (as measured by an observer at infinity) enclosed within a sphere of radius R, exceeds R, the space-time geometry is distorted and a black hole forms. Bekenstein and Hawking [5] made the remarkable observation that one can calculate the entropy of the resulting black hole state in terms of classical properties of the geometry. It is given by one quarter of the area of the horizon of the black hole, measured in Planck units. This is in manifest

[d]More properly, in the Minkowski case we should probably restrict ourselves to linearized waves that we expect to correspond to stable quantum states in the quantum theory.
[e]Recall that in the quantum theory, the classical field corresponding to a single particle has an amplitude which formally goes to zero in the classical limit.

contradiction with local quantum field theory, in which the entropy scales like the volume of the sphere. This is, in some sense, the *reason* that there are no local gauge invariant Green's functions. The region "inside the black hole" only has a space-time description for a very limited proper time, as measured by any observer in this region. We will see that a more fundamental description is in terms of a quantum system with a finite number of states, determined by interpreting the BH entropy as that of a micro-canonical density matrix. The internal Hamiltonian of this system is time dependent and sweeps out the entire Hilbert space of states an infinite number of times as the observer time coordinate approaches the singularity. From the point of view of an external observer this simply means the system thermalizes. The external description can be studied semi-classically and is the basis for Hawking's famous calculation of black hole radiation. Note by the way that Hawking radiation in asymptotically flat space-time removes the asymmetry in our description of the classical phase space. Black hole decay implies that once quantum mechanics is taken into account the final states in scattering amplitudes coincide with the initial states.

At any rate, none of the points in a local Green's function can have a definite meaning, because we cannot isolate something near that point without creating a black hole that envelopes the point. It is easy to see that the most localization we can achieve in a theory of quantum gravity is holographic in nature. That is, if we introduce infinitesimal localized sources on the conformal boundary of an infinite space-time, then straightforward perturbation theory shows that, as long as we aim the incoming beams to miss each other (impact parameter much larger than the Schwarzschild radius corresponding to the center of mass energy, for each subset of sources[f]), there is a non-singular solution of the classical field equations. When these criteria are not satisfied, one can prove that a trapped surface forms [7], and a famous theorem of Hawking and Penrose guarantees that the solution will become singular. The Cosmic Censorship conjecture implies that this singularity is a black hole, with a horizon area bounded from below by that of the trapped surface.

[f]Here we use language appropriate for asymptotically flat space-time. The corresponding scattering theory for asymptotically AdS spaces has been studied in [6].

In quantum field theory, the regime of scattering in which all kinematic invariants are large, is dominated by the UV fixed point. In this regime the differences between different vacuum states disappear. In quantum gravity by contrast, this is the regime in which black holes are formed. In asymptotically flat space, the specific heat of a black hole is negative, which means that at asymptotically high energies, the black hole temperature is very low. Thus, the spectrum of particles produced in black hole production and decay depends crucially on the infrared properties of the system. Different values of the *moduli*, the continuous parameters that characterize all known asymptotically flat string theory models, correspond to different low energy spectra. So in theories of quantum gravity, scattering at large kinematic invariants depends on what some would like to call the vacuum state. This is our first indication that these parameters correspond to different models, not different quantum states of the same system.

Black holes also falsify the claim that the high temperature behavior of the partition function is dominated by a conformal fixed point. In fact, all conformal field theories have positive specific heat and a well defined canonical ensemble. The negative specific heat of black holes in asymptotically flat space-time implies that their entropy grows too rapidly with the energy for the canonical partition function to exist. Although black holes are unstable, they decay by Hawking radiation, and the Hawking temperature goes to zero as the mass of the hole goes to infinity. Thus the high energy behavior of the micro-canonical partition function in asymptotically flat space would appear to be dominated by black holes, and cannot be that of a CFT.

It is interesting to carry out the corresponding black hole entropy calculation in the other two maximally symmetric space-times, with positive or negative values of the c.c. The modified Schwarzschild metric is

$$ds^2 = -(1 - V_N(r) \pm (\frac{r}{R})^2)dt^2 + \frac{dr^2}{(1 - V_N(r) \pm (\frac{r}{R})^2)} + r^2 d\Omega^2,$$

where $V_N(r)$ is the Newtonian potential in d space-time dimensions,

$$V_N(r) = c_d \frac{G_N M}{r^{d-3}},$$

R the radius of curvature of the de Sitter or AdS space, and the

+ sign is for the AdS case. In that case, the horizon radius is the unique zero of g_{tt}. When it is much larger than R it is given approximately by

$$R_S^{d-1} \sim c_d M R^2 L_P^{d-2},$$

where the Planck length is defined by $G_N = L_P^{d-2}$, in units where $\hbar = c = 1$. The area of the horizon is $A_d R_S^{d-2}$, so the BH entropy is

$$B_d (MR)^{\frac{d-2}{d-1}} (\frac{R}{L_P})^{\frac{d-2}{d-1}}.$$

$B_d = A_d c_d^{\frac{d-2}{d-1}}$. Remarkably, this looks like the entropy formula for a conformal field theory in $\mathbf{d-1}$ dimensions, living on a space with volume $\sim R^{d-2}$. In this interpretation, the quantity $(\frac{R}{L_P})^{\frac{d-2}{d-1}}$ plays the role of "the number of independent fields" in the CFT. This formula is one of the key elements of the AdS/CFT correspondence [2] [34] [4]. Note in particular *the dependence of the high energy density of states on the c.c.* In bulk QFT, which motivates the idea of different vacuum states, the c.c. is a low energy property of the theory and the high energy density of states does not depend on it. We will see that the manifold examples of the AdS/CFT correspondence make it abundantly clear that different solutions of the bulk field equations correspond to different quantum Hamiltonians; different models of quantum gravity rather than different states in a given model.

The dS case is even more striking. g_{tt} has two zeroes, the larger of which is the cosmological horizon, which persists even when the black hole mass goes to zero. The sum of the areas of those two horizons is always less than that of the cosmological horizon of "empty dS space", and in fact decreases as the black hole mass increases. There is a maximal mass (Nariai) black hole, whose two horizons have equal area. When combined with the result of Gibbons and Hawking [8], that the dS vacuum state is a thermal state for the local observer in a maximal causal diamond of dS space, this result leads to the conclusion [9] that a quantum theory of a stable dS space must have only a finite number of quantum states.

- The semi-classical theory of quantum tunneling in the presence of gravity begins with the seminal paper of Coleman and De Lucia

[10]. It confirms the picture of different solutions corresponding to different models, rather than different states, although almost all of the literature is couched in the language of vacuum decay. I will use the terms *true and false minima* rather than *true or false vacua* in order to emphasize that the conventional interpretation is wrong. The characteristics of gravitational tunneling depend crucially on the values of the energy density at the true and false minima. Let us begin with the case where the true minimum has negative c.c. One of the most important results in [10] is that in this case, the classical evolution after tunneling does not settle down to the AdS solution with the field sitting at the true minimum. Instead, the geometry undergoes a singular Big Crunch. There is no conserved energy, and as the universe inside the bubble contracts, the energy of the scalar field gets larger and larger. The field explores its entire potential and does not remain near the "true minimum". More importantly, the semi-classical approximation breaks down. Even in quantum field theory, particle production occurs and one might imagine that fluctuations in the energy density could lead to black hole formation. We will reserve to a later section a conjecture about what the real physics of the singularity is. For now we only note that the maximal causal diamond in this crunching geometry has only finite proper time between its past and future tips, as well as a maximal finite area for any space-like $d-2$ surface on its boundary. The main point here is that there is no sense in which this semi-classical approximation describes decay to a well understood ground state. Below, by using the holographic principle, we will find a sensible interpretation of some of these processes (but not as decays) and present arguments that others simply can't occur in well defined models of quantum gravity. This is in stark contrast to the situation in QFT, where of course the value of the potential at its minimum is unobservable. Notice that none of this has anything to do with the AdS solution, which one gets by fixing the scalar field at its true minimum. This solution may or may not represent a sensible model of quantum gravity, but it certainly has no connection to the hypothetical model in which the CDL instanton describes some kind of transition.

When the true minimum has positive c.c., the situation is much better. Classical evolution of the scalar field after tunneling, rapidly brings it to rest at the true minimum. Furthermore, the resulting space-time has an (observer dependent) cosmological horizon. Inside an observer's horizon volume, all fields rapidly approach the empty dS configuration. We will see below that in this case of dS to dS tunneling, more can be gleaned from the nature of the semi-classical CDL solution, and it is all consistent with the idea that the quantum theory of stable dS space has a finite number of of states.

The case of a true minimum with vanishing c.c., whether this is achieved at a finite point in field space, or at asymptotically infinite scalar field, is much more ambiguous. If the falloff of the potential is that found in all asymptotic regions of string theory moduli space[g] then the future causal boundary of the universe is similar to that of Minkowski space: the maximal causal diamond has infinite area holographic screen, and at finite points within that diamond, at late times, the space-time curvature goes to zero, and the scalar field asymptotes to the zero c.c. point. On the other hand, this is NOT an asymptotically flat space. Furthermore, if one takes the analogues of outgoing scattering states for this universe, then most do not extrapolate back to smooth perturbations of the instanton geometry. The meaning of this kind of situation is the central issue in trying to establish the existence of the String Landscape. We will explore these issues, which are far from settled, in section 4 below.

To summarize, CDL tunneling provides abundant evidence for the fact that AdS solutions of gravitational field equations are NOT part of the same model as other stationary points of the same effective action. *One never tunnels to AdS space.* It also suggests that there can be models of quantum gravity with a finite number of states, which describe stable dS space. We will complete that discussion in section 6. Similarly, there is no tunneling to asymptotically flat solutions of the field equations, which again must be regarded as (possibly) defining separate models of quantum gravity. We will argue below that there is no tunneling *from* AdS minima either, and that tunneling from asymptotically flat minima leads to a bizarre picture of the final state.

---

[g] As we will emphasize below, the notion of a potential on string theory moduli space is a problematic one. Nonetheless, if one accepts the validity of the concept one can use the symmetries of string theory to establish bounds on the behavior of the potential at infinity [12].

- Finally, we revisit the question of creating large meta-stable regions of space, which are in "another vacuum". If we start from an asympotically flat, or AdS minimum, and the potential is everywhere much less than the Planck scale and varies on a field space scale $\leq m_P$, then it is easy to find finite energy incoming configurations which move the field into another minimum over a sphere of radius R. However, if there is any potential barrier at all between the asymptotic minimum of the potential and the field value inside R, then the domain wall energy will scale like R^{d-2} and the Schwarzschild radius of the configuration will be $> R$. In other words, a black hole will form before the false vacuum bubble gets too big. Notice that if the false vacuum is a dS space, there will be an additional, volume contribution to the Schwarzschild radius. This guarantees that the black hole ALWAYS forms before the bubble can inflate[h] Thus, while auxiliary minima of a sub-Planckian effective potential do allow the creation of meta-stable states, they are not false vacua. The meta-stable regions that resemble homogeneous vacuum solutions are of limited size. Anything above that size is a black hole, which is to say, a thermodynamic equilbrium state indistinguishable from any other state of the theory that maximizes the entropy within the region $\leq R_S$. Notice also that there is no sense in which the decay of the meta-stable states created here is related to the instanton transitions discussed above. These are localized excitations of the true vacuum state, and will decay back to it by radiating particles off to infinity.

The conclusion is that rather simple classical considerations show that, whatever the theory of quantum gravity is, *it is not a QFT and the QFT concept of a vacuum state does NOT generalize to QG. Different solutions of the same low energy effective gravitational field equations, can*

[h] A lot of confusion is caused by solutions of the equations of GR which describe an arbitrarily large region of dS space or slow roll inflation, connected to an asymptotically flat or AdS region with a small black hole in it. These solutions cannot evolve from data that is non-singular in the past and in particular from incoming scattering data in a space-time with a well defined past conformal infinity. If they represent anything in a real quantum theory of gravity it is an artificially entangled state of two, generally different, quantum Hilbert spaces. In the present discussion we approach localized regions of false minimum by starting from small regions that do not inflate and boosting the incoming energy continuously. In that case the black hole mass is bounded from below by the integrated energy density of the false minimum and the black hole always forms before inflation can occur.

correspond to different models of QG, rather than different states of the same model.

2.3. Matrix Theory and the AdS/CFT Correspondence

Indeed, all of our non-perturbative constructions of quantum gravity have this property in spades. In this section I'll quickly review these constructions, starting with the case of asymptotically flat space.

2.3.1. *Matrix theory*

We have seen that in d dimensional asymptotically flat space, the entropy grows like $E^{\frac{d-2}{d-3}}$, so that conventional constructions of the partition function and the path integral fail. However, at least if $d > 4$, the light-front partition function at fixed longitudinal momentum

$$\mathrm{Tr} e^{-\beta P^-},$$

should be well defined, and we might hope to discover a more or less standard Lagrangian formulation of gravity. The Lagrangian for a single supersymmetric particle (superparticle) in 11 dimensions, is

$$\int dt\ \frac{p}{2}\dot{\mathbf{x}}^2 + i\theta\dot{\theta}.$$

Here $p \geq 0$ is the longitudinal momentum, which is treated as a fixed constant, and the time variable is light front time. $\mathbf{x}$ is a transverse 9-vector and θ_a a 16 component light front spinor. The system is quantized in terms of 9 commuting transverse momentum variables $\mathbf{p}$ and the 16 θ_a, with commutation relations

$$[\theta_a, \theta_b]_+ = \delta_{ab}.$$

The SUSY generators and Hamiltonian are

$$q_a = \theta_a,$$

$$Q_a = (\gamma \cdot \mathbf{p})_{ab}\theta_b,$$

$$P^- = \frac{(\mathbf{p})^2}{2p}.$$

The θ_a don't appear in the Hamiltonian, which describes a single massless relativistic particle. However, they give this massless state a degeneracy, with precisely the spin content of the 11 dimensional SUGRA multiplet.

Notice that this procedure only makes sense when p is strictly greater than zero. Particles with zero longitudinal momentum are non-dynamical. However, when the longitudinal momentum is continuous, the region of low longitudinal momentum becomes singular and one must exercise great care in treating it in order to extract correct results. In QFT this is often done by the method of Discrete Light Cone Quantization (DLCQ), in which the longitudinal direction is formally compactified so that p takes on only the discrete values $\frac{n}{R}$, with n a positive integer. One then studies the limit $R \to \infty$, by considering wave packets made from states including large values of n, so that they are localized in the longitudinal direction and can become independent of R. One convenience of this procedure, often exploited in QFT is that for fixed total longitudinal momentum $\frac{N}{R}$, a multi-particle state can have only a finite number of particles in it, so that in DLCQ field theory is approximated by the quantum mechanics of a finite number of particles.

The word approximated in the previous paragraph has to be stressed. The real system is obtained only in the limit when N is strictly infinite. Thinking about multi-particle states, we see in particular that, at fixed R, only those states with light cone energies $\sim \frac{1}{N}$ will survive in the limit $N \to \infty, R \to \infty$, with $p = \frac{N}{R}$ fixed. This introduces a degree of ambiguity into DLCQ, which can be exploited to simplify the limit. This ambiguity, well known in QFT, has mostly been ignored in the gravitational case, because Seiberg, following work of Sen and Susskind [13] found a particularly compelling form of DLCQ using string dualities.

Matrix Theory proceeds from this kinematical framework, by introducing an alternative to the Fock space treatment of identical particles. Instead, we generalize the variables $\mathbf{x}$ and θ_a to simultaneously diagonalizable $N \times N$ matrices. These can be written as

$$\mathbf{X} = \sum \mathbf{X}^I e_I, \qquad \Theta_a = \sum \theta_a^I e_I,$$

where $e_I^2 = e_I$ and $\mathrm{Tr} e_I = n_I$, $\sum n_I = N$. This representation is redundant if some of the n_I are the same, and we have a gauge symmetry permuting the e_I with equal trace, which is precisely the gauge symmetry of particle statistics. The fact that half integral spin is carried by the anti-commuting variables Θ_a guarantees that the spin-statistics connection is the conventional one. The Lagrangian is

$$L = \frac{1}{R} \mathrm{Tr}\, \left[\frac{1}{2} \dot{\mathbf{X}}^2 + i\Theta\dot{\Theta}\right],$$

and as we run over all possible choices of the e_I we reproduce the Lagrangians for $k \leq N$ supergravitons with all configurations allowed in DLCQ, and total momentum $\frac{N}{R}$.

If we insist on preserving all the SUSY, as well as the $SO(9)$ symmetry of this Lagrangian, there is a unique way of modifying it that allows for interaction between the supergravitons. To see what it is, we note that the Lagrangian we've written down is the dimensional reduction of $N = 4$ super Yang-Mills theory with gauge group $U(1)^N \ltimes S_N$, which is the low energy effective Lagrangian on the maximally Higgsed Coulomb branch moduli space of $U(N)$ SYM theory. The Lagrangian is written in temporal gauge (with the time of the gauge theory identified with light cone time) and the restriction to Bose or Fermi statistics for the particles is just the residuum of the Gauss Law of the non-abelian gauge theory. The full non-abelian Lagrangian is

$$L = \frac{1}{R}\mathrm{Tr}\,\left(\frac{1}{2}(D_t\mathbf{X})^2 - g^2[X^i, X^j]^2 + i\Theta D_t\Theta + g[\gamma\cdot\mathbf{X}, \Theta]\right),$$

where the adjoint covariant derivatives are $D_0 Y = \partial_t Y + g[A_t, Y]$. The constraints are now obtained by varying w.r.t. A_t and then setting $A_t = 0$. We'll describe how the SYM coupling is determined in terms of the Planck length below.

Before doing so, we note that this Lagrangian can also be shown to be the world volume Lagrangian of D0 branes in ten dimensional Type IIA string theory. The excitations on D-branes are open strings satisfying the appropriate mixed Dirichlet/Neumann boundary conditions. For N D-branes, the lowest excitations in open superstring theory have the quantum numbers of the maximally supersymmetric $U(N)$ Yang-Mills multiplet. If all of the spatial boundary conditions are Dirichlet, then the low energy world volume Lagrangian is unique and is given by the above formula. This idea led Seiberg, following Susskind and Sen, to argue that the compactified theory was just given by the D0 brane Lagrangian on the compact space. This conjecture is valid if we preserve at least 16 supercharges. It identifies the correct degrees of freedom, and their Lagrangian is completely determined by symmetries.

The D0 brane picture tells us how to identify the Yang-Mills coupling. Interactions are determined by the open string coupling, g_S, so $g_{YM}^2 = g_S$. On the other hand, Type IIA string theory is the compactification of M-theory (the quantum theory whose low energy limit is 11D SUGRA), on a circle whose radius is small in Planck units. D0 brane charge is

Kaluza-Klein momentum. So we have the identification

$$\frac{l}{l_S g_S} \propto \frac{1}{R},$$

where we've equated the string theory formula for the D0 brane mass to the KK formula. In the duality between M-theory and Type IIA string theory, the string is viewed as an M2 brane wrapped on the small circle, so

$$l_S^{-2} \propto l_P^{-3} R.$$

Combining the two formulae we find $g_S \propto (R/l_P)^{3/2}$. All of these formulae are actually exact consequences of SUSY, so the constants we have omitted can be calculated exactly.

Seiberg's prescription tells us that if we want to find the DLCQ of M-theory compactified on a torus or K3 manifold, we should study the world volume Lagrangian of D0 branes moving on that manifold. If the manifold has size of order the 11D Planck scale, then it is very small in string units, and we should do a T-duality transformation to find a description that is under greater control. For a torus of less than four dimensions, this gives us SYM theory compactified on the dual torus. These are all finite theories and the prescription is unambiguous. Many exact results, including some famous string dualities can be derived from this prescription, and agree with calculations or conjectures that one already had in supergravity or string theory. Other calculations, not protected by supersymmetry non-renormalization theorems are only supposed to be correct when takes the $N \to \infty$ limit, keeping only states whose light cone energy scales like $\frac{1}{N}$.

For a four torus or a K3 manifold, one naively gets the four dimensional SYM theory, which is not renormalizable. However, the T-dual string coupling is large, so we should really be studying the D4 branes (into which the D0 branes are converted by T-duality) in the strong coupling limit. In this limit, D4 branes become M5 branes. The world volume theory on N M5 branes is a maximally superconformally invariant 6 dimensional theory. It is compactified on T^5 or $K3 \times S^1$. Again, the prescription is finite and makes a number of correct exact predictions. It is however more difficult to calculate with since not much is known about the $(2, 0)$ superconformal field theory.

If we add one more circle to either of these constructions, we obtain *little string theory*. This is the world volume theory of N NS5 branes in the zero string coupling limit. Even less is known about this model than about the $(2, 0)$ superconformal field theory, and there have even been questions raised about whether it really exists. With six or more compact dimensions, the Seiberg construction fails and we do not have a working definition of the DLCQ of M-theory with 4 or 5 asymptotically flat dimensions.

Among the most striking features of these constructions is that each different gravitational background gives rise to a different quantum Hamiltonian. Even two versions of M-theory with values of continuous moduli that differ by a finite amount, correspond to the same field theory Hamiltonian on different compactification manifolds. And remember that the canonical variables of this Hamiltonian do not include a gravitational field. The geometry of the compactification manifold is not a dynamical variable in the Matrix theory Hamiltonian.

2.3.2. *The AdS/CFT correspondence*

The correct statement of the AdS/CFT correspondence is that in certain quantum field theories in $d-1$ space-time dimensions, there is a regime of large parameters, in which three important properties are satisfied:

- The high temperature behavior of the partition function on a spatial sphere of radius R is $c(RT)^{d-2}$, with $c \gg 1$.
- The dimension of most operators at the UV fixed point which defines the theory go to infinity.
- The Green's functions of those operators whose dimension remains finite can be computed approximately by solving the classical field equations of a $d+D$ dimensional gravitational Lagrangian, with boundary conditions first outlined by [34]. The space-time metric has a conformal boundary identical to that of $AdS_d \times K$, where K is a compact manifold. If the non-compact space-time is exactly AdS_d then the boundary field theory is conformal.

As a consequence of the last property, we consider such QFTs to be *definitions* of models of quantum gravity, with fixed asymptotic background. The idea that AdS/CFT defines a duality between two independently defined theories, is probably without merit. For a subclass of these theories, one of the large parameters is an integer N which controls the size of the the gauge group of the boundary field theory, and the model has a conventional large N expansion. In this case there is a weak coupling string theory description of the model, which goes beyond the classical gravity expansion described above. In these cases, the models have at least two adjustable parameters. One, N, controls the standard planar expansion of the theory, which can be recast as an expansion in world sheet topology.

The other, loosely called the *'t Hooft coupling*, is continuous (at least in the large N limit). When it is large, the solution of the theory in terms of classical gravitational equations is valid. When the 't Hooft coupling takes on moderate or small values there is a calculation of the correlation functions of all operators whose dimensions are finite in the large N limit, in terms of a world-sheet quantum field theory. In most of the interesting cases[i] the world sheet theory is hard to solve, but enormous progress has been made in establishing the conjecture.

However, even if we were able to calculate everything, including all higher genus contributions in the world sheet theory, this would not constitute an independent definition of the "other side" of the "AdS/CFT duality". String perturbation theory is a non-convergent asymptotic expansion. We know plenty of examples where its existence and finiteness to all orders is *not* a guarantee of the existence of a real quantum model of gravity. Bosonic matrix models related to $1+1$ dimensional string theories are a calculable example [15]. A continuous infinity of other examples is provided by moduli spaces of 4 dimensional compactifications of string theory with $N = 1$ SUSY. These have well defined perturbation expansions. However, general symmetry arguments, as well as many explicit instanton calculations show that there must be a non-perturbative superpotential on this moduli space, if this set of models makes sense at all. This means that all of the perturbation expansions except *perhaps* for a discrete set of points in moduli space do not correspond to well defined models. Furthermore, even if the wildest conjectures about the string theory Landscape are correct, most of these discrete points correspond to space-times with non-zero c.c. This means that the flat space S-matrix elements one calculates in string perturbation theory do not correspond, even qualitatively, to the correct set of observables of the hypothetical underlying model. We will return to this point when we discuss the string Landscape below. Our conclusion here is that the AdS/CFT correspondence is a *definition* of a class of models of QG, in terms of QFTs defined on the conformal boundary of AdS space.

It is important to emphasize that most QFTs fit into neither of these categories, even when they have a large N expansion. All large N models, and many other examples, such as the tensor product of any large collection of mutually non-interacting QFTs (or theories that are small perturbations of such a collection) satisfy the first of our criteria above. Referring back to

[i]The models of [14] are an exception.

the formula for black hole entropy in AdS space, we see that this criterion can be rephrased as: *AdS/CFT gives a rigorous justification of the BH entropy formula for asymptotically AdS space-times.* Comparison of the two formulae leads to the conclusion that the constant c is a measure of the ratio of the AdS curvature radius to the Planck length. Obviously, any classical space-time interpretation of the model will be valid only when this parameter is large, but this is only a necessary condition for the classical gravity approximation to be valid.

To understand better what is going on, let's recall the basic equations of the AdS/CFT correspondence. The Euclidean[j] AdS metric is

$$ds^2 = (1 + \frac{r^2}{R^2})d\tau^2 + \frac{dr^2}{1 + \frac{r^2}{R^2}} + r^2 d\Omega_{d-2}^2.$$

It follows that, at large r, solutions of the Klein-Gordon equation behave like $r^{\lambda_\pm} J(\tau, \Omega)$, with

$$\lambda(\lambda + d - 1) = m^2 R^2.$$

The $\pm$ signs refer to the two roots of these equations. The AdS/CFT prescription is to solve the coupled non-linear Einstein matter equations, with the boundary conditions that the fields behave like the larger root of this equation, and arbitrary source function J. Analogous boundary conditions are imposed on the metric and other higher spin fields. The action as a functional of the source is the generating functional for conformally covariant Green's functions on the boundary.

A consequence of this prescription is that every primary operator in the boundary CFT corresponds to a different field in the bulk. The mass of small fluctuations is related to the dimension of the primary. Thus, the bulk theory will have, generically, an infinite number of fields. The only known way to write an approximately local field theory with an infinite number of fields in AdS space, is to consider field theory with a finite number of fields on $AdS \times K$, where K is a compact manifold. The infinity then corresponds to a complete set of functions on K. The degeneracy of the Laplacian on K for high eigenvalues is power law in the eigenvalue, so this prescription could at most give us a power law growth of the number of fields of mass m, as $m \to \infty$.

[j]The Euclidean rotation familiar from QFT is not valid for QG in asymptotically flat space, because the density of states blows up too rapidly for the finite temperature partition function to be well defined. However, in AdS space the quantum theory is a boundary QFT and the Wick rotation makes sense.

It is well known that the number of primary operators of dimension D grows exponentially with a power of dimension, which implies an exponentially growing number of *fields*, in the approximate local field theory describing fluctuations around the hypothetical $AdS \times K$ background. Kaluza-Klein compactification on K gives rise only to a spectrum of masses that grows like a power of the mass (in $1/R$ units, where R is the radius of curvature of AdS, typically of the same order of magnitude as that of K). In examples where the CFT is dual to a weakly coupled string theory, such an exponential growth *is* seen among string states. So, for a generic CFT, one needs parametrically large entropy in order to claim that the geometrical radii are larger than the Planck length, but also another large parameter to guarantee that geometrical radii are larger than the length defined by the string tension.

It should be emphasized that very few CFTs actually correspond to weakly coupled string theories. The necessary and sufficient condition is that the theory have a conventional *matrix* $1/N$ expansion. This is what is necessary to have both a free string limit, and a topological structure of interactions that corresponds to a sum over world sheet topologies. Neither vector large N limits, nor the topological expansions typical of theories with comparable numbers of flavors and colors, or matter in other large representations of $SU(N)$[k], have a free string interpretation. Thus, for many CFTs, there seems to be no interpretation of their correlation functions as a set of observables corresponding to objects propagating in an AdS space[l].

In all rigorously established examples of the AdS/CFT correspondence the large parameter is an analog of the 't Hooft coupling of a large N gauge theory, a parameter which is continuous in the planar limit. In the two and three dimensional examples the 't Hooft coupling is really a ratio of two large integers, while in four dimensions it is the rescaled Yang Mills coupling. It is important that the theory is conformally invariant for *every* value of the 't Hooft coupling. In the limit when the coupling is large some

[k]The second rank symmetric and anti-symmetric tensor representations of $O(N)$, *do* appear in orientifold projections of large N gauge theories and have a string loop expansion.

[l]Even when the CFT has an entropy and dimension spectrum corresponding to an AdS radius that is large compared to both the "string length" and the Planck length, in the sense described above, it may not have a simple space time interpretation. A simple example is maximally supersymmetric $SU(N) \times SU(M)$ Yang Mills theory with both 't Hooft couplings large, or a perturbation of it by an exactly marginal operator constructed as a product of relevant operators from the individual theories.

dimensions remain of order 1, while others go to infinity. Furthermore, the multiplicity of operators with order 1 dimension grows only like a power of the dimension, consistent with a bulk space-time interpretation on a background of the form $AdS \times K$. All of the examples where this behavior has been established are exactly supersymmetric.

Non-supersymmetric marginal perturbations of these theories *all* lead to models with at most isolated fixed points at 't Hooft coupling of order one. One can also consider orbifolds of the $\mathcal{N} = 4$ SYM theory, whose planar diagrams coincide with the original theory, and are conformally invariant for all values of the 't Hooft coupling. However, the leading non-planar corrections to the beta functions of several couplings are non-zero and depend explicitly on the 't Hooft coupling. The theories will be conformal, if at all, only at 't Hooft coupling of order one. These theories provide interesting analogs of tachyon free non-supersymmetric string theories in flat space-time. Those asymptotically flat models seem completely sensible at string tree level, but the loop diagrams are divergent. If one tries to invoke the Fischler-Susskind mechanism to cancel these divergences, one finds perturbations of the space-time geometry which are singular in either the remote past or future or both. The string perturbation expansion breaks down. There is no evidence that these models really exist. The same is true for the non-supersymmetric orbifold theories. At leading order in the planar expansion, we have a free string theory on an AdS space-time. Finite string coupling corrections destroy this interpretation, except perhaps for a particular AdS radius of order the string scale. The question of whether the model at this particular radius makes sense is the question of whether the leading non-planar beta function has a finite coupling fixed point. In fact, that only guarantees that string perturbation theory in AdS space will make sense at that radius, and one must confront the resummation of the divergent $\frac{1}{N}$ expansion.

2.3.3. *Domain walls and holographic renormalization group flow*

When a flat space QFT has two isolated degenerate vacua, $\phi^i_{\pm}$ it also has domain wall solutions in which the scalar fields vary only in a single coordinate $\phi^i(z)$, and $\phi^i(\pm\infty) = \phi^i_{\pm}$. These solutions are stable and have a finite surface energy density, called the tension of the domain wall. They are limits of meta-stable finite energy states of the field theory with spherical domains of one vacuum inside the other. We have already argued that no such limit exists in theories with gravity. If the spherical domain wall becomes too big it collapses into a black hole.

There are however many examples of infinite hyper-planar domain wall solutions of Lagrangians with gravity, and the AdS/CFT correspondence gives us a novel interpretation of them. Consider a scalar field coupled to gravity with a potential having two stationary points, one a maximum and one a minimum, both with negative c.c. There are AdS solutions corresponding to each of these points, and it is possible for both of them to be stable. Indeed Breitlohner and Freedman showed that tachyonic scalar fields are allowed in AdS_d space, as long as the tachyonic mass satisfies

$$4m^2R^2 > (d-1)^2.$$

Referring to the dictionary relating bulk masses to boundary dimensions, we see that this is the condition for dimensions to be real and that B-F allowed tachyons are dual to *relevant* operators.

The equations for a domain wall solution connecting the two stationary points are

$$\phi''(z) + (d-1)\frac{\rho'}{\rho}\phi'(z) = \frac{dV}{d\phi}.$$

$$2\rho'^{\,2} = \epsilon^2\rho^2(\phi'^{\,2} - V).$$

We have rescaled fields and coordinates so that everything is dimensionless (see the discussion of instanton solutions below) and ϵ is a measure of the rate of variation of the potential V in Planck units. The metric is

$$ds^2 = dz^2 + \rho^2(z)d\mathbf{x}^2,$$

corresponding to a hyper-planar domain wall geometry. Our boundary conditions are that $\phi(\pm\infty)$ be the positions of the two stationary points. ρ then interpolates between the two different AdS geometries.

Near the AdS maximum of the potential, the two solutions of the linearized equation both fall off at infinity, so we only use up one boundary condition by insisting that the solution approach the maximum as $z \to \infty$. The solution then contains both possible power law behaviors and thus, *from the point of view of the AdS/CFT correspondence at the maximum, it corresponds to a perturbation of the Lagrangian of the boundary field theory by a relevant operator.* This is a novelty compared to traditional QFT. The domain wall is not an infinite energy state in the original model, but a perturbation of its Hamiltonian. It becomes clear that we should be trying to view the domain wall solution as the "anti-holographic" representation of a boundary renormalization group flow between two CFTs.

In general there is no such solution. The problem is that one of the linearized solutions of the fluctuation equations around the AdS minimum blows up at infinity. Thus we need two boundary conditions to ensure that the solution approaches the minimum as $z \to \infty$ *and* that its derivative goes to zero there. Having used up one parameter on the other side of the wall, we do not have this freedom. However, we can always find a solution by fine tuning one parameter in the potential, in order to set the coefficient of the growing mode to zero. Thus, the space of potentials with static domain wall solutions connecting two AdS stationary points is co-dimension one in the space of all potentials with two such stationary points[m]. Note that the fact that the second stationary point is a minimum is consistent with, and implied by, the RG interpretation. An RG flow should always approach its IR fixed point along an irrelevant direction in the space of perturbations of that fixed point. The AdS/CFT dictionary translates *irrelevant* as *positive mass squared.*

Having found such an RG flow we are almost ready to declare that we have a self consistent discovery of a new CFT with a large radius AdS dual. However, consistency requires that we check all directions in the bulk scalar field space, to determine if there are any tachyonic modes that violate the B-F bound. One way to guarantee both the existence of the domain wall solution and its B-F stability is to work in SUGRA, and insist that both stationary points preserve some SUSY. A host of solutions of this type have been found, that interpolate between fixed points with different numbers of supercharges in their super-conformal algebra.

Remarkably, when we perturb a supersymmetric CFT with a large radius dual by a relevant operator that violates all supersymmetry, we have yet to find a consistent solution. There are a number of smooth domain wall solutions of this type, but one always finds tachyons that violate the B-F bound in the spectrum of scalar fluctuations of the new minimum. There is, as yet, no theorem that this is *always* the case, but when combined with the failure to find non-supersymmetric large radius AdS spaces by orbifolding one is led to suspect a connection between SUSY and the low curvature of space-time. We will see below that the construction of holographic space-time seems to imply that all consistent theories of

[m]For future reference, we note that the parameter counting remains the same when we search for a domain wall connecting an asymptotically flat minimum to one with negative c.c., although the interpretation of the solution as an RG flow is no longer applicable.

gravity in asymptotically flat space are exactly supersymmetric. Many years of failure to find consistent perturbative string constructions, which violate SUSY in asymptotically flat space-time, have convinced most theorists that no such theories exist.

By contrast, the theory of the String Landscape, to which we will turn in a moment, suggests that there is no particular relation between the size of the cosmological constant and the scale of SUSY breaking. This effective field theory based scenario seeks to identify a huge set of string models with many independent small positive contributions to the effective potential. Adding these to a large negative contribution, one argues that if the number of positive contributions is of order 10^X with X significantly larger than 123, then there will be many of these models with positive c.c.of order the one we observe. One then invokes the "successful" anthropic prediction of the c.c. to explain why we happen to see only this special class of models. As a byproduct, this construction produces a huge set of models with very small negative c.c., without SUSY. Indeed the typical strategy is to find such a negative c.c. AdS solution and then add a single small positive contribution to get a model representing the real world.

It thus seems rather important to determine whether there are in fact non-supersymmetric CFTs with large radius AdS duals. This is a well defined mathematical problem, in stark contrast to the effective potential discussion of the landscape. It's my opinion that more people should be working on it.

2.4. Is there a String Theory Landscape?

The basic idea of the string landscape is easy to state. If one looks at compactifications of string theory to four dimensions, with $N \geq 2$ SUSY, we find moduli spaces of models of quantum gravity, with continuous parameters. The number of such parameters is related to the topological complexity of the compactification manifold. For example, in compactifications of Type IIA string theories on Calabi-Yau manifolds, we find a vector multiplet of N=2 SUSY for every non-trivial $(1,1)$ cycle[n] We find a massless hypermultiplet for every $(2,1)$ cycle. So complicated topologies have high dimensional moduli spaces.

[n]Actually, it's a bit more complicated. One of the vector fields is part of the N=2 SUGRA multiplet. We get a number of non-gravitational vector multiplets equal to $h_{1,1} - 1$.

When we consider compactifications with only N=1 SUSY, for example heterotic strings on CY_3, then we find a similar list of moduli at string tree level and to all orders in perturbation theory. However, there is no non-perturbative argument (in most cases) that these moduli spaces are an exact property of the theory. The fact that there are moduli spaces in perturbation theory is related to a continuous shift symmetry of the superpartner of the dilaton field. There are many non-perturbative effects that violate this symmetry. Thinking in terms of low energy effective field theory, we imagine a non-trivial superpotential on this moduli space, which leads to a non-trivial potential. A generic function on a space of dimension D is expected to have a set of local minima whose number is exponential in D. This is the most naive picture of the string landscape.

As we have described it, the landscape is not under any quantitative control. If one tries to write down a few terms in the expansion of the superpotential around weak coupling, one finds that non-trivial minima always lie at values of the string coupling where the expansion is invalid. A major step in the development of landscape ideas was the notion of *flux compactification* [16]. This was the study of solutions in which field strengths of p-form fields on the internal manifolds are turned on. The Dirac quantization condition tells us that these fluxes obey integer quantization rules, so we anticipate a large discrete lattice of solutions, for an internal manifold of complicated topology. A particularly simple set of solutions was found by [17] using the Lagrangian of Type IIB SUGRA. The internal manifold is conformal to a Calabi-Yau manifold, there are imaginary self dual fluxes of a combination of the Ramond-Ramond and Neveu-Schwarz 3-form fields. The flux superpotential fixes all the complex structure moduli and string coupling. The Kahler moduli remain moduli of these solutions. For some choices of fluxes, the fixed value of the string constant is numerically small, so one claims that one can still trust notions from weak coupling string theory.

One feature of this system that is quite general is the necessity for an orientifold in addition to classical super-gravity fields. All weak coupling string theory approaches to compactification will have to deal with the dilaton field. Apart from Calabi-Yau compactifications with no flux, (for which the Einstein Lagrangian vanishes on shell), there will always be sources for the dilaton field in the compact dimensions. The classical SUGRA contributions to the dilaton source are all positive, so we get an equation

$$\nabla^2 \phi = P.$$

Integrating this equation over a compact manifold, we get a contradiction [18]. The orientifold provides a negative source term, which allows for consistent solutions. Orientifolds are singular and do not belong in effective field theory, but they are certainly innocuous in weakly coupled string theory in flat space. As long as one has the weak coupling string theory formalism at one's disposal, one can imagine that this remains true in curved space. Orientifolds can be defined in a finite manner if one has a world sheet sigma model. We'll discuss this further below.

At tree level, the solutions preserve SUSY in Minkowski space if the value of the flux superpotential at the minimum vanishes. Other choices of fluxes, for which $W_0 \neq 0$ break supersymmetry. These solutions still have vanishing cosmological constant because the Kahler potential of Type IIB SUGRA for the Kahler moduli, has the so called no-scale form. However, quantum corrections to the Kahler potential or superpotential will change this. While the latter is exponentially small in the compactified Kahler moduli one can argue that for small W_0 it can still be the dominant effect at large Kahler moduli. One finds (AdS)-supersymmetric solutions with negative c.c. by tuning fluxes so that W_0 is small.

There are, in my opinion, two related things to worry about in these solutions. The first is the question of what it means for the string coupling to be small, and the second is what to do about the orientifold. The normal meaning of small string coupling is that there is a world sheet expression for observables, and an expansion in powers of the string coupling by summing over world sheet topology. If there were such an expansion, we would have no problem defining orientifolds as finite world sheet field theories. But there cannot be such an expansion in this context, because the string coupling is fixed by competition between different terms in the g_S expansion of the superpotential. So we must view these compactifications as constructs in effective field theory, but the orientifold is problematic in a long wavelength expansion. We know that orientifolds are perfectly finite in flat space perturbative string theory, and many orientifolds are related by dualities to smooth solutions of M-theory. So it is not so much the existence of the orientifold that is at issue, but rather whether its singularity could hide dependence on the fluxes which are the control parameters for these solutions.

Many of these ambiguities are removed, at the expense of a considerable loss in computational power, by looking at F-theory compactifications. F-theory is a rubric for a class of solutions of Type IIB string theory, in which the complex string coupling $\tau = a + i\frac{4\pi}{g_S}$ (a is the RR axion field) varies over a complex 3-fold base space of large volume. The ensemble defines an elliptically fibered CY_4 space, with τ describing the complex structure of the elliptic fiber. The orientifold solutions described above are special limits of F-theory compactifications, which were introduced in order to use weak coupling methods. In a general F-theory compactification the string coupling varies over the 3-fold base and is never weak everywhere. In the orientifold limit the region where the coupling is strong shrinks to the locus of the singular orientifolds. More generally, the only expansion parameter in F-theory is the volume of the 3-fold base in 10 dimensional Planck units. F-theory models with fluxes also exist and have been studied extensively in recent years [19]. While the flux induced superpotential for the complex structure moduli of the base has not been computed explicitly, there seems little doubt that for sufficiently generic fluxes all the moduli will be fixed, leaving only the Kahler moduli. For simplicity we can assume that there is just one Kahler modulus. There is at least one since the overall volume of the compact space will not be determined by the SUGRA action.

Thus it is extremely plausible that on a 3-fold base with large $h_{2,1}$ there will be a large number of smooth solutions of Type IIB SUGRA, with all moduli but the overall volume fixed. Below the Kaluza- Klein scale there will be an effective four dimensional theory with $N = 1$ SUGRA and a single chiral multiplet with a no-scale superpotential. The superpotential W_0 will be a flux dependent constant, and since there are many fluxes, it is plausible that it can be tuned to be much smaller than the KK energy scale, as one finds explicitly for the superpotential computed in the orientifold limit. The use of classical SUGRA is of course predicated on the assumption that the KK radius is much larger than the ten dimensional Planck length. These solutions preserve $N = 1$ SUSY only if the superpotential vanishes. (One way to guarantee this is to search for solutions that preserve a discrete R symmetry. The volume modulus will have R charge 0.) However, as a consequence of the no-scale Kahler potential, all of them will have four flat Minkowski dimensions.

This is not consistent if $W_0 \neq 0$. If it were, there would be a low energy effective action for the modulus in four dimensional $N = 1$ SUGRA, but corrections to the Kahler potential would change the cosmological constant, and there could not be a Minkowski solution. However, it does make sense to postulate the existence of a supersymmetric AdS solution. The condition for supersymmetry is

$$\partial_\rho W - \frac{1}{m_P^2}\partial_\rho K W,$$

where we have parametrized the Kahler modulus as

$$(Rm_{10})^{-4} = \frac{\text{Im } \rho}{m_P}.$$

The real part of ρ is an angle variable, so all corrections to the superpotential must be integer powers of $e^{2\pi i\rho}$[o]. Following KKLT one can then argue that if W_0 is small, the system is self-consistently stabilized at a large value of the imaginary part of ρ, where the corrections to the classical Kahler potential are small.

Our own analysis of this situation differs from that of KKLT in two ways. Rather than considering it a controlled calculation in string theory, we view it as a plausible self consistency check for the existence of a supersymmetric AdS model of quantum gravity, whose low energy Lagrangian and background configuration are those suggested by KKLT. The second difference is that we reject the idea that the weak coupling orientifold calculation is more controlled than the general F-theory set-up. The former has an orientifold singularity, whose effect can only be estimated if we have a systematic world sheet expansion. However, the model fixes the string coupling at a value that is not parametrically small, so no world sheet calculation is likely to exist. The only world sheet calculation one can attempt is an expansion around one of the Minkowski solutions of the classical string equations with the orientifold source. We know that if $W_0 \neq 0$, the string loop expansion leads to divergences in the integral over toroidal moduli space. One can attempt to cancel these divergences with the Fischler-Susskind mechanism [20], but this leads to a time dependent background, which is singular in either the past or the future or both. It does not correspond to the stable supersymmetric model whose existence we are asserting.

[o]If, in F-theory, the cycle associated with this Kahler modulus is wrapped by multiple seven branes, so that the CY_4 is singular on that cycle, then the shift symmetry of the angle variable induces a chiral transformation on the matter fields that couple to the 7-brane gauge group. This chiral symmetry is spontaneously broken by strong coupling gauge theory dynamics, which introduces a new discrete finite variable into the effective field theory, parametrizing the different field theory vacuum states. The result is fractional powers of $e^{2\pi i \frac{\rho}{m_P}}$ in the effective superpotential. Unless the 7-brane group is very large, this does not substantially change our argument.

The only real calculational advantage of the orientifold limit of F-theory is the exact formula for the flux induced superpotential. Rather than pursuing the idea that weak coupling string perturbation theory can be used to calculate some useful property of the hypothetical supersymmetric AdS model, it would seem more profitable to try to find an analogous formula for the superpotential in general F-theory models, or at least to argue that a general model with generic fluxes will indeed stabilize all the complex structure moduli.

The bottom line of this discussion is that F-theory compactifications with generic fluxes seem to stabilize all complex structure moduli at the level of classical SUGRA. We use the phrase seem to because detailed calculations rely on the GVW superpotential, calculated at weak string coupling. Even in the orientifold limit of F-theory, there is no systematic string loop expansion of these models, when $W_0 \neq 0$. Classical solutions in which $W_0 = 0$ as a consequence of an anomaly free discrete R symmetry provide us with moduli spaces of asymptotically flat models of quantum gravity in four dimensions. The Kahler moduli are exact moduli of these models. When $W_0 \neq 0$ we have, at the classical SUGRA level SUSY violating asymptotically flat solutions. The classical SUGRA equations are formally exact in the limit of infinite Kahler moduli, but corrections to this approximation ruin asymptotic flatness. *If we assume the existence of large radius, supersymmetric AdS compactifications*, then effective field theory provides a self consistent solution. Such compactifications should have a low energy effective field theory, and symmetries constrain the superpotential and Kahler potential of that field theory. If $W_0 \ll 1$ (in Planck units) then we find a self consistent solution in which we only need to include the leading correction to the superpotential. For F-theory solutions whose CY_4 has large Betti numbers, of order 100, there are many solutions of this type and it seems plausible that one can find many examples with small W_0. The effective expansion parameter is $|\ln \frac{W_0}{m_{10}^3}|^{-1}$, and the scale of Kaluza-Klein excitations is parametrically larger than the inverse AdS curvature radius. It is believed that the Betti numbers of CY_4-folds are bounded, so the expansion parameter can never be really small.

The KKLT paper can thus be viewed as providing evidence for a large class of large radius supersymmetric AdS compactifications. This conjecture is subject to a rigorous test. One must find 3 dimensional superconformal field theories whose properties mirror those of the conjectured geometries via the AdS/CFT correspondence.

DeWolfe *et al.* [21] have suggested another set of supersymmetric AdS solutions with a tunably small parameter. These are based on solutions of (massive) Type IIA string theory. They again purport to have small string coupling and a parametrically suppressed ratio between the compactification radius and the AdS radius. In these compactifications, the control parameter is a large flux, N. However, in [22] we provided evidence that the ever-present orientifolds in such weak coupling constructions hide a region of the compact manifold where the string coupling is large and the compactification radius scales like the AdS radius. The picture in [22] provides an explanation for the scaling of the entropy with N, which is not available in the weak coupling picture. Again, the real test of all of these conjectures is to find superconformal field theories with the properties implied by these geometries. This is particularly interesting in the Type IIA case, because [23] have exhibited non-supersymmetric versions of these compactifications, which look equally plausible. This implies the existence of a large class of non-supersymmetric fixed points with large radius AdS duals. As I've emphasized above, neither orbifolding nor holographic RG flow, both of which seem like plausible mechanisms for finding examples of such large radius CFTs, actually succeed.

It seems to me that this is a place where progress can be made in assessing the reliability of the effective field theory approach to the String Landscape. There is an apparent conflict between the vast landscape of SUSY violating large radius AdS duals promised by the construction of approximate effective potentials, and our inability to construct even one example of the same from a reliable starting point. Perhaps the most controlled setting for studying this problem is that of AdS_3 models. The effective potential approach to these is quite similar to that for AdS_4, but in 2 dimensions we have a much richer arsenal of tools for studying CFTs without recourse to perturbation theory. This area is relatively unexplored and might repay the attention of young researchers.

I've deliberately avoided discussing the procedure of "uplifting the AdS solutions to meta-stable dS solutions by adding anti-D3-branes". This purports to be a small perturbation of the existing solutions, but it is manifestly not. No one knows how to describe the observables of meta-stable dS states, but it is clear that they have nothing to do with conformal field theories living on the boundary of a $3 + 1$ dimensional AdS space. The procedure of adding anti-branes is perfectly sensible when we are talking about a brane configuration of non-compact codimension 3 or more, embedded in a string model in asymptotically flat space-time. It may also be valid in co-dimension 2. For co-dimension zero the back reaction of branes on the geometry is simply not a small perturbation. If we recall that even a small change in the c.c. changes the high energy spectrum of the theory, we see immediately that one cannot play perturbative or low energy effective field theory games in this situation. We will discuss a possible theory of meta-stable dS spaces below.

2.4.1. *Tunneling in gravitational theories*

The key paper on gravitational tunneling is that of Coleman and De Lucia [10]. I urge every serious student of this subject to study that paper carefully and completely. The study of tunneling in general quantum systems is the study of *instantons*: Euclidean solutions of the classical equations of motion *with appropriate boundary conditions.* In QFT in Minkowski space, the boundary condition is that the scalar fields must rapidly approach their values at some meta-stable minimum of the scalar potential, as the radius goes to infinity. The classical solution is $O(d)$ symmetric in d Euclidean space-time dimensions, and defines a finite "critical bubble". The bubble wall is generically fuzzy, and is defined by saying that the field is closer to the meta-stable minimum than some small parameter ϵ. The derivative of the scalars vanishes at the center of the bubble, and this allows us to analytically continue the bubble geometry to the interior of a forward light cone in Minkowski space. The Euclidean solution provides initial conditions for the propagation of the scalar field inside this light cone. It is easy to see that as one proceeds forward on homogeneous slices of constant negative curvature the scalars smoothly approach their values at the absolute minimum of the potential. One says that *the false vacuum has decayed into the true vacuum.* We will continue to use the terms true and false minimum in the gravitational case even though we have emphasized that the concept of vacuum state does not make any sense in quantum gravity. We will also see that not all instantons describe decay.

In finite temperature field theory, this prescription is modified. The Euclidean time dimension is compactified on a circle, and one searches for periodic Euclidean solutions. The solutions no longer achieve the false minimum, and they describe the decay of a meta-stable thermal ensemble, through a combination of quantum tunneling and thermal hopping over the barrier.

Coleman and De Lucia generalized this prescription to include the dynamics of the gravitational field. Their presentation is oriented towards situations where the gravitational effects are "a small perturbation" of the flat space theory, but they discovered that in many cases this claim is untenable, and the gravitational effects are large. We will not make such a restriction, but it's important to emphasize that CDL discovered examples of every phenomenon we will discuss, within the confines of their restricted approximations. One of the most important features of the CDL analysis is the way in which the nature of gravitational tunneling depends on the cosmological constants at the true and false minima. We will present this as evidence that the nature of the actual quantum theory is in fact quite different in the case of zero, positive and negative c.c.

People often ask me why I place so much confidence in the CDL calculations, since I am always warning that too much reliance on the field theory approximation is dangerous. Indeed, in the proposals I will present below the metric of space-time is not a fluctuating quantum variable, but is instead determined by a rigid set of kinematic constraints on the quantum theory. I believe a reasonable analogy is presented by the Wilson loop variables of large N gauge theory. In the planar limit, the Wilson loop expectation value satisfies a classical field equation in loop space [24] and the $1/N$ expansion can be viewed as a sort of Feynman diagram (string loop) expansion around this classical equation. However, for finite N the Wilson loop operators are not independent canonical variables, and the Hilbert space of the perturbation expansion is too big. The true quantum variables are the gauge potentials in some physical gauge. Nonetheless, Euclidean solutions of the equations for Wilson loops can be used to find tunneling corrections to the $1/N$ expansion. However, the real reason for paying attention to the CDL results is that they can all be related to more fundamental concepts in the theory of QG; concepts like the holographic principle and the AdS/CFT correspondence. We will now proceed to classify gravitational tunneling events according to initial and final values of the c.c.

2.4.2. *No tunneling to or from AdS space*

One of the most annoying aspects of this subject is the tendency of many speakers to talk about tunneling to AdS space. Perhaps the most important point in the CDL paper is the demonstration that this NEVER occurs, except in the thin wall approximation. To understand the result we write the CDL equations for the gravitational field coupled to a scalar via the

Lagrangian

$$\mathcal{L} = \sqrt{-g}[R - \frac{1}{2}(\nabla\phi)^2 - V(\phi)].$$

We work in four dimensions, with a single field, for simplicity, but our conclusions are general. Given a scalar potential

$$V(\phi) = \mu^4 v(\phi/M),$$

the natural space-time scale for motion is $L = \frac{M}{\mu^2}$. If we make a Weyl transformation to dimensionless field variables (we use conventions where coordinates are dimensionless and the metric tensor has dimensions of squared length), and write an $O(4)$ symmetric ansatz:

$$ds^2 = L^2(dz^2 + \rho^2(z)d\Omega^2),$$

$$\frac{\phi}{M} = x(z),$$

where z is a dimensionless radial coordinate and ρ is the dimensionless metric coefficient, then we get Euclidean field equations

$$(\rho')^2 = 1 + \epsilon^2\rho^2[\frac{1}{2}(x')^2 - v(x)],$$

$$x'' + 3\frac{\rho'}{\rho}x' = \frac{dv}{dx}.$$

$\epsilon = \frac{M}{\sqrt{3}m_P}$, where m_P is the reduced Planck mass 2×10^{18} GeV. Note that although μ does not appear explicitly in these equations, it must be less than m_P for the semi-classical approximation to be valid (how much less is a matter of conjecture). Note also that the quantity in square brackets in the first equation is what would have been the "conserved energy" of the second equation in the absence of the friction term.

By convention, the center of the bubble is at $z = 0$ and in the vicinity of this point $\rho = z$. The solution is non-singular only if $x'(0) = 0$. The boundary condition at the upper end of the z interval depends on the c.c. in the false minimum. Our present considerations are independent of that boundary condition. To analytically continue the solution to Lorentzian signature we take $z = it$ and use the Euclidean solution at $z = 0$ as an initial condition for the Lorentzian evolution. The initial conditions are $\dot{x}(0) = 0$, $\rho(0) = 0$ and $x(0)$ a fixed value determined by the boundary conditions at the other end. It must be in the basin of attraction of the true minimum.

The Lorentzian equations are

$$\dot{\rho}^2 = 1 + \epsilon^2 \rho^2 [\frac{1}{2}\dot{\phi}^2 + v],$$

$$\ddot{x} + 3\frac{\dot{\rho}}{\rho}\dot{x} + \frac{dv}{dx} = 0.$$

These equations have an AdS solution in which x is equal to the true minimum of v for all time, and $\rho = \sin(\sqrt{\Lambda}t)$. However, the solution determined by the instanton does not approach this solution, which is unstable to infinitesimal perturbations which are homogeneous and isotropic[p]. Indeed, since the Euclidean solution completely fixes the initial conditions for Lorentzian evolution, $\dot{x}$ will not go to zero as $\rho \to 0$. The kinetic energy of x goes to infinity, because the universe is contracting and we have Hubble anti-friction. x will not stay near the true minimum, but will explore its whole potential surface. This singularity will be reached in a time of order $\frac{M}{\mu^2 \epsilon} \sim \frac{m_P}{\mu^2}$. In a typical particle physics model μ is unlikely to be smaller than a few hundred MeV, so this time is shorter than 10^{-5} sec. For future reference we note that, according to the holographic principle, this implies that an observer trapped in this region can access an entropy that is at most $\sim (\frac{M_P}{\mu})^4 < 10^{80}$, only $\sim 10^{60}$ of which can be in the form of matter and radiation. The actual matter/radiation entropy of our universe is $\sim 10^{80}$. The reader who is confused by these numbers, will be able to go back and check them after we discuss the holographic principle.

The converse of this result is also true: a quantum AdS space cannot decay by tunneling. This follows from the AdS/CFT correspondence. The exact mathematical formulation of CFT requires one to have only unitary highest weight representations of the conformal group in the Hilbert space. It follows that the global Hamiltonian $K^0 + P^0$ is bounded from below. But the Lorentzian continuation of an instanton is always a zero energy solution in which the positive and growing kinetic energy of the expanding bubble is balanced by the increasingly negative potential energy of its interior. It always corresponds to a system which is unbounded from below.

This general argument is exemplified in a beautiful paper by Hertog and Horowitz [25]. These authors found an instanton solution, which seemed

[p]AdS spaces are sometimes stable to small perturbations which fall rapidly at infinity. These are the normalizable fluctuations of the AdS/CFT correspondence. The homogeneous isotropic solutions relevant for instanton physics are not normalizable. Generally they have no extension outside the FRW coordinate patch, as a consequence of the singularity we are discussing.

to indicate a non-perturbative decay of the supersymmetric(!) $AdS_4 \times S^7$ solution of 11 dimensional SUGRA. Upon closer examination, they found that although the perturbation fell off at infinity, it was not a normalizable solution, corresponding to a state in the CFT. Rather, it corresponded to a perturbation of the CFT by a marginal operator that was unbounded from below. The instantons for the putative decays of AdS vacua have also been studied by [26] [27], who provide further evidence that the landscape interpretation of these events is faulty.

The correct way to interpret these facts is to say that if we look at a classical bulk Lagrangian, which has an AdS solution, as well as an instanton which behaves like a normalizable perturbation of this solution at Euclidean infinity, then we will have proven, in the classical approximation, that the AdS Hamiltonian of this system is unbounded from below, and cannot have a CFT dual. It is likely that such a solution is not part of any sensible quantum theory of gravity. Indeed, there is an interesting sidelight on this situation, which already indicates that something serious is wrong with the interpretation of this instanton as a decay of the original AdS space-time.

In ordinary quantum field theory, excitations around the false vacuum are meta-stable only up to some finite energy. If we make the energy density larger than the barrier height the system is simply unstable. Similarly, the thermal ensemble is meta-stable only up to some finite temperature. In quantum gravity in large radius AdS space, we can explore the thermal ensemble by looking at AdS-Schwarzschild solutions of the field equations, which are normalizable and have positive energy of arbitrarily large size. These solutions do not have classical instabilities, indicating that the vacuum decay paradigm of non-gravitational QFT is breaking down.

We can gain more insight into this when we realize that the expanding bubble of the Lorentzian instanton *does not penetrate the interior of a black hole*. The bubble expands only at the speed of light, while the interior geometry expands away from the bubble superluminally. A solution whose initial conditions consist of a space-like separated pair of a black hole and a nucleated critical bubble, has two causally separated future asymptotic regions, both of them space-like singularities. Multiple black holes in the initial state will lead to multiple causally disconnected future regions. Furthermore, single the bubble nucleation probability is *exponentially small* as $\frac{m_P}{\mu}$ goes to infinity, it is easy to see that the black holes can have exponentially larger entropy than the entropy accessible within the bubble. These semi-classical considerations suggest very strongly that there is no sensible quantum mechanical interpretation of AdS solutions that have genuine instanton instabilities. Certainly the interpretation of the instanton as a decay of the original AdS "state" into the system in the interior of the CDL bubble, is completely untenable. This analysis goes through in precisely the same way for CDL "unstable" asymptotically flat space-times, although the existence of Hawking instabilities of black holes in that case, poses further complications.

Our conclusion is that AdS solutions of bulk gravitational field equations never arise as the result of CDL decays, and do not decay in a way that resembles the vacuum decay of a non-gravitational QFT. Some of the solutions are stable, and may well belong to a real theory of QG, which would be defined by a CFT dual. The unstable ones surely belong to a very peculiar quantum theory, if they have any meaning at all. There is thus strong evidence from CDL tunneling, complementing that from the AdS/CFT correspondence, that AdS solutions of gravitational field equations form little isolated models of QG, which have nothing to do with a larger landscape.

2.4.3. *Gravitational tunneling to and from zero c.c. states*

In asymptotically flat space-time, the asymptotic symmetry algebra is the Poincare group. If we do not insist on supersymmetry[q] there is no general argument that the Hamiltonian is bounded from below. However, there is a classical theorem [28] which shows that asymptotically flat solutions of certain Lagrangians do have classically positive energy.

The paper [29] clarified how the space of theories consisting of scalar fields coupled to gravity is divided up by the positive energy theorem. Consider a potential with classically stable Minkowski and AdS solutions and ask whether there is a static domain wall connecting the two solutions. For the AdS/AdS case, we saw that such domain walls with boundary conditions that correspond to normalizable solutions on both sides of the wall, are the holographic representation of RG flows between two fixed points. No such interpretation is possible here, because the analog of the UV fixed point is the Poincare invariant model, which is not a quantum field theory.

[q]It is one of my contentions that we MUST insist on SUSY *i.e.* that every asymptotically flat model of QG is in fact Super Poincare invariant. However, we are exploring more general possibilities in this section, and our explorations lead to important insights for the program based on my conjecture.

The equations determining the domain wall are

$$\rho'^{\,2} = \epsilon^2(\frac{1}{2}x'^{\,2} + v(x)),$$

$$x'' + \frac{3\rho'}{\rho}x' + \frac{dv}{dx} = 0,$$

with boundary conditions

$$x(\pm\infty) = x_{\pm}.$$

$x_{\pm}$ are the false and true maxima of $v(x)$.

As is familiar from linear eigenvalue problems, this system does not have solutions for a generic potential. In the limit in which we model the domain wall as an infinitely thin brane with a given tension, there will only be one value of the tension for which the static solution exists [11]. For tensions below this value there is instead a solution which looks like the asymptotic limit of an expanding bubble wall, corresponding to CDL decay of the Minkowski background (but missing the instability of the previous subsection, for which one must go beyond the thin wall approximation). For tensions above this there is no interpretation of the solution as the limit of an object in the Minkowski background.

More generally, as in any eigenvalue problem, we can find a solution obeying both boundary conditions by tuning a single parameter in the potential. *Thus, the space of all potentials with a Minkowski solution of the field equations contains a co-dimension 1 submanifold, on which a static domain wall connecting Minkowski space to one particular AdS minimum exists, while all for all other AdS minima there are neither domain walls nor expanding bubble solutions.* For a given Minkowski minimum there will generally be only one domain wall, though in supersymmetric situations there may be more [r]. This submanifold in the space of potentials is called *the Great Divide.* By perturbation theory one can show that above the Great Divide the Lagrangian has a positive energy theorem, while below it there are expanding bubble solutions and the ADM energy is unbounded from below.

In [29] we showed that by varying the parameter $\epsilon \sim M/m_P$ in potentials of the form $\mu^4 v(\phi/M)$ we crossed the Great Divide. For $\epsilon \ll 1$, the non-gravitational analysis, which indicates an instability is essentially correct. However, it should be noted that even in this regime, CDL showed

[r] In the AdS to AdS case, static domain walls correspond to relevant perturbations of the CFT for smaller absolute value of the c.c. , which point along RG flows to other fixed points. Such flows are non-generic unless both fixed points are supersymmetric.

that one is above the Great Divide if $|v(x_T)|$, the magnitude of the c.c. at the true minimum, is $\ll 1$. The Great Divide itself is located at $\epsilon = o(1)$ for generic functions $v(x)$. For those models below the Great Divide, the same issues arise as for unstable AdS spaces. Starting from a generic excited state of the Minkowksi solution, we end up with a future that contains multiple causally disconnected space-like singularities, most of whose entropy is contained in black holes. Here however we have to deal with the perturbative Hawking instability of black holes, which returns the degrees of freedom of the black hole to a region causally connected to the expanding bubble. Here we can encounter a paradox: The matter entropy outside the bubble is bigger than that measurable by any observer inside the bubble. One suspects that we are being too naive and neglecting back reaction of all of this matter on the bubble. A possible scenario is that collisions of the bubble wall with a sufficiently large matter density, converts the bubble into a black hole. Indeed, the bare expanding bubble solution has exactly zero energy in empty space. If it collects a finite surface energy density as it passes through a region filled with a uniform density of matter, then it will end up with a mass of order the square of its radius. For large enough radius the Schwarzschild radius of this distribution will be larger than the bubble radius. Thus, a resolution of the apparent paradox of a bubble sweeping up more entropy than any observer inside it can measure, may simply be that in attempting to swallow all of this entropy, the bubble forms a black hole around itself.

The bizarre conclusion of this story would be that, perhaps, below the Great Divide, empty flat space is unstable, but flat space with enough entropy in it nucleates a black hole around the expanding bubble. Of course, another possibility is that there are no actual theories of quantum gravity which contain such meta-stable flat space-time configurations. When we discuss the holographic space-time formalism, we will show that it suggests that all quantum theories of asymptotically flat space-time are exactly supersymmetric. If this is the case then they are automatically Above the Great Divide. This does not yet settle the question of the fate of the asymptotically dS universe, which we appear to inhabit.

2.4.4. *CDL transitions from dS space*

If we take a potential below the Great Divide and add a small positive constant to it, we do not make a significant change in the CDL transition rate. The entire story of the previous section replays with little change.

Above the Great Divide the story is different. With mild assumptions, there is *always* a CDL instanton when a potential has a positive and negative minimum separated by a barrier[s]. However, the transition rate goes to zero like

$$e^{-\pi(RM_P)^2},$$

as R, the dS radius goes to infinity. This suppression by something of order the inverse of the exponential of the dS entropy, is what we would expect for a transition at infinite temperature for a system with a large finite number of states, into a very low entropy state. This is consistent with our previous remark that, according to the holographic principle, the maximal entropy observable in the negative c.c. Big Crunch is a microscopic number. We will see below that the interpretation of dS space as a system with a finite number of states, at infinite temperature, is consistent with all semi-classical evidence about dS space, including its finite Gibbons-Hawking temperature!

Transitions from one dS space to another are also consistent with this picture, and add an extra bit of evidence. Indeed, although we have not emphasized it above, the instantons for transitions out of dS space are compact manifolds, with positive scalar curvature, just like Euclidean dS itself. And like Euclidean dS space they have negative action. The probability interpretation of the instanton calculation comes by subtracting the dS action from the instanton action, which always gives a positive number. In the case of dS to dS transitions, we get two different probabilities, depending on which dS action we subtract. These are interpreted as the probabilities for the forward and reverse transitions

$$P_{1\to 2} = e^{-(S_I - S_1)},$$

$$P_{2\to 1} = e^{-(S_I - S_2)}.$$

The ratio of transition rates is thus

$$e^{-(S_1 - S_2)}.$$

It is a quite remarkable fact (analogous to a result about black holes first discovered by Gibbons and Hawking), that the dS action is exactly the

[s]The exceptions come for potentials in which the maximum is very flat. Analogies with ordinary quantum mechanics lead us to expect a transition from the false minimum to the top of the barrier, which is more or less semi-classical, followed by large quantum fluctuations on the flat top. However, since the system includes gravity, we don't really know how to explore the regime of large quantum fluctuations. It is possible that potentials this flat are simply forbidden in real theories of QG [30] [31] [32].

negative of the dS entropy. This means that these transition rates satisfy the principle of detailed balance appropriate for a system with a finite number of states at infinite temperature. Unlike the case of dS transitions to a negative c.c. Crunch, this semi-classical calculation is under control in both directions. It seems perverse to attach any other meaning to it than what it seems to say: dS space is a system with a finite number of states. Its Hamiltonian is generic and the time evolution of a randomly chosen initial state will sweep out the entire Hilbert space. The dS space with larger c.c. is a low entropy configuration of this system and will be accessed only rarely, in direct proportion to the fraction of the total number of states corresponding to this configuration.

Note that this interpretation meshes perfectly with the one we have proposed for dS to Crunch transitions above the Great Divide. Note further that it does not agree with ANY interpretation of the same transition according to the theory of Eternal Inflation.

As we take the smaller c.c. to zero, the transition rate to the higher c.c. state goes to zero. This makes sense in our interpretation, because the probability of finding a finite entropy subspace of states starting from a random search through an infinite dimensional Hilbert space, is zero. Note however that the limit of zero c.c. is a very subtle one. In the section on stable dS space, we will see that a lot of states must be discarded from the dS Hilbert space in order to describe the Hilbert space of the limiting Poincare invariant theory. The entropy of the latter scales as $(RM_P)^{3/2}$ as the dS radius goes to infinity, while the total entropy of the dS Hilbert space is $\pi(RM_P)^2$. We will see in the next subsection that the required limit for a tunneling solution whose target is a zero c.c. space-time is quite different. The interpretation of such solutions is intertwined with attempts to construct a theory of the String Landscape, and we turn to that problematic subject next.

2.4.5. *Implications for the landscape*

The implications of these results in semi-classical gravity for the idea of a string landscape are profound. Asymptotically flat and AdS models of quantum gravity *are not part of the landscape* and do not communicate with hypothetical landscape states by tunneling. Tunneling only makes sense for meta-stable dS points on an effective potential. These can tunnel to other dS points, to negative c.c. Big Crunches, and to zero c.c. states. None of the physics of these states is encoded in anything like the boundary correlators that string theory has taught us how to compute. *If the landscape exists, the very definition of its observables must be completely different from that of ordinary string theory.*

We have seen that tunneling to negative c.c. crunches falls into two categories. Above the Great Divide, we've provided a plausible quantum interpretation of the CDL tunneling probabilities, in terms of a quantum theory of stable dS space with a finite dimensional Hilbert space. Below the Great Divide, we've argued that these transitions are fraught with interpretational ambiguities. The true endpoint of CDL decay is not a quiescent true vacuum, nor even a single big crunch. The final state depends on which initial excited state of the dS or flat "false vacuum" one begins with. It typically has multiple crunching regions, with different pre-crunch internal geometries, which are causally disconnected from each other.

Nonetheless, many advocates of the landscape insist that any sensible meta-stable model of dS space must be below the Great Divide. The argument is somewhat philosophical, but depends crucially on the fundamental claim that the landscape solves the c.c. problem by invoking the anthropic principle. In order to be certain that this is true, one counts meta-stable landscape points, according to some criterion, and claims that the number is of order 10^{500} or greater. It is important that this number is *much larger* than the ratio between a Planck scale c.c. and the c.c. we observe. One then argues that if generic minima of the potential have a c.c. that is a sum of a such a large number of positive and negative Planck scale contributions[t], there will inevitably be some with c.c. of the value we observe. If anthropic arguments can show that a value bigger than this is incompatible with the existence of intelligent life forms, one has "explained" the small value of the c.c. Note that in order for this counting to work in a way that does not require close scrutiny of each and every minimum of the potential, there must be MANY solutions with a value of the c.c. close to ours. There is no reason for other properties of the low energy world to be similar to those we see. So questions like what the low energy gauge group and representation content are, as well as the value of most low energy parameters, must also be answered anthropically.

I will not spend time here rehashing the futility or experimental implausibility of this claim, but rather emphasize the general picture of string

[t]In the Bousso-Polchinski [33] version of this argument there is one large negative contribution and a large number of smaller positive ones. In KKLT, one argues for a set of AdS solutions, with enough free parameters to make the negative c.c. small, and then adds a small positive contribution.

theory that it implies: string theory has MANY solutions with small c.c. If we are not to regard our own world as simply an accidental consequence of the theory, then we must come up with some argument that makes our conditions more hospitable for observers than other possible meta-stable states. The advocates of these ideas are led to contemplate the question of whether we are *typical* members of the class of observers that the landscape predicts. The answer is that this can only be true if our current dS condition does not last too long.

We know that the universe we observe began in a state of much lower entropy than it has today. This is why we see the second law of thermodynamics in operation. The visible entropy of the universe is dominated by cosmic microwave background photons, and the total entropy by the supermassive black holes in the centers of galaxies. According to modern cosmology, this entropy was created in the post-inflationary history of the universe, through the decay of the inflaton field into radiation, and the gravitational collapse caused by the action of the fluctuations of this field on non-relativistic matter. In the landscape picture, the beginning of this cosmic history is a tunneling event from a higher c.c. meta-stable point, to our own basin of attraction. It is a very low entropy fluctuation.

If the current c.c. dominated era of the universe lasts too long, there is a much more efficient way to make observers than to have a fluctuation that recreates the entire history of the universe. Such a fluctuation, by the CDL calculation, has a probability of order $e^{-10^{123}} = e^{-A/4}$, where $A = 4\pi(RM_P)^2$. On the other hand, in the asymptotic future dS space, the probability to have a random fluctuation that creates a localized mass equal to that of a "single intelligent observer" is $e^{-2\pi Rm_O}$ and the probability that that mass is in the state corresponding to a live intelligent observer is at least $e^{-\frac{A_O}{4}}$, where A_O is the horizon area of a black hole which could enclose the observer. These ridiculously tiny probabilities, are much larger than the probability of the fluctuation that started the universe off. So, either the landscape explanation of the origin of our universe is wrong, or we are far from typical observers, *or the dS state must decay long before all these typical observers can be formed.* This is only possible if our meta-stable dS state is below The Great Divide, which is the choice made by many landscape theorists. As we will see, this claim creates some tension with the only extant proposal for making a true theory of the string landscape.

The attempt to create a true theory of the landscape, analogous to our models of asymptotically flat or AdS spaces has been centered entirely in the

Stanford-Berkeley group. The proposal is that the observables of the theory somehow reside in the causal diamond of a post-tunneling event into a zero c.c. region of the potential, which locally approaches one of the maximally supersymmetric flat space solutions of string theory. The original idea was to construct a sort of scattering theory in the Lorentzian space-time defined by the CDL instanton. It's indeed true that if we consider quantum field theory in such a space-time, one can define scattering states on the past and future boundaries. However, as I suggested in 2004, and was later proven rigorously by Bousso and Freivogel, generic scattering boundary conditions do not lead to small perturbations of the instanton geometry. This can be understood in a heuristic manner. The CDL geometry has a compact throat connecting its past and future regions. If we have an initial or final state with too large an entropy, it will create a black hole of radius larger than the throat. This leads to a space-filling space-like singularity, cutting the future off from the past.

The second proposal was to try to construct an analog of AdS/CFT where the CFT lives on the boundary of the negatively curved space-like slices of the CDL geometry. It is argued that the appropriate boundary conditions for this situation allow quantum fluctuations of the boundary geometry, so that the boundary CFT is coupled to quantum gravity. The hope is that this situation is well defined when the boundary is two dimensional, and leads to a boundary Liouville theory. Two dimensional boundaries are appropriate for 4 dimensional dS spaces, so this proposal relies on the folk theorem that there are no dS solutions of SUGRA above 4 dimensions.

I do not understand the details of this construction or the enthusiasm of its builders, so I will end this section with a list of questions that I think must be answered, if this approach is meaningful.

- What is the probability interpretation of the boundary field theory? Only some of the extant theories of fluctuating two geometries have a quantum mechanical interpretation. In those, the genus expansion is the *divergent* $1/N$ expansion (actually the double scaling limit) of a matrix quantum mechanics. In this context the genus expansion is said to converge. What are the probability amplitudes and what do they have to do with real world measurements? Is the theory quantum mechanics? What are the possible initial states?
- Most of the asymptotically SUSic regions of moduli space are decompactification limits, where the local asymptotically flat

space-time has dimension higher than four. Why are only two dimensional boundaries relevant? One may want to argue that the theory has a two dimensional boundary for all finite FRW times, but the decompactifying dimensions should at least show up as an infinite number of low dimension operators. The formalism has so far restricted attention to massless bulk fields, but surely massive fields whose mass asymptotes to zero must be part of the picture?

- The construction is based on a particular instanton for decay of a particular meta-stable dS point into a particular locally flat geometry. How do all the other instantons fit into the picture? There must be some sort of monstrous duality in which the observables are actually independent of the choice of instanton geometry in the construction?
- Conversely, how does one pick out of the Liouville/CFT observables, the data relevant to our particular universe? This is of course a crucial step in trying to relate these ideas to the real world. Is there any relation between the answer to this question and the practices of those landscape enthusiasts who simply do effective field theory in a particular dS state? Is the answer to this problem computationally effective? That is, can one really hope to separate out the data corresponding to individual members of the 10^{500} strong ensemble?
- The construction purports to be a rigorous definition of what is meant by the phrase *eternal inflation*. What is its prescription for the solution of the *measure problem* in that context? (Some progress has been made on the answer to this question, but not enough to support phenomenological predictions).
- The transition from a dS space with small positive c.c. and one of the zero c.c. egions of the potential, is above the Great Divide. Supposedly one is saved from the problem of fluctuating intelligent observers by much more rapid decays into negative c.c. crunches. We are then left with the bizarre situation in which all of the rigorously defined data about our universe can only be measured in an extremely improbable history for the universe, one in which it lasts long enough for all sorts of fluctuated intelligences to exist.

I will not comment further on this proposal, except to mention that I personally find the challenges of Holographic Space-Time and Cosmological SUSY breaking much less daunting, and their connection to actual observations infinitely more direct. We turn next to an explanation of the Holographic space-time formalism.

2.5. Holographic Space-Time

Having devoted much verbiage to the description of what a theory of QG is *not* we are now ready to propose a general description of what it is. This framework is meant to subsume all of the well defined models we have discovered, which fall under the rubric string/M-theory. That claim has not yet been proven, and I will admit from the beginning that a fully dynamical implementation of the rules of *Holographic space-time* has not yet been found.

All well established models in the string/M-theory menagerie belong to one of two classes. The first corresponds to space-times in dimension $3 \leq d \leq 7$ with AdS asymptotics and an AdS curvature radius that can be taken parametrically large, in the sense that there is a closed set of boundary correlation functions, which can be calculated in a systematic expansion about the GKP/W [34] SUGRA limit. They all have exact AdS SUSY.

The observables in these models are correlation functions on a boundary of the form $R \times S^{d-1}$. In addition, there are many models of asymptotically flat space-time with dimension between 4[u] and 11. The only observable is the S-matrix. All of these models have exact super-Poincare invariance.

In addition there are models which can be viewed as describing certain infinite branes embedded in these spaces. In the AdS case, these are relevant perturbations of the CFT describing the original symmetric model[v] . These models need not be supersymmetric, but they are "supersymmetric in the majority of space-time". In the language of CFT, this means that the high energy, short distance behavior is dominated by a supersymmetric fixed point. Although there are many claims in the literature, there are no well

[u]In 4 dimensions the complete theory of a gravitational S-matrix, is complicated. Fadeev and Kulish [35] have given a prescription, analogous to their treatment of electrodynamics. Not much further work has been done exploring the consequences of their ansatz.

[v]In calling these infinite branes, I am working in the Poincare patch of AdS space, which corresponds to the Hilbert space of CFT in Minkowski space. The corresponding solutions in global coordinates are localized at the center of a global coordinate system. There are also true brane solutions with AdS asymptotics, analogous to D-branes embedded in flat space-time.

established models with nonsupersymmetric fixed points at large curvature radius.

We want to construct a more local description of QG, which will reduce to these supersymmetric models in the infinite volume limit, but which will enable us to describe systems that do not fall into any of these categories, like cosmologies and the real world. In GR, local objects are never gauge invariant, so we should expect our description to be adapted to a certain coordinate system. Indeed, the fundamental postulates of the theory will contain in themselves an explanation for why local physics can never be gauge invariant in QG, a sort of quantum version of the principle of general covariance.

The basic principles of holographic space-time are simple to state:

- The Strong Holographic Principle (Banks-Fischler) - A causal diamond is the intersection of the interior of the backward light-cone of a point P with that of the forward light-cone of a point Q in the causal past of P. The boundary of a causal diamond is a null surface. When we foliate it with space-like $d-2$ surfaces, we find one of maximum area, called the holographic screen. According to the holographic principle, the quantum version of such a causal diamond is a Hilbert space whose dimension is $e^{\frac{A}{4L_P^2}}$, where A is the area of the holographic screen. This formula is asymptotic for large area. The proper quantum concept is the dimension of the Hilbert space, which is of course always an integer.
- Intersections of causal diamonds correspond to common tensor factors in the Hilbert spaces of two diamonds. Geometrically this defines the area of the maximal causal diamond which fits in the intersection. Thus we have

$$\mathcal{H}_1 = \mathcal{O}_{12} \otimes \mathcal{N}_1,$$

$$\mathcal{H}_2 = \mathcal{O}_{12} \otimes \mathcal{N}_2.$$

This encodes the causal structure of the space-time, if we have a rich enough collection of causal diamonds. We ensure this by beginning from a lattice, which encodes the topology of an infinite space-like slice (a Cauchy surface) of the manifold. For each lattice point $\mathbf{x}$ we have a sequence of Hilbert spaces $\mathcal{H}(n, \mathbf{x}) = \otimes \mathcal{P}^n$, where $\mathcal{P}$ is a finite dimensional space we will define below. Geometrically this represents a sequence of causal diamonds whose future tips have larger and larger proper time separation from the initial space-like slice. For a model of a Big Bang space-time we imagine the past tips to lie on the Big Bang hypersurface. This incorporates the idea that the particle horizon is very small near the singularity, but it is clear that nothing singular happens in the quantum theory. For a time-symmetric space-time we take the lattice to lie on a time-symmetric space-like slice, and the past and future tips of the diamonds lie an equal proper time before and after the time-symmetric slice.

- For nearest neighbor points on the lattice, at any n, we insist that the overlap Hilbert space is $\otimes\mathcal{P}^{n-1}$. We interpret these sequences of Hilbert spaces as the sequence of causal diamonds of time-like observers, which penetrate the chosen space-like slice at a given lattice point. The proper time interval between the tips of the nth diamond is a monotonically increasing function of n. Thus, two nearest neighbor sequences of Hilbert spaces, correspond to two time-like observers whose trajectories through space-time are almost identical. The overlaps between other points are constrained by two consistency conditions. Let $d(\mathbf{x}, \mathbf{y})$ denote the minimum number of lattice steps between two points. We require that the overlap not increase as we follow a path of increasing d, starting from $\mathbf{x}$, and that it decrease asymptotically as $d(\mathbf{x}, \mathbf{y})$ goes to infinity.
- The second consistency condition is dynamical. Let $N(\mathbf{x})$ be the maximal value of n at a given lattice point. We prescribe an infinite sequence of unitary operators $U_k(\mathbf{x})$,operating in the Hilbert space $\mathcal{H}(N(\mathbf{x}), \mathbf{x})$, with the property that for $k \leq N(\mathbf{x})$ $U_k = I_k(\mathbf{x}) \otimes O_k(\mathbf{x})$, where I_k is a unitary in $\mathcal{H}(k, \mathbf{x})$ while O_k operates in the tensor complement of this Hilbert space in $\mathcal{H}(N(\mathbf{x}), \mathbf{x})$. This sequence is interpreted as a sequence of *approximations to the S-matrix* in the time symmetric case, and a sequence of *cosmological evolution operators* in a Big Bang space-time. We then encounter the following set of fearsomely complicated consistency conditions. Consider the overlap Hilbert space $\mathcal{O}(m, \mathbf{x}; n, \mathbf{y})$. The individual time evolutions in $\mathcal{H}(N(\mathbf{x}), \mathbf{x})$ and $\mathcal{H}(N(\mathbf{y}), \mathbf{y})$, each prescribe a sequence of density matrices[w] on $\mathcal{O}(m, \mathbf{x}; n, \mathbf{y})$. These two sequences must be conjugate to each other by a sequence of unitary

[w]There is no reason for the state on the overlap to be pure. It is entangled with the other degrees of freedom in each causal diamond.

transformations. A collection of Hilbert spaces with prescribed overlaps, and evolution operators, satisfying all the consistency conditions, is our definition of a quantum space-time.

It's clear from this list, that any quantum space-time, which approximates a Lorentzian manifold when all causal diamonds have large area, will completely prescribe both the causal structure and the conformal factor of the emergent geometry. *We conclude that in this formulation of QG, space-time geometry is not a fluctuating quantum variable.* Given the results of [36] it is likely that any geometry that emerges from this framework will satisfy Einstein's equations with a stress tensor obeying the dominant energy condition. This is because the quantum system will obey the laws of thermodynamics, and those authors claim that this is enough to guarantee Einstein's equations, given the Bekenstein-Hawking connection between area and entropy. Indeed, if we imagine *defining* the stress energy tensor as the right hand side of Einstein's equations, then the only content of those equations is whatever energy conditions we impose. The holographic framework will certainly impose conditions sufficient to prove the area theorem.

This observation is completely in accord with our semi-classical conclusion that different asymptotic behaviors of space-time, even if they are solutions to the same set of low energy field equations, correspond to different models of QG. The holographic construction extends this principle to space-times whose boundaries are not simple conformal transforms of static geometries. One might object that the standard Feynman diagram construction of perturbative QG could not possibly be consistent with such a picture. This is not true. These expansions only describe particles, including gravitons, propagating in a fixed space-time background. Thus, to be consistent with them, one must only require that the quantum variables describe arbitrary scattering states of gravitons in asymptotically flat or AdS space-times (or any other example over which we claim to have good semi-classical control).

Our next task is to introduce just such variables.

2.5.1. *SUSY and the holographic screens*

Consider a "pixel" on a holographic screen. Naively, it's characterized by a null vector and a plane transverse to it, describing the orientation of this pixel in space-time. This is the information content of solutions of the

Cartan-Penrose equation

$$\bar{\psi}\gamma^{\mu}\psi(\gamma_{\mu})^{\alpha}_{\beta}\psi^{\beta} = 0,$$

where ψ is a commuting Dirac spinor. Indeed, this equation implies that $n^{\mu} = \bar{\psi}\gamma^{\mu}\psi$ is a null vector, and that ψ itself is a transverse or null-plane spinor corresponding to this null vector. That is, if $\gamma^{\mu_1\cdots\mu_k}$ are anti-symmetrized products of Dirac matrices with $k \geq 2$ then

$$\bar{\psi}\gamma^{\mu_1\ldots\mu_k}\psi$$

are non-zero only for hyperplanes embedded in a particular $d-2$ plane transverse to n^{μ}. The spinor has only $2^{[\frac{d-2}{2}]}$ independent components. In eleven dimensions this is 16 real components, S_a.

The holographic principle implies that the Hilbert space of a pixel should be finite dimensional, so the only operator algebra we can write down for the S_a, consistent with transverse rotation invariance, is

$$[S_a(n), S_b(n)]_+ = \delta_{ab}.$$

n is a label for the pixel, which we will discuss in a moment. This algebra is the same (up to normalization) as that of a massless superparticle with fixed momentum in 11 dimensions. The smallest representation is the 11D SUGRA multiplet, and all the others correspond to particles that, according to the Coleman-Mandula theorem, cannot have an S-matrix different from 1. If we think about different pixels, they should have independent degrees of freedom, and we would normally ask that the corresponding operators commute. However each of the individual pixel algebras has an automorphism $S_a(n) \to (-1)^{F(n)}S_a(n)$, which we treat as the Z_2 gauge symmetry called $(-1)^F$. We can use this to choose a gauge where spinors corresponding to different pixels anti-commute

$$[S_a(m), S_b(n)]_+ = \delta_{ab}\delta_{mn}.$$

The spin-statistics connection familiar from local field theory is thus built in to the holographic formalism.

Now let us think about the notion of pixel. The holographic principle again requires that a finite area holoscreen should have a finite number of pixels, to each of which we assign a copy of the single pixel algebra. The naive notion of pixel can be thought of as a way to approximate the algebra of functions on the holographic screen by the algebra of characteristic functions of a finite cover of the screen by open sets. This opens the door to more general approximations of the algebra of functions by finite dimensional algebras that are not necessarily commutative. This has numerous advantages. For example, in the case relevant to the real world, a two dimensional holographic screen with $SO(3)$ rotation invariance, we can use the fact that $SU(2)$ has finite dimensional representations of every integer dimension to construct the so called fuzzy sphere. The algebra of $N \times N$ matrices inherits a natural action of $SU(2)$, which contains all integer spins between zero and $N - 1$. It approximates the algebra of functions on the sphere by the usual finite sums of spherical harmonics. The specification of whether we get smooth, continuous, measurable or square integrable functions is encoded in the behavior of the expansion coefficients for large spin.

More generally, if the holographic screen has a Poisson structure, there is a well developed theory of deformation quantization, which, for compact manifolds, leads to a sequence of approximations to the algebra of smooth functions by finite dimensional matrix algebras. In general, this procedure has ambiguities; the analog of the usual ordering ambiguities in quantum mechanics. However, for Kahler manifolds there is much less ambiguity. The space of sections of a holomorphic line bundle over a Kahler manifold is finite dimensional and has a natural Hilbert space structure induced by the Kahler potential. If we take sequences of holomorphic line bundles with dimension going to infinity, we get natural fuzzy approximations to the manifold. Almost all of the manifolds that arise in string compactification are related to Kahler manifolds in some way. Calabi-Yau manifolds are an obvious example, and the Horava-Witten bundles of Calabi-Yau manifolds over an interval are another. It is not known whether general G_2 manifolds have a Poisson structure, but those which exhibit non-abelian gauge groups, are $K3$ fibrations over a sphere or lens space. A choice of Kahler form on the $K3$, combined with the unique $SO(3)$ invariant Poisson structure on S^3 or a lens space, defines a Poisson structure on the entire 7-fold.

Combining these ideas, we obtain a general prescription for compactification of holographic space-time. For compactifications to 4 dimensions we introduce variables satisfying the commutation relations

$$[(\psi^M)_i^A, (\psi^\dagger N)_B^j]_+ = \delta_i^j \delta_B^A Z^{MN} \quad i = 1 \ldots K, \quad A = 1 \ldots K + 1.$$

The operators ψ and $\psi^\dagger$ are $K \times K + 1$ and $K + 1 \times K$ matrices, sections of the two spinor bundles over the fuzzy 2-sphere, the holographic screen for 4 dimensional space-time. The indices M, N can be thought of as either minimal spinor indices in 7 dimensions or $(2, 0)$ or $(1, 1)$ spinors in 6. We know that in string compactifications with 8 or more supercharges, these different interpretations morph into each other as we move around in moduli space. In the interior of moduli space, where we expect the real world to lie, it may be that no particular geometric description is picked out. To be more precise, M and N label a basis in the space of sections of the spinor bundle on the appropriate manifold, appropriately truncated. This gives us a possible new insight into string dualities. It is well known for example that the algebra of $N \times N$ matrices can actually be thought of as a fuzzy approximation to the space of functions on *any* Riemann surface. The topology and geometry of manifolds emerges from fuzzy geometry in the large N limit, by discarding different sets of matrices in the definition of the limiting algebra. In the interior of moduli space in string theory, where the string coupling is not weak, compact manifolds have volumes that are finite in Planck units and should therefore be thought of as finite pixelations. The dual geometry is obtained by taking a different large N limit.

The operators Z^{MN} are sums of $p - forms$ and we may think of them as measuring the charges of branes wrapped around cycles of manifolds. More precisely, each $p - form$ component of Z^{MN} will be a sum of terms, each of which has such an interpretation. Specifying the number of terms in this sum, for each p will tell us the number of independent $p - cycles$ in the manifold. In the string theory literature, the Z^{MN} are often called central charges in the SUSY algebra. However we know that there are interesting examples of singular manifolds, where their algebra is non-abelian, and this gives rise to Yang-Mills gauge potentials in the non-compact dimensions.

We have suppressed another set of matrix indices in the formula for the anti-commutation relations above. Our internal spinors and $p - forms$ are really sections of the corresponding bundles over some fuzzy approximation to the internal manifold. The enumeration of cycles in the previous paragraph is part of the structure of these bundles. The geometry and topology of the manifold are all encoded in the super-algebra of the generators $\psi, \psi^\dagger, Z$. The smallest representation of this super-algebra, for fixed i, A, is the pixel Hilbert space $\mathcal{P}$ referred to above.

An extremely interesting consequence of this method of compactification is that fuzzy manifolds differ from each other discretely. There are no moduli. This is a direct consequence of the holographic principle and has nothing to do with dynamical minimization of potentials. We have noted above that space-time geometry is part of the kinematical framework of holographic space-time. Our discussion of semi-classical gravity and the principle that different solutions of the same gravitational field equations can correspond to different quantum models, rather than different states of the same model, here finds its ultimate justification. Continuous moduli can emerge from the holographic framework when we take the dimension of the function algebra to infinity. There can be different ways to do this, and quantities which go to infinity simultaneously at fixed ratio, define continuous moduli of the limiting geometry.

For example, a fuzzy compactification of a Kahler manifold is provided by the algebra of matrices in the space of holomorphic sections of a line bundle over the manifold. The dimension of this space is fixed by the element of the Picard group, which characterizes the line bundle. These elements are labeled by quantized $U(1)$ fluxes threading two cycles of the manifold and (for ample bundles) the dimension goes to infinity along directions in the Picard group where the fluxes go to infinity. But there are many such directions if the manifold has many two cycles, and the ratios of fluxes through different cycles define continuous Kahler moduli of the limiting manifold.

Note that one cannot really take this kind of limit for a single pixel, or rather if one does so then one has taken the four dimensional Planck length to zero. This would define, at best, a free theory, analogous to free string theory, or at least an interacting subsector that decouples from gravity. The moduli problem of conventional string theory is a result of taking this sort of limit as the starting point of the theory, and then perturbing about it. This remark is even more striking in the context of the theory of stable de Sitter (dS) space that we present in the next section. It follows from the above remark, and the assumption that this theory has a finite number of quantum states, *that it has no moduli.* Furthermore, for a fixed value of the c.c., the volume of the internal manifold in Planck units is severely limited, and the limitation is related to the scale of SUSY breaking! We will deal with this in more detail below, but the essential point is that the dimension of the Hilbert space of the theory is $\pi(RM_P)^2 = K(K+1)\text{lnD}$, where D is the dimension of $\mathcal{P}$, and R the dS radius. Using conventional Kaluza-Klein ideas, we find $\text{lnD} = (M_P/M_D)^2 = (VM_D^{D-4})$. Here D is either 10 or 11, M_D is the D dimensional Planck mass and R the four dimensional dS radius. We will see that the parameter that controls the validity of any four dimensional effective field theory description is $K^{-1/2}$. Thus, a good field theory approximation, for fixed RM_P, requires VM_D^7 to be bounded.

The key restriction on compactifications in this framework is that the algebra of a single pixel should have a representation with precisely one graviton and gravitino in the $K \to \infty$ limit. The classification of such algebras is one of the two central goals of the holographic space-time program. The other is to find equations that determine the scattering matrix. By the way, our focus on four dimensional compactifications is motivated by the search for dS solutions of SUGRA. In the limit (ΛM_P^d) small, a quantum theory of dS space should produce a de Sitter solution of a SUGRA theory. The only known SUGRA Lagrangians that have such solutions, and which also correspond to true compactifications are Lagrangians with minimal SUSY in $d = 4$. Such Lagrangians can have many chiral multiplets, with a relatively unconstrained Kahler potential and superpotential, which can easily have dS minima.

2.6. The Theory of Stable dS Space

The global geometry of dS space is described by the metric

$$ds^2 = -dt^2 + R^2 \cosh^2(t/R) d\Omega_3^2,$$

where $d\Omega_3^2$ is the metric on a unit 3 sphere. As in asymptotically flat or AdS spaces, we can obtain useful information about the quantum theory by investigating perturbations, which do not disturb the asymptotic behavior. Since most ways of foliating this geometry give compact spatial sections, the asymptotic regions to be considered are past and future infinity.

To get an idea of the constraints on such perturbations, consider the exercise of setting small masses m on each point of the sphere, *i.e.* making the "co-moving observers" physical. If we do this at global time T, and space the masses by the particle's Compton wavelength (since in a quantum theory, no particle can be localized more precisely than that), then at $t = 0$ the particle number density is

$$m^3 \cosh^3(T/R),$$

and the 00 component of the stress tensor is exponentially large if $T \gg R$. In other words, long before $t = 0$, the back reaction on the geometry of the test masses becomes important. In order to avoid this, we must make $m \sim \cosh^{-1}(T/R)$ at time T. This strongly suggests that, *if we want to preserve dS asympotics in the future, we must not try to fill the apparently huge volumes of space available in the past with matter.* Rigorous results along these lines have been obtained in [37] [38]. The conclusion of those studies is that if one inserts too much matter in the infinite past, then a singularity forms before $t = 0$. If the singularity can be confined within a marginally trapped surface of radius $< 3^{-\frac{1}{2}}R$, this can be viewed as a black hole excitation of dS space, but if not, the whole space-time experiences a Big Crunch and we are no longer within the class of asymptotically dS space-times.

It is much simpler to understand the finite entropy of dS space, and the arguments that this represents a Boltzmann counting of the total number of quantum states corresponding to the thermodynamic equilibrium state called "the dS vacuum", from the point of view of static coordinates, where

$$ds^2 = -d\tau^2 f(r) + \frac{dr^2}{f(r)} + r^2 d\Omega_2^2,$$

and

$$f(r) = (1 - \frac{R_S}{r} - \frac{r^2}{R^2}).$$

The parameter $R_S d \equiv \frac{2M}{M_P^2}$ is the Schwarzschild radius of a Schwarzschild-de Sitter black hole, and R is the dS radius of curvature. Empty dS space corresponds to $M = 0$. These coordinates cover the maximal causal diamond of a time-like geodesic observer in dS space.

Only the τ translation plus $SO(3)$ rotation generators preserve the static coordinate patch. If we consider quantum field theory on the full dS manifold, then there is an action of the dS group on the field theory Hilbert space, and for free fields, a unique Gaussian state whose two point functions approach those of the Minkowski vacuum at short distances. It has been known for a long time that [39] that this is a thermo-field state for the thermal density matrix in the static patch[x], with temperature $T = \frac{1}{2\pi R}$. Alternatively, this is the state which is chosen by analytic continuation of Euclidean functional integrals on the 4-sphere.

QFT in this geometry actually has an infinite number of states at very low energy, where energy is defined as conjugate to the time τ, at $r = 0$. $f(r)$ vanishes near the horizon, $r = R$, so there is a red shift of finite near horizon frequencies to low frequencies at the origin. If one uses the boundary conditions imposed by the so called Bunch-Davies vacuum on the global dS manifold, one finds an infinite number of states of arbitrarily low energy. It is important to realize that this is exactly the same infinity encountered in global coordinates. At $\tau = 0$ the global geometry has only a finite size and all states are localized in the causal diamond (the other half of the global geometry is just a trick, the thermo-field double trick-for

[x]This is a direct generalization of Israel's discussion [40] of the Hartle-Hawking vacuum in the Kruskal manifold.

computing thermal averages in the causal diamond). As $\tau \to \infty$, nothing falls through the horizon. Rather things get pasted closer and closer to the horizon and they redshift.

Within a causal diamond the infinity is analogous to the infinity of near horizon states of a black hole. And, as in the black hole case, there is a claim that the entropy of dS space is finite and equal to one quarter of the horizon area in Planck units. As with the black hole, we must think of this entropy as representing the maximally uncertain density matrix of the near horizon states, which means that the number of states is finite.

Quantum field theory in a fixed space-time background encourages us to think of dS space as having an infinite number of independent horizon volumes, which are causally disconnected from each other. The thermal entropy of a given horizon is interpreted as a finite entanglement entropy between causally disconnected states of this infinite system. This is supposed to explain the fact that the entropy depends only on the area. We have seen however that the myth of independent horizon volumes is untenable because of gravitational back reaction. Our global considerations suggest a total number of states for an eternal dS space, which is of order the exponential of the Gibbons-Hawking entropy.

I will first outline some general properties of a theory of global dS space, and then a more specific proposal, based on a cartoon of the pixel algebra described in the previous section. In my opinion the correct theory will require us to understand the list of consistent compactifications, which might be quite sparse. It is still within the realm of possibility that there is only *one* consistent answer, and that it describes the real world.

2.6.1. *The two Hamiltonians of Wm. de Sitter*

Our theory of dS space has two Hamiltonians. The first, H, has a random spectrum, distributed in an interval of order $T = \frac{1}{2\pi R}$. Starting from a random initial state, that Hamiltonian will generate expectation values for most operators, which quickly become identical to their thermal averages in the maximally uncertain density matrix. The number of states in the Hilbert space is of order $e^{\pi (RM_P)^2}$, and the average level spacing is $Te^{-\pi (RM_P)^2}$. There will be recurrences on time scales of order $Re^{\pi (RM_P)^2}$.

On time scales less than R, H evolution will not make much of a change in the state. We will postulate another Hamiltonian P_0, which is useful for describing some of the states of the system over these shorter time scales. P_0 will be the operator which approaches the Hamiltonian of a super-Poincare invariant system in the limit $RM_P \to \infty$. It will also be the appropriate operator to identify with approximate descriptions of the system in terms of quantum field theory in a background dS space[y].

In order to understand it, we must first understand which states of the system have such a field theoretic description. A local observer can see only a region of physical size R, so we must ask how many field theory like states can fit in such a region. The density of states of field theory in finite volume grows with energy and the entropy of field theory states in a region of linear size R is of order

$$(RM_c)^3,$$

where M_c is the UV cutoff. The energy of a typical state in this ensemble is

$$E \sim M_c^4 R^3.$$

These estimates are valid as long as the gravitational back reaction is small, a criterion which definitely fails once the Schwarzschild radius E/M_P^2 is of order R. Thus, we must have

$$M_c^4 R^2 < M_P^2,$$

which means that the entropy is of order $(RM_P)^{3/2}$, much less than the total dS entropy. Most of the localized states in the horizon volume are black holes whose radius scales like the horizon volume, and these states do not have a field theoretic description in the horizon volume.

These estimates are valid for any low curvature region, and are similar to the deficit between the entropy of a star and that of a black hole of the same radius. In dS space however we can interpret the extra states as $(RM_P)^{1/2}$ copies of the field theoretic degrees of freedom in a single horizon volume. This allows us to understand the picture of an infinite number of horizon volumes predicted by QFT in curved space-time. As with a black hole, one should postulate a complementarity principle [41], according to which the global description, is a description of the same system as that in static coordinates. In the first the states are interpreted as being localized

[y]This is somewhat confusing since that Hamiltonian is usually associated with the static dS time coordinate. If we look at the action of the corresponding vector field at some interior point of the static observer's causal diamond, the static Hamiltonian converges to the Poincare Hamiltonian. However, they have very different actions on the cosmological horizon. One should identify H with the quantum operator that implements static time translation on the horizon, while P_0 is the corresponding action of the Poincare vector field.

in different regions, while in the static coordinates the same set of states is seen as piled up at the horizon. The time evolution operators corresponding to the two descriptions do not commute with each other. In both the black hole and dS systems, the holographic principle provides an infrared cutoff on the number of states attributed to the system by QFT in curved space-time.

Recall that the Schwarzschild-de Sitter metric is

$$ds^2 = -dt^2 f(r) + \frac{dr^2}{f(r)} + r^2 d\Omega_2^2,$$

where $f(r) = 1 - \frac{R_S}{r} - \frac{r^2}{R^2}$. The black hole mass parameter is given by $2M = M_P^2 R_S$. This metric has two horizons with

$$R^2 = R_+^2 + R_-^2 + R_+ R_-$$

and

$$R_S R^2 = R_+ R_- (R_+ + R_-) = R_+ R_- \sqrt{R^2 + R_+ R_-}.$$

Note that the total entropy of this configuration *decreases* as the black hole entropy $\pi(R_- M_P)^2$ *increases.* There is a maximal black hole mass at which the Schwarzschild and cosmological horizon radii coincide and equal $R_N = \frac{1}{\sqrt{3}} R$. The maximal black hole is called the Nariai black hole.

This entropy formula suggests a model of the system in which the Hilbert space has a finite number of states with logarithm $\pi(RM_P)^2$. Localized states are special low entropy configurations with an entropy deficit, for small R_S[z]

$$\Delta S = 2\pi RM.$$

If we interpret M as the eigenvalue of a Hamiltonian we will call P_0, this relation between the eigenvalue and entropy deficit indicates that the maximally uncertain density matrix is effectively a thermal distribution

$$\rho \propto e^{-2\pi R P_0},$$

for eigenvalues of P_0 much less than the Nariai black hole mass. As a consequence, the Poincare Hamiltonian, a generator acting on localized states in a single cosmological horizon of dS space, which converges to the the Hamiltonian of the super-Poincare invariant limiting theory when the c.c. goes to zero, can be written

$$P_0 = \sum E_n P_n.$$

---

[z]For general M, the entropy deficit is $\Delta S = 2\pi RM(1 + \frac{\Delta S}{S})^{-\frac{1}{2}}$, which indicates that large black holes are present with somewhat larger than thermal probability.

The P_n are commuting orthogonal projection operators, with

$$\text{Tr } P_n = e^{\pi(RM_P)^2 - \delta S_n}.$$

$\Delta S_n = 2\pi E_n R$, when $E_n \ll M_P^2 R$, and near the maximal mass is given by the formula in the previous footnote.

We can summarize the previous few paragraphs by saying that the Bekenstein-Gibbons-Hawking formula for the entropy of black holes in dS space motivates a model for the quantum theory of dS space in which empty dS space is interpreted as the infinite temperature ensemble of a random Hamiltonian H bounded by something of order the dS temperature ($||H|| \leq cT$)[aa]. This implies that localized black hole states are low entropy deformations of the vacuum, and gives a connection between the black hole mass parameter, which is the eigenvalue of another Hamiltonian P_0, and the entropy deficit of its eigen-spaces. This observation leads us to expect what we already know to be true: *the dS vacuum is a thermal state for quantum field theory with a unique temperature* $T = \frac{1}{2\pi R}$, and the present discussion can be viewed as an explanation of that fact from a more fundamental point of view. It is particularly satisfying that this explanation provides a rationale for the uniqueness of the dS temperature.

There is another piece of semi-classical evidence that this picture is valid. The Coleman-DeLucia formalism gives us an unambiguous calculation of the transition rates between two different dS spaces. As discussed above, the CDL formula implies that the ratio of the two transition rates is given by the *infinite temperature limit of the principle of detailed balance.* This is in perfect accord with our model of the dS vacuum as the infinite temperature ensemble in a Hilbert space of finite dimension. Similarly, Ginsparg and Perry [42] and Bousso and Hawking [43] have found instantons for the nucleation of black holes in dS space, and their results are completely consistent with the framework outlined above.

2.6.2. *Towards a mathematical theory of stable dS space*

It is my belief that the theory of dS space only makes sense in 4 dimensions. This follows from the basic principles I've enunciated, plus a knowledge of low energy effective field theory. The basic principle we use is that SUSY is

[aa]The bound on the Hamiltonian should be zero in the classical limit, consistent with the classical notion of a vacuum. This means it is of the form $Tf(T/M_P)$. Since the notion of localized observables in dS space only makes sense when $\frac{T}{M_P} \ll 1$, the linear approximation should be sufficient. So far I have not found any measurable quantity whose value depends on $f(0) \equiv c$.

restored as the c.c. goes to zero, with the gravitino mass going like $m_{3/2} = 10K\Lambda^{1/4}$. We will give two arguments for this behavior below. This formula implies that SUSY breaking must be describable in low energy field theory, which in turn implies that it must be spontaneous, since the gravitino mass and decay constant are much smaller than the Planck scale. Supergravity Lagrangians in 5 or more dimensions do not have de Sitter solutions[bb], while in four dimensions, models with chiral fields and appropriate super and Kahler potentials can have lots of dS solutions. This remains true for dimension less than four. However, the interesting physics of dS space is the localizable physics that is accessible to a local time-like observer. As we will see, this is described by an approximate S-matrix, which approaches that of a super-Poincare invariant model as $RM_P \to \infty$. In 2 and 3 space-time dimensions there can be no such limiting theory, so there is probably no useful model of low dimensional dS space either.

The notion of an approximate S-matrix can be formalized as follows. Consider a causal diamond in dS space whose holographic screen has an area $b\pi(RM_P)^2$, with $b < \frac{1}{2}$. Assume also that $(RM_P) \gg 1$. According to the general principles of holographic space-time there should be an *approximate scattering matrix* $S(b, R)$, which operates on the eigenstates of the Poincare Hamiltonian, relating two bases of eigenstates on the past and future boundaries of the diamond. We do not yet have a prescription for constructing S, but knowledge of effective field theory in dS space leads to the conclusion that this S-matrix becomes insensitive to the dS horizon as $R \to \infty$.

On an intuitive level this sounds obvious, but there is an important subtlety. We define the scattering matrix as the interaction picture evolution operator $U(T, -T)$ in an effective field theory in static coordinates. The time T is chosen such that the causal diamond of the geodesic observer at the origin, between $-T$ and T has holoscreen area $b\pi(RM_P)^2$. The intuitive argument that this S-matrix becomes independent of R as $RM_P \to \infty$ is that the maximal Gibbons-Hawking temperature encountered in that causal diamond is $\frac{(1-b)^{-\frac{1}{2}}}{2\pi R}$, which goes to zero in the limit. The local

[bb]There are solutions of the form $dS \times K$, where K is a negatively curved manifold. If K is compact, there is no control over the amount of SUSY breaking, because there are large corrections to the classical bulk solutions, and both the dS and compact radii of curvature are naturally of order the cutoff. For the dS radius this is just the fine tuning of the c.c. in effective field theory, but the compact radius is an additional fine tuning. Some of the literature considers non-compact K, but throws away all but the constant mode on K. The meaning of these papers is completely obscure to me.

geometry also approaches Minkowski space. If we consider a configuration space Feynman diagram contributing to the S-matrix, then all parts of it within the causal diamond converge to their flat space values as the dS radius goes to infinity.

As we approach the horizon, field theory in static coordinates encounters an infinity. The coefficient of $d\tau^2$ vanishes, which means that the norm of the Killing vector field $\frac{\partial}{\partial \tau}$, goes to infinity. As a consequence, very high frequency modes of the field, localized near the horizon, are low energy states as viewed from the origin. As we approach the horizon, we appear to see an infinite number of modes, all of which "our friend at the origin" considers low energy.

A general relativist will attribute this to our insistence on using "bad coordinates". The message of the holographic principle is that the pileup of states near the horizon is just the the holographic image of all physical excitations which have fallen through the horizon in coordinate systems that are regular there. It also instructs us to cut off the infinity, so that the total entropy of these states is finite. The latter instruction cannot be understood in terms of quantum field theory, but must be built in to the quantum theory of dS space we are trying to construct. In the next section, we will describe how thinking about Feynman diagrams with internal lines that penetrate the horizon leads to a relation between the gravitino mass and the c.c. We'll derive that relation from different considerations in this section.

According to our general formalism, all the states in dS space are accounted for in the irreducible representation of the pixel algebra

$$[(\psi^M)_i^A, (\psi^{\dagger\ N})_B^j = \delta_i^j \delta_B^A Z^{MN},$$

where M and N run over a basis of sections of the spinor bundle over the fuzzy compactification. For each pixel, the irrep has dimension D and we have

$$\pi(RM_P)^2 = K(K+1)\ln\, D.$$

In terms of Kaluza-Klein language, $\ln\, D = V$, the volume of the compact dimensions in higher dimensional Planck units. We also have the K-K relation $V = (M_P/M_D)^2$.

Particle states localized within our causal diamond are described by considering the algebras of block diagonal matrices, with block sizes K_i, with $\sum K_i = K$ [44]. The spinor bundle over such an algebra is the direct sum of the set of $K_i \times K_i + 1$ matrices (and their $K_i + 1 \times K_i$ conjugates), each tensored with the internal spinor bundle. If, as $K \to \infty$, the representation space of the pixel algebra approaches a direct sum of supersymmetric particle state spaces, then the block diagonal construction, with $K_i \to \infty$ and $\frac{K_i}{K_j}$ fixed, approaches the Fock space of that collection of supermultiplets, with the correct Bose/Fermi gauge equivalence (particle statistics). We must of course include block decompositions with an arbitrary number of blocks. Indeed, a direct sum of algebras always has a permutation gauge symmetry, when we view it as constructed from block diagonal matrices.

If K is fixed and very large, only some of these block diagonal constructions really resemble particles. If K_i is too small, then the would-be particle will not be localizable on the holographic screen, whereas if K_i is too large there will not be any multi-particle states. The compromise, which maximizes the entropy, while still retaining particle-like kinematics, is to take each K_i of order $\sqrt{K}$. The total entropy in such states is of order $(RM_P)^{3/2}$, which is the same scaling we derived by heuristic consideration of particle states in dS space, which do not form black holes.

There are 3 important remarks to make about this construction.

- By considering off-diagonal bands in the block diagonalization of the algebra of $K \times K$ matrices[cc], we see of order $\sqrt{K}$ identical copies of the highest entropy particle states. These may be considered *particle states in other horizon volumes* and we see how we can reproduce the claim of QFT in curved space-time, in the $K \to \infty$ limit[dd]. However, thinking in terms of the static coordinates, all but one of these collections of particles should be lumped together into the states on a particular observer's holographic screen. There are of course of order $(RM_P)^2$ such states.
- The fixed ratios between the K_i should be interpreted as the ratios of magnitudes of the longitudinal momenta of the different particles. Those familiar with Matrix Theory will recognize this rule. We can motivate it by the following remarks. The conformal group of the two sphere is the spin-Lorentz group $SL(2, C)$ and the spinor bundle contains solutions of the conformal Killing spinor equation $D_z s = \gamma_z s$, where z is a holomorphic coordinate on the sphere and γ_z is the pullback of the two dimensional Dirac matrices by the

[cc]This means the ith upper off diagonal band, completed by the $K - i$th lower off diagonal.

[dd]The transformations that map one off diagonal band into the next should be thought of as discrete analogs of the dS boosts, which change one static observer into another.

zweibein. The solutions of the conformal Killing spinor equation transform as a Dirac spinor q^a_α under $SL(2,C)$. The requirement that the representations of the pixel algebra are supermultiplets in the large K_i limit implies in particular that there are generators that converge to

$$S_a(\Omega_0) = S_a\delta(\Omega, \Omega_0),$$

where S_a are two component real spinors under $SO(2)$, which satisfy a Clifford algebra. These operators are a "basis" for the space of sections of the spinor bundle. They should be thought of as operator valued measures on the space of sections. When we integrate them against the conformal Killing spinors we get

$$Q_\alpha(\Omega_0) = \int S_a(\Omega_0, \Omega) q^a_\alpha(\Omega) = S_a q^a_\alpha(\Omega_0).$$

If

$$[S_a, S_b]_+ = p\delta_{ab},$$

then

$$[Q_\alpha, Q_\beta] = (\gamma^0\gamma^\mu)_{\alpha\beta}P_\mu,$$

$$P_\mu = p(1, \Omega).$$

In deriving the continuous generators from the fuzzy sphere, the normalization p arises in the usual way. The discrete generators differ from the continuous one by an infinite normalization proportional to K_i, so the ratios of p_i are the ratios of K_i.

The precise super-particle spectrum that comes out in the limit depends on the details of the rest of the pixel algebra representation. The classification of pixel algebras whose limit gives rise to a super-particle spectrum containing the $N = 1$ SUGRA multiplet is the analog in this formalism of classifying all supersymmetric compactifications of string theory with minimal SUSY in 4 dimensions. However, if we keep the pixel algebra fixed and take $K \to \infty$, as is appropriate for a theory that is the limit of stable dS space, then we only obtain models with no moduli. Other supersymmetric models, which can be described in terms of perturbative string theory, come from more elaborate limits in which we take both K and the size of the pixel algebra to infinity at the same time, obtaining continuous moduli. These are not related to dS models.

The control parameter that governs the restoration of super-Poincare symmetry is the typical particle momentum, K_i, which scales like $\sqrt{K}$. Rotational symmetry is of course exact, while the Lorentz group is realized as the conformal group of the two sphere. The accuracy with which it can be represented is limited by the total number of spherical harmonics available, which scales like K On the other hand, we can expect the violation of the super-Poincare relation

$$[Q_\alpha, P_0] \sim K^{-\frac{1}{2}}.$$

For a theory with spontaneously broken SUSY, the superpartner of any state is that state plus one gravitino, so we get the estimate

$$m_{3/2} = K^{-\frac{1}{2}} M_P.$$

Taking into account the relation between the dS entropy, K and $\ln D$ we get

$$m_{3/2} = c(\ln D)^{1/4} \Lambda^{1/4} = 10c\Lambda^{1/4}.$$

The last estimate incorporates Witten's idea [45] that the volume of extra dimensions is the explanation for the ratio of 100 between the reduced Planck scale and the unification scale. We might expect c to be of order 1 but we cannot say that we've accounted for all factors of 2π correctly. If c is of order 1 then we get a gravitino mass of order 10^{-2} eV and a gravitino decay constant $F \sim 30(\mathrm{TeV})^2$.

2.7. Implications for Particle Phenomenology

I'll begin this section with an alternative derivation between the gravitino mass and the cosmological constant, based on the notion of Feynman diagrams with internal lines going through the horizon. We want to consider a dS space with very large RM_P. Low energy physics is approximately the same as it is in the limiting super-Poincare invariant model. The latter is described by an $N = 1$ SUGRA Lagrangian, with a super-Poincare invariant vacuum. In order to ensure that the cosmological constant is self-consistently zero, we impose a discrete R symmetry on the low energy Lagrangian. We want to compute the leading correction to this supersymmetric Lagrangian, which leads to the SUSY violation we expect in dS space.

This is computed, as effective Lagrangians always are, in terms of Feynman diagrams, and the new effects of dS space obviously have to do with diagrams in which internal lines go out to the horizon. They cannot lead to explicit violation of SUSY, and renormalization of parameters in the effective Lagrangian will not violate SUSY. However, interactions with the horizon *can* violate R symmetry. If we consider a diagram whose external legs are localized near the origin, then lines going out to the horizon are extended over space-like intervals of geodesic length R. If we assume that the gravitino is the lightest R charged particle in the model, the leading R violating diagrams will have two gravitino lines leading out to the horizon and will have an exponential suppression $e^{-2m_{3/2}R}$. It does not make sense to neglect the gravitino mass in this formula, but the rest of the diagram is evaluated in the $\Lambda = 0$ theory. Recalling that the horizon has a huge number (infinite in the field theory approximation) of very low energy states, of order $e^{\pi(RM_P)^2}$, we can write the contribution of this diagram as

$$\delta\mathcal{L} \sim e^{-2m_{3/2}R} \sum | < 3/2|V|s > |^2,$$

where V is the operator representing emission from and absorption of the gravitino by the horizon.

The horizon is a null surface and the massive gravitino can only propagate near it for proper time of order its Compton wavelength. As a quantum particle it does a random walk, and we take the proper time step to be the Planck scale. Thus, the area in Planck units that it covers is of order $\frac{M_P}{m_{3/2}}$, and we take this as an estimate of the logarithm of the number of states for which the matrix element is of order 1. The total contribution is thus of order

$$\delta\mathcal{L}e^{-2m_{3/2}R+b\frac{M_P}{m_{3/2}}}.$$

This formula can be self consistent only for one behavior of the vanishing gravitino mass in the $RM_P \to \infty$ limit. If we assume the gravitino mass goes to zero too rapidly, for example like the naive SUGRA prediction $m_{3/2} \sim \sqrt{\Lambda}/m_P$, then the formula predicts exponentially large corrections to the effective Lagrangian. If we assume it goes to zero too slowly the effective Lagrangian is exponentially small, which is inconsistent with the assumption. In effective field theory, it is this correction the the Lagrangian that is responsible for the gravitino mass. For self-consistency, the exponential dependence on R must cancel exactly

$$m_{3/2}\sqrt{\frac{bM_P}{2R}}.$$

This is the same scaling we found in the previous section, but we learn less about the coefficient.

We conclude that the low energy Lagrangian of stable dS space has the form

$$\mathcal{L}_0 + \mathcal{L}_{\Delta R}.$$

The full Lagrangian must predict a dS solution, and implement the relation between the gravitino mass and the c.c. An example of such a Lagrangian would be

$$\mathcal{L}_{\Delta R} = \int d^2\theta \, (W_0 + FG),$$

with G a single chiral superfield, the goldstino multiplet, which we assume is the only low energy matter field. $\mathcal{L}_0$ would have a discrete R symmetry, which forbade both of these terms. In order that there be no SUSY vacuum in low energy effective field theory, we have to assume that G has R charge 0. However, the demands of the underlying theory are not so strict. We could for example insist only that the R symmetry forbid terms up to cubic order in G and that the natural scale in $\mathcal{L}_0$ is just the Planck scale. Then there might be SUSY minima at $S \sim m_P$, but the Lagrangian could be above the Great Divide, and consistent with the underlying finite dimensional model for dS space.

While this model satisfies the basic consistency conditions, it is not our world. In the real world, we must couple the SUSY violating order parameter to standard model supermultiplets. In particular, gaugino masses would result from terms of the form

$$\int d^2\theta \, f_i(G/M) W_\alpha^{i\ 2}$$

and would be given by

$$m_{1/2}^1 = f_i'(G/M)(F/M).$$

Since $F \sim 30(\mathrm{TeV})^2$, M cannot be larger than a few TeV if we are to obey the experimental bounds (there are factors of standard model fine structure constants in f_i). This indicates that there must be a new strongly coupled gauge theory with confinement scale M, which contains fields transforming under the standard model. The Goldstino field G must be an elementary field with renormalizable couplings to the new gauge system, a composite field from that system, or a combination of both.

Given these couplings, squarks and sleptons will get mass via gauge mediation. We are forced, by the low maximal scale of SUSY violation, to consider a model of direct gauge mediation. Such models are notorious for having problems with coupling unification. One must have complete representations of the unified gauge group, with low multiplicities, which means that the hidden sector gauge group must be small and the representations of the new chiral matter of low dimension. So, for example, if the unified gauge group is $SU(5)$, we can, when two loop corrections are taken into account, tolerate at most 4 $5 + \bar{5}$ pairs in the hidden sector.

While I have not done a definitive survey, all examples I've studied of hidden sectors that satisfy these constraints contain light fields with standard model quantum numbers, which are ruled out by experiment. Simple unification appears incompatible with direct gauge mediation. One appears forced to utilize Glashow's *trinification* scheme, in which the standard model is embedded in

$$SU_1(3) \times SU_2(3) \times SU_3(3) \rtimes Z_3,$$

where Z_3 cyclically permutes the three $SU(3)$ groups. The standard model chiral superfields are embedded in 3 copies of

$$(1, \bar{3}, 3) + (3, 1, \bar{3}) + (\bar{3}, 3, 1),$$

as the 15 states that transform chirally under the standard model. There is a nice embedding of this in E_6, but that would put us back in the forbidden realm of simple unification. More interesting is the way that this structure, *including the prediction of the number of generations*, arises from 3 D3-branes at the Z_3 orbifold in Type IIB string theory. We also note that the vector-like spectrum of this model contains 3 copies of the conventional SUSY Higgs fields. However, at least in the orbifold construction the implied structure of standard model Yukawa couplings comes out wrong.

We can add a hidden sector to trinification, without ruining standard model coupling unification, by postulating an $SU(N)$ gauge theory, with $N = 3, 4$ and chiral fields T_i and $\tilde{T}_i$ in the $(\bar{N}, 3_i) + (\bar{3}_i, N)$. These models have a pyramidal quiver diagram and are called the Pyramid Schemes [46]. At the level of the orbifold construction the new fields come from $D7$-branes and one can think of the model as an F-theory solution with an orbifold singularity in its base.

There is no room here to go into the intricate details of model building, but the Pyramid Schemes throw new light on the strong CP problem, the little hierarchy problem, the origin of the μ term in the MSSM, the nature of dark matter, *etc.*. They have a rich phenomenology and can easily be ruled out at the LHC. It is not clear whether the LHC energy is high enough to reveal the complete structure of these models.

What I would like to emphasize is that the theory of stable dS space we have adumbrated gives rather detailed predictions for Terascale physics. Thus, despite its rather abstract origins, and the incomplete nature of the theory of holographic space-time, we may hope in the near future for experimental input that could encourage us to continue to work on this set of ideas, or convince us to abandon them.

2.8. Appendix: Exercises on CDL Tunneling

In my lectures, I asked the students to work out some of the theory of gravitational tunneling for themselves, because there are so many erroneous notions in the community about the results of Coleman and De Luccia. Much of my second lecture was an extended "recitation section", in which I outlined the solution of these problems. There are also some exercises on black hole solutions.

1. Show that the metric

$$ds^2 = -f(r)dt^2 + \frac{dr^2}{f(r)} + r^2 d\Omega^2,$$

with

$$f(r) = 1 - c_d \frac{M}{r^{d-3} M_P^{d-2}} \pm \frac{r^2}{R^2},$$

solves the d space-time dimensional Einstein equations with cosmological constant Λ , where $R^{-1} = b_d \sqrt{\Lambda}/M_P^{d/2-1}$ is the Hubble scale associated with the c.c. Work out the necessary constants for all d. The Einstein equations are

Show that positive c.c. corresponds to the choice of negative sign in $f(r)$. These are the Schwarzschild black hole solutions for all possible maximally symmetric background space times.

2. Show that for positive c.c. $f(r)$ has two zeroes, corresponding to the two positive roots of a cubic equation

$$(r - R_+)(r - R_-)(r + R_+ + R_-).$$

$R_{\mp}$ is the position of the black hole (cosmological) horizon. Show that both R_{pm} are $< R$ and the entropy deficit

$$\pi(R^2 - R_+^2 - R_-^2)M_P^2$$

is always positive and is approximately

$$\Delta S \simeq 2\pi RM$$

when $R_- << R$. Find the maximal black hole mass in de Sitter space and argue that it has the smallest total entropy. This little exercise shows that localized states are low entropy excitations of the dS vacuum, which we have argued should be modeled by an infinite temperature density matrix on a Hilbert space with a finite number of states.

3. The Coleman De Lucia (CDL) equations for gravitational tunneling are the equations for a scalar field coupled to Euclidean Einstein gravity, with $SO(4)$ symmetry. This is the Euclidean analog of FRW cosmology: a four dimensional space-time with a maximally symmetric 3 dimensional subspace. The equations are

$$\phi'' + 3\frac{\rho'}{\rho}\phi' = \frac{dU}{d\phi},$$

$$\rho'2 = 1 + \frac{\rho^2}{3m_P^2}(\frac{1}{2}\phi'^{\ 2} - U)$$

The metric is

$$ds^2 = dz^2 + \rho^2(z)d\Omega_3^2.$$

A. If the potential has the form

$$U = -\mu^4 v(x),$$

where $x = \phi/M$, and we make everything dimensionless using the space-time scale $z = \frac{M}{\mu^2}\tau$, show that the equations take the form

$$\ddot{x} + \frac{\dot{a}}{a}\dot{x} = -\frac{dv}{dx},$$

$$\dot{a} = 1 + \epsilon^2 a^2(\frac{1}{2}\dot{x}^2 + v),$$

where $\epsilon^2 = \frac{M^2}{3m_P^2}$.

If we are tunneling *from* a solution with non-positive c.c.,then the Euclidean space-time is infinite. We insist that, at infinity, the solution approach the Flat or Hyperbolic space solution. By convention we set the field value at infinity to zero, and $v(0) = c$, so that the c.c. is $-c\mu^4$. The other maximum of v is called x_T. The solution for the metric at infinity is

$$a = \sinh(\epsilon c\tau)/\epsilon c \to \tau.$$

The last limit is $c \to 0$. The exact solution will also have a point where $a = 0$. This is the center of the vacuum bubble.

The equations (but not the metric) have the form of an FRW *EUniverse* with a scalar field. The "Big Bang" is the center of the bubble, which we conventionally call $\tau = 0$ and we must have $\dot{x} = 0$ there in order to have a regular solution. We must choose the value of $x(0)$ in order to satisfy the boundary conditions at infinity. We choose $\dot{a}(0) = 1$. This expanding Euniverse condition just says that we are following the Euclidean configuration to larger radius spheres. In Euclidean space, this analog Big Bang is not a singularity. Since $c \geq 0$, the real, Lorentzian signature c.c. is non-positive, but the analog EUniverse has non-negative c.c. at the maximum of v at $x = 0$. The equations correspond to motion under a complicated frictional force, plus the force derived from the potential v. Show that the EEnergy,

$$EE = \frac{1}{2}\dot{x}^2 + v,$$

will be monotonically decreasing as long as $\dot{a}$ remains positive, as will the speed.

B. In the non-gravitational case ($\epsilon = 0$) $\dot{a}$ is always positive. Then, it's clear that the friction term goes to zero at large t. Argue that there are values of $x(0)$, such that $x(t)$ will undershoot the maximum of v at $x = 0$. Argue that by starting close to x_T we can find solutions that overshoot $x = 0$. Argue that continuity implies there is a solution, which asymptotes to 0 at $\tau = \infty$. Argue that this analysis remains valid when $\epsilon \ll 1$, as long as v is of order one and $v(x_T) - v(0)$ is of order one.

C. When ϵ is of order 1 a new behavior sets in. The sign of $\dot{a}$ can change in the region where v is negative. If this happens, the solution never reaches infinite radius. The radius shrinks to zero and the $\dot{x}$ does not go to zero. Friction turns to anti-friction as the radius shrinks and the velocity actually goes to infinity. We can think of this as a Big Crunch of the EUniverse. Such solutions exist when ϵ is very small, but they are confined to a small range of values of $x(0)$ near the minimum of v. The transition between overshooting and undershooting happens at a larger negative value of $x(0)$ so the instanton always exists. Argue that as ϵ is increased, the onset of Big Crunch solutions moves to more negative values of x. Eventually it crosses the transition point between overshoot and undershoot solutions, and no instanton exists.

D. When $c > 0$, even when the instanton exists, its interpretation is not that of an unstable bubble that can appear as a state in the AdS space.

Argue this as follows: Near infinity the instanton $x(\tau)$ becomes small, and is well approximated by a solution of the equations for small fluctuations around Euclidean AdS space. Argue that, because of the boundary conditions on the instanton at $\tau = 0$ it is a linear combination of both the normalizable and non-normalizable solutions of the linearized equations. As you will learn in other lectures on the AdS/CFT correspondence, this means that it corresponds to adding an operator to the Hamiltonian. States in the model with the original Hamiltonian correspond to purely normalizable solutions at infinity. In all cases of the AdS/CFT correspondence where such instantons have been found, the operator that is added is unbounded from below.

E. The overshoot solutions are those for which $x(0)$ is near x_T. Thus, as ϵ is raised, the point at which instantons disappear is the point at which $x(0)$ is forced to x_T in order to avoid a crunch. However, this is no longer an instanton, because if we start a solution at x_T with zero velocity, it stays there. What happens instead is that the point recedes in geodesic distance, and the solution becomes infinite in both the $\tau = 0$ and $\tau = \infty$ limits. Show that the interpretation of this solution is as a static domain wall between two AdS regions (or an AdS and Minkowski region). Show more generally that the existence of such a static domain wall always requires the fine tuning of one parameter in the potential. We summarize this in the statement that the sub-manifold in the space of potentials, on which a static domain wall solution exists, *has co-dimension one.* This sub-manifold is called The Great Divide. On one side of the Great Divide instantons exist, while on the side we call Above the Great Divide, they don't exist. There is a connection between this and the positive energy theorem in General Relativity, which I will explain in the lectures.

F. Show that the Euclidean continuation of dS space is a 4 sphere, and that it has negative Euclidean action. In fact, in an echo of the Gibbons-Hawking [8] result for Euclidean black holes, the action is just equal to minus the entropy of dS space. Correspondingly, instantons for the "decay" of dS are compact 4 manifolds with negative Euclidean action. We make a probability formula that is ≤ 1 by subtracting the negative dS action[ee] of the initial "decaying" state.

$$P_{12} = e^{-(S_I - S_{dS_1})}.$$

If the state 2 to which dS_1 "decays" is also a dS space, then we can form

[ee]The rule of subtracting off the action of the initial configuration is motivated by quantum field theory, where we can *prove* that this is the right thing to do.

the reverse probability

$$P_{21} = e^{-(S_I - S_{dS_2})}.$$

This leads to

$$\frac{P_{12}}{P_{21}} = e^{-\Delta\ Entropy}.$$

Argue that this is the infinite temperature form of the *principle of detailed balance*. Consider two finite collections of states, such that the transition amplitude for any state in collection 1 to any state in collection 2 is the same. Show that unitarity implies that there are reverse transitions, and that the probabilities for the two ensembles to decay into each other are related by the above equation. The CDL formula thus provides evidence for the picture expounded in the lectures, in which dS space is modeled as a system with a finite number of states. Notice that the instanton transition for the lower c.c. dS space, is here interpreted, not as an instability, but as a temporary sojourn of a large system in a very low entropy configuration, like the air in a room collecting in a little cube in the corner.

More controversial is the contention, also expounded in the lectures, that the same interpretation is valid *above The Great Divide* for dS "decays" into negative c.c. Big Crunches. The holographic principle shows that the latter are low entropy states, and we should expect rapid transitions back from them to the equilibrium dS configuration. These reverse transitions, cannot be modeled by instantons, because the initial configuration is not classical in any way.

2.9. Appendix: Potentials in String Theory

In tree level string theory, one can only add sources to the system if they correspond to vertex operators for asymptotic states of the system in a fixed space-time background. In asymptotically flat space, this means that one can only add constant sources, as in the definition of the field theoretic 1PI potential, for massless particles. They correspond to rather singular limits of genuine scattering amplitudes, but, *so long as the particle remains massless for all values of the source*, they seem sensible. The italicized phrase means that the analog of the effective potential can only be defined when it is exactly zero.

In non-supersymmetric string theory, even when there are no tachyons, the perturbation expansion is singular at one loop. Fischler and Susskind

[20] argued that these singularities could be removed by changing the background space-time. This procedure leads to time dependent solutions, and general considerations show [12] that the time dependence is singular[ff].

Fischler and Susskind tried to argue that their procedure gave a method for computing quantum corrections to the effective potential in string theory. They showed that there was a Lagrangian, at the appropriate order in string coupling, which reproduced the modified background solution they had found. Students who have studied the rest of these lectures, will know that such a demonstration says nothing whatever about the existence of other solutions of the same equations of motion, as bona fide theories of quantum gravity. This argument is independent of the question dealt with above, to the effect that the Fischler Susskind solution itself does not provide evidence for the existence of a model of quantum gravity based on their modified background.

Another attempt to define effective potentials in string theory tries to define a String Field Theory [47]. Open String Field Theory is an elegant construction, which reproduces tree-level open string amplitudes. However, at the loop level it is singular, because of the familiar fact that open string loops imply closed strings. *Any* regularization of that singularity forces us to introduce an independent closed string field. Closed String Field Theory is *not* a non-perturbative definition of theory. Its Lagrangian must be corrected at each order in perturbation theory, in order to reproduce the correct loop amplitudes. Furthermore, the series that defines the string field action is divergent. Much has been made of the fact that the open string field theory "contains closed strings automatically", and it's been proposed that this gives a non-perturbative definition of the theory. In fact, the appearance of closed strings is ambiguous and the relevant open string diagrams are singular. When one tries to regulate the singularities, one finds that one must introduce an independent closed string field, with the difficulties noted above.

The upshot of this is that there is no indication in any perturbative string theory calculation, that there is a beast like the mythical effective potential, whose minima classify different consistent theories of quantum gravity. Every non-perturbative definition of string theory leads to precisely the opposite conclusion, as we have sketched in the main lectures.

[ff]In a systematic F-S expansion, the time dependence appears linear but at large times this expansion breaks down. One can try to do a more exact solution of the low energy field equations, but this leads to singular cosmological solutions. There does not appear to be a way to make the F-S mechanism into a controlled expansion.

References

[1] T. Banks, W. Fischler, S. H. Shenker and L. Susskind, "M theory as a matrix model: A conjecture," Phys. Rev. D **55**, 5112 (1997) [arXiv:hep-th/9610043].

[2] J. M. Maldacena, "The large N limit of superconformal field theories and supergravity," Adv. Theor. Math. Phys. **2**, 231 (1998) [Int. J. Theor. Phys. **38**, 1113 (1999)] [arXiv:hep-th/9711200].

[3] S. S. Gubser, I. R. Klebanov and A. M. Polyakov, "Gauge theory correlators from non-critical string theory," Phys. Lett. B **428**, 105 (1998) [arXiv:hep-th/9802109]. ;E. Witten, "Anti-de Sitter space and holography," Adv. Theor. Math. Phys. **2**, 253 (1998) [arXiv:hep-th/9802150].

[4] O. Aharony, S. S. Gubser, J. M. Maldacena, H. Ooguri and Y. Oz, "Large N field theories, string theory and gravity," Phys. Rept. **323**, 183 (2000) [arXiv:hep-th/9905111].

[5] J. D. Bekenstein, "Black Holes And The Second Law," Lett. Nuovo Cim. **4**, 737 (1972). ; J. D. Bekenstein, "Black Holes And The Second Law," Lett. Nuovo Cim. **4**, 737 (1972). ; S. W. Hawking, "Particle Creation By Black Holes," Commun. Math. Phys. **43**, 199 (1975) [Erratum-ibid. **46**, 206 (1976)].

[6] J. Polchinski, "S-matrices from AdS spacetime," arXiv:hep-th/9901076. ; L. Susskind, "Holography in the flat space limit," arXiv:hep-th/9901079. S. B. Giddings, The boundary S-matrix and the AdS to CFT dictionary, Phys. Rev. Lett. 83, 2707 (1999) [arXiv:hep-th/9903048]; M. Gary and S. B. Giddings, "The flat space S-matrix from the AdS/CFT correspondence?," Phys. Rev. D **80**, 046008 (2009) [arXiv:0904.3544 [hep-th]] ; M. Gary, S. B. Giddings and J. Penedones, "Local bulk S-matrix elements and CFT singularities," Phys. Rev. D **80**, 085005 (2009) [arXiv:0903.4437 [hep-th]].

[7] K. A. Khan and R. Penrose, "Scattering of two impulsive gravitational plane waves," Nature **229**, 185 (1971). ;T. Banks and W. Fischler, "A model for high energy scattering in quantum gravity," arXiv:hep-th/9906038. ;H. J. Matschull, "Black hole creation in 2+1-dimensions," Class. Quant. Grav. **16**, 1069 (1999) [arXiv:gr-qc/9809087] ; D. M. Eardley and S. B. Giddings, "Classical black hole production in high-energy collisions," Phys. Rev. D **66**, 044011 (2002) [arXiv:gr-qc/0201034].

[8] G. W. Gibbons and S. W. Hawking, "Cosmological Event Horizons, Thermodynamics, And Particle Creation," Phys. Rev. D **15**, 2738 (1977).

[9] W. Fischler, Taking de Sitter seriously, Talk given at Role of Scaling Laws in Physics and Biology (Celebrating the 60th Birthday of Geoffrey West), Santa Fe, Dec. 2000; T. Banks, "Cosmological breaking of supersymmetry or little Lambda goes back to the future. II," arXiv:hep-th/0007146. ;T. Banks, "Cosmological breaking of supersymmetry?," Int. J. Mod. Phys. A **16**, 910 (2001).

[10] S. R. Coleman and F. De Luccia, "Gravitational Effects On And Of Vacuum Decay," Phys. Rev. D **21**, 3305 (1980).

[11] M. Cvetic, S. Griffies and H. H. Soleng, "Local and global gravitational aspects of domain wall space-times," Phys. Rev. D **48**, 2613 (1993) [arXiv:gr-qc/9306005] ; M. Cvetic, S. Griffies and H. H. Soleng, "Nonextreme and ultraextreme domain walls and their global space-times," Phys. Rev. Lett. **71**, 670 (1993) [arXiv:hep-th/9212020] ; M. Cvetic, S. Griffies and S. J. Rey, "Static domain walls in N=1 supergravity," Nucl. Phys. B **381**, 301 (1992) [arXiv:hep-th/9201007].
[12] T. Banks and M. Dine, "Dark energy in perturbative string cosmology," JHEP **0110**, 012 (2001) [arXiv:hep-th/0106276].
[13] L. Susskind, Unpublished, 1997; A. Sen, "D0 branes on T(n) and matrix theory," Adv. Theor. Math. Phys. **2**, 51 (1998) [arXiv:hep-th/9709220] ; N. Seiberg, "Why is the matrix model correct?," Phys. Rev. Lett. **79**, 3577 (1997) [arXiv:hep-th/9710009].
[14] A. Giveon, D. Kutasov and N. Seiberg, "Comments on string theory on AdS(3)," Adv. Theor. Math. Phys. **2**, 733 (1998) [arXiv:hep-th/9806194].
[15] E. Brezin, V. A. Kazakov and A. B. Zamolodchikov, "Scaling violation in a field theory of closed strings in one physical dimension," Nucl. Phys. B **338**, 673 (1990). ; D. J. Gross and I. R. Klebanov, "One Dimensional String Theory on a Circle," Nucl. Phys. B **344**, 475 (1990). ; G. W. Moore, "Double scaled field theory at c = 1," Nucl. Phys. B **368**, 557 (1992). ; J. Polchinski, "Classical Limit Of (1+1)-Dimensional String Theory," Nucl. Phys. B **362**, 125 (1991) ;J. Polchinski, "On The Nonperturbative Consistency Of D = 2 String Theory," Phys. Rev. Lett. **74**, 638 (1995) [arXiv:hep-th/9409168].
[16] K. Dasgupta, G. Rajesh and S. Sethi, "M theory, orientifolds and G-flux," JHEP **9908**, 023 (1999) [arXiv:hep-th/9908088] ; R. Blumenhagen, L. Goerlich, B. Kors and D. Lust, "Magnetic flux in toroidal type I compactifications," Fortsch. Phys. **49**, 591 (2001) [arXiv:hep-th/0010198].
; K. Becker, M. Becker, M. Haack and J. Louis, "Supersymmetry breaking and alpha'-corrections to flux induced potentials," JHEP **0206**, 060 (2002) [arXiv:hep-th/0204254] ; P. K. Tripathy and S. P. Trivedi, "Compactification with flux on K3 and tori," JHEP **0303**, 028 (2003) [arXiv:hep-th/0301139].
; M. R. Douglas and S. Kachru, "Flux compactification," Rev. Mod. Phys. **79**, 733 (2007) [arXiv:hep-th/0610102], and references therein.
[17] S. B. Giddings, S. Kachru and J. Polchinski, "Hierarchies from fluxes in string compactifications," Phys. Rev. D **66**, 106006 (2002) [arXiv:hep-th/0105097].
[18] D. Z. Freedman, G. W. Gibbons and P. C. West, "Ten Into Four Won't Go," Phys. Lett. B **124**, 491 (1983).
[19] C. Beasley, J. J. Heckman and C. Vafa, "GUTs and Exceptional Branes in F-theory - I," JHEP **0901**, 058 (2009) [arXiv:0802.3391 [hep-th]] ; C. Beasley, J. J. Heckman and C. Vafa, "GUTs and Exceptional Branes in F-theory - II: Experimental Predictions," JHEP **0901**, 059 (2009) [arXiv:0806.0102 [hep-th]].
[20] W. Fischler and L. Susskind, "Dilaton Tadpoles, String Condensates And Scale Invariance," Phys. Lett. B **171**, 383 (1986). ; W. Fischler and L. Susskind, Phys. Lett. B **173**, 262 (1986).

[21] O. DeWolfe, A. Giryavets, S. Kachru and W. Taylor, "Type IIA moduli stabilization," JHEP **0507**, 066 (2005) [arXiv:hep-th/0505160].

[22] T. Banks and K. van den Broek, "Massive IIA flux compactifications and U-dualities," JHEP **0703**, 068 (2007) [arXiv:hep-th/0611185].

[23] M. Dine, A. Morisse, A. Shomer and Z. Sun, "IIA moduli stabilization with badly broken supersymmetry," JHEP **0807**, 070 (2008) [arXiv:hep-th/0612189].

[24] Yu. M. Makeenko and A. A. Migdal, "Exact Equation For The Loop Average In Multicolor QCD," Phys. Lett. B **88**, 135 (1979) [Erratum-ibid. B **89**, 437 (1980)].

[25] T. Hertog and G. T. Horowitz, "Towards a big crunch dual," JHEP **0407**, 073 (2004) [arXiv:hep-th/0406134] ; T. Hertog and G. T. Horowitz, "Holographic description of AdS cosmologies," JHEP **0504**, 005 (2005) [arXiv:hep-th/0503071].

[26] J. L. F. Barbon and E. Rabinovici, "Holography of AdS vacuum bubbles," JHEP **1004**, 123 (2010) [arXiv:1003.4966 [hep-th]].

[27] D. Harlow, "Metastability in Anti de Sitter Space," arXiv:1003.5909 [hep-th].

[28] R. Schon and S. T. Yau, "On the Proof of the positive mass conjecture in general relativity," Commun. Math. Phys. **65**, 45 (1979). ; E. Witten, "A Simple Proof Of The Positive Energy Theorem," Commun. Math. Phys. **80**, 381 (1981).

[29] A. Aguirre, T. Banks and M. Johnson, "Regulating eternal inflation. II: The great divide," JHEP **0608**, 065 (2006) [arXiv:hep-th/0603107].

[30] T. Banks, M. Dine, P. J. Fox and E. Gorbatov, "On the possibility of large axion decay constants," JCAP **0306**, 001 (2003) [arXiv:hep-th/0303252].

[31] N. Arkani-Hamed, L. Motl, A. Nicolis and C. Vafa, "The string landscape, black holes and gravity as the weakest force," JHEP **0706**, 060 (2007) [arXiv:hep-th/0601001].

[32] T. Banks, M. Johnson and A. Shomer, "A note on gauge theories coupled to gravity," JHEP **0609**, 049 (2006) [arXiv:hep-th/0606277].

[33] R. Bousso and J. Polchinski, "Quantization of four-form fluxes and dynamical neutralization of the cosmological constant," JHEP **0006**, 006 (2000) [arXiv:hep-th/0004134].

[34] S. S. Gubser, I. R. Klebanov and A. M. Polyakov, "Gauge theory correlators from non-critical string theory," Phys. Lett. B **428**, 105 (1998) [arXiv:hep-th/9802109]. E. Witten, "Anti-de Sitter space and holography," Adv. Theor. Math. Phys. **2**, 253 (1998) [arXiv:hep-th/9802150].

[35] P.P. Kulish, L.D. Fadeev, Theor. Math. Phys. 4 (1970), p. 745; L.D. Faddeev, in: New developments in relativistic quantum field theory and its applications. Lecture in VIII Winter School of Theoretical Physics in Karpacz. - Acta Univer. Wratislaviensis, 1972, vol.1, N 164, 69 - 78; P.P. Kulish, Theoretical and Mathematical Physics, 1971, vol.6, N 1, 28 - 35 (In Russian); Infrared divergences of quantum gravitational field, by P.P. Kulish, (in Russian). Zapiski Nauchnykh Seminarov LOMI, vol.77 (1978) 106 - 123. (English translation: Jour. Soviet Mathematics, vol.22, N 5 (1983).

[36] T. Jacobson, "Thermodynamics of space-time: The Einstein equation of state," Phys. Rev. Lett. 75, 1260 (1995); T. Padmanabhan, "The Holography of Gravity encoded in a relation between Entropy, Horizon area and Action for gravity," Gen. Rel. Grav. **34**, 2029 (2002) [arXiv:gr-qc/0205090] ;T. Padmanabhan, "Is gravity an intrinsically quantum phenomenon? Dynamics of gravity from the entropy of spacetime and the principle of equivalence," Mod. Phys. Lett. A **17**, 1147 (2002) [arXiv:hep-th/0205278] ;T. Padmanabhan, "Gravity from Spacetime Thermodynamics," Astrophys. Space Sci. **285**, 407 (2003) [arXiv:gr-qc/0209088] ; T. Padmanabhan, "Gravitational entropy of static spacetimes and microscopic density of states," Class. Quant. Grav. **21**, 4485 (2004) [arXiv:gr-qc/0308070] T. Padmanabhan, "Gravity and the thermodynamics of horizons," Phys. Rept. **406**, 49 (2005) [arXiv:gr-qc/0311036] ;T. Padmanabhan, "Why gravity has no choice: Bulk spacetime dynamics is dictated by information entanglement across horizons," Gen. Rel. Grav. **35**, 2097 (2003). ;T. Padmanabhan, "Gravity: A new holographic perspective," Int. J. Mod. Phys. D **15**, 1659 (2006) [arXiv:gr-qc/0606061] ;D. Kothawala, S. Sarkar and T. Padmanabhan, "Einstein's equations as a thermodynamic identity: The cases of stationary axisymmetric horizons and evolving spherically symmetric horizons," Phys. Lett. B **652**, 338 (2007) [arXiv:gr-qc/0701002] ;T. Padmanabhan and A. Paranjape, "Entropy of Null Surfaces and Dynamics of Spacetime," Phys. Rev. D **75**, 064004 (2007) [arXiv:gr-qc/0701003] ;T. Padmanabhan, "Gravity as an emergent phenomenon," Int. J. Mod. Phys. D **17**, 591 (2008). ;T. Padmanabhan, "Entropy density of spacetime and thermodynamic interpretation of field equations of gravity in any diffeomorphism invariant theory," arXiv:0903.1254 [hep-th] ;T. Padmanabhan, "Equipartition of energy in the horizon degrees of freedom and the emergence of gravity," Mod. Phys. Lett. A **25**, 1129 (2010) [arXiv:0912.3165 [gr-qc]] ;T. Padmanabhan, "Surface Density of Spacetime Degrees of Freedom from Equipartition Law in theories of Gravity," Phys. Rev. D **81**, 124040 (2010) [arXiv:1003.5665 [gr-qc]] ;T. Padmanabhan, "Entropy density of space-time and gravity: A conceptual synthesis," Int. J. Mod. Phys. D **18**, 2189 (2009). ; S. Kolekar and T. Padmanabhan, "Holography in Action," arXiv:1005.0619 [gr-qc]. "On the Origin of Gravity and the Laws of Newton," arXiv:1001.0785 [hep-th].

[37] G. Horowitz, *Unpublished*

[38] R. Bousso and B. Freivogel, "Asymptotic states of the bounce geometry," Phys. Rev. D **73**, 083507 (2006) [arXiv:hep-th/0511084].

[39] G. W. Gibbons and S. W. Hawking, "Cosmological Event Horizons, Thermodynamics, And Particle Creation," Phys. Rev. D **15**, 2738 (1977).

[40] W. Israel, "New Interpretation Of The Extended Schwarzschild Manifold," Phys. Rev. **143**, 1016 (1966).

[41] L. Susskind, L. Thorlacius and J. Uglum, "The Stretched Horizon And Black Hole Complementarity," Phys. Rev. D **48**, 3743 (1993) [arXiv:hep-th/9306069] ; L. Susskind, "String theory and the principles of black hole complementarity," Phys. Rev. Lett. **71**, 2367 (1993) [arXiv:hep-th/9307168].

[42] P. H. Ginsparg and M. J. Perry, "Semiclassical Perdurance Of De Sitter Space," Nucl. Phys. B **222**, 245 (1983).
[43] R. Bousso and S. W. Hawking, Phys. Rev. D **54**, 6312 (1996) [arXiv:gr-qc/9606052].
[44] T. Banks, B. Fiol and A. Morisse, "Towards a quantum theory of de Sitter space," JHEP **0612**, 004 (2006) [arXiv:hep-th/0609062].
[45] E. Witten, "Strong Coupling Expansion Of Calabi-Yau Compactification," Nucl. Phys. B **471**, 135 (1996) [arXiv:hep-th/9602070].
[46] T. Banks and J. F. Fortin, "A Pyramid Scheme for Particle Physics," JHEP **0907**, 046 (2009) [arXiv:0901.3578 [hep-ph]].
[47] T. Kugo, "String field theories," Prog. Theor. Phys. Suppl. **188**, 1 (2011), and references cited therein.

LHC Physics

Chapter 3

Fundamentals of LHC Experiments

Jason Nielsen

Santa Cruz Institute for Particle Physics and Department of Physics, University of California, Santa Cruz, CA 95064, USA

nielsen@scipp.ucsc.edu

Experiments on the Large Hadron Collider at CERN represent our furthest excursion yet along the energy frontier of particle physics. The goal of probing physical processes at the TeV energy scale puts strict requirements on the performance of accelerator and experiment, dictating the awe-inspiring dimensions of both. These notes, based on a set of five lectures given at the 2010 Theoretical Advanced Studies Institute in Boulder, Colorado, not only review the physics considered as part of the accelerator and experiment design, but also introduce algorithms and tools used to interpret experimental results in terms of theoretical models. The search for new physics beyond the Standard Model presents many new challenges, a few of which are addressed in specific examples.

3.1. Introduction

Experimental results combined with theoretical considerations imply the existence of new physics beyond the Standard Model, at energies no greater than 1 TeV. Although this has been known for a while [1], the possibility of accessing this energy scale, known as the "terascale," has now been realized in current-day hadron colliders and the experiments that use them. These notes provide a brief overview of the experimental considerations and design needed to measure particle interactions at the terascale.

To probe directly the physics at the 1 TeV scale, we need a momentum transfer Q^2 of approximately 1 TeV between the initial state particles. This direct requirement assumes we are most interested in producing real (on-shell) new particles. There is still a role, of course, for precision experiments on the intensity frontier that measure the effects of off-shell new particles in loop diagrams, but that lies outside the current discussion of the energy frontier.

3.2. Proton Beams for Terascale Physics

To achieve the goal of beams for terascale physics, we consider three possibilities. First, we could pursue collisions of high-energy beams on stationary targets, but the center-of-mass energy in such collisions scales only as $\sqrt{E_{\text{beam}}}$, so this option seems impractical. Second, we might investigate a lepton-antilepton collider, in which the center-of-mass energy scales with E_{beam}. Third, we might pursue a hadron-hadron collider, either with proton-proton or proton-antiproton collisions. Because the hadrons are composite particles with varying constituent momenta, this approach is somewhat less convenient than a lepton-antilepton collider. Nevertheless, we shall see that other considerations favor the hadron collider concept as implemented in the Large Hadron Collider.

Synchrotron radiation limits the choices of lepton-antilepton colliders, since the energy radiated by a charged particle in each turn of a circular accelerator is proportional to γ^4/ρ. For example, radiation losses per turn were $0.2\,\mu$W for each electron in the LEP collider 45 GeV beams. Even though the losses can be overcome, replacing the energy through acceleration after each turn is prohibitively inefficient. For these reasons we expect high-energy colliders either to accelerate more massive charged particles (protons or muons) or to push the ring radius ρ to its ultimate limit in a linear collider.

There are two caveats relating to linear colliders. First, a linear collider seems to require full accumulation of beam particles and full acceleration for each shot, unless the beam particles can be recycled. Second, accelerating beams to 500 GeV (the energy required for a 1 TeV lepton collider to probe the "terascale") requires very high acceleration gradients over a very long linear path. Given that the state of the art in radiofrequency gradients is about 30 MV/m, accelerating a single beam to 500 GeV would take about 15 km. Work is ongoing to improve those gradients for a high-energy linear electron-positron collider.

The rest of our discussion will focus on circular proton or antiproton colliders. Whereas accelerating non-relativistic particles requires only a bending magnetic field fixed to $B = p/q\rho$ and a constant accelerating cyclotron frequency $\omega_c = qB/m$, accelerating relativistic particles requires an accelerating frequency $\omega_c = qB/\gamma m$. During acceleration, then, ramping up the B field bends the beams of increasing momentum within a fixed bending radius while allowing a fixed accelerating frequency to be used. In MKS units, $p = 0.3B\rho$, where the momentum p has units of GeV/c, B has units of Tesla, and ρ has units of meters. There are obvious limits to both B and ρ, some physical and some geographical.

If our goal is to have a proton beam of at least 3.5 TeV in order to access the terascale physics at $Q^2 > 1$ TeV, we must work within these limitations. To maximize the bending radius, we might choose the large underground ring at CERN (an octagon with alternating straight accelerating and bending sections), which has a 4.3 km average radius. In this ring, a 3.5 TeV beam requires an average bending field of 4.2 T, while a 7 TeV beam requires 5 T. (Since the bending magnets are not distributed everywhere around the ring, the peak requirements are somewhat higher, up to 8.3 T for the 7 TeV case.) These fields are far above the domain saturation cutoff for regular ferromagnets, but superconducting magnets can achieve close to 10 T in the steady state, limited by intercable and interfilament forces. A heavy laboratory-industrial partnership has developed NbTi "Rutherford cable" needed to bend beams of the required momentum in the CERN ring now occupied by the Large Hadron Collider.

Given the possibility of accelerating and bending high-energy beams in a large enough ring, what beam intensities are needed for studies of the rare interactions of the terascale? The answer depends on several variables, including the beam size, particle spacing, and potential effects on the experiments. The small interaction rates expected at $Q^2 = 1$ TeV require instantaneous luminosities of $10^{34}\,\text{cm}^{-2}\text{s}^{-1}$ to collect enough events to study. With the design parameters of the LHC, this corresponds to 3×10^{14} protons per beam. (These required luminosities are much greater than those available at the Fermilab Tevatron, where the proton-antiproton collisions have instantaneous luminosities of $4 \times 10^{32}\,\text{cm}^{-2}\text{s}^{-1}$.)

Up to this point we have kept open the possibility of proton-antiproton or proton-proton collisions, but now we have to make a choice. Antiproton-proton collisions have some advantage, as they enhance the $q\bar{q}$ initial states due to the dominant parton distributions at high parton momentum. Unfortunately it seems unfeasible to pack 3×10^{14} antiprotons into a single beam. At the Tevatron, 25 years of experience producing and accumulating antiprotons has resulted in a maximum accumulation rate of 3×10^{10} antiprotons per hour, giving a maximum antiproton count of 3×10^{12} in a beam. This is a factor of 100 too low of the value needed to produce events in the rarest interactions. As a result, to investigate rare processes with

$Q^2 > 1\,\text{TeV}$ occurring at low rates, the proton-proton collider is the only choice at present.

The Large Hadron Collider is the final stage of the CERN accelerator chain shown in Fig. 3.1. Protons are taken from a single bottle of hydrogen gas, accelerated in the linac and Proton Synchrotron to 26 GeV, and injected into the Super Proton Synchrotron. After being accelerated in the SPS to 450 GeV, protons are injected into the twin rings of the LHC. The staged injection energies for each accelerator minimize the required operational range for the bending magnets.

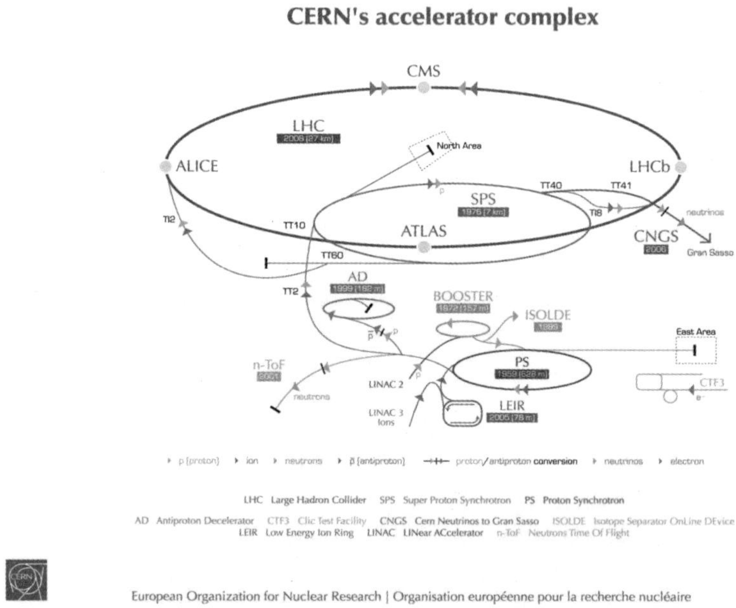

Fig. 3.1. Overview of the CERN accelerator chain, showing the relationship between low-energy accelerators and the Large Hadron Collider with experiments. Image credit: CERN

The LHC itself is composed of two rings, one for each proton beam direction. Because the existing CERN tunnel has a diameter of just 3.7 m, a "twin-bore" design first proposed by Blewett is used. In this design both rings are contained in a single superconducting cold mass and cryostat, but the magnetic dipole fields point in opposite directions to provide Lorentz bending force toward the center of the ring. An important part of the magnet design is the total dipole length of 15 m, chosen to reduce the number of inter-cryostat connections.

The high magnetic field and large volume of these dipoles imply an enormous amount of stored magnetic energy in each magnet. A simple calculation per magnet of

$$E = \frac{1}{2}LI^2 = \frac{1}{2}(0.099\,\mathrm{H})(11.8\,\mathrm{kA})^2 = 6.9\,\mathrm{MJ} \tag{3.1}$$

demonstrates the need for a quench protection system. The superconducting cable quenches and becomes a regular ohmic conductor if the temperature or magnetic field rise above critical values. If the current is not carried away through shunt resistances, it has the potential to boil off liquid helium explosively.

The protons beams themselves are accelerated with radio frequencies of 400 MHz, giving rise to synchroton oscillations that group protons into RF "buckets." The collision rates in the interaction regions are proportional to $N_{\mathrm{p}}^2 n_b$, where N_{p} is the number of protons per bunch and n_b is the number of bunches. The LHC design goal is 2808 bunches. Within the beams, the proton population is distributed in both position space and momentum space, and the "emittance" is a measure of that spread. The longitudinal emittance relates to the bunch length, while the transverse emittance relates to the bunch (beam) width. The LHC beams are squeezed as they approach the interaction points, with the "strength" of the squeezing gradient given by $\beta^\star$. Formally, $\beta^\star$ is the distance from the interaction point where the beam width doubles, and a lower value of $\beta^\star$ means the beam is smaller at the interaction point. During early LHC operations a typical value of $\beta^\star$ has been 1.5 m.

To calculate the expected rate of collisions in the LHC, we define the instantaneous luminosity L, measured in units of area^{-1}time^{-1}. Ultimately phase 1 of the LHC will reach instantaneous luminosities of $10^{34}\,\mathrm{cm}^{-2}\mathrm{s}^{-1}$ at full intensity. We can also write L in units of cross section, using the definition $1\,\mathrm{barn} = 10^{-24}\,\mathrm{cm}^2$ or $1\,\mathrm{nb} = 10^{-33}\,\mathrm{cm}^2$, to estimate the rate of specific physics interactions. For example, with $L = 10^{33}\,\mathrm{cm}^{-2}\mathrm{s}^{-1} = 1\,\mathrm{nb}^{-1}\mathrm{s}^{-1}$ and inelastic proton-proton interaction cross section $\sigma = 70\,\mathrm{mb}$ we expect a rate of $L\sigma = 70 \times 10^6$ interactions per second. Since the experiments collect data over an extended period of time, the integrated luminosity

$$\mathcal{L} = \int L dt \tag{3.2}$$

is defined with units area^{-1}, and then the dimensionless product $\mathcal{L}\sigma$ is the total number of interactions.

> *Exercise 1:* Compare the center-of-mass energy in electron-positron collisions for (a) head-on collision of particles, each with energy E_{beam}, and (b) collision of a beam particle with energy E_{beam} on a fixed target particle of mass m_e.
>
> *Exercise 2:* Starting from the classical formula for the radiated power from an accelerated electron, show that the loss per turn due to synchrotron radiation is $\Delta E = 4\pi e^2\gamma^4/3\rho$, where ρ is the ring radius and γ is the Lorentz factor of the electron.

3.3. Particle Detectors for the Energy Frontier

With the necessary beams designed and constructed to produce rare interactions sensitive to TeV-scale physics, we turn to the challenge of measuring the high-energy particles in the final states of those interactions. It is especially instructive to consider how particle detectors work from the point of view of the fundamental interactions. This allows us to predict how existing detectors would respond to new unusual particles in theories beyond the Standard Model.

In a nutshell, the ultimate goal of particle detection is to measure the 4-momentum of each final state particle. This can be accomplished by measuring (p_x, p_y, p_z, m), (p_x, p_y, p_z, E), or (p_T, η, ϕ, E). This last form makes use of the definitions for transverse momentum $p_T^2 = p_x^2 + p_y^2$ and pseudorapidity $\eta = -\ln[\tan(\theta/2)]$. The relativistic momentum can be measured in a magnetic spectrometer, while the velocity in the lab frame can be measured in certain cases with precision timing circuits. Particle mass can be inferred from the energy loss through ionization, and the scalar energy can be measured in a shower of large cross section interactions.

It is clear that there must be *some* interaction of the final state particles with the detector medium, and we make use of the two highest-rate interactions – the electromagnetic and strong interactions. These two interactions affect the particles on two very different energy scales, since the electromagnetic interaction occurs on the atomic length scale (eV energy scale) and the strong interaction occurs on the nuclear length scale (GeV energy scale) [2].

A magnetic spectrometer is at the heart of every LHC experiment. The Lorentz force causes charged particles to move in helical trajectories in a solenoidal field. Measuring the sagitta of the projected helix gives a clean estimate of the particle's transverse momentum, using the relation

$$s = \frac{0.3}{8} \frac{L^2 B}{p_T} \tag{3.3}$$

where L is the total arc length over which the sagitta is measured. In general, the uncertainty on the transverse momentum scales as $\sigma_{pT}/p_T \sim p_T$. Requirements on momentum resolution, dictated by physics goals, translate directly to requirements on sagitta resolutions. As an example, the ATLAS collaboration set a goal of 10% momentum resolution for 1 TeV muons expected from some new physics signatures. This implies a resolution of 50 μm on the sagitta measurement for trajectories that are nearly straight in the muon spectrometer. Increasing the magnetic field B or the arc length L (which is effectively the detector radius for straight tracks) increases the sagitta and leads to smaller relative uncertainties. This motivates large detectors like the ATLAS muon spectrometer, which has a maximum radius of 20 m.

The interaction underlying most non-destructive measurements is the electromagnetic interaction, whether a particle scatters elastically off charges in material or loses energy through ionization of atomic electrons. In the former case, scattering from multiple charge centers results in an uncertainty on the particle's original momentum vector. This uncertainty scales with the square root of the number of scatterers – $\langle \theta^2_{\text{MS}} \rangle = N \langle \theta^2 \rangle$, where θ is the deflection expected from a single scatter – and it is a powerful motivation for limiting the amount of material in the particles' path. In the latter case, the energy loss follows the Bethe-Bloch formula, which in one form for a singly-charged particle looks like

$$\frac{dE}{d\rho x} \sim \left(\frac{N_{\text{Avo}} Z}{A} \right) \left(\frac{\alpha^2 \hbar}{m_e c} \right) \left(\frac{1}{\beta^2} \right). \tag{3.4}$$

This energy loss translates directly to a number of ionization electrons along the particle trajectory.

Charged particle detectors used controlled electric fields to collect ionization electrons and cations. In some cases, most notably in detectors with gaseous media, a central sense wire lies at the center of a radial electric field. As ionization electrons drift toward the center, the strongest part of the electric field, they are accelerated and induce an avalanche of additional ionization. This multiplication factor makes it possible to observe the passage of a particle in a relatively low-density medium. The position resolution of such detectors depends on exact knowledge of the drift time and is limited by placement accuracy of the central wires. Another detector type uses solid-state semiconductors, usually silicon, as the interaction medium. Particles passing through the semiconductor ionize valence electrons and create electron-hole pairs, which drift under an applied electric field to strip or pixel readout elements patterned on the surface. These detectors benefit from increased ionization energy loss (roughly 400 keV/mm), but there is no amplification of collected charge, and special low-noise readout electronics are required to detect the signal. A major advantage of these detectors is that the readout element patterns can be microns apart, yielding extremely good position resolution. The disadvantages with respect to gaseous detectors are cost (gas is much cheaper than lithographed silicon) and processing requirements. Any defects in the semiconductor crystal trap the charge carriers and prevent them from reaching the readout elements. Because of their high cost and precision position resolution, solid-state detectors are most often placed near the interaction point, where they measure the first points of the particle trajectories before any scattering can take place.

By measuring the energy of a particle and either γ or β, we can use the relation $E = \gamma m$ to determine the particle mass and therefore the particle type. Measurements of γ and β come from energy loss (dE/dx) as collected by readout electronics, direct velocity measurements via precision timing, or radiation emitted by charged particles in a dielectric medium (Cherenkov or transition radiation).

Energy measurements in calorimeters are by nature destructive measurements, since their aim is to fully contain and collect the energy of the incoming particle. There are two different mechanisms by which a particle entering the dense material of the calorimeter showers into a large number of second particles. The first mechanism, the electromagnetic shower, proceeds by alternating electron bremsstrahlung to photons and subsequent conversion to electron-positron pairs. Only photons and electrons participate in the electromagnetic shower, since both bremsstrahlung and pair production are maximized for low mass particles. The second mechanism, the hadronic shower, is due to nuclear interactions. For both mechanisms, the cascade undergoes exponential branching until the energy of individual particles falls below some critical energy and ionization takes over.

Because of the exponential growth of the shower, the total number of secondary particles is proportional to the energy of the primary particle that entered the calorimeter. The energy resolution of the calorimeter, then, is given by

$$\frac{\sigma_E}{E} \sim \frac{\sigma_N}{N} \sim \frac{1}{\sqrt{E}} \tag{3.5}$$

since the fluctuations in the large number of particles follow a Gaussian distribution. The secondary particles in the shower can be collected in a dedicated active medium distinct from the absorber material (as in a non-homogeneous calorimeter) or in the same medium that serves as absorber (in a homogeneous calorimeter). This distinction has a substantial effect on the energy resolution; for example, the electromagnetic energy resolution is $\sigma_E/E = 10\%/\sqrt{E}$ in ATLAS but $2.7\%/\sqrt{E}$ in CMS. In the LHC experiments, two distinct calorimeters are deployed, one representing a small number of nuclear interaction lengths but large number of electromagnetic radiation lengths, and one with a large number of nuclear interaction lengths. These are the electromagnetic and hadronic calorimeters, respectively.

Particles that do not interact strongly, such as muons and neutrinos, penetrate the calorimeters without showering. Muons are detected in standalone external spectrometers or chambers integrated in the magnet yoke. Neutrinos do not interact with any of the detector material, and any missing momentum in the collision is attributed to them. Figure 3.2 shows in detail the experimental signature of the Standard Model particles, as detected in the CMS experiment.

The high interaction rates required to search for new physics at the TeV scale present extra challenges for the LHC experiments. Charged particle tracking algorithms are designed to function with detector occupancies of up to $\mathcal{O}(1\%)$. These algorithms, which work by stringing together significant energy deposits ("hits"), start with the highest-granularity silicon detectors near the interaction point and work outward, accounting for energy loss in each detector layer encountered. The experiments have been designed to meet the requirements of low occupancy even in particle-dense environments like boosted jets from high-mass resonances or Higgs boson decays.

Certain examples of physics beyond the Standard model give rise to striking experimental signatures, and it is worth looking at a few such examples. First, exotic charged massive particles (CHAMPs) typically have low velocities, with β significantly less than 1. This causes them to lose greater amounts of energy through ionization, and they may even stop in the middle of the detector if the energy loss is great enough. Since the

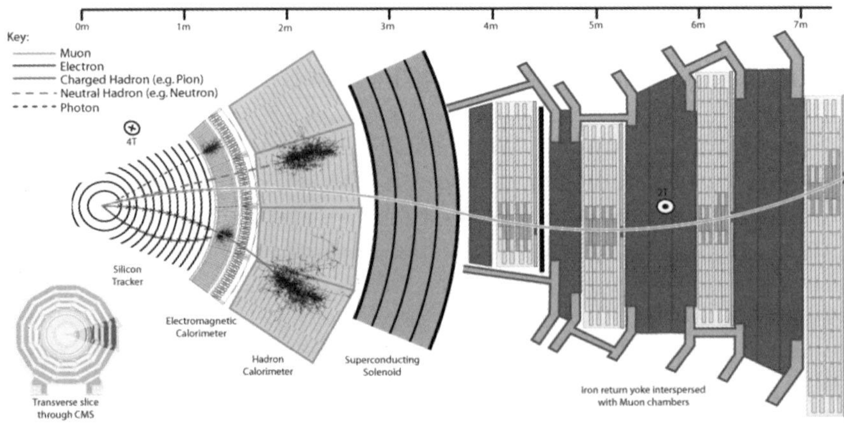

Fig. 3.2. Transverse slice through the CMS detector, showing the individual detector subsystems and particle signatures in each. The particle type can be inferred by combining the detector response in the different subdetectors. Image credit: CERN

particles are massive they do not shower in the electromagnetic calorimeter. R-hadrons, stable particles with heavy colored constituents, can have similar signatures. Doubly-charged particles lose $q^2 = 4$ times the normal ionization energy loss in the tracking detectors, a unique signature for any experiment with dE/dx sensitivity. Second, metastable particles, such as long-lived neutralinos in theories of supersymmetry breaking, may have $c\tau$ values of order 1 m, and the decay products do not point back to the primary interaction vertex. Even though it is challenging to reconstruct particles originating in the middle of the detector, successful reconstruction makes it possible to measure the lifetime of the parent particle. Third, exotic "quirks," predicted in new theories of strong interactions, can appear as a mesoscopic bound state, with properties similar to a doubly-charged particle, or as a state that oscillates like a macroscopic string with two charge endpoint particles.

These and other particle interactions can be modeled with dedicated computer programs of differing complexity. On one end lies GEANT4 [3, 4], a simulation toolkit with detailed lists of high- and low-energy interactions. Each particle is tracked step-by-step through a custom model of the detector. On the other end lie fast simulation programs like PGS [5] and Delphes [6], which use parameterized detector response to each type of particle.

Careful consideration of the physical interactions behind the experimental techniques allows one to extrapolate detector behavior to the most unusual new possibilities! Unfortunately experiments do not have large enough computing budgets to store the detector response to each bunch crossing, so they use a multi-level trigger system to decide immediately which events should be saved. This bears repeating: unless the new physics is selected by a trigger algorithm, it will be lost forever.

The LHC experiments use a combination of low-level hardware and high-level software triggers to filter events for further study. In the hardware triggers, individual trigger objects (jets, electromagnetic clusters, muons) are identified using data from fast readout detectors. These are passed to streamlined versions of offline software algorithms to be reconstructed more fully. For example, an electron trigger might require a Level-1 electromagnetic cluster with $E_T > 20\,\text{GeV}$, followed in the High-Level Trigger with a set of track-matching requirements and further cluster shape cuts to reject jets and π^0 mesons. If the rate of electron events becomes too great, then some filter has to be tightened to reduce the overall rate.

Exercise 3: Beginning from the Lorentz force law, prove the relation in Eqn. 3.3 between the sagitta and transverse momentum of a charged particle.

Exercise 4: In silicon strip detector with strip pitch (spacing) d, a hit on a strip means a particle passed somewhere in the $\{-d/2, +d/2\}$ range centered on the strip. Show that the variance of this uniform distribution is the same as that of a Gaussian distribution with width $d/\sqrt{12}$.

3.4. Physics Studies with Hadronic Jets

Many Standard Model and new physics signatures include hadronic jets. These jets present many challenges for the LHC experiments, and since they are the result of peturbative and non-perturbative QCD effects they also test our theoretical framework. Because the jets are measured chiefly in the calorimeters, experimentalists develop special energy calibrations to account for the effects of hadronization and contributions from pileup.

Even though we do not expect a one-to-one correspondence between reconstructed jets and partons (except possibly for the highest energy partons) we do expect that the relation between the two should be calculable on average. In the past decade experimentalists have become more sophisticated in defining jets and comparing measurements with theoretical predictions. Such comparisons are always made at the "particle-level" or "hadron-level," after the parton shower and hadronization but before the detector interaction occurs. This means that theoretical predictions must apply a showering and hadronization model to parton-level results, and experiments must unfold the effects of jet reconstruction in the detectors.

We have first to define exactly what is meant by "jets," and there are three main considerations. First, what particles should be clustered, or what inputs will be given to the algorithms? Second, which particles should be combined into each jet, based on proximity and energy? Third, how should the input 4-momenta be combined? The goal is to define jet clustering algorithms that are fast, robust under particle boosts, and able to deal with collinear and infrared radiation. This problem has been studied in great detail [7], but it is useful to summarize some of the solutions.

The answer to the first question is that the inputs can in fact be any kind of 4-momentum, possibly associated with truth particles (to give truth jets) or calorimeter cells (reconstructed jets). The answer to the third question seems to have been decided in favor of adding 4-momenta vectorially. There are many answers to the second question of how to choose the particles to combine into each jet!

Perhaps the simplest way to cluster particles is to use a cone algorithm, in which a distance $\Delta = \sqrt{(\Delta\eta)^2 + (\Delta\phi)^2}$ is defined with respect to a seed particle (typically the highest-p_T particle). All particles satisfying $d < R$ lie in a circle of radius R in the $\eta - \phi$ plane of the calorimeter, and these particles are combined to form a jet 4-momentum. This algorithm is well-defined geometrically, but the choice of seed particle is not stable when collinear radiation is considered.

A more sophisticated class of algorithms combine nearest particles first, effectively reversing the branching of the parton shower, instead of fixing a seed particle and cone. These "sequential recombination" algorithms cluster particles with smallest d_{ij} first, where

$$d_{ij} = \min\left(k_{ti}^{2p}, k_{tj}^{2p}\right) \frac{\Delta_{ij}^2}{R^2} \tag{3.6}$$

for particles i, j and beam 4-vector t. (Particles whose closest neighbor is the beam are considered stable jets.) In this general formulation, different values of the exponent p give very different algorithms. For $p = 0$ (Cambridge-Aachen algorithm), particles near each other in η, ϕ coordinate space are clustered first, whereas for $p = 1$ (kT algorithm) lower-momentum particles are clustered first.

Cacciari *et al.* pointed out recently that the case $p = -1$ also defines a useful algorithm, the anti-kT algorithm [8]. This choice means that the particles around the hardest particle are clustered first. The anti-kT algorithm guarantees a cone-like geometry with well-defined jet borders around the highest momentum particles (see Fig. 3.3), but it maintains the infrared safety and collinear safety of the sequential recombination family. This algorithm has become a preferred jet algorithm for LHC experiments, along with a modified stable infrared-safe cone algorithm called SISCone [9].

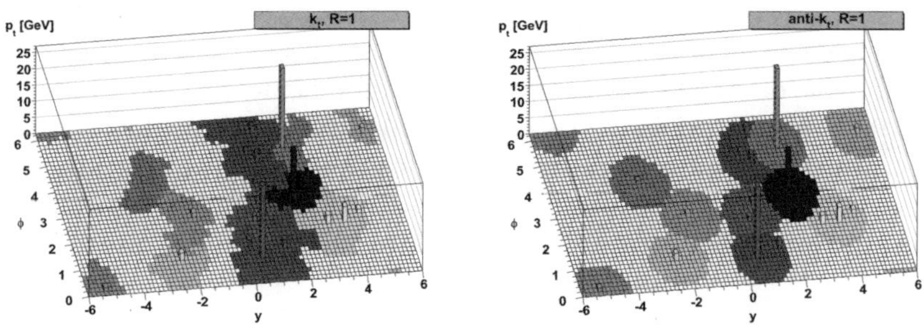

Fig. 3.3. Comparison of jet catchment areas for two different jet clustering algorithms [8]. All particles within a jet's catchment area will be clustered into the jet.

Total energy measured for a jet at the detector level must be corrected to match the energy at the particle level, and the calibration of the jet energy scale is a major experimental uncertainty in signatures with hadronic jets [10]. In short, the following effects are included in the calibration: varying detector response due to non-linearities or uninstrumented regions, mixed electromagnetic and hadronic showers in the same calorimeter, overall absolute energy scale calibration (assuming differences in relative response have been treated), and loss or gain of particles in the region defined by the jet area. These effects are estimated using calibration data samples in the jet-jet or γ+jet signatures, where the true jet energy can be estimated from the other object's recoil. Because there are no sufficiently large data samples of jets at the highest energies, jet calibration at those energies is based on extrapolation or on Monte Carlo simulation.

Certain jets, specifically those associated with heavy quarks (b, c), have several special properties due to the quarks themselves. The heavy quarks stand out for their long lifetimes (due to CKM suppression), large mass with respect to their decay products, and high multiplicity decays. These properties give rise to a distinct decay geometry, shown in Fig. 3.4. The momentum vectors of decay products from the B point to a secondary vertex, not the primary interaction vertex, and the distance between the two vertices depends on the b lifetime, if we use the spectator model approximation for the decay of the heavy quark hadron. Typical values before boost factors are $c\tau_b = 500\,\mu\text{m}$, $c\tau_c = 500\,\mu\text{m}$, and $c\tau_\tau = 90\,\mu\text{m}$. Because finding a common vertex for two or more tracks is one of the most challenging problems in tracking, the impact parameter of a track is used as a proxy to determine if it is consistent with having come from the primary interaction vertex. If a large number of tracks in a jet are inconsistent with the primary vertex, then it is likely that there is a heavy flavor hadron in the jet. One of the primary motivations for developing precision solid-state tracking detectors, which are even sometimes named "vertex detectors," is to measure track parameters precisely enough to allow for vertexing. The same parameters are also use for exact impact parameter measurements with 10% precision on values of $\mathcal{O}(300\,\mu\text{m})$.

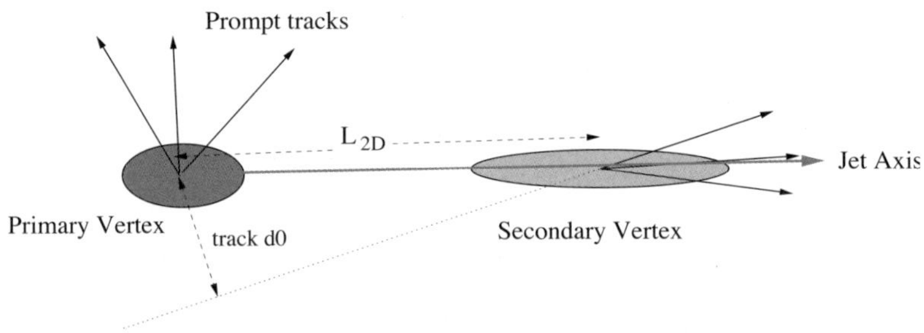

Fig. 3.4. Displaced secondary vertex from decay of a long-lived particle. Tracks from the decay are not expected to point back to the primary vertex as prompt tracks do. Flavor tagging (b-tagging) algorithms are designed to identify tracks with significant impact parameter d_0 and a vertex with significant decay length L_{2D}.

Flavor tagging algorithms are not limited to decay lengths and impact parameters. Identifying medium-p_T leptons from semileptonic heavy flavor decays provides an independent tagging mechanism. Since charm quarks also have long lifetimes and semileptonic decays, we use discriminating variables based on the mass of all particles in the secondary vertex to distinguish b-jets from c-jets. Ultimately, multivariate techniques combine information from all of these measurements to give powerful separation between b, c and light quark jets. Tighter requirements on b-jets reduce c and light quark jet contamination but also reduce the b-tagging efficiency, as shown in Fig. 3.5.

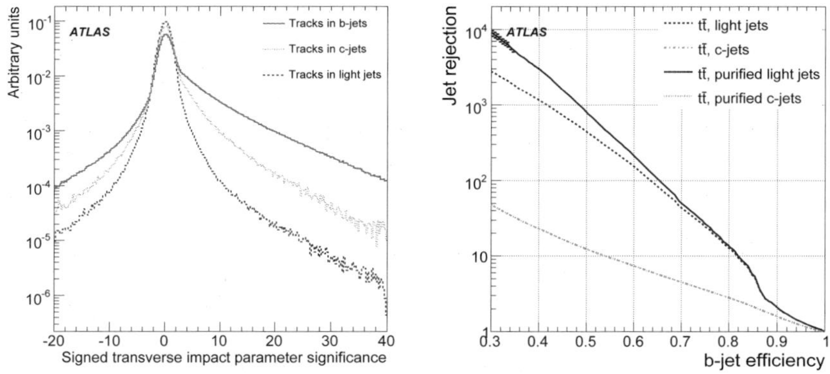

Fig. 3.5. Impact parameter significance distribution (left) and flavor tagging pefor-mance (right) from the ATLAS experiment [11]. Flavor tagging algorithms exploit the difference in the impact parameter distributions to select jets from bottom quarks while rejecting jets from light quarks (u, d, s) and, to a lesser extent, jets from charm quarks.

In the Standard Model, the proton-proton collision at a fixed center-of-mass energy is in fact a parton-parton collision between partons of unknown energy. As a result, the longitudinal momentum of the initial state is completely unknown, and this complicates the final state reconstruction. The key is to realize that the initial transverse momentum is well known; it is essentially 0 because of the small horizontal emittance of the beams and the low energy scale of Λ_{QCD}. The final state transverse momentum is expected therefore to also equal 0, and any deviation can be interpreted as missing transverse momentum or "missing transverse energy" (E_T^{miss} or MET), presumably due to non-interacting particles produced in the interaction but not detected.

The LHC experiments define the missing transverse energy as the opposite of the vector transverse sum of all detected particles. Such a measurement is only relevant if the detectors are nearly hermetic to both charged and neutral particles; this has put strict requirements on the hermeticity of the experiments. Corrections are applied for the detector response to muons and jets, and in some cases information from tracking and calorimetry is combined to optimize the missing energy reconstruction. The missing energy scale and resolution are calibrated using events known to have specific missing energy, e.g., $Z(\to \nu\bar{\nu})$+jets events. It is important to calibrate several different points to enable extrapolation to the higher values of $E_T^{\rm miss}$ we expect in new physics signatures. Like all energy measurements, the absolute resolution on the missing transverse energy scales as $\sqrt{E}$; early measurements of the ATLAS resolution yielded $\sigma = 0.57\sqrt{E\,(\mathrm{GeV})}$.

Exercise 5: The cone algorithm for jet clustering has its drawbacks, but it does have one redeeming quality. Use the definition of the pseudorapidity η to show that a cone size of $\Delta R \equiv \sqrt{(\Delta\eta)^2 + (\Delta\phi)^2}$ is invariant under boosts along $\hat{z}$.

Exercise 6: Use Fig. 3.4 to show that, for a typical track from a B hadron decay, $d_0 \sim \langle c\tau_B \rangle$, assuming the kick transverse to the jet axis is due to the large mass of the B hadron. Hint: consider the angle θ the track makes with respect to the b-jet axis. (This relation shows that the B mass is not as important as the lifetime when we consider track impact parameters, and it explains why tracks from charm decay have d_0 values of similar magnitude.)

3.5. Searches for Higgs Bosons

The search for the Higgs boson, whether in the Standard Model or beyond, is a key goal for understanding the physics of the terascale. The W and Z boson and t quark masses near the 100 GeV scale give quantitative constraints on the Standard Model Higgs boson mass, the only unknown parameter in the electroweak sector. In particular, precision measurements of m_W, m_Z, m_t constrain Higgs loop contributions and favor low Higgs masses, below 200 GeV. Direct experimental searches at LEP and Tevatron rule out $m_H < 114$ GeV and $158 < m_H < 175$ GeV, respectively [12, 13].

Standard Model Higgs boson production cross sections and branching ratios depend only on the Higgs mass. As seen in Fig. 3.6, the gluon fusion production mechanism dominates, but other production mechanisms are important for signatures with small background contributions. The branching fractions change quickly with increasing m_H as phase space for new decay channels opens. One notable feature is the dominant WW decay even above $2m_Z$; this is explained by the Standard Model Lagrangian term

$$\mathcal{L} \sim (2M_W^2 H W_\mu^+ W^{-\mu} + M_Z^2 H Z_\mu Z^\mu). \tag{3.7}$$

A successful Higgs boson search would show not only a discrepancy in the data with respect to the Standard Model prediction (without Higgs), but also consistency with expected Higgs production. We choose a specific

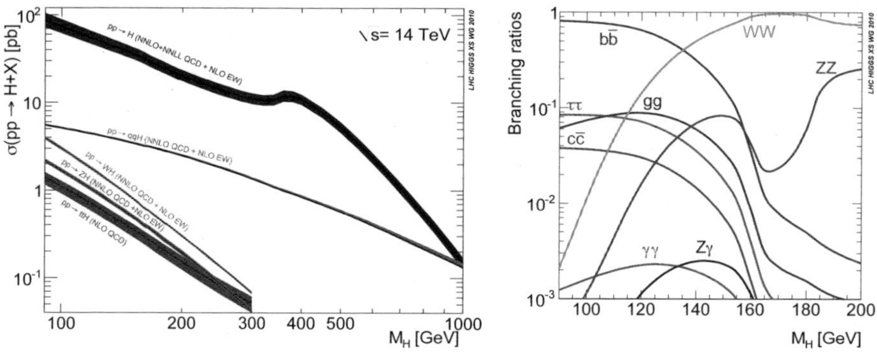

Fig. 3.6. Calculations of the Standard Model Higgs boson production pp cross section (left) and branching fractions (right) [14].

experimental signature or "channel" and develop an event selection to reject backgrounds from SM physics processes. The data sample may include contributions from background processes and putative signal events, but the goal is to measure the background contributions directly from data, if possible, to avoid any bias or errors in simulation.

A vital, if controversial, part of the Higgs search is the statistical interpretation of observed results in the context of a Higgs boson production hypothesis. What is the probability that the observed dataset is consistent with background-only production? With signal plus background production? Suppose 40 background (SM) events are expected in a search for a model that predicts 10 events from a new physics signal. Can the new physics model be excluded definitively if 40 events are observed? Can the Standard Model be excluded if 50 events, or even 60 events, are observed?

Much work in the past decade has focused on bringing sophisticated statistical tools to bear on this question in particle physics. The most common shorthand for presenting results is a re-interpretation in terms of a Gaussian distribution. If the observed experiment is considered as one of many possible experiments, given a certain model, then it is possible to calculate where the observed experiment lies in the distribution of the "pseudo-experiments" and convert its percentile to a number of "sigma." That is, if α is the probability to measure a less likely value than the observed experiment, then $\alpha = 0.3173$ corresponds to a 1σ deviation. (It is important to be aware of the distinction, sometimes overlooked, between one-sided and two-sided definitions of α [15]. Two-sided definitions are typical for measurements, while one-sided is often used for counting events when the signal is unknown.)

Usually we are interested in testing two complementary hypotheses, that of background-only production ($s = 0$) and that of signal+background production ($s > 0$). If the data favor the latter hypothesis and strongly disfavor the former, then we have a discovery. Experiments often report results in terms of a likelihood ratio $LR = L_{s+b}/L_b$ and by asking the following: how often can certain values of the LR be expected from an experiment in the presence of signal? For a discovery we talk about excluding the background hypothesis at $> 5\sigma$, which is $P_b < 10^{-7}$.

Most of the progress in the past five years has been on the treatment of systematic uncertainties, which reflect the inherent uncertainty in the number of background and signal expected in the datasets. Poorly constrained backgrounds can doom a Higgs search just as surely as low integrated luminosity. An overview of some methods, with technical results somewhat beyond the scope of these lectures, is given in Ref. 16.

Now it is instructive to introduce four of the main Standard Model Higgs boson searches at the LHC and to have a peek at one that may be important in the future.

For low mass Higgs bosons ($m_H < 130\,\text{GeV}$), one might expect to search for Higgs resonances produced in gluon fusion and decaying to $b\bar{b}$ pairs. Unfortunately, non-resonant $gg \to b\bar{b}$ production has an enormous production rate at the LHC, about 6 orders of magnitude greater than Higgs production! Fortunately, there are two other possibilities.

Higgs decays to tau lepton pairs (approximately 10% branching fraction) do not suffer from the gg background because the tau is not colored. The dominant background process is $Z \to \tau^+\tau^-$, but the Z mass resonance is well below the search region. After two tau lepton candidates have been reconstructed in an event, using algorithms that identify fully leptonic decay or hadronic decay, the missing transverse energy is used to estimate the energy taken by two or more neutrinos. With no other information, the kinematic system would be underconstrained, and mass reconstruction of the $\tau\tau$ system would be impossible. If one assumes the tau leptons from Higgs decays are highly boosted, then the neutrinos in the tau decay all have the same momentum as their sister decay products. With this trick, it is possible to reconstruct the full tau momentum and calculate the invariant mass of the Higgs candidate (see Exercise 8).

Higgs decays to two photons are a tiny fraction of all decays, but there are no diphoton resonances in the Standard Model above 1 GeV, so any sign of a resonance would be a clear indication of new physics. The LHC experiments, particularly ATLAS and CMS, have been designed to have excellent resolution for both photon energy and direction, and significant effort has gone into rejecting fake photons (misidentified electrons or jets) to reduce background. The non-resonant diphoton background can be measured as a falling distribution in data, as shown in Fig. 3.7 compared to the signal expectations for Standard Model Higgs production. This search is limited only by the number of events that can be collected, given the small branching ratio.

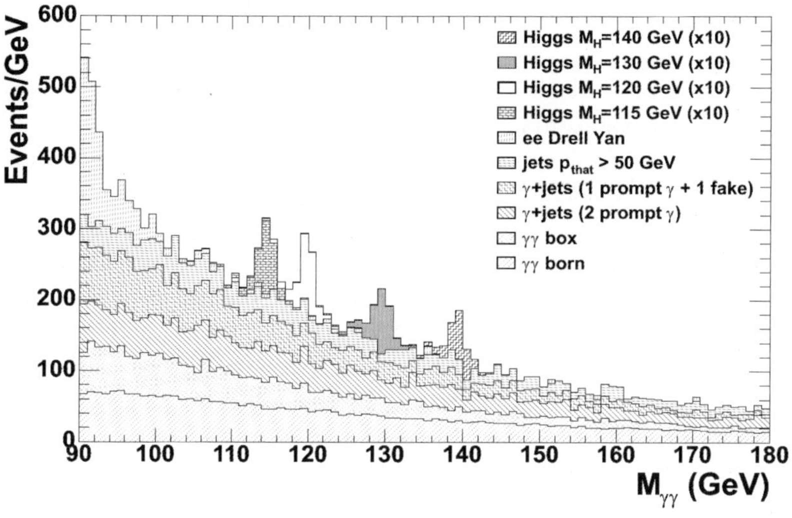

Fig. 3.7. Invariant mass spectrum of diphoton candidates selected in the low-mass Higgs boson search [17]. The significance of the diphoton resonances depends on the mass resolution of the experiment and the level of background underneath the signal peak.

For high mass Higgs bosons ($m_H > 170\,\text{GeV}$), the $ZZ \to \ell^+\ell^-\ell^+\ell^-$ decay channel offers another clean signature. Even though there is non-resonant ZZ Standard Model production, the reconstructed Higgs resonance would stand out clearly, as shown in Fig. 3.8. Again, the LHC experiments are designed to make precise measurements of the high-p_T electrons and muons, even for large Higgs masses; this defines the target resolution for the muon spectrometers.

For medium mass Higgs bosons ($130 < m_H < 170\,\text{GeV}$), the dominant decay is $H \to W^+W^-$. This signal process has a large rate using all production mechanisms, and the decay to a dilepton signature is clean. There are

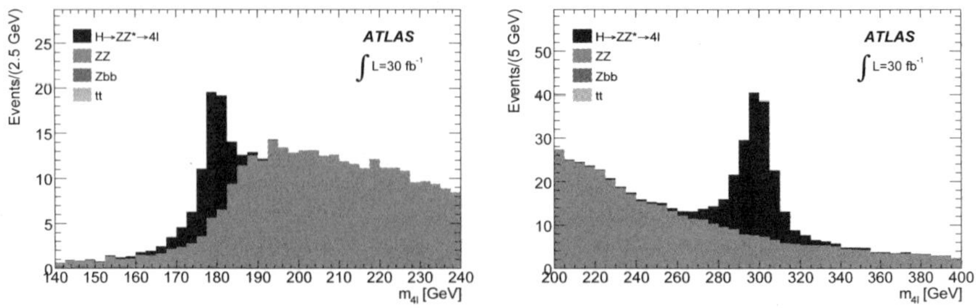

Fig. 3.8. Reconstructed 4-lepton mass for signal and background processes, assuming $m_H = 180\,\text{GeV}$ (left) and $m_H = 300\,\text{GeV}$ (right) [11].

two challenges for searches in this channel. First, the presence of two high-p_T neutrinos means there is no invariant mass peak for the reconstructed Higgs. (Since the W bosons are not boosted, the reconstruction trick from the tau channel cannot be re-used.) Second, direct WW production and top quark pairs have similar dilepton+E_T^{miss} signatures, but even here there is one extra trick for selecting WW pairs from Higgs decay. Because the Higgs is a scalar boson with spin 0, the two W bosons from Higgs decay must have opposite spin, and the leptons from W decay tend to be closer in direction than in the $t\bar{t}$ case. The angle between leptons is one of several input variables for multivariate tools that separate Higgs signal from SM background. This technique has already been used at the Tevatron to exclude certain Higgs mass hypotheses between 158 and 175 GeV [13].

The peek into the future regards the dominant $b\bar{b}$ decay channel for low-mass Higgs bosons. Since electroweak fits favor low Higgs masses, this decay channel would seem to be of prime importance in the Higgs search, but the $b\bar{b}$ jet background limits its sensitivity. (If the $b\bar{b}$ mass resolution were as good as the $\gamma\gamma$, a peak might still be resolved, but the jet energy resolution is limited by fluctuations in the hadronic shower.) One way to reduce the $b\bar{b}$ background is to require associated production of W, Z or $t\bar{t}$, and all of these associated production channels are being studied. A novel idea focuses the WH/ZH search to the region of phase space where the vector boson has large transverse momentum. Events in this region have a highly-boosted Higgs boson and $b\bar{b}$ decay products observed in a single fat jet. By shrinking the jet clustering radius until the two b subjets are resolved (as in Fig. 3.9), it is possible to compare the masses of the subjets to the mass of the boosted Higgs jet [18]. The large mass drop from the parent Higgs jet to both daughter b subjets is nearly unique;

only the WZ decay to $b\bar{b}$ has a similar reduction. The background from two-body $gg \to b\bar{b}$ decay is greatly reduced, and the dominant background becomes W/Z+light jet production. The benchmark calibration for this new technique is the reconstruction of the Z peak in Standard Model WZ and ZZ production.

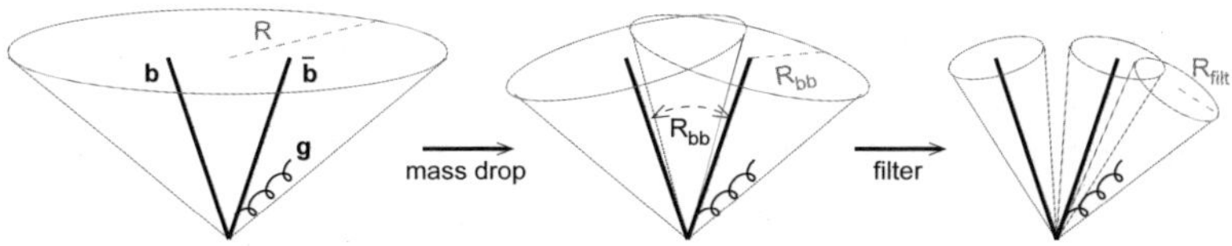

Fig. 3.9. Description of jet algorithm to identify boosted jets and substructure within [18]. The algorithm is designed to identify cleanly two b-jets from $H \to b\bar{b}$ decay, but it can be used for any decay of a particle to products having much lower mass.

Because the Higgs production cross section is small, several promising channels must be combined in the low-mass region to ensure sensitivity. When added to the powerful high-mass channels (WW, ZZ), these searches guarantee the LHC experiments will have something definitive to say about the Standard Model Higgs boson in the near future, perhaps with several fb^{-1} of 7 TeV data.

Exercise 7: There is a quick way to estimate the "number of sigma" significance of a signal observed above the background expectation. Use Poisson probabilities to define a likelihood ratio $L = P_{s+b}/P_b$ and show that to good approximation

$$\sigma \equiv \sqrt{2 \ln L} = \sqrt{2\left[(s+b)\ln\left(1+\frac{s}{b}\right) - s\right]},$$

where s and b are the expected numbers of signal and background events, respectively.

Exercise 8: (suggested by M. Strassler): Consider an event $pp \to Z + \gamma$, where the Z decays to $\tau^+\tau^-$. Use the fact that, although the neutrinos carry off energy, they do not significantly alter the directions of the tau leptons' other decay products, since the taus are highly boosted. Show that the Z boson mass can be reconstructed using only the photon momentum, the observed missing p_T , and the tau momentum directions (not their energies). This technique can also be used to measure the Higgs boson mass in $H \to \tau^+\tau^-$ decays.

3.6. Searches for Physics beyond the Standard Model

Searches for physics beyond the Standard Model (BSM) follow one of two approaches. The search strategy may focus on a specific model, or it may target any discrepancy from the Standard Model expectation. Both of the strategies are used in the LHC experiments, which hunt for general features in data that may correspond to a wide range of BSM models. Results in this section are a sampling of techniques used in new physics searches.

Perhaps the most straightforward signature shared by BSM models is the total event energy. New physics related to the terascale often has energies near this scale. The outgoing events in the hard scatter set the event scale, which is near the mass of the heavy new particles. Whatever the exact nature of their subsequent decay, the event energy scale is roughly preserved. As a result, the total event energy is a good estimate for the mass of new particles produced in pairs. To make this connection, it is best to focus on a robust definition of the total events energy.

There are at least three common calculations corresponding to total event energy. The first is the simple $\sum E_T$, for which all calorimeter energy is summed. This definition does not cover non-interacting particles, such as neutrinos or weakly interacting massive particles, nor does it account for extra calorimeter activity due to pileup events. The second is the oft-misunderstood H_T, usually defined as the scalar sum of missing transverse energy and the transverse energies of identified jets and leptons. This definition works well if there are one or two non-interacting particles, but it still suffers from pileup contamination and depends on the identification of the physics objects. The third is the effective mass M_{eff}, usually defined as the scalar sum of transverse energies of the four hardest identified jets and the missing transverse energy. This definition suppresses low-energy contributions from pileup events, but it does not capture the leptonic parts of the new particles decay chains. All three of these definitions are used in various channels by the LHC experiments, where they show good discriminating power between Standard Model background and the new physics signal.

Many arguments have been advanced for higher-mass versions of SM particles, and some of these correspond to resonances (invariant mass peaks) of simple objects, such as leptons or jets. Examples include Kaluza-Klein towers of particles confined in extra dimensions, Z'/W', and t', a 4th-generation up-type quark. Reconstruction of these resonances is straightforward, if the 4-momenta of all decay objects are known. The sensitivity to resonances is limited by background contamination and invariant mass resolution, both of which are high priorities for the LHC experiments.

If some of the decay products from the new resonance are invisible (non-interacting), a simple invariant mass calculation will not capture the signal. In these cases, the transverse mass

$$m_T^2 = 2E_{T1}E_{T2}\left[1 - \cos(\Delta\phi)\right] \tag{3.8}$$

can be used, using as E_T the missing transverse energy. If two particles decay to invisible daughters, as in cascade decay chains to lightest supersymmetric particles, it is still possible to apportion correctly the missing transverse energy. For example, in the supersymmetric decay chain

$$pp \to X + \tilde{\ell}_R^+\tilde{\ell}_R^- \to X + \ell^+\ell^-\chi_1^0\chi_1^0 \tag{3.9}$$

the final state neutralinos both appear as missing transverse energy. If the χ_1^0 mass is known, then the mass equivalence of the slepton mothers gives enough constraints to construct a kinematic variable whose distribution endpoint gives the slepton mass [19].

For a more concrete example, consider the following decay in theories of gauge-mediated supersymmetry breaking:

$$\tilde{\chi}_2^0 \to \tilde{\ell}^\pm\ell^\mp \to \tilde{\chi}_1^0\ell^\mp\ell^\pm \to \tilde{G}\gamma\ell^\pm\ell^\mp \tag{3.10}$$

Assuming that the background in this dilepton channel can be estimated using an opposite-flavor sample, the background-subtracted invariant mass distribution (Fig. 3.10) shows a sharp edge, indicating a kinematic limit [20]. In this case, the location of the edge is determined by the relation between the neutralino and slepton masses. The minimum mass for the $\ell\ell\gamma$ system has a similar endpoint, given by the difference of the neutralino masses. By using these and other kinematic solution endpoints, it is possible to reconstruct all of the masses in the decay chain.

An alternative to targeted searches has emerged in the last decade. Instead of developing event selections and kinematic variables designed for each of many different models and decay chains, some experiments have deployed general search strategies. These programs, count events in each of several high-p_T object classes ($1\mu1$jet, $1e2$jet, etc.) and compare the results to SM expectations [21–23]. The challenge is to describe completely the SM backgrounds for all signatures at once! Several discriminant distributions are considered for each class, including the scalar p_T sum of all objects, the invariant mass (or transverse mass) of all objects, and the missing transverse

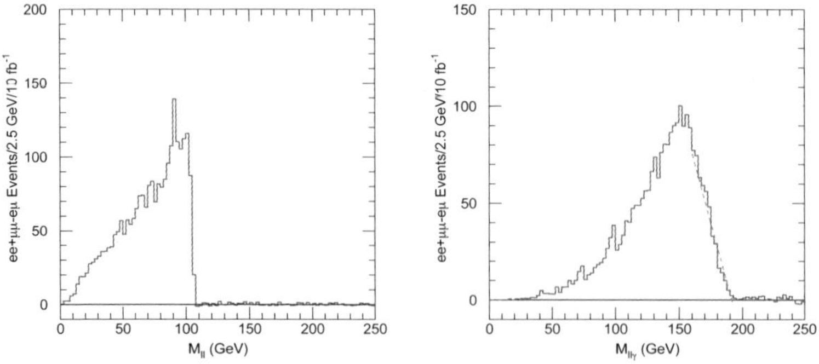

Fig. 3.10. Invariant dilepton mass distribution (left) and minimum dilepton+photon mass distribution (right) with background subtraction for GMSB signal events [20]. The endpoints of the background-subtracted distributions can be used to determine the masses of intermediate particles in the new physics decay chain. These figures represent data corresponding to an integrated luminosity of $10\,\mathrm{fb}^{-1}$.

energy. Any discrepancies between observed data and expectations are flagged for further studies.

In conjunction with the rise in general searches, new emphasis has been placed on simplified phenomenological models, which include the gross effects of new particle mass spectra and decay chains without focusing on the details of particle couplings and spin effects. These models have been successful in identifying experimental signatures that may have been overlooked [24], and they offer hope of helping match experimental observations with consistent theoretical models of new physics [25].

3.7. Conclusion

The Large Hadron Collider and associated experiments have been designed and constructed to answer questions about physics at the 1 TeV scale. The size, scope, and details of the experiments stem directly from the physics goals and requirements on measurements at that energy scale.

The focus in the last few sections on Higgs boson searches and other specific searches led naturally to a concentration on results from the general-purpose ATLAS and CMS experiments, but the other LHC experiments have been designed to pursue different physics goals that are no less interesting. All of the detector interactions, many of the design considerations, and some of the analysis techniques are being brought to bear on those goals as well.

A basic understanding of detector physics and practical limitations makes interpretation of experimental results more exciting and engaging, and the next few years may even bring news of exotic new particles with unexpected signatures.

Acknowledgments

My discussions with Matthew Strassler played an important role in the development of this material, and I thank him for his suggestions. I also thank the TASI 2010 organizers – Tom Banks, Michael Dine, and Subir Sachdev – for their efforts and patience, and the local hosts K.T. Mahantappa, Tom DeGrand, and Susan Spika for their hospitality.

References

[1] E. Eichten, I. Hinchliffe, K. D. Lane, and C. Quigg, Super Collider Physics, *Rev.Mod.Phys.* **56**, 579–707, (1984). doi: 10.1103/RevModPhys.56.579,10.1103/RevModPhys.56.579.
[2] D. Green, *The Physics of Particle Detectors.* (Cambridge University Press, Cambridge, 2000).
[3] S. Agostinelli et al., GEANT4: A Simulation toolkit, *Nucl.Instrum.Meth.* **A506**, 250–303, (2003). doi: 10.1016/S0168-9002(03)01368-8.
[4] J. Allison, K. Amako, J. Apostolakis, H. Araujo, P. Dubois, et al., Geant4 developments and applications, *IEEE Trans.Nucl.Sci.* **53**, 270, (2006). doi: 10.1109/TNS.2006.869826.
[5] J. Conway. *PGS4: Pretty Good Simulation of high energy collisions*, (2009). `http://physics.ucdavis.edu/~conway/research/software/pgs/pgs4-general.htm`.
[6] S. Ovyn, X. Rouby, and V. Lemaitre, DELPHES, a framework for fast simulation of a generic collider experiment. (2009).
[7] G. C. Blazey et al., Run II jet physics. (2000).
[8] M. Cacciari, G. P. Salam, and G. Soyez, The anti-k_t jet clustering algorithm, *JHEP.* **04**, 063, (2008). doi: 10.1088/1126-6708/2008/04/063.
[9] G. P. Salam and G. Soyez, A practical Seedless Infrared-Safe Cone jet algorithm, *JHEP.* **05**, 086, (2007). doi: 10.1088/1126-6708/2007/05/086.
[10] A. Bhatti et al., Determination of the jet energy scale at the collider detector at Fermilab, *Nucl. Instrum. Meth.* **A566**, 375–412, (2006). doi: 10.1016/j.nima.2006.05.269.
[11] G. Aad et al., *Expected Performance of the ATLAS Experiment - Detector, Trigger and Physics.* (CERN, Geneva, 2009).
[12] R. Barate et al., Search for the standard model Higgs boson at LEP, *Phys. Lett.* **B565**, 61–75, (2003). doi: 10.1016/S0370-2693(03)00614-2.
[13] Tevatron New Physics and Higgs Working Group, Combined CDF and D0

Upper Limits on Standard Model Higgs- Boson Production with up to 6.7 fb^{-1} of Data. (2010).

[14] LHC Higgs Cross Section Working Group, S. Dittmaier, C. Mariotti, G. Passarino, and R. Tanaka (Eds.), Handbook of LHC Higgs Cross Sections: 1. Inclusive Observables, *CERN-2011-002*. (CERN, Geneva, 2011).

[15] K. Nakamura et al., Review of particle physics, *J.Phys.G.* **G37**, 075021, (2010). doi: 10.1088/0954-3899/37/7A/075021.

[16] E. Gross. LHC Statistics for Pedestrians. In eds. L. Lyons, H. Prosper, and A. de Roeck, *PHYSTAT-LHC Workshop on Statistical Issues for LHC Physics*, pp. 205–212, Geneva, (2008). CERN.

[17] G. Bayatian et al., CMS physics technical design report, Volume II: Physics performance, *Journal of Physics G: Nuclear and Particle Physics*. **34**(6), 995, (2007). URL `http://stacks.iop.org/0954-3899/34/i=6/a=S01`.

[18] J. M. Butterworth, A. R. Davison, M. Rubin, and G. P. Salam, Jet substructure as a new Higgs search channel at the LHC, *Phys. Rev. Lett.* **100**, 242001, (2008). doi: 10.1103/PhysRevLett.100.242001.

[19] C. G. Lester and D. J. Summers, Measuring masses of semiinvisibly decaying particles pair produced at hadron colliders, *Phys. Lett.* **B463**, 99–103, (1999). doi: 10.1016/S0370-2693(99)00945-4.

[20] I. Hinchliffe and F. E. Paige, Measurements in gauge mediated SUSY breaking models at LHC, *Phys. Rev.* **D60**, 095002, (1999). doi: 10.1103/PhysRevD.60.095002.

[21] B. Abbott et al., Search for new physics in e muon X data at D0 using Sleuth: A Quasi model independent search strategy for new physics, *Phys. Rev.* **D62**, 092004, (2000). doi: 10.1103/PhysRevD.62.092004.

[22] A. Aktas et al., A general search for new phenomena in e p scattering at HERA, *Phys. Lett.* **B602**, 14–30, (2004). doi: 10.1016/j.physletb.2004.09.057.

[23] T. Aaltonen et al., Model-Independent and Quasi-Model-Independent Search for New Physics at CDF, *Phys. Rev.* **D78**, 012002, (2008). doi: 10.1103/PhysRevD.78.012002.

[24] D. Alves et al., Simplified Models for LHC New Physics Searches. (2011).

[25] N. Arkani-Hamed et al., MARMOSET: The Path from LHC Data to the New Standard Model via On-Shell Effective Theories. (2007).

Chapter 4

Theoretical Particle Physics at Hadron Colliders: An Introduction

Matthew J. Strassler*

New High Energy Theory Center,
Department of Physics and Astronomy, Rutgers University,
Piscataway, NJ, 08854, USA
strassler@physics.rutgers.edu

These lectures give a brief overview of the topics of which every student of modern hadron collider physics must be aware, and which are not easily found organized elsewhere. They begin with a basic review of the Higgs mechanism and discussion of the associated unitarity and hierarchy problems. Then the subject turns to the basics of a proton-proton collider: the rates for various standard model and new-physics processes at the LHC, the nature of triggers and reconstruction, and the kinematical variables used at hadron colliders. Two discussions — one preliminary, one advanced — of how QCD impacts hadron colliders is given, touching on confinement and the running coupling, parton distribution functions and luminosities, hadrons, and jets thereof. The role of and limitations of theory in calculations of backgrounds is covered, as is the experimental role of the third generation of fermions. Finally, the lectures conclude with an extremely and overly brief (and, I hope, interestingly idiosyncratic) review of new physics to look for at the LHC, and the challenges involved.

4.1. Brief Introduction

There are two natural approaches to a set of introductory lectures on LHC physics. The first option is to focus on the exciting possibilities for new physics: the ideas and concepts behind the models that extend the Standard Model, and the structure of their predictions. The second option is to emphasize the critical issues without which LHC physics cannot be understood: what the experimental detectors can and can't do, what the particles

*On leave.

and interactions of the Standard Model do and don't do, and what these considerations make experimentally possible and impossible. It is easier to find lectures following the first approach, both on-line and in books. But the key to understanding the first approach is understanding the second, and it is much harder to find summarized in the literature. For this reason I'm taking the second route.

Also, given that I have five lectures and want to cover many topics, I will be short on technicalities (especially those that can easily be found elsewhere) and long on concepts, with the aim of providing a comprehensive theoretical introduction. You should read Jason Nielsen's lectures for the experimental flip side of this discussion.

I must also apologize to you and to my colleagues for the extremely limited referencing. I simply ran out of time to do a decent job before the deadline for completion of the published lectures. The last two sections suffer from a lack of figures, for similar reasons. I have suggested some additional reading material at the end of the lectures.

4.2. The Standard Model of Particle Physics

4.2.1. *The basics*

Let me remind you of what I will assume you have seen before and more or less know. The Standard Model (SM) of Particle Physics consists of three gauge groups, $SU(3)_C \times SU(2)_W \times U(1)_Y$, with gauge bosons (8 gluons g_μ, 3 weak bosons W_μ and 1 hypercharge boson X_μ), and three (running) coupling "constants" g_3, g_2, g_1, or $\alpha_3, \alpha_2, \alpha_1$, where $\alpha_i \equiv g_i^2/4\pi$. (Recall g_3, g_2, g_1 are sometimes called g_s, g, g'; similarly $\alpha_s \equiv \alpha_3$ and $\alpha_W \equiv \alpha_2$. Note $\alpha \equiv \alpha_{QED}$ is not α_1; see below.) The most important physical scales in the SM are the QCD confinement scale $\Lambda_{QCD} \equiv \Lambda \sim 300$ MeV, and the electroweak breaking scale $v = 246$ GeV $= 1/\sqrt{G_F}$, about which I will say more later. There are three generations of matter, each containing the following *2-component left-handed Weyl fermions*:

- a lepton ℓ_i and neutrino ν_i, forming a doublet of $SU(2)$;
- an antilepton $\bar{\ell}_i$, a singlet of $SU(2)$;
- an up-type and down-type quark doublet u_i, d_i of $SU(2)$;
- an up-type antiquark $\bar{u}_i$ and a down-type antiquark $\bar{d}_i$, both singlets of $SU(2)$.

Finally there is a "Higgs sector", consisting of at least one $SU(2)$-doublet scalar (fundamental or composite), which gets a vacuum expectation value ("vev") and breaks $SU(2)_W \times U(1)_Y \to U(1)_{EM}$, giving masses to the W and Z bosons while leaving the photon massless, and giving mass to the fermions. The full set of charges of the matter and Higgs fields is shown in Table 4.1.

Table 4.1. The transformations under the SM gauge group of the matter and Higgs fields; here the index $i = 1, 2, 3$ labels the three generations.

	$SU(3)_C$	$SU(2)_W$	$U(1)_Y$
$L_i = \begin{pmatrix} \nu_i \\ \ell_i \end{pmatrix}$	$\mathbf{1}$	$\mathbf{2}$	$-\frac{1}{2}$
$\bar{\ell}_i$	$\mathbf{1}$	$\mathbf{1}$	1
$Q_i = \begin{pmatrix} u_i \\ d_i \end{pmatrix}$	$\mathbf{3}$	$\mathbf{2}$	$\frac{1}{6}$
$\bar{u}_i$	$\overline{\mathbf{3}}$	$\mathbf{1}$	$-\frac{2}{3}$
$\bar{d}_i$	$\overline{\mathbf{3}}$	$\mathbf{1}$	$\frac{1}{3}$
$H = \begin{pmatrix} H^+ \\ H^0 \end{pmatrix}$	$\mathbf{1}$	$\mathbf{2}$	$\frac{1}{2}$

Let me remind you of the details of Electroweak Symmetry Breaking (EWSB). This phase transition leaves a massless photon, coupling to matter with strength e (or $\alpha = e^2/4\pi$) via electric charge $Q = T_3 + Y$ (where T_3 is the third component of weak-isospin $SU(2)_W$, and Y is hypercharge). Note that, for example,

- for the electron $T^3 = -\frac{1}{2}$ and $Y = -\frac{1}{2}$,
- for the up quark $T^3 = +\frac{1}{2}$ and $Y = \frac{1}{6}$,
- for the up antiquark $T^3 = 0$ and $Y = -\frac{2}{3}$.

Meanwhile the W and Z bosons become massive (so that the weak force, which actually has $\alpha_W > \alpha$ and is thus not very weak at short distances, becomes the very weak force we know at long distances.) This occurs via the Higgs mechanism, which I now review.

4.2.2. *The Higgs mechanism*

We start with the Lagrangian for the Higgs boson doublet, with its kinetic and potential terms.

$$\mathcal{L} = (D_\mu H)^\dagger D^\mu H - V(H^\dagger H) \tag{4.1}$$

Here, defining $\mathbf{1}$ and τ^a as 2-by-2 unit and Pauli matrices, the covariant derivatives are

$$D_\mu = \partial_\mu \mathbf{1} - \frac{i}{2}\left(g_1 X_\mu \mathbf{1} + g_2 \sum_{a=1}^{3} W_\mu^a \tau^a\right). \tag{4.2}$$

Suppose the potential $V(H^\dagger H)$ has a minimum away from zero — for instance, if

$$V(H^\dagger H) = -\mu^2 H^\dagger H + \frac{1}{2}\lambda (H^\dagger H)^2 \tag{4.3}$$

with $\mu^2 > 0$, $\lambda > 0$, and there are minima at $H^\dagger H = \frac{1}{2}v^2$, where $v^2 = 2\mu^2/\lambda$. One such minimum would be at

$$\langle H \rangle = \begin{pmatrix} 0 \\ \frac{v}{\sqrt{2}} \end{pmatrix} \tag{4.4}$$

and any $SU(2) \times U(1)$ rotation of this example would also be a minimum, since the potential has $SU(2) \times U(1)$ symmetry. Then we can redefine fluctuations of H as follows

$$H \equiv \langle H \rangle + \delta H = \begin{pmatrix} 0 \\ \frac{v}{\sqrt{2}} \end{pmatrix} + \begin{pmatrix} \mathcal{G}^+ \\ \frac{1}{\sqrt{2}}(h^0 + i\mathcal{G}^0) \end{pmatrix}. \tag{4.5}$$

Consequently

$$D_\mu H \to -\frac{iv}{2\sqrt{2}}\left[(g_1 X_\mu - g_2 W_\mu^3)\begin{pmatrix} 0 \\ 1 \end{pmatrix} + g_2(W_\mu^1 + iW_\mu^2)\begin{pmatrix} 1 \\ 0 \end{pmatrix}\right] + D_\mu(\delta H) \tag{4.6}$$

and so $(D_\mu H)^\dagger D^\mu H$ contains the terms

$$M_W^2 W_\mu^+ W^{-\mu} + \frac{1}{2} M_Z^2 Z_\mu Z^\mu \tag{4.7}$$

where

$$M_W = \frac{1}{2} g_2 v \ , \ M_Z = \frac{1}{2}\sqrt{g_1^2 + g_2^2}\, v, \tag{4.8}$$

$$W_\mu^+ = \frac{1}{\sqrt{2}}(W_\mu^1 + iW_\mu^2) \ , \ Z_\mu = \frac{1}{\sqrt{g_1^2 + g_2^2}}(g_1 X_\mu - g_2 W_\mu^3). \tag{4.9}$$

These terms make the Ws and Z massive, leaving the photon

$$A_\mu = \frac{1}{\sqrt{g_1^2 + g_2^2}}(g_2 X_\mu + g_1 W_\mu^3) \tag{4.10}$$

massless.

Exercise: Verify these formulas, and re-express them in terms of $\sin\theta = g_1/\sqrt{g_1^2+g_2^2}$, $\cos\theta = g_2/\sqrt{g_1^2+g_2^2}$, *etc. From the measured masses of the* W *and* Z, *calculate the value of* $\tan\theta = g_1/g_2$. *Which of the two couplings is weaker? What happens in the limit* $g_1 \to 0$*?*

There's a loose end here. Massive spin-one bosons have three degrees of freedom, while massless spin-one bosons have only two. The W and Z were originally massless when $v = 0$; where did the third degree of freedom for the massive W and Z come from? The answer is that they arise from $\mathcal{G}^{\pm}$ and $\mathcal{G}^0$, the would-be "Nambu-Goldstone bosons" (NGBs) of δH. In the absence of the gauge interactions, the potential has a classical symmetry $SU(2)\times U(1)$ which is broken to $U(1)$ by $\langle H\rangle$, and this would give rise to 3 NGBs $\mathcal{G}^{\pm}, \mathcal{G}^0$. (Note, unlike the pions in QCD, which are pseudo-NGBs of a explicitly though weakly broken $SU(2)\times SU(2)$, these are *exact* massless NGBs of an *exact* spontaneously-broken symmetry.) But in the presence of the gauge interactions, the symmetry is a fake, and there aren't NGBs. There is no symmetry there to spontaneously break, despite the somewhat unfortunate terminology "Electroweak Symmetry Breaking", which is conceptually very misleading. But the degrees of freedom for the Higgs are still real. And in the phase where the vev of $H^\dagger H$ is substantial and non-zero, they become absorbed into the W and Z through the terms in $(D_\mu\langle H\rangle^\dagger)D_\mu\delta H$ that are of the form $(W_\mu^- v)(\partial_\mu\mathcal{G}^-)$, etc. Through these mixing terms the would-be NGBs become the *longitudinal polarization modes* of the massive W, Z bosons.

Our loose end has a loose end. We disposed of three of the degrees of freedom of H. What happens to the fourth one h^0? This is nothing but the physical Higgs boson of the SM. Notice that it has unusual interactions.

- Like any charged scalar, it has $hhVV$ interactions with gauge bosons

$$(D_\mu H)^\dagger D^\mu H \to \frac{1}{4}g_2^2 W_\mu^- h^0 W^{+\mu} h^0 + \frac{1}{8}\sqrt{g_1^2+g_2^2} Z_\mu^0 h^0 Z^{0\mu} h^0. \quad (4.11)$$

- Unlike charged scalars, it does NOT have an hhV interaction.
- *Unlike any other type of scalar*, it does have a tree-level hVV interaction, proportional to its vev v.

$$(D_\mu H)^\dagger D^\mu H \to g_2 h^0 (M_W W_\mu^+ W^{-\mu} + \frac{1}{2} M_Z Z_\mu^0 Z^{0\mu}). \quad (4.12)$$

Exercise: CHECK THIS FORMULA! These interactions are a direct sign of electroweak symmetry breaking!
- There are also $h^0(\mathcal{G}^{\pm}W^{\mp}$ and $h^0\mathcal{G}^0 Z^0)$ interactions.

- There is *no* tree-level $h^0 A_\mu A^\mu$ interaction.
- Small (but important) loop-induced interactions, of the form $h^0 F_{\mu\nu} F^{\mu\nu}$, are present for both photons and gluons.

We will make use of all of these facts in the LHC search for and study of the Higgs boson, and checking all of these facts will be a key test that any scalar discovered at the LHC is really participating in EWSB.

The Higgs also gives mass to SM fermions, *none of which can have mass until the Higgs has a vev.* Any mass terms for SM fermions in the absence of EWSB would violate gauge invariance! Masses for the charged leptons (electron e, muon μ and tau τ) come from "Yukawa couplings":

$$\mathcal{L} = \sum_{i,k=1}^{3} y_{ik}^{(\ell)} (L_i \cdot H^\dagger)\bar{\ell}_k = \sum_{i,k=1}^{3} y_{ik}^{(\ell)} (\nu_i H^- + \ell_i H^{0*})\bar{\ell}_k \to \sum_{i,k=1}^{3} m_{ik}^{(\ell)} \ell_i \bar{\ell}_k \tag{4.13}$$

where

$$m_{ik}^{(\ell)} = \frac{1}{\sqrt{2}} y_{ik} v. \tag{4.14}$$

Thus the masses of the leptons are proportional to their coupling strength to the Higgs boson. More precisely, the masses are related to the eigenvalues of the Yukawa coupling matrix $y_{ik}^{(\ell)}$. Similarly, down-type and up-type quark masses, come from $(Q_i \cdot H^\dagger)\bar{d}_k = (u_i H^- + d_i H^{0*})\bar{d}_k$ and $(Q_i \times H)\bar{u}_k = (u_i H^0 - d_i H^+)\bar{u}_k$ terms. The masses of all these particles are known, and in the SM the top quark couples most strongly to the Higgs boson. The up and down quarks and the electron barely couple at all (as in Table 4.2.)

Table 4.2. The Yukawa couplings of the Standard Model.

	u_i	d_i	ℓ_i
1st generation	~0.00002	~0.00004	0.00000294
2nd generation	~0.0069	~0.0006	0.006899
3rd generation	~0.99	~0.25	0.0102

Neutrino masses are different. They come from a dimension-five (non-renormalizable) interaction $\frac{f_{ij}}{M_0}(L_i \times H)(H \times L_j)$. Here we define M_0 so that the f couplings are not larger than about 1. Presumably there is new physics at or below the scale M_0. The neutrino masses have not yet been measured; only some differences of masses-squared are known at present, and there is an upper bound on the sum of the masses. I will not say much about them; unless there are new right-handed neutrinos accessible at the LHC, the neutrinos may just as well be, as far as the LHC is concerned, be three massless Weyl fermions.

Now I have just told you a long story. Which part of this story is known? And which part is still conjecture?

The existence of the vev v is already certain. There is no other way to give masses to W and Z bosons, or the fermions, that would be consistent with experiment. We have even seen, through the properties of its decay, that the top quark appears to couple as required to the $\mathcal{G}^{\pm}$ Goldstone modes.

But the existence of the particle H, and more phenomenologically h^0, is less certain. For instance, there might be several $SU(2)$-doublet Higgs bosons H_r, $r = 1, 2, \ldots, n$, with $v = \sqrt{\sum 2\langle H_r\rangle^2}$, which would lead to $4n-3$ physical Higgs scalars, $2n-1$ of them real and neutral, and n each of charge $+1$ and -1. (Note there would still generally be only three would-be NGBs, turning into the longitudinal modes of the W and Z.) There could also be additional $SU(2) \times U(1)$ singlet Higgs bosons, which would give additional neutral scalars. However, there might be no physical Higgs boson at all; there might just be a very wide and unobservable resonance, or some other kind of analytic structure in an amplitude.

But we know this: *the SM without the Higgs boson, and with nothing else in its place, is inconsistent.* This is because it can be shown rigorously that the SM with EWSB but without the Higgs, treated in perturbation theory, violates unitarity.

4.2.3. *Problem: Without a Higgs, the SM cannot predict beyond 1 TeV*

In particular, $W^+W^- \to W^+W^-$, $W^+W^- \to Z^0Z^0$, $W^+W^- \to t\bar{t}$, and many other processes grow violently with energy when calculated in perturbation theory with the Higgs diagrams dropped. They violate unitarity bounds as the energy approaches a TeV or so.

There are several possible solutions to this problem. The obvious one is to put the Higgs (or Higgses) back in. Unitarity is restored through the presence of a Higgs particle (or multiple Higgs particles) which through its (or their) coupling to masses give contributions that cancel off the energy growth. These Higgs particles must have mass near or below 1 TeV if the unitarity problem is to be avoided.

Another possibility is that unitarity is restored through the presence of other classes of particles (such as a tower of W, Z, and t copies) which through carefully adjusted couplings give contributions that cancel off the energy growth. The lightest of these particles must have mass near or below 1 TeV.

A third is that perturbative field theory calculations become invalid at or around 1 TeV, because the interactions of heavy particles — of W and Z bosons and at least t quarks — become strong, due to new dynamics (new forces and particles) at or below that scale.

A fourth is that quantum field theory itself somehow breaks down at around 1 TeV.

I should note these possibilities are not all mutually exclusive. Any or all of them might show up at the LHC.

Let me reemphasize: *The unitarity problem of the Higgs-less Standard Model is an extremely serious problem.* It must be solved, either through the simple one-doublet SM Higgs boson or an extension of the SM. And this is why a TeV-scale collider is needed, to explore a regime where the particles and forces we know so far form an inconsistent whole. No matter what such a machine finds around or below the 1 TeV range, it is "new physics", where the theory of the known particles, which is fully perturbative around the energy scale of 100 GeV, loses its ability to make any predictions.

4.2.4. *Problem: The hierarchy*

There is a *second* problem, known as the "hierarchy problem". The problem is the following. In any quantum field theory, we expect dimensional analysis to hold unless there is a very good reason. But in the Standard Model, dimensional analysis fails.

Why should dimensional analysis generically hold? Because even if you tried to adjust the classical (or better, the "bare") interactions of a quantum field theory to be smaller than you would expect from dimensional analysis, quantum corrections would generally raise the effective strength of the interaction that you measure to be what you expected from dimensional analysis. Said another way, *quantum corrections generally respect dimensional analysis.* When they don't, there's always a good reason.

Actually, we don't come across dimensional analysis arguments much in the SM, precisely because all the interactions *except* the Higgs mass *do* have a good reason to violate dimensional analysis! The gauge and Yukawa couplings are all dimensionless, and so expected to be of order 1. However the electron Yukawa coupling y_e is 10^{-5}. Is that reasonable? Yes. If y_e were zero, there would be a new symmetry (a chiral symmetry rotating the e^+ field). Therefore quantum corrections must satisfy $\delta y_e \propto y_e$, so if y_e is small for some reason, quantum corrections leave it small. The only other quantities you might worry about are explicit masses for the gauge bosons (not those obtained from EWSB, just direct $m^2 A_\mu A^\mu$ terms.) These cannot arise quantum mechanically because a spin-one particle with only two degrees of freedom simply can't quantum mechanically get a third degree of freedom from nowhere, as it would have to if it were to become massive.

However, the W and Z masses that arise from EWSB are an issue; we must ask how large, on dimensional grounds, we would expect v to be. We know $v^2 = |\mu^2|/\lambda$ in the SM. To get this result, we assumed perturbation theory (and thus a classical analysis) was a good approximation. This implies λ is at most 3 or so, so that $\lambda^2/4\pi \sim 1$ or less. So $|\mu^2|$ cannot be larger than about $(1\ \text{TeV})^2$.

But by dimensional analysis, the natural value for $|\mu^2|$ would be the largest physical mass scale in particle physics: $\delta\mu^2 \propto M^2_{max}$. And (naively at least) M_{max} is at least as large as the Planck scale, 10^{18-19} GeV. And therefore we have an apparent problem: the natural value for $|\mu|$ is of order the Planck scale, so that the natural value for v is either around the Planck scale (if the potential is concave down near zero) or zero (if it is concave up near zero.) Remarkably, the actual value of v is 246 GeV, definitely non-zero but 16 orders of magnitude below the Planck scale. How is this possible?

I don't know the answer, nor does anyone else. This is the hierarchy problem. It is an unexplained failure of dimensional analysis. Note, however, that unlike the perturbative unitarity problem, which is really serious, *the "hierarchy problem" of the standard model is a disturbing issue, but it is not an inconsistency of the theory.* It is merely an observation that there is something unnatural about the theory. (Indeed it is called a "naturalness" problem.) This might turn out just be an issue of perspective.

Many attempts have been made to solve the hierarchy problem, by changing the theory to either (a) alter the dimensional analysis itself, or (b) create a reason why dimensional analysis is not satisfied by μ^2, or (c) change quantum field theory altogether. All (most?) of these solutions produce new particles at the TeV scale, and should also be detectable by a TeV-scale collider.

Both the serious perturbative unitarity problem and the disturbing hierarchy problem tell us that looking at the TeV scale is a good idea. *There will definitely be new physics; there may be a LOT of new physics.*

4.2.5. *Why the Large Hadron Collider?*

So we definitely want a TeV-scale machine. Why a Large (27 km in circumference) Collider of Hadrons (bound states of quarks, antiquarks and gluons)? Why a machine with a maximum beam energy of 7 TeV, and center-of-mass energy $\sqrt{s} = 14$ TeV? (Sometimes I will talk about reduced energy of 7 TeV, which is where the machine is operating at present, but in these lectures all numbers, unless otherwise specified, refer to 14 TeV collisions.)

The goals of such a machine should be to

- Explain the EWSB scale $v = \langle h \rangle = 2m_W/g_W = 246$ GeV;
- Find the Higgs boson(s) if it is (they are) present;
- Find other particles or phenomena if present below a few TeV;
- Help resolve the hierarchy problem; and
- Perhaps shed light on dark matter.

Back in the mid-1980s, when the LHC was proposed, this implied that the collider and its detectors had to be designed to find anything in the SM and anything that might lie Beyond the Standard Model (BSM). The machine had to be able to observe and measure

- Higgs boson decay modes, in particular $h \to ZZ \to \mu^+\mu^-e^+e^-$, $e^+e^-e^+e^-$, $\mu^+\mu^-\mu^+\mu^-$; $h \to \gamma\gamma$, $h \to b\bar{b}$, $h \to \tau^+\tau^-$;
- Effects of supersymmetry (SUSY), for instance heavy gluinos $\tilde{g}$ decaying to quarks and invisible neutralinos, making events with large energy released in quarks (which manifest themselves as "jets" of hadrons, for reasons we will learn later) but with momentum carried off invisibly by new undetectable particles;
- Effects of technicolor, through alteration of WW, ZZ and WZ scattering, a process identified by accompanying forward and backward jets near the beam, and requiring detection of the decay products of the W and Z bosons.

These in turn implied that the LHC detectors had to be able to

- Identify isolated photons, electrons, muons and taus;

- Detect isolated high-energy colored "partons" — a term referring inclusively to quarks, antiquarks and gluons — as "jets";
- Measure momenta of all visible particles precisely, so to be able to infer the presence of invisible particles through $\sum \vec{p}_{visible} \neq 0$;
- Be sufficiently hermetic so as not to miss anything potentially observable.

These requirements determined the technical specifications of the accelerator and the design of the detectors, which are covered in Jason Nielsen's lectures.

Since 1985, there have been many new models of Beyond-the-Standard-Model physics, including large and warped extra dimensions, Little Higgs theories, and other variants of SUSY. But only a few have qualitatively different requirements. The ones listed above are fairly (but not completely) generic, and so the accelerator and detectors are designed to find almost anything nature might throw at us. They are also versatile enough to search for many things for which they were not specifically designed.

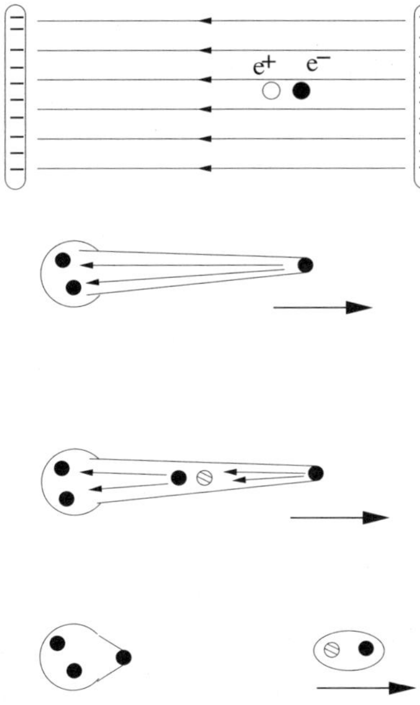

Fig. 4.1. In precise analogy to e^+e^- pair-production in a electric field, an attempt to pull a quark out of a proton creates a flux tube with high enough chromo-electric to produce a new quark-antiquark pair, thus creating a $q\bar{q}$ meson that can be separated cleanly from the qqq baryon that is left behind.

4.3. LHC Fundamentals

4.3.1. *The proton*

Since the LHC is mainly a proton-proton collider, we'd better learn what a proton really is. So let's test your knowledge. Here are some simple statements about the proton. Are they true or false?

- *There are three quarks in a proton.*
- *Each quark carries on average 1/3 the momentum of the proton.*
- *There are only u and d quarks in a proton (not s, c, b, t).*
- *The quarks are confined inside the proton and can never escape it.*

Actually all of these statements are false. In fact,

- A proton contains two u quarks, a d quark, and innumerable gluons and quark-antiquark pairs. (It still has the same quantum numbers as uud, which is why it is often mistakenly described as a uud bound state.)
- The pairs come in all flavors $u\bar{u}$, $d\bar{d}$, $s\bar{s}$, $c\bar{c}$, $b\bar{b}$ (t is exceedingly rare because of its large mass.)
- The gluons carry a substantial fraction of the proton's momentum.
- Quarks, if given a kick, can break free of their parent proton through quark-antiquark pair production, leading to at least one new hadron in the process, as in for example Fig. 4.1.

In general these are good things; otherwise the only collisions we would have are quark-quark collisions, and these are some of the worst ones for making new physics. $q\bar{q}$, qg, gg and even $b\bar{b}$ can all be better. Also, if quarks were completely confined, then when we scattered quarks or produced new ones, we would never see clear evidence that we had done so.

4.3.2. *A simple process, first pass*

Let's discuss a classic LHC process, in which two strongly interacting particles collide, annihilate, and form weakly-interacting ones: $u\bar{d} \to W^+ \to e^+\nu_e$. How do we calculate the rate for this process at a proton-proton collider?

EXERCISE: Note that there are other initial states which are suitable for making a W^+. What are they?

First we need the initial u and $\bar{d}$, which we obtain via the use of the parton model and parton distribution functions. In the parton model, we can define the probability $f_u^p(x)$ to find, inside a fast proton with momentum $P_\mu \approx (E, 0, 0, E)$, a u with momentum $p_\mu \approx xP_\mu$ $(0 < x < 1)$. That we can do so sensibly is *not* obvious! This is both an oversimplified model (which we will unsimplify later) and a highly nontrivial consequence of very special properties of QCD, not shared by all theories.

We then define the differential cross-section for $u\bar{d} \to W^+ \to e^+\nu_e$

$$d\hat{\sigma}_{u\bar{d}\to W\to e\nu}(p_u^\mu = x_1 P_1^\mu, p_{\bar{d}}^\mu = x_2 P_2^\mu, p_e^\mu, p_\nu^\mu). \tag{4.15}$$

We can calculate this (naively) using Feynman diagrams.

EXERCISE: Calculate the amplitude for the above process. Then square it, averaging over initial spins and colors, and summing over final spins, to get a simple expression which depends only on dot-products of the momenta. Assume the W boson is off-shell. Ignore all quark and lepton masses (but not the W mass!) Be careful to remember the parity-violating couplings of the W.

Then naively (and almost correctly!)

$$\begin{aligned} &d\sigma_{pp\to(W^+\to e^+\nu_e)+\ldots}(P_1^\mu, P_2^\mu, p_e^\mu, p_\nu^\mu) \\ &\quad = \int_0^1 dx_1 \int_0^1 dx_2\ f_u(x_1) f_{\bar{d}}(x_2)\ d\hat{\sigma}_{u\bar{d}\to W\to e\nu}(x_1 P_1^\mu, x_2 P_2^\mu, p_e^\mu, p_\nu^\mu) \end{aligned} \tag{4.16}$$

plus another term where the u comes from proton number 2 and $\bar{d}$ from proton 1 *(plus terms for other initial quark-antiquark states; see the earlier exercise.)* I have dropped the superscript p on f_u^p, *etc.*, since we'll be considering protons throughout these lectures.

In this and all classic processes, we take a partonic cross-section (which we calculate using Feynman diagrams), and convolve it with the parton distribution functions (which we measure in "deep inelastic scattering" and elsewhere) to obtain a proton-proton cross-section.

Note that we are not studying $pp \to e^+\nu_e$. (For one thing, that would violate charge conservation!) The dots in the subscripts to $d\sigma$ in Eq. (4.16) represent all the debris from the two disrupted protons (the "Underlying Event.") We can't easily measure that debris precisely, and in fact we're not going to try. So far our theories haven't suggested that new physics will be found there. (Attention: Theory Bias!) We will drop this notation below, but remember events at hadron colliders are never pristine: the protonic debris of the underlying event is always there, mostly forgotten by theorists, but a serious nuisance for experimentalists that one must not ignore.

Experimentally, it's even worse than this. Just in order to get enough collisions, the accelerator physicists arrange their beams — which for technical reasons come in bunches — so that whenever two bunches cross, the average number of simultaneous proton-proton collisions is greater than one. So every interesting collision is bathed in additional debris from 5 to 10 very boring proton-proton collisions. This is called "Pile-up". It adds a certain layer of usually featureless noise to the experimental data, but affects precision measurements and cannot be ignored either.

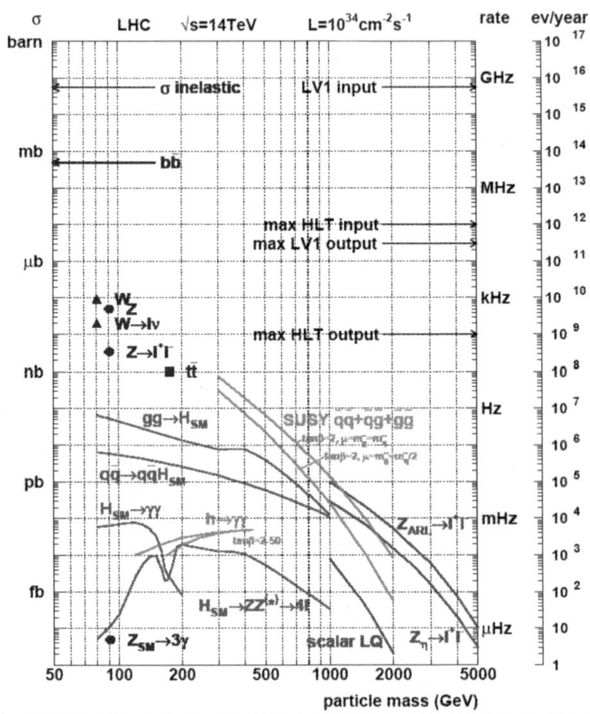

Fig. 4.2. Cross sections (in barns) and production rates (in Hertz) for various important processes at the LHC. Also shown are CMS trigger data rates (see Sec. 4.3.4), with "LV1" meaning the first level of the trigger and "HLT" meaning the higher level(s) of the trigger.

4.3.3. *Rates and luminosities*

Production of W bosons is a classic process. But that does not mean it is *typical.* Look at Fig. 4.2, and be afraid.

Let's define some terms, in order that we may interpret this figure. First, cross section σ has units of area: cm^2, or 1 barn $= 1$ b $= 10^{-24}\ \mathrm{cm}^2 = (10^{-12}\ \mathrm{cm})^2$. Notice that the scale on Fig. 4.2 runs from millibarns to femtobarns (mb, μb, nb, pb, fb.) There are many orders of magnitude here.

The radius of a proton is 10^{-13} cm, of order Λ^{-1}, so its naive geometric cross-section is about $\pi r_p^2 \sim 30$ mb. And indeed the total cross section $\sigma_{pp}^{tot} \sim 2\sigma_{pp}^{inelastic} \sim 100$ mb (which slowly grows with energy as a result of effects known collectively as the Pomeron, something you should learn about in the context of "Regge theory" but which I will not discuss here.)

The weak scale, on the other hand, involves length-scales of order 10^{-16} cm, or area 80 nb. And that is why

$$\sigma_{pp\to W\to e\nu} \sim 20 \text{ nb} \sim 2\times 10^7 \text{ fb} \sim 10^{-7}\ \sigma_{pp}^{total}. \tag{4.17}$$

In short, weak processes are rare because the ratio of v to Λ is about 10^3, and that ratio is squared in cross-sections.

But notice also that the Higgs decay mode to two photons, very important if the Higgs is lightweight, appears yet again much further down:

$$\sigma_{pp\to h\to\gamma\gamma} \sim 10^2 \text{ fb} \sim 10^{-12}\ \sigma_{pp}^{total} \tag{4.18}$$

The lesson is that searching for a light Higgs boson involves looking for a needle in a haystack.

Actually a light Higgs (lighter than 130 GeV) decays dominantly via $h \to b\bar{b}$, so why don't we look there? As you can see in the figure,

$$\sigma_{gg\to h} \sim \sigma_{pp\to h\to b\bar{b}} \sim 30 \text{ pb} = 10^{-7}\sigma_{pp\to b\bar{b}} \tag{4.19}$$

The signal is completely drowned in QCD-generated background, mostly from two gluons in the proton colliding to make a b and a $\bar{b}$. A two-photon signal, though much smaller, still wins, because backgrounds to the two-photon process are, relatively speaking, smaller — and also, because photons can be more precisely measured than bottom quarks, making the mass resonance in two photons sharper and easier to observe over its smooth background. (If this isn't clear, don't worry; we'll say more about mass resonances later.)

Here are some additional definitions that you should be familiar with, and some exercises to check your understanding. The Luminosity $\mathcal{L}$ is the number of collisions per time per area, such that if you multiply it by the cross section σ for a particular process, you get the number of collisions per second producing that process, *i.e.* the rate for that process. If you integrate the luminosity over time you get the Integrated Luminosity $\int \mathcal{L}(t)\, dt$ over a period of time, and $\sigma \int \mathcal{L}(t)\, dt$ gives you, for a particular process with cross-section σ, the average number of events expected during the corresponding period.

The units of Luminosity are $\text{cm}^{-2}\ \text{sec}^{-1}$. A very rough estimate for how luminosity will increase over the coming years at the LHC is

- 2010: $10^{30-32}\ \text{cm}^{-2}\ \text{sec}^{-1}$
- 2011-2012: $10^{33}\ \text{cm}^{-2}\ \text{sec}^{-1}$
- 2015-2018: $10^{33-34}\ \text{cm}^{-2}\ \text{sec}^{-1}$
- 2019: $10^{35}\ \text{cm}^{-2}\ \text{sec}^{-1}$

but this is subject to very significant revision depending upon discoveries and machine performance. (Indeed it changed very much between the time when these lectures were delivered and their final written version!)

Meanwhile, the units of Integrated Luminosity are cm^{-2} or pb^{-1} (in words, "inverse picobarns") or fb^{-1}. And how much Integrated Luminosity do we expect? Well, how long is a year? 3×10^7 sec, but no accelerator runs, or runs efficiently, every day. Typically a good year of running is 10^7 sec, so $\int\ \mathcal{L}(t)\, dt$ for a year is $\approx 10^7\ \text{sec} \times \mathcal{L}$, conventionally, in all planning documents. This gives, as a very rough estimate,

- 2010: $0.1\ \text{fb}^{-1}$ at 7 TeV
- 2011: $2-5\ \text{fb}^{-1}$ at 7 TeV
- 2012: $5-10\ \text{fb}^{-1}$ at 7 TeV
- 2013-2014: shutdown for repairs and upgrades
- 2015: $5-20\ \text{fb}^{-1}$ at (or near) full energy
- 2016-2018: $50-100\ \text{fb}^{-1}$ per year at full energy
- 2019: $100-1000\ \text{fb}^{-1}$.

But again, this is likely to change significantly.

Let's put this into practise. How many $W \to e\nu$ events[a] are expected in 2011? From Fig. 4.2, we see $\sigma_{pp\to W\to e\nu} \approx \sigma_{pp\to W}\text{Br}(W \to e\nu) \sim 20\ \text{nb} = 2 \times 10^7$ fb at 14 TeV. This drops to about 5 nb at 7 TeV, where the LHC is running in 2011. Now we multiply by 5 inverse fb $= 5 \times 10^6$ inverse nb to obtain the number of these events: 2.5×10^7. How many in 2016? The

[a] I will use some shorthand, dropping charges and bars on particle labels, when the expressions are unambiguous.

cross-section increases by 4 when we go from 7 to 14 TeV, and with $50-100$ inverse fb we expect at least 10^9.

A billion Ws decaying to electrons. Back in the 1980s, every W was precious. Now we're swimming in them.

EXERCISE: List the decay modes of the W boson. Estimate (theoretically) the "branching ratio" for $W^+ \to e^+\nu$ (i.e., what fraction of W^+ decays go to positron and neutrino?) [Hint: either ignore quark mixing, or (if you know how) account for it – you will get essentially the same answer. Why?] How many W bosons, independent of how they decay, will be produced in 2016?

4.3.4. *The trigger*

Fig. 4.2, and the simple estimates that we've just learned how to do, lead to an obvious, and critical, problem. The LHC will produce — *must* produce, if it is to provide us with a substantial sample of Higgs bosons, supersymmetric particles, etc. — 10^{12-16} events per year. Simply put, the data storage and management requirements for that many events exceed what is possible and reasonable for even a very large scientific project.

To deal with this, the data flow rates must be reduced by up to 10^6 in order to bring the rate to 10^3 Terabytes per year $\sim 10^{9-10}$ events per year. The only way to do this is to throw away, real time, irretrievably, as much as 99.9999% of the collisions!

Obviously this requires an automated system, called the "trigger." It is the *sine qua non* of a hadron collider. The idea is to keep only events that "look" interesting — interesting to the automated system, of course. The system must be smart enough to throw out 99.9999% of the hay and very few of the needles. It needs to be selective so as to enhance rare processes and reduce common and boring ones. Clearly, if we make bad choices and our automated system is badly or unwisely programmed, we will throw away the new physics. So the trigger had better be designed with care! And theory — both in the understanding of the SM and in its suggestions for BSM physics — plays a large role in guiding these choices (Attention: Theory Bias!) *If theorists are insufficiently imaginative, it is possible we could miss the new physics altogether because the trigger throws it away.*

The basic principles of a trigger are simple: select events which are relatively rare in the SM. The most common events contain either just some proton debris, or debris plus two back-to-back relatively low-energy jets. Rare processes includes ones with

- isolated electroweak particles: γ, e, μ, τ, even with fairly low energy (as low as 5–20 GeV),
- very high-energy jets (from quarks or gluons) with a few hundred GeV or more,
- though with a lower-energy requirement for b quarks (which are special jets, as we'll see),
- and/or invisible particles (apparent missing momentum.)

But in selecting events in this way, the trigger introduces other problems. All measurements are distorted by the trigger, since it involves a rather complicated set of selection criteria. Nothing a theorist calculates is ever measured; the trigger changes it to a greater or lesser degree. Any measurement must account for the "efficiency" of the trigger (what fraction of a given type of events are selected?) and the resulting distortion of observable distributions. To do this, the effect of the trigger must be measured, a tricky technical challenge. And any prediction at a hadron collider requires that the effect of the trigger be simulated. This must always be done by the experimentalists, as it is detector-dependent and very complicated. So a theorist must always have this in mind when calculating something.

4.3.5. *Drowning in data*

Now this is not the only problem. Even after the trigger cuts away all but one in a million events, there are still 10^{9-10} events recorded each year. Many new-physics signatures will show up in only $10 - 1000$ of these events. How are we going to find them? Well, with 10^3 physicists, we are not going to find the 1 in 10^6 events that are special by paging through event pictures! And besides, pictures wouldn't tell the full quantitative story anyway. Instead, yet another automated system is needed to do interpretation of data: initial ("reconstruction", automated) and detailed ("analysis", semi-automated.) And again (Attention: Theory Bias!) there are all sorts of assumptions that go into these systems.

The basic strategy in looking for new physics is based on some very robust assumptions. First, SM physics is well understood. All heavy SM particles decay microscopically; i.e., they travel less (in some cases vastly less) than a millimeter before decaying. Only a few lightweight particles (γ, e, μ, ν_i, and a few hadrons, such as π^+, K^+, K^0, p, n) can travel more than a millimeter (and most of these travel meters or more.) Second, in most BSM models, new heavy particles are produced, decaying microscopically to SM particles and possibly new stable invisible particles (which we can't detect directly.) So based on these assumptions, one builds a detector to detect $\gamma, e, \mu, \pi^+, K^+, K^0, p, n$ emerging *from the collision point* (or very near) and traveling *through the detector.* Then one writes reconstruction software to recognize these "objects" from the electronic signals they leave as they traverse the detector.

But if there's some process that looks different from what we typically expect, software might miss it, *even* in cases where the human eye might notice it. (Attention: Theory Bias)

The problem with looking for something rare, as rare as new physics is relative to SM physics — 8 to 14 orders of magnitude more rare — is that, as we all know, everything that can go wrong will go wrong sometimes. Even things that happen in only one in every 10^6, or even 10^{12}, events can be significant challenges. For instance, an important background to $h \to \gamma\gamma$ comes from fake isolated photons, which are as common in the relevant energy range as real ones. When one observes missing momentum, it is often as likely to be fake, due to a detector problem, as real. Fake τs, strange-looking jets, are almost as common as real ones. And a variety of very strange effects can arise: from "beam halo" muons, created by protons that stray from the beam and hit something upstream from the detector, from "beam-gas collisions," in which a proton hits a rare gas molecule in the evacuated beam pipe, and from "secondary scattering", a collision of an outgoing particle with detector material, which produces a spray of particles. Cosmic rays, too, can produce odd phenomena. The experimentalists have to deal with all of these issues with great care.

4.3.6. *The role of theory*

The worst problem — for theorists — is that theory can barely predict anything reliably. All processes are sensitive to the treatment of the strong interactions. Perturbation theory is technically difficult and barely works, and simple perturbative amplitudes need to be supplemented by resummation of infinite classes of Feynman graphs. Nonperturbative processes are always present and occasionally are very important. Worst of all, experiments simply can't measure what theorists can calculate easily, and theorists can't calculate what experiments can measure easily. We are always struggling in this field... though things have recently been getting a lot better.

To match to any experimental measurement, theory must be passed through extensive and not-entirely-reliable software ("Monte Carlo" programs, some for simulation of the physics of the collision and some for simulation of the detector's response to what is produced.) Only via simulation can one really make predictions for what experiments actually measure. Furthermore, theory that predicts SM backgrounds often has significant uncertainties, and its predictions must be verified in the data (which is not always easy). In fact searches for new physics typically try to compare data with data, not data with theory, since theory is often so unreliable.

In general, if there's anything that could possibly make a theorist humble, it's a hadron collider!

Because of this complexity, and the limitations of theory, a critical part of understanding how to do physics at the LHC is to understand, in detail, how the experiments are done. So you should *read and understand* Jason Nielsen's lectures at this school; you may find them challenging, but you will not regret it! No great hadron collider theorist ever was ignorant of basic experimental techniques.

4.4. Basic Kinematics

Now let me give some kinematical definitions that we will need, along with some essential formulas. (See the particle data book, at pdg.lbl.gov, in the chapter called "Kinematics.")

Since the beampipe defines an axis, with many detectors arranged with approximate cylindrical symmetry around this axis, the natural coordinates for describing the physics are cylindrical around the beampipe, with θ the polar angle and ϕ the azimuthal angle. But it turns out θ is not as convenient as a different quantity called "pseudorapidity" η. It is a monotonic function of θ. If we choose θ so that $\theta = 0$ is forward, π backward, and $\pi/2$ perpendicular to the beampipe, then define

$$\eta \equiv -\log\tan(\theta/2) = \log\cot(\theta/2). \tag{4.20}$$

Note that

$$\sinh\eta = \cot\theta \ ; \ \cosh\eta = 1/\sin\theta \ ; \ \tanh\eta = \cos\theta \ . \tag{4.21}$$

The variable η runs from minus infinity to infinity, with $\eta \to -\eta$ if we reflect the beampipe around the perpendicular plane, and with

- $\eta \approx \pi/2 - \theta$ for $\theta \sim \pi/2$;
- $\eta \approx \log(2/\theta)$ for $\theta \ll 1$;

- $\eta \approx \log(2/[\pi - \theta])$ for $\pi - \theta \ll 1$.

EXERCISE: What angles θ correspond to $\eta = 0, \pm 1, \pm 2, \pm 3, \pm 4$? Obtain these as numbers and sketch them on a diagram.

It is useful to define a "distance" between two directions

$$\Delta R = \sqrt{(\Delta\eta)^2 + (\Delta\phi)^2}. \tag{4.22}$$

Why is this a nice quantity? Well, let's see.

Rapidity y is a different quantity from pseudorapidity. It is not a function of angle but a function of E and p_z of a particle:

$$y = \frac{1}{2}\log\frac{E + p_z}{E - p_z} = \tanh^{-1}(p_z/E) = \tanh^{-1} v_z. \tag{4.23}$$

But for relativistic particles $y = \eta$!

Now here are three easy but crucial exercises, which will clarify why η is used instead of θ.

EXERCISE: Prove that under a boost in the z direction with velocity β, $y \to y + \tanh^{-1}\beta$ [thus demonstrating that a rapidity difference Δy is Lorentz-invariant under boosts along the beam direction.]

EXERCISE: Prove that for a massless particle, $y = \eta$ [thus demonstrating that for massless or nearly massless particles moving in the detector, ΔR is Lorentz-invariant under boosts along the beam direction.] In particular, show

$$\eta = \frac{1}{2}\log\frac{|\vec{\mathbf{p}}| + p_z}{|\vec{\mathbf{p}}| - p_z} = \tanh^{-1}(p_z/|\vec{\mathbf{p}}|). \tag{4.24}$$

EXERCISE: Prove that for small $|\eta|$, $\Delta R \approx \sqrt{(\Delta\theta)^2 + (\Delta\phi)^2}$.

Why is it so good to use a quantity R that is invariant under boosts in the z direction? It is because the lab frame is not typically the relevant frame for the physical processes. In general the partonic energy fractions x_1 and x_2 of the colliding partons are not equal. Therefore *the center-of-mass frame of the parton-parton collision is not the lab frame, but is instead in motion along the beampipe.* This, combined with the facts that you should by now have proved in the previous exercises, motivate the use of η and R —

which are the same in the (often unreconstructed) parton-parton collision frame as they are in the lab frame — rather than θ, which transforms under z-boosts in a nasty way.

To fully appreciate this, we should try to understand naive parton-parton kinematics. For particles with $m \ll E$,

$$(k_1 + k_2)^2 \approx 2k_1 \cdot k_2 = 2E_1E_2 - 2\vec{p}_1 \cdot \vec{p}_2 = 2E_1E_2(1 - \cos\theta_{12}). \quad (4.25)$$

Take the two proton momenta to be $P_1^\mu = E_b(1,0,0,1)$, $P_2^\mu = E_b(1,0,0,-1)$, where E_b is the beam energy, so the center of mass energy of the collider in the lab frame is $E_{pp} = 2E_b$. We can write

$$s \equiv (P_1 + P_2)^2 = (2E_b)^2 = E_{pp}^2 = 2m_p^2 + 2P_1 \cdot P_2 \approx 2P_1 \cdot P_2. \quad (4.26)$$

Then, if we have parton momenta $p_i = x_iP_i$, the square of the *partonic* center-of-mass energy, $\hat{E}$, is given by

$$\hat{s} \equiv (x_1P_1 + x_2P_2)^2 = \hat{E}^2 \approx 2x_1P_1 \cdot x_2P_2 \approx x_1x_2s. \quad (4.27)$$

Thus

$$\hat{E}/E_{pp} = \sqrt{x_1x_2}. \quad (4.28)$$

In Table 4.3 we compare quantities describing the two partons and their kinematics, expressing them in the center-of-mass frame of the two colliding partons and also in the lab frame (the center-of-mass frame of the colliding protons:) Here $\hat{v}_z, \hat{y}, \hat{\gamma}$ refer to the motion of the partonic center-of-mass

Table 4.3. Important properties of the parton-parton collision as viewed in the parton-parton center-of-mass frame and in the lab frame.

parton c.o.m frame	lab frame
$E^{tot} = \hat{E} = \sqrt{\hat{s}} = \sqrt{x_1x_2}\sqrt{s}$	$E^{tot} = (x_1 + x_2)E_b = \frac{1}{2}(x_1 + x_2)\sqrt{s}$
$p_z^{tot} = 0$	$p_z^{tot} = (x_1 - x_2)E_b = \frac{1}{2}(x_1 - x_2)\sqrt{s}$
$\vec{p}_T^{\,tot} = 0$	$\vec{p}_T^{\,tot} = 0$
$\hat{v}_z = 0$	$\hat{v}_z = \frac{x_1 - x_2}{x_1 + x_2}$
$\hat{y} = 0$	$\hat{y} = \log\sqrt{\frac{x_1}{x_2}}$ $(e^{\hat{y}} = \sqrt{x_1/x_2})$
$\hat{\gamma} = 0$	$\hat{\gamma} = \frac{x_1 - x_2}{x_1 + x_2}$

frame along the z axis. Note also that

$$x_1 = \sqrt{\frac{\hat{s}}{s}} e^{\hat{y}} \; ; \; x_2 = \sqrt{\frac{\hat{s}}{s}} e^{-\hat{y}}. \tag{4.29}$$

EXERCISE: Prove the formulas in the table and the formula above.

EXERCISE: Show that for fixed $\hat{s}$ and fixed s, collisions can only have rapidity satisfying $|\hat{y}| \leq \log \sqrt{\frac{\hat{s}}{s}}$.

In the specific case that the process of interest involves $2 \to 2$ parton scattering $p_1, p_2 \to p_3, p_4$, we can go further. Define the partonic Mandelstam variables in the standard way.

$$\begin{aligned} \hat{s} &\equiv (p_1 + p_2)^2 = (p_3 + p_4)^2 \; ; \\ \hat{t} &\equiv (p_3 - p_1)^2 = (p_4 - p_2)^2 \; ; \\ \hat{u} &\equiv (p_4 - p_1)^2 = (p_3 - p_1)^2 \; . \end{aligned} \tag{4.30}$$

(The equalities follow from momentum conservation.) It is well-known that $\hat{s} + \hat{t} + \hat{u} = m_1^2 + m_2^2 + m_3^2 + m_4^2$ (*which you should verify if you've never seen it before, or have forgotten.*) If the particles are all effectively massless then

$$\hat{t} = -2E_1 E_3 (1 - \cos\theta_{13}) = -(\hat{s}/2)(1 - \cos\hat{\theta}) \tag{4.31}$$

since $E_1 = E_2 = E_3 = E_4$ in the parton center-of-mass frame. Similarly,

$$\hat{u} = -\hat{s} - \hat{t} = -2E_1 E_4 (1 - \cos\theta_{14}) = -(\hat{s}/2)(1 + \cos\hat{\theta}). \tag{4.32}$$

Note that the *transverse momentum* $\vec{p}_{i,T} \equiv (p_{i,x}, p_{i,y})$ is invariant under z-boosts; it is the same in the lab frame and in the parton center-of-mass. frame. To be precise, in the latter frame,

$$p_T = \frac{\sqrt{\hat{s}}}{2} \sin\hat{\theta} = \frac{\sqrt{\hat{t}\hat{u}}}{2\hat{s}} \; . \tag{4.33}$$

The final expression in this equation is built from Lorentz invariants, and so is true in terms of the lab frame variables too.

Let us summarize the list of quantities that are invariant under Lorentz-boosts along the beampipe, in the case of "massless" observed particles (which will include γ, e, μ, and most observed hadrons, but not always jets, which often have a non-negligible mass.) The z-boost-invariant quantities we have encountered are

$$\hat{s}, \hat{t}, \hat{u} \; ; \phi, \Delta\eta \to \Delta R; p_T \; . \tag{4.34}$$

Why is this so important? Consider two collisions that can occur in proton-proton collisions that have exactly the same parton center-of-mass frame quantities, and differ only in the boost of the event as viewed in the lab frame, namely x_1/x_2. For instance, consider one collision that creates, from a u and a $\bar{d}$, a W boson at rest in the lab frame, and another that similarly creates a W boson with a rapidity of 0.5 in the lab frame. Now, these are certainly different collisions. The remnants of the protons left behind, and thus the underlying event, are completely different in the two collisions. But theoretically and even experimentally we don't care about that physics, which is low-energy muck that we barely measure and mostly try to avoid. If we are just interested in counting W bosons — as we are when using the LHC to study high-energy phenomena generated by high-energy parton-parton collisions — we can treat these two collisions as representing the same high-energy phenomenon. Thus, effectively integrating over the low-energy junk that can accompany these collisions, we combine these collisions in our experimental samples.

In short, collisions that have the same parton-center-of-mass phenomena, and differ only in the z-boost of the partonic frame relative to the lab frame, can be combined together into a single sample representing a single phenomenon. That this is possible and sensible is not instantly obvious! It follows non-trivially from QCD, as we will discuss later. But because it is true, we will characterize events in terms of their z-boost-invariant quantities, because we will always be combining events that share those quantities. (There are of course — as always — subtleties with such statements. But consider this a general, though imperfect, principle.)

Finally, let me mention a couple of event-wide z-boost-invariant quantities. One is a simple measure of the scale of visible p_T:

$$S_T \text{ or } H_T \equiv \sum_{\text{objects } i} |\vec{p}_{i,T}| \ . \tag{4.35}$$

(Unfortunately, exactly which objects to include in the sum varies; there is no agreed-upon convention right now. You need to read the definition carefully if you see an article or plot using this quantity.) Another is a simple measure of invisible $\vec{p}_T$:

$$\vec{\not{p}}_T \equiv - \sum_{\text{all visible}} \vec{p}_{i,T} \ . \tag{4.36}$$

Clearly this is the "missing transverse momentum"; $|\vec{\not{p}}_T|$ is conventionally (though inaccurately) called the "missing energy" or "missing transverse energy" or "MET" or $\not{E}_T$. If the detector were perfectly hermetic then

$$\vec{\not{p}}_T \equiv + \sum_{\textbf{all invisible}} \vec{p}_{i,T} \ . \tag{4.37}$$

Obviously we must sum over everything visible, or everything invisible, to do correct accounting.

Question: why don't we look for missing p_z or missing E? *Answer:* The scattered partons carried E and p_z but little or no $\vec{p}_T$. That means that the debris from the underlying event, the remnants from the shattered protons, has total energy $E < 14$ TeV and nonzero p_z, but little or no $\vec{p}_T$. Typically the many hadrons produced in this debris each have low p_T, but many have very high E and p_z. Quite a few of them therefore are at very low angles relative to the beam, and remain inside or very near the beampipe until they have traveled outside of the detector. These hadrons we cannot measure. So it is impossible, for technical reasons, to do a proper accounting of all the E or p_z carried by all the final state particles, which would be necessary to see if any is missing. Fortunately these particles carry very little transverse momentum, so we can still do a largely accurate accounting of p_T, and look to see if any is missing as a result of invisible particles, such as neutrinos, being produced.

4.5. QCD at the LHC

We will be referring to several critical special features of QCD. a theory of $N_c = 3$ colors, confining at a strong coupling scale $\Lambda_{QCD} \equiv \Lambda \sim 300$ MeV, and with $N_f = 6$ flavors of quarks, of which $N_f^{light} = 3$ have $m < \Lambda$ (u, d, s) and $N_f^{heavy} = 3$ have $m \gg \Lambda$ (c, b, t). Here are some important facts about QCD that we need to recall (or learn, as the case may be.)

4.5.1. *Beta functions, confinement, asymptotic freedom and fixed points*

QCD FACT 1: this theory is asymptotically free in the ultraviolet, "confining" in the infrared.

This is not true of gauge theories in general! Consider a generalization of QCD, to a theory with N_c colors and N_f flavors of quarks and antiquarks in the $\mathbf{N_c} + \bar{\mathbf{N}}_\mathbf{c}$ representations. For small N_f, the gauge coupling α_s has a beta function, $\beta_{\alpha_s} \propto -(\frac{11}{3}N_c - \frac{2}{3}N_f)$ in the one-loop approximation, that is negative at all scales $\mu \gg \Lambda$. Thus the coupling is small in the UV and large in the IR. But if N_f is very large, then $\beta > 0$, and the reverse is true, and the theory is *not* asymptotically free.

If, for an $SU(N_c)$ gauge theory, the running coupling $\alpha_s N_c \ll 1$, then the gauge coupling satisfies

$$\alpha_s(\mu) N_c \approx \frac{2\pi}{(\frac{11}{3} - \frac{2}{3}\frac{N_f}{N_c}) \log(\Lambda/\mu)}. \tag{4.38}$$

For $\beta < 0$ ($\beta > 0$) this is valid for $\mu \gg \Lambda$ ($\mu \gg \Lambda$.) (This is strictly correct only if N_f does not change with μ. It does of course change whenever μ crosses a mass threshold of a quark, but this detail is not critical for us at the moment.) Perturbation theory is a good approximation when $\alpha_s N_c \ll 1$ — which, please note, is a stronger constraint than $\alpha_s \ll 1$.

For $\beta < 0$, as $\mu \to 0$, α_s grows large. But *confinement does not logically follow.* In fact, if N_f is large but not too large, so that $\beta < 0$ at one-loop, higher-loop corrections may cause $\beta(\alpha_s)$ to vanish at a finite α_s. In real-world QCD, N_f is too small and this does not happen. But it may well happen for $N_f \sim 10$–15.

Conversely, confining theories need not be asymptotically free. Even if $\alpha_s \to \infty$ as $\mu \to 0$, it may be, as $\mu \to \infty$, that $\alpha_s \to$ a nonzero constant, again as a result of a breakdown in Equation (4.38) due to higher loop corrections.

Thus, coexistence of UV freedom and IR slavery is not automatic. It is crucial that $N_f \ll \frac{11}{2} N_c$ at all scales. In fact, we do not, strictly, know that QCD is UV-free. If there are in nature a sufficient number of heavy undiscovered colored particles, we will someday learn that it is not. All we know now is that QCD is has a negative beta function up to a few hundred GeV.

EXERCISE: Look up the formula for the beta function of a QCD-like theory at two loops. Show that there is a zero of the beta function when $0 < 11N_c - 2N_f \ll N_c$. *Bonus question, if you know about the large-*N_c *expansion: show that for large* N_c *this is stable against higher loop corrections. [Hint: what value does* $\alpha_s N_c$ *take at the beta-function zero?]*

4.5.2. *Parton distribution functions*

QCD FACT 2: "Factorization": Scattering of confined partons can be described by convolving process-independent (universal!) parton distribution functions (pdfs) with process-dependent short-distance ($\hat{s}, \hat{t}, \hat{u} \gg \Lambda^2$) perturbative amplitudes.

This isn't trivial either. Clearly, without $\alpha_s N_c \ll 1$ at $\mu \gg \Lambda$, it would be false, and without confinement, it would be irrelevant. A highly nontrivial proof does exist in some special cases, but a general proof of this statement is still lacking. The best argument, right now, is that it has worked at previous colliders and there is no known reason why it should break down for the processes of greatest interest at the LHC.

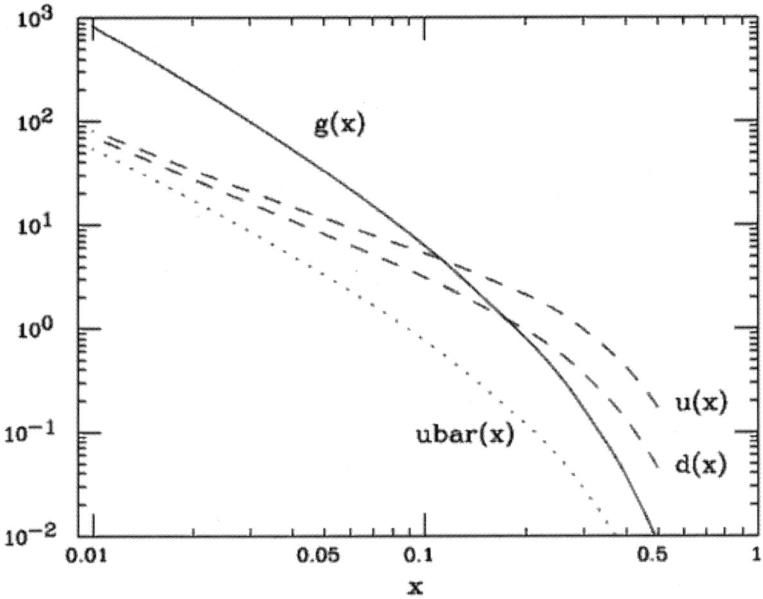

Fig. 4.3. Log-Log plot of parton distribution functions.

Figures 4.3 and 4.4 show some parton distribution functions (pdfs) and parton luminosities. (Caution: don't get too attached to these until we've reached the end of the course.) Observe the following features in Fig. 4.3 (where $f_u(x)$ is called $u(x)$, etc.):

- Gluons dominate at $x < .1$;
- u, d quarks dominate at $x > .1$;
- $u \sim 2d \gg \bar{u}$ for $x \sim .3$;
- $u \sim d \sim \bar{u}$ for $x \ll .1$.

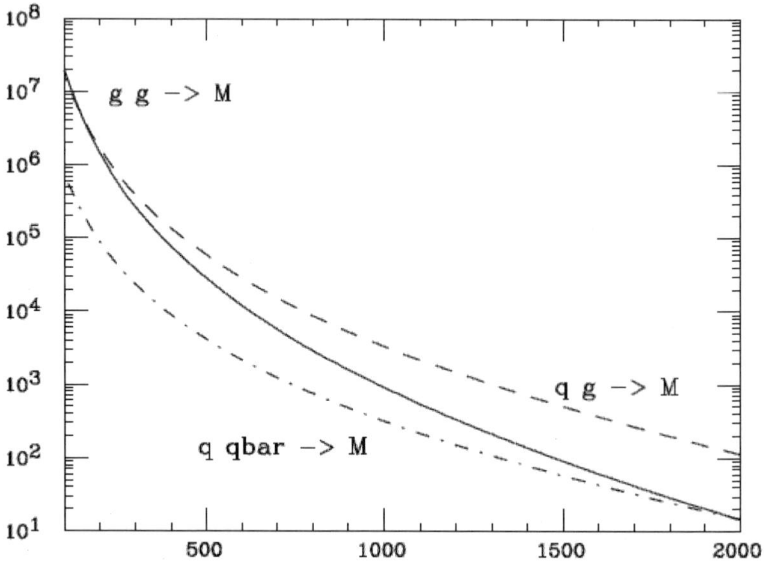

Fig. 4.4. Log-Linear plot of parton luminosities.

Keep in mind that for a proton at $E_b = 7$ TeV, a quark with $x = .1$ is carrying $E = 700$ GeV of energy, so it can have a collision at the TeV scale.

Figures 4.5–4.8 show more of the pdfs, sometimes weighted by x or x^2 to make the graphs readable — look carefully at the captions. Here we see that

- $\bar{d}, \bar{u}, \bar{s}$ are all pretty similar at all x;
- $s = \bar{s}, c = \bar{c}$ to a good approximation at all x;
- $\bar{q} < g$ at all x;
- $u, d, s, c, \bar{u}, \bar{d}, \bar{s}, \bar{c}$ are all pretty similar at very small x.

How do we interpret these features?

To a good approximation, we can divide the u and d quark distributions into "valence quark" and "sea quark" distributions, $u_v = u - \bar{u}$, $u_s = \bar{u}$. The valence quark distributions are most significant, and much larger than the sea distributions, at higher values of x. Most non-valence partons are gluons, which rise rapidly from small values at large x to dominate at small x. The remainder of the partons form a sea of quark-antiquark pairs, from which all the antiquarks and the non-valence quarks are drawn: $\bar{u}, u_s = \bar{u}, \bar{d}, d_s = \bar{d}, \bar{s}, s$, etc. The gluons are dominantly emitted from the valence quarks or from each other, and grow very rapidly in number at small x. The sea quarks and antiquarks are emitted by the gluons; they are pulled up with the gluons at small x, but they can never catch up.

Notice for $x > .01$, f_g always falls much faster than $1/x$, $f_{u,d}$ falls just a bit faster than $1/x$. This is true until $x \sim .3$, when both begin to drop off more rapidly as they approach zero at $x = 1$.

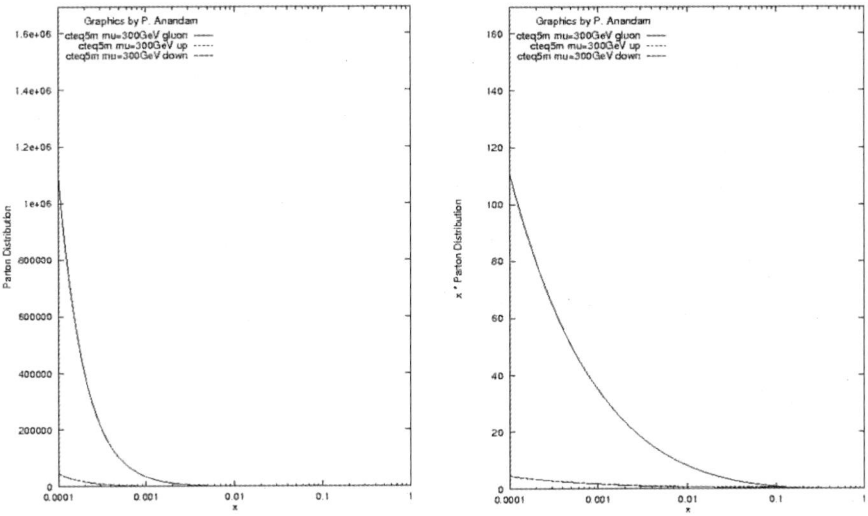

Fig. 4.5. The g, u and d parton distribution functions. Left: the functions $f_i(x)$. Right: the functions $xf_i(x)$. At very low x, gluons dominate, $f_u \approx f_d$ at low x, and all three functions grow faster than $1/x$ as $x \to 0$.

4.5.3. *Parton luminosity*

We saw

$$d\sigma_{pp\to(W^+\to e^+\nu_e)+\ldots}(P_1, P_2, p_e, p_\nu)$$

$$= \int_0^1 dx_1 \int_0^1 dx_2 \; f_u(x_1) f_{\bar{d}}(x_2) \; d\hat{\sigma}_{u\bar{d}\to W\to e\nu}(x_1 P_1, x_2 P_2, p_e, p_\nu) \tag{4.39}$$

plus a term with $f_{\bar{d}}(x_1) f_u(x_2)$, plus some similar terms with $u\bar{d} \to u\bar{s}, c\bar{d}$, *etc.*. As we discussed earlier, $d\hat{\sigma}$ depends on $\hat{s}, \hat{t}, \hat{u}$ but not on $\hat{y}$! That is,

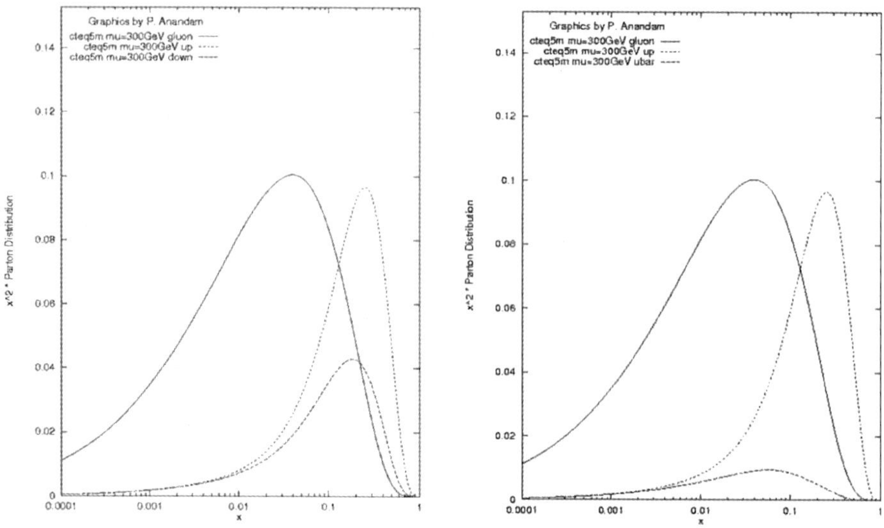

Fig. 4.6. Left: The g, u and d parton distribution functions, multiplied by x^2. In contrast to Fig. 4.5, the structure at larger x is visible. The u, d quarks dominate above $x = .1$, where $f_u \sim 2f_d$; below $x = 0.1$, gluons dominate, and gradually $f_u \to f_d$. Recall that $x = 0.1$ for a 7 TeV proton corresponds to a parton with 700 GeV of kinetic energy. Right: the g, u and $\bar{u}$ parton distribution functions, times x^2. Note $f_u \gg f_{\bar{u}}$ at large x, but $f_u \sim f_{\bar{u}}$ at small x.

the differential cross-section depends on the Lorentz-invariant kinematics of the scattering partons, but not on how fast the parton center-of-mass frame is moving in the lab frame. So we may rewrite the integral in terms of $\hat{s}$ and $\hat{y}$.

$$\int dx_1 \int dx_2 = \frac{1}{s} \int d\hat{s} \int d\hat{y} \tag{4.40}$$

and thus the total cross section at the collider is

$$\sigma_{pp \to (W^+ \to e^+ \nu_e) + \dots} = \int_0^s d(\hat{s}/s) \hat{\sigma}_{u\bar{d} \to W \to e\nu}(\hat{s}, \hat{t}, \hat{u}) L_{u\bar{d}}(\hat{s}/s) \tag{4.41}$$

where

$$L_{u\bar{d}} = \int_{-\infty}^{\infty} d\hat{y}\ f_u\left(\sqrt{\frac{\hat{s}}{s}} e^{\hat{y}}\right) f_{\bar{d}}\left(\sqrt{\frac{\hat{s}}{s}} e^{-\hat{y}}\right) \tag{4.42}$$

is the $u\bar{d}$ *parton luminosity*, which depends only on $\hat{s}$. It tells us the probability of finding a left-moving u and a right-moving $\bar{d}$ with a particular

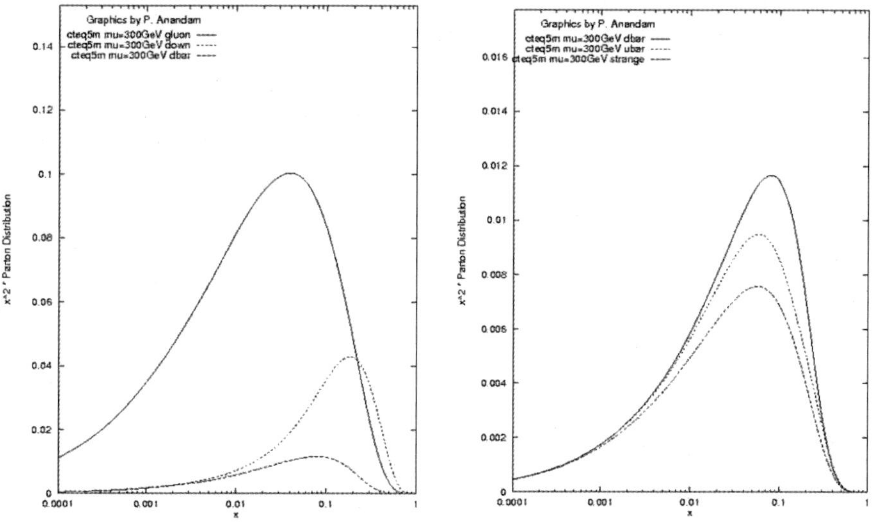

Fig. 4.7. Left: The g, d and $\bar{d}$ parton distribution functions, multiplied by x^2. Compare to Fig. 4.6. Right: the $\bar{d}$, $\bar{u}$ and s parton distribution functions, showing they differ at medium x by up to 20%, but not at much lower x.

fixed $\hat{s}/s$. (Since we could also take a u from the right-moving proton, the total $u\bar{d}$ parton luminosity is typically defined with a factor of 2, but this is a matter of convention.)

EXERCISE: Derive the last three equations.

In Fig. 4.4 are shown the parton luminosities, summed over quark flavor, for a 14 TeV LHC. Note that (using an obvious shorthand)

- qg is as big as or bigger than gg for $\sqrt{\hat{s}} > 200$ GeV;
- qg is 10 times larger than gg for $\sqrt{\hat{s}} \sim 2$ TeV;
- qg is always of order 10 times larger than $q\bar{q}$;
- $q\bar{q}$ and gg are about equal at $\sqrt{\hat{s}} > 2$ TeV.

Not shown is that $ug \sim 2dg$ for moderate to large $\sqrt{\hat{s}}$.

You may have heard it said that "the LHC is essentially a gluon-gluon collider," but you can see that this is too naive. The $q\bar{q}$ and qg initial states are crucial to understanding the phenomenology. From Fig. 4.4 we also see that

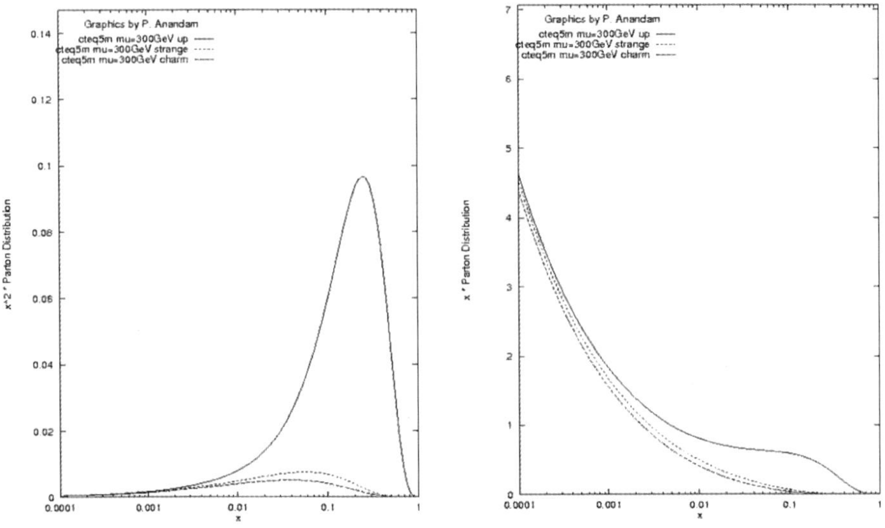

Fig. 4.8. The u, s and c parton distribution functions. Left: $x^2 f_i(x)$; Right: $x f_i(x)$. There are almost as many $s\bar{s}$ and $c\bar{c}$ pairs inside the proton as $u\bar{u}$ pairs, but none at high x, where u and d quarks dominate. Recall that $x = 0.001$ for a 7 TeV proton corresponds to a quark with 7 GeV of kinetic energy.

- $qg, q\bar{q}$ fall a bit slower than $\hat{E}^{-4}$
- gg falls $\hat{E}^{-(3-5)}$.

The rapid fall-off of the luminosities with $\hat{E} = \sqrt{\hat{s}}$ is crucial in almost every aspect of LHC phenomenology.

Let's return to Fig. 4.2, and interpret it more carefully. The total cross section is dominated by elastic scattering and by glancing blows that create additional hadrons. These events typically only create hadrons with $p_T <$ 1 GeV. If we choose bunch crossings at random, and look at the ones where we see a little bit of energy in the detector, we are imposing the *minimum amount of bias* in our selection ... so these are called "min. bias" events, and the trigger which selects (a random subset of) them is called the "min. bias trigger."

With some small probability, two partons may strike each other a little more strongly and scatter with $p_T > 1$ GeV. Once p_T is large enough (strictly, once $\hat{s}, |\hat{t}|, |\hat{u}| > 1$ GeV^2) then perturbation theory applies and this process can be computed, at least roughly. The most likely event involves parton i elastically scattering off parton j, in the t channel. That's because QCD, like electromagnetism, is (on the distance scales of 1/100 GeV^{-1}) a long-range force, and so scattering is peaked at small angles. Less likely, but still common, is that partons i and j annihilate in the s channel and make partons i' and j'; this is true quantum scattering. An example is $gg \to b\bar{b}$, which shows up in Fig. 4.2 as $\sigma_{pp\to b\bar{b}} \sim 300\ \mu\text{b}$.

Because of the rapidly falling parton luminosities, the probability to make partons with transverse momentum p_T in any given process drops roughly like p_T^{-6} or more. More precisely, $L_{ij} \sim \hat{E}^{-(3-4)}$, and in QCD $\sigma(\hat{s}) \sim 1/\hat{s}$ for an s-channel process, so rates for most interesting processes are scaling like $\hat{E}^{-5}$ or faster.

Let's now apply these observations to learn a bit more about physics at the LHC, looking again at Fig. 4.2.

W or Z bosons are somewhat rare, compared to the total cross sections, because several prices must be paid relative to QCD processes. First, weak rather than strong interactions are needed, and the coupling α_W is somewhat weaker (not that much though) than α_S. Second, the initial state requires an antiquark, and that's rare even at small x compared to the dominant gluons. And you need an accidentally color-neutral initial state. Finally, of course, at least one of the quarks must have have $x > m_W/2E_b \sim .006$.

Compare this with $t\bar{t}$ production. To make $t\bar{t}$ (mostly $gg \to t\bar{t}$) does use the strong interactions, and initial-state gluons, which are not at all rare. And this partially compensates for the larger energy required: $2m_t \sim 4m_W$. But two-body production amplitudes is smaller, relatively speaking, than resonance production; and moreover rates fall fast with energy. So the cross-section for $t\bar{t}$, even with its advantages, is still 100 times smaller than W production.

Supersymmetric particles are also produced in pairs, and the colored ones are produced through strong interactions, as in $t\bar{t}$. Note that the lines that indicate supersymmetric particle production, if they were extrapolated down to the top quark mass scale, would like a bit above $t\bar{t}$. That is simply because there are many squarks and because the gluino, as a color octet, has even stronger interactions than the color-triplet top quark.

We'll come back to the Higgs boson and its production, and some of the other lines in Fig. 4.2, later.

A word of caution is merited here. Many of the events corresponding to a given process are either unobservable or drowned in background, so the actual number of usable events may be much smaller than shown in Fig. 4.2. In fact the fraction of events that actually can be separated from

SM processes will often be a percent or a tenth of a percent. That's part of why we need such large data sets to make discoveries at the LHC.

4.6. Hadrons

4.6.1. *The light hadrons of QCD*

QCD FACT 3: QCD has some exceptionally light hadrons. Pion and kaon and eta mesons are significantly lighter than the other hadrons, and lighter, at least in the case of pions, than any reasonable definition of the confinement scale. Consequently, all other mesons are able, very rapidly, to decay to them.

Now why is this? The light masses $m_\pi < \Lambda$, $m_K \sim \Lambda$ arise because of "spontaneous chiral symmetry breaking". The theory has $N_f^{light} = 3$ light quarks and antiquarks. If the quarks were massless, the theory would have a $SU(3) \times SU(3)$ global symmetry. In the real world this is *both* explicitly broken (by the quark masses m_q) *and* spontaneously broken (by $\langle q\bar{q}\rangle \sim \Lambda^3$). The non-zero quark masses assure there are no exact NG bosons, because the "spontaneous breaking" from the $q\bar{q}$ vev breaks a "symmetry" that the quark masses have already explicitly done away with. But because there *would be* a symmetry if m_q were zero, there are exceptionally light particles, pseudo-NG (or PNG) bosons, with $m_\pi \sim \sqrt{m_q\Lambda}$.

We cannot take these states for granted. For instance, in a theory with $N_f^{light} = 0$ there are no global symmetries and no light pions. The theory at scales near Λ has only gluons (and is called"Pure Yang-Mills Theory") and has a very different spectrum with many stable hadrons (called "glueballs.") Meanwhile, for $N_f^{light} = 1$ and N_c small, there are also no approximate symmetries (the axial $U(1)$ present classically is famously broken, explicitly, by a quantum anomaly,) so again there are no light hadrons. (The η' can become light at large N_c; see Witten's famous treatment of the η' mass.)

EXERCISE: Consider a QCD-like theory with $N_c = 3$ and $N_f^{light} = 2$ **massless** *quarks. What is the spectrum of the pion-like states and the nucleon-like states? How does it change if $N_c = 4$? How does it change if $N_f^{light} = 4$?*

The presence of these light states, and their associated long lifetimes, determines much of what we actually see in LHC detectors. They are so light, in fact, that essentially every hadron decays to light mesons (and/or protons and neutrons) eventually. Pions are therefore the most common particle emerging in the LHC detectors, with kaons and etas not far behind.

Table 4.4. The charge assignments, under the electroweak gauge group, of the Higgs, and the Higgs potential.

	$SU(2)_W$	$U(1)_Y$	$V(H)$
$H = \begin{pmatrix} H^+ \\ H^0 \end{pmatrix}$	**2**	$\frac{1}{2}$	$-\mu^2 H^\dagger H + \frac{1}{2}\lambda(H^\dagger H)^2$

The equations that describe the pions are almost the same as those for the Higgs boson. In a world with massless quarks, one would simply replace the Higgs boson and its potential, shown in Table 4.4, with a quark bilinear field Σ and an associated potential, Table 4.5.

Table 4.5. The charge assignments under the u, d flavor group of the quark bilinear field Σ, and the potential for Σ.

	$SU(2)_L$	$SU(2)_R$	$V_0(\Sigma)$
$\Sigma = \begin{pmatrix} (\bar{u}u) & (\bar{u}d) \\ (\bar{d}u) & (\bar{d}d) \end{pmatrix}$	**2**	**2**	$-A\text{tr}\left(\Sigma^\dagger\Sigma\right) + B\text{tr}\left(\Sigma^\dagger\Sigma\right)^2$

The details of the potential, which is generated quantum mechanically by the strong QCD interactions, don't matter much; what matters is that symmetries are broken. The vev $\langle\Sigma\rangle = \chi\mathbf{1}$ in the QCD vacuum breaks $SU(2)_L \times SU(2)_R \to SU(2)_V$ (sometimes called $SU(2)_D$, where V is for "vector", D for "diagonal.") The three broken generators lead to NGBs, which are massless pions.

In the real world, masses for $\bar{u}u$ and $\bar{d}d$ break $SU(2) \times SU(2)$ *explicitly* down to up-number and down-number, and so the potential for Σ takes the form

$$V = V_0 + m_u \bar{u}u + m_d \bar{d}d + c.c. = V_0 + \frac{1}{2}\text{tr}\left(\mathcal{M}\Sigma^\dagger\right) + \text{c.c.} \qquad (4.43)$$

where $\mathcal{M}$ is a diagonal matrix with nonzero entries m_u, m_d. Famously, this means that the NGBs become PNGBs, with $m_\pi^2 \propto m_u + m_d$ times a factor of order a few hundred MeV. *(Why is the $\pi^\pm$ a bit heavier than the π^0?)*

EXERCISE: Substitute $\Sigma = f e^{i\pi^a \tau^a / f}$*, where* τ^a *are the Pauli matrices* $(a = 1, 2, 3)$ *to show that the pion mass is indeed proportional to* $\sqrt{m_q}$*, not* m_q*. The overall normalization cannot be determined, but on dimensional grounds must be of order* Λ*.*

The generalization of this story to $SU(3) \times SU(3)$ explains why the K's and η are light also. But they are heavier than the pions, because the larger strange quark mass explicitly breaks $SU(3) \times SU(3)$ more dramatically than the small up and down masses break $SU(2) \times SU(2)$,

4.6.2. *Hadron decays*

Though in principle there could be others, in practice there are three types of bound states of quarks, antiquarks and gluons observed: mesons (with the quantum numbers of a quark and an antiquark,) baryons (with the quantum numbers of three quarks,) and antibaryons. Baryon number is conserved in the SM (to an excellent but very slightly imperfect approximation — see 't Hooft) so the lightest baryon is stable.

Among these, from lowest mass to highest mass, are three classes. (Actually four; the proton is special, of course, since it is stable for all practical purposes.) The first class involves metastable hadrons, which can only decay via electroweak interactions, with lifetimes $\tau \gg \Lambda^{-1}$. Above them lie unstable hadrons ("resonances"), decaying via strong interactions to other hadrons, with lifetimes τ somewhat less than Λ^{-1}. Not far above them the hadrons become highly unstable states that can barely be defined, with $\tau \sim m^{-1}$. The number of hadrons is infinite, but most are in the last category; only a finite number of states (dozens) will ever be identified. Beyond the mesons, baryons and antibaryons, all other types of hadrons (such as "glueballs" and other "exotics") are presumably all in the last category; it's not that they don't exist, but that they are such wide resonances that we can't pick them out.

We may categorize the hadrons differently, based on their appearance in an LHC detector. A typical hadron, when produced at the LHC, may carry energy much larger than its mass; often it will have a Lorentz boost of order 10 or even 100. Accounting for this, let us divide the hadrons into classes by where they decay. We refer to hadrons as decaying "promptly" if the quantity $\gamma c \tau$ (γ the Lorentz boost factor, c the speed of light, τ the lifetime) is less than 300 μm, which, for $\gamma \sim 10$, implies $\tau \sim 0.1 ps$. Such hadrons are not observed directly, only inferred from their decay products. Hadrons are "effectively stable" at the LHC if $\gamma c\tau \gg 1\text{m}$; in this case they largely or completely traverse the detector. In between, certain hadrons may decay inside the bulk of the detector,creating at a detectably displaced vertex at which two or more daughter particles emerge. We will often refer to them as "metastable" — but notice the potential confusion of language. In particular, note that a π_0, which decays to two photons in 10^{-17} seconds ($c\tau = 3$ nm), is long-lived (metastable) from the point of view of the strong interactions (*i.e.*, as in the previous paragraph), but short-lived (prompt) from the point of view of the detector (*i.e.*, as in the current paragraph.) These two points of view are both useful. Which one is being used should be clear from context.

Here is a table — not by any means an exhaustive list — of the most experimentally important hadrons in QCD, and some of their properties. This table may look a bit bewildering. But every element in the table is (up to a factor of at most two) understood. A well-educated theorist can explain and/or estimate every entry! You should aspire to that level of knowledge. The Particle Data Book (see pdg.lbl.gov) is a handy reference for all metastable and unstable hadrons known.

In this context it is also useful to mention the properties of the two unstable charged leptons. The muon has mass of 105 MeV, a lifetime of 2×10^{-6} sec ($c\tau \sim 600m$) and decays mainly to $e\nu\bar{\nu}$. The tau has mass of 1777 MeV, a lifetime of 3×10^{-13} sec, and decays as follows:

- 17% of the time, to $e^-\nu\bar{\nu}$;
- 17% of the time, to $\mu^-\nu\bar{\nu}$;
- 50% of the time, to 1 charged hadron plus ≥ 0 neutral hadrons;
- 15% of the time, to 3 charged hadrons plus ≥ 0 neutral hadrons;

Note that 85% of τ decays are to a single charged particle, the remainder to three charged particles.

EXERCISE: Why is the ratio of the tau lifetime to the muon lifetime so small? Estimate the ratio as a formula, and then see if you get the right answer when you plug in numbers.

From the experimentalist's perspective, different lifetimes correspond to different locations in the detector, and different experimental challenges and techniques. Recall that $10^{-12}\text{ sec} \times c \sim 0.3$ mm (or that one foot is the distance traveled by light in a nanosecond,) that the radius of the beampipe is about ~ 3 cm, and that time dilation effects can be important. In this

Table 4.6. Properties of important hadrons.

Name	Mass(MeV)	Lifetime (sec)	Dominant Decay	"Flavor"
π^+	140	3×10^{-8}	$\mu^+\nu_\mu$	$u\bar{d}$
π^0	135	8×10^{-17}	$\gamma\gamma$	$u\bar{u}, d\bar{d}$
K^+	494	1×10^{-8}	$\mu^+\nu_\mu, \pi^+\pi^0$	$u\bar{s}$
K^0_S	498	9×10^{-11}	$\pi^+\pi^-, \pi^0\pi^0$	$d\bar{s}, s\bar{d}$
K^0_L	498	5×10^{-8}	$\pi\pi\pi, \pi\ell\nu$	$d\bar{s}, s\bar{d}$
η	548	6×10^{-19}	$\gamma\gamma$,$3\pi^0$	$u\bar{u}, d\bar{d}, s\bar{s}$
ρ^+	770	4×10^{-24}	$\pi^+\pi^0$	$u\bar{d}$
ρ^0	770	4×10^{-24}	$\pi^+\pi^-$	$u\bar{u}, d\bar{d}$
ω	782	8×10^{-23}	$\pi^+\pi^-\pi^0$	$u\bar{u}, d\bar{d}$
K^{*+}	892	1×10^{-23}	$K^+\pi^0, K^0\pi^+$	$u\bar{s}$
K^{*0}	896	1×10^{-23}	$K^+\pi^-, K^0\pi^0$	$d\bar{s}$
η'	958	3×10^{-21}	$\pi^+\pi^-\eta, \ldots$	$u\bar{u}, d\bar{d}, s\bar{s}$
p	938	$> 10^{42}$	–	uud
n	940	887	$pe^-\bar{\nu}$	udd
ϕ^0	1020	1×10^{-22}	$K^+K^-, K^0_L K^0_S$	$s\bar{s}$
Λ	1115	2×10^{-10}	$p\pi^-, n\pi^0$	uds
Σ^+	1189	8×10^{-11}	$p\pi^0, n\pi^+$	uus
Σ^0	1193	7×10^{-20}	$\Lambda\gamma$	uds
Ξ^0	1314	3×10^{-10}	$\Lambda\pi^0$	uss
Ξ^-	1321	2×10^{-10}	$\Lambda\pi^-$	dss
Ω^-	1672	8×10^{-11}	$\Lambda K^-, \Xi^0\pi^-$	sss
D^+	1869	1×10^{-12}	$K + \cdots$	$c\bar{d}$
D^0	1864	4×10^{-13}	$K + \cdots$	$c\bar{u}$
B^+	5279	2×10^{-12}	$D + \cdots$	$b\bar{u}$
B^0	5279	2×10^{-12}	$D + \cdots$	$b\bar{d}$

context, we may divide up the hadrons yet again, into those which decay promptly, those that decay with a detectable displacement but before leaving the vacuum inside the beampipe, those that decay decay after passing through the beampipe wall but inside the detector, and those that pass all the way through the detector. In this table, wherever a particle name is unaccompanied by a specific charge, all particles with this name have comparable lifetimes (*e.g.* the $\rho^\pm$ and ρ^0 have similar lifetimes.)

A cautionary remark: remember that you live in a quantum world. Lifetimes are averages; boosts vary, and decays are stochastic. Commonly a particle will decay earlier than shown, and more rarely a particle (especially a high-energy one) will decay late. These effects can be important.

Table 4.7. Hadrons categorized by experimentally-relevant lifetimes.

Prompt	Displaced (<1 cm)	Displaced (>1 cm)	$\sim$ Stable
π^0, η, η'		K_S	$\pi^\pm, K^\pm, K_L^0$
ρ, ω, ϕ, K^*			
Σ^0		$\Sigma^+, \Lambda, \Xi, \Omega$	p, n
B^*, D^*	$B, D, \Lambda_c, \Lambda_b$	extreme-p_T B	
τ	extreme-p_T τ		μ, e, γ

EXERCISE: Consider π^+ particles with $p_T = 140$ GeV and $\eta = 0$. [I do mean GeV, not MeV.] What fraction of them will decay at a distance of less than 1 meter?

EXERCISE: Consider the B^0 meson. What is its average travel distance if it has $p_T = 10$ GeV and $\eta = 0$? At what energy do typical B^0 mesons traveling perpendicular to the LHC beam exit the LHC beampipe before decaying?

4.6.3. *Detecting hadrons*

The detectors obviously can only directly detect particles that live long enough to actually enter and traverse a substantial portion of the detector. A detector is actually a set of subdetectors, crudely (and not entirely accurately) of the following four types: a tracker for measuring charged particle tracks, an electromagnetic calorimeter for measuring the energies of electromagnetically interacting particles (photons and electrons), a hadronic calorimeter for measuring the energies of strongly interacting particles (though note that many charged hadrons will also leave some energy in the electromagnetic calorimeter), and a muon system for detecting charged particles that make it through the whole detector. A magnetic field in the tracker and one in the muon system causes charged-particle tracks to curve, allowing particle momenta to be measured.

These four classes of subdetectors can be used to distinguish photons, electrons, muons, charged hadrons, and neutral hadrons, through the following simple table.

In the table I have emphasized that promptly decaying particles like π^0, η can leave unusual and therefore sometimes identifiable clusters of stable particles. In fact, τs do this as well. We'll return to τ's, and also D and B mesons, later in the lectures.

Table 4.8. The experimental signatures of various particles.

Name	Track	ECAL	HCAL	μ system
γ	N	Y	N	N
$e^{\pm}$	Y	Y	N	N
$\mu^{\pm}$	Y	N	N	Y
$\pi^{\pm}, K^{\pm}, p, \bar{p}$	Y	(Y)	Y	N
$K^0, n, \bar{n}$	N	N	Y	N
$\pi_0 \to \gamma\gamma, \eta \to \gamma\gamma$	N	YY	N	N

At the CMS and ATLAS detectors, stable hadrons of the same charge are generally not distinguished. (This is not true at the LHCb and ALICE detectors, however, partly because of their differing science goals.)

Now why do we care about these hadrons? They are the most common particles seen at hadron colliders. They appear in the underlying event — when the protons fracture into many pieces, each piece forms into a hadron — and in jets, the manifestations of hard quarks, antiquarks and gluons.

4.6.4. *Jets, a naive view*

When a quark, antiquark or gluon is kicked out of the proton, it creates a "jet" of (generally excited) hadrons, more or less collimated in angle. Note that most of these hadrons are unstable (in the QCD sense) and decay down immediately via the strong interactions to multiple metastable (in the QCD sense) hadrons — mostly $\pi^{\pm}$s, stable on detector timescales, and π^0s, which are unstable on detector timescales and are observed as photon pairs.

Jets are ubiquitous at the LHC. How and why do they form? Naively, partons are confined (or "cloaked") by QCD interactions, so (as in Fig. 4.1) at least one hadron should be produced when a parton is kicked out of a proton. If a quark is kicked out with a lot of energy, then there is plenty of energy available to make lots of hadrons. But this logic is deeply flawed. It does not explain why $p_{parton} \sim p_{jet}$, which is crucial in our efforts to use jets to tell us what the underlying partons are doing. In fact it does not explain why the hadrons emerge in a jet at all. We will return to these unanswered questions later.

Ignoring these subtleties for now, and assuming that, as is in the end roughly true, a quark of momentum p^{μ} turns into a jet with momentum $\sim p^{\mu}$, consider the process $u\bar{d} \to W \to u\bar{s}$. (To make things simpler, I changed the d quark flavor to an s quark in the final state, so that this process is dominated by an s-channel amplitude, with no t-channel contribution.)

This process is similar to Z boson decay to $q\bar{q}$, a colorless source decaying to two jets of energy $\sim$ 45 GeV, whose properties are studied in Fig. 4.9. From the measurement of this process at LEP, we know that such jets have $\sim$ 10 tracks (up to 20), which implies $\sim$ 15 stable hadrons, mostly $\pi^{\pm}$ and π^0. But these hadron multiplicities are averages, and they have large fluctuations.

From Fig. 4.10., we see that track multiplicity is gradually growing with jet momentum. Many of these tracks are at low energy, and will curl in the strong magnetic field of the detector. Some very nearly overlap. That means it is easy to miss one.

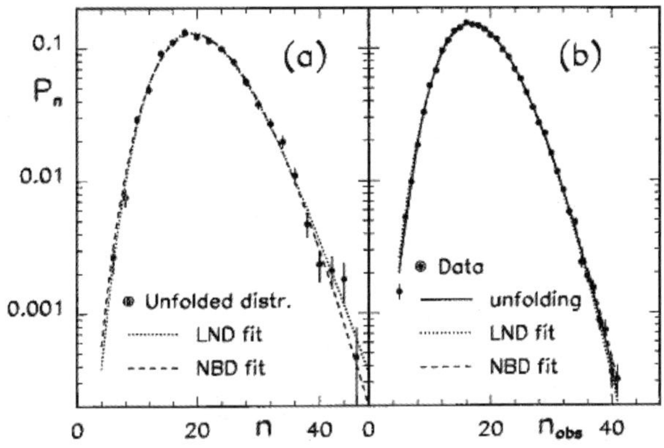

Fig. 4.9. Charged-hadron multiplicity in $Z \to q\bar{q}$ events.

And that in turn implies that the measured charge even of a gluon jet may be nonzero. This is an important experimental limitation. We cannot easily distinguish u, d, $\bar{u}$, $\bar{d}$ and g jets from one another simply by looking at the total charge of all the hadrons in the jet. The correlation between parton charge and total jet charge is very weak. Nor can looking at the types of hadrons in a jet help. Pions, along with a few kaons, dominate all of these jets.

That is unfortunate. It means that a process such as $u\bar{d} \to W \to u\bar{s}$ has a huge background from $gg \to gg$ that cannot be reduced experimentally. And that means that it is virtually unobservable.

This said, gluon jets are a little fatter and have higher multiplicity on average than quark jets. We'll see why later. Fluctuations around the average are large, though, so this tendency doesn't help enormously. Still,

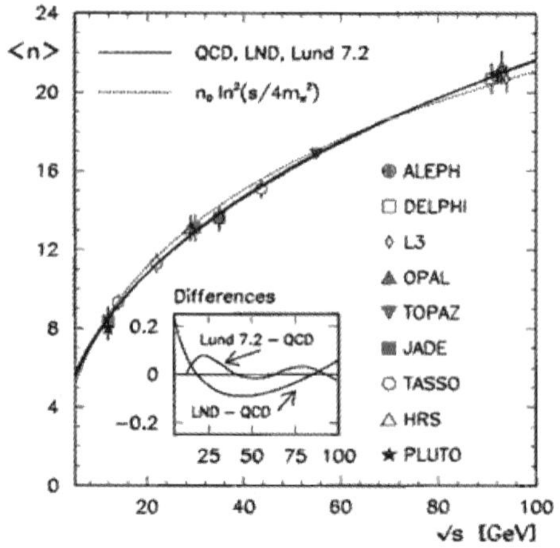

Fig. 4.10. Total charged-hadron multiplicity in $e^+e^- \to$ dijet events as a function of the center-of-mass energy $\sqrt{s}$.

it is likely that the ability, statistically speaking, to distinguish quarks from gluons may be useful in some contexts.

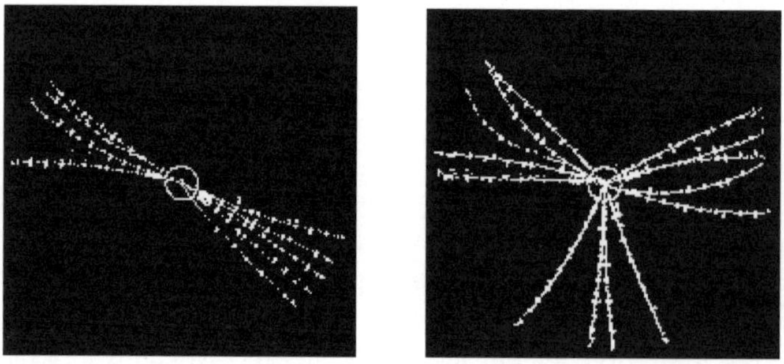

Fig. 4.11. Typical dijet and trijet events in 20 GeV e^+e^- collisions, showing only the tracks left by the charged hadrons.

4.6.5. *Jets, a less naive view: Partons and hadrons*

QCD FACT 4: "Parton-Hadron duality" — that high-energy partons turn into relatively narrow jets of hadrons with $p^\mu_{jet} \sim p^\mu_{parton}$ — should not be

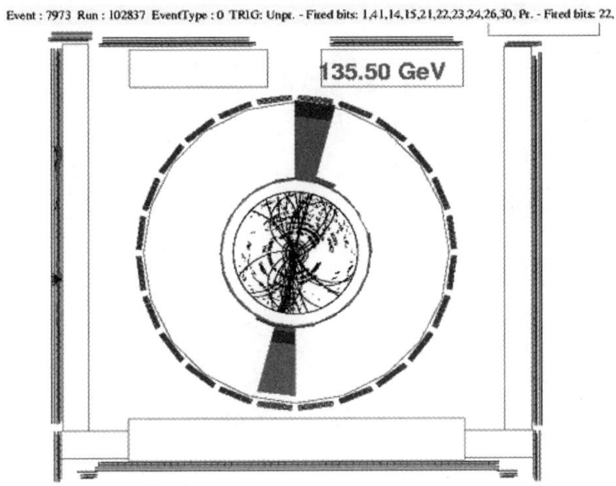

Fig. 4.12. A dijet event at CDF. The blotches show where the energy is deposited. Note the straighter tracks from the high-p_T hadrons in the jets are accompanied by curling tracks — lower-p_T hadrons from the jets and from the underlying event. Note also the many hits that are not able to be associated with reconstructed tracks.

taken for granted! It need not be true in principle, and in some theories, it is almost certainly false.

Whether jets form, and how, depends on the size and behavior of the strong coupling constant α_s. The angular width of a jet is related to the quantity $\alpha_s N$ (N the number of colors.) We can calculate, and observe, that lower-energy jets are broader than higher-energy jets, because of the running of the coupling. So in a theory where $\alpha_s N_c$ never becomes small in the ultraviolet (such as a theory with a non-trivial ultraviolet fixed point,) jets are likely too broad to be observed as such.

As we've discussed, we should not even take the existence of hadrons for granted; many theories don't have them. But even when you have them, the process of hadronization — by which partons turn into hadrons — occurs with a rate $\propto N_f^{light}/N_c$, or, if $N_f^{light} = 0$, with a rate $\propto 1/N_c^2$. If the time for hadronization is rather slow, then there will be time for momentum to be redistributed via the complex interactions among the quarks and gluons. This would have the effect of making $p_{jet} \neq p_{parton}$, and also broadening the jets. If hadronization is sufficiently slow, any jetty structure may be lost altogether through violent rearrangement of energy. Only if the hadronization time is very fast, and the process smooth, can a parton in the short-distance collision possibly appear in the detector as a

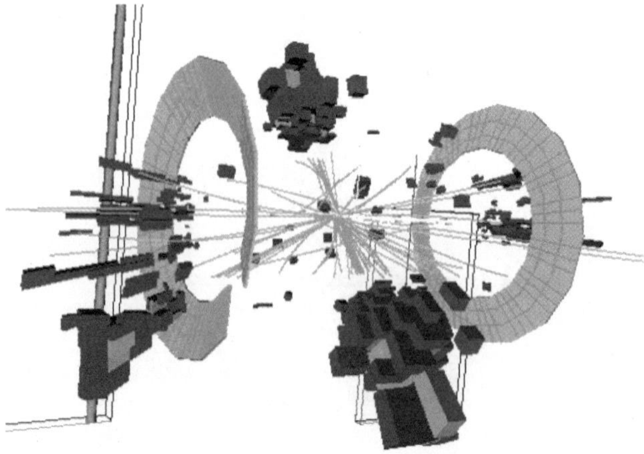

Fig. 4.13. A dijet event at D0. Note the tracks that are not pointing at the two jets; these are from the underlying event.

long-distance jet with roughly the same momentum. To justify these claims is the goal of this section.

The easiest of these statements to understand is the importance of N_f^{light}. The hadronization process, if N_f^{light} isn't zero, proceeds by a flux tube splitting through quark-antiquark pair production in the strong chromo-electric field of the flux tube, as in Fig 4.1. It scales with the number of quarks of low mass. For a low mass quark with $m_q \ll \Lambda$, the process is fast, since the confined chromo-electric field $\sim \Lambda^2$, by dimensional analysis. But pair-production is exponentially suppressed for $m_q \gg \Lambda$; the chromo-electric field is simply too low to produce the heavy quarks.

More subtle is the $1/N_c$ of the hadronization. I'm going to give you a heuristic argument, which is enough to get you the right intuition.

Strong coupling — the point at which perturbation theory breaks down — occurs not at $\alpha_s \sim 1$ but at $\alpha_s N_c \sim 1$. What I want you to do is temporarily think (inaccurately, but still usefully) of a flux tube as a chain of gluons, with color and anticolor $c_i, \bar{c}_{i-1}$ connecting a quark of color c_1 and anticolor $\bar{c}_n$

$$(\bar{c}_n)(c_n\bar{c}_{n-1})\dots(c_4\bar{c}_3)(c_3\bar{c}_2)(c_2\bar{c}_1)(c_1). \tag{4.44}$$

Here each color c_i can take values between 1 and N_c. To this chain I can always add a gluon, at no cost. The replacement

$$(c_3\bar{c}_2) \to \sum_{c_0} (c_3\bar{c}_0)(c_0\bar{c}_2) \tag{4.45}$$

costs a factor of the coupling constant, but on the other hand the process can occur for any choice of color c_0, which runs from one to N. The total cost is therefore

$$\alpha_s \sum_{c_0} 1 = \alpha_s N_c \sim 1 \; ; \tag{4.46}$$

in other words, no cost at all.

Now suppose we fragment the flux tube into two pieces through the pair production of a light quark and antiquark. We have to split, say, the $c_3\bar{c}_2$ gluon: $g \to d\bar{d}$, where d has color c_3, $\bar{d}$ anticolor $\bar{c}_2$.

$$(\bar{c}_n)(c_n\bar{c}_{n-1})\dots(c_4\bar{c}_3)(c_3) \qquad (\bar{c}_2)(c_2\bar{c}_1)(c_1). \tag{4.47}$$

Note how the flux tube is broken in two. The cost is

$$\sim \alpha_s \sim \frac{\alpha_s N_c}{N_c} \sim \frac{1}{N_c}. \tag{4.48}$$

This time we do not have a color index to sum over — the initial gluon provides the colors for the final quark and antiquark. But we could substitute a $u\bar{u}$ or $s\bar{s}$ pair for the above $d\bar{d}$ pair — in other words, there is a flavor index to sum over. So the rate for splitting by $g \to q\bar{q}$ goes as N_f^{light}/N_c. As mentioned earlier, heavy quark pair-production is strongly suppressed because the energy density isn't large enough.

Now if there are no light quarks, then to break the flux tube requires a non-planar graph. Non-planar graphs are suppressed by $1/N_c^2$, so hadronization is much slower in this case. For example, this can happen through the exchange of a $(c_3\bar{c}_1)$ gluon between the first and third gluon from the right, leaving

$$(\bar{c}_n)(c_n\bar{c}_{n-1})\dots(c_4\bar{c}_1)(c_3\bar{c}_2)(c_2\bar{c}_3)(c_1) \to (\bar{c}_n)(c_n\bar{c}_{n-1})\dots(c_4\bar{c}_1)(c_1)$$
$$+(c_3\bar{c}_2)(c_2\bar{c}_3). \tag{4.49}$$

A closed loop has broken off from the flux tube. This required two vertices, but the color and anticolor of the gluon had to be precisely chosen, so the cost of this process is

$$\alpha_s^2 = \frac{(\alpha_s N_c)^2}{N_c^2} \sim \frac{1}{N_c^2} \tag{4.50}$$

Note the very big difference, for large N_c, between $N_f^{light} = 1$ and $N_f^{light} = 0$. There is a jump in the $1/N_c$ dependence between these two cases.

Where is QCD itself in this story? Well, in the real world, $N_f^{light} = N_c = 3$, so there are no suppression factors for splitting flux tubes via quark-antiquark pair production. Presumably this is what allows parton-hadron duality to work so well in the context of jets.

Now that we have some reason to believe that confinement in QCD may not badly impact jets, we can trust a *perturbative* treatment of jet formation. We will return to this treatment later.

4.7. Data and Prediction

4.7.1. *The simplest observations: Z and W bosons*

It's time to start thinking in more detail about how measurements are made. For example, we discussed $u\bar{d} \to W \to e\nu$; how would we observe this process?

It is easier to know that we made a Z boson, so let's start with that. In particular, let's consider $pp \to Z + \cdots \to e^+e^- + \cdots$. The trigger will fire on events with 2 "electrons" of moderate p_T. (Note that experimentalists, as a short hand, use "electron" for electrons and positrons, "muon" for muon and antimuon, except when it is necessary to specify the particle's charge — and they often use "lepton" to refer to electrons and muons only, since taus decay so quickly that they are not observed directly.) More precisely, the trigger records events with two isolated moderate-p_T tracks that leave energy in the electromagnetic calorimeter. Whether they turn out to be an electron or a positron is determined later, during reconstruction of the event.

We are interested in those events with

$$m_{e^+e^-} = \sqrt{(p_+ + p_-)^2} \approx m_Z. \tag{4.51}$$

I use an approximate sign both because the experimental measurement isn't perfect and, even more important in this case, $\Gamma_Z \approx 2.5$ GeV. Now $m_{e^+e^-}$ is a strict Lorentz-invariant quantity, so this measurement is especially easy. In each collision, we get to measure this quantity no matter what the momentum of the Z itself happens to be in a particular collision.

We should note, however, that e^+e^- pairs don't just come from decaying Z bosons. We can get them from off-shell photons and Z's (written as γ^*,

Z^*, where the asterisk means "off-shell"). The photon and Z diagrams interfere.

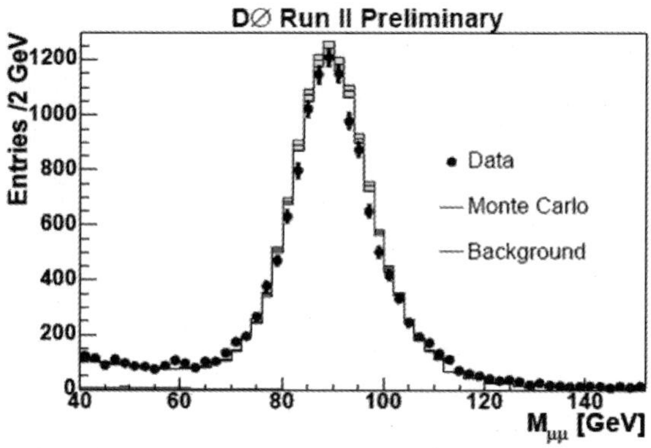

Fig. 4.14. The Z peak in the $\nu^+\mu^-$ invariant mass spectrum; the same is seen for e^+e^-.

So what we'll do is collect all the collisions in our data set that have an electron and a positron, and plot the invariant mass m_{ee}. We'll see two peaks: one a nice isolated peak at $m_{ee} = m_Z$, and a second rise toward $m_{ee} = 0$. The measurement at small m_{ee} is affected by many subtleties, including fake electrons at low p_T, the trigger, which requires the electrons already have moderate p_T (of order 10–20 GeV), and other sources of low-mass e^+e^- pairs, such as photons converting to a pair within the beampipe material. But out where the Z boson peak sits, the measurement is very clean.

What about W bosons decaying leptonically? In this case we need to look at events that have an $e^\pm$ or $\mu^\pm$ and nothing else. (Well, not nothing. As always, we expect a few dozen tracks and H_T of 50-200 GeV from the underlying event. And often there's a random jet or two also.) So we need the trigger to fire on events with, say, an electron that has p_T greater than some threshold. From these events, we would select the ones where there is a high-p_T electron and no other high-p_T tracks. But we don't know where the neutrino went — we know its $\vec{p}_T$ should equal the missing transverse momentum, but we don't know its p_z — so we can't compute the invariant mass of the e and ν. What should we plot instead?

How about $p_{e,T}$? Suppose first the W were at rest in the lab frame. Then the electron and neutrino would carry $p_T = \frac{1}{2} m_W \sin\theta$, where θ is the angle of either one with the beampipe. If the W decayed isotropically on average (true if its spin direction were random) then

$$\frac{d\hat{\sigma}}{dp_T} = \frac{d\hat{\sigma}}{d\cos\hat{\theta}}\frac{d\cos\hat{\theta}}{dp_T} \propto \frac{d\cos\hat{\theta}}{d\sin\hat{\theta}} = \tan\hat{\theta} = \frac{p_T}{\sqrt{m_W^2 - 4p_T^2}} \tag{4.52}$$

and thus the electron p_T would actually peak at its maximum of $m_W/2$. This function is called a "Jacobian peak" (because it arises from a change-of-variables) at $p_T = m_W/2$. So although we can't learn the W mass from just a few events (as we can from the Z peak in e^+e^-,) we have a hope of learning it from a large ensemble of events.

Now in reality the W bosons at the LHC are moving. If we only had to consider their motion along the beampipe, the formula above, which is z-boost-invariant, would remain unchanged. And so we would get the same distribution in the above variable, even combining many W decays where the W's have different z boosts.

But unfortunately the Ws also generally have some p_T of their own, and that spoils this simple formula. The initial-state q and $\bar{q}$ often will radiate one or more low-p_T ("soft") gluons ("initial state radiation," or "ISR"). This typically gives $p_{W,T} \sim 10$ GeV and is often much larger, ruining our calculation and smearing out the peak at $m_W/2$. (Two other smaller but non-negligible sources of smearing are imperfect measurement of $p_{e,T}$ and the finite W width, $\Gamma_W \sim 2$ GeV.) It turns out that a better variable to use in determining the W mass is what is known as the "Transverse Mass," m_T.

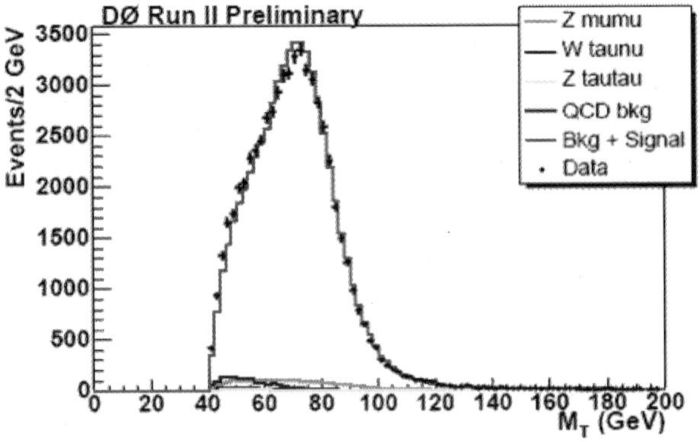

Fig. 4.15. The Jacobian peak from W bosons in the M_T variable.

The way to understand this variable is to imagine compressing the event into the two-dimensional plane transverse to the beampipe. Define a *2+1-dimensional massless three-vector* with $\tilde{p}_e = (|\vec{p}_{e,T}|, \vec{p}_{e,T})$, with two-momentum $\vec{p}_{e,T}$, and $\tilde{p}_e^2 = 0$ in the 2+1-dimensional sense. Then the transverse mass is the *2+1-dimensional invariant mass* built from this vector and a similar vector with momentum $\tilde{\not{p}}_T = (|\vec{\not{p}}_T|, \vec{\not{p}}_T)$:

$$m_T^2 = (\tilde{p}_e + \tilde{\not{p}}_T)^2 = 2(|p_{e,T}||\vec{\not{p}}_T| - \vec{p}_{e,T} \cdot \vec{\not{p}}_T) = 2|p_{e,T}||\vec{\not{p}}_T|(1 - \cos\phi_T) < m_W^2 \,. \tag{4.53}$$

Since m_T is manifestly invariant under z-boosts, we may analyze it in the frame where $y_W = 0$. There it is easy to show that if the W is not at rest, but has a kick sideways and picks up a velocity v, then $m_T \to m_T +$ order(v^2). Thus the measurement of the W mass using this variable is rather stable against the effects of nonzero $p_{W,T}$, and therefore against theoretical error on the distribution of the W's p_T.

Experimentally, one doesn't use all the events in the data set. It is better to cut away events with small $p_{e,T}$ and large $|\eta_e|$ to reduce the effect of the trigger, tracking inefficiencies, fakes, etc. What remains is relatively clean. And data agrees with theory. But it isn't so easy to obtain that agreement. Let's see why.

4.7.2. *What does theory say?*

Let's ask three simple (no, just simple-looking) questions about W production:

- How many SM $W \to e\nu$ events will be produced at the LHC?
- What will be the p_T distribution of the electrons from SM W production at the LHC?
- How many of these events will be *observed* at the LHC?

These questions are in increasing level of difficulty.

We can almost answer the first, though it requires calculating corrections to the tree-level matrix element. *(These are large! $qg \to qW$ is comparable to $q\bar{q} \to W$. Why might you expect this?!)*

The second requires us to combine the W production process with the decay distributions. The W has spin so the e^- distribution is not isotropic in the W rest frame. And the W is not at rest in the lab frame. If the W has only p_Z and no p_T, this does not affect the p_T of the e^-. *Why?* But when the W recoils against a jet (see above) it picks up p_T also, and this significantly affects the high-p_T tail of the e^- distribution.

The third question is very difficult, because it requires knowing and combining many different types of knowledge. These include the geometric structure of the detector (in particular the η range), the p_T cut on the electron required by detector and trigger effects, the kinematic distribution in p_T and η of the e^- from W decay, and the probability that the electron is isolated.

Calculations in the context of such complexity cannot be done analytically! Even with exact analytic expressions for the differential distribution for the electron momentum, the integrals over η, ϕ, x_1, and x_2 would have to be done with a computer. And the other constraints on the data — such as the isolation of the electron from any jets in the event — are very complicated constraints on the integrations.

Therefore, for LHC physics, we must have computer programs for dealing with the kinematical distributions. And this brings us to a new subject.

4.7.3. *Monte Carlo event generators*

Rather than do the integrals directly, these "Monte Carlo" programs work by producing simulated events with correct probability weights, and thus do the integrals by sampling.

This is powerful and practical because it makes it possible to account for additional effects (uninstrumented or less-instrumented gaps in the detector, inefficiencies of the detector in detecting electrons, scattering of electrons off detector material) and to change the kinematic restrictions (such as the p_T cut) without redoing the integrals from scratch.

Again, it is important to emphasize that the use of these programs is not optional! They are essential, for practical reasons, for all predictions for physics at the LHC.

Famous ones are PYTHIA, HERWIG, SHERPA, MadGraph, ALPGEN, CompHEP, MC@NLO, etc.; they are different programs, intended in some cases to do different things at different levels of accuracy, yet often overlapping in their applications.

Also many of these programs implement models of incalculable (or essentially incalculable) but experimentally-relevant physical phenomena. We already need them for W/Z decay to leptons.

For example, PYTHIA, HERWIG and SHERPA include models that produce a simulated version of the underlying event. Since the "soft" particles (mostly pions, and photons from π^0s) can affect the measured energies or general appearance of the "harder" leptons or jets we care about, they must be accounted for, at least statistically. These models are very crude, since the underlying event is largely non-perturbative and cannot be calculated from first principles.

Also, these programs contain routines for computing the effects of "*Initial State Radiation*" (ISR), in which a quark or gluon is created during the collision prior to the formation of the W/Z, and "*Final State Radiation*" (FSR), extra quarks or gluons produced in W and Z decays.

One of the most important things that PYTHIA, HERWIG and SHERPA do is implement a routine for turning a quark or gluon into a jet, using techniques based as on a combination of QCD theory and phenomenology extracted from data. This requires three steps:

- Showering: semiclassical radiation of (mainly) gluons off the parent quark/gluon (jet formation);
- Hadronization: confinement of all the quarks and gluons in the shower into hadrons;
- Decays: decay of all unstable hadrons to more stable hadrons or leptons.

Importantly, most aspects of jets are determined in the first stage, which is under very good theoretical control.

PYTHIA and HERWIG mainly contain routines for $2 \to 2$ scattering processes with $2 \to 1$ and a few $2 \to 3$ processes as well. Particle decays are simulated, with varying degrees of detail. All calculations are done at tree-level (or in a few cases where the tree-level vanishes, at first non-vanishing order.) A number of new-physics processes are included.

SHERPA, ALPGEN, MadGraph, CompHEP and others allow calculations of much more complex processes, including all tree-level SM processes (in principle) and many new-physics processes (if the Feynman rules are entered by the user, which can be done in an automated way.) However, 1-loop-level event generation is coming! A huge advance in 1-loop calculations has recently been made; this will be fully automated in the next several years.

4.8. Comments on the Third Generation

As the heaviest fermions, with Yukawa couplings ~ 1 (for t, and possibly for b and τ), the 3rd generation may have especially strong couplings to new phenomena. Not only the Higgs but many other new particles may decay mostly to this generation.

Quarks and charged leptons in the 2nd and 3rd generation decay by emitting a W boson. The W is virtual, except for the top, which decays to a real W and a b quark. For example, the tau may decay to a 3-body leptonic final state:

$$\tau^- \to \nu_\tau (W^-)^* \to \nu_\tau e^- \bar{\nu}_e \tag{4.54}$$

and the b quark can similarly decay to $c\mu^-\nu_\mu$ or to $cu\bar{d}$.

In the standard model, all decays via a Z boson or a photon are absent at tree level, and are generally small even when quantum effects are accounted for. This absence of FCNCs (flavor changing neutral currents) in the SM and in the data is a very strong constraint on any type of new physics, which could give new sources of FCNCs. Thus we do not observe $\tau^- \to \mu^-\gamma$ or $\tau \to e^-e^+e^-$, and $b \to s\gamma$ and $b \to s\mu^+\mu^-$ are very rare (though observed.)

So almost all decays in the SM occur through flavor-changing *charged* currents. But even these are not simple. For example, we observe both $b \to c\mu\bar{\nu}$ and $b \to d\mu\bar{\nu}$. We also observe $s \to u$ and $c \to s$ and $c \to d$ transitions. In fact in the SM all possible transitions via a W are allowed, with probabilities that are proportional to the squares of the entries in a *unitary* matrix, due to Cabibbo (in its 2-by-2 form) and to Kobayashi and Maskawa:

$$V_{CKM} = \begin{pmatrix} V_{ud} & V_{us} & V_{ub} \\ V_{cd} & V_{cs} & V_{cb} \\ V_{td} & V_{ts} & V_{tb} \end{pmatrix} \approx \begin{pmatrix} .974 & .226 \pm .001 & .0036 \pm .0002 \\ .226 \pm .001 & .973 & .0415 \pm .0010 \\ .0087 \pm .0003 & .041 \pm .001 & .999 \end{pmatrix}. \tag{4.55}$$

(There is a similar matrix in the lepton sector, but it probably has no experimental impact at the LHC, because ν_τ, ν_μ and ν_e are all equally invisible to the LHC detectors.)

Thus the branching fractions satisfy relations such as

$$\frac{\mathrm{Br}(b \to u\mu\bar{\nu})}{\mathrm{Br}(b \to c\mu\bar{\nu})} = \frac{|V_{ub}|^2}{|V_{cb}|^2} \tag{4.56}$$

to a pretty good approximation. (Among other things, since $m_c \gg m_u$, the final-state kinematics are different in the numerator and denominator, so this relation isn't exact as written.) We also have relations such as

$$\frac{\mathrm{Br}(b \to cd\bar{c})}{\mathrm{Br}(b \to cs\bar{c})} = \frac{|V_{cd}|^2}{|V_{cs}|^2} \tag{4.57}$$

to a moderately good approximation. (The more quarks appear in the decay, the more subtleties we have to account for from the fact that quarks are "confined" into hadrons.)

Because of the approximately diagonal structure of the CKM matrix, the most common quark-flavor decays are $t \to b$, $b \to c$, $c \to s$ and $s \to u$. (We also have $d \to u$ in some cases, for example in neutron beta decay, but because of phase space this is very slow. There is also $u \to d$ in some nuclear decays.)

From the decay diagram, it is clear that the decay of a particle decaying via a virtual W quark will be proportional to $|1/m_W^2|^2$. Then, by dimensional analysis, the width of quark q_i is proportional to

$$\Gamma_i \sim g_2^4 \frac{m_q^5}{m_W^4} |V_{ij}|^2 \tag{4.58}$$

(times factors of order 2, π *etc.*) where V_{ij} is the largest CKM matrix element involving quarks q_j that are lighter than quark q_i (so that the decay is kinematically allowed.) For leptons the formula is the same, but all ν_i are light enough to contribute, and do so with equivalent kinematics. Therefore $|V_{ij}|^2$ is replaced, for muon decay, by $\sum_j |V_{\mu\nu_j}|^2 = 1$.

From this formula we learn

- Heavier particles have *much* shorter lifetimes than particles of the same charge:

  (a) $m_\tau \sim 10\ m_\mu$, but $\Gamma_\tau \sim 10^5 \Gamma_\mu$.
- Since $|V_{us}|^2 \sim .1$ and $|V_{cb}|^2 \sim 10^{-5}$, the lifetime of strange and bottom quarks are anomalously long compared to the charm quark, which has $|V_{cs}|^2 \sim 0.95$. (In fact, historically, this long lifetime is what made "strange" particles strange.)

All the quarks except the top quark live so long that complicated QCD interactions are important in their behavior. In particular, these particles always form hadronic bound states. But the top quark has a very short lifetime, with a width well over 1 GeV. Its lifetime is so short that it decays before strong QCD physics has a chance to play any role; it never has time to form a bound state, and its decays can be described using perturbation theory.

4.8.1. *Taus and tau identification*

Taus will be very important in LHC searches for Higgs, SUSY, etc. They decay 34% of the time to e or μ, 50% to one charged hadron (plus neutral hadrons), and 15% to 3 charged hadrons (plus neutral hadrons). The decay is always to $W^*\nu_\tau$, and since the neutrino can only be indirectly detected, the energy of the τ, E_τ, can never be directly measured! Moreover, a leptonic tau decay looks, to the detector, just like any old e or μ. The only hint that there's a tau is some amount of missing momentum. But most events with taus have other neutrinos (on their own, or from other taus) and so identifying a μ or e as coming from a tau is usually impossible. Hadronically-decaying taus, on the other hand, look like special jets: they have 1 or 3 tracks (typical jets have $\sim 5-20$ tracks), and the invariant mass of their decay products is always less than $m_\tau \sim 1.8$ GeV (whereas the invariant mass of the particles in a typical jet is of order $m_{jet} \sim 0.1 - 0.2\, p_T^{jet}$). So to find hadronic taus, we can look for ultra-narrow low-track-number jets (see Fig. 4.16). (Warning: such jets do occur in QCD. It is rare, but there are very many jets at the LHC, and consequently hadronic taus have a rather high fake rate.)

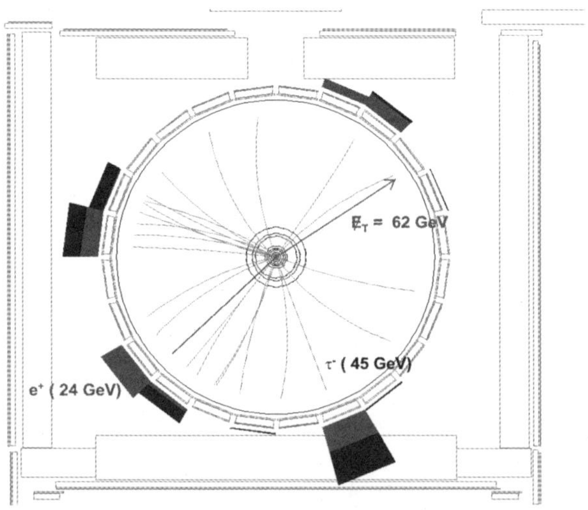

Fig. 4.16. A top quark pair production event at CDF with a one-prong tau (indicated on the figure as a single isolated straight track pointing at an energy deposit in the calorimeter) in the final state.

4.8.2. *Bottom quarks and heavy flavor tagging*

Bottom quarks (and charm quarks) make jets much like other quarks and gluons, but with a single B (or D) meson [or baryon] which tends to carry a large fraction of the energy. When a B meson decays, it will typically travel (E_B/m_B) 450 μm before decaying, and typically decay to $D\ell\nu$ or

D plus other hadrons. Meanwhile, a D meson has a 200 μm lifetime and decays to kaons K plus leptons or hadrons.

Thus a b quark turns into a jet which typically has a displaced vertex (or two), which implies that it often has ≥ 2 tracks missing the original collision point ("primary vertex") by 200-1000 μm . And it also has a μ or e embedded in it, about $\sim$ 10% of time per lepton. Note these are *non-isolated* leptons, as there are many pion tracks and photons near them.

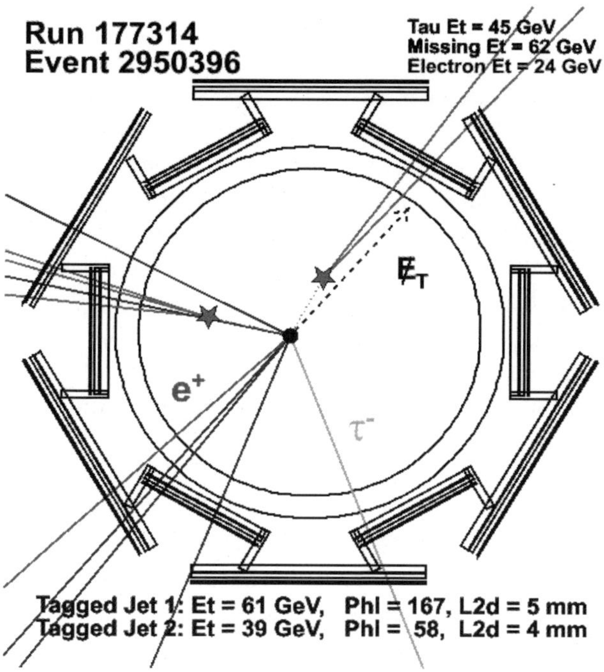

Fig. 4.17. Closeup of the tracking in the previous event, showing the two displaced vertices from the decaying b and $\bar{b}$ quarks (more precisely, the two B mesons) produced in the t and $\bar{t}$ decays.

For a c quark, the situation is similar, but compared to a b quark it is less likely, because of its shorter lifetime, to have detectably-displaced tracks and a vertex. Also, the lower mass of the D meson implies it decays to tracks that are more collimated.

Typical "heavy-flavor tagging" algorithms (the numbers depend on technique used) allow 50–70% (15–20%) of b (c) jets to be identified, or "tagged", as "heavy flavor". It is hard to tell b jets from c, though there

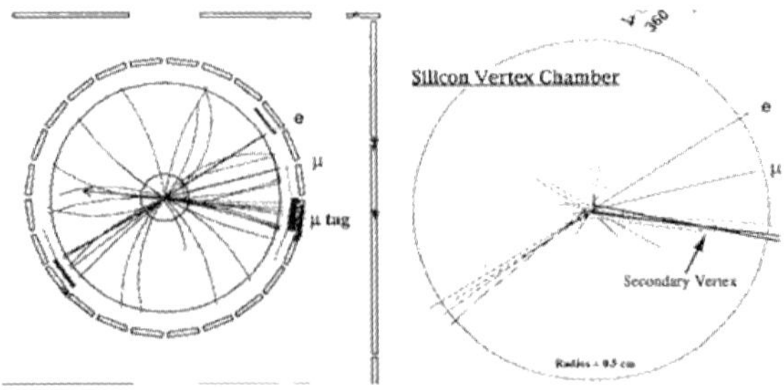

Fig. 4.18. Closeup of the tracking in another CDF $t\bar{t}$ event.

are some methods that work some of the time (such as looking for the B meson decay chain.) Meanwhile, 1–2% of ordinary g jets get tagged due to a real physical effect in QCD: gluon splitting via $g \to c\bar{c}, b\bar{b}$. In these cases the jet contains real D, B mesons, and tagging some of these jets is unavoidable. Finally, 0.5–1% of other jets are erroneously tagged, just due to small errors in tracking of the hadrons in the jet. The details depend on experimental parameter choices; by setting parameters differently, one can adjust the compromise between a pure but limited sample and a larger but less pure sample.

Now, a few comments. First, don't forget mistagging or gluon splitting. For some measurements, these effects are a large source of background. Also, though it is hard to tell b from $\bar{b}$, in those cases with a non-isolated lepton in the jet the embedded lepton's charge is correlated (imperfectly) with the b charge. Finally, some words about isolation of leptons. We distinguish μ's and e's from heavy flavor (or μs from π decay) from more interesting leptons by their "isolation" from other particles. But sometimes b's *can* create jets where the lepton is the only high-momentum particle, thus providing a source of fake isolated leptons.

4.8.3. *Top quarks and their role at LHC*

The LHC, by any measure, is a top factory. The production rate is huge: 10^6–10^8 per year. Production is dominated by $gg \to t\bar{t}$ followed by top quark decay to bW^+ (almost 100% of time) and then $W \to q\bar{q}$ (2/3), $\tau\nu$ (1/9), $\mu\nu$ (1/9), $e\nu$ (1/9). Thus (ignoring q vs. $\bar{q}$, and defining $\ell \equiv e, \mu$), $t\bar{t}$ events can be divided into those that are "fully hadronic" $t\bar{t} \to qqqqb\bar{b}$, 44% of the events, with a very large QCD background; "semi-leptonic"

$t\bar{t} \to qq\ell^{\pm}\nu b\bar{b}$, 30% of the events, and fully reconstructable (as explained below); and "dilepton" events, $t\bar{t} \to \ell^+\nu\ell^-\bar{\nu}b\bar{b}$ 5% of the events, with low background but low rate. See Fig. 4.18 for an example. Finally, 21% of the events contain one or two τs. These are more complicated and we won't discuss them here, but see Figures 4.16 and 4.17 for an example.

Understanding top quark physics in detail is a key goal for the early LHC. Tops provide almost everything: jets (including b jets), electrons, muons, taus and MET. This makes them an excellent place to test out whether the detector is working properly in all respects. The production mechanism is simple (relatively!) since QCD, the top mass, and the main top quark decay mode are essentially known. This makes it relatively easy to use the top quark and its various decays as a set of "standard candles" against which calibration of the detector can be carried out. Of course, before doing this one must make sure that indeed the top quark is behaving as expected. Potentially we will see new physics early in top quark decays or production. And we have to understand it even if its behavior is just as expected in the SM, because it is a huge background to many new physics signals.

The easiest events to obtain and study are the semileptonic ones. One can use the lepton for triggering. The lepton, $\not{E}_T$ and tagging of the b jets (or even the occasional c-jet tag) can be used to reduce QCD and W+jets background. Two of the (usually) untagged jets will have $m_{jj} \sim m_W$. A third (possibly-tagged) jet can be combined with the hadronically-decaying W such that $m_{jjj} \sim m_t$. And finally, for the remaining (possibly-tagged) jet, one may check that $m_{j\ell} < 155$ GeV, as required by kinematics.

EXERCISE: Check this last claim. Show that if we have two 2-body decays — $A \to BC$ and $B \to DE$ — then m_{CD} is bounded from above. Determine the upper bound in terms of m_A, m_B, m_E. This is especially useful when E is invisible. In our case, show that if $A = t$, $B = W$, $C = b$, $D = e$ and $E = \nu$, then $m_{be} < 155$ GeV. (The precise number depends on what you take for m_t – just try to get close.)

If needed, we can fully reconstruct these events. We can find the neutrino momentum assuming $m_{e\nu} = m_W$ and $\vec{p}_{\nu,T} = \not{\vec{p}}_T$. (The required equation for $p_{\nu,z}$ is quadratic, so there are two discrete solutions. This usually doesn't cause a problem. Sometimes one solution is obviously more plausible.) In this process, the successful reconstruction of the hadronically-decaying W and t are key tools for calibrating jet measurements at the LHC experiments.

The next easiest events are those with two leptons. It is easy to trigger when you have two leptons; backgrounds are small. But there are 2 neutrinos, so the events are hard to reconstruct. There are many new physics signals with two leptons and missing momentum, so this channel is a large background to new $\ell^+\ell^-\not{E}_T$ signals!!

Finally, let us say a few words about fully hadronic events. These have large 6-jet QCD backgrounds. There are challenges at the level of the trigger (the p_T threshold for triggering is much higher for jets than for leptons) and there is a combinatorial challenge (if you have six jets, how do you reliably choose the correct jet triplets? Some progress is possible if $p_{t,T} \gg m_t$, because then the three jets in the top quark decay will be often quasi-collimated into one fat jet . This has recently become a well-known tool in the experimentalists tool box.

4.9. Advanced QCD: Evolution of PDFs and Emergence of Jets

There are two crucially important effects we have not discussed yet. One is that the parton distribution functions are not just functions of x but of an arbitrary scale μ that should be chosen to be near to the scale of interest. In other words, pdfs are not $f(x)$ but $f(x, \mu^2)$; partonic initial states run — or "evolve" — with scale.

The second is that jets emerge because partonic final states themselves run, or evolve, in a similar way. Both of these effects are independent of confinement and arise from the quasi-conformal behavior of high-energy QCD interactions. Gluon emission dominates! The large-N_c limit captures most of the physics in the structure of jets. Meanwhile, the order-$N_f/N_c \sim 1$ effects from $q\bar{q}$ pairs play a limited role in perturbative jet structure, but protect that structure from hadronization.

The evolution of pdfs is an easier topic, so we'll do it first. Consider a u quark in a proton. Here is some intuition (strictly true in light-cone gauge.) As it travels it constantly emits and absorbs gluons, virtual processes occurring at all scales of quark "virtuality" q^2. (All scales because, for $q^2 \gg \Lambda^2$, there is no mass-scale in the problem and the physics is nearly scale-invariant.) The time and distance scale corresponding to a virtuality q^2 is of order $1/\sqrt{q^2}$; if we look at a quark on a very short time scale of order t, we will find that, although its average transverse momentum p_T is zero, its quantum mechanically fluctuating p_T is of order $1/\sqrt{t}$.

If the quark scatters, through a process in which energy-momentum of order Q^2 is transferred, and thus over a time scale of order $\frac{1}{\sqrt{Q^2}}$, the scattering interrupts the ongoing absorptions and emissions for $q^2 < Q^2$. Some of the emitted partons will not be reabsorbed, because the quark, having been scattered, is not there to reabsorb them. These orphaned partons become *real.* As we noted, the scattering quark will have *non-negligible* p_T of order $\sqrt{q^2}$, the virtuality of the last emitted gluon before the scattering. This last emitted gluon, which carries off balancing p_T, becomes real, and can turn into a soft jet. More generally, many gluons are emitted. But only the highest-p_T gluons typically are observable experimentally as jets. These radiated gluons of low to moderate p_T are called "initial state radiation", or "ISR". (Note, however, that separating ISR from other radiation is not gauge invariant. It only makes approximate sense for radiation for which the opening angle of the emission is small, in which regime the ISR-like diagram dominates over other diagrams that might throw radiation into the same region of phase space. Since one diagram is much larger than the others, it dominates the amplitude and must on its own be essentially gauge-invariant.)

What is true for the quark is also true for any antiquark or gluon. The same logic applies.

Although long before the scattering process the parton initially carries $p_z = x_0 E_b$, ISR also steals some of its p_z as well, leaving it with $q_z = xE_b < x_0 E_b$. Thus, the probability of finding a parton with momentum fraction x is changed as a result of QCD radiation, and depends in q^2, which, in turn, cannot exceed the Q^2 of the scattering process. In short, when we go from the parton model, where the partons are free, to QCD, where the partons can emit gluons etc, the parton distribution functions go from simple functions of x to functions of both x and the scale Q^2 of the scattering. This is not unlike the fact that the coupling constant α_s itself, which is constant classically, becomes a scale-dependent quantity quantum mechanically.

Intuitively, if a parton which, averaging over times $\sim \Lambda^{-1}$, has $p = xP$ in the proton, then it will be at a lower x when probed on a shorter time $1/Q$ by the short-distance scattering process. Evolution is one-way: emission of gluons only lowers the energy of the initial parton. (Again, this intuition is only strictly accurate line-by-line in light-cone gauge.) Slow, predictable "evolution" of the pdfs through the process of "parton splitting" is a fundamental prediction of QCD, and it was precisely tested in the 1970s and 1980s.

The emission, and therefore evolution, looks α_s-suppressed, but this is not so! Consider a quark of momentum p_μ emitting a gluon of momentum k_μ; define $p_\mu = k_\mu + q_\mu$; $p^2, k^2 \ll q^2 = -2p \cdot k < 0$. Then defining also $E_0 = x_0 E_b$,

$$-\frac{1}{q^2} \approx \frac{1}{2p \cdot k} = \frac{1}{E_0 E_g (1 - \cos\theta_g)} \tag{4.59}$$

which diverges when $\cos\theta_g \to 0$ (the "collinear" singularity) or $E_g \to 0$ (the "soft" singularity.) A log enhancement arises when we integrate this expression over phase space, canceling the apparent α_s suppression.

EXERCISE: Check the above formulas.

Now we need to study the process of parton splitting in more detail. What is the amplitude $\mathcal{M}$ for a quark to emit a gluon before scattering? It is simpler to compute first a *scalar* quark entering a subamplitude $\mathcal{A}$:

$$\mathcal{M} \sim \frac{\epsilon \cdot (p+q)}{q^2} \mathcal{A} = \frac{\epsilon \cdot p}{p \cdot k} \mathcal{A}. \tag{4.60}$$

Here, we used $\epsilon \cdot k = 0$ to write $\epsilon \cdot p = \epsilon \cdot q$. For now we ignore the color structure; it is simple to add it back in. We square, sum over gluon helicities, and use

$$\sum_h \epsilon_\mu^{(h)} \epsilon_\nu^{(h)*} = -g_{\mu\nu} + \frac{n_\mu k_\nu + n_\nu k_\mu}{n \cdot k} \tag{4.61}$$

where $n = (1, 0, 0, -1)$ defines the light-cone gauge $n^\mu \cdot \epsilon_\mu = 0$. (Remember, the physics is gauge invariant, but the intuitive picture agrees with diagrams only in light-cone gauge. In other gauges, there are other diagrams that have to be considered as well.) The result is proportional to

$$\frac{k \cdot p \; n \cdot p}{n \cdot k (p \cdot k)^2} \propto \frac{1}{p \cdot k} \tag{4.62}$$

and thus $\sum_h |\mathcal{M}^2|$ has the same singularities as $\mathcal{M}$. And this result is also true for spin-$\frac{1}{2}$ quarks and spin-1 gluons that split by gluon emission.

The soft singularity (from the long-range chromo-electric field) is not observable. The quark barely moves — $q^\mu \approx p^\mu$ — and the pole cancels against a virtual contribution from a loop diagram in which the quark emits and reabsorbs a photon of momentum k^μ. But the collinear singularity significantly reduces the x of the emitting particle, and is in this sense directly observable. In the collinear limit we can approximate $\theta \ll 1$, so then

$$\begin{aligned} k_\mu &= (1-z)E_0(1, \sin\theta, 0, \cos\theta) \approx (1-z)x_0 E_b(1, \theta, 0, 1) \ ; \\ -q^2 &= 2p \cdot k \approx E_0 E_g \theta^2 = (1-z)x_0^2\theta^2 \ . \end{aligned} \tag{4.63}$$

There is then a phase space integral to do. For *fixed* z, we will have

$$\int \frac{d\cos\theta}{1-\cos\theta} \approx \int \frac{d\theta^2}{\theta^2} = \int \frac{dq^2}{q^2}. \tag{4.64}$$

So the emission is order $\alpha_s \ln Q_2^2/Q_1^2 \sim 1$ if $Q_2 \gg Q_1$. This tells us we do not have a pure expansion in α_s. Instead, we have an expansion in $\alpha_s \ln Q_2^2/Q_1^2$, which we must *resum*, combining numerous gluon emissions into a complete expression. This result then in turn gets order-α_s corrections, with no logarithmic enhancements.

To resum these large logarithms, we must integrate these effects over Q^2. But before we do this, we should stop and be more careful. We need to distinguish two things. The first is the physical scale of some particular scattering, Q. That's physics. The second is an arbitrary scale that we will use to separate the radiation that we will treat as ISR (resumming the large logarithms using the collinear approximation) from the radiation that we will account for as part of the hard scattering process (calculating carefully in a Feynman graph.) That scale we will call the "factorization scale" μ_F. It is a conceptually separate scale from Q; Q is from the kinematics of the physical process, while μ_F, like a renormalization scale, is something *we* introduce to make calculations more accurate. There is no best choice for μ_F; it is an artifact of the calculational technique, and has no effect in a perfect all-orders calculation. Though arbitrary, it should be chosen to be of order Q, so that our first order in perturbation theory will not differ too much from the true answer, giving us a convergent expansion. We do not want logarithms of Q/μ_F, which will appear in our calculations, to be too large. In these respects μ_F is very similar to a renormalization scale.

The point is that what we mean by a pdf depends on μ_F. For example, consider the process $uu \to uug$, where the g is at a small angle from the beampipe. Should we compute all the Feynman graphs for $uu \to uug$? Or should we compute the graphs for $uu \to uu$ and treat the g as ISR? If we did a calculation to all orders, it wouldn't matter, but in an approximate calculation we should choose something that will give us the best possible approximation. If the partonic center-of-mass energy is Q, and the gluon has $p_T \ll Q$, then treating it as ISR is probably a good approximation, but if $p_T \sim Q$ that is surely not so. So we want a dividing line — an arbitrary one — between the harder and softer emissions. That dividing line is μ_F. And the large logarithms we saw above are really not of $\ln Q_2^2/Q_1^2$. The question is what happens if we change μ_F from μ_1 to $\mu_2 > \mu_1$; then we must include emissions in this range, which gives us new contributions of order $\alpha_s \ln \mu_2^2/\mu_1^2$ (and not, as we said naively earlier, $\ln Q_2^2/Q_1^2$.) It is these effects which we need to resum.

For example, the evolution of the up-quark pdf is given by[b]

$$f_u(x, \mu_2^2) = f_u(x, \mu_1^2) + \int_x^1 dx_0 \; f_u(x_0, \mu_1^2) \times G[x, x_0; \mu, \mu_2] \tag{4.65}$$

where $G[x, x_0, \mu_1, \mu_2]$ is the probability, as $q^2 \to \mu_2^2$, for a u quark of virtuality $q^2 = \mu_1^2$ to cascade down from x_0 to x_1 by g emission.

$$G[x, x_0; \mu_1, \mu_2] = \int_{\mu_1^2}^{\mu_2^2} d\mu^2 \int_0^1 dz \; \tilde{P}_{qg\leftarrow q}(x_0, x, z; \mu^2) \; \delta(x - zx_0) \; . \tag{4.66}$$

Here $\tilde{P}_{qg\leftarrow q}$ is the differential probability for a quark of virtuality μ^2 to emit a gluon such that its x decreases from x_0 to $x = zx_0$. [In what follows I will relabel the subscript $qg \leftarrow q$ by qq; note that in general the order of the subscripts matters.]

The function $\tilde{P}$ is given by a QCD tree graph. It is scale-invariant except for the overall running $\alpha_s(\mu^2)$. To make the rest of the formula scale invariant, it must be that $\tilde{P}(x, x_0, z, \mu^2) = \frac{\alpha_s(\mu^2)}{\mu^2} P(x, x_0, z)$. The diagram describing the splitting is independent of E_b, so it only can depend on the ratio $x/x_0 = z$. Therefore $P(x, x_0, z) = P(z)$. And finally,

$$\begin{aligned} G[x, x_0; \mu_1, \mu_2] &= \frac{\alpha_s}{2\pi} \int_{\mu_1^2}^{\mu_2^2} \frac{d\mu^2}{\mu^2} \int_0^1 dz \; P_{qq}(z) \; \delta(x - zx_0) \\ &= \frac{\alpha_s}{2\pi} \int_0^{\log \mu_2^2/\mu_1^2} d \log \mu^2 \int_0^1 \frac{dz}{x_0} \; P_{qq}(z) \; \delta(z - x/x_0) \; . \end{aligned} \tag{4.67}$$

[b]Note that Peskin and Schroeder sticks with the notation Q for μ_F. This leads, in my opinion, to many confusions. Do not confuse the arbitrary factorization scale, in terms of which the evolution equations are expressed, with a scale in a physical process! The whole point of the pdfs is that they are process-independent and know nothing about what scattering process is underway.

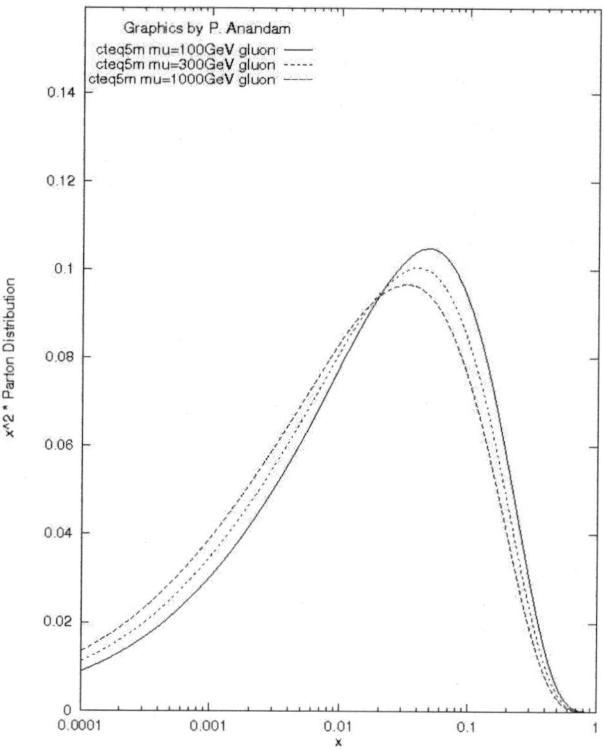

Fig. 4.19. Running of the gluon parton distribution function with changing μ_F.

Note the change in the argument of the delta function, and the corresponding Jacobian. Now, we have

$$f_u(x,\mu_2^2) = f_u(x,\mu_1^2) + \frac{\alpha_s}{2\pi}\int_x^1 \frac{dx_0}{x_0}\, f_u(x_0,\mu_1^2)\int_0^{\log\mu_2^2/\mu_1^2} d\log\mu^2\; P_{qq}(x/x_0) \tag{4.68}$$

and so

$$\frac{\partial f_u(x,\mu_2^2)}{\partial\log\mu^2} = \frac{\alpha_s}{2\pi}\int_x^1 \frac{dx_0}{x_0} P_{qq}(x/x_0)\, f_u(x_0,\mu_1^2). \tag{4.69}$$

This is not quite complete. We can also get a u quark starting from a

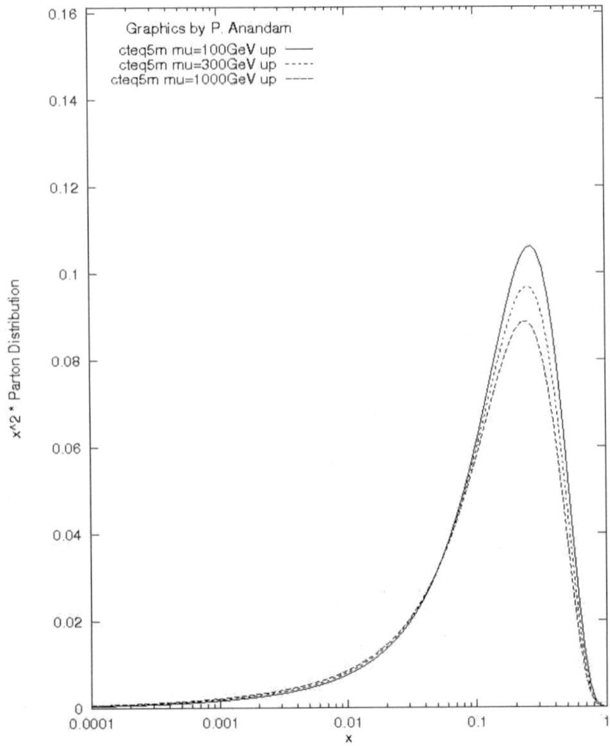

Fig. 4.20. Running of the up-quark parton distribution function with changing μ_F.

g that splits to a $u\bar{u}$ pair. So in fact

$$\frac{\partial f_u(x,\mu_2^2)}{\partial \log \mu^2} = \frac{\alpha_s}{2\pi}\int_x^1 \frac{dx_0}{x_0}\ \left[P_{qq}(x/x_0)\ f_u(x_0,\mu_1^2) + P_{qg}(x/x_0)\ f_g(x_0,\mu_1^2)\right] \tag{4.70}$$

where

$$P_{qq} = \frac{h_1(z)}{(1-z)}\ ;\ P_{gq} = \frac{h_2(z)}{z}\ ;\ P_{gg} = \frac{h_3(z)}{z} + \frac{h_4(z)}{1-z}\ ;\ P_{qg} = h_5(z). \tag{4.71}$$

All the $h_i(z)$ are calculable linear or quadratic polynomials that have a constant term; see Peskin and Schroeder for their precise form.

Note that $P_{qq}(z) = P_{gq}(1-z) \propto \frac{1}{z}$. This soft singularity — a high probability for soft gluon emission — is *observable* after all, despite what I said earlier. It leads to the very large $f_g(x)$ for $x \ll 1$.

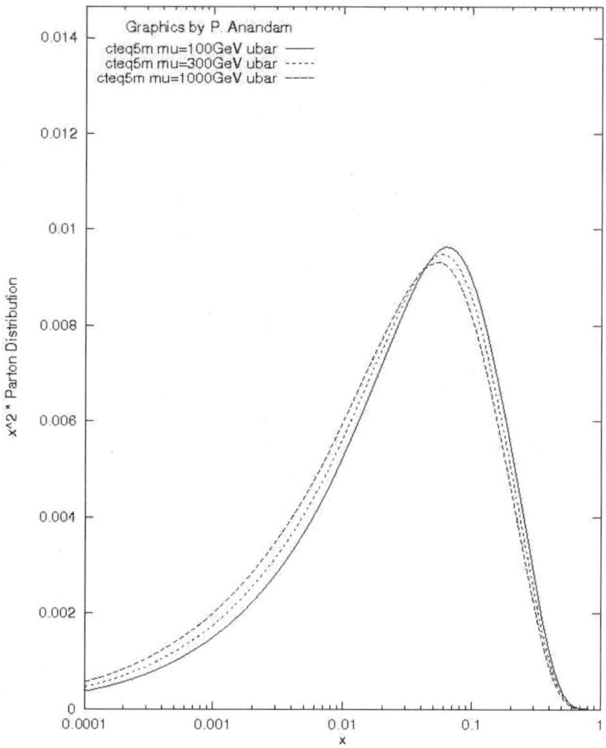

Fig. 4.21. Running of the up-antiquark parton distribution function with changing μ_F.

To summarize the situation, all parton splittings show a collinear singularity. Gluon emission from any parton also has a soft singularity, while quark emission (*i.e.* in $g \to q\bar{q}$) has no soft singularity. As a result, as μ^2 increases, the gluons win big: the $1/z$ emission enhancement leads to a very large $f_g(x)$ at small x. The gluons then can split to q and $\bar{q}$ pairs, leading to large $f_q(x), f_{\bar{q}}(x)$ at small x — large but still much smaller than $f_g(x)$. The N_f enhancement of $g \to q\bar{q}$ is ineffectual; gluons are still more likely to split to other gluons than to quarks.

The evolution of the pdfs $f_g, f_u, f_{\bar{u}}$ with scale is shown in Figures 4.19–4.21. Note how the valence distributions of the u (and also d, not shown) quarks decrease while the gluons and the sea quarks increase as μ_F increases.

4.9.1. *Jet evolution*

Now we turn to the physics of jet formation. There are some key physical similarities to, as well as important differences from, what we have just encountered with pdfs.

If we have a scattering process, such as $uq \to uq$ that occurs at a scale $Q_0^2 \gg \Lambda^2$, it typically doesn't produce an on-shell final-state quark. Instead it makes a quark with virtuality q^2, with $0 < \Lambda^2 < q^2 < Q^2$.

Nothing prevents this interacting quark from emitting collinear and soft radiation as it does prior to the collision. But there are important small changes. First, the initial quark is more off-shell than the final one; $|q^2|$ *decreases* as we move away from the collision. The reverse is true in initial-state radiation. Second, in contrast to the initial state, where we measure the effect of the scattering quark, and perhaps one or two of its ISR particles, but most of the emission goes down the beampipe, in the final state we measure much more. Most of the radiation is observed in the detector. Therefore in the final state we must consider the evolution not only of the original emitting particle but also of those particles emitted, which themselves become additional emitters. This branching process of emission and emission into the final state is called a *parton shower*. (Actually, it also occurs in the initial state, but I ignored it there because most of it is not observed, and because it clutters up the original discussion.)

4.9.1.1. *Collinear emission*

Collinear emission follows the same equations as before (at leading order in α_s only.) Emission at large q^2 has large k_T (where k_T is the transverse momentum of an emitted particle relative to its emitter, not to the beampipe) and is visible as final-state radiation ("FSR.") But most of the emission is collinear with the original quark's direction, and *forms the quark's jet.* Note that this is all long before hadronization — the jet is forming from the collinear singularity structure of QCD, in a perturbative, computable regime.

To describe the final state, we need to model the emission fully as a classical probabilistic branching process. And in fact the same approach is needed if we want to describe ISR in detail, as we do when simulating events on a computer. That this is possible (at a certain approximation) is a non-trivial, key result, allowing event simulation with pdfs and jets (using programs such as PYTHIA, HERWIG, SHERPA, *etc.*...) It is far from immediately obvious that a quantum mechanical process can be approximated well by an classical one.

We would want this approximation to be extendable, in a controllable way, to arbitrary orders in perturbation theory. How to do this remains an active research area.

4.9.1.2. *Soft emission, and dipoles*

It is clear why collinear emission creates a jet of partons traveling all in the same direction. But why doesn't soft emission destroy the jet structure and throw hadrons everywhere? The reason is that emission occurs not so much from quarks and gluons as from quark-antiquark pairs — dipoles — or more generally color lines and anticolor lines that are color-correlated. This simple picture becomes increasingly accurate in the large N_c limit – and in real-world QCD, for which $1/N_c^2 \gg 0.11$, this picture is not too bad.

Consider gluon emission from a quark. Let's follow the color quantum numbers as a quark with color c_1 emits two gluons, ending up as a quark with color c_3 preceded by two gluons with color $(c_i\bar{c}_{i+1})$.

$$c_1 \to (c_1\bar{c}_2)c_2 \to (c_1\bar{c}_2)(c_2\bar{c}_3)c_3 \to \ldots . \tag{4.72}$$

For $N_c^2 \gg 1$, colors are independent; forces between an object with color c and another with anticolor $\bar{c}'$ are large only if $c = c'$. That means that when we consider two random lines with uncorrelated color, they don't talk to each other. The c_1 color of the first gluon only feels the anticolor $\bar{c}_3$ of the second gluon at order $1/N_c^2$.

By contrast, the $\bar{c}_2$ and c_2 charges form a dipole. Soft emission will only occur *inside* this dipole, in physical space. Compare this situation with the ordinary electric field $\vec{E}$ of an e^+e^- pair with small opening angle ϕ. We know that the interference between emission off the e^+ and e^- must be destructive for $\theta > \phi$ — otherwise, far from this particle pair, we would get a doubled monopole $\vec{E}$ field, whereas what we should get is a dipole $\vec{E}$ field, with the positive and negative monopoles canceling. The largest electric fields are found between the two particles; that's where the any soft emission will be found.

What this implies for soft emission is a form of "angular ordering". Each emitted soft gluon is typically emitted at an angle comparable to or smaller than the angle between its two emitters; at larger angles there is a destructive interference. The new gluon forms two new dipoles, its anticolor pairing with the color of its emitter to one side, and its color pairing with the anticolor of its emitter on the other side. The opening angles of these new dipoles are typically smaller than that of the original dipole. So any future emission occurs at even smaller angles. And the process repeats in a scale-free manner.

In short, when a new gluon is emitted in this process,

$$(c_1\bar{c}_2)(c_2\bar{c}_3)c_3 \to (c_1\bar{c}_4)(c_4\bar{c}_2)(c_2\bar{c}_3)c_3 \to \dots \tag{4.73}$$

the gluon is both "inside" in color-space and "inside" in physical angular space. Think about this for a moment. This correlation between the color-space configuration and the momentum-space configuration builds a string! This is the string which arose in my heuristic argument for hadronization. I called it a flux tube at the time, but really, the string of gluons makes sense in perturbation theory. It pre-exists *before* confinement! Even as perturbation theory is beginning to break down, this string is already beginning to fracture into colorless pieces through $g \to q\bar{q}$ processes that occur along it, as in Fig. 4.1. These processes split the string into gauge-invariant string-segments, which are small in physical space and so have invariant mass $\sim \Lambda$, not Q. Then they turn into hadrons, with mass not much larger than Λ. These hadrons are either stable on QCD time-scales or decay immediately to those which are (such as pions and kaons.)

This pre-clustering, in which the string rapidly breaks (with a rate proportional to N_f/N_c) into hadron-sized pieces, is crucial for jet survival! The result is that the string is cut into little string bits before it has time to drag on itself. The combination of collinear radiation, angular ordering of soft radiation, and rapid breakup of the string are all essential to why the momentum of the parton ends up as the momentum of a jet, to a decent approximation. In particular, in QCD hadronization, there's no time for momentum to be moved around significantly.

To summarize, at small $\alpha_s N_c$ jets *always* form through a perturbative, quasi-scale-invariant process. This is called the "parton shower"; it involves the resummation of many graphs. But in principle this perturbative information might be lost in the complexities of hadronization. In particular, were N_f/N_c too small — and especially if N_f were zero — the time for hadrons to form would be slow, and all the momentum carried by the showers of partons would be reorganized as the long-lived flux tubes wiggled around in space. Fortunately for hadron colliders like the LHC, N_f/N_c is big enough that the flux tubes break long before they get a chance to wiggle. Thus hadronization lets the jets through, with $p^\mu_{jet} \sim p^\mu_q$. And this provides the critical link between the observed hadronic physics and the high-energy quarks and gluons. Without this link, we could not reconstruct hadronic decays of W, t, $\tilde{g}$, etc.

Here is some further reading on all aspects of collider QCD. The best LHC-QCD/SM lectures available, in my opinion, are those of the expert M.L.Mangano, home.cern.ch/mlm. A QCD book with many technical details is *QCD and Collider Physics*, by Ellis, Stirling and Webber.

4.10. What Lies Beyond the Standard Model?

Finally, armed with all of this SM information, we are prepared to ask the question of what lies beyond it. What follows is the briefest possible summary, enough to get you started, and with some important and possibly controversial comments to make it worth your while.

4.10.1. *The Higgs boson of the Standard Model*

I consider the Higgs boson, in its simplest form, part of the SM. But it is new physics from the point of view of LHC, so let's quickly review it. In the SM, the Higgs boson is produced dominantly in four processes. From the highest rate process to the lowest rate one, these are

- $gg \to h$ via loops of colored particles (mainly t in the SM; the loop is sensitive to new physics);
- $qq \to hqq$ via intermediate W, Z bosons: (this is called Vector Boson Fusion, or VBF; it tests the all-important WWh, ZZh couplings);
- $q\bar{q} \to Wh, Zh$ (which again tests WWh, ZZh couplings);
- $t\bar{t}h$ (a very difficult measurement which is the only way to measure y_t, the top Yukawa coupling).

Note that any scalar can have a coupling to gg (if it interacts with quarks directly, or with any other unknown colored particle) but recall that *only a scalar with an EWSB vev* can have tree-level, and therefore rather large, couplings to WW and ZZ! So measuring these processes is very important in testing whether a new scalar is really the Higgs boson.

Meanwhile, the SM Higgs decay modes depend on its mass. A light Higgs is narrow and decays to $b\bar{b}, \tau^+\tau^-, gg, \gamma\gamma, \ldots$. The di-photon decay is the cleanest; it arises through loops of charged massive particles, such as the W boson and the t quark. A heavy Higgs is much wider, and decays to W^+W^-, ZZ (and only a bit of $t\bar{t}$, despite the large coupling.)

EXERCISE: Explain, in physical terms, why the WW, ZZ decays of the Higgs dominate the $t\bar{t}$ coupling for a very massive Higgs boson. Hint: what happens if you turn off the gauge coupling? what do you have to do to make the Higgs massive holding the vev v fixed?

That's the SM Higgs. But what will be different if we are faced with the very likely possibility that the Higgs sector is not the simplest possible one? The answer is not a short one.

4.10.2. *Non-minimal Higgs sectors*

What could lie in a BSM Higgs sector? Without being too specific, there are a number of general observations one can make. There can be multiple neutral and charged Higgs bosons in the Higgs sector, with different masses, production rates and decay modes. The Higgs production and decay mechanisms can be significantly affected. The WWh and ZZh couplings can be shared among several bosons, each of which has a vev smaller than v. The ggh and $\gamma\gamma h$ couplings can be bigger or smaller than in the SM, and the bbh and $\tau\tau h$ couplings are easily affected. Even the $\mu^+\mu^-h$ coupling can become large enough to be important. In fact, the dominant decay of the most easily produced Higgs bosons can be to pairs of light, rarely produced scalars (e.g. $h \to aa$, where a is a CP-odd scalar decaying to light SM fermions.) Or the Higgs may decay invisibly almost all of the time. That would make life hard. Conversely, the Higgs may be copiously produced in the decays of heavy new particles, which (if colored) may have large cross-sections, making the Higgs much easier to discover.

So discovery of a BSM Higgs sector can be a lot easier or a lot harder than of the SM Higgs. But in any case, experimental life is a challenge, because with such a huge range of possibilities, a vast array of analysis strategies must be pursued!

In the SM and most SUSY models, if $m_h < 140$ GeV, $h \to b\bar{b}$ is common, and $h \to \tau\tau, \gamma\gamma$ are helpful channels to look in. But consider the *Minimally Non-Minimal Standard Model* (MNMSM), the SM plus a single real scalar S. I will now show you how just a single new particle can change everything.

4.10.2.1. *MNMSM 1: An Invisible Higgs Sector*

Take the scalar potential

$$V(H,S) = -\mu^2|H|^2 + \lambda|H|^4 + \eta S^2|H|^2 + \frac{1}{2}M_s^2S^2 + \kappa S^4. \qquad (4.74)$$

For some choices of parameters $\langle H\rangle = v$, $\langle S\rangle = 0$. In this case, write $H = v + h$, and you will find

$$V(h,S) = \frac{1}{2}m_h^2 h^2 + \frac{1}{2}m_s^2 S^2 + 2\eta v h S^2 + \text{quartic terms.} \tag{4.75}$$

What phenomena arise?

- If $m_h > 2m_s$, then $h \to SS$ decay becomes possible.
- If $\eta v^2 \gg m_b m_h$, the rate could be $\sim 100\%$!
- The Z_2 symmetry $S \to -S$, which is unbroken, implies S is stable and invisible!

So if η is not small, this Higgs sector is largely invisible. All visible decays of the h, such as $h \to b\bar{b}$, are dramatically reduced by a large common factor; the vast majority of the time, $h \to SS$ and vanishes. Such a Higgs is not impossible to detect indirectly, but it is very challenging to do so, and harder still to study it in any detail.

The lesson? *One new particle can ruin your whole day.*

4.10.2.2. *MNMSM 2: Two SM-Higgs-like bosons, and possibly novel decays*

Alternatively, for other choices of parameters $\langle S\rangle \equiv v' \neq 0$. Let $S = v' + s$. In that case there is a non-zero hs term in the Lagrangian in addition to the h^2 and s^2 terms — and so the two fields mix, and the two mass eigenstates are scalars ϕ_1 and ϕ_2 that are mixtures of h and s,

$$\phi_1 = h\cos\theta + s\sin\theta\ ;\ \phi_2 = -h\sin\theta + s\cos\theta\ . \tag{4.76}$$

Let's define ϕ_1 to be the lighter of the two. If it is sufficiently light, $m_1 < 140$ GeV or so, ϕ_1 may decay like a light SM-h, as in $\phi_1 \to b\bar{b}$. The heavier ϕ_2, however, if heavier than 160 GeV or so, may decay like a heavy SM-h, as in $\phi_2 \to ZZ$. Or maybe, via the hss and hhs couplings, we may even have $\phi_2 \to \phi_1\phi_1 \to b\bar{b}\,b\bar{b}$. And so, with one new particle, we get significantly altered phenomenology, and possibly a completely new signature.

4.10.2.3. *MNMSM 3: A Higgs decaying to long-lived particles*

Here's one more possibility to think about. Perhaps the Z_2 symmetry $S \to -S$ is explicitly broken, by tiny effects. (Note: this is not fine-tuned! small explicit symmetry-breaking effects are common, and are protected against large quantum corrections by the symmetry that would be restored if they were absent. An example is the electron Yukawa coupling, which is small but not fine-tuned.) This would allow a tiny $S|H|^2$ term, which would induce Sh mixing even with $\langle S \rangle = 0$. The hSS term allows $h \to SS$, possibly with 100% branching fraction. The tiny mixing allows $S \to b\bar{b}$ with a very long lifetime. And this means that, depending on the mixing angle, the S may decay anywhere. Perhaps it decays mainly outside the detector (giving an invisible Higgs.) Or perhaps it decays promptly (giving a difficult $b\bar{b}b\bar{b}$ signal.) Or, if the coupling is right, decays can occur right in the middle of the tracking volume. This last signature has no SM background! But it does have a huge detector background from secondary interactions. The Tevatron analysis searching for such a signal took more than 2 years, and the constraints are still weak. The challenge at the LHC is that this signature has a very low trigger rate, at least for standard triggers.

4.10.2.4. *Summary and implications*

This was for one new particle, the real scalar S. Imagine the range of possibilities if we allow 2 (or 5!) new particles. (Four new particles arise in the minimal supersymmetric theories described below.) Should we investigate all these possibilities?! It seems an infinite task, filled with unmotivated models...

Well, you've heard of Occham's Razor – an elegance criterion. I hereby present to you Strassler's Machete: a usefulness criterion. *A theory is important experimentally if it predicts a non-obvious signature which has not been previously (or sufficiently) considered, could be observed, and is unlikely to be discovered by accident using existing methods.* We will use this criterion later.

4.10.3. *Models beyond the SM*

Construction of BSM models are mainly driven by the hierarchy problem, so it is useful to recall this problem and pose it more formally. On the one hand the SM Higgs has mass-squared term $\mu^2 \sim (100\ \text{GeV})^2$. On the other hand, Yukawa couplings y_i range from $\sim 10^{-5}$ to as large as ~ 1.

$$\mathcal{L} = \mu^2 H H^\dagger + y_i F H f + \dots . \tag{4.77}$$

(Recall my notation that F and f represents respectively $SU(2)$ doublets and singlets.) The hierarchy problem can be phrased as a tension between these two types of terms.

The mass-squared term argues for the Higgs being composite, strongly-interacting, or in some other way very complicated. This is because of naturalness. If the Higgs is a fundamental field, then $\mu^2 \ll M_{max}^2$ is unnatural. (See Sec. 4.2.4.) But, for instance, if the Higgs is a composite object at a scale Λ_H, then it is natural that $|\mu^2| \sim \Lambda_H^2$. (Note this relation is true of scalar hadrons [though not of pseudoscalar PNGBs, which are naturally light] in QCD.)

But the Yukawa couplings — especially the top quark coupling, which is ~ 1 — want the reverse: they argue for a fundamental Higgs boson. If the Higgs is a fundamental field, then the Yukawa couplings are dimensionless and can easily be set to be of order 1. But if the Higgs is composite, then Yukawa interactions have dimension > 4 above the compositeness scale Λ_H. Their natural size is then no more than of order Λ_H/M_{Planck} to some positive power, which is (no matter what the power) inconsistent with $y_t \sim 1$.

Let us say this somewhat more formally. The operator H that creates the Higgs boson has some dimension d_H; this dimension is 1 at a given scale if the Higgs is effectively fundamental at that scale, but could be larger than 1 at other scales if the operator is composite and/or strongly interacting at those scales. Consider the proton operator; at 1 GeV it creates a proton and has dimension 3/2, but at much higher scales it creates three quarks and has dimension 9/2. Similarly, the operator H might have a large anomalous dimension $d_H - 1$ at scales well above the weak scale, with a transition to $d_H \sim 1$ at the scale Λ_H. That would good for naturally having a Higgs with mass of order Λ_H. The operator $H^\dagger H$ will have dimension of order $2d_H$ (though it may itself have a large anomalous dimension away from this naive value, something we'll disregard here for a moment.) For instance, if $d_H = 3$, and $d_{H^\dagger H} \sim 6$, then $H^\dagger H$ is an *irrelevant operator*, and its coefficient μ^2 shrinks as the renormalization scale decreases, just the way four-fermion operators in the SM become unimportant at low energy. That means that the Higgs mass is going to be no larger than the scale Λ_H where d_H becomes of order 1.

But if d_H is far from 1, then the natural size of the Yukawa couplings is of order Λ_H/M_{Planck} to the power $(d_H - 1)$. And that is a problem. Though maybe not such a terrible problem. Note that most of the Yukawa couplings are small, so actually a composite or strongly interacting Higgs might even be ok... were it not for the top quark, with $y_t \sim 1$. So one way around this is to make the top quark composite too, so that its couplings do not run perturbatively. However, it's not so easy. Such an approach risks causing problems with flavor physics, in particular with unobserved FCNCs.

What are the existing approaches to solving, adjusting, coping with, or otherwise evading the hierarchy problem? Here are a few:

- Supersymmetry (SUSY): Quantum corrections to μ^2 vanish in the limit of unbroken SUSY; in the real world SUSY is spontaneously broken, feeding down to SM particles at the TeV scale, and giving $\mu^2 \sim \text{TeV}^2$.
- Technicolor: The Higgs is composite (and possibly the top quark as well) at $\Lambda_H \sim 1$ TeV.
- Warped Extra Dimensions: Similar phenomenology to models in which (some) SM fields are composite, and for a good reason, because of "AdS/CFT" duality. [Some working models are actually dual forms of technicolor.] There is a lot of fine print here in this statement; see below.
- Little Higgs: the Higgs is a PNGB, composite at 10 TeV; its mass is suppressed well below 1 TeV by "collective symmetry breaking".
- Large Anomalous Dimensions: an idea, but with no realistic model; the Higgs and SM fermions are charged at high energy under a strong gauge group, such that the $H^\dagger H$ operator has a large anomalous dimension while FHf operators do not.
- Large-Volume Flat Extra Dimensions: $M_{Max} = 1$ TeV, and Quantum Gravity (perhaps String Theory?) is found at the TeV scale.
- Standard Model Plus Denial: *There Is No Hierarchy Problem.* Perhaps the hierarchy arises from a selection effect — maybe any region of the universe without such a hierarchy is uninhabitable? Or perhaps there is some misunderstanding of how QFT operates when quantum gravity is present?

Now I am going to give you a taste of all of these, and some others.

4.10.4. *Supersymmetry*

In supersymmetry (SUSY), every particle has a "sparticle" partner of opposite statistics. Also the structure of the theory generally requires at least two Higgs $SU(2)$ doublets. This seems like a huge number of extra as-yet-unobserved particles, but the theory does solve the hierarchy problem naturally. The Higgs h^0 is a fundamental scalar, and it remains light because of cancellations — positive and negative contributions from bosons and partner fermions — to the Higgs mass-squared μ^2. The cancellation isn't perfect, but the failure to cancel is of order the mass splittings within supermultiplets. We need the Higgs to be below the 1 TeV mass scale (so that v naturally can be below that scale) so all of these new particles must be at or below 1 TeV. The minimal model (Minimal SUSY SM, or "MSSM") has over 50 new particles and over 100 new parameters. And why should we expect nature to present us the minimal version of this model?

The thing about SUSY is that it must be broken (spontaneously) or we would long ago have seen the sparticle partner to the photon and electron, among others. But how is SUSY broken? There must be some dynamics that causes it. The SM sparticles must learn about the breaking — how may this happen? It may happen because they participate directly in the SUSY-breaking dynamics (unlikely, for technical reasons), or indirectly, through gauge boson effects ("gauge mediation"), Planck-scale-suppressed effects, (called "gravity mediation", but this is a misnomer), or through an anomaly ("anomaly mediation.") There are a number of other possibilities.

In the "MSSM," R-parity (+1 for SM particles, -1 for their sparticles) is set to be conserved. (It is the slickest way to avoid catastrophic baryon-number violation that would be present if R-parity were completely absent.) Consequently,

- SUSY particles must be produced in pairs;
- A SUSY particle must decay to SM particles plus one other SUSY particle (or more generally an odd number of sparticles);
- The lightest R-parity-odd particle (LSP) is stable;
- And therefore the final state of every SUSY event contains two LSPs!

Let us call the lightest sparticle for any SM particle (all the quarks, leptons, gauge bosons and Higgs bosons) the "LSMSP':: the Lightest Standard Model Sparticle. Note we do not include here the gravitino, the partner of the graviton. A key issue for the MSSM is the mass $m_{3/2}$ of the gravitino.

In minimal versions of Gravity Mediation, the gravitino is relatively heavy, and the LSMSP is also the LSP, the lightest of all sparticles. But in gauge mediation, the gravitino is the LSP, the LSMSP is the NLSP, the next-to-lightest R-parity-odd particle. The difference has a huge effect on the phenomenology.

We know some things about stable particles. Light stable charged particles or colored particles would have been found already at colliders. And heavier stable charged particles would show up as unusual atoms and screw up Big Bang Nucleosynthesis , so they are strongly excluded. Thus the LSP, since it is stable, must be neutral and colorless, and interacting as weakly as a neutrino, or weaker.

In some cases this object is in fact a good dark matter candidate! This gets people very excited about SUSY. Note, however, that the existence of a dark matter candidate is not a consequence of SUSY, despite what you will read. It is a consequence of R-parity — a new global symmetry — and not SUSY! (R-parity without SUSY has a dark-matter candidate; SUSY without R-parity does not.) In fact, any new global symmetry (continuous or discrete) will give a new stable particle. So when you hear someone say that SUSY is better than other theories because it gives us a dark matter candidate... well, you should think for yourself.

Do we know anything about the spectrum of the sparticles? Unfortunately it is not enough to make a simple list of the sparticles and their masses. While the gluino and the first two generations of squarks/sleptons are pretty simple, the top squark and (conjugate of the) anti-top squark mix, giving two different scalar mass eigenstates t_1, t_2. The same is often true for bottom squarks. Meanwhile the electroweak gauge boson spartners mix with the Higgs boson spartners. In particular, the $\tilde{W}^+$ and $\tilde{H}^+$ mix to form two "charginos" $\tilde{\chi}_i^{\pm}$, $i = 1, 2$. And the $\tilde{W}^0$ ("Wino"), $\tilde{X}^0$ [often called "$\tilde{B}^0$", or "Bino"], and $\tilde{H}_u^0, \tilde{H}_d^0$ ("Higgsinos") mix to form four "neutralinos" $\tilde{\chi}_i^0$, $i = 1, 2, 3, 4$. I should also mention that there are five Higgs bosons: h^0, H^0, A^0 and $H^{\pm}$, mixtures of the eight real components of two complex doublets H_u, H_d (with the other three components forming the longitudinal polarizations of the W and Z.)

Think about what this is going to mean experimentally. When it comes time, if SUSY is part of nature, to figure out the parameters in the Lagrangian, we need to invert all of this mixing! We will need the masses *and* mixing angles of *all* the charginos and neutralinos, stops, sbottoms, and Higgses. This is very difficult to do at the LHC alone! We will be working with partial information for a very long time, and other experiments will surely be needed.

In the meantime, we have to worry that SUSY might cause problems with the physics of flavor. There will be large FCNCs in a SUSY model unless the spectrum of squarks and sleptons is special. The spectrum might be such that all (or at least the first two generations) of squarks/sleptons with a definite charge are very nearly degenerate. Or their masses might be tightly aligned with those of the quarks/leptons. It is possible that only the third generation of squarks is light. And there is even the possibility that R-parity is replaced with a $U(1)_R$-symmetry, with only dimension-6 flavor violation.

As I mentioned, there are lots of unfixed parameters. But in many models, the SUSY masses of sparticles are rather similar at high scales. The running of masses, due to interactions with gauge bosons, are largest for colored particles. The colored particles then end up heavier than the uncolored ones (by a factor of 5–10,) but often still light enough that they are more often produced than the uncolored ones. And their more spectacular decays are often easier to separate from backgrounds than those of uncolored sparticles.

So — typically, meaning in a wide variety of SUSY models (though watch out for lots of fine print,) the dominant mechanism for sparticle production is pair production of gluinos and/or squarks. Each of these sparticles decays to high-p_T jets plus a neutralino or chargino. The neutralinos/charginos decay to the LSMSP plus lower p_T leptons/neutrinos or jets. And thus every event gives two "cascade decays," producing jets, maybe leptons and/or neutrinos, and two LSMSPs.

Note, however, that this structure is true of many classes of non-SUSY models, for similar reasons. Therefore, the same search strategies are also sensitive to many non-SUSY models also. Conversely, discovery of such a signal does not necessarily imply SUSY. Much more information will be needed.

4.10.4.1. *SUGRA versus GMSB*

Now how do Gravity Mediation ("SUGRA") and Gauge Mediation ("GMSB") typically differ? In the mass of the gravitino, which is much lighter in gauge mediation. This has important implications for the behavior and nature of the LSMSP.

In Gravity Mediation the LSMSP is the LSP, so it must be neutral, colorless, and thus invisible to the detector. This means the only options in the MSSM are a sneutrino (which happens rather rarely in explicit models) or a neutralino. And therefore Gravity Mediation in the *minimal* SUSY SM gives large MET signals, and no reconstructible new resonances. (Note, however, this may not be true in non-minimal gravity mediation! See below.)

But in gauge mediation, the initial breaking of supersymmetry, which occurs at a scale called M_{SUSY} for historical reasons, is communicated to the standard model by gauge interactions, not by interactions of gravitational strength. This allows the mass splitting of the graviton and gravitino to be much smaller than the splitting between standard model particles and their superpartners. The gravitino mass is $M_{SUSY}^2/\sqrt{3}M_{Planck} \sim$

(1 eV)$(M_{SUSY}/100\text{ TeV})^2$, while the masses of the superpartners (which must be of order a TeV) are parametrically of order $(\alpha_i/4\pi)M^2_{SUSY}/M_X$, where X is a messenger field that couples the supersymmetry breaking sector to the Standard Model.

Since the LSMSP is *not* the LSP, it is unstable, and decays to the gravitino. And therefore it does *not* need to be neutral and colorless. In minimal gauge mediation the colored sparticles are heavy. So here we expect either the Bino or the $SU(2)$-singlet stau is the LSMSP. Again, however, in non-minimal gauge mediation there are other possibilities, including the Wino and Higgsino. In fact, in more general models, even the stop and gluino have shown up as LSMSPs!

Let's start with the case of a neutralino $\tilde{\chi}^0$ LSMSP. Then, in GMSB, typically $\tilde{\chi}^0 \to \tilde{G}$ plus γ or Z or h^0 (where $\tilde{G}$ is the gravitino). What is its lifetime? It is highly variable, with lifetime (in distance units)

$$c\tau \sim 10^{-2}\text{ cm}\left(\frac{M_{SUSY}}{100\text{ TeV}}\right)^4\left(\frac{100\text{ GeV}}{m_{\tilde{\chi}}}\right)^5. \tag{4.78}$$

If M_{SUSY} is very large, the lifetime is very long, the neutralino is invisible, and we get something almost identical to a gravity-mediation signal. If the neutralino decays are prompt, however, each event will display visible consequences of the two LSMSP $\to$ LSP decays (and somewhat reduced MET compared to the long-lived case.) This can make life very easy experimentally, if for example there are two photons in each event, or somewhat harder if there are two Z or h bosons in every event. In between, the decays of the neutralinos are highly displaced, and the events can look really crazy. There can be photons, pairs of leptons, or pairs of jets that seem to appear from nowhere, well away from the collision point. These possibilities range from challenging to very challenging experimentally.

If instead the $\tilde{\tau}$ is the LSP, then typically one has a chain like

$$\tilde{g} \to q\bar{q}\tilde{\chi}^0 \to q\bar{q}\tau\tilde{\tau} \to q\bar{q}\tau\tau\tilde{G} \tag{4.79}$$

so if the decay is prompt there are four τs in every SUSY event! This is potentially striking. Meanwhile, if the $\tilde{\tau}$ is long-lived, it gives something even more striking: a CHAMP (Charged Heavy Massive Particle). This is a heavy muon-like particle. When such a particle travels slowly, it would give a very high-momentum track that nevertheless has a large energy-loss rate in matter (dE/dx) typical of a slow particle, and can have a measurably-slowed time-of-flight across the detector. This signature has almost no background.

The full richness of the possible phenomenology could take us a week. There are some old reviews and a serious of recent long (and still not exhaustive!) papers on the subject.

What is the lesson? Even within the MSSM, the predictions for what we might observe vary enormously. We have to be prepared for many different possibilities.

As a corollary we see that the map between models and signatures is extremely nonlinear! and this is true not just in supersymmetry. One model can give many signatures; one signature can arise from many models.

Once we discover some process with certain characteristics that *might* be consistent with SUSY, we've just begun the battle. First, we have to figure out what we are looking at! What are the masses and quantum numbers of the particles being produced and appearing in the cascade decays? Then, we have to show that this process is actually consistent with one or more SUSY models. This involves both testing whether SUSY is really true, and determining the parameters in the SUSY Lagrangian. Now, this process might take a *long* time (extending beyond the LHC) and could be very ambiguous and tortured. That's because many non-SUSY models *also* predict jets, leptons and MET. And many non-minimal SUSY models differ from the minimal one by a great deal. But if we are lucky, there will be a natural, simplest-possible guess, and if we are really lucky, it will be true.

4.10.5. *Composite Higgs and gauge/string duals thereof*

Now let me say something about models that have a composite Higgs boson. All of these models have an essential problem: how can one generate fermion Yukawa couplings $yFHf$ in a reasonable way? and without FCNCs?

4.10.5.1. *Technicolor*

Technicolor, or TQCD (analogous to QCD) is a very general idea. Its simplest realizations don't work at all, but it is so simple that perhaps a variant of it is right.

Technicolor is scaled-up QCD with a subgroup of its flavor group gauged by the electroweak gauge group. Just as QCD breaks global $SU(2)_L \times SU(2)_R \to SU(2)_V$ at 300 MeV through strong dynamics, perhaps TQCD breaks gauged $SU(2)_W \times U(1)_Y \to U(1)_{EM}$ at 246 GeV through strong dynamics. The pions $\pi^{\pm}, \pi^0$ of QCD would be replaced by technipions, NGBs $\mathcal{G}^{\pm}, \mathcal{G}^0$ that form the longitudinal modes of the W, Z bosons. The wide, heavy, and almost undetectable σ resonance of QCD would become a wide, heavy Higgs boson h^0. And the dramatic $\rho^\pm$, ρ^0, ω resonances at 800 MeV would become dramatic $\rho_T^\pm$, ρ_T^0, ω_T resonances at the TeV scale. (Let's call their mass-scale M_{TC}.) Thus technicolor predicts new TeV-mass resonances and a (possibly unobservable) very heavy Higgs.

Technically, this is done by using the equations for the pions of QCD, but replacing the quarks u, d with techniquarks U, D, and replacing $SU(3)_C \times SU(2)_L \times SU(2)_R$ with $SU(N)_{TC} \times SU(2)_W \times SU(2)_R \supset U(1)_Y$, and finally gauging the $SU(2)_W \times U(1)_Y$ as the electroweak forces.

In QCD the σ, ρ, ω resonances couple strongly to $\pi\pi$, and so in TQCD the corresponding resonances couple strongly to $\mathcal{G}\mathcal{G}$ — the longitudinal W and Z bosons. They do not couple strongly to light quarks or gluons — which means they are difficult to produce! One must study scattering of longitudinal W, Z bosons using Vector Boson Fusion. Or one can perhaps look for rare processes such as $gg \to \omega_T \to WWZ$. The problem is that there are large backgrounds at the LHC from scattering of *transverse* W and Z bosons! (The Superconducting SuperCollider, canceled in the 90s, would have been a much safer bet for this process.)

In some variants of technicolor, $\bar{t}$ and perhaps $Q_3 = (t\ b)$ are also composite. In this case the techni-resonances often decay to $t\bar{t}$, adding new opportunities for discovery and study.

But the essential problem with Technicolor is that it is difficult to make this idea consistent with existing data. First, the mechanism has to somehow not mess up precision electroweak measurements. Naively, if there is new strong dynamics with a big resonance at M_{TC}, one would expect corrections to the SM of order $(100\text{ GeV}/M_{TC})^2$, and this would push M_{TC} up to something of order 3 TeV, which is uncomfortably high. (There is more to this story, known as the problem of the "S parameter"; see Peskin and Schroeder's last chapter.) Second, a way must be found to generate FHf Yukawa couplings with new interactions that somehow do not generate flavor-violating $FFFF$ or $FFff$ or $ffff$ terms that would lead to large (and unobserved) FCNCs. Naively, these four fermion operators would have coefficients of order $1/$ (few TeV)^2; but the smallness of K–$\bar{K}$ meson mixing suggests these coefficients must be more like $1/(100\text{ TeV})^2$, at least for the first two generations. B meson physics puts additional constraints. Theorists have trouble solving these problems. But technicolor involves strong dynamics, and if those dynamics are not like those of QCD, they are little understood. Perhaps there is strong dynamics that can evade these problems. In any case, technicolor cannot be rigorously excluded.

4.10.5.2. *Warped extra dimensions a la Randall-Sundrum*

Randall-Sundrum (RS) models involve the idea that the world is actually 4 + 1 dimensional (or possibly 4+1+5, where the extra five dimensions make the space 10-dimensional, convenient for superstring theory) where the 4+1 form a part of an approximately Anti-de-Sitter ("AdS") space. Through the "AdS/CFT correspondance", this 4+1-dimensional (or 4+1+5-dimensional) theory may be either exactly or roughly a "dual" (*i.e.*, quantum mechanically equivalent) description of a 3+1–dimensional gauge theory, with coordinates x^μ.

The coordinates of the new space are infinite Lorentzian coordinates x^μ and a "radial" coordinate $r > 0$. We'll consider only the case with cutoffs r_{max} and r_{min} at large and small r. You should think of the r_{min} as dual to an infrared momentum cutoff, perhaps from a confining scale (but not necessarily.) And you should think of r_{max} as dual to new ultraviolet physics, perhaps representing a scale where gravity and a string compactification become important.

When the AdS/CFT equivalence applies, everything *gauge-invariant* about the CFT must have a corresponding representation in the theory that lives on the AdS space. For instance,

- a *global* current in the CFT is dual to a bulk *gauge* field living in *AdS*;
- hadrons created by a global current in an confining CFT are dual to a tower of Kaluza-Klein (KK) modes of the bulk gauge field in *AdS*;
- a *gauged* current in the CFT corresponds to a bulk gauge field in *AdS* *plus* a 3+1-dimensional mode at $r \to \infty$ that couples to it
- adding a UV cutoff r_{max} to the last case allows the boundary mode to be replaced by a normalizable zero-mode of the gauge field in the bulk.

In typical versions of these models, then, we could put $SU(2)\times U(1)$ (at least) as gauge interactions in the bulk. This gives us W_μ^a and X_μ at the bottom of a tower of composite spin-one bosons. (This is similar to how the electromagnetic current creates the fundamental photon but also the tower of ρ mesons.) Thus we would have our W boson and (as in QCD) a ρ_T, above which is a ρ'_T, a ρ''_T, etc. Then physics at r_{min} must somehow encode the breaking of $SU(2) \times U(1)$. This can be done with a bulk Higgs-like boson which gets an r-dependent vev and breaks $SU(2) \times U(1) \to U(1)$ at small r, or through boundary conditions r_{min}.

Qualitatively, Technicolor and this type of Warped Extra Dimension model are dual to one another. But in detail they are not. Traditional Technicolor is QCD-like, and is excluded. But through AdS/CFT, an RS model is at best dual to a theory at large N and g^2N — large numbers of colors and large 't Hooft coupling — a regime which is not very QCD-like. So one might ask, is this variant of techicolor any better at solving the problems listed earlier? Unfortunately, the answer appears to be no. It largely shares the same problems as technicolor, with constraints forcing the new resonances into the multi-TeV range.

There are various attempts at solving the problems of Technicolor, These can be tried in the extra-dimensional context too. And they work more-or-less as well. But they aren't really better, either; the first resonances are still generally pushed to a few TeV, not the 1 TeV that would be more natural. So, even with the extra-dimensional version, there still really isn't an entirely satisfying model of Technicolor.

There are several other versions of RS, some not so similar to Technicolor, and arguably with fewer problems. They deserve more discussion than I can give here. Still, the first resonance often has to be at around 3 TeV.

Warped extra dimensions have some additional possible predictions to look for. First, there are spin-two resonances: Kaluza-Klein graviton resonances, from the compactification of 4+1 dimensional gravity on the finite interval for r. Such states are dual to a spin-two glueball of a Technicolor theory. In QCD glueballs fall apart very rapidly to $\pi\pi$ and are too wide to observe. We might expect this in TQCD too. But in some variants of RS models (are they realistic?) there are narrower spin-two particles, and these can be produced in gg collisions and decay often to $\gamma\gamma$. (Couplings to light fermions such as e^+e^- are often mentioned, though this requires a definite model of flavor that avoids flavor-changing neutral currents — an issue often ignored in the literature.)

Kaluza-Klein gluon resonances are also noteworthy. If there is also a tower of states above the gluon (as may well occur if the $\bar{t}$ or $(t\ b)$ are composite) then there may be a resonance easily produced in gg collisions and decaying for example to $t\bar{t}$. How would we reconstruct such a resonance? The trick is that the top quarks will be highly boosted, and can be separated from background using the fact that they look like unusual fat jets with three subjets. Techniques for reducing QCD background have been developed successfully by experimentalists and by theorists in the last few years.

4.10.5.3. *Higgsless models*

A variant of these theories is called a "Higgsless Model." The idea here is to take AdS space really seriously, and solve the unitarity problem of the SM through a tower of KK W, Z bosons, in a theory with no Higgs boson at all. This is dual, essentially, to a limit where H is a very high dimension operator and the Higgs is never produced. (Is this realistic for flavor?) In any case, it predicts a tower of KK W bosons. (But these look like a tower of techni-ρ resonances, so can this easily be distinguished from the technicolor case?) This model is formally interesting because of its unusual solution to the unitarity problem. But it is far from clear it is realistic, particularly in its ability to generate realistic fermion Yukawa couplings.

Experimentally, all of these models with towers of states unfortunately put only the lowest level (and possibly the second) within the LHC reach. It may be quite difficult to distinguish the different models from one another. The couplings of the new resonances to SM particles will be the critical quantities to measure in diagnosing the new dynamics.

4.10.6. *Little Higgs*

In Technicolor, the Higgs is like the σ of QCD: heavy, wide, and decaying to WW, ZZ. The Higgs mass and vev, and the ρ_T mass, are naturally of the same order, namely Λ_{TC} to within a factor of 3 or so.

In the Little Higgs model, the Higgs is like a lightweight Kaon of QCD — a PNGB, so it is exceptionally light — but with a potential that causes it develop a vev. This allows the scale of new dynamics Λ_{LH} to be closer to 10 TeV, where it causes no problems with electroweak precision measurements, and yet, because the Higgs is a PNGB, it can naturally be light and have a small vev.

For instance: if we take $SU(3)_L \times SU(3)_R$ and embed $SU(2) \times U(1)$ so that $SU(2)_L$ rotates $u \leftrightarrow d$, then (K^+, K^0) is a doublet under $SU(2) \times U(1)$, just like H is supposed to be.

$$\begin{bmatrix} \frac{\pi^0}{\sqrt{2}} + \frac{\eta}{\sqrt{6}} & \pi^+ & K^+ \\ \pi^- & -\frac{\pi^0}{\sqrt{2}} + \frac{\eta}{\sqrt{6}} & K^0 \\ K^- & \bar{K}^0 & -2\frac{\eta}{\sqrt{6}} \end{bmatrix}. \tag{4.80}$$

This is true in QCD. Now let's imagine an analogous dynamics where the above-mentioned particles are composites of some new force; let's label the new particles the same was as in QCD, but with a hat on top. Then all we need to do for EWSB to happen is to arrange for some additional dynamics to cause $\langle \hat{K}^0 \rangle$ to be nonzero, in particular to equal $v = 246$ GeV, and then we will break $SU(2) \times U(1) \to U(1)$, as we need. The $\hat{K}^{\pm}$ and the imaginary part of $\hat{K}^0$ become the longitudinal modes of the W and Z; the real part of $\hat{K}^0$ becomes the Higgs boson h^0. That would be neat.

Now, gauging $SU(2) \times U(1)$ explicitly breaks the $SU(3) \times SU(3)$, so the "h^0" in the $\hat{K}^0$ field becomes a PNGB. But to avoid screwing up electroweak precision measurements, we need the new dynamics to sit far above 1 TeV. This isn't good. A loop effect from the $SU(2)$ gauge boson corrects the mass of the h^0:

$$m_h^2 \sim \frac{g_2^2}{4\pi} \Lambda_{LH}^2. \tag{4.81}$$

This puts the Higgs mass less than a factor of 10 below the strong coupling scale, which is too heavy.

But there's a trick, called "Collective Symmetry Breaking," that solves this problem. To understand this phenomenon, let's simplify the discussion by first ignoring hypercharge and looking at a toy example. Imagine some kind of larger global symmetry structure G, with two copies of a global $SU(2)$ — that is $SU(2) \times SU(2)$ — embedded in G. Break G *spontaneously* so that $SU(2) \times SU(2) \to SU(2)_W$, the weak $SU(2)$ group (which we are still imagining is global, not gauged.) And imagine the theory has exact NGBs that form a complex scalar doublet $(\hat{K}^+, \hat{K}^0)$ of $SU(2)_W$ after the breaking.

Now change the theory a little bit. Break G *explicitly* as well, by turning on gauge couplings g_2, g_2' for $SU(2) \times SU(2)$. Again imagine breaking the (now gauged) group spontaneously, to the diagonal $SU(2)_W$. Now suppose G were such that if *either* $g_2 = 0$ or $g_2' = 0$, "h^0" (the real part of $\hat{K}^0$) is still an exact NGB. In other words, imagine both g_2 and g_2' are needed to "collectively" break the symetry that protects the h^0 mass. Then it must be that $m_h^2 \propto g_2 g_2'$ to some power, so that if either coupling vanishes, m_h vanishes too. But there is no one loop graph that goes like $g_2 g_2'$. *Try drawing one!* A one-loop graph either has a gauge boson from the first $SU(2)$ or from the second, and these are proportional to g_2^2 and $(g_2')^2$ respectively. So the h^0 mass can only be generated at two-loop order:

$$m_h^2 \sim \frac{(g_2 g_2')^2}{(4\pi)^2} \Lambda_{LH}^2. \tag{4.82}$$

This suppresses the Higgs mass by one order in perturbation theory. Note this is unlike SUSY, which suppresses the Higgs mass to all orders and allows us to raise the ultraviolet cutoff as far as we want. Here, the suppression by one loop factor allows us to raise the UV cutoff Λ_{LH} to 10 TeV, as opposed to the 1 TeV which is natural in technicolor.

This seems rather magical. How does it work? What suppresses the usual one-loop corrections to the Higgs mass from W, Z, and t loops? The trick is that the model, by its very construction, automatically contains W', Z', t', and H' partners for some of the SM particles, which have the *same spin* as W, Z, t, H (unlike SUSY) but naturally couple to the Higgs with coefficients equal and *opposite in sign* to the couplings of the W, Z, t and H itself. This causes all the one-loop graphs to naturally cancel, as they do in SUSY. In our toy example above, the required W' appears automatically when the gauge group was broken from $SU(2) \times SU(2)$ to $SU(2)_W$; it is a necessary consequence of the collective symmetry breaking.

As in SUSY, the W and its partner W' don't have equal masses, so their two contributions at one loop don't cancel exactly. Suppressing m_h requires the mass of the W' be less than or of order 1 TeV, and that puts it, and the other same-spin partners, within reach of the LHC (although the strong dynamics that produced the Higgs particle as a composite lies out of reach at 10 TeV.) Thus Little Higgs models generally predict that the LHC will observe W', Z', t' (perhaps b'), and H' partners with the *same* spin as their SM partners. However, the details are quite different in different models. This is too large a subject to cover here.

4.10.7. *Breakdown of quantum field theory?!*

Could the TeV scale be mark the final demise of our incredibly successful framework of quantum field theory? This doesn't seem especially likely... wouldn't we have seen signs by now?

But here's one famous example of how this could happen. If there are large extra dimensions that are flat (or at least, relatively flat, compared to the warped case) then full-fledged higher-dimensional quantum gravity could await us at the LHC. (Flat extra dimensions are *not* dual — that is, quantum mechanically equivalent — to gauge theory; they are very different, for many reasons, from the case with a single warped extra dimension.) If there are D extra dimensions with a large volume Ω, then the true Planck scale differs from the one we infer from Newton's constant and relativistic quantum mechanics by

$$M_{Planck}^2 = \Omega \times (M_{Planck}^{(true)})^{D+2}. \tag{4.83}$$

So if the volume is big, the true Planck scale is much smaller than what we infer, and could be as low as a few TeV.

Could this really be true? Skepticism is warranted. But at least we know enough about gravity to have some idea of what to expect. If string theory is right, and strings are not too strongly interacting, we expect to see resonances which correspond to the stringy excitations of the known particles. With sufficient energy, we expect black holes (at the highest energies the LHC can reach) which decay, via Hawking Radiation, into quasi-spherical, quasi-thermal distributions of particles.

Some other, more subtle, breakdown in QFT seems more likely to me. String theory has revealed some difficult-to-understand dynamics, such as non-commutative field theory, non-geometric compactifications and Little String Theory. Perhaps one of these, or one yet undiscovered, awaits us? And perhaps it resolves the hierarchy and flavor problems in a fashion that we have not yet imagined?

4.10.8. *Beyond the minimal versions of models*

You've noticed I've talked mostly about a simple version of each class of models. In the literature the tendency is to stick with minimal or near-minimal versions of theories. Why is that? Well, as theorists, we just don't care much about non-minimal models. If you have a minimal version of a model that solves the hierarchy problem, a non-minimal version is simply more complicated and makes fewer predictions. But this bias of ours is somewhat dangerous, because the impact of non-minimal models on experimentalists is much more dramatic. A small change in a Lagrangian or mass spectrum can completely change the experimental signatures of a model. If the experimentalists only prepare for minimal versions of a model, they are not truly prepared. In other words, a theoretical bias toward minimal models risks delaying discoveries by years, perhaps even a decade or more.

A little history reveals that the SM itself is/was not minimal. It has three generations, not one ("who ordered that?!") It has neutral currents ($SU(2) \times U(1)$ instead of just $SU(2)$.) And it has neutrino masses. Minimalists scoffed at these ideas; they were wrong.

And a little investigation reveals that many non-minimal models, though they fail Occham's razor, pass Strassler's machete.

So we'd better think about this. We have already seen that adding just one particle can completely change Higgs physics. What about adding a sector as rich as the SM itself?

An enormous class (by no means exhaustive) of experimentally allowed and theoretically reasonable models involve adding a hidden sector to your favorite minimal model. For example: consider Gravity-Mediated SUSY plus a sector of neutral particles, not charged under $SU(3) \times SU(2) \times U(1)$ SM interactions, but with their own interactions, and coupled to the SM in some other way, somewhere in the 1 GeV to 100 TeV scale.

Let's call the lightest superpartner in the hidden sector the LHSP (lightest hidden sparticle.) If the LSMSP is heavier than the LHSP, then (even with conserved R-parity) the LHSP is the LSP. And it follows that the LSMSP will decay (eventually) into the hidden sector!

Now, this can make a gravity-mediated model look like a gauge-mediated model, with $\tilde{\chi}^0 \to \gamma/Z/h$ plus something invisible. In particular, just as in gauge mediation, the LSMSP no longer need be neutral; it could be a stau, stop or gluino. And just as in gauge mediation, the LSMSP lifetime can be short, medium or long. In a class of models called "hidden valleys", some of the hidden sector particles produced in the decay of the LSMSP can themselves decay back to the SM. This makes the final state potentially much more complicated, and can dramatically reduce the average observed MET. In this case, interpreting the data as a SUSY model will be tied up with understanding the hidden sector, a process which clearly will be confusing for a time. And the range of possible hidden sectors – and thus final states of SUSY events – is mind-boggling.

The possible signatures of Hidden Valleys are wide-ranging, but include several novel features — all of which, though they can arise in certain other models, are usually thought of as rare, whereas here they are generic. One has to be prepared for everything from odd and unusual clusterings of jets or leptons, many new long-lived particles decaying in the detector, large numbers of low-p_T $b\bar{b}$ pairs, new light or ultra-light ("dark photon" or "dark vector boson") e^+e^- or $\mu^+\mu^-$ resonances, several short-lived particles all emerging in the decay of a long-lived LSMSP... This is by no means a complete list.

Are we, as a community, prepared for all these posssibilities? It really isn't obvious. Could we even have missed new physics at this type at the Tevatron? Have we asked the wrong questions? Is new physics on the verge of slipping by the trigger? the reconstruction software? the analysis techniques? These worries are not idle. We must work hard, collectively, to make sure that no new physics slips by the LHC.

4.11. Final Remarks

The LHC is a difficult experimental endeavor, with many unknowns and many challenges. It is the only near-term hope for breaking through the bottleneck that has paralyzed our field for over 30 years. If it is a failure... Well, we can't even think afford to think about that. Enormous efforts are necessary to ensure its success. Theorists' role is to help with understanding backgrounds and assisting with search strategies. But in this effort, theoretical bias is a concern, and we need to be on the lookout for blind spots in our thinking.

Wish us luck! Hopefully at the next TASI, or the next, we'll be talking about interpretation of LHC discoveries.

4.12. Suggested Reading

A lot of the information a student will need for LHC physics is in basic quantum field theory texts such as that of Peskin and Schroeder, and to a certain extent those of Srednicki and of Weinberg.

One should never neglect the Particle Data Book, pdg.lbl.gov, which has a huge amount of useful information summarized in it, and generally has a good bibliography for each section. The mini-reviews of various topics are very useful and a quick way to get started learning a subject. I recommend that you look through the list of chapters and read a couple that don't immediately appear interesting to you. You are likely to change your mind, or at least to close the book with a deeper appreciation of an important issue.

For many issues in QCD and in collider physics more generally, I recommend "QCD and Collider Physics" by R. Keith Ellis, W. J. Stirling and B. R. Webber, which covers many important topics.

Some issues in confinement were discussed in the early part of my Trieste 2001 lectures,
www.ictp.trieste.it/~pub_off/lectures/lns007/Strassler/Strassler.pdf

For the theory of pions, also called the "chiral Lagrangian", needed both in QCD and in similar theories that arise in Technicolor and in Little Higgs models, I recommend "Weak Interactions and Modern Particle Theory" by Howard Georgi.

You will also find the various lectures by Michelangelo Mangano on many basic aspects of LHC physics enormously valuable. You can find them currently on home.cern.ch/mlm.

For new physics models, The old Supersymmetry Primer, arxiv:hep-ph/9709356, of Steve Martin is still one of the best around. An old review of gauge mediation is Theories with Gauge-Mediated Supersymmetry Breaking, G.F. Giudice, R. Rattazzi, arXiv:hep-ph/9801271. For Technicolor and Warped Extra Dimensions, try TASI 2004 Lectures: To the Fifth Dimension and Back, by Raman Sundrum, arXiv:hep-th/0508134, as well as TASI Lectures on Electroweak Symmetry Breaking from Extra Dimensions, by Csaba Csaki, Jay Hubisz, Patrick Meade, hep-ph/0510275. Little Higgs models were discussed in 2004 TASI lectures by Martin Schmaltz, with the lovely title "Little Higgs Goes to TASI." Hidden Valleys were recently reviewed by me at a recent Blois conference, confs.obspm.fr/Blois2010/Strassler.pdf.

String Model Building, Landscape and Phenomenology

Chapter 5

TASI Lectures: Particle Physics from Perturbative and Non-perturbative Effects in D-Braneworlds

Mirjam Cvetič[1,2] and James Halverson[1]

[1]*Department of Physics and Astronomy, University of Pennsylvania, Philadelphia, PA 19104-6396, USA*

[2]*Center for Applied Mathematics and Theoretical Physics, University of Maribor, Maribor, Slovenia*

In these notes we review aspects of semi-realistic particle physics from the point of view of type II orientifold compactifications. We discuss the appearance of gauge theories on spacetime filling D-branes which wrap non-trivial cycles in the Calabi-Yau. Chiral matter can appear at their intersections, with a natural interpretation of family replication given by the topological intersection number. We discuss global consistency, including tadpole cancellation and the generalized Green-Schwarz mechanism, and also the importance of related global $U(1)$ symmetries for superpotential couplings. We review the basics of D-instantons, which can generate superpotential corrections to charged matter couplings forbidden by the global $U(1)$ symmetries and may play an important role in moduli stabilization. Finally, for the purpose of studying the landscape, we discuss certain advantages of studying quiver gauge theories which arise from type II orientifold compactifications rather than globally defined models. We utilize the type IIa geometric picture and CFT techniques to illustrate the main physical points, though sometimes we supplement the discussion from the type IIb perspective using complex algebraic geometry.

5.1. Introduction

Progress in our understanding of quantum field theory and particle physics over the last fifty years has given us a truly remarkable model in the Standard Model of particle physics. To high accuracy and precision, based on all experimental evidence to date, it appears to be the correct low energy effective field theory for all particle interactions below the weak scale. By now there is much excitement about the results which will come from the Large Hadron Collider (LHC) over the next few years. Most expect that the Higgs boson will be found, at which point all particles of the Standard Model will have been discovered.

Compared to other possibilities offered by non-abelian gauge theory, the Standard Model is rather complicated: there are three families of quarks and leptons which transform under the gauge group $SU(3)_C \times SU(2)_L \times U(1)_Y$ and 26 parameters which make up the particle masses, mixing angles, and gauge coupling constants. There are large hierarchies in the parameters, as masses of the lightest and heaviest fermions in the theory differ by over ten orders of magnitude. Furthermore, though the up-flavor quarks (for example) transform in the same way with respect to the symmetries of the quantum theory, there is a hierarchy of about five orders of magnitude between the masses of the u-quark and t-quark. These rather striking experimental facts necessitate a theoretical explanation, and the Standard Model, though very successful, does not provide one.

Any underlying theoretical framework should be able to explain the origin of the Standard Model gauge group, particle representations, and parameters. If it is to be the fundamental theory of nature, that framework should also give a sensible theory of quantum gravity. The best candidate for such a framework is superstring theory, which has been shown to naturally give rise to all of these ingredients.

Identifying particular string vacua which could give rise to our world is often difficult, though. One difficulty is that superstring theory requires ten-dimensional spacetime, and therefore six of those dimensions must be compact and of very small size to evade experimental bounds on extra dimensions. For the sake of $\mathcal{N} = 1$ supersymmetry, the standard type of manifold for compactification is a Calabi-Yau threefold, of which there are at least thirty thousand[a], and almost certainly many more. In addition to other ingredients one can choose in defining a string theory, this choice of compactification manifold gives rise to a vast number of string theories whose four-dimensional effective theory might possibly give rise to the particle physics seen in our world. These vacua

[a]This is being as conservative as possible. In [1], 473,800,776 four (complex) dimensional toric varieties were identified which have a Calabi-Yau threefold hypersurface by Batyrev's construction. These threefolds have 30,108 distinct pairs of Hodge numbers, giving the lower bound on the number of topologically distinct threefolds.

are conjectured to be local minima of a potential on the moduli space of string theory, a notion which is often referred to as the landscape [2; 3; 4].

In these lectures, we focus on a corner of the string landscape which addresses many of the fundamental questions in particle physics in an illuminating geometric fashion. We focus on type II string theory (IIa in concrete examples), where spacetime-filling stacks of D-branes wrapping non-trivial cycles in the Calabi-Yau give rise to four-dimensional gauge theories. We emphasize that all of the ingredients necessary to construct the Standard Model (or MSSM) are present in these compactifications, which are able to realize:

- Non-abelian gauge symmetry with $G = U(N)$, $SO(2N)$, or $Sp(2N)$ for each brane.
- Chiral matter at brane intersections, with natural family replication.
- Hierarchical masses and mixing angles, dependent on geometry in the Calabi-Yau.

In type IIa, all of these effects are described by geometry, and one can often employ CFT techniques for their calculation. In the presence of orientifold planes, which are needed for globally consistent supersymmetric models, these theories can realize all of the representations present in a standard Georgi-Glashow SU(5) grand unified theory (GUT). Such theories are often know as D-braneworlds or type II orientifold compactifications[b].

Despite their success in realizing gauge symmetry and chiral matter content, initial studies of D-braneworlds did not give rise to some important parameters in particle physics. In particular, in *all* weakly coupled D-braneworlds, both the Majorana neutrino mass term $M_R\, N_R\, N_R$ and the Georgi-Glashow top-quark Yukawa coupling $10\,10\,5_H$ are forbidden in string perturbation theory by global $U(1)$ symmetries. In many concrete realizations these same global symmetries also forbid other Yukawa couplings, giving rise to massless families of quarks or leptons, in direct contradiction with experiment.

In [13; 14; 15], it was shown that euclidean D-brane instantons in D-braneworlds can generate non-perturbative corrections to superpotential

[b]In this framework, the first globally consistent models with chiral matter were presented in [5; 6; 7] and the first supersymmetric globally consistent models with chiral matter were presented in [8; 9]. For comprehensive reviews and further information, see [10; 11; 12] and references therein.

couplings involving charged matter. Thus, all of the mentioned coupling issues can, in principle, be ameliorated by non-perturbative effects. The instanton corrections are exponentially suppressed, with the factor depending on the volume of the cycle wrapped by the instanton in the Calabi-Yau, and therefore can account for the large hierarchies seen in nature. Taking into account D-instantons, then, the global $U(1)$ symmetries which forbid couplings in perturbation theory are a virtue of these compactifications, rather than a drawback. For a comprehensive review of D-instantons in type II string theory, see [16] and references therein.

As there are comprehensive reviews of both generic aspects of D-braneworlds [10; 11; 12] and D-instantons [16], we intend these notes to be a short review of both topics. For the sake of brevity, we often omit in-depth derivations, choosing instead to give the reader an intuition for the geometry of these models and the corresponding particle physics. We hope that they are sufficient to prepare the reader to read either the existing literature or the in-depth reviews.

These lectures are organized as follows. In section 5.2 we give a rudimentary introduction to D-branes and explain how they give rise to gauge theories. Based on scales in the theory, the possibility of large extra dimensions is discussed. In section 5.3 we discuss important aspects of chiral matter. We begin by briefly reviewing conformal field theory techniques and then use open string vertex operators to derive the supersymmetry condition as a function of angles between branes. We introduce the notion of the orientifold projection, and discuss the appearance of particular representations in this context. In section 5.4 we discuss conditions required for global consistency of type II orientifold compactifications. We derive the conditions on homology necessary for Ramond-Ramond tadpole cancellation and present the generalized Green-Schwarz mechanism, as well as the constraints they impose on chiral matter. In section 5.5 we discuss the basics of toroidal orbifolds and present a Pati-Salam model. In section 5.6 we discuss the appearance of Yukawa couplings in string perturbation theory via CFT techniques and present examples of two important couplings forbidden in perturbative theory. In section 5.7, we present the basics of D-instantons. We discuss details of gauged axionic shift symmetries which are important for superpotential corrections. We also discuss details of charged and uncharged zero modes in terms of both CFT and sheaf cohomology, and present a concrete example of the instanton calculus. Finally, in section 5.8 we discuss the advantages and disadvantages of the "bottom-up" approach, which involves studying quiver gauge theories rather than fully defined orientifold compactifications.

5.2. Generic Background: D-branes, String Parameters, and Scales

Early attempts to understand the physics of superstring theory involved the study of the Ramond-Neveu-Schwarz (RNS) action, a (1+1)-dimensional superconformal worldsheet action with spacetime as the target space, which must be ten-dimensional to cancel the worldsheet conformal anomaly. Rather than describing particle worldlines, the RNS action describes the physics of string "worldsheets" embedded in ten-dimensional spacetime. There are only two possibilities for the topology of the superstring: either an S^1 or the interval, describing closed and open strings, respectively. The closed strings therefore have no special points, but the open strings do, and it is important to ask what boundary conditions must be imposed at their endpoints.

Very early in the history of string theory, it was realized that the open string equations of motion allow for two types of boundary conditions, known as Neumann and Dirichlet. As an open string propagates in ten-dimensional spacetime, it can have either type of boundary condition in each dimension of spacetime. An open string with $(p+1)$ Neumann dimensions and $(9-p)$ Dirichlet dimensions has boundary conditions given by

$$\begin{aligned} \mu &= 0,\ldots,p & \partial_\sigma X^\mu|_{\sigma=0,\pi} &= 0 \\ \mu &= p+1,\ldots,9 & \partial_\tau X^\mu|_{\sigma=0,\pi} &= 0, \end{aligned} \tag{5.1}$$

where σ is the worldsheet coordinate along the string and τ is the worldline proper time in the particle limit. A look at the definition of the Dirichlet condition might worry the reader, as the τ-derivative vanishing means that the ends of the string are "stuck" at particular points in spacetime, which would seem to break Poincaré invariance. For this very reason early work on open strings did not consider the possibility of Dirichlet boundary conditions.

It was realized, however, that there are issues with ignoring the possibility of Dirichlet boundary conditions. In particular, it was soon realized that type IIa and type IIb superstring theory are T-dual to one another. The simplest statement of the duality is that type IIb compactified on an S^1 of radius R gives the same physics as type IIa compactified on an S^1

of radius $\frac{\alpha'}{R}$, where $\alpha' = l_s/(2\pi)^2$ depends on the string length l_s. A basic fact about the duality is that it exchanges Neumann and Dirichlet boundary conditions of the open strings on the circular dimension, and thus if the duality is to hold, Dirichlet and Neumann boundary conditions must be on the same footing.

The key insight which solved the issue about Poincaré invariance [17] was that open strings end on objects which themselves carry energy, providing an object to which the momentum at the ends of the open string can escape. These objects, given the name D-branes based on the importance of Dirichlet boundary conditions, source Ramond-Ramond charge from the closed string sector. A Dp-brane is an object on which an open string with $(p+1)$ Neumann dimensions and $(9-p)$ Dirichlet dimensions can end. The endpoints of the open string can only move in the Neumann dimensions, and therefore the strings are confined to the Dp-brane.

A massless open string which starts and ends on the same D-brane is interpreted as a gauge boson, since string quantization shows that it transforms in the adjoint of some group G, usually $U(N)$, $SO(2N)$, or $Sp(2N)$. One interesting question is whether there is anything to be learned from the fact that a Yang-Mills theory with gauge group G is confined to some submanifold of the total spacetime, whereas the closed string (gravitational) sector propagates in the full spacetime. Indeed, ignoring small constant factors for the time being[c], the spacetime effective action S contains a gauge term for the Dp-brane and a gravitational term as

$$S \supset (\frac{M_s}{g_s})^{p-3} \int_{\mathbb{R}^{3,1}\times\pi_a} F_{ab}F^{ab} + \frac{M_s^8}{g_s^2} \int_{\mathbb{R}^{3,1}\times\mathcal{M}} \mathcal{R}_{10d}, \tag{5.2}$$

where the string mass $M_s = (\frac{1}{2\pi\alpha'})^{1/2}$ is the natural mass scale in the theory, π_a is the $(p-3)$-cycle in the Calabi-Yau manifold $\mathcal{M}$ wrapped by the Dp-brane, and g_s is the string coupling constant. Dimensionally reducing to four dimensions, we obtain

$$S \supset (\frac{M_s}{g_s})^{p-3}\, \mathcal{V}_{\pi_a} \int_{\mathbb{R}^{3,1}} F_{\mu\nu}F^{\mu\nu} + \frac{M_s^8}{g_s^2}\, \mathcal{V}_{\mathcal{M}} \int_{\mathbb{R}^{3,1}} \mathcal{R}_{4d} \tag{5.3}$$

from which we can read off $M_p^2 \sim \frac{M_s^8}{g_s^2}\mathcal{V}_{\mathcal{M}}$ and $\frac{1}{g_{\text{YM}}^2} \sim \frac{M_s^{p-3}}{g_s}\mathcal{V}_{\pi_a}$. If the geometry is factorizable such that $\mathcal{V}_{\mathcal{M}} = \mathcal{V}_{\pi_a}\mathcal{V}_t$, where $\mathcal{V}_t$ is the volume of the dimensions in $\mathcal{M}$ transverse to the Dp-brane, it immediately follows

[c]One can be more precise, of course, by expanding the DBI action to leading order in α' to obtain the standard gauge kinetic term with appropriate factors. Here, we are just interested in consequences of dimensional analysis.

that

$$M_p^2\, g_{YM}^2 \sim \frac{M_s^{11-p}}{g_s}\,\mathcal{V}_t. \tag{5.4}$$

This relation between known parameters on the left-hand side and string-theoretic parameters on the right-hand side allows for the following important observation: in the braneworld scenarios, a low string scale M_s allows for the possibility of large extra dimensions transverse to the brane, whose volume is $\mathcal{V}_t$. The possibility of large extra dimensions was explored in [18; 19], which looked in particular at the possibility of two large extra dimensions.

Being a bit more precise, the dynamics of massless open string modes in the worldvolume are given by the Dirac-Born-Infeld (DBI) action plus the Wess-Zumino (WZ) action.

$$S_{\text{eff}} = S_{\text{DBI}} + S_{\text{WZ}}.$$

Together they form the relevant worldvolume Lagrangian to leading order in the string coupling and derivatives. Looking to the subset a, b of the ten-dimensional spacetime indices M, N along the worldvolume of the D-brane, the actions are given by[d]

$$\begin{aligned} S_{\text{DBI}} &= -\mu_p \int d^{p+1}x\; e^{-\phi}\, \sqrt{-det(G_{ab}+B_{ab}+2\pi\alpha' F_{ab})} \\ S_{\text{WZ}} &= -\mu_p \int d^{p+1}x\;\; \text{tr}\, e^{2\pi\alpha'\mathcal{F}} \wedge \sqrt{\frac{\hat{A}(\mathcal{R}_T)}{\hat{A}(\mathcal{R}_N)}} \wedge \bigoplus_q C_q \end{aligned} \tag{5.5}$$

where the Ramond-Ramond charge of a p-brane is $\mu_p = 2\pi l_s^{-p-1} = (2\pi)^{-p}(\alpha')^{(-p-1)/2}$ and the fields are defined $G_{ab} \equiv \partial_a X^M \partial_b X^N g_{MN}$ and $B_{ab} = \partial_a X^M \partial_b X^N B_{MN}$. $\hat{A}(x)$ is the A-roof genus, given in terms of the Pontryagin classes p_i of a real bundle x as $\hat{A}(x) = 1 - \frac{1}{24}p_1 + \frac{1}{5760}(7p_1^2 - 4p_2) + \ldots$, and $\mathcal{R}_T$ and $\mathcal{R}_N$ are the curvature forms of the tangent bundle and normal bundle of the brane worldvolume, respectively.

The key physics to note is that the DBI action describes the coupling of the open string gauge field modes in F to the massless NSNS sector, that is, the dilaton, metric and two-form. The WZ action, on the other hand, describes the coupling of the Ramond-Ramond forms C_q which charge the D-branes to the gauge field F. As we will see in section 5.4, the WZ action is a useful tool for deriving and understanding global consistency conditions of type II orientifold compactifications.

[d] We realize we are very brief, and refer the reader to [12] for more details.

5.3. Massless Spectrum and Conformal Field Theory

Having introduced D-branes and the fact that gauge theories are confined to them in section 5.2, in this section we present details related to those gauge theories. Specifically, using conformal field theory techniques we will discuss the appearance of gauge bosons and chiral matter, as well as details related to the presence of orientifolds, whose presence often allows for more interesting four-dimensional particle physics.

A particularly natural arena in which to discuss the massless spectrum of D-braneworlds is that of conformal field theory, where quantization techniques allow us to directly identify interesting properties of massless superstrings. For brevity, we present only the details relevant for our presentation, and refer the interested reader to [20; 21; 22; 23; 24; 25] and references therein for more details on the BRST quantization of superstrings.

We consider open strings attached to the D-brane, with Dirichlet boundary conditions transverse to the D-brane and Neumann boundary conditions along the D-brane worldvolume. In the conformal field theory description, we have a concrete representation of massless states, as they are given by vertex operators. The worldsheet fermions appearing in vertex operators have two possibilities for the boundary conditions, Neveu-Schwarz and Ramond, differentiated by a sign in ψ upon going around the spatial direction of the closed superstring, with the Ramond sector carrying the minus sign.

5.3.1. *Non-abelian gauge symmetry*

In the NS sector with superconformal ghost number $\phi = 0$, the vertex operator for the gauge boson is given by

$$V_{A_\mu} = \xi_\mu \partial_z X^\mu e^{ik\cdot X}, \tag{5.6}$$

where $\mu \in \{0, 1, 2, 3\}$ and ξ_μ is the polarization vector in the target space, making this a spin 1 field. Here we employ radial quantization by mapping the Euclidean worldsheet coordinates (τ_E, σ) to complex coordinates $z, \overline{z}$ as

$$\begin{aligned} z &= e^{\tau_E + i\sigma} \\ \bar{z} &= e^{\tau_E - i\sigma}. \end{aligned} \tag{5.7}$$

Since these are open strings, there is a worldsheet boundary condition $0 \leq \sigma \leq \pi$, which is equivalent to $\operatorname{Im} z \geq 0$. Instead of considering two sets of

Virasoro generators defined on the upper-half plane $\mathrm{Im}\, z \geq 0$, we employ the doubling trick[e] in order to consider one set of Virasoro generators on the *whole* complex plane with holomorphic coordinate z.

The NS vertex operator in the (-1) ghost pictures is given by:

$$V_{A_I}(-1) = \xi_\mu e^{-\phi} \psi^\mu e^{ik\cdot X}, \tag{5.8}$$

where the conformal dimensions of the fields appearing in the vertex operator are

$$[e^{\alpha\Phi}] = -\frac{\alpha(\alpha+2)}{2} \qquad [\psi] = \frac{1}{2} \qquad [e^{i\,k\cdot X}] = \frac{\alpha' k^2}{2}. \tag{5.9}$$

Calculating the conformal dimension of (5.8), we obtain $[V_{A_I}] = \frac{1}{2} + \frac{1}{2} + \frac{\alpha' k^2}{2}$. The requirement that this be equal to one shows that this vertex operator corresponds to a massless field, in particular a massless spin-1 field confined to the worldvolume of the D-brane. It therefore has a natural interpretation as a gauge boson.

Thus far, the vertex operators discussed correspond to the massless degrees of freedom in a pure $U(1)$ gauge theory, living on the D-branes. As is well known from the theory of open strings, there is a generalization which corresponds to adding degrees of freedom at the endpoints of the strings. These degrees of freedom are allowed because they break no symmetries (conformal, Poincaré, etc) of the worldsheet theory and are known as Chan-Paton factors. Though this is a trivial generalization of the worldsheet theory, it has profound implications for spacetime physics. In particular, for $a \in \{1, \ldots, N\}$, the Chan-Paton factor Λ_a associated with one end of an open string corresponds to that end being confined to a stack of N coincident D-branes. Adding a Chan-Paton factor for each end of the massless open string associated with (5.8), the generalized vertex operator is

$$V_{A_\mu} = \xi_\mu e^{-\phi} \psi^\mu e^{ik\cdot X} (\Lambda_a \otimes \bar{\Lambda}_b) \tag{5.10}$$

with $a, b \in \{1, \cdots N\}$, Λ_a the fundamental, and $\overline{\Lambda}_b$ the antifundamental. In the absence of orientifolds, the N^2 degrees of freedom in the Chan-Paton factors transform in the adjoint of $U(N)$. Thus, a stack of N coincident D-branes has a $U(N)$ gauge theory living on its worldvolume.

[e]The conformal field theory references should contain more details, for the reader unfamiliar with the doubling trick.

5.3.2. *Orientifold projection*

In addition to D-branes, type II string theory also allows for the presence of another type of object which carries Ramond-Ramond charge. These objects have a fixed negative tension and are known as orientifold planes,

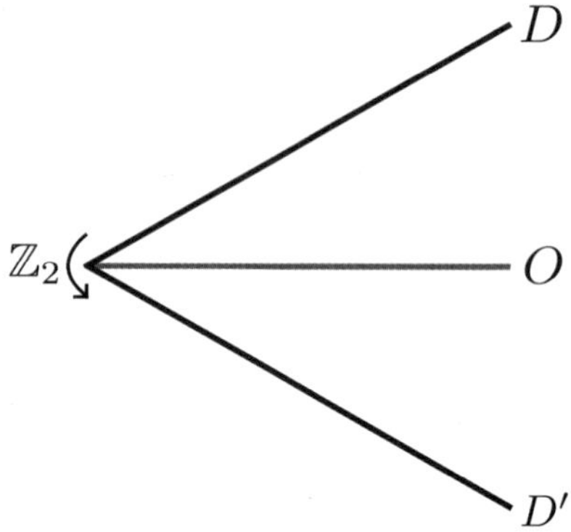

Fig. 5.1. Graphical depiction of a brane D and its orientifold image D', as well the the orientifold O, which is the fixed point locus of the specified $\mathbb{Z}_2$ action.

and it is this negative tension which allows for supersymmetric globally consistent models, as we will see in section 5.4. To specify an orientifold compactifications in type II, one must provide an orientifold projection, in addition to a Calabi-Yau manifold. The orientifold projection is a combination of three actions:

- $\Omega : \sigma \mapsto -\sigma$, the worldsheet parity operator
- $(-)^F$, an action on worldsheet fermions
- a $\mathbb{Z}_2$ involution on the Calabi-Yau.

The $\mathbb{Z}_2$ involution, which must be antiholomorphic in type IIa and holomorphic in type IIb, generically has a non-trivial fixed point locus, where the orientifold planes sit.

The involution acts on non-trivial cycles in the Calabi-Yau, and therefore it also acts on any D-brane wrapping a non-trivial cycle. Associated to any D-brane in an orientifold compactification, therefore, is an image brane, as depicted in Figure 5.1. The $\mathbb{Z}_2$ involution fixes the homology of the O-planes, which via global consistency has profound implications for particle physics. We postpone this detailed discussion until section 5.4.

In addition to implications for the homology of D-branes and O-planes, the orientifold projection imposes constraints on the physical states of the theory. If the cycle wrapped by a stack of N D-branes is not invariant under the orientifold, then there are no additional conditions on the physical state, and both that stack and its image stack give rise to $U(N)$ gauge symmetry. If the cycle is orientifold invariant, one must distinguish between two cases:

- the cycle is pointwise fixed
- the cycle is fixed, but only in homology.

In both cases the orientifold projection imposes a constraint on the Chan-Paton factors, given by $(\Lambda_a \otimes \bar{\Lambda}_b) = \pm(\Lambda_a \otimes \bar{\Lambda}_b)^T$, where the different signs are for the different cases in homology. The extra constraint changes the gauge theory on the D-brane from $U(N)$ to $SO(2N)$ or $Sp(2N)$, depending on sign, where the details are dependent upon whether one is in type IIa or type IIb. We refer the reader to section 2.2.8 of [12] for more details.

5.3.3. *Chiral matter and representations*

Now that we have introduced the very basics of orientifolds, we have the requisite background for describing generic matter content in weakly coupled type II string theory. The basic idea is that an open string could, in general, end on two different stacks of D-branes, a and b. We say that such a string is in the ab-sector, and therefore the adjoint representation of the a-stack and b-stack correspond to the aa-sector and the bb-sector, respectively. Orientifolds allow for further generality, as now each D-brane has an image brane, allowing for the extra possibilities of strings in the aa'-sector, ab'-sector, and bb'-sector.

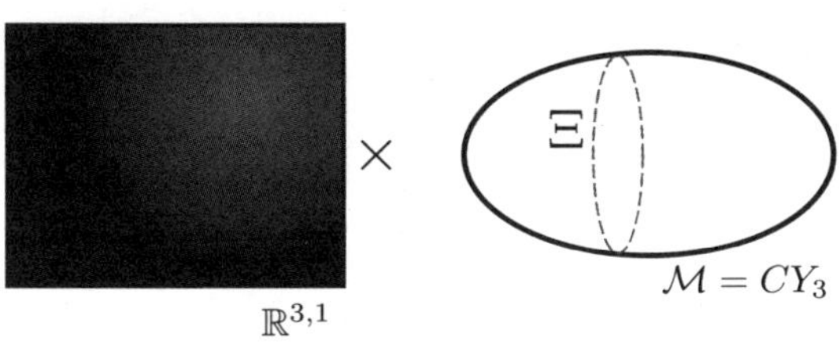

Fig. 5.2.

Fig. 5.3. A target space view of a spacetime filling gauge D-brane, which wraps a non-trivial cycle in the Calabi-Yau.

A natural question to ask is whether there is a geometric way to see the appearance of matter from the Higgsing of the adjoint representation of some higher gauge group. Considering a stack of $(N+M)$ D-branes on a generic cycle, called a, the only brane sector is the aa-sector, which corresponds to the adjoint of $U(N+M)$ and has $(N+M)^2 = N^2 + M^2 + 2NM$ degrees of freedom. Performing a geometric deformation where one unfolds the stack of $(N+M)$ D-branes into stacks of N and M D-branes, called b and c, the degrees of freedom from the aa-sector now lie in the bb, cc, or the bc sector. These sectors respectively give rise to the adjoints of $U(N)$ and $U(M)$, as well as bifundamentals $(N, \overline{M})$ and $(\overline{N}, M)$. This is a concrete geometric interpretation in string theory for how the adjoint of $U(N+M)$ breaks to the adjoint of $U(N)$, $U(M)$ and bifundamentals. In this simple example, the matter fields in the bifundamental representations live at the intersection of two branes.

Having motivated the notion that matter lives at the intersection of different stacks of D-branes, we focus on intersecting D6-branes in type IIa for a concrete discussion of the quantization of an open string between two stacks of intersecting branes. Calling the three-cycles wrapped by two stacks of D6-branes $\pi_a, \pi_b \in H_3(\mathcal{M}, \mathbb{Z})$, we write

$$\begin{aligned} \pi_a &= N_{aI} A^I + M_a^J B_J \\ \pi_b &= N_{bI} A^I + M_b^J B_J, \end{aligned} \tag{5.11}$$

where we have used the symplectic basis of three-cycles A^I and B_I. They satisfy

$$A^I \circ B_J = \delta^I_J, \qquad \int_{A^I} \alpha_J = \delta_{IJ}, \qquad \int_{B_I} \beta_J = -\delta^{IJ} \tag{5.12}$$

where α_J and β^J are the dual basis of three-forms on the Calabi-Yau. From this it is straightforward to calculate the topological intersection number

$$\pi_a \circ \pi_b = N_{aI} M_b^I - M_a^I N_{bI}. \tag{5.13}$$

This already has interesting implications for physics: if, for example, $\pi_a \circ \pi_b = 3$, then these D-branes intersect at three points in the Calabi-Yau, giving three copies of whatever quantized matter lives at a single intersection. This is quite naturally interpreted as family replication.

The quantization of open strings living at the intersection of two D-branes[f] utilizes the standard open string quantization, together with the

[f]We recommend appendix A of [26] for all the details.

non-trivial boundary conditions

$$\begin{aligned}\partial_\sigma X^{2I-1} = X^{2I} &= 0 \quad \text{at} \quad \sigma = 0 \\ \partial_\sigma X^{2I-1} + (\tan \pi\theta_I)\partial_\sigma X^{2I} &= 0 \quad \text{at} \quad \sigma = \pi \\ X^{2I} - (\tan \pi\theta_I) X^{2I-1} &= 0 \quad \text{at} \quad \sigma = \pi\end{aligned}$$

where $I = \in \{2,3,4\}$[g]. Going to complexified notation for the worldsheet bosons and quantizing, with boundary conditions taken into account, the expansion in terms of oscillators is given by

$$Z^I = X^{2I-1} + iX^{2I} = \sum_{n\epsilon\mathbb{Z}} \frac{\alpha^I_{n-\theta_I}}{n-\theta_I} z^{-n+\theta_I} + \sum_{n\epsilon\mathbb{Z}} \frac{\tilde{\alpha}^I_{n+\theta_I}}{n+\theta_I} \bar{z}^{-n-\theta_I}, \qquad (5.14)$$

and we note that since $\tilde{\alpha}^{I\dagger}_{n+\theta_I} = \alpha^I_{n+\theta_I}$, $Z \mapsto \overline{Z}$ under $\theta_I \mapsto -\theta_I$. The only non-vanishing commutator is $[\alpha^I_{n\pm\theta}, \alpha^{I'}_{m\mp\theta}] = \pm m\, \delta_{n+m} \delta^{II'}$. We can also complexify worldsheet fermions, giving

$$\begin{aligned}\text{Neveu-Schwarz Sector:} \qquad \Psi^I &= \psi^{2I-1} + i\psi^{2I} = \sum_{r\in\mathbb{Z}+\frac{1}{2}} \psi_{r-\theta_I} z^{-r-\frac{1}{2}+\theta_I} \\ \text{Ramond Sector:} \qquad \Psi^I &= \psi^{2I-1} + i\psi^{2I} = \sum_{r\in\mathbb{Z}} \psi_{r-\theta_I} z^{-r-\frac{1}{2}+\theta_I},\end{aligned} \qquad (5.15)$$

where the only non-vanishing anticommutator is $\{\psi^I_{m-\theta_I}, \psi^{I'}_{n+\theta_I}\} = -\delta_{mn}\, \delta^{I,I'}$. We refer the reader to [26] for more details on oscillator quantization for D-branes at intersecting angles, including zero point energies and mass formulae.

Instead, we show the equivalent physics using the vertex operator formalism of CFT. In the vertex operator formalism, it is the presence of bosonic twist fields σ_θ [23] that ensure the boundary conditions (5.14) for D6-branes intersecting at non-trivial angles θ_I. As one might expect, we will have vertex operators for both fermions and bosons living at the intersections of two branes. From studying their conformal dimensions, we will extract mass formulae and show that, though the fermion is always massless, the mass of the boson depends on the angles of intersection.

Recall that in quantizing the superstring, one often chooses to bosonize the worldsheet fermions Ψ^M with $M \in \{0, \cdots, 9\}$, rather than working

[g]Here we depart from conventions elsewhere in the literature, which often use θ_1, θ_2, and θ_3 for the angles in each of the three complex internal dimensions. We use the labeling $0, \ldots, 9$ for real dimensions and $0, \ldots, 4$ for complexified dimensions, thus $2, 3, 4$ for the angles between branes on the internal space.

with the fermions directly. Usually each of the five complexified worldsheet fermions is bosonized as

$$\begin{aligned} \text{Neveu-Schwarz Sector:} \qquad & \Psi^M \cong e^{iH_M} \\ \text{Ramond Sector:} \qquad & \Psi^M \cong e^{i(1\pm\frac{1}{2})H_M}, \end{aligned} \tag{5.16}$$

where the half integer in the last line is present in order to take care of the Ramond boundary conditions on the worldsheet. The $\pm$ ambiguity corresponds to each complexified worldsheet fermion having spin $\pm\frac{1}{2}$, and the 2^5 sign choices reflect that fact that the Ramond sector ground state is a 32-dimensional Dirac spinor in ten dimensions.

The key difference between a standard open superstring and an open superstring at the intersection of two D-branes is the boundary conditions, which must be taken into account. As they only apply in the internal dimensions, they only change three of the complexified worldsheet fermions, which become

$$\begin{aligned} \text{Neveu-Schwarz Sector:} \qquad & \Psi^I \cong e^{i\theta_I H_I} \\ \text{Ramond Sector:} \qquad & \Psi^I \cong e^{i(\theta_I\pm\frac{1}{2})H_I}, \end{aligned} \tag{5.17}$$

with $I \in \{2,3,4\}$, where the sign in the last line depends crucially on how the angles are defined. Here we choose the conventions $0 < \theta_I < 1$ for $I = 2,3$ and $-1 < \theta_4 \leq 0$. The two complexified worldsheet fermions which are not subject to boundary conditions form a two-component Weyl spinor in four dimensions, which we write as S^α in the vertex operators.

Having discussed the relevant ingredients, we would like to explicitly write two vertex operators, in the NS-sector and R-sector, for open strings stretched between spacetime filling D6-branes with non-trivial intersection in the Calabi-Yau. Omitting Chan-Paton factors, since they aren't immediately relevant to the discussion, the vertex operators are

$$\begin{aligned} V_{-1} &= e^{-\phi} \prod_{I=2}^{3} \sigma_{\theta_I} e^{i\theta_I H_I} \sigma_{1+\theta_4} e^{i(1+\theta_4)H_4} e^{ik\cdot X} \\ V_{-\frac{1}{2}} &= u_\alpha e^{-\frac{\phi}{2}} S^\alpha \prod_{I=2}^{3} \sigma_{\theta_I} e^{i(\theta_I-\frac{1}{2})H_I} \sigma_{1+\theta_4} e^{i(\frac{1}{2}+\theta_4)H_4} e^{ik\cdot X}, \end{aligned} \tag{5.18}$$

respectively. To calculate the mass of these states, one must know the conformal weights of the fields appearing in the vertex operators, which are given by

$$[e^{\alpha\phi}] = -\frac{\alpha(\alpha+2)}{2} \qquad [\sigma_\theta] = \frac{\theta(1-\theta)}{2} \qquad [e^{iaH_I}] = \frac{a^2}{2}$$

$$[e^{ik\cdot X}] = \frac{\alpha' k^2}{2} \qquad [S^\alpha] = [e^{i(\pm\frac{1}{2}H_1 \pm \frac{1}{2}H_2)}] = 2\frac{(\pm\frac{1}{2})^2}{2} = \frac{1}{4}. \quad (5.19)$$

Knowing these, it is straightforward to calculate the conformal weights of the vertex operators (5.18) to be

$$\begin{aligned}[V_{-1}] &= \frac{1}{2} + \sum_{I=2}^{3}\left(\frac{\theta_I(1-\theta_I)}{2} + \frac{1}{2}\theta_I^2\right) + \frac{(\theta_4+1)(-\theta_4)}{2} + \frac{(\theta_4+1)^2}{2} \\ &= \frac{1}{2} + \frac{1}{2}\sum_{I=1}^{3}\theta_I + \frac{1}{2} + \frac{\alpha' k^2}{2} \end{aligned} \quad (5.20)$$

$$\left[V_{-\frac{1}{2}}\right] = \frac{3}{8} + \frac{1}{4} + 3\frac{1}{8} + \frac{\alpha' k^2}{2}. \quad (5.21)$$

The mass formulae [23; 20; 21; 27; 28] are derived from the requirement that the conformal dimension be one, yielding

$$\alpha' m^2 = \sum_{I=1}^{3}\theta_I \qquad \text{and} \qquad \alpha' m^2 = 0, \quad (5.22)$$

for the spacetime bosons and fermions, respectively. Therefore, if the sum of the three angles is negative, zero, or positive then the boson is tachyonic, massless, and massive, respectively. In the case where the boson is massless, the NS-sector boson V_{-1} becomes massless and forms a supermultiplet with the R-sector fermion $V_{-\frac{1}{2}}$. Thus, $\sum_I \theta_I = 0$ is the local condition for intersecting branes to give rise to supersymmetric matter.

The angle condition is a local picture of mutually supersymmetric branes. The global picture is that for a D-brane to give rise to a supersymmetric gauge theory, it must wrap a supersymmetric cycle, given by a special Lagrangian in type IIa or a holomorphic divisor in type IIb. For the effective four-dimensional theory to be supersymmetric, the branes must preserve the *same* supersymmetry. It has been shown that the global conditions for two D6-branes on special Lagrangians to preserve the same supersymmetry reduces locally to the condition on angles.

The vertex operators (5.18) also generically include Chan-Paton factors, which might satisfy further constraints due to the orientifold projection.

Table 5.1. Representations and multiplicities for chiral matter at the intersection of two D6-branes.

Representation	Multiplicity
$\begin{smallmatrix}\Box\\ \Box\end{smallmatrix}_a$	$\frac{1}{2}\left(\pi_a \circ \pi'_a + \pi_a \circ \pi_{O6}\right)$
$\Box\Box_a$	$\frac{1}{2}\left(\pi_a \circ \pi'_a - \pi_a \circ \pi_{O6}\right)$
$(\Box_a, \overline{\Box}_b)$	$\pi_a \circ \pi_b$
$(\Box_a, \Box_b)$	$\pi_a \circ \pi'_b$

The generic case we would like to discuss is the structure of Chan-Paton factors at the intersection of two gauge D-branes, in which case the factors are a tensor product of some combination of fundamentals and antifundamentals. That is, for gauge branes with $U(N_a)$ and $U(N_b)$ gauge symmetry, the possibilities are $(\Box_a, \Box_b)$, $(\Box_a, \overline{\Box}_b)$, $(\overline{\Box}_a, \Box_b)$ and $(\overline{\Box}_a, \overline{\Box}_b)$, where the choice between fundamental and antifundamental depends on the direction of the string and whether or not the string ends on a brane or its orientifold image. Thus, the most common possibility is that chiral matter appearing at the intersection of two D-branes is in the bifundamental representation. The possible representations and multiplicities of chiral matter are listed in Table 5.1.

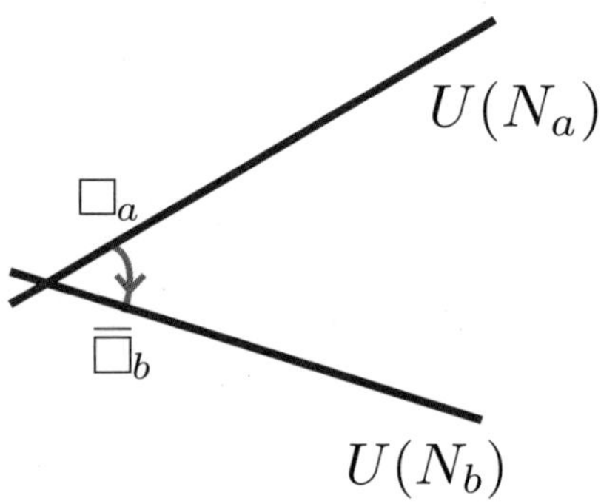

Fig. 5.4. A bifundamental string beginning on a $U(N_a)$ brane and ending on a $U(N_b)$ brane, with appropriate Chan-Paton factors at the endpoints.

There are two special cases which are interesting to discuss, one of which arises in the table. First, we revisit the possibility that a string begins and ends on the same $U(N_a)$ brane. In such a case, the Chan-Paton factors take the form $\Box_a \otimes \overline{\Box}_a = \mathrm{Adj}_a \oplus 1$. That is, due to the decomposition into a direct sum, the string beginning and ending on the same brane can transform in the adjoint representation, and is therefore a gauge boson. Second, one might wonder about the properties of a string beginning on a brane and ending on its orientifold image. In such a case the Chan-Paton factors take the form $\square_a \otimes \square_a = \square\square_a \oplus \begin{smallmatrix}\square\\\square\end{smallmatrix}_a$, and therefore chiral matter can be in the symmetric or antisymmetric representation of the gauge group. This is of great importance, for example, in $SU(5)$ GUT models. There the 10 representation is needed, which can be realized as $10 = \begin{smallmatrix}\square\\\square\end{smallmatrix}_5$ in type II D-braneworlds.

Thus, we see that D-braneworlds can give rise to all of the ingredients necessary for realizing the gauge symmetry and matter content of the Standard Model. In particular, a stack of multiple D-branes give rise to non-abelian gauge symmetry, with the possibility of chiral matter living at the intersection of two D-branes. We have shown the presence of chiral supermultiplets locally at the intersection of two D-branes. Upon taking into account global aspects, namely the need for compactification, the fact that branes can intersect multiple times in the internal space gives a geometric reason for family replication.

5.4. Global Consistency

Since string theory gives rise to low energy gauge theories in a variety of ways, it has been important throughout its history to address whether or not string theory gives rise to *consistent* gauge theories. For instance, the first superstring revolution was sparked in [29] when Green and Schwarz showed that the type I string in ten dimensions with $SO(32)$ gauge symmetry is anomaly free. Over time, anomaly cancellation has been shown to arise naturally in many corners of the landscape. In all corners, the conclusion thus far has been the same: the natural ingredients arising in a string theory ensure the consistency of the low energy effective theory.

In this section, we will present known results for how this occurs in type II orientifold compactifications. We begin with tadpole cancellation, which amounts to conditions on the homology of spacetime filling D-branes and O-planes that ensures the necessary cancellation of Ramond-Ramond charge on the internal space. These global conditions on homology impose constraints on chiral matter which are *necessary* for tadpole cancellation. We will show that a subset of these constraints on chiral matter are precisely the conditions for the cancellation of non-abelian anomalies. We will also show that the presence of Chern-Simons couplings of Ramond-Ramond forms to $U(1)$ field strengths gives rise to a generalized Green-Schwarz mechanism which cancels abelian and mixed anomalies. Couplings of this type also generically give a Stuckelberg mass to the corresponding $U(1)$ gauge bosons, whose corresponding global symmetries impose phenomenologically important selection rules on superpotential couplings.

5.4.1. *Ramond-Ramond tadpole cancellation*

Historically, by studying amplitudes arising in CFT descriptions, it was shown that certain one-loop cylinder, Mobius strip, and Klein-bottle diagrams have infrared divergences due to the presence of massless Ramond-Ramond tadpoles, which are required to cancel for consistency of the theory. With the advent of D-branes, a geometric picture of tadpole cancellation became clear in the works of [30; 31; 32]. Following those works and working in type IIa, we examine the RR seven-form kinetic term of the ten-dimensional supergravity Lagrangian, along with relevant Wess-Zumino terms (5.5) of the D-brane effective action

$$S \supset -\frac{1}{4\kappa^2}\int_{\mathbb{R}^{3,1}\times\mathcal{M}} H_8\wedge *H_8 \;+\; \mu_6\sum_a N_a\int_{\mathbb{R}^{3,1}\times\pi_a} C_7$$
$$+\;\mu_6\sum_a N_a\int_{\mathbb{R}^{3,1}\times\pi'_a} C_7 \;-\; 4\mu_6\sum_a N_a\int_{\mathbb{R}^{3,1}\times\pi_{O6}} C_7, \tag{5.23}$$

where $H_8 = dC_7$ is the field strength of the Ramond-Ramond seven-form which couples to D6-branes and O6-planes, π_a is the three-cycle wrapped by a D6-brane and π'_a is wrapped by its orientifold image, and π_{O6} is the three-cycle wrapped by the O6-plane. The ten-dimensional gravitation coupling is $\kappa^2 = \frac{1}{2}(2\pi)^7(\alpha')^4$. Given this action and the Poincaré dual $\delta(\pi_a)$ of π_a, the equation of motion is

$$\frac{1}{\kappa^2}\,d(*H_8) = \mu_6\sum_a N_a\left(\delta(\pi_a)+\delta(\pi_{a'})\right) \;-\; 4\mu_6\,\delta(\pi_{O6}) \tag{5.24}$$

and we see from the left-hand side that the right-hand side is an exact form, and is therefore trivial in $H^3(\mathcal{M},\mathbb{Z})$. But Poincaré duality is an *isomorphism* between cohomology and homology, and therefore the Poincaré dual of the right hand side must be trivial in homology, yielding

$$\sum_a N_a\left([\pi_a]+[\pi_{a'}]\right) = 4\,[\pi_{O6}], \tag{5.25}$$

where $[\pi_a]$ is the homology class of the three-cycle π_a.

This is the D6-brane tadpole cancellation condition in type IIa orientifold compactifications. It is a condition on the homology of the cycles which the D6-branes and O6-planes wrap. Qualitatively, satisfying this condition ensures that the Ramond-Ramond charge is canceled on the internal space, which is necessary since the spacetime filling D6-branes and O6-planes source Ramond-Ramond charge and the directions transverse to them are a submanifold of the compact Calabi-Yau $\mathcal{M}$. That is, the condition must be satisfied, as otherwise the flux lines would have nowhere to go in a compact manifold.

The condition on homology is the necessary and sufficient for cancellation of homological RR tadpoles, but the condition on homology induces constraints on chiral matter which are necessary for tadpole cancellation. These are interesting in their own right. Using Table 5.1 and intersecting (5.25)[h] with another three-cycle π_a wrapped by a D6-brane (in the case where orientifolds are absent), we obtain

$$0 = \pi_a \circ \sum_b N_b \ \pi_b = \sum_b N_b \ I_{ab} = \sum_b N_b(\#(a,\bar{b}) - \#(\bar{a},b)). \tag{5.26}$$

This can be rewritten as $\#a = \#\bar{a}$, which is precisely the condition for non-abelian anomaly cancellation if $N_a > 2$. Generalizing to the case with orientifolds and image branes, the full condition is

$$N_a \geq 2: \qquad \#a - \#\bar{a} + (N_a + 4)\ \#\square\!\square_a + (N_a - 4)\ \#\begin{smallmatrix}\square\\\square\end{smallmatrix}_a = 0$$

$$N_a = 1: \qquad \#a - \#\bar{a} + (N_a + 4)\ \#\square\!\square_a = 0 \mod 3, \tag{5.27}$$

where the mod 3 condition for the $N_a = 1$ case comes from the fact that there is no antisymmetric representation of a $U(1)$. For more details of this derivation, we refer the reader to [33].

Thus, we see that type II string theory provides a beautiful geometric picture for the existence of anomaly cancellation: the Ramond-Ramond charge of spacetime filling branes must be canceled on the internal space, which yields a condition on homology[i] that induces necessary constraints on chiral matter. These constraints on chiral matter happen to include the cancellation of non-abelian anomalies, but also include some genuinely stringy constraints.

---

[h]With indices switched. That is, the sum is over b, rather than a.

[i]It is important to note that, in addition to the necessary cancellation of homological Ramond-Ramond charge, one must also cancel K-theory charges, due to the fact that D-branes are classified by K-theory groups [34] rather than homology groups. We refer the interested reader to [10; 11; 12] for more details.

5.4.2. *Generalized Green-Schwarz mechanism*

While the constraints on chiral matter corresponding to the cancellation of non-abelian anomalies are satisfied immediately if the homological tadpole cancellation condition is satisfied, tadpole cancellation does not cancel the abelian, mixed abelian-non-abelian, and mixed abelian-gravitational anomalies. In [30] it was shown that there is a generalization of the Green-Schwarz mechanism [29] to the case of intersecting branes. The mechanism cancels the anomalies [30; 31] by the gauging of axionic shift symmetries associated with the Ramond-Ramond forms.

We now address some of the details. Again for concreteness we work with the type IIa supergravity action. Expanding the exponential of the field strengths of the gauge fields in the Wess-Zumino action (5.5), each stack of $D6$-branes, indexed by a, has Chern-Simons couplings of the form

$$\int_{\mathbb{R}^{3,1}\times\pi_a} C_3\wedge \mathrm{Tr}(F_a\wedge F_a), \qquad \int_{\mathbb{R}^{3,1}\times\pi_a} C_5\wedge \mathrm{Tr}(F_a), \tag{5.28}$$

where F_a is the gauge field strength on the brane. As we are concerned with mixed anomaly cancellation in the effective four-dimensional gauge theory, we expand the Ramond-Ramond forms in basis (β^I,α_J) Poincaré dual to the integral basis of three-cycles (A^I,B_J) defined in section 5.3.3 as

$$C_3=\Upsilon^I\,\alpha_I+\tilde{\Upsilon}_I\,\beta^I \qquad \text{and} \qquad C_5=\tilde{\Delta}^I\,\alpha_I+\Delta_I\,\beta^I \tag{5.29}$$

where the coefficients of α and β are the four-dimensional axions and two-forms

$$\Upsilon^I=\int_{A^I} C_3 \qquad \tilde{\Upsilon}_I=-\int_{B_I} C_3$$
$$\tilde{\Delta}^I=\int_{A_I} C_5 \qquad \Delta_I=-\int_{B^I} C_5. \tag{5.30}$$

Upon dimensional reduction of $\mathcal{N}_a$ D6-branes on $\pi_a=Na_I\,A^I+M_a^I\,B_I$, the generic Chern-Simons couplings (5.28) can be written in terms of axionic couplings of the form

$$\mathcal{N}_a\int_{\mathbb{R}^{3,1}}(N_{aI}\Upsilon^I-M_a^I\tilde{\Upsilon}_I)\wedge\mathrm{Tr}(F_a\wedge F_a), \quad \mathcal{N}_a\int_{\mathbb{R}^{3,1}}(N_{aI}\tilde{\Delta}^I-M_a^I\Delta_I)\wedge\mathrm{Tr}(F_a). \tag{5.31}$$

The axions and two-forms come in four-dimensional Hodge dual pairs $(d\Upsilon^I,-d\Delta_I)$ and $(d\tilde{\Upsilon}_I,d\tilde{\Delta}^I)$, which can be derive from the ten-dimensional Hodge duality $dC_3=-\star_{10}dC_5$. One can show that the axions transform as

$$\Upsilon^I \mapsto \Upsilon^I + \mathcal{N}_a\, M_a^I \Lambda_a \qquad \text{and} \qquad \tilde{\Upsilon}_I \mapsto \tilde{\Upsilon}_I + \mathcal{N}_a\, N_{aI} \Lambda_a \tag{5.32}$$

under $U(1)_a$, which clearly leaves the four-dimensional couplings (5.31) invariant. It was shown in [31] that this gauging of the axionic shift symmetry precisely cancels the abelian and mixed anomalies.

Another important fact is that the couplings of the form $\int \Delta \wedge \mathrm{Tr}(F_a)$ and $\int \tilde{\Delta} \wedge \mathrm{Tr}(F_a)$ generically give rise to a Stuckelberg mass term for the $U(1)_a$ gauge bosons. Since we are dealing with orientifold compactifications, there are also image branes on π'_a with field strength $-F_a$, so that the relevant couplings take the form

$$\mathcal{N}_a \int_{\mathbb{R}^{3,1}} (N_{aI} - N'_{aI})\tilde{\Delta}^I \wedge \mathrm{Tr}(F_a) \;-\; \mathcal{N}_a \int_{\mathbb{R}^{3,1}} (M_a^I - M'^I_a)\Delta_I \wedge \mathrm{Tr}(F_a). \tag{5.33}$$

Though the $U(1)_a$ gauge bosons receive a mass, no symmetries of the action are broken, and so the gauge symmetry selection rules associated with the $U(1)_a$ gauge symmetry survive in the low energy effective action as global selection rules. These are precisely the global $U(1)$ symmetries which forbid superpotential terms in string perturbation theory.

However, as is fortunate for phenomenological purposes, it is often the case that some linear combination of the $U(1)_a$ gauge symmetries remains massless. As a condition on homology, this means that a linear combination $\sum_x q_x\, U(1)_x$ is massless when

$$\sum_x \mathcal{N}_x\, q_x\, (\pi_x - \pi'_x) = 0. \tag{5.34}$$

As with the case of the condition on homology for tadpole cancellation, this can be intersected with a cycle π_a wrapped by a D6-brane to give constraints on the allowed forms of chiral matter. Using Table 5.1, these constraints are given by

$$-q_a \mathcal{N}_a\, (\#(\square\!\square_a) + \#(\begin{smallmatrix}\square\\\square\end{smallmatrix}_a)) + \sum_{x \neq a} q_x \mathcal{N}_x\, (\#(a,\overline{x}) - \#(a,x)) = 0, \tag{5.35}$$

which becomes

$$-q_a\, \frac{\#(a) - \#(\overline{a}) + 8\#(\square\!\square_a)}{3} + \sum_{x \neq a} q_x \mathcal{N}_x\, (\#(a,\overline{x}) - \#(a,x)) = 0, \tag{5.36}$$

for the special case $\mathcal{N}_a = 1$. Any linear combination of $U(1)$'s which satisfies these conditions is an anomaly-free $U(1)$ with a massless gauge boson. When trying to realize the standard model in D-braneworlds, there must be such a $U(1)$ which allows for an interpretation as hypercharge. The particular linear combination corresponding to hypercharge, sometimes called a "hypercharge embedding", has important implications for the realization of MSSM matter fields, and thus also for the structure of couplings.

5.5. Concrete Model-Building Example: Toroidal Orbifolds

The beautiful geometric picture of particle physics offered by type IIa intersecting braneworlds has encouraged much work in model building. Many of these examples are compactified on various toroidal orbifolds [j], which offer two distinct advantages over more generic Calabi-Yau backgrounds. In particular, the homology cycles on a toroidal orbifold are particularly easy to visualize, which makes model building a bit more intuitive. Furthermore, toroidal orbifolds offer a CFT description, and therefore all of the power of vertex operator formalism can be brought to bear.

Generically, the toroidal orbifold is a six-torus modded out by a discrete group Γ, so that $\mathcal{M} = T^6/\Gamma$. We think of a factorizable six-torus $T^6 = T^2 \times T^2 \times T^2$. One might wonder about the simplest possibility, where Γ is trivial and therefore we simply have the type IIa string compactified on an orientifold of T^6. Unfortunately, due to simple considerations from the supersymmetry condition, these models cannot realize the MSSM. In the literature, therefore, Γ is non-trivial and is usually of the form $\mathbb{Z}_N$ or $\mathbb{Z}_N \times \mathbb{Z}_M$.

Before discussing the effects of Γ-action on T^6, it is necessary to mention another detail or two about the orientifold. Introducing complex coordinates $z^i = x^i + i\,y^i$ on each of the T^2's, the anti-holomorphic involution acts as $\overline{\sigma} : z^i \mapsto \overline{z}^i$. On each T^2 there are exactly two different choices for the complex structure which are consistent with the involution. They correspond to the torus and the tilted torus. The twofold basis of one-cycles for the torus and tilted torus are $([a^i],[b^i])$ and $([a'^i],[b^i])$, where $[a'^i] = [a^i] + \frac{1}{2}[b^i]$. Since the six-torus is factorizable, the three-cycles can be written as a product of three one-cycles as

$$\pi_a = \prod_{i=1}^{3}(n_a^i[a^i] + \tilde{m}_a^i[b^i]), \tag{5.37}$$

---

[j] For recent work on global toroidal orbifold models and other recent work, see [35] and references therein. For systematic work on the landscape of string vacua for particular toroidal orbifolds, see [36].

where $\tilde{m}_a^i = m_a^i$ for untitled tori and $\tilde{m}_a^i = m_a^i + \frac{1}{2}n_a^i$ for tilted tori. Using the fact that the only non-vanishing intersection of one-cycles is $[a^i] \circ [b^i] = -1$, it is straightforward to calculate

$$I_{ab} = \prod_{i=1}^{3}(n_a^i \tilde{m}_b^i - \tilde{m}_a^i n_b^i) = \prod_{i=1}^{3}(n_a^i m_b^i - m_a^i n_b^i). \tag{5.38}$$

One should make careful note that the intersection number *does not* depend on the choice of tilted or untilted tori. This makes sense, of course, because topological quantities such as I_{ab} must not depend on metric-related issues, such as complex structure moduli.

Since we have specified a manifold on which to compactify, it is useful to recast the generic tadpole cancellation conditions (5.25) in terms of the wrapping numbers (n, m). Independent of the tilt on each T^2, the O6-plane is wrapping the cycle $2[a^i]$, so that the entire three-cycle reads $\pi_{O6} = 8[a^1][a^2][a^3]$. The action of $\overline{\sigma}$ on a generic cycle is simply $(n^i, \tilde{m}^i) \mapsto (n^i, -\tilde{m}^i)$. Parameterizing the cycles in terms of wrapping numbers as in (5.37), the RR tadpole cancellation conditions become

$$\begin{aligned} &[a^1][a^2][a^3]: \qquad \sum_{a=1}^{K} N_a \prod_i n_a^i = 16 \\ &[a^i][b^j][b^k]: \qquad \sum_{a=1}^{K} N_a n_a^i \tilde{m}_a^j \tilde{m}_a^k = 0 \qquad \text{with} \qquad i \neq j \neq k. \end{aligned} \tag{5.39}$$

One might wonder why there are no equations for the three-cycle basis components of the form $[b][b][b]$ or $[a][a][b]$. This is because $\tilde{m}^i \mapsto -\tilde{m}^i$ under $\overline{\sigma}$ kills any contribution to a component with an odd number of b's.

As type IIa compactified on an orientifold of T^6 cannot realize the MSSM, it is important to examine the possibility of non-trivial Γ. Here we consider a well-studied choice for the orbifold group, where $\Gamma = \mathbb{Z}_2 \times \mathbb{Z}_2$. The generators of the $\mathbb{Z}_2 \times \mathbb{Z}_2$ are given by ω and θ, defined to be

$$\omega : (z^1, z^2, z^3) \mapsto (-z^1, -z^2, z^3) \qquad \theta : (z^1, z^2, z^3) \mapsto (z^1, -z^2, -z^3). \tag{5.40}$$

Since Γ also acts on the homology cycles of T^6, the simplification of the tadpole conditions (5.25) in terms of wrapping numbers must take this into account. In fact, this can be done for any Γ. We refer the interested readers to the reviews [10; 11; 12] for the derivation and expressions of the orbifold tadpole conditions.

As an example, we present the wrapping numbers for a globally consistent model of [37] in Table 5.2. This model is a type IIa orientifold

Table 5.2. Toroidal orbifold wrapping numbers for a Pati-Salam model with $U(4)_a \times U(2)_b \times U(2)_c \times U(1)_1 \times U(1)_2 \times U(1)_3 \times U(1)_4$ gauge symmetry. N is twice what one might expect, due to the orbifold action.

Stack	N	$(n^1, m^1) \times (n^2, m^2) \times (n^3, m^3)$
a	8	$(0, -1) \times (1, 1) \times (1, 1)$
b	4	$(3, 1) \times (1, 0) \times (1, -1)$
c	4	$(3, -1) \times (0, 1) \times (1, -1)$
1	2	$(1, 0) \times (1, 0) \times (2, 0)$
2	2	$(1, 0) \times (0, -1) \times (0, 2)$
3	2	$(0, -1) \times (1, 0) \times (0, 2)$
4	2	$(0, -1) \times (0, 1) \times (2, 0)$

on $T^6/(\mathbb{Z}_2 \times \mathbb{Z}_2)$ with intersecting D6-branes and Pati-Salam $SU(4)_C \times SU(2)_L \times SU(2)_R$ gauge symmetry after the Green-Schwarz mechanism has given masses to $U(1)$ gauge bosons. From the point of view of bifundamental matter under the Pati-Salam group, the chiral spectrum is particularly nice, as it contains three families of $(4, \overline{2}, 1)$ and $(\overline{4}, 1, 2)$. However, in addition to the Pati-Salam gauge symmetry, which arises from stacks a, b, and c, "filler" branes labeled with integers are needed to satisfy the tadpole conditions. This gives rise to many chiral exotics arising at the intersection of a filler brane with a Pati-Salam brane. The appearance of chiral exotics at intersections with filler branes occurs somewhat frequently in type II orientifolds, and often spoils the phenomenology.

5.6. Perturbative Yukawa Couplings

To this point, we have reviewed how low energy effective theories with particular gauge symmetry and chiral matter arise in the context of type II orientifold compactifications. While these effects are the most important if a string vacuum is to realize the particle physics of our world, it is also crucial that the couplings of the low energy theory reproduce the structure in the Standard Model. This is a particularly important detail to investigate in the context of intersecting brane models, as the gauge symmetries whose gauge bosons are given a Stuckelberg mass via the Green-Schwarz mechanism impose global selection rules on couplings, forbidding crucial superpotential terms in string perturbation theory. If a model is to realize such a forbidden, but desired, coupling, it must be due to a non-perturbative effect.

5.6.1. *Yukawa couplings from string amplitudes*

Before we address non-perturbative effects, we address those Yukawa couplings which *are* present in string perturbation theory, and thus give the leading order effects. To determine the structure of a given Yukawa coupling, one must calculate the relevant correlation function in conformal field theory. We take the example of an up-flavor quark Yukawa coupling, which appears in the superpotential as $H_u\, Q_L\, u_R$. It is phenomenologically preferred that at least one such Yukawa coupling be present in string perturbation theory, as the top-quark Yukawa coupling is $\mathcal{O}(1)$, which is difficult to obtain via a non-perturbative effect. In terms of the vertex operators presented in section 5.3, the relevant correlator is

$$\langle V_{-1}^{H_u}\; V_{-1/2}^{Q_L}\; V_{-1/2}^{u_R}\rangle, \tag{5.41}$$

where the NS and R sector vertex operators are chosen so that we are using the bosonic component of the Higgs supermultiplet, and the fermionic components of the quarks.

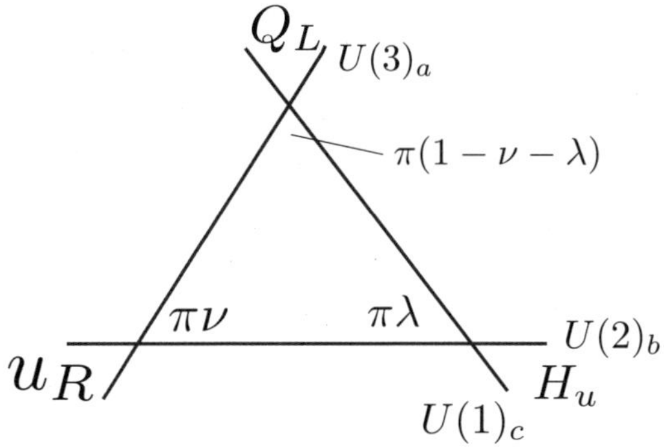

Fig. 5.5. A two-dimensional slice of the Calabi-Yau graphically depicting, an up-flavor quark Yukawa coupling. The area of the triangle (worldsheet instanton) affects the scale of the coupling.

Before being concerned about the precise structure of the correlator, the most coarse thing that one can do is determine whether or not the operator is forbidden by symmetries in string perturbation theory. This depends entirely on how the H_u, Q_L, and u_R fields are represented in the brane stacks. For example, consider the case of a three stacks of D-branes with $U(3)_a \times U(2)_b \times U(1)_c$ gauge symmetry, and that the linear combination $U(1)_Y = \frac{1}{6}U(1)_a + \frac{1}{2}U(1)_c$ is left massless by the Green-Schwarz mechanism, which we identify with hypercharge. Now suppose that the fields are realized as

$$H_u \sim (b,c) \qquad Q_L^1 \sim (a,b) \qquad Q_L^2 \sim (a,b) \qquad u_R \sim (\overline{a},\overline{c}), \tag{5.42}$$

where two families of the left-handed quark doublets are realized as Q_L^1 and one as Q_L^2. Then the possible up-flavor quark Yukawa couplings have $U(1)$ structure

$$H_u\, Q_L^1\, u_R: \ \ (0,2,0) \qquad \text{and} \qquad H_u\, Q_L^2\, u_R: \ \ (0,0,0), \tag{5.43}$$

the first of which has non-zero global $U(1)$ charge and is therefore forbidden in string perturbation theory. In this case we have one family of up-flavor quarks perturbatively allowed and two disallowed, which gives a nice explanation of the large top-quark mass. However, if the model is to be phenomenologically viable, non-perturbative effects *must* generate Yukawa couplings for the other two families, otherwise the up-quark and charm-quark will be massless.

Since the vertex operators for each of the relevant fields is known explicitly, the correlator (5.41) can be calculated explicitly using CFT techniques [23; 38; 39; 40; 41]. The non-trivial aspect of the calculation of this correlation function involves calculating the three-point correlator of the bosonic twist fields which take into account the boundary conditions associated with the angles between branes. Having the picture of a toroidal orbifold in our head, the target space picture of this Yukawa coupling for one of the tori looks like Figure 5.5, where the corresponding bosonic twist field amplitude that needs to be calculated is

$$\langle \sigma_\nu(z_1)\, \sigma_{-\nu-\lambda}(z_3)\, \sigma_\lambda(z_4) \rangle. \tag{5.44}$$

Calculation of this correlator can be performed by calculating the four point disk correlator

$$\langle \sigma_\nu(z_1)\, \sigma_{-\nu}(z_2)\, \sigma_{-\lambda}(z_3)\, \sigma_\lambda(z_4) \rangle \tag{5.45}$$

and extracting (5.44) in the limit where $z_2 \mapsto z_3$. The spacetime picture of the four point correlator is the blue trapezoid in Figure 5.6, and the geometric picture of the $z_2 \mapsto z_3$ limit is to take the uppermost brane north, past the point of convergence of the dotted red lines. For technical details on the calculation of these correlators, we refer the reader to [38].

The twist field correlator determines the angular dependence of the Yukawa coupling, which is often referred to as the "quantum" part of the coupling, due to its dependence on CFT quantum correlators. The explicit structure of a Yukawa coupling in a $T^6 = T^2 \times T^2 \times T^2$ background is given by

$$h = \sqrt{2}\, g_s\, 2\pi \prod_{I=1}^{3} \Big(\frac{16\pi^2\, \Gamma(1-\nu^I)\Gamma(1-\lambda^I)\,\Gamma(1-\nu^I-\lambda^I)}{\Gamma(\nu^I)\Gamma(\lambda^I)\Gamma(\nu^I+\lambda^I)}\Big)^{\frac{1}{4}} \sum_m \exp(\frac{-A_I^m}{2\pi\alpha'}), \tag{5.46}$$

where the indices $I \in \{1,2,3\}$ are one for each two-torus and A_I^m is the area of the m-th triangle (worldsheet instanton [42; 7]) on the j-th two-torus. The factor involving the worldsheet instantons is often called the classical factor.

5.6.2. *Coupling issues, exemplified*

In our simple example in the previous section, we saw that two of the families of up-flavor quark Yukawa couplings were forbidden in string perturbation theory, as $H_u Q_L^1 u_R$ had non-zero global $U(1)$ charge. In certain scenarios non-perturbative effects can generate the missing couplings, but in this case the issue could be avoided entirely if all three families of left-handed quark doublets appear as Q_L^2 instead of Q_L^1, as in that case all of the up-flavor quark Yukawa couplings would be perturbatively allowed. This depends heavily on how the chiral matter in a given orientifold compactification is realized at the intersection of two branes, as that determines the structure of the global $U(1)$ charges. There are important phenomenological couplings which are *always* forbidden in string perturbation theory, though, so that if a weakly coupled type II orientifold compactification is to realize them, it *must* be at the non-perturbative level. In this section, we discuss the non-perturbative generation of the always forbidden Majorana neutrino mass term and its role in the seesaw mechanism. In addition, we discuss the non-perturbative generation of the always forbidden $10\,10\,5_H$ Yukawa coupling, which gives mass to the top-quarks in Georgi-Glashow GUTs.

Example One: The neutrino masses

Consider a single Dirac neutrino mass coupling $h_\nu\ H_u\, L\, N_R$, which can be calculated as a function of moduli using the conformal field theory techniques above. Then, after electroweak symmetry breaking, the Dirac mass term is $m_{D_\nu} = h_\nu\, \langle H_u \rangle$. Comparing to a generic quark mass $m_q = h_Q\, \langle H_u \rangle$, we see that the quark masses and the neutrino masses are generically of the same order, unless the Yukawa couplings are tuned such that $h_\nu \ll h_Q$.

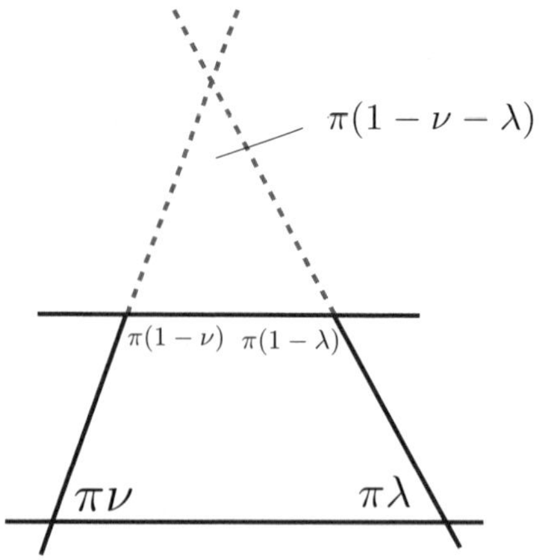

Fig. 5.6. A spacetime picture of a four-point coupling. The three-twist correlator, relevant for Yukawa couplings, is extracted from the four-twist correlator appearing in this coupling.

While worldsheet instantons are able to account for the standard model fermion hierarchies to some degree, it is only in very small regions of moduli space where they could account for the hierarchy $m_Q \sim 1\,GeV$ and $m_\nu \sim 10^{-3}\,eV$. As in the particle theory literature, we would prefer to have some mechanism to account for this hierarchy, rather than attributing it to some miraculous result of moduli stabilization.

One popular field theoretic mechanism which accounts for the small neutrino masses is the type I seesaw mechanism. In this mechanism, in addition to the Dirac type neutrino mass term $h_\nu\, H_u\, L\, N_R$, there is a Majorana neutrino mass term of the form $M_R N_R N_R$. The neutrino mass matrix takes the form

$$\begin{pmatrix} 0 & h_\nu\ \langle H_u\rangle \\ h_\nu\ \langle H_u\rangle & M_R \end{pmatrix}$$

giving rise to mass eigenvalues M_R and $\frac{h_\nu^2 \langle H_u\rangle^2}{M_R}$ in the limit of large M_R. Thus, one of the mass eigenvalues has been "seesawed" to a very small value by the large Majorana mass, giving (in this mechanism) the reason for the very small neutrino masses observed in nature.

While this mechanism is nice from the point of view of field theory, there is an important difficulty which arises when attempting to realize a Majorana mass term in string theory. As the right-handed neutrinos N_R will have global $U(1)$ quantum numbers in a type II orientifold compactification, the Majorana mass term $M_R N_R N_R$ will also be charged with respect to the global $U(1)$'s, and is therefore forbidden in string perturbation theory. Therefore, in type II orientifold compactifications it is difficult to account for the smallness of the neutrino masses in string perturbation theory: in the absence of extreme fine-tuning of the moduli, m_{D_ν} and m_Q are of the same order, and the seesaw mechanism cannot be realized, as the Majorana mass term is forbidden.

Example Two: The $10\,10\,5_H$ in $SU(5)$ GUTs

There is another well-known coupling problem that arises when trying to realize Georgi-Glashow GUTs in weakly coupled type II orientifold compactifications. In these models, the $SU(5)$ gauge theory is realized by a stack of five spacetime-filling D-branes wrapping a non-trivial cycle in the Calabi-Yau, and chiral matter charged under the $SU(5)$ factor is realized at the intersection of this five-stack with some other D-brane. It is the possibility of symmetric and antisymmetric matter representations, in addition to the bifundamental, which allows for the realization of the standard $SU(5)$ GUT particle representations. Specifically, the 10 representation of $SU(5)$ can be realized as $\begin{array}{|c|}\hline \\ \hline \\ \hline\end{array}_5$ at the intersection of the five-stack with its orientifold image.

Though the proper spectrum can be realized, there is an immediate problem at the level of couplings. Since the 10 is realized as an antisymmetric, it comes with charge 2 under the $U(1)$ of the five-stack. In addition, the 5_H comes with charge 1, since it a fundamental of $SU(5)$, and the top-quark Yukawa coupling $10\,10\,5_H$ has charge 5 and is therefore *always* forbidden in string perturbation theory. On the other hand, the bottom-quark Yukawa coupling $10\,\overline{5}\,\overline{5}_H$ can be present in string perturbation theory, giving a massive bottom-quark and a massless top-quark. This inverts the standard hierarchy and is a major phenomenological pitfall that must be remedied if one hopes to realize realistic Georgi-Glashow GUTs in weakly coupled type II orientifold compactifications.

5.7. Non-perturbative Superpotential Corrections: D-instantons

We have now seen that it would be phenomenologically useful if some non-perturbative effect were able to generate superpotential couplings which are forbidden in string perturbation theory. Doing so would require that the non-perturbative physics somehow cancels the excess $U(1)$ charge associated with a perturbative Yukawa coupling. A the reader familiar with the KKLT [43] scenario (for example) might be concerned that D-instantons cannot serve this purpose, as there a euclidean D3-instanton in type IIb generated a non-perturbative correction without charged matter which was responsible for stabilizing a Kähler modulus. However, it was shown in [13; 14; 15] that in the presence of spacetime filling gauge D-branes the axionic shift symmetries which are gauged by the Green-Schwarz mechanism cause the axions to be charged with respect to the global $U(1)$ symmetries. These axions appear in instanton corrections and can cancel the net $U(1)$ charge of perturbatively forbidden couplings, giving rise to these couplings at the non-perturbative level.

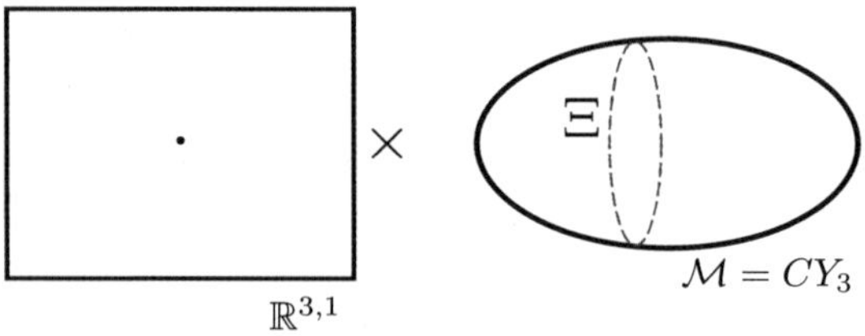

Fig. 5.7.

Fig. 5.8. A target space view of a euclidean D-brane instanton, which is pointlike in spacetime and wraps a non-trivial cycle Ξ in the Calabi-Yau.

Consider a type II orientifold compactification with euclidean D-instantons in the background. The instantons are pointlike in spacetime and wrap a non-trivial cycle in the Calabi-Yau. The instanton action is of the form

$$S_{inst} = S_{cl}^E + S(\mathcal{M}, \Phi) \tag{5.47}$$

where $\mathcal{M}$ are the set of instanton zero modes and Φ are the set of charged matter fields present in the low energy theory. The instanton correction to the low energy effective theory in four dimensions takes the form

$$S_{np}^{4d}(\Phi) = \int [\mathcal{DM}] \; e^{-S_{inst}}, \tag{5.48}$$

where the structure of the correction is determined by $\mathcal{M}$ and Φ, and its magnitude is set by the classical instanton suppression factor, which depends on the volume wrapped by the instanton in the Calabi-Yau.

5.7.1. *Instanton heuristics*

For the sake of concreteness, we again work in type IIa, where the instantons are euclidean D2 branes which are point-like in spacetime and wrap non-trivial three-cycles in the Calabi-Yau. Taking a three-cycle Ξ, the classical action of the instanton is

$$S_{cl}^E = T_{E2}[\frac{1}{g_s}\int_\Xi \sqrt{\det G} - i\int_\Xi C_3], \tag{5.49}$$

where the first term comes from the Born-Infeld action and the second from the Wess-Zumino action. Here $e^{-Re\, S_{cl}^E}$ is a real suppression factor that sets the scale of superpotential corrections. Its value is set by

$$Re\, S_{cl}^E = \frac{T_{E2}}{g_s} V_\Xi = \frac{8\pi^2}{g_a^2}\frac{V_\Xi}{V_{\pi_a}}, \tag{5.50}$$

and can lead to very large suppression due to its exponential nature. This is phenomenologically very relevant, as (for example) it allows one to realize neutrino masses of the correct order with only a highly suppressed Dirac term LH_uN_R [44], without resorting to the seesaw mechanism. Another alternative to the seesaw mechanism includes the generation of the Weinberg operator LH_uLH_u by a D-instanton [45].

One might recall from section 5.4.2 that dimensional reduction of the Ramond-Ramond three-form C_3 on three cycles gives rise to four-dimensional axions Υ^I and $\tilde{\Upsilon}_I$, which transform under $U(1)_a$ due to the generalized Green-Schwarz mechanism. These axions enter the classical instanton action as

$$Im\, S_{cl}^E = T_{E2}\int_\Xi C_3 = T_{E2}\,(N_{\Xi I}\Upsilon^I - M_\Xi^I\tilde{\Upsilon}_I). \tag{5.51}$$

The gauge field one-form associated with $U(1)_a$ transforms as $A \mapsto A+d\Lambda_a$, and the axions transform as

$$\Upsilon^I \mapsto \Upsilon^I + \mathcal{N}_a\, M_a^I\Lambda_a \qquad \text{and} \qquad \Upsilon_I \mapsto \Upsilon_I + \mathcal{N}_a\, N_{a,I}\Lambda_a \tag{5.52}$$

from which it can be seen that the classical instanton action transforms as

$$e^{-S_{cl}} \mapsto e^{-S_{cl}+i\,Q_\Xi^a\,\Lambda_a}, \tag{5.53}$$

with $Q_\Xi^a = \mathcal{N}_a\;\Xi\circ\pi_a$.

Taking the orientifold and image branes into account induces extra shifts in the classical instanton action, so that in full generality

$$Q_\Xi^a = \mathcal{N}_a\,\Xi\circ(\pi_a - \pi_a'). \tag{5.54}$$

This is precisely the net charge carried microscopically by the charged instanton zero modes in the path integral measure, as will be discussed in section 5.7.2. If a subset of matter fields $\phi_i \in \Phi$ has charges Q_i such that $Q^a_\Xi + \sum_i Q^a_i = 0 \ \forall a$, an instanton wrapped on Ξ can, in principle, generate a superpotential coupling of the form

$$e^{-S^{E2}_{cl}} \prod_i \phi_i, \tag{5.55}$$

since the non-trivial transformation of the axions cancels the global $U(1)$ charge of $\prod_i \phi_i$.

5.7.2. *Fermionic zero modes*

The arguments of the previous section heuristically showed that the gauging of shift symmetries by the Green-Schwarz mechanism make it possible for couplings of the form (5.55) to be gauge invariant, depending on the $U(1)$ charges of the matter fields. Whether or not such a a term is actually generated by an instanton depends heavily on the microscopic properties of the instanton, in particular its fermionic zero modes, which correspond to massless open strings.

The importance of fermionic zero modes for determining non-perturbative corrections is well known in other areas of the landscape. For example, in [46] Witten argued that an M5-instanton wrapped on a 6-cycle Ξ_{M5} must satisfy

$$\chi(\Xi_{M5}, \mathcal{O}_{\Xi_{M5}}) = \sum_{i=0}^{3} h^i(\Xi_{M5}, \mathcal{O}_{\Xi_{M5}}) = 1 \tag{5.56}$$

if it is to contribute to the superpotential. This is a constraint on particular (uncharged) fermionic zero modes of the instanton, which are counted by the Hodge numbers $h^i(\Xi_{M5}, \mathcal{O}_{\Xi_{M5}})$. Similar constraints exist for the uncharged modes in type II, as, for example, a euclidean D3-instanton wrapped on a holomorphic divisor D must satisfy $\chi(D, \mathcal{O}_D) = 1$ if it is to contribute to the superpotential.

The influence of the fermionic modes on non-perturbative corrections is easy to see: the path integral integral is over all fermionic zero modes, so if these modes are not lifted or the instanton action does not have appropriate terms for soaking them up, then the Grassman integral evaluates to zero. Suppose (in an unrealistic but illustrative example) that $S(\mathcal{M}, \Phi) = a\,\xi + b\,\eta + c\,\xi\eta$, where the Greek variables are fermionic zero modes. Then from

(5.48) the non-perturbative correction would be

$$e^{-S_{cl}^E}\int[d\xi][d\eta]\;e^{-(a\,\xi+b\,\eta+c\,\xi\eta)}=e^{-S_{cl}^E}\int[d\xi][d\eta]\;(1-(a\,\xi+b\,\eta+c\,\xi\eta))$$
$$=c\,e^{-S_{cl}^E}. \tag{5.57}$$

It is easy to see that if the third term in $S(\mathcal{M},\Phi)$ were not present, then the correction would be absent.

One might intuitively think that, since the basic branes in our theory are gauge D-branes and euclidean D-instantons, there exist two types of fermionic zero modes living in the instanton worldvolume, corresponding to a string from the instanton to itself and a string from the instanton to a gauge brane. Indeed, this is the case, as can be shown (for example) by CFT techniques. The strings from the instanton to itself are known as uncharged zero modes, and are the modes counted by $h^i(D,\mathcal{O}_D)$ in type IIb[k] which contribute to the holomorphic genus. The strings from the instanton to a gauge brane are charged under the gauge group of the D-brane, and thus are known as charged modes.

5.7.2.1. *Uncharged zero modes*

Perhaps the most crucial of the uncharged modes are the ones associated with the breakdown of supersymmetry. Recall that type II string theory compactified on a Calabi-Yau manifold gives rise to $\mathcal{N}=2$ supersymmetry in four dimensions, an $\mathcal{N}=1$ subalgebra of which is preserved by the orientifold, with supercharges Q^α, $\overline{Q}^{\dot\alpha}$. A spacetime-filling D-brane wrapping a 1/2-BPS cycle[l] might preserve the same supercharges $Q^\alpha, \overline{Q}^{\dot\alpha}$, in which case the D-brane is supersymmetric with respect to the orientifold. The orthogonal complement to the $\mathcal{N}=1$ algebra preserved by the orientifold, which has supercharges Q'^α and $\overline{Q}'^{\dot\alpha}$, then gives four Goldstinos associated with the four broken supersymmetries.

The key point is that, due to localization in the four extended dimensions, a 1/2-BPS D-instanton does not preserve the four supercharges Q^α, $\overline{Q}^{\dot\alpha}$ preserved by a gauge D-brane and the orientifold, but instead the combination Q'^α and $\overline{Q}^{\dot\alpha}$. There are then four Goldstinos in the instanton worldvolume associated with the breakdown of supersymmetry: two chiral modes

[k]For a beautiful explanation of how these modes lift to modes of vertical M5 instantons in F-theory, see [47], and for a lift to F-theory of an instanton generating the $10\,10\,5_H$ see [48]. For a generic review of ED3 zero modes from the point of view of sheaf cohomology, see section 2 of [49].

[l]These are special Lagrangians for D6-branes, holomorphic divisors for D7-branes.

θ^α associated with the breaking of Q^α, and two anti-chiral modes $\overline{\tau}^{\dot\alpha}$ [50; 51] associated with the breaking of $\overline{Q}'^{\dot\alpha}$. With the θ mode identified as the θ mode for $\mathcal{N}=1$ theories in four dimensions, the instanton might contribute a superpotential correction if the $\overline{\tau}$ modes are somehow saturated or lifted.

There are numerous ways for this to happen. One common possibility is that the instanton wraps an orientifold invariant cycle, in which case the $\overline{\tau}$ mode is projected out. In the case of a type IIa orientifold compactification, this can be seen directly in the CFT formalism, where the vertex operators associated with the θ and the $\overline{\tau}$ mode are given by

$$\begin{aligned} V^{\theta}_{-\frac{1}{2}} &= \theta_\alpha\, e^{-\frac{\phi}{2}}\, S^\alpha(z)\, \Sigma_{\frac{3}{8},\frac{3}{2}}(z) \\ V^{\overline{\tau}}_{-\frac{1}{2}} &= \overline{\tau}^{\dot\alpha}\, e^{-\frac{\phi}{2}}\, S_{\dot\alpha}(z)\, \Sigma_{\frac{3}{8},-\frac{3}{2}}(z), \end{aligned} \tag{5.58}$$

and the Σ fields are spin fields describing fermionic degrees of freedom on the internal space. The subscripts of Σ give the conformal dimension and worldsheet $U(1)$ charge. The orientifold projection induces extra constraints on the structure of Chan-Paton factors, which the $\overline{\tau}$ modes do not satisfy and therefore they are projected out.

From the point of view of type IIb compactified on a generic Calabi-Yau with a euclidean D3 instanton wrapping a holomorphic divisor D, the presence or absence of θ and $\overline{\tau}$ is counted by the Hodge number $h^{0,0}(D) \cong h^0(D,\mathcal{O}_D)$. The presence of the holomorphic $\mathbb{Z}_2$-action σ associated with the orientifold allows for a decomposition of ordinary sheaf cohomology into a sum of $\mathbb{Z}_2$-equivariant sheaf cohomology as

$$H^i(D,\mathcal{O}_D) \cong H^i_+(D,\mathcal{O}_D) \oplus H^i_-(D,\mathcal{O}_D), \tag{5.59}$$

reflecting the fact that each of the zero modes transform with a sign under σ. The $\overline{\tau} \in h^0_-(D,\mathcal{O}_D)$ mode is the one which transforms with a $-$ sign, and is therefore absent in the case of an orientifold invariant divisor. Such an instanton is called an $O(1)$ instanton.

These are not the only uncharged zero modes, however. For example, in addition to the θ and $\overline{\tau}$ Goldstino modes associated with the breakdown of supersymmetry due to the localization of the instanton in four-dimensional spacetime, there are additional uncharged modes corresponding to the localization of the instanton on submanifolds of the Calabi-Yau. These deformation modes also admit both a CFT description, when available, and a description in terms of cohomology. The latter can be seen in type IIb as deformations of a holomorphic divisor D, which are given by global sections of the normal bundle, and thus are counted by $h^0(D, N_{D|\mathcal{M}})$[m]. If a cycle has no deformation modes, it (as well as an instanton wrapping it) is said to be rigid.

In order for an instanton to contribute to the superpotential, it must realize the θ mode and none of the other uncharged modes[n], as the superpotential appears in the $\mathcal{N} = 1$ spacetime action as $\int d^4x\, d^2\theta\, W(\Phi)$. There are at least two possible reasons for the absence of a zero mode. First, it may be absent to begin with due to being projected out. This is the reason for the absence of the $\overline{\tau}$ mode in the case of an $O(1)$ instanton. The second possibility is that the extra zero modes are "saturated" or "soaked up". This is the case in (5.57), where the χ and η integrals evaluate to one when integrating over the $c\,\chi\eta$ term.

5.7.2.2. *Charged zero modes*

The charged zero modes are massless strings from a euclidean D-instanton to a gauge D-brane, which are therefore charged under the gauge group of the D-brane. They are the microscopic modes that carry the global $U(1)$ charges which compensate for the overshoot in $U(1)$ charge of perturbatively forbidden couplings. The form of the superpotential corrections involving charged matter depends heavily on these charged modes and is calculated using the instanton calculus presented in [13] and reviewed in [16]. In short, it tells one how to determine the structure of $S(\mathcal{M}, \Phi)$ in the instanton action based on CFT disc diagrams involving charged matter fields and charged zero modes. For the sake of brevity, we refer the reader to those sources for a general discussion of instanton calculus, and instead we present the CFT basics of charged zero modes and an illustrative example that makes the physics and basics of the methods very clear.

The Ramond sector open string vertex operator corresponding to a charged matter mode between a brane $\mathcal{D}_a$ and an instanton $\mathcal{E}$ is given

[m]Deformation modes, as well as other uncharged modes, are often said to be counted by $h^i(D, \mathcal{O}_D)$ in the literature, which may not be the most illuminating presentation. To help motivate this, note that in a Calabi-Yau manifold a holomorphic divisor has $K_D = N_{D|\mathcal{M}}$, so that by Serre duality we have $H^0(D, N_{D|\mathcal{M}}) \cong H^2(D, K_D \otimes N^*_{D|\mathcal{M}}) \cong H^2(D, \mathcal{O}_D)$. Therefore, by some simple isomorphisms, the intuitive notion of deformation modes as normal bundle sections is recast as $\mathcal{O}_D$ sheaf cohomology.

[n]In type IIb, the precise statement in cohomology is $h^0_+(D, \mathcal{O}_D) = 1$ and all others zero. This necessary and sufficient constraint automatically satisfies the necessary constraints $\chi(D, \mathcal{O}_D) = 1$, as it must.

by

$$V^{\lambda_a^i}_{\frac{1}{2}}(z) = \lambda_a^i e^{-\frac{\phi}{2}} \Sigma^{D_a,\mathcal{E}}_{\frac{3}{8},-\frac{1}{2}}(z)\sigma_{h=1/4}(z), \tag{5.60}$$

where i is the gauge index of the brane $\mathcal{D}_a$ and the σ fields are the 4D spin fields arising from the twisted 4D worldsheet bosons carrying half-integer modes. Due to the four Neumann-Dirichlet boundary conditions between the brane and the instanton in $\mathbb{R}^{3,1}$, the zero point energy in the NS-sector is shifted by $L_0 = 1/2$. This makes all NS-sector states massive, so that the only charged zero modes come from the Ramond sector. We emphasize that the net $U(1)$ charge of these zero modes, as dictated by Table 5.3, is precisely equivalent to the charge of the classical instanton action (5.54).

Table 5.3. Representations and multiplicity of charged modes appearing at the intersection of an instanton $\mathcal{E}$ and a gauge brane $\mathcal{D}$.

Zero Mode	Representations	Number
$\lambda_a \equiv \lambda_{\mathcal{E},\mathcal{D}}$	$(-1_{\mathcal{E}}, \square_{\mathcal{D}})$	$I^+_{\mathcal{E},\mathcal{D}}$
$\overline{\lambda}_a \equiv \lambda_{\mathcal{D},\mathcal{E}}$	$(1_{\mathcal{E}}, \overline{\square}_{\mathcal{D}})$	$I^-_{\mathcal{E},\mathcal{D}}$

We examine the up-flavor quark sector of the model presented in Table 1 of [52][o]. That model is a four-stack quiver with $U(3)_a \times U(2)_b \times U(1)_c \times U(1)_d$ gauge symmetry which becomes $SU(3)_C \times SU(2)_L \times U(1)_Y$ due to the Green-Schwarz mechanism, with the Madrid hypercharge embedding $U(1)_Y = \frac{1}{6}U(1)_a + \frac{1}{2}U(1)_c + \frac{1}{2}U(1)_d$. The fields relevant for up-flavor quark Yukawa couplings are realized as

$$H_u:\ (b,c) \qquad Q_L:\ (a,\overline{b}) \qquad u_R^1:\ 1\times(\overline{a},\overline{c}) \qquad u_R^2:\ 2\times(\overline{a},\overline{d}), \tag{5.61}$$

so that the Yukawa couplings have global U(1) charge

$$H_u\, Q_L\, u_R^1 : (0,0,0,0) \qquad\qquad H_u\, Q_L\, u_R^2 : (0,0,1,-1). \tag{5.62}$$

Since u_R^2 has multiplicity two, two families are perturbatively forbidden, while one family is perturbatively realized, as one might hope given the hierarchy of the top-quark mass relative to the up and the charm.

To generate the missing $H_u Q_L u_R^2$ couplings in type IIa, an instanton $\mathcal{E}$ would have to exhibit intersection numbers

$$I_{\mathcal{E}a} = 0 \qquad I_{\mathcal{E}b} = 0 \qquad I_{\mathcal{E}c} = 1 \qquad I_{\mathcal{E}d} = -1, \tag{5.63}$$

[o]In fact this model is a quiver, not a globally defined orientifold compactification. See section 5.8 for more information.

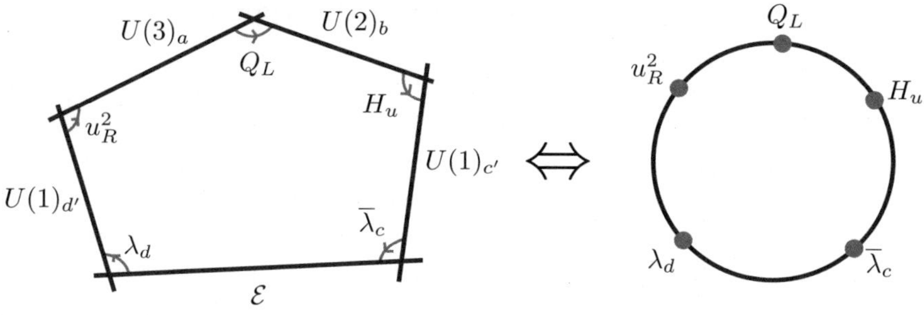

Fig. 5.9.

Fig. 5.10. A spacetime view of a non-perturbatively generated up-flavor quark Yukawa coupling, along with the corresponding disc diagram which contributes to the instanton action.

which gives rise to two charge modes, $\overline{\lambda}_c$ and λ_d. One can heuristically "see" that this cancels the excess global $U(1)$ charge in Figure 5.9 by the fact that spacetime picture is closed and the arrows point in in a consistent direction. The corresponding disk diagram, also drawn in the figure, contributes the instanton action.

If, in a global embedding which realizes this spectrum, a rigid O(1) instanton exists with this intersection pattern, then one can perform the instanton calculus with the mentioned disk contribution. The path integral (5.48) takes the form

$$\begin{aligned}
&\int d^4x\, d^2\theta\, d\overline{\lambda}_c\, d\lambda_d\ e^{-S^{\mathcal{E}}_{cl}+Y^J\,\overline{\lambda}_c\,H_uQ_Lu_R^{2,J}\,\lambda_d} \\
&= e^{-S^{\mathcal{E}}_{cl}} \int d^4x\, d^2\theta\, d\overline{\lambda}_c\, d\lambda_d\ Y^J\,\overline{\lambda}_c\,H_uQ_Lu_R^{2,J}\,\lambda_d \\
&= e^{-S^{\mathcal{E}}_{cl}} \int d^4x\, d^2\theta\ Y^J\,H_uQ_Lu_R^{2,J}, \qquad (5.64)
\end{aligned}$$

where J runs across the two family indices for u_R^2. In such a case the up-quark and charm-quark masses are suppressed by a factor of $e^{-S^{\mathcal{E}}_{cl}}$ relative to the top-quark mass. In principle Y^J can give a hierarchy between the up-quark and charm-quark, since it depends on worldsheet instantons, but this generically depends heavily on the details of moduli stabilization.

5.8. Braneworld Quivers: The Bottom-Up Approach

To this point we have discussed how all of the basic ingredients of real-world particle physics can be realized in the context of weakly coupled type II orientifold compactifications. In particular, gauge symmetry lives on the worldvolume of spacetime filling D-branes with possible gauge groups $U(N)$, $Sp(2N)$, or $SO(2N)$. Chiral matter appears at the intersection of two stacks of D-branes, with the type and amount of chiral matter dictated by the topological intersection numbers of the D-branes. Finally, the presence or absence of superpotential couplings depends crucially on the charge of couplings under the $U(1)$ symmetries associated with the $U(N)$ branes.

D-braneworlds offer a beautiful geometric picture which suggests the possibility of arranging branes in such a way that something very similar to the MSSM is obtained. A top-down approach would first require specifying a Calabi-Yau manifold $\mathcal{M}$ together with a $\mathbb{Z}_2$ involution on the space, which would allow for the identification of O-planes and follow with an investigation of the types of arrangements of D-branes allowed by tadpole cancellation. Perhaps with the specification of further data (e.g. fluxes in type IIb, for chirality), the massless spectrum can be calculated and the global $U(1)$ charges of the matter can be determined, allowing for the determination of perturbative superpotential couplings. One can then perform a scan of possible instanton cycles to determine which might have the proper fermionic zero mode structure for superpotential contribution. Needless to say, this quickly becomes rather involved.

Though a "top-down" model is necessary if string theory is to provide the correct description of particle physics in our world, this does not necessarily mean that the best way to identify promising models is by taking a top-down approach to each vacuum. Recently, a "bottom-up" approach [53; 54; 55] has emerged which suggests looking at certain subsets of data associated with a string vacuum, with the hope that one can say non-trivial things across broader patches of the landscape, despite the fact that certain details have been ignored. This approach is only good to the extent that the ignored details don't destroy the physics determined by the subset of vacuum data of interest.

We already saw an example of this approach in section 5.3, when looking at the up-quark Yukawa couplings $H_u\, Q_L^1\, u_R$ and $H_u\, Q_L^2\, u_R$, where the fields are realized by three-stacks of D-branes. Notice that we specified neither a Calabi-Yau manifold nor a $\mathbb{Z}_2$ involution in this example, and yet were able to make statements about couplings based on assumptions about how chiral MSSM matter is realized at the intersection of various brane stacks. This information would be a subset of the information associated with the quiver realized in a type II orientifold compactification.

In fact, much has been learned recently about particle physics in type II by studying quivers, which are a subset of the data defining a type II orientifold compactification. Generically, a quiver is made of nodes and edges between them, where the nodes represent gauge D-branes and an edge represents matter at the intersection of the corresponding D-branes. Thus, a quiver encodes the gauge symmetry and matter content in a type II orientifold compactification, including global $U(1)$ charges. We emphasize that these *are not* globally consistent string compactifications. They cannot be, as quivers are only a subset of the data associated with a string vacuum. However, a given quiver can be shown to be *compatible* with global consistency if it satisfies the necessary conditions (5.27) and (5.35), which do contain genuinely string constraints that are not already present in field theory.

An example of a consistent type II quiver which realizes the exact MSSM spectrum is given in Figure 5.11, which corresponds to MSSM matter being realized as

$$Q_L:\ 2\times(a,\overline{b}),\ 1\times(a,b) \qquad u_R:\ 3\times(\overline{a},\overline{c}) \qquad d_R:\ 3\times \begin{smallmatrix}\Box\\ \Box\end{smallmatrix}_a$$

$$L:\ 3\times(b,\overline{c}) \qquad E_R:\ 3\times \Box\!\Box_c \qquad H_u:\ 1\times(b,c) \qquad H_d:\ 1\times(\overline{b},\overline{c}),$$

where the gauge symmetry is $U(3)_a \times U(2)_b \times U(1)_c$ and the massless hypercharge is realized as $U(1)_Y = \frac{1}{6}U(1)_a + \frac{1}{2}U(1)_b$. There are a few details required to understand the quiver diagram properly. First, since we are in the framework of type II orientifold compactifications, there is an image brane associated with each $U(N)$ brane, which is all of the branes, in this case. Rather than doubling the number of nodes, we double the number of arrows on the edges, with an arrow coming out of a node representing a fundamental, and an arrow going into a node representing an antifundamental. Therefore, there are four options for arrow orientation on each edge, representing all possible bifundamental representations between those two branes and their orientifold images. Second, edges from a node to itself implicitly correspond to a string stretching between a brane and its image, which yields a symmetric or antisymmetric representation. In the $U(N)$ case, in addition to being symmetric or antisymmetric of $SU(N)$, there is a choice of $U(1)$ charge ± 2, and so we label these edges with either an A or an S and either a $+$ or a $-$.

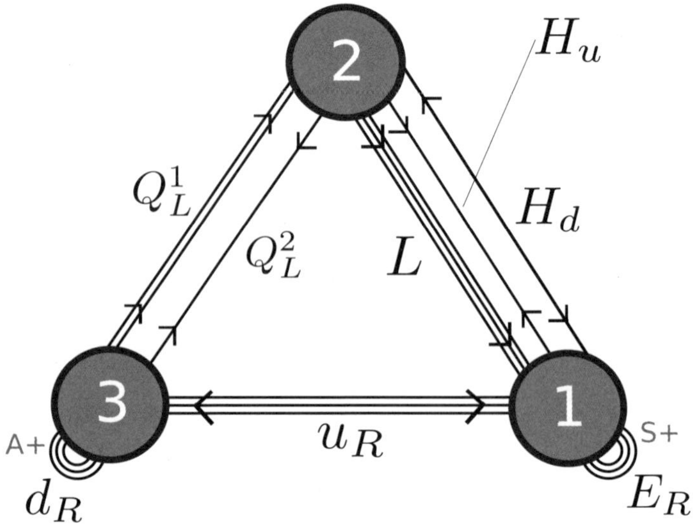

Fig. 5.11.

Fig. 5.12. A type II quiver which realizes the exact MSSM spectrum and satisfies the necessary constraints (5.27), (5.35), and (5.36) for tadpole cancellation and a massless hypercharge.

This quiver contains enough information to say a great deal about the structure of the Yukawa couplings. Specifically, two of the families of up-quarks and all of the leptons have perturbatively realized Yukawa couplings, while one of the up-quark families and all of the down-quark families are forbidden in string perturbation theory. Note that the perturbatively forbidden up-flavor quark Yukawa coupling has global $U(1)$ charge $(0, 2, 0)$, as does the R-parity violating coupling LLE_R. Therefore, any D-instanton which generates the missing up-flavor Yukawa coupling will also generate an R-parity violating operator at a very high level. For this reason, and perhaps others, any type II orientifold compactification giving rise to this quiver is phenomenologically ruled out.

Any given quiver can, in principle, be realized in many different global embeddings, allowing one to make statements across broader patches of the landscape. For example, in addition to the quiver presented in Figure 5.11, there are 23 other three-stack quivers with $U(1)_Y = \frac{1}{6}U(1)_a + \frac{1}{2}U(1)_b$ and the exact MSSM spectrum that also satisfy the necessary constraints (5.27) and (5.35). There is one other linear combination for the hypercharge which might realize the MSSM spectrum with three-stacks, which also gives rise to a total of 24 quivers with the exact MSSM spectrum that satisfy (5.27) and (5.35). One can therefore make the following statement: if the exact MSSM is to be realized in a type II orientifold compactification at the intersections of three stacks of D-branes with $U(3) \times U(2) \times U(1)$ gauge symmetry, that compactification's corresponding quiver must be one of the mentioned 48, the couplings of which we can say a great deal about based on their global charges.

Popular avenues for studying quivers include systematic study of hypercharge embeddings and MSSM quivers [56], as well as the study of particular quivers at the level of couplings [57]. Additionally, systematic work has been done along these lines at the level of couplings [33] and the mass hierarchical structure of MSSM quarks and leptons have been investigated in [58; 52]. The basic strategy in the systematic works was to study the phenomenology of MSSM quivers, possibly extended by three right-handed neutrinos or a singlet S which can give rise to a dynamical μ-term. All presented quivers satisfy the necessary constraints on the chiral spectrum for tadpole cancellation and a massless hypercharge, and thus it is not ruled out that these quivers can be embedded in a consistent type II orientifold compactification. Beyond these necessary constraints, the quivers also satisfy a host of phenomenological constraints. In particular, for a quiver to be semi-realistic, one has to require that D-instanton effects generate enough of the forbidden Yukawa couplings to ensure that there are no massless quark or lepton families. In doing so, however, a D-instanton which generates a Yukawa coupling might also generate a phenomenological drawback, such as an R-parity violating coupling, a dimension five proton decay operator, or a μ-term which is far too large. Such a quiver would be ruled out.

5.9. Conclusion and Outlook

In these lectures, we have presented the basic perturbative and non-perturbative physics of type II orientifold compactifications. This corner of the string landscape is particularly nice for understanding aspects of four-dimensional particle physics, as spacetime filling D-branes give rise to four-dimensional gauge symmetry and chiral matter can appear at their intersections. The importance of the non-perturbative D-instanton effects cannot be overstated in these compactifications, as in their absence a type II orientifold compactification often gives rise to massless families of quarks or leptons, due to global $U(1)$ selection rules that forbid their Yukawa couplings. Indeed, they are *necessary* for some aspects of particle physics, as, for example, the Majorana neutrino mass term and the $10\,10\,5_H$ top-quark Yukawa coupling are *always* perturbatively forbidden in weakly coupled type II. In addition to generating phenomenologically desirable couplings, instanton effects must be taken into account because they could also generate couplings which spoil the physics. Furthermore, instantons generate the leading superpotential contributions for Kähler moduli in type IIb, and thus play an important role in their stabilization.

Though the picture of particle physics in these models is beautiful, one still has to decide how to deal with the enormity of the landscape. It is a useful fact that coupling issues can be studied at the quiver level, which specifies how chiral matter transforms under the gauge groups of the D-branes, and thus the global $U(1)$ charges of matter, while postponing the issue of global embeddings to a later date. We believe this approach to be of great use in identifying promising quivers for global embeddings, as it is not worth trying to realize a global embedding for the sake of particle physics if it can already be seen at the quiver level that a model is ruled out phenomenologically. In addition to phenomenological constraints on quivers, string consistency conditions on chiral matter greatly constrain the possibilities, so it is not true that "anything goes".

From the point of view of globally consistent type II orientifold compactifications, both type IIa and type IIb have their advantages and disadvantages. On one hand, the appearance of chiral matter at the intersection of D6-branes in type IIa is purely dependent on geometric issues, giving an intuitive picture of particle physics. In addition, many useful CFT techniques have been developed for the type IIa string compactified on a toroidal orbifold. In type IIb, on the other hand, the appearance of chiral matter depends on the choice of worldvolume flux on the D7-branes, and therefore depends on more than geometry. The major advantage of type IIb, however, is that much more is known about moduli stabilization and that the full power of complex algebraic geometry can be utilized, since the 1/2-BPS gauge branes and euclidean D-instantons wrap holomorphic divisors rather than special Lagrangians. In addition, type IIb offers a description as the $g_s \mapsto 0$ limit of F-theory, which has been of great interest in string phenomenology over the last few years[p].

The interplay between type IIb D-braneworlds and F-theory compactifications runs deep, as one might expect. In particular, a major motivating factor for the study of the appearance of Georgi-Glashow GUTs in F-theory

[p]For work on F-theory GUTs, see [59; 60; 61]. For lectures and further information, see [62] and references therein.

is the absence of the $10\,10\,5_H$ Yukawa coupling in type IIb string perturbation theory. Though this coupling can be generated by an instanton in type II [63; 64], it is exponentially suppressed by the classical action of the instanton and therefore still has trouble explaining the top-quark hierarchy, unless the Kähler moduli are stabilized such that the Yukawa coupling is $\mathcal{O}(1)$. In the F-theory lift of a type IIb GUT, the $10\,10\,5_H$ occurs at a point of D_6 enhancement, which is the lift of where the orientifold, $U(5)$ brane, and $U(1)$ brane intersect in type IIb. These objects also can intersect at a point of E_6 enhancement in F-theory, which gives a perturbative ($\mathcal{O}(1)$) contribution to $10\,10\,5_H$. Though this is one of the phenomenological advantages of F-theory GUTs over type II, instantons still play important roles in other aspects of the physics. This is still an active area of research, and in particular the microscopic description of charged modes in F-theory needs clarification.

Acknowledgments

We would like to acknowledge Iñaki Garćia-Etxebarria, Paul Langacker, Robert Richter and Timo Weigand for recent collaborations. We thank I.G.E. and R.R. in particular for discussions related to the content of the lectures. We thank the TASI organizers for providing a wonderful school, and the participants for lively discussions. This work was supported in part by the National Science Foundation under Grant No. NSF PHY05-51164, DOE under grant DE-FG05-95ER40893-A020, NSF RTG grant DMS-0636606, the Fay R. and Eugene L. Langberg Chair, and the Slovenian Research Agency (ARRS).

References

1. M. Kreuzer and H. Skarke, Complete classification of reflexive polyhedra in four-dimensions, *Adv.Theor.Math.Phys.* **4**, 1209–1230, (2002).
2. L. Susskind, The Anthropic landscape of string theory. (2003).
3. F. Denef and M. R. Douglas, Distributions of flux vacua, *JHEP.* **0405**, 072, (2004). doi: 10.1088/1126-6708/2004/05/072.
4. A. Schellekens, The Emperor's Last Clothes? Overlooking the String Theory Landscape, *Rept.Prog.Phys.* **71**, 072201, (2008). doi: 10.1088/0034-4885/71/7/072201.
5. C. Angelantonj, I. Antoniadis, E. Dudas, and A. Sagnotti, Type I strings on magnetized orbifolds and brane transmutation, *Phys.Lett.* **B489**, 223–232, (2000). doi: 10.1016/S0370-2693(00)00907-2.

6. R. Blumenhagen, L. Gorlich, B. Kors, and D. Lust, Asymmetric orbifolds, noncommutative geometry and type I string vacua, *Nucl.Phys.* **B582**, 44–64, (2000). doi: 10.1016/S0550-3213(00)00270-4.
7. G. Aldazabal, S. Franco, L. E. Ibanez, R. Rabadan, and A. Uranga, Intersecting brane worlds, *JHEP.* **0102**, 047, (2001).
8. M. Cvetič, G. Shiu, and A. M. Uranga, Three family supersymmetric standard - like models from intersecting brane worlds, *Phys.Rev.Lett.* **87**, 201801, (2001). doi: 10.1103/PhysRevLett.87.201801.
9. M. Cvetič, G. Shiu, and A. M. Uranga, Chiral four-dimensional N=1 supersymmetric type 2A orientifolds from intersecting D6 branes, *Nucl.Phys.* **B615**, 3–32, (2001). doi: 10.1016/S0550-3213(01)00427-8.
10. R. Blumenhagen, M. Cvetič, P. Langacker, and G. Shiu, Toward realistic intersecting D-brane models, *Ann.Rev.Nucl.Part.Sci.* **55**, 71–139, (2005). doi: 10.1146/annurev.nucl.55.090704.151541.
11. F. Marchesano, Progress in D-brane model building, *Fortsch.Phys.* **55**, 491–518, (2007). doi: 10.1002/prop.200610381.
12. R. Blumenhagen, B. Kors, D. Lust, and S. Stieberger, Four-dimensional String Compactifications with D-Branes, Orientifolds and Fluxes, *Phys.Rept.* **445**, 1–193, (2007). doi: 10.1016/j.physrep.2007.04.003.
13. R. Blumenhagen, M. Cvetič, and T. Weigand, Spacetime instanton corrections in 4D string vacua: The Seesaw mechanism for D-Brane models, *Nucl.Phys.* **B771**, 113–142, (2007). doi: 10.1016/j.nuclphysb.2007.02.016.
14. L. Ibanez and A. Uranga, Neutrino Majorana Masses from String Theory Instanton Effects, *JHEP.* **0703**, 052, (2007). doi: 10.1088/1126-6708/2007/03/052.
15. B. Florea, S. Kachru, J. McGreevy, and N. Saulina, Stringy Instantons and Quiver Gauge Theories, *JHEP.* **0705**, 024, (2007). doi: 10.1088/1126-6708/2007/05/024.
16. R. Blumenhagen, M. Cvetič, S. Kachru, and T. Weigand, D-Brane Instantons in Type II Orientifolds, *Ann.Rev.Nucl.Part.Sci.* **59**, 269–296, (2009). doi: 10.1146/annurev.nucl.010909.083113.
17. J. Polchinski, Dirichlet Branes and Ramond-Ramond charges, *Phys.Rev.Lett.* **75**, 4724–4727, (1995). doi: 10.1103/PhysRevLett.75.4724.
18. N. Arkani-Hamed, S. Dimopoulos, and G. Dvali, The Hierarchy problem and new dimensions at a millimeter, *Phys.Lett.* **B429**, 263–272, (1998). doi: 10.1016/S0370-2693(98)00466-3.
19. I. Antoniadis, N. Arkani-Hamed, S. Dimopoulos, and G. Dvali, New dimensions at a millimeter to a Fermi and superstrings at a TeV, *Phys.Lett.* **B436**, 257–263, (1998). doi: 10.1016/S0370-2693(98)00860-0.
20. D. Friedan, E. Martinec, and S. Shenker, Covariant quantization of superstrings, *Physics Letters B.* **160**(1-3), 55–61, (1985).
21. D. Friedan, E. Martinec, and S. Shenker, Conformal invariance, supersymmetry, and string theory, *Nuclear Physics B.* **271**(3-4), 93–165, (1986).
22. V. Knizhnik, Covariant fermionic vertex in superstrings, *Physics Letters B.* **160**(6), 403–407, (1985).

23. L. Dixon, D. Friedan, E. Martinec, and S. Shenker, The conformal field theory of orbifolds, *Nuclear Physics B.* **282**, 13–73, (1987).
24. J. Polchinski, String theory, vols. 1 and 2, *Cambridge University Press.* (1998).
25. M. Peskin, Introduction to string and superstring theory ii, *TASI Lectures.* (1986).
26. M. Cvetič and R. Richter, Proton decay via dimension-six operators in intersecting D6-brane models, *Nucl.Phys.* **B762**, 112–147, (2007). doi: 10.1016/j.nuclphysb.2006.10.029.
27. M. Berkooz, M. R. Douglas, and R. G. Leigh, Branes intersecting at angles, *Nucl.Phys.* **B480**, 265–278, (1996). doi: 10.1016/S0550-3213(96)00452-X.
28. C. Bachas, A Way to break supersymmetry. (1995).
29. M. B. Green and J. H. Schwarz, Anomaly cancellation in supersymmetric d=10 gauge theory and superstring theory, *Physics Letters B.* **149**, 117–122, (1984).
30. G. Aldazabal, S. Franco, L. E. Ibanez, R. Rabadan, and A. Uranga, D = 4 chiral string compactifications from intersecting branes, *J.Math.Phys.* **42**, 3103–3126, (2001). doi: 10.1063/1.1376157.
31. R. Blumenhagen, V. Braun, B. Kors, and D. Lust, Orientifolds of K3 and Calabi-Yau manifolds with intersecting D-branes, *JHEP.* **0207**, 026, (2002).
32. R. Blumenhagen, V. Braun, B. Kors, and D. Lust, The standard model on the quintic. (2002).
33. M. Cvetič, J. Halverson, and R. Richter, Realistic Yukawa structures from orientifold compactifications, *JHEP.* **0912**, 063, (2009). doi: 10.1088/1126-6708/2009/12/063.
34. E. Witten, D-branes and K theory, *JHEP.* **9812**, 019, (1998).
35. S. Forste and G. Honecker, Rigid D6-branes on T6/[Z(2)xZ(2M)xOmegaR] with discrete torsion. (2010).
36. R. Blumenhagen, F. Gmeiner, G. Honecker, D. Lust, and T. Weigand, The Statistics of supersymmetric D-brane models, *Nucl.Phys.* **B713**, 83–135, (2005). doi: 10.1016/j.nuclphysb.2005.02.005.
37. M. Cvetič, T. Li, and T. Liu, Supersymmetric Pati-Salam models from intersecting D6-branes: A Road to the standard model, *Nucl.Phys.* **B698**, 163–201, (2004). doi: 10.1016/j.nuclphysb.2004.07.036.
38. M. Cvetič and I. Papadimitriou, Conformal field theory couplings for intersecting D-branes on orientifolds, *Phys.Rev.* **D68**, 046001, (2003). doi: 10.1103/PhysRevD.68.046001,10.1103/PhysRevD.70.029903.
39. S. Abel and A. Owen, Interactions in intersecting brane models, *Nucl.Phys.* **B663**, 197–214, (2003). doi: 10.1016/S0550-3213(03)00370-5.
40. M. Bertolini, M. Billo, A. Lerda, J. F. Morales, and R. Russo, Brane world effective actions for D-branes with fluxes, *Nucl.Phys.* **B743**, 1–40, (2006). doi: 10.1016/j.nuclphysb.2006.02.044.
41. D. Lust, P. Mayr, R. Richter, and S. Stieberger, Scattering of gauge, matter, and moduli fields from intersecting branes, *Nucl.Phys.* **B696**, 205–250, (2004). doi: 10.1016/j.nuclphysb.2004.06.052.
42. D. Cremades, L. Ibanez, and F. Marchesano, Computing Yukawa couplings

from magnetized extra dimensions, *JHEP*. **0405**, 079, (2004). doi: 10.1088/1126-6708/2004/05/079.
43. S. Kachru, R. Kallosh, A. D. Linde, and S. P. Trivedi, De Sitter vacua in string theory, *Phys.Rev.* **D68**, 046005, (2003). doi: 10.1103/PhysRevD.68.046005.
44. M. Cvetič and P. Langacker, D-Instanton Generated Dirac Neutrino Masses, *Phys.Rev.* **D78**, 066012, (2008). doi: 10.1103/PhysRevD.78.066012.
45. M. Cvetič, J. Halverson, P. Langacker, and R. Richter, The Weinberg Operator and a Lower String Scale in Orientifold Compactifications, *JHEP*. **1010**, 094, (2010). doi: 10.1007/JHEP10(2010)094.
46. E. Witten, Nonperturbative superpotentials in string theory, *Nucl.Phys.* **B474**, 343–360, (1996). doi: 10.1016/0550-3213(96)00283-0.
47. R. Blumenhagen, A. Collinucci, and B. Jurke, On Instanton Effects in F-theory, *JHEP*. **1008**, 079, (2010). doi: 10.1007/JHEP08(2010)079.
48. M. Cvetič, I. Garcia-Etxebarria, and J. Halverson, Global F-theory Models: Instantons and Gauge Dynamics. (2010).
49. M. Cvetič, I. Garcia-Etxebarria, and J. Halverson, On the computation of non-perturbative effective potentials in the string theory landscape: IIB/F-theory perspective. (2010).
50. R. Argurio, M. Bertolini, S. Franco, and S. Kachru, Metastable vacua and D-branes at the conifold, *JHEP*. **0706**, 017, (2007). doi: 10.1088/1126-6708/2007/06/017.
51. R. Argurio, M. Bertolini, G. Ferretti, A. Lerda, and C. Petersson, Stringy instantons at orbifold singularities, *JHEP*. **0706**, 067, (2007). doi: 10.1088/1126-6708/2007/06/067.
52. M. Cvetič, J. Halverson, and R. Richter, Mass Hierarchies from MSSM Orientifold Compactifications, *JHEP*. **1007**, 005, (2010). doi: 10.1007/JHEP07(2010)005.
53. I. Antoniadis, E. Kiritsis, and T. Tomaras, A D-brane alternative to unification, *Phys.Lett.* **B486**, 186–193, (2000). doi: 10.1016/S0370-2693(00)00733-4.
54. G. Aldazabal, L. E. Ibanez, F. Quevedo, and A. Uranga, D-branes at singularities: A Bottom up approach to the string embedding of the standard model, *JHEP*. **0008**, 002, (2000).
55. I. Antoniadis, E. Kiritsis, and T. Tomaras, D-brane standard model, *Fortsch.Phys.* **49**, 573–580, (2001).
56. P. Anastasopoulos, T. Dijkstra, E. Kiritsis, and A. Schellekens, Orientifolds, hypercharge embeddings and the Standard Model, *Nucl.Phys.* **B759**, 83–146, (2006). doi: 10.1016/j.nuclphysb.2006.10.013.
57. L. Ibanez and R. Richter, Stringy Instantons and Yukawa Couplings in MSSM-like Orientifold Models, *JHEP*. **0903**, 090, (2009). doi: 10.1088/1126-6708/2009/03/090.
58. P. Anastasopoulos, E. Kiritsis, and A. Lionetto, On mass hierarchies in orientifold vacua, *JHEP*. **0908**, 026, (2009). doi: 10.1088/1126-6708/2009/08/026.
59. R. Donagi and M. Wijnholt, Model Building with F-Theory. (2008).

60. C. Beasley, J. J. Heckman, and C. Vafa, GUTs and Exceptional Branes in F-theory - I, *JHEP.* **0901**, 058, (2009). doi: 10.1088/1126-6708/2009/01/058.
61. C. Beasley, J. J. Heckman, and C. Vafa, GUTs and Exceptional Branes in F-theory - II: Experimental Predictions, *JHEP.* **0901**, 059, (2009). doi: 10.1088/1126-6708/2009/01/059.
62. T. Weigand, Lectures on F-theory compactifications and model building, *Class.Quant.Grav.* **27**, 214004, (2010). doi: 10.1088/0264-9381/27/21/214004.
63. R. Blumenhagen, M. Cvetič, D. Lust, R. Richter, and T. Weigand, Non-perturbative Yukawa Couplings from String Instantons, *Phys.Rev.Lett.* **100**, 061602, (2008). doi: 10.1103/PhysRevLett.100.061602.
64. R. Blumenhagen, V. Braun, T. W. Grimm, and T. Weigand, GUTs in Type IIB Orientifold Compactifications, *Nucl.Phys.* **B815**, 1–94, (2009). doi: 10.1016/j.nuclphysb.2009.02.011.

Chapter 6

Supergravity and String Vacua in Various Dimensions

Washington Taylor

Center for Theoretical Physics, Department of Physics,
Massachusetts Institute of Technology,
77 Massachusetts Avenue,
Cambridge, MA 02139, USA
wati@mit.edu

These lectures describe some classes of known supersymmetric string vacua in various dimensions greater than 4, and how these sets of theories fit into the larger space of possible quantum supergravity theories. Anomalies and other quantum consistency conditions, combined with supersymmetry, place strong limitations on the space of physical theories which can be coupled to gravity in space-times with a large number of dimensions. There is a unique supersymmetric theory of gravity in eleven dimensions, which is realized in the regime of string theory known as "M-theory". As the dimensionality of space-time decreases, the range of possible supergravity theories expands, as does the set of string vacua. We focus on theories in 10, 8, and 6 dimensions, and develop the basic technology for describing a variety of string vacua, including heterotic, intersecting brane, and F-theory constructions. The resulting picture of the space of supergravity and string vacua provides a number of lessons for understanding the space of theories in four dimensions.

6.1. Introduction

6.1.1. *Motivation*

Quantum field theory is an incredibly successful framework for describing the fundamental processes underlying most observable physics. Quantities such as the anomalous magnetic dipole moment of the electron can be computed from first principles in quantum electrodynamics (QED), giving results that agree with experiment for up to 10 significant digits. While in quantum chromodynamics (QCD) many physically interesting questions cannot be addressed using perturbative calculations, the framework of field theory itself is believed to be adequate for describing most observed phenomena involving strong nuclear interactions. The difficulty in the case of QCD arises from the strong coupling constant, which necessitates nonperturbative treatment by methods such as lattice simulation. The standard model of particle physics, which underlies essentially all observed phenomena outside gravity, is itself a quantum field theory. And it is likely that whatever phenomena are discovered at the LHC will also be describable in terms of a quantum field theory, whether in terms of a Higgs scalar field, supersymmetry, new strongly coupled physics, or other more exotic possibilities.

Quantum field theory itself, however, does not place stringent constraints on what kinds of theories can be realized in nature. Field theories can be constructed with virtually any gauge symmetry group, and with a wide range of possible matter content consisting of particles transforming in various representations of the gauge group. While the absence of quantum anomalies in gauge symmetries, as well as other macroscopic consistency conditions, place some simple constraints on what is possible within a quantum field theory, the space of apparently consistent field theories is vast.

The one physically observed phenomenon that cannot be directly described in terms of quantum field theory is gravity. As discussed in other lectures in this school, in some cases quantum gravity has a dual description in terms of a quantum field theory in a lower dimension. Attempting to describe a diffeomorphism invariant theory in space-time of dimension 4 or greater in terms of a field theory in the same space-time, however, runs into well-known difficulties. Quantum field theory is defined in a fixed background space-time, and allowing the space-time metric and topology itself to fluctuate takes us outside the regime of applicability of standard field theory methods.

Constructing a consistent quantum theory of gravity has proven to be substantially more difficult than identifying a quantum theory describing the other forces in nature. String theory unifies gravity with quantum physics. From the point of view of the low-energy field theory, however, string theory requires an infinite sequence of massive fields to smooth the divergences encountered when the theory is coupled to gravity. While there are many different ways in which the extra dimensions of string theory can be "compactified" to give different low-energy field theories coupled to gravity in four dimensions, the space of such models still seems much smaller than the full space of 4D quantum field theories.

This leads to a fundamental question, which can be taken as motivation for these lectures:

Motivating question: *Does the inclusion of gravity place substantial constraints on the set of consistent low-energy quantum field theories?*

By low-energy here, we mean below the Planck scale.

To address this question we will attempt to give a global characterization of two general classes of theories, one of which is contained within the other.

$$\begin{aligned} \mathcal{G} &= \{\text{apparently consistent low-energy field theories coupled to gravity}\} \\ \mathcal{V} &= \{\text{low-energy theories arising from known string constructions}\} \end{aligned} \tag{6.1}$$

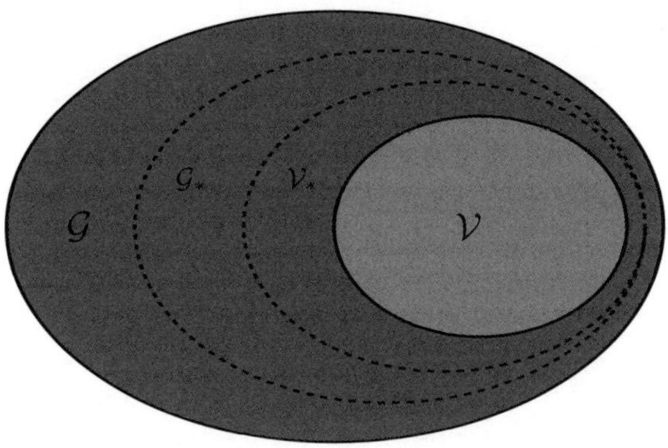

Fig. 6.1. Venn diagram of supergravity theories. The set $\mathcal{G}$ of apparently consistent quantum gravity theories (largest set, red + green regions) contains the set $\mathcal{V}$ of known string vacua in any particular dimension (smallest set, green region). Intermediate sets $\mathcal{G}_*$, $\mathcal{V}_*$ denote the mathematically complete sets of consistent gravity theories and string vacua respectively. Set inclusions satisfy $\mathcal{G} \supseteq \mathcal{G}_* \supseteq \mathcal{V}_* \supseteq \mathcal{V}$.

By "apparently consistent", we mean that there is no known obstruction to constructing a consistent UV-complete quantum theory with the desired properties. By "known string constructions", we mean a compactification of string theory in any of the regimes in which it is currently understood (*i.e.*, heterotic, type I/II, M-theory, F-theory, *etc.*). While some of these vacua cannot yet be described even perturbatively through a sigma model, and no background-independent fundamental formulation of string theory yet exists, we assume that there is a consistent quantum theory underlying string constructions. Pragmatically, this means that we assume that a few high-dimensional supergravity theories have consistent UV completions through string/M-theory, and consider the network of models that can be constructed by compactifying and adding branes and other features.

Assuming that string theory is indeed a consistent theory of quantum gravity, we have the inclusion

$$\mathcal{G} \supseteq \mathcal{V}\,. \tag{6.2}$$

The sets $\mathcal{G}$ and $\mathcal{V}$ as defined above are dependent upon our state of knowledge. We denote by $\mathcal{G}_*$ the space of *actually* consistent gravity theories, and $\mathcal{V}_*$ the complete space of possible string vacuum constructions (including those not yet discovered). The sets $\mathcal{G}_*$ and $\mathcal{V}_*$ are presumed to admit a mathematically precise definition, which has not yet been determined; these sets do not depend upon our state of knowledge. We then have the series of inclusions

$$\mathcal{G} \supseteq \mathcal{G}_* \supseteq \mathcal{V}_* \supseteq \mathcal{V}\,. \tag{6.3}$$

By discovering new quantum consistency constraints, the space $\mathcal{G}$ can be reduced, and by discovering new string vacuum constructions the space $\mathcal{V}$ can be expanded[a]. As we discuss in these lectures, focusing on theories $x \in \mathcal{G} \setminus \mathcal{V}$ can be a useful way of advancing knowledge. If $x \in \mathcal{V}_*$ then there is a new string construction including x. If $x \notin \mathcal{G}_*$ there is a physical constraint that makes x an inconsistent theory. And if $x \in \mathcal{G}_* \setminus \mathcal{V}_*$ then there is a constraint particular to string theory that is violated by x. This discussion is closely related to Vafa's notion of the "swampland",[1,2] which consists of gravity theories that appear to be consistent but which are not realized in string theory. In the notation defined above, the swampland is the set $\mathcal{G}\setminus\mathcal{V}_*$. We will sometimes refer to the set $\mathcal{G}\setminus\mathcal{V}$ as the "apparent swampland". Part of the motivation for the approach taken here is the notion that we can learn a lot by trying to remove theories from the apparent swampland.

When the dimension of space-time is sufficiently large, and we restrict attention to gravity theories with supersymmetry, the constraints on $\mathcal{G}$ are quite strong. In particular, as we discuss below in more detail, in 11 and 10 space-time dimensions the constraints are sufficiently strong that all supergravity theories not realized in string theory are known to be inconsistent.

[a] It is of course possible that some string vacuum constructions currently viewed as plausible by some parts of the community may be inconsistent, even if string theory is a fine theory. In this case we define $\mathcal{V}$ to be the subset of hypothesized string vacua that are actually in $\mathcal{V}_*$; in addition to the goals outlined in the text, the program described here can also be of assistance in determining which hypothetical string vacuum constructions are inconsistent (for example if a constraint on $\mathcal{G}$ can be identified that is violated by some proposed vacua).

Thus, for the space of 11- and 10-dimensional supergravity theories we have $\mathcal{G} = \mathcal{V}$. It follows that

$$\mathcal{G}_* = \mathcal{V}_* \quad \text{for 10D and 11D supergravity}. \tag{6.4}$$

We say that *string universality* holds for supersymmetric theories of gravity in these dimensions, meaning that every theory not known to be inconsistent is realized in string theory.[3]

In these lectures we focus attention on the discrete data characterizing the structure of supergravity theories: namely the field content and symmetries of the theory. The statement of string universality in (6.4) has been demonstrated in 10 and 11 dimensions at the level of these discrete structures. Any complete supergravity theory will contain a range of additional detailed structure encoded in continuous parameters such as coefficients of higher-derivative terms, the metric on moduli space, etc. Determining the uniqueness of such additional structure would refine our understanding of the relationship between the spaces $\mathcal{G}$ and $\mathcal{V}$. While this represents a very interesting class of questions and challenges, we do not discuss these issues here.

In dimensions 10 and 11, as we shall describe in more detail below, there are only a handful of possible theories. As the dimension of space-time and the amount of supersymmetry decreases, the constraints on possible theories become weaker, and the range of possible string constructions increases. We focus here on the classes of gravity theories with minimal supersymmetry in even dimensions. As the dimension decreases, the range of interesting physical phenomena also expands. In 8 dimensions, there are supergravity theories coupled to gauge theories with many different gauge groups. In 6 dimensions there can also be matter fields living in many different representations of the gauge group. In 4 dimensions, couplings such as Yukawa-type interactions arise, which increase the complexity of the theories significantly. These lectures focus primarily on dimensions 10, 8, and 6. By following possible string constructions through the decreasing range of dimensions, we can systematically introduce the new features and mechanisms giving rise to the phenomena that arise in each dimension. Thus, this approach gives a convenient pedagogical framework for introducing many fundamental aspects of different approaches to the construction of string vacua.

We do not discuss odd-dimensional supergravity theories here. While there are also interesting questions in odd dimensions, the global picture of theories in dimensions such as 9D is similar to the story described here for

8D theories. In even dimensions, there is also a somewhat richer structure both from the point of view of supergravity, where anomaly constraints can be stronger, and string theory, where F-theory provides a powerful nonperturbative framework for describing the space of string vacua. A systematic analysis of the discrete classes of string vacua with minimal supersymmetry in dimensions 9, 8, and 7 can be found in the work of de Boer *et al.*[4]

Another aspect of the theme of global structure of the space of theories is the connectivity of the space of theories. While different string theories, and different string vacua, may seem physically distinct, most of these theories are connected in various ways. Perturbative and nonperturbative duality symmetries relate different string constructions. In eight dimensions and six dimensions, most or all of the wide range of possible supersymmetric string vacua are different branches of a single theory, living in a continuous moduli space that is described in different regimes by different string constructions. In each of these dimensions we describe how the connectivity of this set of spaces arises, and the sense in which the diversity of string models fit into a single overarching theory.

We conclude this introductory section with a brief outline of the material presented in these lectures. In each of the dimensions 10, 8, and 6, we characterize the supergravity theories and introduce string vacuum constructions, then we compare what is known about the spaces $\mathcal{G}$ and $\mathcal{V}$. Ten dimensions provides a good starting point for a systematic discussion of both supergravity theories and string vacua. After a brief summary of 11-dimensional supergravity, we describe the basic supergravity theories in 10 dimensions from which the lower-dimensional theories descend. On the supergravity side, anomaly constraints and the Green-Schwarz anomaly cancellation mechanism provide powerful tools for understanding the space $\mathcal{G}$. On the string side, basic objects such as strings and branes arise naturally and can be most easily understood in 10 dimensions. As mentioned above, in 10 dimensions $\mathcal{G} = \mathcal{V}$. Going down to 8 dimensions, we introduce the heterotic and F-theory approaches to string compactification. We describe how different gauge groups can be realized in these two types of constructions, and how the constructions are related. In each case, 8 dimensions provides a natural domain in which to introduce the essential features of these classes of string vacua. In 8 dimensions we believe that we can identify $\mathcal{V}$ with $\mathcal{V}_*$, but many theories still lie in $\mathcal{G} \setminus \mathcal{V}$. In six dimensions, as in 10 dimensions, anomaly constraints provide a powerful tool for understanding the space $\mathcal{G}$ of gravity theories without known inconsistencies. We discuss additional complexities in the heterotic and F-theory vacuum constructions in 6D, and introduce intersecting brane models as another class of string vacua. F-theory vacua provide the largest set of 6D vacua constructions. Mathematical structure appearing in the anomaly cancellation conditions for any six-dimensional supergravity theory corresponds very closely to topological data for F-theory constructions. This enables us to identify a "bottom-up" map from low-energy theories to candidate string vacua in six dimensions. This map gives us a characterization of the embedding $\mathcal{V} \subset \mathcal{G}$ and allows us to identify general features of models in the "apparent swampland" $\mathcal{G} \setminus \mathcal{V}$. F-theory provides a framework in which known supersymmetric 6D string vacua fit together into a single theory, with a moduli space connected through continuous deformations and phase transitions that can be understood in terms of F-theory geometry. We conclude the lectures with some comments on four-dimensional theories. While in four dimensions the spaces $\mathcal{G}$ and $\mathcal{V}$ are much larger and less well understood than in higher dimensions, some lessons from higher-dimensional supergravity and string constructions may be helpful in characterizing global aspects of the space of possible theories.

6.1.2. *Background*

These lectures are intended for an audience with some knowledge and experience with the basic principles and tools of quantum field theory. Some familiarity with elementary aspects of string theory is also helpful, though little specific technical knowledge is assumed. In the early part of the lectures, a number of concepts related to supersymmetry, supergravity, and perturbative string theory are reviewed briefly. The reader interested in more background on these topics should consult the textbooks of Green, Schwarz, Witten[5,6] and Polchinski.[7,8] The lectures generally follow the notation and conventions of Polchinski. A comprehensive overview of early work on supergravity theories in various dimensions can be found in the two-volume compilation by Salam and Sezgin.[9]

The material covered in these lectures has evolved somewhat since the lectures were given, as some new results have clarified parts of the story. These written lecture notes integrate developments up to the time of writing (March 2011).

6.1.3. *Supersymmetry*

In these lectures we will restrict attention to theories with supersymmetry. Supersymmetry is a symmetry that relates bosons to fermions. We are interested here in theories where supersymmetry can be described as an extension of the Poincaré symmetry group that characterizes field theories in Minkowski space. At the level of the algebra of generators of the symmetry group, supersymmetry extends the Poincaré algebra by a set of fermionic generators $Q_\alpha, \bar{Q}_\alpha$ satisfying the anticommutation relations

$$\{Q_\alpha, \bar{Q}_\beta\} = 2P_\mu \Gamma^\mu_{\alpha\beta}, \tag{6.5}$$

where α, β are spinor indices. The way in which bosonic and fermionic fields transform under an infinitesimal supersymmetry transformation parameterized by a spinor ϵ takes the schematic form

$$\delta\phi \sim \bar{\epsilon}\psi, \qquad \delta\psi \sim \Gamma^\mu \epsilon \partial_\mu \phi. \tag{6.6}$$

Some theories have multiple supersymmetries, with generators Q^A_α, parameterized by an index $A = 1, \ldots \mathcal{N}$, where $\mathcal{N}$ is the number of supersymmetries in the theory. In this case (6.5) generalizes to

$$\{Q^A_\alpha, \bar{Q}^B_\beta\} = 2\delta^{AB} P_\mu \Gamma^\mu_{\alpha\beta}. \tag{6.7}$$

The supersymmetry algebra (6.7) can be extended by central charges that identify topological charges in the theory; for example, (6.7) can be extended to

$$\{Q^A_\alpha, \bar{Q}^B_\beta\} = 2\delta^{AB} P_\mu \Gamma^\mu_{\alpha\beta} + Z^{AB}\delta_{\alpha\beta}, \tag{6.8}$$

where the central charge Z^{AB} commutes with all other generators.

A description of supersymmetry and spinors in various dimensions is given in Appendix B of Volume 2 of Polchinski's text on string theory.[8] We will not use too many detailed aspects of supersymmetry and spinors in these lectures, but will assume some basic facts for which more detailed explanations can be found in that reference.

For a quantum field theory in flat space-time, supersymmetry is generally a global symmetry of the theory. If a theory of gravity has supersymmetry, however, then the symmetry becomes local. Just as translation symmetry in flat space-time becomes a symmetry under local diffeomorphisms in a generally covariant theory of gravity, in a supersymmetric generally covariant theory the spinor parameter ϵ itself becomes a general space-time dependent function. A theory of gravity with local supersymmetry is called a *supergravity* theory. Each supergravity theory contains $\mathcal{N}$ massless spin 3/2 *gravitino* fields $\psi^A_{\mu\alpha}$ that are partners of the graviton $g_{\mu\nu}$.

A primary reason for considering only supersymmetric theories of gravity in these lectures is that supersymmetry imposes additional structure that makes both gravity and string theory easier to analyze and to understand. Indeed, there are unresolved foundational and conceptual issues associated with global aspects of non-supersymmetric quantum gravity theories with positive cosmological constant, some of which are discussed in the lectures of Tom Banks at this school.[10] There are, however, physical reasons to be interested in supersymmetric theories, both from the point of view of low-energy phenomenology, and from the top-down point of view of string theory.

From the phenomenological point of view, supersymmetry has a number of desirable features. Supersymmetry modifies the renormalization group equations so that the strong, weak and electromagnetic couplings appear to unify at a high scale. Supersymmetry protects the mass of the scalar Higgs, giving a possible solution to the "hierarchy" problem. Supersymmetry also provides a natural candidate for dark matter.

From the string theory point of view, supersymmetry plays a crucial role in removing tachyonic instabilities from the theory at the Planck scale. While it is possible that some intrinsically non-supersymmetric versions of string theory can be made mathematically consistent, string theory is best understood as a supersymmetric theory of quantum gravity in 10 dimensions.

While understanding the space of supersymmetric theories is a rewarding enterprising in its own right, which may also give new insights into the structure of non-supersymmetric theories, the most optimistic reason to study supersymmetric theories is the possibility that supersymmetry is manifest in our world at an experimentally accessible energy scale. The simplest framework in which this occurs can be analyzed following the assumption that physics can be split into two different energy scales. Under this assumption, at and below some scale Λ, all of the relevant physics describing our world can be characterized by a supersymmetric quantum field theory coupled to (classical) gravity. Above the scale Λ, supergravity and/or string theory are needed to describe physics at energies up to the Planck scale. If this assumption is correct, it means that quantum gravity will play a phenomenological role primarily in determining which supersymmetric QFT's can arise at the intermediate scale Λ. If, on the other hand, this assumption is incorrect and supersymmetry is broken at the Planck scale, making any progress in understanding the connection between quantum gravitational consistency and low-energy physics will be extremely challenging. From a theoretical point of view, one of the most exciting consequences of the discovery of supersymmetry at the LHC would be the confirmation that this separation of scales exists, guaranteeing that supersymmetry can be used in efforts to understand physics up to the Planck scale.

6.1.4. *Supergravity in 11 dimensions*

We are interested, therefore, in understanding supersymmetric theories of gravity (supergravity), which may also include gauge fields and various kinds of matter. In higher dimensions, supergravity theories are quite constrained. For dimensions $D > 11$, any representation of the Clifford algebra associated with gamma matrices of the relativistic symmetry group has dimension 64 or greater. This leads to massless particles related to the graviton by supersymmetry that have spin greater than 2. No interacting theories of this kind are known in dimensions above 11, and it is believed (though perhaps not rigorously proven) that the highest dimension in which a supersymmetric theory of gravity can exist is 11 dimensions.

In 11 dimensions, there is a unique supersymmetric theory of gravity. This theory has one supersymmetry ($\mathcal{N} = 1$), and 32 supercharges Q_α carrying an index in the 32-dimensional spinor representation of $SO(1, 10)$. The massless fields in the theory describe particles that are in the *supermultiplet* of the graviton; *i.e.*, states that are related to the graviton by acting with the supersymmetry generators. These fields include:

$g_{\mu\nu}$: the graviton (quantum of fluctuation in the space-time metric)
The graviton is symmetric and traceless, with $(9 \times 10)/2 - 1 = 44$ degrees of freedom.

$C_{\mu\nu\lambda}$: an antisymmetric 3-form field (analogous to a gauge field A_μ but with more indices)
The 3-form field has $(9 \times 8 \times 7)/6 = 84$ degrees of freedom.

$\psi_{\mu\alpha}$: the gravitino, with 128 degrees of freedom.

In each case the number of degrees of freedom can be understood by considering the appropriate representation of the $SO(9)$ little group for massless states in $SO(1, 10)$, in a fashion directly analogous to the standard analysis of states in 4D QFT.

The low-energy action for the bosonic fields of 11-dimensional supergravity is given by

$$S = \frac{1}{2\kappa_{11}^2} \left[\int \sqrt{g}(R - \frac{1}{2}|F|^2) - \frac{1}{6} \int C \wedge F \wedge F \right], \tag{6.9}$$

where κ_{11} is the 11-dimensional Newton constant, and the field strength F

is a 4-form given by

$$F^{(4)} = dC^{(3)} . \tag{6.10}$$

This is a theory of pure supersymmetric gravity, with no conventional gauge symmetries or matter content (though the 3-form field C does have a higher-index version of an abelian gauge symmetry). For further details and references on this theory, the reader is referred to Polchinski's text.[8]

From the point of view of these lectures, the significant feature of supergravity in 11 dimensions is that there is only one possible theory, at least at the discrete level of field content and symmetries. Thus,

$$\mathcal{G}^{(11)} = \{M_{11}\} \tag{6.11}$$

where M_{11} is the supergravity theory with bosonic action (6.9). Although this theory does not itself contain strings, it can be viewed as the strong coupling limit of a ten-dimensional string theory.[11] Thus, this theory is included in the general space $\mathcal{V}$ of "known string vacua", and in 11 dimensions we have

$$\mathcal{G}^{(11)} = \mathcal{V}^{(11)} \quad \Rightarrow \quad \mathcal{G}_*^{(11)} = \mathcal{V}_*^{(11)} . \tag{6.12}$$

So string universality holds in 11 dimensions. The quantum theory of 11-dimensional supergravity is often referred to as "M-theory". As we discuss below, M-theory can also be described in the light-cone gauge through quantization of the membrane in the 11-dimensional theory, or alternatively in terms of pointlike branes in a 10D theory.

6.2. Supergravity and String Vacua in Ten Dimensions

We now consider supergravity theories in ten dimensions, and their UV completions through string theory. In ten dimensions, there are theories with one or two supersymmetries. We begin with a brief summary of the gravity theories with two supersymmetries, before considering the theories with one supersymmetry in more detail. We then discuss the string realization of these theories and the relationship between $\mathcal{G}$ and $\mathcal{V}$. Again, further details on most of the material in this section can be found in Polchinski.[8]

In 10 dimensions we can define an 11th gamma matrix

$$\Gamma^{11} = \prod_{\mu=0}^{9} \Gamma^{\mu} \tag{6.13}$$

analogous to γ^5 in four dimensions. As in four dimensions, Γ^{11} has eigenvalues ± 1 corresponding to Weyl spinors of fixed chirality. In dimensions of the form $D = 4k + 2$, the Weyl representations are self-conjugate. In ten dimensions it is also possible to impose a Majorana (reality) condition. The Lorentz group in 10D thus has two distinct real 16-dimensional representations $\mathbf{16}, \mathbf{16}'$ corresponding to Majorana-Weyl chiral spinors. The different supergravity theories in 10D have different choices of spinor representations for the supersymmetry generators Q^A_α.

6.2.1. $\mathcal{N} = 2$ *supergravity in ten dimensions*

6.2.1.1. *Type IIA supergravity*

The type IIA $\mathcal{N} = 2$ supergravity theory in ten dimensions has two supersymmetries of opposite chirality, generated by $Q^1 \in \mathbf{16}, Q^2 \in \mathbf{16}'$. The theory has bosonic fields

$$g_{\mu\nu}, B_{\mu\nu}, \phi \tag{6.14}$$

living in an $\mathcal{N} = 1$ supersymmetry multiplet. The field $B_{\mu\nu}$ is an antisymmetric two-form field, and the field ϕ is a scalar (the *dilaton*). The IIA theory has additional bosonic fields described by a 1-form and 3-form

$$A_\mu, C_{\mu\nu\lambda}\,. \tag{6.15}$$

From the nature of the stringy origin of these fields, they are referred to as *Ramond-Ramond*, or *R-R* fields.

Counting degrees of freedom, the bosonic fields in the IIA theory have 128 components. In accord with supersymmetry, this is the same as the number of fermionic degrees of freedom, which are contained in a pair of Majorana-Weyl gravitinos ($\mathbf{56} + \mathbf{56}'$) and a pair of Majorana-Weyl spinors[b] ($\mathbf{8} + \mathbf{8}'$). Because there is one supersymmetry of each type, the theory is non-chiral, with one spinor and one gravitino of each chirality.

The IIA theory is directly related to 11-dimensional supergravity by *compactification* on a circle S^1. Wrapping the 11D theory described by the action (6.9) on a circle of radius R, as R becomes small the momentum (Kaluza-Klein) modes on the extra circle become very massive, and in the low-energy limit the zero modes of the theory combine into the fields of

[b]Note that the Majorana-Weyl spinors characterizing the supersymmetry generators transform under the relativistic symmetry group $SO(1,9)$, while the on-shell degrees of freedom in a massless spinor field transform under the little group $SO(8)$; this explains the discrepancy between the 16 and 8 real degrees of freedom in these representations.

the IIA supergravity theory. The dilaton, for example, comes from the component of the metric tensor in 11D with both indices wrapped in the compact direction $g_{11\,11} \to \phi$. The correspondence between the 11D and 10D degrees of freedom for all the bosonic fields is given by

$$\text{IIA}: \quad \begin{array}{cccccc} \text{11D SUGRA} & g^{(11)}_{\mu\nu} & g^{(11)}_{\mu\,11} & g^{(11)}_{11\,11} & C^{(11)}_{\mu\nu\lambda} & C^{(11)}_{\mu\nu\,11} \\ \downarrow (S^1) & \downarrow & \downarrow & \downarrow & \downarrow & \downarrow \\ \mathbb{R}^{1,9} & g_{\mu\nu} & A_\mu & \phi & C_{\mu\nu\lambda} & B_{\mu\nu} \end{array} \tag{6.16}$$

6.2.1.2. *Type IIB supergravity*

Type IIB supergravity in ten dimensions is similar to the type IIA theory in some ways, but is distinguished by having two supersymmetries of the same chirality, $Q^1, Q^2 \in \mathbf{16}$. The type IIB theory again contains the $\mathcal{N} = 1$ multiplet of fields $g_{\mu\nu}, B_{\mu\nu}, \phi$, as well as a Ramond-Ramond axion χ, two-form $\tilde{B}_{\mu\nu}$, and self-dual 4-form field $D^+_{\mu\nu\lambda\sigma}$ whose massless states transform in the chiral **70** representation of the little group $SO(8)$. In this theory there are two gravitinos with the same chirality ($2\times$ **56**), and two spinors of identical chirality ($2 \times \mathbf{8}'$). The self-dual field D^+ makes it difficult to write a local Lagrangian for this theory, but classical equations of motion for the fields can be written in an unambiguous fashion. The type IIB supergravity theory has a classical global symmetry under $SL(2, \mathbb{R})$, under which the two-form fields $B, \tilde{B}$ rotate into one another as a doublet representation.

The type IIA and IIB supergravity theories are uniquely fixed by the supersymmetries of the theory, at least in terms of the field content and low-energy equations of motion.

6.2.2. $\mathcal{N} = 1$ *supergravity in ten dimensions*

We now turn to ten-dimensional supergravity theories with only one supersymmetry, $Q_\alpha \in \mathbf{16}$. There are two kinds of supersymmetry multiplet that can appear in such theories; in addition to the gravity multiplet, there is a vector multiplet containing a vector field and a spinor called the *gaugino* in the **8** representation of SO(8). The field content of the multiplets, and the associated numbers of bosonic and fermionic degrees of freedom, can be summarized as

$$\text{10D } \mathcal{N} = 1 \text{ multiplets} \quad \begin{cases} \text{SUGRA} & g_{\mu\nu}, B_{\mu\nu}, \phi, \psi_{\mu\alpha}, \zeta_\alpha & [64 + 64 \text{ DOF}] \\ \text{vector} & A_\mu, \lambda_\alpha & [8 + 8 \text{ DOF}] \end{cases} \tag{6.17}$$

Note that the spinor ζ_α in the gravity multiplet is in the $\mathbf{8}'$ of SO(8) and has the opposite chirality to any gauginos.

Classically, $\mathcal{N} = 1$ supergravity can be coupled to a set of vector fields A^a_μ realizing any abelian or nonabelian gauge group. The action for such a theory is

$$S \sim \frac{1}{2\kappa_{10}^2} \int d^{10}x\sqrt{g} \left[e^{-2\phi} \left(R + 4\partial_\mu\phi\partial^\mu\phi - \frac{1}{2}|H|^2 \right) - \frac{e^{-\phi}}{g_{\mathrm{YM}}^2} F^a_{\mu\nu} F^{a\,\mu\nu} \right] \tag{6.18}$$

where g_{YM} is the Yang-Mills coupling constant, F is the usual field strength of the abelian or nonabelian gauge group, and the three-form field strength H is related to the two-form B field through

$$H = dB - \omega_Y \qquad d\omega_Y = \sum_a F^a \wedge F^a \,. \tag{6.19}$$

The additional contribution of the Chern-Simons term ω_Y to H is a consequence of supersymmetry;[12,13] this term produces a coupling of the form BF^2 that is related through supersymmetry to the gauge field kinetic term. We now discuss the conditions for quantum consistency of supergravity theories in 10D.

6.2.3. *Anomalies in ten-dimensional supergravity theories*

Not every theory that admits a classical Lagrangian formulation is quantum mechanically consistent. Classical symmetries can be broken by quantum anomalies. The most well-known example is the chiral anomaly in four-dimensional gauge theories, in which the current associated with a chiral symmetry develops a quantum correction at one loop. The non-vanishing of the divergence of the chiral current $\partial_\mu j^{5\,\mu} \sim F \wedge F$ amounts to a quantum breaking of the chiral symmetry. The anomaly appears because the quantum theory cannot be regulated or made UV complete without breaking the symmetry. Anomalies can be understood in terms of one-loop corrections where the failure of a regulator to preserve the symmetry is manifest. Alternatively, anomalies can be understood in terms of a failure of the measure in the path integral to respect the classical symmetry of the theory

$$\int d\psi d\bar{\psi} \neq \int d\psi' d\bar{\psi}' \,. \tag{6.20}$$

In general, anomalies have a topological origin and can be related to an appropriate mathematical index theorem. For a good introduction to anomalies and more details on aspects relevant to these lectures, see Harvey's 2003 TASI lectures.[14]

If a global symmetry in a theory becomes anomalous, such as the chiral symmetry in some 4D gauge theories, then it simply means that the symmetry is broken quantum-mechanically. The theory can still be a consistent quantum theory. If, however, a local symmetry is anomalous, it is generally impossible to find a consistent quantum completion of the theory. For example, if a theory contains a U(1) gauge symmetry that is anomalous, then the theory cannot be fixed unless an additional degree of freedom (such as a Stückleberg field) combines with the $D-2$ degrees of freedom of the massless gauge field to complete the necessary set of $D-1$ degrees of freedom for a massive gauge field.

Similarly, if local diffeomorphism invariance is broken by a quantum anomaly, there is no way to make the theory consistent as a quantum theory of gravity. As mentioned above, Weyl representations are self-conjugate in dimensions of the form $D = 4k + 2$. This means that particles and antiparticles have the same chirality, so that chiral fermions can contribute to gravitational and mixed gauge-gravitational anomalies as well as to purely gauge anomalies. The detailed form of such anomalies in ten dimensions was worked out in a classic paper by Alvarez-Gaume and Witten.[15] In ten dimensions, anomalies arise from hexagon diagrams with external gauge fields or gravitons (see Figure 6.2). The anomaly structure of any theory

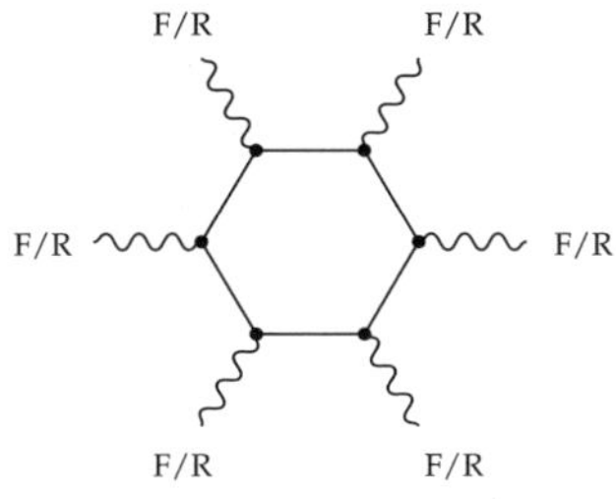

Fig. 6.2. Hexagon diagrams give rise to gravitational, gauge, and mixed gauge-gravitational anomalies in ten dimensions.

in D dimensions can be encoded in a $(D+2)$-form *anomaly polynomial* $\hat{I}$. Alvarez-Gaume and Witten showed that the respective contributions of n gaugino spinor fields (**8**), the gravitino field (**56**), and a self-dual 4-form field (**70**) to the 10D anomaly polynomial are

$$\hat{I}_{(8)} = -\frac{\mathrm{Tr}(F^6)}{1440} + \frac{\mathrm{Tr}(F^4)\mathrm{tr}(R^2)}{2304} - \frac{\mathrm{Tr}(F^2)\mathrm{tr}(R^4)}{23040} - \frac{\mathrm{Tr}(F^2)[\mathrm{tr}(R^2)]^2}{18432} + \frac{n\,\mathrm{tr}(R^6)}{725760} + \frac{n\,\mathrm{tr}(R^4)\mathrm{tr}(R^2)}{552960} + \frac{n\,[\mathrm{tr}(R^2)]^3}{1327104} \quad (6.21)$$

$$\hat{I}_{(56)} = -495\frac{\mathrm{tr}(R^6)}{725760} + 225\frac{\mathrm{tr}(R^4)\mathrm{tr}(R^2)}{552960} - 63\frac{[\mathrm{tr}(R^2)]^3}{1327104} \quad (6.22)$$

$$\hat{I}_{(70)} = 992\frac{\mathrm{tr}(R^6)}{725760} - 448\frac{\mathrm{tr}(R^4)\mathrm{tr}(R^2)}{552960} + 128\frac{[\mathrm{tr}(R^2)]^3}{1327104}. \quad (6.23)$$

In the expression for $\hat{I}_{(8)}$, the terms containing F arise from the coupling of the gaugino spinor to the gauge field. The notation Tr indicates that these traces are evaluated in the adjoint representation of the gauge group. Note that for abelian vector fields, these terms must vanish, as the adjoint representation of $U(1)$ is trivial.

The anomaly contributions from eqs. (6.21-6.23) must be summed over all chiral fields in a theory to determine the total anomaly polynomial. The condition that the total anomaly must cancel is a very strong condition on 10D gravity theories.

Let us first consider the anomaly contributions from the two $\mathcal{N} = 2$ theories. The type IIA supergravity theory is non-chiral; each chiral field has a counterpart with the opposite chirality. So in this theory all anomalies cancel.

The type IIB theory, on the other hand, has a more complicated set of contributions. The two gravitinos contribute two factors of $\hat{I}_{56}$. The self-dual four-form field D^+ contributes one factor of $\hat{I}_{70}$. The two spinors in the $\mathbf{8}'$ contribute as two gauginos in (6.21), but with the opposite sign and with no gauge contribution. The total anomaly for the IIB theory is then

$$-2\hat{I}_8(F \to 0, n \to 1) + 2\hat{I}_{56} + \hat{I}_{70} = 0\,. \quad (6.24)$$

Thus, the type IIB supergravity theory can be a quantum consistent theory through a rather intricate cancellation.

6.2.4. *The Green-Schwarz mechanism*

From the above analysis, it seems that all $\mathcal{N} = 1$ theories must be anomalous. Consider in particular the terms in $\hat{I}_{(8)}, \hat{I}_{(56)}$ proportional to $\mathrm{tr}R^6$ and $[\mathrm{tr}(R^2)]^3$. Since the spinor ζ_α in the gravity multiplet is in the $\mathbf{8}'$ and contributes to the R^6 term with the opposite sign of a gaugino, the trR^6 terms can only cancel if $n = 496$. So the gauge group must have dimension 496. But by a similar argument, the $[\mathrm{tr}(R^2)]^3$ terms cannot cancel unless $n = 64$. So both terms cannot cancel simultaneously.

This seems to doom all $\mathcal{N} = 1$ theories of supergravity in ten dimensions. But, at the time of Alvarez-Gaume and Witten's analysis it was already known that a string theory, known as the type I string, exists and corresponds at low energies to an $\mathcal{N} = 1$ supergravity theory in ten dimensions with gauge group SO(32). Green and Schwarz analyzed this string theory carefully and showed that it evades the apparent anomaly constraint through a feature now known as the *Green-Schwarz mechanism* for anomaly cancellation.[16]

The key to the Green-Schwarz mechanism is the realization that the anomalous gauge variation associated with 1-loop diagrams can be canceled by tree-level diagrams when higher-order terms that are themselves not gauge invariant are added to the action. To implement this mechanism in the case of a nonabelian gauge group , the field strength (6.19) must be enhanced at higher orders in the derivative expansion by a Chern-Simons term in the spin connection

$$H = dB - \omega_Y + k\omega_R, \qquad d\omega_R = \mathrm{tr}R^2\,, \tag{6.25}$$

with k a constant factor. The two-form B must then transform as $\delta B = \mathrm{Tr}(\Lambda F) - \mathrm{tr}(\Theta R)$ where $\delta A = d\Lambda$, and the connection 1-form transforms through $\delta\omega_1 = d\Theta$. The anomaly can now be cancelled by a tree diagram (see Figure 6.3) in which a B-field is exchanged, when a "Green-Schwarz" term

$$\Delta S \sim \int B \wedge X_8(F, R) \tag{6.26}$$

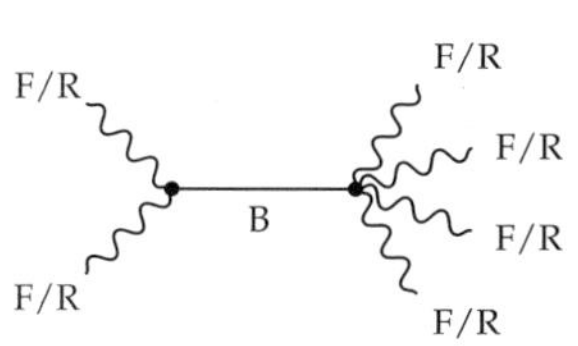

Fig. 6.3. The Green-Schwarz mechanism: A tree diagram describing exchange of a B field can cancel the anomalous part of the one-loop hexagon diagram in ten dimensions in special cases.

is added to the action such that the anomaly can be expressed in the factorized form

$$\hat{I} = Y_4(F,R)X_8(F,R), \qquad Y_4 = \mathrm{tr}R^2 - \frac{1}{30}\mathrm{Tr}F^2\,, \tag{6.27}$$

where the constant k has taken the value $k = 30$.

For a general nonabelian gauge group of dimension n, the total anomaly can be rearranged to take the form

$$\begin{aligned}\hat{I}_{12} = &\frac{1}{1440}\left(-\mathrm{Tr}F^6 + \frac{1}{48}\mathrm{Tr}F^4\mathrm{Tr}F^2 - \frac{(\mathrm{Tr}F^2)^3}{14400}\right)\\ &+(n-496)\left(\frac{\mathrm{tr}R^6}{725760} + \frac{\mathrm{tr}(R^4)\mathrm{tr}(R^2)}{552960} + \frac{[\mathrm{tr}(R^2)]^3}{1327104}\right)\\ &-\frac{Y_4X_8}{768}\,,\end{aligned} \tag{6.28}$$

where

$$X_8 = \mathrm{tr}R^4 + \frac{(\mathrm{tr}R^2)^2}{4} - \frac{(\mathrm{Tr}F^2)(\mathrm{tr}R^2)}{30} + \frac{\mathrm{Tr}F^4}{3} - \frac{(\mathrm{Tr}F^2)^2}{900}\,. \tag{6.29}$$

Thus, the Green-Schwarz mechanism can cancel the anomaly for a nonabelian group of dimension 496 precisely when the first term vanishes. There are exactly two nonabelian groups for which $\mathrm{Tr}F^6$ can be expressed as the necessary combination of $\mathrm{Tr}F^4\mathrm{Tr}F^2$ and $(\mathrm{Tr}F^2)^3$. These two groups are $SO(32)$ and $E_8 \times E_8$.

Now consider a gauge group with abelian factors. Because supersymmetry requires (6.19) to contain a contribution from the Chern-Simons term of each abelian factor, there is a coupling in the action of the form BF^2 for each $U(1)$ factor in the theory. For the pure gravitational anomaly to vanish, the Green-Schwarz term (6.26) must be added to the action. But the R^4 part of X_8 in this coupling combines with the BF^2 term required by supersymmetry to give a tree diagram contribution of the form F^2R^4. And as mentioned above such terms do not appear in hexagon diagrams for abelian gauge factors. Thus, in any theory with an abelian gauge group factor the Green-Schwarz mechanism cannot cancel the anomaly in a way that is simultaneously compatible with supersymmetry. As a result, there are no consistent $\mathcal{N} = 1$ supergravity theories in ten dimensions with gauge groups having abelian factors. The details of this argument ruling out $U(1)$ factors were given in a paper with Adams and DeWolfe.[17] This shows, in particular, that the theories with gauge groups $U(1)^{496}$ and $E_8 \times U(1)^{248}$, which were for some time in the "apparent swampland," cannot be consistent supersymmetric quantum theories of gravity.

Summary: we have shown that there are only four possible consistent ten-dimensional supergravity theories

$$\mathcal{G}^{(10)} = \{\text{IIA},\ \text{IIB},\ \text{SO}(32), \text{E}_8 \times \text{E}_8\}\,. \tag{6.30}$$

6.2.5. *String theories in ten dimensions*

In this section we give a short overview of how quantization of strings and other extended objects can lead to consistent frameworks for describing quantum gravity. For a detailed pedagogical introduction to string theory and branes, the reader should consult Polchinski.[7,8]

The fundamental excitations in any quantum field theory are pointlike quanta associated with localized particles in space-time. While directly quantizing an interacting theory of gravity in more than three dimensions using quantum field theory methods proves problematic, greater success has been realized in quantizing gravity using extended objects. As we have seen, supersymmetric theories of gravity in higher dimensions often contain antisymmetric p-form fields. For example, all supergravity theories in ten dimensions contain an antisymmetric two-form field $B_{\mu\nu}$ in the gravity multiplet, and 11-dimensional supergravity contains an antisymmetric three-form field $C_{\mu\nu\lambda}$. These antisymmetric tensor fields can couple to extended charged objects in the same way that the electromagnetic vector potential A_μ couples to charged particles. The coupling of a pointlike object with charge q under a vector field A_μ is described by a contribution to the action given by an integral along the particle world-line $x^\mu(\tau)$

$$S_0 = q \int A_\mu dx^\mu\,. \tag{6.31}$$

Similarly, the field $B_{\mu\nu}$ couples to charged stringlike excitations with one direction of spatial extent through

$$S_1 = \int B_{\mu\nu} dx^\mu dx^\nu\,, \tag{6.32}$$

and a 3-form field couples to a membrane with two spatial dimensions in a similar fashion, etc. Just as gravity coupled to electromagnetism in four dimensions admits classical solutions with mass and charge (Reissner-Nordström black holes), gravitational theories in higher dimensions with $(p + 1)$-form fields admit extended charged "black brane" solutions. In supersymmetric theories, the central charge in the supersymmetry algebra places an upper bound on the charge/mass ratio possible for such brane solutions. Solutions which saturate this bound are known as *extremal* solutions. Extremal brane solutions will be quantized in a quantum theory of gravity. By considering the quantum field theory on the world-volume of the branes, a theory of quantum gravity in the bulk space-time emerges in many cases.

The principle of *brane democracy*[18] states that any of the branes appearing in a theory can in some sense be treated as fundamental degrees of freedom. Indeed, to the extent it is possible to quantize branes in any supergravity theory, the resulting quantum theory seems to give at least a limited description of a consistent quantum gravity theory. The p-form fields and associated extended objects appearing in 11D and 10D supergravity theories are listed in the following table

Theory	Field	Brane
11D	$C_{\mu\nu\lambda}$	M2-brane (+ dual M5)
$\mathcal{N} = 1, 2$ 10D	$B_{\mu\nu}$	(F) string (+ NS5-brane) (6.33)
IIA	$A_\mu, C_{\mu\nu\lambda}$	(D)-branes
IIB	$\tilde{B}_{\mu\nu}, D^+_{\mu\nu\lambda\sigma}$ [RR-fields]	(D)-branes

For each $(p+1)$-form field, there is a corresponding p-brane that is electrically charged, in the sense that the brane couples to the field in a fashion analogous to (6.31) and (6.32). There is also a dual $(D-p-4)$-brane that couples magnetically, in the sense that there is a flux for the field strength through a space-like $(p+2)$-sphere surrounding the magnetic charge. For example, there is a dual 5-brane charged magnetically under $B_{\mu\nu}$ in every 10D supergravity theory, so that a 3-sphere surrounding the 5-brane carries nonzero flux $\int_{S^3} H$.

The branes that can be quantized in the most well-understood fashion are the "fundamental" strings appearing in all ten-dimensional supergravity theories as objects charged under $B_{\mu\nu}$. The perturbative approach to string theory begins by considering the theory on the world-sheet of these strings, described by a sigma model mapping the world-sheet into space-time. This approach leads to five tachyon-free superstring theories in ten dimensions: the type IIA and type IIB theories, whose low-energy limits are the type IIA and type IIB supergravity theories; the heterotic $SO(32)$ and $E_8 \times E_8$ theories, with the associated $\mathcal{N} = 1$ supergravity theories as low-energy limits; and type I string theory, which gives a different description of the $SO(32)$ theory. These theories give perturbative descriptions of the quantum supergravity theory in each case, and can be directly formulated in terms of a sigma model for backgrounds solving the classical equations of motion with nontrivial geometry, dilaton, and B fields. The perturbative description is less clear when the background contains nonzero Ramond-Ramond fields, but recent progress has been made in this direction.[19,20]

Just as fundamental strings form a natural route to quantizing ten-dimensional supergravity theories, the membrane carrying charge under the three-form field C is a natural candidate for quantizing 11-dimensional supergravity. Although a generally covariant quantization of the membrane has not been found, by restricting to light-cone gauge the theory on the membrane can be regulated in such a way that the resulting theory is a simple matrix quantum mechanics theory. This matrix quantum mechanics theory, known as "M(atrix) theory" gives a nonperturbative description of M-theory in light-cone gauge.[21–23]

A major breakthrough in understanding quantum gravity and string theory was made when Polchinski observed that the supergravity brane solutions charged under the R-R fields of type II string theory could be identified with dynamical branes in space-time located on hypersurfaces where open strings end with Dirichlet boundary conditions ("D-branes"). This insight provides two ways of describing the dynamics of such branes: analyzing fluctuations around the classical soliton solution, or quantizing open strings ending on the brane. Quantizing strings connecting N coincident branes leads to a world-volume $SU(N)$ gauge theory; in an appropriate near-horizon limit the gauge theory becomes supersymmetric Yang-Mills theory. The celebrated AdS/CFT correspondence is the correspondence between this Yang-Mills theory and the dual gravity theory in the vicinity of the brane.[24,25] Again, quantizing fluctuations around a brane has led to a theory of gravity. In this case, the gravity theory is in a space of higher dimensionality than the field theory on the brane world-volume, and the correspondence gives a nonperturbative definition of quantum gravity in terms of a field theory. The AdS/CFT correspondence underlies many of the talks at this TASI school. One other situation where quantizing the theory on branes leads to quantum gravity is realized by quantizing the theory on N D0-branes in type IIA string theory. This gives an alternate route to the M(atrix) model of M-theory mentioned above.[22]

The upshot of this discussion is that theories of quantum gravity can be studied by quantizing p-dimensional extended objects coupled to dynamical $(p+1)$-form fields in supergravity theories. String theory, the AdS/CFT correspondence, and M(atrix) theory are all examples of this general principle. In later parts of these lectures we will also use branes as a tool in constructing string vacua in dimensions less than 10.

6.2.6. *Summary: String universality in ten dimensions*

We showed above that there are only four distinct theories of supergravity in ten dimensions that do not suffer from inconsistencies due to quantum anomalies. Since each of these theories can be realized in string theory, we have string universality for supergravity theories in ten dimensions

$$\mathcal{G}^{(10)} = \mathcal{G}_*^{(10)} = \mathcal{V}_*^{(10)} = \mathcal{V}^{(10)} = \{\text{IIA, IIB, SO(32)}, \text{E}_8 \times \text{E}_8\} . \quad (6.34)$$

Of course, as mentioned above, we have only considered the discrete field content and symmetries of the theory, and have not proven that each of these quantum theories is perturbatively and nonperturbatively unique. Trying to prove, for example, that all higher-derivative terms in the theory are uniquely determined by supersymmetry and quantum consistency is a further interesting enterprise on which some initial progress has been made.[26]

Note that the strategy of analyzing theories in the "apparent swampland" $\mathcal{G} \setminus \mathcal{V}$ has led to several steps of progress enroute to this result. After Green and Schwarz identified their anomaly cancellation mechanism, the $E_8 \times E_8$ theory was in the apparent swampland. Identifying a string theory underlying this supergravity theory became a clear challenge, which was met when the "Princeton string quartet" of Gross, Harvey, Martinec and Rohm successfully constructed the $E_8 \times E_8$ heterotic theory.[27] Thereafter, for some time the theories with $U(1)^{496}$ and $E_8 \times U(1)^{248}$ gauge groups remained in the apparent swampland, motivating the eventual demonstration that these theories are inconsistent.

As discussed above, there are actually two string theory realizations of the $\mathcal{N} = 1$ 10D supergravity theory with $SO(32)$ gauge group, the type I and heterotic $SO(32)$ string theories. While there is no proof that these theories are equivalent, strong evidence suggests that there is a nonperturbative *duality* symmetry which identifies these two apparently distinct string theories.[11,28] From the discussion so far, it seems that there are a number of disconnected theories in ten dimensions. These theories are all connected, however, through a network of duality symmetries.[11,29] In particular, once the theories are compactified to lower dimensions, it becomes clear that apparently different 10D supergravity/superstring theories give rise to different descriptions of the same continuous space of gravity theories in the lower dimension. We now turn to eight dimensions, where we see an explicit example of such a duality symmetry for lower-dimensional theories.

6.3. Supergravity and String Vacua in Eight Dimensions

We begin our consideration of eight-dimensional theories with a brief introduction to eight-dimensional supergravity. We then discuss some general aspects of the compactification of supergravity theories from 11D and 10D to lower dimensions. This sets the stage for introducing two approaches to constructing 8D superstring vacua: heterotic compactifications on a two-torus, and F-theory compactifications on an elliptically fibered K3 surface. We show that these two rather different string constructions give rise to the same set of 8D theories, providing an example of a duality symmetry relating ostensibly very different string constructions.

6.3.1. *Supergravity in eight dimensions*

As in 10 dimensions, gravity theories can be constructed in 8 dimensions with either one or two supersymmetries. The $\mathcal{N} = 2$ 8D supergravity theory contains only the supergravity multiplet and thus has a uniquely determined field content. This supergravity theory can be realized through dimensional reduction of 11D or 10D supergravity on a torus, and thus has a natural string theory realization. In each dimension we focus on the supergravity theories with minimal supersymmetry, where novel phenomena emerge. In 8D $\mathcal{N} = 1$ theories, the new feature present is a wider range of possible gauge groups relative to the highly constrained set of 10D $\mathcal{N} = 1$ supergravity theories.

Just as in ten dimensions, the minimal $\mathcal{N} = 1$ supersymmetry in eight dimensions has 16 supercharges, and the multiplets of interest consist of the supergravity and vector multiplets

$$\text{8D } \mathcal{N} = 1 \quad \begin{cases} \text{SUGRA} & g_{\mu\nu}, B_{\mu\nu}, 2A_\mu, \sigma; \psi_{\mu\alpha}, \chi_\alpha \quad [48 + 48 \text{ DOF}] \\ \text{vector} & A_\mu, 2\phi, \lambda_\alpha \quad [8 + 8 \text{ DOF}] \end{cases} \tag{6.35}$$

In eight dimensions, the graviton multiplet contains two vector fields, often referred to as *graviphotons*. Spinors in eight dimensions, as in four dimensions, can be Majorana or Weyl, but not both. Classically, the supergravity multiplet can be coupled to any number of vector multiplets.[30,31]

8D theories can in principle have pure gauge or mixed gauge-gravitational anomalies from pentagon diagrams.[15] The gauge factor is always of the form $\mathrm{tr}_R F^3$ or $\mathrm{tr}_R F^5$ where the trace is taken in the representation R under which the charged matter transforms; these terms vanish in the adjoint representation, which is the only representation possible for the charged fermions in a supersymmetric theory, so there are no local anomalies in 8D supergravity theories. In our current state of knowledge regarding 8D theories, then, there are no restrictions from anomalies on the gauge group and

$$\mathcal{G}^{(8D,\mathcal{N}=1)} = \mathcal{G} = \{\text{SUGRA + YM for any } G\}. \tag{6.36}$$

This is an infinite set of theories with distinct gauge groups. We discuss at the end of this section how this set of theories may be further constrained.

To physicists familiar with supersymmetry in 4 dimensions, or even 10 or 6 dimensions, it may be surprising that the supergravity multiplet in 8 dimensions has a number of degrees of freedom (96) that is not a power of 2. To understand this it is helpful to briefly review the construction of supermultiplets.[8,32] In any dimension D, a massless supermultiplet is formed by starting with a representation of the corresponding little group $SO(D-2)$ and acting with raising operators associated with the Γ matrices representing the supersymmetry algebra. For example, the $\mathcal{N} = 1$ supergravity multiplet in 10 dimensions is formed by acting on the 8-dimensional vector representation $\mathbf{8}_v$ of the little group $SO(8)$ by the set of supersymmetry raising operators that combine into the $\mathbf{8}_v + \mathbf{8}'$ representation of $SO(8)$. The tensor product of these representations gives

$$\mathbf{8}_v \times (\mathbf{8}_v + \mathbf{8}') = \mathbf{35} + \mathbf{28} + \mathbf{1} + \mathbf{56} + \mathbf{8}' \tag{6.37}$$

which corresponds to the set of degrees of freedom in the components in the supergravity multiplet. In eight dimensions, the little group is $SO(6)$. The 16 supersymmetry raising operators break up into representations $\mathbf{6} + 2 \times \mathbf{1} + \mathbf{4} + \bar{\mathbf{4}}$ of $SO(6)$, and the supergravity multiplet is formed from the massless $\mathbf{6}$ representation, giving

$$\mathbf{6} \times (\mathbf{6} + 2 \times \mathbf{1} + \mathbf{4} + \bar{\mathbf{4}}) = \mathbf{20} + \mathbf{15} + \mathbf{1} + 2 \times \mathbf{6} + \mathbf{20}_\psi + \bar{\mathbf{20}}_\psi + \mathbf{4} + \bar{\mathbf{4}}, \tag{6.38}$$

which are the degrees of freedom in the fields of the 8D supergravity multiplet.

6.3.2. *Compactification of supergravity theories*

We have described above the various supergravity theories in 11 and 10 dimensions. As we have seen, the 10-dimensional type IIA theory can be related to a compactification of the 11D supergravity theory on a circle S^1. More generally, a variety of supergravity theories in lower dimensions can be constructed by compactifying the 11D and 10D supergravity theories on various geometries. For example, considering a D-dimensional supergravity theory on a manifold of the form $X_k \times \mathbb{R}^{1,D-k-1}$, where X_k is a k-dimensional compact manifold, gives rise to a supergravity theory in $D-k$ dimensions. To have a vacuum solution in the lower-dimensional theory, the compactification must be constructed in such a way that the equations of motion of the D-dimensional supergravity are satisfied. The massless spectrum in $D-k$ dimensions then comes from the zero modes of the massless fields in the higher-dimensional theory. If the manifold X_k is a torus T^k, then such a compactification amounts to a periodic identification of all fields in the theory (with some choice made for boundary conditions on fermions and higher spin fields). In general such a toroidal compactification preserves all of the supersymmetries of the theory, giving a supergravity theory in the smaller dimension with the same number of supercharges. The spectrum in the toroidally compactified theory has the same number of degrees of freedom as the higher-dimensional theory, organized according to the relativistic symmetry group of the lower-dimensional theory. More complicated geometries can be chosen that break some of the supersymmetry, giving a wider variety of theories in lower dimensions.

In addition to purely geometrical compactifications, additional features such as branes and fluxes can be included that increase the range of possibilities. The inclusion of branes gives constructions such as *intersecting brane models*, where branes wrapping cycles on the compactification manifold break some of the supersymmetry and give rise to gauge groups and matter. *F-theory* models, as we discuss later in this section, can be thought of as compactifications of the type IIB theory with branes, although they also have a natural interpretation in terms of pure geometry. Including fluxes for the various p-form fields in the supergravity theories can also break some supersymmetry and increase the variety of possible low-energy theories. We do not discuss compactifications with fluxes much in these lectures. Although they are a rich source of structure in lower-dimensional string vacua, there are no interesting classes of flux compactifications known in dimensions 8 or 6.

For now, we focus on compactifications on purely geometrical spaces, without branes or fluxes. We assume then that space-time is of the form

$$\mathcal{M}_{10} = \mathbb{R}^{1,D-k-1} \times X_k \,, \tag{6.39}$$

where $D = 10$ or $D = 11$. We assume that we are expanding the theory around a background where all fields other than the metric (and dilaton when $D = 10$) vanish, with a constant value for the dilaton when $D = 10$. To preserve supersymmetry there must be a supersymmetry transformation under which the variation of the gravitino vanishes. Thus, in particular, there must be a spinor η on the space-time so that $\delta\psi_\mu = D_\mu\eta = 0$. Such a covariantly constant spinor can exist on a manifold of dimension $k = 2N$ only when parallel transport on the manifold, or *holonomy*, leaves a component of the spinor unchanged. This is possible when the holonomy around all curves lies in a subgroup $SU(N)$ of the structure group $SO(2N)$. A manifold of dimension $2N$ has holonomy in $SU(N)$ if and only if it is a complex Kähler manifold with a nowhere vanishing holomorphic N-form. Such manifolds are known as *Calabi-Yau* manifolds. We will not need to deal with many technical aspects of Calabi-Yau manifolds in these lectures, but we will learn a few things about some of the simpler Calabi-Yau spaces in the context of various string compactifications.

The simplest example of a Calabi-Yau manifold is a torus of even dimensionality T^{2N}. For a torus the holonomy is trivial around any curve since the connection vanishes. For compactification on a torus, all supersymmetries of the higher-dimensional theory are retained after compactification. For complex dimension 1, the two-torus is the only Calabi-Yau manifold. For complex dimension 2, besides the torus T^4, there is one other topological class of Calabi-Yau manifold; this manifold is known as the *K3 surface*. Compactification on a K3 surface breaks half of the supersymmetries of the higher-dimensional theory. We will describe this manifold in more detail and study some of its features below in our discussion of 8-dimensional F-theory vacua. The number of topologically distinct Calabi-Yau manifolds in complex dimension three is much larger; it is not even known if this number is finite. Compactification on a Calabi-Yau threefold reduces the number of supersymmetry generators by a factor of 4.

There are a few other situations in which supersymmetry can be preserved under purely geometrical compactifications. If a 7-manifold has holonomy in the group $G_2 \subset SO(7)$, then a supersymmetry is preserved.[33] For example, 11-dimensional supergravity compactified on a G_2-holonomy 7-fold gives a theory with one supersymmetry in four dimensions, reducing the number of supersymmetry generators by a factor of 8. Similarly, compactification on an 8-fold with holonomy in Spin(7) or Sp(2) gives theories with supersymmetry reduced by a factor of 1/16 or 3/16.[34–36]

A table of the primitive geometric compactifications that admit a covariantly constant spinor, and hence supersymmetry in the dimensionally reduced theory, is given below. Products of these spaces will, of course, also give a supersymmetric dimensionally reduced theory; for example, compactification of a 10D theory on K3 $\times T^2$ gives a 4D theory with supersymmetry reduced by a factor of 1/2. In this table, CY3, CY4 refer to Calabi-Yau threefolds and fourfolds with holonomy precisely equal to $SU(N)$; product spaces have reduced holonomy and preserve more supersymmetry. For reductions on compact spaces of dimension up to 8, the manifolds in this table and products thereof are the only smooth geometries which preserve some supersymmetry. A more detailed discussion of the mathematics behind this assertion and further references can be found in the book by Gross, Huybrechts and Joyce[37] and in Morrison's TASI notes.[35]

(real) dimension	manifold type	holonomy	SUSY
k	T^k	$\{1\}$	1
4	K3 (CY2)	SU(2)	1/2
6	CY3	SU(3)	1/4
7	G_2	G_2	1/8
8	Sp(2)	hyper-Kähler	3/16
8	CY4	SU(4)	1/8
8	Spin(7)	Spin(7)	1/16

We can now ask the question: "How can we get a supergravity theory in eight dimensions by a geometric compactification?". Given what we have learned so far, it would seem that this can only be done by compactifying a 10-dimensional supergravity theory on a two-torus T^2. Indeed, compactifying IIA and IIB supergravity/string theory on T^2 gives rise to two (dual) descriptions of the same $\mathcal{N} = 2$ supergravity in eight dimensions, while compactifying $\mathcal{N} = 1$ supergravity on T^2 gives $\mathcal{N} = 1$ supergravity in 8D. In the following section we describe the compactification of the heterotic theories to 8D in more detail. We then describe another approach to constructing eight-dimensional supergravity theories through *F-theory*. F-theory can be thought of as a compactification of the type IIB theory on a curved manifold with branes. The branes give rise to an effective

12-dimensional geometric picture, so that $\mathcal{N} = 1$ supergravity in eight dimensions can be thought of as arising from F-theory on a K3 surface. As we now discuss in more detail, the two approaches of heterotic on T^2 and F-theory on K3 give dual descriptions of the same set of 8D supergravity theories.

$$\begin{array}{ccc} \text{het} = \text{I (10D) [16 Q]} & & \text{F-theory (“12D”) [32 Q]} \\ \Big\downarrow T^2 & \overset{\text{dual}}{\Longleftrightarrow} & \Big\downarrow \text{K3} \\ \text{8D [16 Q]} & & \text{8D [16 Q]} \end{array}$$

6.3.3. *Heterotic string vacua in eight dimensions*

6.3.3.1. *Compactification of 10D* $\mathcal{N} = 1$ *theories on* T^2

We now consider the theories that can be realized by compactifying the $SO(32)$ and $E_8 \times E_8$ theories from ten dimensions to eight dimensions on a torus T^2. From the point of view of the 10D $\mathcal{N} = 1$ supergravity theory, we can deduce the field content of the 8D theory by considering how the representations of the 10D little group $SO(8)$ for massless fields decompose when reduced to the 8D little group $SO(6)$. Operationally, we can perform the reduction of fields by choosing for each 10D Lorentz index $(0, \ldots, 9)$ either an 8D Lorentz index ($\mu = 0, \ldots, 7$) or an index in one of the compact directions ($c = 8, 9$). For example, the 10D vector field (with massless states transforming in the $\mathbf{8}_v$ of $SO(8)$) reduces to an 8D vector field A_μ (with states in the $\mathbf{6}$ of $SO(6)$) and two scalar fields $\phi = A_8, \phi' = A_9$. Thus, the 8D vector multiplet is simply the dimensional reduction of the 10D vector multiplet. The dimensional reduction of the 10D $\mathcal{N} = 1$ supergravity multiplet gives the 8D supergravity multiplet and two vector multiplets ($128 = 96 + 2 \times 16$). The fact that the supersymmetry multiplet is reducible after dimensional reduction comes from the decomposition of the $\mathbf{8}_v$ multiplet on which the SUSY generators act in (6.37) through $\mathbf{8}_v \rightarrow \mathbf{6} + 2 \times \mathbf{1}$. In the anomaly-free 10D $\mathcal{N} = 1$ supergravity theories, there are 496 gauge bosons. The compactification of the bosonic fields of such a supergravity theory on T^2, assuming all gauge fields vanish, gives an 8D theory with one supergravity multiplet and 498 vector multiplets.

$$g_{\mu c}, B_{\mu c}, A_\mu \qquad\qquad \phi, A_c, g_{cc'}, B_{cc'}$$

$$\downarrow \qquad\qquad\qquad\qquad \downarrow$$

$$500\ A_\mu \quad [U(1)^4 \times G] \qquad\qquad \sigma + \ 996\ \phi \ \text{(scalars/moduli)}$$

The scalar σ in the supergravity multiplet and the two scalars in each vector multiplet combine to give 997 scalar fields in the 8D theory.

From the discussion so far, it would seem that the gauge group of the dimensionally reduced theory must be given by $G \times U(1)^4$, where $G = SO(32)$ or $E_8 \times E_8$ is the nonabelian gauge group of the 10D theory. In either case, the rank of the 8D gauge group is 20. The gauge group in the reduced theory is generically broken to a smaller group, however, by *Wilson lines* around the circles of the compactification space. From the point of view of the low-energy theory this corresponds to Higgsing the gauge group by turning on expectation values for any of the 992 scalars arising from the dimensional reduction of A_8, A_9, which transform in the adjoint representation of the gauge group. The gauge group can also be increased to a larger group by a stringy gauge enhancement mechanism. In either case the rank of the group stays fixed at 20. We now consider each of these possibilities in turn.

The possibility of nontrivial Wilson lines for the gauge fields A_μ around each of the two directions in the torus can be understood in a straightforward fashion. A constant nonabelian gauge field can be turned on in the directions A_8 and A_9 without breaking supersymmetry, as long as the curvature $F = [A_8, A_9] = 0$ vanishes. (Note that nonzero curvature gives an energy density $E \sim F^2$, which is incompatible with the vanishing vacuum energy required by supersymmetry in Minkowski space.) In the presence of Wilson loops, the 10D gauge group is broken to $H \subset G$ where $H = \{h : [h, A_8] = [h, A_9] = 0\}$. Since A_8, A_9 can always be chosen as part of a maximal torus, the rank of the resulting 8D gauge group is always 20. This gives a large family of possible unbroken gauge groups for the 8D theory. For generic commuting Wilson lines A_8, A_9, the nonabelian gauge group will be completely broken to

$$G_{\text{generic}} = U(1)^{20} . \tag{6.40}$$

In such a generic vacuum, two of the $U(1)$ factors are graviphotons. Associated with the 16 unbroken generators of the original nonabelian gauge group there are 32 scalar fields. Four additional scalars are associated with the shape and size of the compactification torus. There is one more scalar coming from the 10D dilaton, for a total of 37 scalar fields. These fields parameterize the space of supersymmetric vacua of the theory. There is no potential for these massless scalar fields, which are known as *moduli* of the theory. This gives us a picture of the 8D supergravity theory resulting from the heterotic compactification as having a continuously connected 37-dimensional moduli space of vacua; in lower-dimensional subspaces of the moduli space the gauge group is enhanced.

From the preceding discussion it seems that the maximum enhancement of the gauge group can be to one of the original heterotic groups. As mentioned above, however, string theory also provides a mechanism for enhancing the gauge group further beyond $U(1)^4 \times SO(32)$ or $U(1)^4 \times E_8 \times E_8$. To understand this mechanism, it will be helpful to briefly review some basics of the mathematics of lattices. Lattices are relevant here because the k-dimensional torus can be thought of as a quotient $\mathbb{R}^k/\Gamma$, where Γ is a k-dimensional lattice. Lattices will also play an important role in many other constructions in these notes.

6.3.3.2. *Interlude on lattices*

A k-dimensional lattice is defined to be the subset of $\mathbb{R}^k$ given by integral linear combinations of a set of k linearly independent basis vectors e_i

$$\Gamma = \{n_i e_i, n_i \in \mathbb{Z}\} . \tag{6.41}$$

As in much of the mathematical literature, we are interested in lattices that carry an integral symmetric bilinear inner product

$$v \cdot w \in \mathbb{Z} \ \forall v, w \in \Gamma . \tag{6.42}$$

We assume that every lattice discussed in these notes carries such a structure. We present here only some basic aspects of the theory of lattices. For a more thorough introduction to the subject see the text by Conway and Sloane.[38]

For a given choice of basis, a lattice Γ can be represented by the integral matrix $e_i \cdot e_j$. We will sometimes use Γ to denote this matrix as a convenient shorthand for describing a given lattice. A lattice Γ is *Euclidean* if

$$v \cdot v > 0 \ \forall v \in \Gamma . \tag{6.43}$$

This corresponds to the condition that the associated integral matrix is positive definite. More generally, we will be interested in lattices of indefinite signature (p, q).

A lattice is said to be *even* if

$$v \cdot v \in 2\mathbb{Z} \;\; \forall v \in \Gamma\,. \tag{6.44}$$

The *dual* Γ^* of a lattice Γ is defined to be the set of dual vectors whose inner product with all elements of Γ is integral

$$\Gamma^* = \{v : v \cdot w \in \mathbb{Z} \;\; \forall w \in \Gamma\}\,. \tag{6.45}$$

A lattice is *self-dual* or *unimodular* if it is equal to its dual

$$\Gamma = \Gamma^*\,. \tag{6.46}$$

This is equivalent to the condition that the associated matrix has unit determinant, $\det \Gamma = \pm 1$. Alternatively, the self-duality condition is equivalent to the condition that the basis vectors e_i span a unit cell of volume ± 1. An important theorem due to Milnor states that any unimodular lattice has signature satisfying

$$p \equiv q \pmod 8 \qquad (\Gamma \text{ unimodular})\,. \tag{6.47}$$

Example: As an example of an even unimodular lattice consider the lattice

$$U = \begin{pmatrix} 0 & 1 \\ 1 & 0 \end{pmatrix}, \qquad e_1^2 = e_2^2 = 0, \quad e_1 \cdot e_2 = 1\,. \tag{6.48}$$

This lattice has signature $(1,1)$. A sublattice of U is spanned by the vectors with respect to which U is diagonalized

$$(e_1 + e_2)^2 = 2,\; (e_1 - e_2)^2 = -2,\; (e_1 + e_2) \cdot (e_1 - e_2) = 0\,. \tag{6.49}$$

A particularly noteworthy class of even lattices are the *root lattices* of the algebras associated with the simply-laced Lie groups $(SU(N), SO(2N), E_6, E_7, E_8)$. For such an algebra the simple roots r_i form a basis for the lattice. The Cartan matrix is formed from the inner products of the simple roots $a_{ij} = r_i \cdot r_j$, where the roots are normalized such that $a_{ii} = 2$. For simply-laced algebras, all off-diagonal entries in the Cartan matrix are either 0 or -1. These algebras and the associated matrices are often conveniently described in terms of *Dynkin diagrams*. The Dynkin diagram for a given algebra/Cartan matrix is given by drawing a node for each simple root r_i, and connecting with a line each pair of nodes i, j satisfying $r_i \cdot r_j = -1$. The number of nodes in the Dynkin diagram corresponds to the rank of the associated Lie algebra, which is the same as the dimension of the corresponding lattice. The lattice and Dynkin diagram associated with the algebra of $SU(N)$ is denoted A_{N-1}, while that associated with the algebra of $SO(2N)$ is denoted D_N.

Example: The lattice A_2 is described by the Cartan matrix associated with the Lie algebra of $SU(3)$

$$(a_{ij}) = \begin{pmatrix} 2 & -1 \\ -1 & 2 \end{pmatrix}. \tag{6.50}$$

This lattice is even but not unimodular. The associated Dynkin diagram is

Example: The lattice E_8 is described by the Cartan matrix associated with the E_8 group

$$(a_{ij}) = \begin{pmatrix} 2 & 0 & -1 & 0 & 0 & 0 & 0 & 0 \\ 0 & 2 & 0 & -1 & 0 & 0 & 0 & 0 \\ -1 & 0 & 2 & -1 & 0 & 0 & 0 & 0 \\ 0 & -1 & -1 & 2 & -1 & 0 & 0 & 0 \\ 0 & 0 & 0 & -1 & 2 & -1 & 0 & 0 \\ 0 & 0 & 0 & 0 & -1 & 2 & -1 & 0 \\ 0 & 0 & 0 & 0 & 0 & -1 & 2 & -1 \\ 0 & 0 & 0 & 0 & 0 & 0 & -1 & 2 \end{pmatrix} \tag{6.51}$$

This lattice is both even and unimodular. The associated Dynkin diagram is

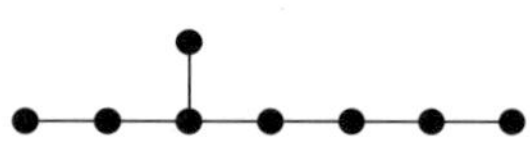

As a consequence of (6.47), a Euclidean lattice can only be self-dual if it has a dimension $p \equiv 0 \pmod 8$. E_8 is the unique Euclidean self-dual lattice of dimension 8. There are two even self-dual Euclidean lattices of dimension 16:

$$E_8 \oplus E_8, \text{ and } \Lambda_{16}, \tag{6.52}$$

where Λ_{16} is the *Barnes-Wall* lattice. This lattice is closely related to the root lattice D_{16} of $SO(32)$. Λ_{16} includes D_{16} as a sublattice, and contains an additional set of points corresponding to the weights of a spinor representation of $SO(32)$ (essentially another copy of D_{16}, but shifted by an offset moving each lattice point to one of the biggest "holes" in the original lattice).

For lattices of indefinite signature (p, q), there is a unique even self-dual lattice with any given signature when $p \equiv q \pmod 8$. This lattice is denoted

$\Gamma^{p,q}$, and is given by

$$\Gamma^{p,q} = U \oplus \cdots \oplus U \oplus (\pm E_8) \oplus \cdots \oplus (\pm E_8) \tag{6.53}$$

where the number of factors of U is min (p, q) and the number of factors of E_8 is $(p - q)/8$.

6.3.3.3. *Toroidal compactification and enhanced symmetries*

We can now describe enhanced symmetries of the toroidally compactified heterotic string in terms of the language of lattices. We begin by recalling the quantization of the simple bosonic string on a compact circle of radius R, following the notation of Polchinski.[7] In terms of world-sheet coordinates σ, τ, and setting $\alpha' = 1$, the position of the string in the compact direction is given by

$$X = x + w\sigma + p\tau + \text{oscillators}(\alpha_n, \tilde{\alpha}_n) \tag{6.54}$$

$$= x + \frac{1}{\sqrt{2}} l_L(\tau + \sigma) + \frac{1}{\sqrt{2}} l_R(\tau - \sigma) + \cdots \tag{6.55}$$

where the winding number and momenta are quantized through

$$w = mR, \qquad p = n/R. \tag{6.56}$$

T-duality is the symmetry that exchanges $w \leftrightarrow p$ through $m \leftrightarrow n$ and $R \leftrightarrow 1/R$ ($R \to \alpha'/R$ when α' is reinstated). The left- and right-moving momenta $l_{L,R}$ are related to winding number and momenta through

$$l_L = \frac{p + w}{\sqrt{2}}, \quad l_R = \frac{p - w}{\sqrt{2}}. \tag{6.57}$$

These momenta live on the even lattice U from (6.48), with inner product

$$l \cdot l = l_L^2 - l_R^2 = 2pw = 2nm \in 2\mathbb{Z}. \tag{6.58}$$

Restoring units, physical momenta are given by

$$k = \sqrt{2/\alpha'}\, l, \tag{6.59}$$

and the mass shell condition for string states is

$$M^2 = k_\mu k^\mu = \frac{2}{\alpha'} \left[l_L^2 + 2(N - 1) \right] = \frac{2}{\alpha'} \left[l_R^2 + 2(N - 1) \right]. \tag{6.60}$$

There are then two different classes of massless vector fields arising from closed string states. There are states with both left and right moving oscillator number $N = \tilde{N} = 1$ and vanishing left and right momenta

$$\alpha_{-1}^\mu \tilde{\alpha}_{-1}^* | l_L = 0; l_R = 0 \rangle \pm \alpha_{-1}^* \tilde{\alpha}_{-1}^\mu | l_L = 0; l_R = 0 \rangle, \tag{6.61}$$

where $*$ denotes the compact direction. These states are the string states associated with the vector bosons $g^{\mu *}, B^{\mu *}$ in the dimensionally reduced theory. There are additional massless vector states with only one nonzero oscillator number, namely

$$\tilde{\alpha}^{\mu}_{-1}|l_L; l_R\rangle, \; l_R = 0, \; l_L^2 = l \cdot l = 2 \tag{6.62}$$

and the analogous states with a single α^{μ}_{-1}. These massless states occur in the spectrum precisely when $R = \sqrt{\alpha'}$, the self-dual radius for the circle.

The upshot of this analysis is that when there are points in the lattice with $l \cdot l = 2$ and either $l_R = 0$ or $l_L = 0$ then there are additional massless vectors in the theory. In the case of compactification on a single circle S^1, these massless vectors appear at the self-dual radius, and enhance the generic $U(1)^2$ gauge symmetry to $SU(2)^2$. Note that the condition $l \cdot l = 2$ is always satisfied by a set of elements of the momentum lattice, while the existence of nonzero lattice vectors satisfying the condition $l_R = 0$ or $l_L = 0$ depends upon the values of the scalar moduli — in this case the compactification radius.

Generalizing to a compactification on a higher-dimensional torus T^D, the momenta l take values in a signature (D, D) lattice

$$\Gamma^{D,D} = U \oplus \cdots \oplus U \,. \tag{6.63}$$

Again, the generic gauge group arising from compactification is $U(1)^{2D}$, but enhanced symmetries arise for states with $l \cdot l = 2$ and either $l_R = 0$ or $l_L = 0$.

The moduli space of theories associated with compactification on T^D can be parameterized by the embeddings of the lattice $\Gamma^{D,D}$ in the momentum space l_L, l_R of the compact directions with inner product $l_L^2 - l_R^2$. Given one embedding of the lattice in these coordinates, any other embedding can be realized by acting with a transformation in the group $SO(D, D)$ that preserves the inner product. Two embeddings that are related by a transformation in $SO(D) \times SO(D)$ that separately preserves l_L^2 and l_R^2 are physically equivalent, so the moduli space is locally given by $SO(D, D)/SO(D) \times SO(D)$. There is a further equivalence between embeddings related by a transformation in $G_D = SO(D, D; \mathbb{Z})$, the discrete (T-duality) group of transformations that map the lattice $\Gamma^{D,D}$ to itself. The global space of toroidal compactifications is thus given by the Narain moduli space

$$SO(D, D; \mathbb{Z}) \backslash SO(D, D)/SO(D) \times SO(D) \,, \tag{6.64}$$

where the discrete duality group acts on the left, independently from the $SO(D) \times SO(D)$ symmetry acting on the right. As a simple example, for the type IIA and IIB theories compactified on a single circle ($D = 1$), the T-duality group is the $\mathbb{Z}_2$ group whose nontrivial element exchanges momentum and winding, giving $n \leftrightarrow m$ in (6.58). This T-duality transformation gives a duality symmetry relating type IIA string theory compactified on a circle of radius R and the IIB theory compactified on a circle of radius α'/R.

For the toroidal compactifications just discussed, the unimodular form of the lattice (6.63) follows automatically from the construction. More generally, however, we can consider a world-sheet theory with p chiral left-moving bosons ϕ^a and q chiral right-moving bosons $\tilde{\phi}^b$. The operators in such a theory are of the form

$$\mathcal{O}_{l,\tilde{l}} = \exp\left(il_a\phi^a + i\tilde{l}_b\tilde{\phi}^b\right) . \tag{6.65}$$

The closure of the set of operators under operator products implies that $(l, \tilde{l})$ lie in a lattice Γ of dimension $p + q$. The single-valued nature of the operator product implies that

$$l_a l_a - \tilde{l}_b\tilde{l}_b \in \mathbb{Z} , \tag{6.66}$$

so that the lattice is integral and has signature (p, q). For a consistent string theory, the world-volume theory must be modular invariant. As shown by Narain[39,40] (see also Polchinski[7] for more details), modular invariance of the world-volume theory implies that the lattice Γ must be even and self-dual. For $p = q$, then, the lattice (6.63) arising from toroidal compactification is the only lattice possible for a consistent string theory.

6.3.3.4. *Lattices and compactification of the heterotic string*

We can apply the preceding general discussion to the heterotic string. The world-sheet degrees of freedom on the heterotic string are 10 chiral left-moving bosons and 26 chiral right-moving bosons. (Note that we use the opposite convention from Polchinski for the right-left splitting of degrees of freedom to match the dominant negative signature convention for the associated lattices.) In ten dimensions, ten of the right-moving bosons correspond to space-time momenta, so the compactification lattice is an even unimodular lattice $\Gamma^{0,16}$. From (6.52) we know that there are only two possibilities for 16-dimensional Euclidean even unimodular lattices,

$$(-)\Gamma^{0,16} = E_8 \oplus E_8, \text{ or } (-)\Gamma^{0,16} = \Lambda_{16} . \tag{6.67}$$

These are precisely the heterotic theories giving $\mathcal{N} = 1$ 10D supergravity with the gauge groups $E_8 \times E_8$ and $SO(32)$. Since Λ_{16} is not exactly the root lattice of $SO(32)$, the correct symmetry group for the latter theory is actually $\mathrm{Spin}(32)/\mathbb{Z}_2$, though it is generally referred to as the "$SO(32)$" theory. The group Spin(32) is the simply connected cover of $SO(32)$ just as $SU(2)$ is the simply connected cover of $SO(3)$. Spin(32) has a center $\mathbb{Z}_2^2$, and the group SO(32) is given by the quotient $SO(32) = \mathrm{Spin}(32)/\mathbb{Z}_2$ by a particular $\mathbb{Z}_2$ subgroup of the center. The quotients by the other two $\mathbb{Z}_2$ factors both give the same group, which is different from SO(32) and generally denoted $\mathrm{Spin}(32)/\mathbb{Z}_2$. Further details related to the discrete quotient are discussed in the references.[6,41]

Now let us consider compactification of the heterotic string to dimensions below 10. Consider first the compactification on S^1 to 9 dimensions. In this case, one of the left-moving bosons and 17 of the right-moving bosons correspond to internal degrees of freedom associated with the compactification, and the resulting lattice is

$$\Gamma^{1,17} = U \oplus (-E_8) \oplus (-E_8)\,. \tag{6.68}$$

Note that this lattice is the unique even unimodular lattice of signature $(1, 17)$. It follows that compactification of both the $E_8 \times E_8$ and $\mathrm{Spin}(32)/\mathbb{Z}_2$ theories on a circle S^1 give the same moduli space of theories in 9 dimensions.

Compactifying to eight dimensions, the lattice is

$$\Gamma^{2,18} = U \oplus U \oplus (-E_8) \oplus (-E_8)\,. \tag{6.69}$$

The moduli space for this theory is the 36-dimensional space

$$SO(2, 18; \mathbb{Z})\backslash SO(2, 18; \mathbb{R})/SO(2) \times SO(18)\,, \tag{6.70}$$

along with the single scalar in the gravity multiplet (the dilaton). The generic gauge group is $U(1)^{20}$, as discussed above. It is possible to generate an enhanced gauge group by choosing moduli where there are additional massless gauge bosons. This can produce gauge groups even larger than the original $E_8 \times E_8$ or $SO(32)$ of the uncompactified theory. In the heterotic string the ground state for the left-moving oscillators is massless, like the type II string, and only the ground state for right-moving oscillators is tachyonic. So enhanced gauge symmetries only arise when $l_R^2 = 2, l_L^2 = 0$. Since there are 18 right-moving bosons, the symmetry group is determined by the sublattice of $\Gamma^{2,18}$ lying in the negative-signature 18-dimensional space spanned by the l_R. The possible enhanced nonabelian gauge symmetry groups of the theory are therefore precisely the set of groups whose root lattices G admit an embedding into $\Gamma^{2,18}$.

$$G: \; -\Lambda_G \hookrightarrow \Gamma^{2,18} . \tag{6.71}$$

Such groups can have rank up to 18. For example, there is an embedding

$$-D_{18} \hookrightarrow \Gamma^{2,18} \tag{6.72}$$

from which it follows that the group $SO(36)$ can be realized in the 8D heterotic theory.[8] Similarly, it was shown by Ganor, Morrison and Seiberg[42] that $-A_{17}$ can be embedded in $\Gamma^{2,18}$, associated with a theory having gauge group $SU(18)/\mathbb{Z}_3$.

Determining which groups do or do not admit an embedding of the form (6.71) is a nontrivial problem. Fortunately, some powerful theorems on lattice embeddings were proven by Nikulin[43] that state necessary and sufficient conditions for such an embedding to be possible. In many cases it is also possible to prove that the embedding, when it exists, is unique up to lattice automorphisms. The precise statement of these theorems is somewhat intricate; in many situations, however, the following simplified theorem, which is a corollary of the stronger theorems, is sufficient.

Theorem [Nikulin]

Let S be an even lattice of signature (s_+, s_-) and let T be an even, unimodular lattice of signature (t_+, t_-). There exists an embedding of S into T provided the following conditions hold:

(1) $t_+ \geq s_+$ and $t_- \geq s_-$
(2) $t_+ + t_- - s_+ - s_- > l(S^*/S)$

where the lattice quotient S^*/S is a finite abelian group, and $l(S^*/S)$ is the minimum number of generators of this group □

It follows from this theorem, for example, that the rank 18 group $SU(19)$ can be realized as the gauge group of a heterotic string compactification in 8D. The associated lattice $S = -A_{18}$ has $s_+ = 0, s_- = 18$, so satisfies the first condition for $T = \Gamma^{2,18}$. For A_{n-1} the discrete group S^*/S is $\mathbb{Z}_n$[38] (an easy exercise for the reader is to check this for A_2). Thus, for $S = A_{18}$, $S^*/S = \mathbb{Z}_{19}$, which is a cyclic group with one generator, so the second condition becomes $20 - 18 > 1$ and is satisfied. The structure of the discrete group S^*/S also affects the global structure of the gauge group, for example leading to the $\mathbb{Z}_3$ quotient in the case of A_{17} mentioned above.[42]

The stronger theorems proven by Nikulin[43] use more detailed aspects of the number-theoretic structure of S^*/S. The proofs are carried out using p-adic analysis.

This concludes our brief introduction to heterotic compactifications to 8D.

6.3.4. *F-theory vacua in eight dimensions*

We now turn to another approach to string compactifications, known as *F-theory.*[44–46] F-theory can be thought of in several ways: as a limit of a class of type IIA or M-theory vacua, or as a framework for characterizing nonperturbative type IIB vacua[c]. We will primarily approach the subject from the latter perspective in these lectures. Other pedagogical introductions to F-theory can be found in the Les Houches notes of Denef[47] and in the TASI notes by Morrison.[35]

6.3.4.1. *7-branes and geometry*

As described in Section 6.2, type IIB supergravity has a classical $SL(2,\mathbb{R})$ symmetry. In the quantum theory, the spectrum of string excitations charged under the B fields is quantized, and this symmetry is broken to $SL(2,\mathbb{Z})$. Just as the two-form fields $B, \tilde{B}$ transform as a doublet under this symmetry, fundamental strings and Dirichlet strings (D1-branes) also transform as a doublet under this discrete group. The group $SL(2,\mathbb{Z})$ thus gives a group of nonperturbative duality symmetries for the IIB string. The axion and dilaton combine into a complex *axiodilaton*

$$\tau = \chi + ie^{-\phi} . \tag{6.73}$$

The imaginary part of the axiodilaton is positive, so τ lives in the upper half-plane of $\mathbb{C}$. The axiodilaton transforms under $SL(2,\mathbb{Z})$ as

$$\tau \to \frac{a\tau + b}{c\tau + d}, \qquad ad - bc = 1 . \tag{6.74}$$

This group is generated by the transformations

$$T : \ \tau \to \tau + 1 \tag{6.75}$$

$$S : \ \tau \to -1/\tau \tag{6.76}$$

[c]Most of what I know about F-theory I learned from David Morrison, my guru in the way of F-theory. Credit for any clear and correct insights in the F-theory portions of these notes should go to him, while I of course am responsible for all errors and confusion.

and is precisely the group of transformations on the upper half-plane corresponding to modular transformations on a two-torus parameterized by the complex structure parameter τ. It is thus natural to interpret the set of possible values for the axiodilaton as the modular parameter for a two-torus. In the language of algebraic geometry, a two-torus equipped with a complex structure is an *elliptic curve.*

A D7-brane in the IIB theory is magnetically charged under the axiodilaton. This means that the value of the axion field undergoes a monodromy around a small circle surrounding the D7-brane in $\mathbb{R}^9$, given by the T modular transformation

$$\tau \to \tau + 1\,. \tag{6.77}$$

Acting on a D7-brane with an $SL(2,\mathbb{Z})$ transformation produces a more general class of (p,q) 7-branes, with monodromies conjugate to (6.77). A supergravity background containing a configuration of (p,q) 7-branes can be geometrized by interpreting τ at each point in space-time as parameterizing a torus locally fibered over space-time. This leads to the essential geometric picture of F-theory, a space that is *elliptically fibered* over space-time. While this geometric picture suggests a 12-dimensional enhancement of space-time, it is important to emphasize that there is no underlying 12-dimensional supergravity theory associated with this picture. The presence of different kinds of 7-branes can produce an axiodilaton configuration that may be strongly coupled in any duality frame, and which is therefore essentially nonperturbative. While F-theory captures many nonperturbative aspects of the moduli space of supersymmetric string vacua, there are also limitations to this approach as it is currently understood. F-theory does not at this time have an intrinsic definition including a dynamical principle. Thus, explicit computations, for example of the metric on the compact space or the effect of internal fluxes, depend upon finding a dual perturbative description in which the desired physics can be formulated and computed more precisely.

An alternative to the type IIB picture of F-theory is through a limit of compactification of M-theory. Consider a compactification of the 11-dimensional M-theory on a torus with modular parameter τ. Treating one cycle of the torus as compactification of M-theory to type IIA, we have the type IIA theory on a circle of radius R parameterizing the other cycle of the original torus. By T-duality, as discussed in Section 6.3.3.3, this is equivalent to type IIB on a circle of radius α'/R. In the limit where the size of the original torus becomes small, $R \to 0$ and this becomes the type IIB theory in an infinite flat space with axiodilaton parameter τ. In principle, this construction can be fibered over any base space, reproducing F-theory as a limit of M-theory compactified on a torus shrunk to 0 size; for a more detailed pedagogical description of this picture of F-theory, see Denef's Les Houches notes.[47]

An F-theory compactification is associated with an elliptic fibration over a compact base space B of complex dimension d, giving a space-time theory in dimension $D = 10 - 2d$. For the theory to be supersymmetric, the total space of the elliptic fibration must be a Calabi-Yau manifold. Thus, for example, F-theory vacua in eight dimensions are described by an elliptically fibered Calabi-Yau manifold with two complex dimensions, F-theory vacua in 6 dimensions come from elliptically fibered Calabi-Yau threefolds, and F-theory vacua in 4 dimensions come from elliptically fibered Calabi-Yau fourfolds. We denote the total space of the elliptic fibration by X, characterizing the fibration through the following diagram

$$\begin{array}{ccc} T^2 & \longrightarrow & \mathrm{X} \\ & & \downarrow \\ & & \mathrm{B} \end{array}$$

Away from 7-brane sources, the axiodilaton varies smoothly, describing a smooth elliptic fibration over the base space. At the position of a 7-brane, the elliptic fibration becomes singular. This does not mean, however, that the total space X is necessarily singular. As a rough analogy, consider the two-sphere $S^2 = \{(x, y, z) : x^2 + y^2 + z^2 = 1\}$ as a circle fibration over a line segment parameterized by $z \in [0, 1]$. At every $z \neq 0, 1$ there is a copy of the circle that varies smoothly with z. The fibration is singular at the endpoints, where the circle degenerates to a point. But the total space S^2 is still smooth. A similar thing happens in the vicinity of a (p, q) 7-brane in F-theory. The total space X is smooth, though the structure of the fibration becomes singular.

We now focus specifically on compactifications of F-theory to 8 dimensions, the simplest class of F-theory vacua. First we discuss the structure of the base B and elliptically fibered Calabi-Yau complex two-fold X. Like any codimension two massive object, 7-branes produce localized curvature in space-time that leads to a deficit angle in the geometry away from the brane. For a single 7-brane, the metric at large distances becomes asymptotically flat, but the total angle around the brane is reduced by $\pi/6$. This can be seen from the ratio between the circumference and diameter of a large circle, which approaches $11\pi/6$ (instead of 2π) as the circle becomes large. For F-theory to be defined on a compact space, we need enough 7-branes to close the transverse 2D space into a compact geometry. With 12 7-branes, the deficit angle becomes 2π, so that the asymptotic space becomes a cylinder (no change in circumference at large radius). Putting together two such configurations, we can construct a topological sphere from a space-time containing 24 7-branes. This can also be understood from the Gauss-Bonnet theorem, which states that the total curvature of a sphere is 4π, where each 7-brane locally contributes $\pi/6$ to the curvature.

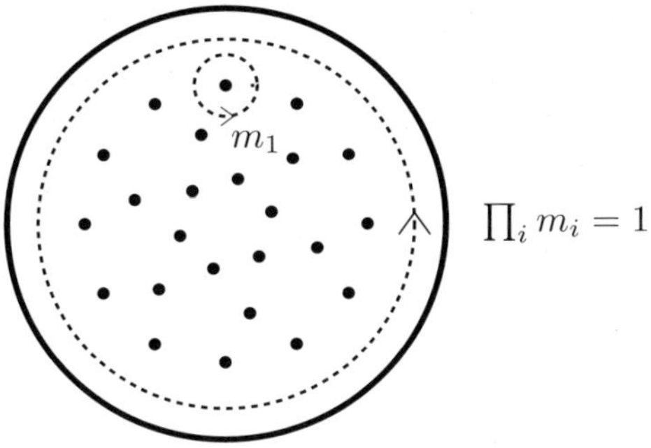

Fig. 6.4. K3 as an elliptic fibration over the sphere S^2. The fibration is singular at 24 points. The product of the monodromies around the individual points must be the identity, as the associated curve is contractable.

Thus, a compactification of F-theory to eight dimensions is associated with a collection of 24 7-branes on a base space that is topologically a two-sphere $B = S^2$. Around each 7-brane there is a monodromy m_i acting on τ. The product of the monodromies must be trivial, $\prod_i m_i = 1$, since a curve surrounding all 24 7-branes can be contracted to a point on the sphere (see Figure 6.4). Note that not all the 7-branes can be D7-branes, since if each m_i gives the modular transformation (6.77) then $\prod_i m_i$ would give the modular transformation $T^{24} : \tau \to \tau + 24 \neq \tau$. The total space X of the elliptic fibration must be a Calabi-Yau two-fold, and is not a torus if there is only one supersymmetry in 8 dimensions. X is therefore a K3 surface. While the presentation of K3 as an elliptic fibration over S^2 is the principal characterization of K3 that we will use in studying F-theory, it is useful to briefly digress on the structure of the K3 surface as seen from other perspectives.

6.3.4.2. *The K3 surface*

We now introduce some basic features of the K3 surface. For a much more detailed introduction to the mathematics and physical applications of K3 surfaces, see the lecture notes by Aspinwall on K3.[48]

The K3 surface is the only topological type of Calabi-Yau manifold of complex dimension two besides the four-torus T^4, as mentioned above. K3 surfaces are simply connected, so they have no first homotopy (π_1) or homology (H_1) structure. K3 surfaces have a 22-dimensional second homology group $H_2(K3, \mathbb{Z})$. The intersection pairing on this homology group gives the lattice $\Gamma^{3,19}$, which as discussed above must take the form

$$H_2(K3, \mathbb{Z}) = \Gamma^{3,19} = U \oplus U \oplus U \oplus (-E_8) \oplus (-E_8) . \tag{6.78}$$

One of the simplest ways to view the K3 surface is in a particular limit where all the curvature is concentrated at points and the surface can be viewed as an orbifold of a torus. Consider the torus T^4, considered as a product of two 2-tori with modular parameters $\tau_1 = \tau_2 = i$, so that coordinates on the two 2-tori are subject to the usual identifications on each 2-torus $z_i \sim z_i + 1 \sim z_i + \tau$. We now take an orbifold of the space by imposing the additional identification of points related through the $\mathbb{Z}_2$ action

$$\rho : (z_1, z_2) \to (-z_1, -z_2) . \tag{6.79}$$

The resulting space is flat everywhere except at 16 orbifold points that are locally of the form $\mathbb{C}^2/\mathbb{Z}_2$ (see Figure 6.5). This singular limit of the K3 surface can be continuously deformed into a smooth K3 surface by *blowing up* the 16 singular points. The blowing-up procedure, which will be described in more detail below, essentially involves replacing a singular

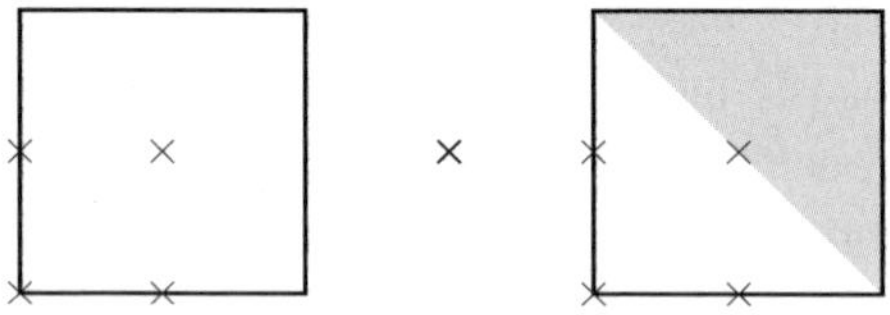

Fig. 6.5. An orbifold limit of the K3 surface can be viewed as the space $T^4/\mathbb{Z}_2$, which is locally flat except at 16 orbifold points (marked with "x"'s). The product of the regions on the two 2-tori that have not been marked in gray gives a fundamental domain for the orbifold space.

point in a surface with a complex projective space $\mathbb{CP}^1$ consisting of the set of limiting lines approaching the singular point (note that all projective spaces discussed in these lectures are complex projective spaces; from here on we denote $\mathbb{P}^n$ for $\mathbb{CP}^n$.). For a singularity of the form $\mathbb{C}^2/\mathbb{Z}_2$, blowing up the singular point at the origin produces a smooth space. From this picture we can see fairly clearly how the 22 homology cycles arise. There are six homology classes $\bar{\pi}_{ij}$ in the original T^4 corresponding to cycles wrapped around the ij coordinate axes (where the real coordinate axes are labeled by $i = 1, 2, 3, 4$). In the T^4 these have intersection numbers

$$\bar{\pi}_{ij} \circ \bar{\pi}_{kl} = - \epsilon_{ijkl} \,. \tag{6.80}$$

In the orbifold space, there are closely related cycles π_{ij}. These lift in the covering space to two copies of $\bar{\pi}_{ij}$, with 4 intersections in the covering space. The intersection in the orbifold space is then

$$\pi_{ij} \circ \pi_{kl} = -2 \, \epsilon_{ijkl} \,. \tag{6.81}$$

Another 16 cycles come from the $\mathbb{P}^1$'s formed by blowing up the 16 singular points. We denote these cycles e_{ij}, where the values of the indices $i, j \in \{1, \ldots, 4\}$ correspond to the points $(0,0), (1/2, 0), (0, \tau/2), (1/2, \tau/2)$ on the two toroidal factors T^2. The inner product on these cycles is $e_{ik} \cdot e_{jl} = -2 \, \delta_{ij}\delta_{kl}$. The 22 cycles π_{ij}, e_{ij} span a 22-dimensional lattice. This is not quite the complete homology lattice of K3, however. This is a sublattice, known as the Kummer lattice, of the full homology lattice. To complete the full lattice additional fractional cycles must be added, such as

$$\gamma_{13} = \frac{1}{2}\pi_{13} + \frac{1}{2}(e_{11} + e_{12} + e_{21} + e_{22}) \,. \tag{6.82}$$

These cycles are topologically spheres, which can be seen from $\gamma_{13} \cdot \gamma_{13} = -2 = 2g - 2$ where $g = 0$ is the genus of the surface[d]. Including a set of such fractional cycles gives a complete set of generators for the homology lattice (6.78) of K3, as reviewed in more detail in several references.[48–50]

6.3.4.3. *Elliptically fibered K3 surfaces*

Now we return to our discussion of F-theory. As described above, F-theory on a K3 surface is described geometrically by an elliptic fibration over the sphere S^2 with 24 singularities, where each singularity has a monodromy

[d]This formula relating the intersection form of a curve with itself to the genus of the curve is the special case of a more general formula (6.90), which we derive later, for a curve in flat or toroidal space.

conjugate to $\tau \to \tau + 1$. Locally, an elliptic fibration can be described by a modular parameter τ varying holomorphically over the base $\mathbb{P}^1$. Generically, the total space of such a fibration is a smooth K3 surface.

A convenient algebraic-geometric description of an elliptic curve is as the set of points in the projective plane satisfying a cubic equation. More specifically, consider the projective space $\mathbb{P}^{2,3,1}$ defined by the set of coordinates (x, y, z) with the equivalence

$$(x, y, z) \sim (\lambda^2 x, \lambda^3 y, \lambda z), \ \forall \lambda \in \mathbb{C} \setminus \{0\} . \tag{6.83}$$

The equation

$$F = -y^2 + x^3 + fxz^4 + gz^6 = 0 \tag{6.84}$$

defines a complex curve of genus one, *i.e.*, an elliptic curve, for general complex values of f and g. We can take a local coordinate chart leaving out the points where $z = 0$ by using eq. (6.83) to set $z = 1$ so that the curve in the local (x, y) coordinate chart is defined by

$$F = -y^2 + x^3 + fx + g = 0 . \tag{6.85}$$

This is known as the *Weierstrass* form of the elliptic curve. One easy way to see that this equation should define a genus one curve is to think of x as giving a local coordinate chart on $\mathbb{P}^1$. The vanishing of F then gives y as a double-valued function of x, with branch points at the points where the cubic $x^3 + fx + g$ vanishes. Generically there are 3 such branch points, and another branch point at $x = \infty$ ($z = 0$), so (6.84) describes a branched cover of $\mathbb{P}^1$ with 4 branch points, which is topologically a torus or genus one curve. Note that while there are two complex parameters f, g defining the Weierstrass model, the moduli space of elliptic curves is only one-dimensional; rescaling $f \to \lambda^4 f, g \to \lambda^6 g, x \to \lambda^2 x, y \to \lambda^3 y$ gives equivalent elliptic curves.

We can now give a more precise mathematical description of a K3 surface realized as an elliptic fibration over a complex one-dimensional base space such as $B = \mathbb{P}^1$. Consider w as a local coordinate on B. An elliptic fibration is given locally by a choice of elliptic curve at each point on B. We can parameterize such a family of elliptic curves in the Weierstrass form by taking f, g to be functions of w

$$F = -y^2 + x^3 + f(w)x + g(w) = 0 . \tag{6.86}$$

This gives a local Weierstrass description of the elliptic fibration as a hypersurface in $\mathbb{C}^3$. To give the global description of the surface, we can reinstate the projective variable z in the fiber, and a projective variable v in the base. The Weierstrass form for an elliptic fibration selects a point at ∞ on each elliptic curve (the point at $z = 0$). This choice of a point on each fiber defines a global section of the fibration. Thus, F-theory is defined by a compactification on an elliptically fibered Calabi-Yau *with section.*

A (complex) codimension one space defined as the vanishing locus in $\mathbb{C}^n$ of a single function such as (6.85) is singular when all derivatives of the function simultaneously vanish. Such singularities are treated rigorously in the mathematical framework of algebraic geometry, but we can understand this simply by noting that level surfaces of the function define locally smooth complex analytic (and algebraic) sets as long as the gradient of the function is nonvanishing. Thus, the elliptic curve defined by (6.85) is singular at a point (x, y) when

$$F = \frac{\partial F}{\partial y} = \frac{\partial F}{\partial x} = -2y = x^3 + fx + g = 3x^2 + f = 0\,, \tag{6.87}$$

Combining these equations, we have $3x^3 + 3fx + 3g = 2fx + 3g = 0 \Rightarrow x = -3g/2f$, so the curve is singular when the *discriminant* Δ vanishes

$$\Delta = 4f^3 + 27g^2 = 0\,. \tag{6.88}$$

Considering again the elliptic fibration over the base $\mathbb{P}^1$, we take $f(w)$ and $g(w)$ to be functions on the base, and the K3 surface is locally defined by (6.86). At a singularity where the discriminant (6.88) vanishes the partial derivative of F with respect to w, $\partial F/\partial w = f'x + g'$, is generically nonzero, so at such points the total space of the K3 surface is locally nonzero even though the fiber degenerates. We generically expect 24 singularities in the elliptic fibration as discussed above. Thus, the discriminant should be a degree 24 polynomial in the coordinate w on the base. We therefore expect that $f(w)$ has degree 8 and $g(w)$ has degree 12.

It is helpful to view f, g, and Δ from a more global perspective. For this we need a little more mathematical machinery, which will be extremely helpful in understanding F-theory compactifications on Calabi-Yau manifolds of higher dimension. While these notes are intended to be relatively self-contained, the reader interested in understanding the concepts outlined in the next few paragraphs more thoroughly may find it useful to study some basic aspects of algebraic geometry. For the material presented here, the text of Perrin[51] is a good starting point. The book by Barth, Hulek, Peters, and Van de Ven[52] also contains a great deal of useful material on compact complex surfaces, including aspects of the structure of elliptic fibrations.

While locally f and g are functions of the coordinate w taking values in $\mathbb{C}$, when these functions are considered on the complete base $B = \mathbb{P}^1$ they are actually sections of *line bundles* over $\mathbb{P}^1$. Complex line bundles over any complex space X are characterized by their first Chern class $c_1(X)$, and are thus associated with an element of $H^2(X, \mathbb{Z})$. In particular, for $X = \mathbb{P}^1$, the first Chern class is in $H^2(\mathbb{P}^1, \mathbb{Z}) = \mathbb{Z}$.

We will freely move between the cohomology representative of a line bundle on a space X and the Poincaré dual homology class, which is a *divisor* on X. A divisor on X is a linear combination (over $\mathbb{Z}$) of irreducible algebraic hypersurfaces of the form $\sum_i n_i H_i$. There is a one-to-one correspondence between divisors and line bundles over any complex space X. The divisor associated with a line bundle $\mathcal{L}$ can be thought of as the linear combination of the homology class containing the zero locus of a section of $\mathcal{L}$ and the negative of the homology class containing the poles of the section. We will refer to the line bundle associated with a divisor D as $\mathcal{O}(D)$. On $\mathbb{P}^1$ the only irreducible algebraic hypersurface is (the homology class of) a point P, so every divisor is characterized as a multiple nP. For example, the function z on $\mathbb{P}^1$ can be extended to a section of the line bundle $T\mathbb{P}^1 = \mathcal{O}(2P)$, and vanishes at two points (0 and $z = \infty \to w = 1/z = 0$). On the other hand, a section dz of the cotangent bundle goes to $dz = -dw/w^2$ on the chart around $z = 1/w = \infty$, and has a double pole at $w = 0$, so $T^*\mathbb{P}^1 = \mathcal{O}(-2P)$. We can take the product of two line bundles by multiplying the transition functions; this corresponds to adding the corresponding divisors $\mathcal{O}(D) \otimes \mathcal{O}(F) = \mathcal{O}(D + F)$.

On any complex manifold there is a special class known as the *canonical class*, K. On a complex manifold of dimension d, the canonical class corresponds to the line bundle associated with the dth power of the cotangent bundle, and is given locally by the maximum antisymmetric power of the holomorphic differential, $dz_1 \wedge \cdots \wedge dz_d$. The canonical class is essentially a measure of the total curvature of a space. On $\mathbb{P}^1$ the canonical class is

$$K = c_1(T^*) = -2P\,. \tag{6.89}$$

A manifold in any complex dimension is Calabi-Yau if and only if[e] the manifold is Kähler and $K = 0$.

Given the machinery just defined, we can now understand a formula that will be useful in studying several relevant aspects of curves on surfaces. For

[e]Technically, manifolds with K in a torsion class, $nK = 0$, are sometimes classified as Calabi-Yau, but we do not worry about such issues here.

a smooth curve C on a surface S with canonical class K_S, we have

$$(K_S + C) \cdot C = 2g - 2\,. \tag{6.90}$$

This statement follows from a fairly straightforward three-line argument that uses some important theorems from elementary algebraic geometry. The first step is the *adjunction formula*, which determines the canonical class on C in terms of K_S, $\mathcal{O}(K_S) \otimes \mathcal{O}(C)|_C = \mathcal{O}(K_C)$, where $|$ denotes the restriction of the line bundle to C. The second step is relating this restriction to the intersection form using $\mathcal{O}(D)|_C = D \cdot C$. And the third step uses the *Riemann-Roch* theorem, which essentially says that on a smooth curve C, $\mathcal{O}(K_C) = 2g - 2$. In flat space, or on a torus or K3 surface, $K_S = 0$ so (6.90) becomes $C \cdot C = 2g - 2$, which we used in discussing the K3 surface earlier.

As another simple application of the adjunction formula, consider a hypersurface D in projective space $\mathbb{P}^{d+1}$ defined by a degree $d + 2$ homogeneous polynomial. There is a single irreducible hypersurface H (up to linear equivalence) on $\mathbb{P}^{d+1}$, associated with the vanishing of any coordinate function z_i. The canonical class of $\mathbb{P}^{d+1}$ is $K = \mathcal{O}(-(d + 2)H)$, while the divisor D associated with a degree $d+2$ polynomial is $D = (d+2)H$, so the canonical class on D is $\mathcal{O}(K_D) = \mathcal{O}(K + D)|_D = 0$. Thus, any such D is a Calabi-Yau. This gives an alternative proof that the cubic on $\mathbb{P}^2$ gives an elliptic curve. Similarly, a quartic on $\mathbb{P}^3$ gives a K3 surface, and a quintic on $\mathbb{P}^4$ gives a Calabi-Yau threefold.

Returning to our elliptically fibered K3 surface, in terms of the canonical class, f and g are sections of the line bundles $\mathcal{O}(-4K)$ and $\mathcal{O}(-6K)$, and Δ is a section of the bundle $\mathcal{O}(-12K)$. Characterizing Δ by the curvature class of the associated bundle, the condition for an elliptic fibration to describe an elliptically fibered Calabi-Yau surface over B can be written as

$$-12K = \Delta\,. \tag{6.91}$$

We will refer to this condition as the "Kodaira condition". It is a special case of the relation proven by Kodaira between the canonical class of the total space of an elliptic fibration and the canonical class of the base. In the case $B = \mathbb{P}^1$, the Kodaira condition states that $\Delta = 24P$, which is just the statement above that the singularity locus consists of 24 points on the sphere S^2.

The space of elliptically fibered K3 surfaces is parameterized by the coefficients of the polynomials (sections) $f(w), g(w)$. The coefficients are moduli that parameterize the complex structure of the K3 surface just as

τ parameterizes the complex structure of T^2. The number of coefficients is $9 + 13 = 22$. This parameterization is, however, redundant. Just as constants f, g provide one extra parameter for a single elliptic curve with a redundancy under scaling, the homogeneous functions $f(w, v), g(w, v)$ have four redundant degrees of freedom through general linear transformations on the homogeneous coordinates w, v. Thus, the number of complex degrees of freedom for the set of elliptically fibered K3 surfaces (with section) is $22 - 4 = 18$, in agreement with the number of degrees of freedom found in the heterotic construction. One additional scalar field is associated with the volume of the base $B = \mathbb{P}^1$.

6.3.4.4. *Gauge groups and singularities*

As discussed above, the generic elliptic K3 is smooth. The resulting F-theory vacuum is an 8D supergravity theory with an abelian gauge group $U(1)^{20}$ (where as usual two of the $U(1)$'s lie in the gravity multiplet.) The derivation of the rank of the gauge group from the F-theory picture is actually somewhat subtle. The abelian gauge group factors arise from global structure of the compactification, unlike nonabelian gauge group factors, which, as we discuss below, arise from the geometry in a simpler local fashion. Naively it might seem that the gauge group should be $U(1)^{24}$, with each 7-brane associated with a singularity giving a separate $U(1)$ factor. Global constraints, however, reduce the rank of the gauge group to 20. This cannot be understood easily from a simple supergravity picture, since the 7-branes cannot all be described perturbatively in the same duality frame. The constraint to rank 20 can be understood from the topology of the global K3, or from the point of view of probe branes on the base, which interact with the 7-branes through open strings and string junctions;[53,54] more geometrically this corresponds to the structure of rational sections of the fibration (Mordell-Weil group),[46,55] which encode part of the topology of the global K3 space.

Now let us consider the non-generic situation in which several of the 7-branes on the spherical base B coincide. This leads to a more complicated singularity in the elliptic fibration and often to a singularity in the total space of the K3. At such a singularity of the K3, one or more two-cycles may shrink to a point. When a two-cycle shrinks to a point, the theory develops an extra massless gauge boson. This gives rise to an enhanced gauge group in the 8D supergravity theory. In the M-theory/IIA picture, the extra gauge boson comes from a membrane/D2-brane wrapped on the vanishing cycle.[11] It was shown by Kodaira that the possible types of singularity in a complex surface that can be resolved by a succession of blow-ups can be systematically classified.[56,57] For each type of singularity there is a corresponding Dynkin diagram, encoding the intersection form on the shrunk two-cycles. In the F-theory picture, these Dynkin diagrams precisely characterize the nonabelian gauge group arising in that singular limit of the K3 surface. Note the chronic sign difference between the intersection form on the two-cycles and the Cartan matrix; this arises from a difference in conventions between the algebraic-geometric and Lie algebra frameworks.

In the following section we work through an explicit example of an F-theory singularity of type A_3. In general, the singularity type depends upon the degree of vanishing of f, g and Δ at the singular point. For example, when f and g are nonvanishing, and Δ vanishes to order n, there is a singularity of the form A_{n-1}, associated with nonabelian gauge group $SU(n)$. The complete list of Kodaira singularity types is given in the following table.

ord (f)	ord (g)	ord (Δ)	singularity	nonabelian symmetry
≥ 0	≥ 0	0	none	none
0	0	n	A_{n-1}	$SU(n)$
≥ 1	1	2	none	none
1	≥ 2	3	A_1	$SU(2)$
≥ 2	2	4	A_2	$SU(3)$
2	3	$n+6$	D_{n+4}	$SO(8+2n)$
≥ 2	≥ 3	6	D_4	$SO(8)$
≥ 3	4	8	E_6	E_6
3	≥ 5	9	E_7	E_7
≥ 4	5	10	E_8	E_8

It is an educational exercise to work through the blow-up procedure for some of the different singularities in the table, and to verify the appearance of the stated Dynkin diagram in the intersection form of the blown-up $\mathbb{P}^1$'s. When a (complex) codimension one singularity occurs which has degrees higher than any allowed in the Kodaira table, the geometry cannot be resolved to a space which is locally Calabi-Yau. In Section 6.4.7, we discuss non-Kodaira singularities of higher codimension, which can lead to physically interesting transitions in the space of theories.

We now have a general picture of how F-theory describes 8D supergravities through compactification on elliptically fibered K3 surfaces. The elliptically fibered K3 surface is described through a Weierstrass equation of the form (6.86), where f, g are degree 8 and degree 12 polynomials respectively on the base $\mathbb{P}^1$ (sections of $\mathcal{O}(-4K), \mathcal{O}(-6K)$ respectively in the global picture). Generically the gauge group is $U(1)^{20}$, though as the moduli vary the singularities associated with vanishing of Δ from (6.88) can coincide, giving more complicated singularity types and enhancing the 8D gauge group to include nonabelian factors.

As a global example of an F-theory vacuum on an elliptically fibered K3 with large symmetry group, we describe the theory with $E_8 \times E_8$ gauge symmetry, following Morrison and Vafa.[46] Choose

$$f = \alpha z^4, \qquad g = z^5 + \beta z^6 + z^7 . \tag{6.92}$$

This gives

$$\Delta = 27z^{10} + \cdots + 27z^{14} . \tag{6.93}$$

This Weierstrass model has E_8 singularities at $z = 0, \infty$, as can be verified from the Kodaira table above.

We can now address the question of which gauge groups G can be realized in eight dimensions through an F-theory compactification on an elliptically fibered K3. The condition that the K3 be elliptically fibered and have a section identifies two cycles f, s on the total space with intersection products $f \cdot f = 0, f \cdot s = 1, s \cdot s = -2$. The linear combinations $f, s+f$ then have intersection products in K3 given by a copy of U. The elliptic fibration thus removes a factor of U from $H_2(K3; \mathbb{Z}) = \Gamma^{3,19}$, giving $\Gamma^{2,18}$.[42] In this remaining lattice, we can shrink any combination of two-cycles satisfying $c \cdot c = -2$ to get nonabelian gauge bosons. The set of nonabelian gauge groups that can be realized is then, just as in the heterotic theory, the set of G such that

$$G : \ -\Lambda_G \hookrightarrow \Gamma^{2,18} . \tag{6.94}$$

Again, as in the heterotic theory, the rank of the gauge group when all $U(1)$ factors is included is always 20.

6.3.4.5. *Example: A_3 singularity*

It is helpful to study the geometry of a particular case to understand the general principles of the Kodaira classification. Consider an elliptic fibration over a local patch in $\mathbb{C}$ parameterized by w, given by the Weierstrass form (6.86) with

$$f = -\frac{1}{3} - w^2 \tag{6.95}$$

$$g = \frac{2}{27} + \frac{1}{3}w^2 .$$

While f and g are both nonvanishing at $w = 0$, a small calculation shows that the discriminant vanishes to order w^4

$$\Delta = 4f^3 + 27g^2 = -w^4 - 4w^6 . \tag{6.96}$$

At $w = 0$, $F = -y^2 + x^3 - x/3 + 2/27$, which has a singularity at $x = -1/3, y = 0$. To simplify the analysis, it is convenient to change coordinates

$$x \to x + \frac{1}{3} \tag{6.97}$$

to move the singularity to $x = 0$. The Weierstrass equation then becomes

$$F = -y^2 + x^3 + x^2 - w^2 x = 0 . \tag{6.98}$$

This gives a local equation for the complex surface described by the elliptic fibration in coordinates $(x, y, w) \in \mathbb{C}^3$. The surface has a singularity at $x = y = w = 0$. This singularity can be resolved by blowing up the codimension one singularity repeatedly until the space is smooth. We can do this by working in a sequence of coordinate charts containing the various blow-ups. We refer to the local chart where the surface is defined through (6.98) as Chart 0.

Chart 1:

To resolve the singularity in chart 0, we blow up the singular point. We replace the point $(0, 0, 0)$ with a $\mathbb{P}^2$ given by the set of limit points described by homogeneous coordinates $[x : y : w]$ along curves approaching $(0, 0, 0)$. The new space produced by the blow-up process can be described as a subspace of $\mathbb{P}^2 \times \mathbb{C}^3$, where homogeneous coordinates $[u : v : t]$ on the $\mathbb{P}^2$ satisfy the relations

$$uy = vx, \; uw = tx, \; vw = ty . \tag{6.99}$$

When one or more coordinates x, y, w are nonzero a unique point in the $\mathbb{P}^2$ is determined, $[u : v : t] \sim [x : y : w]$. At the point $(x, y, w) = (0, 0, 0)$, however, (6.99) imposes no conditions on the coordinates u, v, t.

This blow-up process can be described in different coordinate charts on the $\mathbb{P}^2$. We choose, for example, a local chart where $w \neq 0$. In this chart we cannot have $t = 0$ (or $u = v = 0$ also from (6.99)), so we fix the homogeneous coordinates on $\mathbb{P}^2$ using $t = 1$. We then have

$$x = uw, \quad y = vw\,. \tag{6.100}$$

Relabeling $u \to x_1, v \to y_1$, this chart (Chart 1) is realized by changing coordinates to

$$(x, y, w) = (x_1 w_1, y_1 w_1, w_1)\,. \tag{6.101}$$

For $w_1 \neq 0$, there is a unique point (x, y, w) in the original chart (chart 0) corresponding to each point (x_1, y_1, w_1) in the new chart. For $w_1 \to 0$, however, there is a set of points $(x_1, y_1, 0)$ in the new chart that all correspond to $(0, 0, 0)$ in Chart 0. This gives a local patch on the new $\mathbb{P}^2$ formed from the blow-up, containing the points $[x_1 : y_1 : 1]$. In the coordinates of Chart 1, the local equation (6.98) becomes

$$F = (-y_1^2 + x_1^2 + x_1^3 w_1 - w_1 x_1) w_1^2 = 0\,. \tag{6.102}$$

The $\mathbb{P}^2$ that is added through the blow-up process is known as the *exceptional divisor* associated with the blow-up. The resulting equation (6.102) is reducible and contains two copies of the exceptional divisor at $w_1 = 0$. Factoring out the overall w_1^2 from (6.102) (*i.e.*, removing the copies of the exceptional divisor), we have the equation for the *proper transform* of the original space

$$F_w = -y_1^2 + x_1^2 + x_1^3 w_1 - w_1 x_1 = 0\,. \tag{6.103}$$

This equation describes the surface in chart 1 after the original singularity at $(x, y, w) = (0, 0, 0)$ has been blown up. (We use the subscript on F to denote the coordinate chart used for the blow-up.) The intersection of the space defined through (6.103) with the exceptional divisor at $w_1 = 0$ gives the exceptional divisor on the surface, which is generally a curve or set of curves associated with blowing up the point at $w_1 = 0$ within the surface defined by F. At $w_1 = 0$, (6.103) becomes $-y_1^2 + x_1^2 = 0$

$$y_1 = \pm x_1\,. \tag{6.104}$$

This defines a pair of curves in $\mathbb{P}^2$, which we call $C_1^{\pm}$. The equation (6.103) still contains a singularity at the point $(x_1, y_1, w_1) = (0, 0, 0)$, where the curves $C_1^{\pm}$ cross. So we must again blow up the singularity to produce a smooth space.

Chart 2:

We replace the singular point in Chart 1 with another exceptional divisor $\mathbb{P}^2$, this time using the local coordinates

$$(x_1, y_1, w_1) = (x_2, y_2 x_2, w_2 x_2) = 0\,. \tag{6.105}$$

After removing two copies of the exceptional divisor $x_2 = 0$ we get the new local equation

$$F_{wx} = -y_2^2 + 1 + x_2^2 w_2 - w_2 = 0\,. \tag{6.106}$$

This gives another exceptional curve C_2, associated with the intersection of (6.106) with the exceptional divisor $x_2 = 0$

$$C_2 = \{(x, y, w) : x = 0, w = 1 - y^2\}\,. \tag{6.107}$$

Since (6.106) has no further singularities, we have completely resolved the local singularity and have a smooth space in coordinate chart 2.

From the way in which the exceptional curves $C_1^{\pm}, C_2$ intersect, we identify the A_3 form of the singularity found by Kodaira. To compute the intersections, we write (6.104) in terms of coordinates in chart 2

$$y_2 x_2 = \pm x_2 \;\Rightarrow\; y_2 = \pm 1\,, \tag{6.108}$$

which combined with $w_1 = w_2 x_2 = 0$ gives the points $[1 : \pm 1 : 0]$ in homogeneous coordinates on $C_2 = \mathbb{P}^2$, showing that $C_1^{\pm}$ each intersect C_2 at a single point but do not intersect one another. The intersections of these curves are shown graphically in Figure 6.6, along with the associated Dynkin diagram. In principle, we would need to check all coordinate charts at each step to be sure we have not missed any additional singularities. Here we have chosen charts that completely resolve the original singularity.

Fig. 6.6. Exceptional curves resolving an A_3 singularity, depicted as (a) geometry of the exceptional divisors, which take the form of $\mathbb{P}^1 = S^2$'s with associated intersections, (b) associated Dynkin diagram.

The example just described is an A_3 singularity. The shrinking two-cycles in the K3 describe an extra set of massless gauge bosons that extend the gauge group to a nonabelian group. In particular, the cycles $C_1^{\pm}, C_2$ form the simple roots of the Lie algebra $SU(4)$. By taking linear combinations of these cycles we can produce all the nonzero roots of $SU(4)$, each of which corresponds to a curve in the K3 with $c^2 = -2$. When this type of singularity occurs at a point in a global elliptically fibered K3 compactification of F-theory, the resulting gauge group for the 8D theory has an $SU(4)$ factor. The rank of the group stays unchanged in such an enhancement, so the total rank is still 20.

6.3.5. *The space of 8D supergravities*

We have now seen that two very different string-theoretic constructions give rise to the same space of 8D supergravity theories. For both the heterotic and F-theory constructions, there is a low-energy 8-dimensional theory with one supergravity multiplet and vector multiplets forming a rank 20 gauge group. At a generic point in this moduli space the gauge group is $U(1)^{20}$, but at special loci in the moduli space the gauge group is enhanced. The possible nonabelian gauge groups that can be realized in each description are precisely those whose root lattices can be embedded (with a factor of -1) into $\Gamma^{2,18}$. The moduli space of theories has 37 dimensions, parameterized by the scalars in the gravity and vector multiplets. In both the heterotic and F-theory descriptions, this moduli space of theories is a connected space.

The hypothesis that the heterotic string compactified on the torus T^2 and F-theory compactified on K3 give rise not only to the same low-energy supergravity theory, but to the same physical theory at the nonperturbative level, asserts the existence of a duality symmetry between these two classes of string vacua. This duality symmetry, first formulated by Vafa,[44] underlies many duality symmetries relating lower-dimensional theories. While, as we have seen, this duality symmetry holds at the level of the field content and symmetries of the theories, a complete mathematical proof of this and all other string duality symmetries is still lacking—in large part because there is as yet no mathematically complete definition of string theory. Nonetheless, the duality between these two ostensibly very different realizations of 8D supergravity theories has been explored in a number of ways beyond what we have described here. An explicit correspondence between the moduli has been identified for some classes of heterotic compactifications.[46,58] The duality between F-theory on K3 and type I string theory on T^2 was explored in detail by Sen.[59] Protected one-loop corrections to the effective action on the heterotic side have also been matched from the geometry of F-theory,[60–63] suggesting some deeper geometric structure underlying these theories. All evidence suggests that the heterotic and F-theory compactifications are physically equivalent; this suggests in turn that there may be a unique quantum supergravity theory in 8 dimensions with minimal supersymmetry.

Classically, we can couple the 8D supergravity theory to any number of vector multiplets, and the restriction to a rank 20 gauge group that embeds through eq. (6.94) is not apparent. As for ten-dimensional $\mathcal{N} = 1$ supergravity, we would like to understand whether quantum consistency restricts the gauge group in a way that matches the range of models realized in string theory. There is at this time no complete argument from the point of view of the macroscopic supergravity theory that restricts the gauge group of the theory in this way. It does seem plausible, however, that quantum consistency of the supergravity theory may impose additional constraints that have not yet been fully elucidated. One possibility is that extra consistency conditions on the gauge group may arise from supersymmetric constraints on the geometric structure of the moduli space.

While space-time anomalies in the 8D theories are not restrictive, we can also consider anomalies on the world-volume of solitonic string excitations of the theory; such anomalies may impose additional constraints on the theory. Every 8D supergravity theory has an antisymmetric two-form field $B_{\mu\nu}$ in the spectrum. There are classical solutions describing stringlike black brane excitations of the theory that couple to the B field. While there is no proof that quantum excitations of this kind of string soliton must be included in the complete quantum gravity theory for such an 8D supergravity, it seems likely that this is the case. Locally, a classical solution describing a small loop of such string represents a small deformation from the flat space-time background. Some general arguments for the conclusion that quanta of any object carrying an allowed conserved charge must be present in any supergravity theory were recently stated by Banks and Seiberg.[64]

A hint for how world-volume anomaly conditions on a solitonic string may place constraints on the set of allowed theories can be seen in 10 dimensions. Anomaly cancellation in the world-volume theory of N D-strings in the type I description of the $SO(32)$ supergravity theory was analyzed by Banks, Seiberg and Silverstein.[65] This world-volume theory has $(0, 8)$ supersymmetry in 2 dimensions. The theory is chiral and carries a $Spin(N)$

gauge symmetry in its world-volume, which would be inconsistent without cancellation of the gauge anomaly. The theory contains a vector multiplet with 8 left-moving chiral fermions in the adjoint representation of the gauge group. There is a matter multiplet containing 8 bosons corresponding to transverse fluctuations of the string and 8 right-moving fermions in the symmetric tensor representation of the gauge group. There are also 32 left-moving fermions in the singlet representation of the gauge group, which can be associated in the string theory picture with open strings stretching from the D-string to the space-filling type I D9-branes that generate the $SO(32)$ gauge group (these D9-branes and the type I picture are discussed in more detail in the following section). For the gauge anomaly to cancel we must have

$$\sum x_R A_R = 0 \tag{6.109}$$

where x_R denotes the number of chiral left-moving fermions in the representation R of the gauge group (right-moving fermions entering with the opposite sign), and A_R is the constant of proportionality between the trace of F^2 in representation R and the fundamental representation

$$\mathrm{tr}_R F^2 = A_R \mathrm{tr} F^2 \,. \tag{6.110}$$

Here, as previously, tr_R is the trace in representation R, with the absence of index on the trace on the RHS indicating the fundamental representation. For $Spin(N)$, we have $A_{\mathrm{adj}} = N - 2$ and $A_{\mathrm{sym}} = N + 2$. Denoting the number of left-moving fermions in the singlet representation by $2r = 32$ where r is the rank of the space-time gauge group, the anomaly cancellation condition is

$$8(N-2) + 2r - 8(N+2) = 0 \,, \tag{6.111}$$

which vanishes precisely when $r = 16$. This confirms the consistency of the $SO(32)$ $\mathcal{N} = 1$ theory in 10 dimensions, and can be interpreted as a constraint on the rank of the space-time gauge group. While this argument is formulated in the language of type I string theory, it should be possible to reproduce the analysis from the point of view of the low-energy theory on the D1-brane itself. By analyzing fluctuations around the solitonic string solution, the rank of the space-time gauge group should thus be fixed to be 16 by cancellation of the world-volume anomaly. Note that this argument may become more subtle for more general space-time gauge groups, in particular for the $E_8 \times E_8$ theory.

It has been suggested by Uranga[66] that this kind of argument can be applied in supergravity theories with 16 supercharges in fewer dimensions. If correct, this could lead to a demonstration that the world-volume theory of the solitonic string would be anomalous in any D-dimensional supergravity theory with 16 supercharges unless the rank of the space-time gauge group is fixed to be $16 + 2 \times (10 - D) = 36 - 2D$. In particular, in 8D the gauge group would need to be rank 20. While plausible in schematic form, the details of this argument have not been worked out. In particular, as we have seen, in eight dimensions, the 18 vector multiplets in the Cartan algebra of the gauge group appearing in the theory are on an equal footing, and should all play the same role in the world-volume theory of the solitonic string solution. Thus, we expect a cancellation between the extra vector multiplets and the $U(1)$ factors in the supergravity multiplet. Such a cancellation is plausible since the graviphoton has different properties from the other vector fields; for example, the couplings of the BF^2 terms for the $U(1)$ factors in the different multiplets have opposite signs. But a detailed proof that this works out is still lacking, and is left as a challenge for future work. Proving by such an argument that the rank of the gauge group in eight dimensions is constrained is also not enough to prove that the set of consistent theories is precisely those given by string theory. There are groups, such as $SU(2)^{18} \times U(1)^2$ that are of rank 20 and yet cannot be embedded into $\Gamma^{2,18}$ as in (6.71). A proof of the stronger embedding constraint from the point of view of the solitonic strings, combined with the known string constructions of supergravity theories with 16 supercharges through toroidal compactification of the 10D theory, would amount to a proof of string universality for this class of supergravity theories. This could lead to similar conclusions for 4D theories with $\mathcal{N} = 4$ supersymmetry and 6D theories with $\mathcal{N} = (1,1)$ supersymmetry as well as the 8D theory with $\mathcal{N} = 1$ supersymmetry, and would show that the rank and number of possible nonabelian gauge groups for these theories is finite in each case.

The story just outlined must be incomplete in at least some respects. One complication that must be addressed is the existence of orbifold string compactifications that give rise to theories with 16 supercharges and gauge groups of lower rank. For example, the CHL string[41,67,68] gives a theory in eight dimensions with only 10 vector multiplets, and another class of heterotic orbifolds gives rise to an 8D theory with 2 vector multiplets.[4] A complete argument for a bound on the rank of the gauge group in 8D supergravity theories would need to be compatible with the presence of these other discrete structures for the spectrum. Another interesting question is whether these discrete families of string vacua can be smoothly connected in some way to the moduli space of 8D supergravity theories with rank 20 gauge group.

6.4. Supergravity and String Vacua in Six Dimensions

We now turn to supergravity theories in six dimensions. We again focus on theories with the minimum amount of supersymmetry, where the most interesting new phenomena arise. In six dimensions the SUSY generators are chiral, and there are theories with $(2,2), (2,0), (1,1)$, and $(1,0)$ supersymmetry. The (2, 2) theory with maximal supersymmetry arises from compactification of 10D type II supergravity on a torus, and the field content of the theory is uniquely constrained by the supersymmetry structure. The theories with $(1,1)$ supersymmetry are in the class of theories with 16 supercharges discussed at the end of the previous section. Theories with $(2,0)$ supersymmetry are strongly constrained by anomalies and correspond to the theories realized through compactification of the type II theory on a K3 surface.[69] Theories with $(1,0)$ supersymmetry have the richest structure. In particular, these theories can contain matter fields that transform in a variety of representations of the gauge group. These are the 6D theories on which we focus in these lectures. These supersymmetric theories can all be formulated in 6D Minkowski space. While in four dimensions, there are supersymmetric models in AdS space, and gauged supergravity theories with stable supersymmetric backgrounds, such models do not occur in six dimensions,[70] so all the supersymmetric 6D theories of interest admit Minkowski vacua.

We begin by describing the constraints from supergravity and then consider string constructions. There is a much wider range of possible string constructions for 6D $\mathcal{N} = 1$ supergravity theories than for 8D supergravities. We consider several approaches here, adding intersecting brane models on a K3 compactification of type IIB to to our repertoire, and explaining the additional complications involved in heterotic and F-theory constructions beyond those encountered in 8D compactifications.

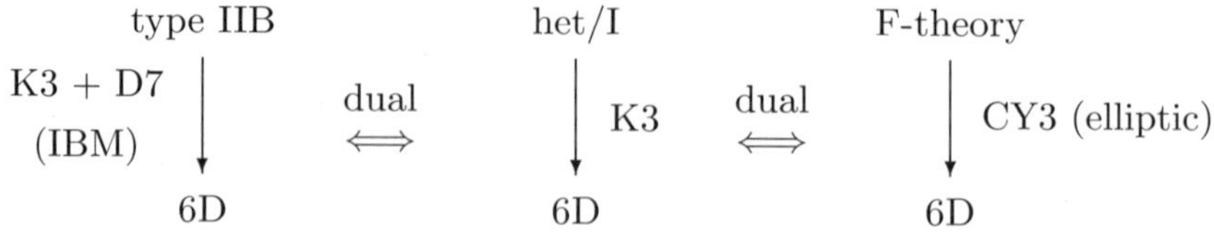

In Section 6.4.1 we describe the basic structure of 6D supergravity theories with minimal supersymmetry, and in Section 6.4.2 we characterize the space of theories that satisfy the 6D Green-Schwarz anomaly cancellation conditions. We then introduce some basic aspects of orientifolds in Section 6.4.3, which we use to describe intersecting brane models in Section 6.4.4 and magnetized brane models, which are equivalent to compactifications of the heterotic string, in Section 6.4.5. We briefly describe a variety of additional approaches to heterotic and type I/II constructions that have been used for 6D vacua in Section 6.4.6. We describe F-theory constructions in six dimensions in Section 6.4.7. We show in Section 6.4.8 that the close relationship between a lattice determined by anomaly cancellation in the low-energy theory and the mathematical structure of F-theory allows us to use the data from a low-energy theory to characterize the topological structure of any corresponding F-theory compactification. This is helpful in characterizing the global structure of the set of possible theories. We summarize the current state of knowledge regarding 6D gravity theories with minimal supersymmetry in Section 6.4.9.

6.4.1. *Six-dimensional gravity with $\mathcal{N} = (1,0)$ supersymmetry*

We focus on the massless spectrum of six-dimensional supergravity theories. There are four massless supersymmetry multiplets that appear in $\mathcal{N} = (1,0)$ theories with 8 supercharges. These multiplets are summarized in the following table.

Multiplet	Matter Content
SUGRA	$(g_{\mu\nu}, B^+_{\mu\nu}, \psi^-_\mu)$
Tensor (T)	$(B^-_{\mu\nu}, \phi, \chi^+)$
Vector (V)	(A_μ, λ^-)
Hyper (H)	$(4\varphi, \psi^+)$

The supergravity multiplet contains, in addition to the metric, a bosonic self-dual two-form field $B^+_{\mu\nu}$. There are also tensor multiplets that contain anti-self-dual two-form fields $B^-_{\mu\nu}$ as well as a single scalar field. In general an $\mathcal{N} = 1$ supergravity theory can have any number T of tensor multiplets, although the theory only has a Lagrangian description when $T = 1$. The two-form fields $B^{\pm}$ transform under an $SO(1,T)$ action that also transforms the scalar fields in the tensor multiplets. These scalar fields parameterize a T-dimensional moduli space $SO(1,T)/SO(T)$ that is closely analogous to the moduli space for toroidal compactifications (6.64). (A further discrete quotient by a duality symmetry group must be taken in the quantum theory, as discussed further below.) As in higher dimensions, the vector multiplet contains the 6D gauge field and a chiral gaugino field. The gauge group of the theory in general takes the form

$$G = G_1 \times G_2 \times \cdots \times G_k \times U(1)^n/\Gamma \tag{6.112}$$

where G_i are simple nonabelian gauge group factors, and Γ is a discrete group. The matter hypermultiplets in 6D supergravity theories live in a manifold with a quaternionic Kähler structure. These hypermultiplets can transform in an arbitrary representation (generally reducible) of the gauge group.

To summarize, the discrete data characterizing the field content and symmetries of a 6D $\mathcal{N} = 1$ supergravity theory consist of the following:

T: the (integer) number of tensor multiplets

G: the gauge group of the theory; we denote by V the number of vector multiplets in the theory.

$\mathcal{M}$: the representation of G characterizing the matter content of the theory. We denote by H the number of hypermultiplets (including uncharged multiplets) in the theory.

A complete description of the theory would involve further information such as the metric on the scalar moduli space and higher-derivative terms in the action. We do not address this more detailed structure in these lectures. Understanding the extent to which this structure is uniquely determined by supersymmetry and quantum consistency is an interesting direction for future research. The Lagrangian for 6D supergravity theories with one tensor multiplet ($T = 1$) was originally described by Nishino and Sezgin,[71,72] and the field equations for models with multiple tensors were developed by Romans.[73]

The question we now want to address is: What combinations of T, G, and $\mathcal{M}$ are allowed in a consistent 6D supergravity theory? *i.e.*, what is the space $\mathcal{G}^{6D,\mathcal{N}=1}$? To begin to answer this question we consider the known quantum constraints on this class of theories.

6.4.2. *Anomalies and constraints on supergravity in 6D*

The structure of quantum anomalies in six dimensions is very similar to that in ten dimensions. The chiral fields of the theory that contribute to anomalies are the self-dual and anti-self-dual two-form fields $B^{\pm}_{\mu\nu}$, the gravitino ψ^{-}_{μ}, gauginos λ^{-}, and the chiral fermions χ^{+} and ψ^{+} from the tensor and hyper multiplets. The anomaly is characterized by an 8-form anomaly polynomial $I_8(R, F)$. The anomaly arises from one-loop "box" diagrams with 4 external gauge bosons or gravitons. The Green-Schwarz mechanism again comes into play to cancel anomalies through tree diagrams mediated by an exchange of B fields[74,75] (See Figure 6.7). The story in six dimensions is complicated, however, by the presence of multiple B fields. The

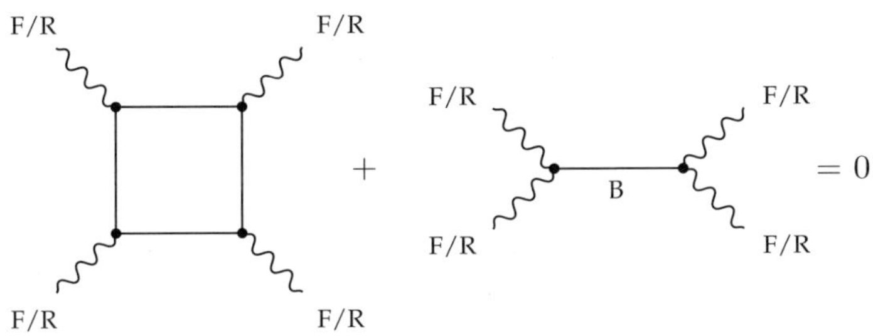

Fig. 6.7. Green-Schwarz anomaly cancellation in six dimensions cancels one-loop anomalies through tree-level diagrams mediated by self-dual and anti-self-dual B fields.

generalization of the Green-Schwarz mechanism including multiple B fields was worked out by Sagnotti. The 6D gravitational, nonabelian gauge, and mixed gauge-gravitational anomalies cancel when the 8-form I_8 factorizes in the form[76–78]

$$I_8 = \frac{1}{2}\Omega_{\alpha\beta}X_4^{\alpha}X_4^{\beta} \tag{6.113}$$

where

$$X_4^{\alpha} = \frac{1}{2}a^{\alpha}\mathrm{tr}R^2 + \sum_i b_i^{\alpha}\left(\frac{2}{\lambda_i}\mathrm{tr}F_i^2\right). \tag{6.114}$$

Here, $\Omega_{\alpha\beta}$ is a signature $(1, T)$ inner product, a^α and b_i^α are vectors in $\mathbb{R}^{1,T}$, and λ_i are normalization constants for the simple group factors G_i appearing in G, where for example $\lambda_{SU(N)} = 1, \lambda_{E_8} = 60, \ldots$. We have not written the anomaly conditions for $U(1)$ gauge factors. These take a similar but slightly more complicated form.[79–81] We do not treat $U(1)$ factors systematically in these lectures; they add some technical complications to the story but do not play an important rule in the main points we wish to emphasize here.

It may be helpful to consider a special case of the anomaly conditions where the factorization takes a particularly simple form. When $T = 1$, there is always a basis for $\mathbb{R}^{1,T}$ where

$$\Omega_{\alpha\beta} = \begin{pmatrix} 0 & 1 \\ 1 & 0 \end{pmatrix} \qquad a = (-2, -2), \quad \text{and } b = \frac{1}{2}(\alpha, \tilde{\alpha})\,. \tag{6.115}$$

The anomaly factorization condition can then be written as

$$I_8 = X^1 X^2 = (\mathrm{tr}R^2 - \sum_i \alpha_i \mathrm{tr}F_i^2)(\mathrm{tr}R^2 - \sum_i \tilde{\alpha}_i \mathrm{tr}F_i^2). \tag{6.116}$$

This form of the anomaly cancellation condition appears in much of the literature on 6D models with $T = 1$.

The anomaly cancellation conditions are not the only constraints that a 6D supergravity theory must satisfy for consistency. As observed by Sagnotti,[76] there is also a constraint coming from the condition that the gauge fields have kinetic terms with the proper (negative) sign. Wrong-sign gauge field kinetic terms would lead to an instability in the theory. As in 10D, the gauge kinetic term is related by supersymmetry to the BF^2 coupling that plays a role in the Green-Schwarz mechanism. The gauge kinetic term is thus proportional to $-j \cdot b\, \mathrm{tr}F^2$. The sign constraint on this gauge kinetic term gives the condition that $j \cdot b > 0$. This constraint plays an important role in restricting the set of possible consistent supergravity theories in six dimensions. We discuss other quantum consistency constraints later in this section and in Section 6.4.9.

Anomaly cancellation through (6.113) imposes a set of conditions that relate the field content of the theory and the vectors a, b_i appearing in the anomaly polynomial. Each type of term in the anomaly polynomial (R^4,

$F^2R^2, \ldots$) must cancel separately. This gives the following conditions

$$R^4: \qquad H - V = 273 - 29T \tag{6.117}$$

$$F^4: \qquad 0 = B^i_{Adj} - \sum_R x^i_R B^i_R \tag{6.118}$$

$$(R^2)^2: \qquad a \cdot a = 9 - T \tag{6.119}$$

$$F^2R^2: \qquad a \cdot b_i = \frac{1}{6}\lambda_i \left(A^i_{Adj} - \sum_R x^i_R A^i_R \right) \tag{6.120}$$

$$(F^2)^2: \qquad b_i \cdot b_i = \frac{1}{3}\lambda_i^2 \left(\sum_R x^i_R C^i_R - C^i_{Adj} \right) \tag{6.121}$$

$$F_i^2F_j^2: \qquad b_i \cdot b_j = 2\sum_{R,S} x^{ij}_{RS} A^i_R A^j_S, \qquad i \neq j. \tag{6.122}$$

In these expressions, A_R is the set of group-theory coefficients for each representation defined by (6.110), and similarly B_R and C_R are defined through

$$\mathrm{tr}_R F^4 = B_R \mathrm{tr} F^4 + C_R (\mathrm{tr} F^2)^2 \,. \tag{6.123}$$

The numbers x^i_R represent the number of matter fields in the R representation of G_i, and similarly x^{ij}_{RS} is the number of matter fields transforming in the $R \times S$ representation of $G_i \times G_j$. Values of the group theory coefficients A_R, B_R, C_R are straightforward to compute using elementary group theory methods. These values for a few representations of $SU(N)$ are given in the following table. The values in the table hold for $N \geq 4$. For $SU(2)$ and $SU(3)$, A_R are the values given in the table, while $B_R = 0$ and C_R is computed by adding the terms from the table with $N = 2, 3$ in the combination $C_R + B_R/2$.

Rep.	Dimension	A_R	B_R	C_R
$\Box$	N	1	1	0
Adjoint	$N^2 - 1$	$2N$	$2N$	6
$\begin{array}{c}\Box\\\Box\end{array}$	$\frac{N(N-1)}{2}$	$N - 2$	$N - 8$	3
$\Box\Box$	$\frac{N(N+1)}{2}$	$N + 2$	$N + 8$	3
$\begin{array}{c}\Box\\\Box\\\Box\end{array}$	$\frac{N(N-1)(N-2)}{6}$	$\frac{N^2-5N+6}{2}$	$\frac{N^2-17N+54}{2}$	$3N - 12$

More extensive lists and methods for deriving these factors can be found in various relevant papers.[79,82,83]

The relations (6.119-6.122) determine the inner products of the vectors a, b_i in terms of the matter content of the theory. It can be proven using group theory identities[78] that these anomaly cancellation conditions[f] automatically lead to integral values for the inner products $a \cdot a, a \cdot b_i, b_i \cdot b_j$. Thus, the vectors a, b_i define an integral lattice

$$\Lambda \subset \mathbb{R}^{1,T}\,. \tag{6.124}$$

We refer to Λ as the *anomaly lattice* of the 6D theory. (Note that the basis chosen in (6.115) for $T = 1$ is not always the basis in which the basis vectors for the lattice are integral.)

Example:

Consider a theory with one tensor multiplet ($T = 1$), nonabelian gauge group

$$G = SU(N) \tag{6.125}$$

and charged matter content

$$\text{matter} = 2 \times \begin{smallmatrix}\square\\[-2pt]\square\end{smallmatrix}(\mathbf{N(N-1)/2}) + 16 \times \mathbf{N}. \tag{6.126}$$

For each representation $\mathbf{R}$ of G listed, the matter content contains one complex scalar field in representation $\mathbf{R}$ and a corresponding field in the conjugate representation $\bar{\mathbf{R}}$. Together these fields form the quaternionic structure needed for the scalar moduli space. (Note that for special representations like the $\mathbf{2}$ of $SU(2)$, the representation is itself quaternionic, so that the conjugate need not be included. In cases like this the field is often referred to as a "half hypermultiplet"). The total number of charged hypermultiplets in the theory is $H = N^2 + 15N$ and the number of vector multiplets is $V = N^2 - 1$, so from (6.117) we have

$$H - V = H_{\text{neutral}} + H_{\text{charged}} - V = H_{\text{neutral}} + 15N + 1 = 244\,. \tag{6.127}$$

It follows that $N \leq 16$. From the table above of values of A_R, B_R, C_R, it is easy to verify that (6.118) is also satisfied for any N, and the matrix of inner products defining the anomaly lattice for the theory is given by

$$\Lambda = \begin{pmatrix} a \cdot a & -a \cdot b \\ -a \cdot b & b \cdot b \end{pmatrix} = \begin{pmatrix} 8 & 2 \\ 2 & 0 \end{pmatrix}. \tag{6.128}$$

---

[f] for gauge groups such as $SU(2)$ and $SU(3)$ with no fourth order invariant, cancellation of global anomalies[84,85] is also needed[78]

Example:

Now consider a theory with one tensor multiplet ($T = 1$), gauge group

$$G = SU(5) \times SU(6) \tag{6.129}$$

and matter content

$$\text{matter} = 2 \times (\mathbf{10}\,(\text{⊟})\,, \mathbf{1}) + 1 \times (\mathbf{5}, \bar{\mathbf{6}}) + 10 \times (\mathbf{5}, \mathbf{1}) + 7 \times (\mathbf{1}, \mathbf{6}) + 161 \times (\mathbf{1}, \mathbf{1})\,. \tag{6.130}$$

The total number of hypermultiplets is $H = 303$ and the number of vector multiplets is $V = 59$, so $H - V = 244 = 273 - 29T$, and (6.117) is satisfied. It is easy to verify that (6.118) is also satisfied, and the matrix of inner products defining the anomaly lattice for the theory is given by

$$\Lambda = \begin{pmatrix} a \cdot a & -a \cdot b_1 & -a \cdot b_2 \\ -a \cdot b_1 & b_1 \cdot b_1 & b_1 \cdot b_2 \\ -a \cdot b_2 & b_1 \cdot b_2 & b_2 \cdot b_2 \end{pmatrix} = \begin{pmatrix} 8 & 0 & 2 \\ 0 & -2 & 1 \\ 2 & 1 & 0 \end{pmatrix}. \tag{6.131}$$

The lattice $\Lambda \subset \mathbb{R}^{1,T}$ is two-dimensional; we see from (6.131) that the matrix of inner products is degenerate, and there is a linear relation between the vectors $-a = 2b_1 + 4b_2$. In the basis (6.115), we have $a = (-2, -2)$, $b_1 = (1, -1)$ and $b_2 = (0, 1)$, reproducing the desired inner products.

The anomaly cancellation conditions for 6D supergravity theories place strong constraints on the range of possible consistent theories. It has been proven that when $T < 9$, the set of possible gauge groups G and matter representations $\mathcal{M}$ charged under the nonabelian gauge group factors is finite.[78,82] This bound is valid for theories with nonabelian and abelian gauge group factors, although for theories with abelian factors the charges of the matter fields under the $U(1)$ factors are not constrained to a finite range of possible values by anomaly cancellation.[81] The proof of the finiteness result is slightly technical, but relies at its core on the relation (6.117), which bounds the number of matter multiplets for a given number of tensor fields and gauge content. Roughly, if the gauge group factors are all of limited dimension ($\dim(G_i) < D\ \forall i$), then it can be shown that almost all pairs of gauge group factors share hypermultiplets with charge under both factors. This leads to a situation where $H \sim \mathcal{O}(k^2)$ and $V \sim \mathcal{O}(k)$, where k is the number of distinct nonabelian factors in the gauge group. As $k \to \infty$, the constraint (6.117) must be violated. For the details of this argument the reader is referred to the original papers.[78,82] Anomaly cancellation alone does not rule out the possibility of a finite number of gauge group factors with unbounded dimension. It was shown by Schwarz[86] that there are infinite families of theories with $T = 1$ that satisfy the anomaly factorization conditions. For example, the theories with gauge group $G = SU(N) \times SU(N)$ and matter content $2 \times (\mathbf{N}, \bar{\mathbf{N}})$ satisfy the anomaly cancellation conditions and have $H - V = 2$ for any N. There are 5 infinite families of models that satisfy the factorization conditions. For each such family, however, it is possible to prove that everywhere in the moduli space at least one gauge group factor must have the wrong sign on the kinetic term.[78] For example, for the theories just mentioned that were found by Schwarz, it is easy to check that the anomaly conditions give

$$a \cdot b_1 = a \cdot b_2 = 0, \qquad b_1^2 = b_2^2 = -2, \qquad b_1 \cdot b_2 = 2\,. \tag{6.132}$$

From this it follows that, when $a^2 > 0$,

$$a \cdot (b_1 + b_2) = 0 \;\; \& \;\; (b_1 + b_2)^2 = 0 \;\; \Rightarrow \;\; b_1 + b_2 = 0 \;\; \Rightarrow \;\; j \cdot b_1 = -j \cdot b_2\,. \tag{6.133}$$

Thus, at least one of the gauge group factors has the wrong sign on the kinetic term for any vector j. This rules out families of theories with gauge group factors of unbounded dimension, as long as $T < 9$. When $T \geq 9$, however, the norm of the vector a is no longer positive definite, (recall $a^2 = 9 - T$), so the above argument no longer works. In fact, there are infinite families of models with $T = 9$ and greater that satisfy the anomaly cancellation equations and have proper-sign kinetic terms.[78]

This gives a basic outline of the set of 6D $\mathcal{N} = 1$ supergravity theories that satisfy the anomaly cancellation and gauge kinetic term sign constraints. It was recently shown that a further constraint can be placed on the set of 6D supergravity theories that can give consistent quantum theories. The set of allowed charges for objects that couple to the B fields of the theory form a lattice Γ of signature $(1, T)$. This lattice must be integral, from the Dirac quantization condition.[87] The consistency of the dimensional reduction of the theory to 2D or 4D requires that the lattice is furthermore self-dual.[88] This conclusion is compatible with the general mathematical framework for treating (anti)self-dual p-form fields that has recently been under development.[89–92] The charge lattice is invariant under a discrete duality group $G^{1,T} \subset SO(1, T)$. This reduces the part of the moduli space of the theory parameterized by the scalars ϕ in the tensor multiplets to

$$G^{1,T} \backslash SO(1, T)/SO(T)\,. \tag{6.134}$$

The vectors b_i in the anomaly lattice are associated with gauge dyonic strings[93] associated with instantons in each gauge group factor, and thus represent vectors in the charge lattice Γ. There must therefore be an embedding of the lattice spanned by the vectors b_i into the lattice Γ. The vector a should also give a vector in the charge lattice Γ, though this is not rigorously proven. The condition that an embedding of the anomaly lattice Λ into Γ is possible imposes further constraints on the set of allowed 6D theories. For example, consider the theory with $T = 2$, gauge group

$$G = SU(N) \times SU(N)\,, \tag{6.135}$$

and charged matter content

$$2N \times (\mathbf{N}, \mathbf{1}) + 2N \times (\mathbf{1}, \mathbf{N})\,. \tag{6.136}$$

The anomaly lattice for this model is spanned by vectors $-a, b_1, b_2$ with inner products

$$\Lambda = \begin{pmatrix} 7 & 0 & 0 \\ 0 & -2 & 0 \\ 0 & 0 & -2 \end{pmatrix}\,. \tag{6.137}$$

This lattice does not admit an embedding into any unimodular lattice Γ. If it did, the unit cell of Λ would have a volume given by an integer times the volume of the unit cell of Γ. The determinant of the matrix Λ would then be a perfect square. But the determinant is 28, which is not a perfect square, so the lattice does not admit a unimodular embedding and the theory is not a consistent theory. This criterion cuts down further the space of allowed 6D supergravity theories, though it does not eliminate some known infinite families of models that obey all known consistency constraints.

Thus, we have characterized the set of theories $\mathcal{G}^{6D,\mathcal{N}=1}$ with no currently known inconsistencies. We now turn to string constructions of 6D theories to see what subset of the apparently consistent supergravity theories can be realized in string theory.

6.4.3. *Branes and orientifolds*

Six dimensions is an excellent playground for systematically developing the methods of string compactification. The range of possibilities for string constructions in 6D is still much simpler than in 4D, but provides for a rich range of possible gauge groups and matter structure. The strong constraints of anomaly cancellation provide a useful mechanism for checking the internal consistency of different approaches to string vacuum constructions, as well as providing a natural framework for relating different constructions.

We describe in some detail three distinct ways of constructing 6D string vacua. We expand upon the development in Section 6.3 of the heterotic and F-theory approaches, illustrating some of the new features that must be included for compactifications of these types to 6 dimensions. We also introduce the basic structures needed for another class of compactification, the *intersecting brane model* (IBM) construction. We consider intersecting brane models associated with compactifications of type IIB on an orbifold limit of the K3 surface, where D7-branes added to the geometry realize the gauge groups and matter content of the theory. A more detailed introduction to the physics of intersecting brane model constructions, with a focus on the phenomenology of four-dimensional vacua, can be found in the lecture notes by Cvetic and Halverson from this school.[94] We give here a much more rudimentary introduction to the subject of IBM's, focusing on how the models generated through this construction fit into the broader set of 6D string vacua.

To understand intersecting brane models and heterotic/type I compactifications in 6D, we must first review some basic aspects of D-branes and orientifolds. These objects are explained clearly and pedagogically in Polchinski's text,[7,8] but we summarize the basic structures involved in order for these lectures to be somewhat self-contained.

The essential feature of a D-brane is that it is a locus in space-time where open strings end. A Dp-brane in the flat ten-dimensional space-time of type IIA or IIB string theory is a fluctuating hypersurface of spatial dimension p, whose motion is described by the quantized open strings ending on the brane. When N Dp-branes are coincident, the D-branes carry a world-volume gauge group $U(N)$; the dynamics of this gauge group and the scalars describing transverse fluctuations of the brane are captured by the physics of the open strings. One approach to describing theories in fewer than 10 dimensions with chiral matter is to use systems of intersecting branes. When two different branes intersect, the strings stretching between the branes give rise to chiral matter fields in the dimensionally reduced theory.

Consider a pair of flat D-branes in $\mathbb{R}^{1,9}$ that are parallel and coincident in all but 4 spatial dimensions. For example, two D7-branes that are both extended in dimensions 0-5, but each extended in a different two-plane in the four-dimensional space 6789. We can view these branes as intersecting in two different coordinate planes; for example the branes may each extend on different lines in the 67 plane and the 89 plane (See Figure 6.8). The branes preserve a common supersymmetry if the angles between the branes

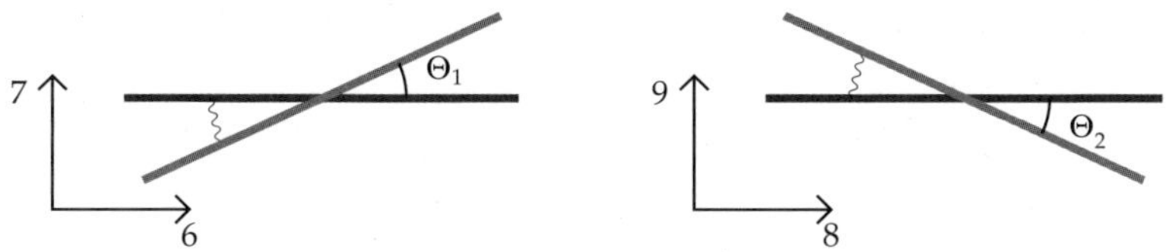

Fig. 6.8. Brane stacks intersecting in two planes preserve some supersymmetry if the intersection angles are equal and opposite in the two planes, $\theta_1 = -\theta_2$. Strings stretching between the branes live in the bifundamental representation $(N, \bar{M})+(\bar{N}, M)$ of the gauge groups on the two brane stacks.

in the two planes are equal and opposite[95]

$$\theta_1 + \theta_2 = 0\,. \tag{6.138}$$

If there are stacks of N, M coincident branes intersecting, then the open strings connecting the branes produce fields living on the brane intersection locus that transform in the bifundamental representation $(N, \bar{M})+(\bar{N}, M)$. If we compactify the theory down to six dimensions, this will give matter in the low-energy 6D theory transforming in the same representation.

The obstacle to considering compactifications with D-branes is that the D-branes act as sources for the Ramond-Ramond fields. A D-brane that extends in all noncompact dimensions of space-time will produce a flux in the transverse compact directions. D-branes that preserve the same supersymmetry will generally carry the same kind of charge. With net D-brane charge, the flux lines will have nowhere to end on the compact space, and the geometry will not admit a solution of the equations of motion. The situation is like a set of positive charges in a compact space; the electric field lines point away from the positive charges, and without negative charges to collect the field lines, a compact solution is impossible. Thus, we need something to cancel the R-R charges of the D-branes in order to use the branes to generate the low-energy gauge group and matter content. One solution to this problem is the *orientifold.* Orientifolds are similar to D-branes, but carry negative tension and negative charge. When orientifolds are included, compactifications with D-branes can preserve SUSY and give rise to interesting low-energy theories[g].

[g]Note that the 7-branes on S^2 in the F-theory description of 8D vacua described in Section 6.3.4.1 evade this problem in a different way; the condition that the product of monodromies cancels on the F-theory base is equivalent to the condition that the charges cancel, which can be done in a supersymmetric way when the total space of the elliptic fibration defining the F-theory model is a Calabi-Yau. Like everything else in string theory, these apparently different constructions are related through duality.[59]

To understand orientifolds, it is easiest to begin with the space-filling orientifold that relates the type IIB theory to the type I string theory. Consider the transformation Ω that acts by a reflection on the internal coordinate of a string

$$\Omega : \sigma \to -\sigma\,. \tag{6.139}$$

This transformation exchanges the right- and left-moving string degrees of freedom $\Omega : x_L \leftrightarrow x_R$, $\Omega : \alpha \leftrightarrow \tilde{\alpha}$. This is a symmetry of the type IIB string theory. If we consider only string states that are invariant under this symmetry (essentially gauging the symmetry Ω by taking the orbifold of the theory by Ω), then the string spectrum will be simplified. For example, the B field of the type IIB theory is associated with the first-quantized string states

$$\left[\alpha^{\mu}_{-1}\tilde{\alpha}^{\nu}_{-1} - \tilde{\alpha}^{\mu}_{-1}\alpha^{\nu}_{-1}\right]|0\rangle\,. \tag{6.140}$$

This set of states is projected out by the orientifold action. The theory realized by the orientifold of the type IIB theory is the type I theory, which has no B field, only the R-R field $\tilde{B}$. The type I theory has only one supersymmetry in 10 dimensions. As we know, an $\mathcal{N} = 1$ theory in 10D must have a gauge group of dimension 496 to be anomaly-free. Thus, some additional structure must provide a gauge group for the type I theory. The projection under Ω produces an unoriented string theory. String diagrams with unoriented topology, such as the Möbius strip and Klein bottle, must be included. From the analysis of such diagrams, the type I orientifold background has been shown to be inconsistent without the addition of 16 space-filling D9-branes, whose world-volume theory (after including the orientifold projection) is precisely the gauge group $SO(32)$ needed for an anomaly-free theory.[8] This can be understood in terms of the presence of a space filling "orientifold 9-plane," with negative tension and D9-brane charge -16, which cancels the charge of the 16 D9-branes.

By combining the world-sheet symmetry Ω with a space-time reflection ρ, orientifold planes of lower dimension can be produced. One-loop string diagrams show that such an orientifold plane of codimension d carries D-brane charge $16/2^d$. One easy way to understand this result is by analyzing toroidal compactification and T-duality for the type I theory. Recall that for a string compactified on a circle S^1 of radius R, the spectrum of momentum and winding modes is given by

$$M^2 = \frac{n^2}{R^2} + \frac{m^2 R^2}{\alpha'}\,. \tag{6.141}$$

As discussed in Section 6.3.3.3, T-duality is the symmetry that exchanges

$$\text{T}-\text{duality}:\ n \leftrightarrow m\,. \tag{6.142}$$

T-duality also exchanges $x_R \leftrightarrow -x_R$. It follows that T-duality exchanges a D-brane wrapped in the compact direction with one that is unwrapped ($D_p \leftrightarrow D_{p\mp1}$); the gauge field component A_μ on the Dp-brane wrapped in the compact direction μ becomes the transverse scalar field X^μ of the unwrapped brane

$$T:\ A_\mu \leftrightarrow X^\mu\,. \tag{6.143}$$

The relation (6.143) can be understood from the point of view of the string world-sheet,[7] or from the world-volume theory of a brane on a compact space.[96] Just as T-duality can reduce or increase the dimension of a D-brane, in a similar fashion T-duality reduces the dimension of a space-filling orientifold plane. The action of $\Omega : x \to x$ on $x = x_R + x_L$ before T-duality becomes the reflection $\Omega : x \to -x$ on $x = -x_R + x_L$ after T-duality. On the dual circle, this action has two fixed points, at $x = 0, x = \pi R$. Thus, under T-duality of the type I theory on d circles in the torus T^d the 16 D-branes are distributed across 2^d orientifold $(9-d)$-planes. This confirms the statement above that each orientifold $(9-d)$-plane carries D-brane charge $-16/2^d$.

This brief summary of the physics of D-branes and orientifold planes gives us enough background to construct intersecting brane models on compactifications of the type IIB theory to six dimensions.

6.4.4. *Intersecting brane models in 6D*

Intersecting brane models (IBM's) provide a rich range of examples of string compactifications with a variety of gauge groups and chiral matter. Semi-realistic IBM's in four dimensions have been the subject of much study.[94,97] Here we will just introduce the basics of the construction in six dimensions to show how some 6D supergravity theories can be realized from this approach. A more detailed analysis of 6D intersecting brane models was given by Blumenhagen, Braun, Körs, and Lüst;[49] the presentation here follows the analysis and notation of that work. Additional features of these models have been developed in work with Nagaoka.[98]

While in principle, intersecting brane models on smooth Calabi-Yau manifolds give rise to a very general class of string compactifications, the analysis is greatly simplified by working on a toroidal orbifold. We will use here the orbifold limit of K3 described in Section 6.3.4.2. Thus, we consider the orbifold of T^4 by the symmetry

$$\rho : z_i \to -z_i \,. \tag{6.144}$$

Here, as before, z_1, z_2 are complex coordinates on two T^2 factors of T^4. We assume that both T^2 factors are rectangular, and have moduli $\tau_j = iR_j$, $j = 1, 2$.

In order to include D-branes we must also include orientifolds. We include an orientifold 7-plane (O7-plane) defined through the orientifold action $\Omega\sigma$, where σ gives the reflection

$$\sigma : z_i \to \bar{z}_i \,. \tag{6.145}$$

This gives an O7-plane in the 68 directions. Since we are taking the orbifold separately by ρ and $\Omega\sigma$, we must also include the combined orientifold action $\Omega\rho\sigma$; this gives another O7-plane in the 79 directions. As discussed above, there are really 4 copies of each O7-plane in the covering space T^4, at antipodal points on the perpendicular tori. After taking the orbifold action, the combined set of cycles where the orientifold is wrapped becomes

$$\pi_{O7} = 2(\pi_{68} - \pi_{79}) \,. \tag{6.146}$$

Now consider D7-branes that are wrapped on the T^4 as a product of one-cycles on the two T^2 factors with winding numbers

$$(n_1, m_1; n_2, m_2) \,. \tag{6.147}$$

For example, a D7-brane wrapped on the cycle $(1, 1; 1, -1)$ is depicted in Figure 6.9. The homology class of a brane with winding numbers (6.147) is

$$\pi = n_1 n_2 \, \pi_{68} + n_1 m_2 \, \pi_{69} + m_1 n_2 \, \pi_{78} + m_1 m_2 \, \pi_{79} \,. \tag{6.148}$$

Associated with each such brane there is an orientifold image with winding numbers (reflected across the horizontal axes) $(n_1, -m_1; n_2, -m_2)$ and associated homology class π'. We consider a general D7-brane configuration to be composed of stacks of N_a coincident branes with winding numbers $(n_1^a, m_1^a; n_2^a, m_2^a)$. The total D-brane charge for all branes extended in space-time and wrapped around any particular cycle of the compact space must vanish as discussed above. Each O7-plane carries -4 units of D7-brane charge. Combining the branes and their orientifold images, the condition for the charges to vanish on the compact space is then

$$\sum_a N_a n_1^a n_2^a = 8, \qquad \sum_a (-N_a m_1^a m_2^a) = 8 \,. \tag{6.149}$$

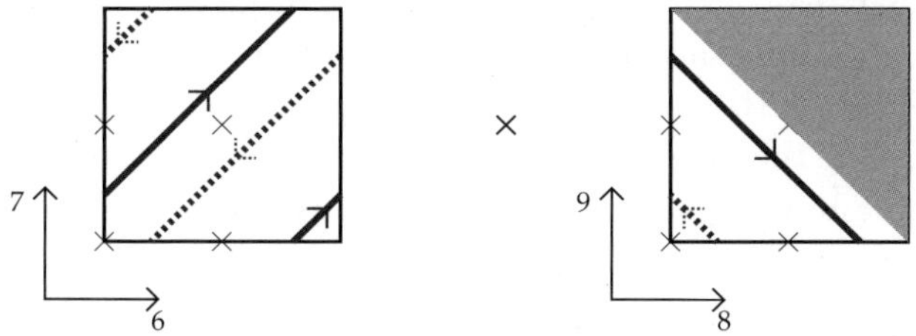

Fig. 6.9. A diagonal D7-brane wrapped on the cycle $(1, 1)$ on the first T^2 and the cycle $(1, -1)$ on the second T^2. The dotted part of the brane is associated with the piece that lifts under the orbifold action $\rho : z_i \to -z_i$ to the shaded part of the covering space outside the fundamental domain.

These equations are often referred to as "tadpole cancellation" conditions since they are needed for cancellation of tadpoles in the R-R fields of the theory; similar conditions arise in most brane model constructions. For 6D models, these conditions can be related to the anomaly cancellation conditions of the associated supergravity theory.

If we assume that the branes all preserve a common supersymmetry, then the SUSY condition (6.138) implies that for all branes in the system

$$\frac{m_1}{n_1} = -\alpha \frac{m_2}{n_2} \tag{6.150}$$

for a common value of α that parameterizes the moduli of the two T^2 factors.

A stack of N diagonal branes that do not coincide with their orbifold counterpart (*i.e.*, branes that are not parallel to the orientifold plane and that do not pass through the orbifold points) carry a $U(N)$ gauge group. Each such brane is wrapped on a cycle that is topologically a torus and will have a single matter field in the adjoint representation, which describes transverse motions of the brane. As discussed in the previous section, open strings between intersecting branes produce bifundamental fields transforming under the gauge groups on each brane. Adjoint, symmetric, and antisymmetric two-index tensor representations of the group $SU(N)$ for a particular brane stack will also be produced by intersections of the brane with itself and its orientifold image. A detailed analysis of the matter structure for a general brane configuration determines the number of matter fields of each type in terms of the number of brane intersections.[49,98] Other types of branes can be included, such as branes parallel to the orientifold planes, fractional branes, and branes intersecting the orbifold points;[49,98]

these brane types can include other gauge group factors such as $Sp(N)$ and introduce additional interesting features that we do not explore here. Another complication that we do not treat in detail here is associated with the $U(1)$ factors in the gauge group. In general some of the $U(1)$ factors in an intersecting brane model will be anomalous, and will acquire a mass through the Stückleberg mechanism. As above, we primarily focus here on the nonabelian part of the gauge group, and do not treat the $U(1)$ factors carefully.

To summarize, in the basic class of 6D IBM models just described, we have a set of stacks of N_a branes, with each stack carrying a gauge group factor $U(N_a)$. The winding numbers for each stack satisfy (6.150), and the tadpole constraint associated with the total charge of the branes and orientifold plane (on each homology class) is satisfied through the conditions (6.149). The set of models that satisfy these conditions is fairly limited. We mention here only the simplest example.

Example: 6D IBM on K3

The simplest example is to take a stack of 8 branes with winding numbers $(1, 1; 1, -1)$ as depicted in Figure 6.9. This model clearly satisfies the SUSY and tadpole cancellation conditions. The supergravity theory associated with this model has gauge group

$$G = U(8)\,. \tag{6.151}$$

From (6.81) it follows that $\pi \cdot \pi' = \pi \cdot o_7 = 8$. All the intersections between the brane stack and its orientifold image coincide with the orientifold plane, which means that each such intersection gives an antisymmetric two-index matter field.[49] The matter content of the theory is thus

$$\text{matter} = 1 \times (\mathbf{63}) + 8 \times (\mathbf{28})\,. \tag{6.152}$$

Considering only charged hypermultiplets, we have $H_{\text{charged}} - V = 224$. This leaves 20 uncharged hypermultiplets to saturate the gravitational anomaly constraint, which is precisely the number of scalar hypermultiplets that come from the closed string sector. The F^4 anomaly equation (6.121) for an $SU(N)$ gauge group factor under which there are f fundamental matter representations, D adjoint representations, and A antisymmetric representations is

$$f = 2N - 2ND - A(N - 8)\,, \tag{6.153}$$

which is satisfied for this matter content. In all 6D intersecting brane models of this type, there is a single tensor multiplet in the 6D theory

($T = 1$); the anti-self-dual two-form from the tensor and the self-dual two-form from the gravity multiplet combine to form the two-form field that descends from the 10D $B_{\mu\nu}$. Connecting to the general framework for 6D supergravities, we can compute the anomaly lattice of this theory, which is spanned by vectors $-a, b$ with inner product matrix

$$\Lambda = \begin{pmatrix} 8 & 8 \\ 8 & 8 \end{pmatrix}. \tag{6.154}$$

This lattice is degenerate, and represents a 1D lattice spanned by the generator $b = -a$. We show in Section 6.4.8 how this data can be used to construct the equivalent F-theory model.

6.4.5. *Magnetized brane models and heterotic bundles in 6D vacua*

We now consider 6D theories associated with compactifying $\mathcal{N} = 1$ supergravity on a K3 surface. The general compactification of this type is described by a nonabelian instanton configuration on K3. We primarily consider a simple subclass of these models, where the instantons are characterized by $U(1)$ fluxes on cycles in the K3 surface. The models in this class were the first 6D compactifications of string theory.[74] We can think of such compactifications from the string point of view either as heterotic or type I string compactifications. We begin from the supergravity point of view and then connect to the type I picture, which in spirit is like a T-dual of the intersecting brane models just considered. As for the intersecting brane models on the orbifold limit of K3 discussed above, all models constructed by compactification on a smooth K3 have $T = 1$, with the anti-self-dual and self-dual two-forms combining to form the dimensionally reduced B field from 10D.

6.4.5.1. *D-brane charges on branes and orientifolds*

We begin from the point of view of supergravity. In compactifying an $\mathcal{N} = 1$ 10D theory to eight dimensions on a torus, the only additional structure available was the possibility of nontrivial Wilson lines around the cycles of the torus. Compactifying on a non-toroidal Calabi-Yau requires the further introduction of a gauge bundle carrying instantons. To see why this is the case, recall from (6.25) that the 3-form H contains Chern-Simons contributions from the Yang-Mills and space-time curvatures. The Bianchi identity in a general background is then generalized from $dH = 0$ to

$$dH = \text{tr}R \wedge R - \text{Tr}F \wedge F\,. \tag{6.155}$$

Since the left-hand side is an exact form, the right-hand side must vanish in cohomology, implying a nonzero instanton number for a compactification on K3

$$\int_S c_2(F) = \frac{1}{8\pi^2}\int_S \text{Tr}F \wedge F = \frac{1}{16\pi^2}\int_S \text{Tr}R \wedge R = -\frac{1}{2}\int_S p_1(R) = 24\,, \tag{6.156}$$

where c_2 and p_1 are the second Chern class and first Pontryagin class, respectively[h]. For compactification of an $\mathcal{N} = 1$ 10D theory on K3, then, we must have a background gauge field configuration with instanton number 24.

An alternative perspective on the conclusion that the gauge bundle on K3 must have 24 instantons comes from the type I picture, which can be used when the 10D gauge group is $SO(32)$ (really $\text{Spin}(32)/\mathbb{Z}_2$). Recall that in the type I picture the gauge group comes from 16 D9-branes superimposed on an orientifold O9-plane. An important aspect of D-brane physics is the presence of Chern-Simons couplings in the world-volume action between powers of the field strength F and the Ramond-Ramond p-form potentials.[99] For D9-branes, there is a coupling to the 6-form potential of the IIB and I theories proportional to

$$\int F \wedge F \wedge \hat{C}_6\,. \tag{6.157}$$

($\hat{C}_6$ is the dual to the R-R two-form $\tilde{B}$.) This means that an instanton on a D9-brane carries the charge of a D5-brane.[100] This statement, as well as the more general statement that a system of Dp-branes can carry charge associated with a D$(p \pm 2k)$-brane can be understood easily from T-duality on the D-brane world-volume.[101] By taking the world-volume derivative of (6.143) we have

$$T: \;\; \partial A_\mu \leftrightarrow \partial X^\mu\,. \tag{6.158}$$

Thus, T-duality can relate, for example, flux $F = \partial A$ on a D9-brane to the slope ∂X of a tilted D8-brane in a T-dual picture. The T-dual of a diagonal D8-brane on a torus has both D9-brane and D7-brane charge; the D7-brane

[h]Note the factor of 2 in moving between the vector and adjoint representations of $SO(1,3)$, denoted respectively by tr and Tr. Note also that we have changed normalization relative to (6.25) so that the instanton number is an integer when expressed in conventional form.

charge can be associated with the flux encoded in F on the D7-brane. Generalizing this picture to multiple T-dualities shows that $F \wedge F$ on a Dp-brane carries D$(p-4)$-brane brane charge, etc. (A further generalization to multiple Dp-branes, replacing $\partial \to \partial + iA_\mu$, leads to the T-dual statement that $[X, X]^k$ on a stack of multiple Dp-branes carries charge associated with higher-dimensional D$(p+2k)$-branes.[102–104]) Through the geometric picture of T-duality as relating fluxes to tilting on branes, we see that intersecting diagonal D7-branes on a torus T^4 can be T-dual to a system of D9-branes carrying D7-brane and D5-brane charges through world-volume fluxes. This relates the intersecting brane models discussed in the previous section to the type I models discussed here. There are some subtleties, however, in this correspondence, since we are really working on K3 and not a torus. Also, both the IBM and magnetized brane models are described in the supergravity approximation where the volume of the compactification space is large, and T-duality maps a large compactification torus to a small dual torus, where the supergravity approximation is no longer valid.

In the type I picture, the need for a gauge field configuration with instanton number 24 follows from structure of the D-brane world-volume theory related to the Chern-Simons couplings (6.157). From supersymmetry, there are similar terms of the form $R \wedge R \wedge \hat{C}_6$ in the D9-brane world-volume action,[105] and analogous couplings occur for the orientifold plane.[106,107] The quantization of the coupling is such that each D9-brane wrapped on a K3 carries -1 units of D5-brane charge, while the O9-plane wrapped on K3 carries -8 units of D5-brane charge. Combining one O9-plane and 16 D9-branes gives a total deficit of 24 units of D5-brane charge, or instanton number 24, needed to cancel the $\hat{C}_6$ tadpole. The term "magnetized brane" refers to $U(1)$ fluxes F on the D9-brane world-volume that can be used to realize the needed instanton.

6.4.5.2. *Abelian instantons on K3*

To understand a general 6D compactification of the 10D $\mathcal{N} = 1$ theory, then, we must consider the moduli space of instantons for a nonabelian gauge theory on K3. This leads into a rich and interesting mathematical story, which leads beyond the scope of these lectures. Here we will focus on the simple class of "magnetized brane" models where the instanton structure is completely encoded in fluxes within a commuting set of $U(1)$'s in the SO(32) gauge group. This is the class of 6D models originally studied by Green, Schwarz, and West.[74] By describing the structure of these theories in terms of the homology lattice of K3 with associated intersection product, following work with Kumar,[50] we make contact with the general theme of lattice embeddings that we have already encountered in the 8D context, and put these models into a framework that connects with the other classes of 6D vacua.

We begin the discussion with a review of some basic facts about K3. Like any complex surface, K3 has a choice of complex structure

$$\Omega \in H^{2,0}(S) \subset H^2(S) \tag{6.159}$$

that is fixed up to a scale factor. The complex structure satisfies

$$\int \Omega \wedge \Omega = 0, \quad \int \Omega \wedge \bar{\Omega} \propto \mathrm{Vol}(S) > 0\,. \tag{6.160}$$

Writing $\Omega = x + iy$, it follows that

$$x \cdot y = 0, \; x \cdot x = y \cdot y > 0\,, \tag{6.161}$$

where in this section we will use $\cdot$ to denote the wedge product in cohomology or the equivalent intersection form in homology, freely moving back and forth using Poincaré duality. There is a Kähler structure associated with a Kähler form $J \in H^{1,1}(S)$ that satisfies

$$\int J \wedge \Omega = 0 \quad \Rightarrow \quad J \cdot x = J \cdot y = 0 \tag{6.162}$$

$$\int J \wedge J \propto \mathrm{Vol}(S) > 0 \quad \Rightarrow \quad J \cdot J > 0. \tag{6.163}$$

It follows that (x, y, J) define a positive-definite 3-plane in $H^2(S;\mathbb{R}) = \mathbb{R}^{3,19}$.

Now, let us consider fluxes of an $SO(32)$ gauge theory on K3. Fluxes wrapped around nontrivial two-cycles are quantized, and can be normalized to $F = 2\pi i f$ where

$$f \cdot f \in 2\mathbb{Z}, \; f \in H^2(K3, \mathbb{Z}) = \Gamma^{3,19}\,. \tag{6.164}$$

We thus naturally identify such a flux with an element of the lattice $\Gamma^{3,19}$.

The group $SO(32)$ contains 16 mutually commuting $U(1)$ factors (associated with $SO(2)$ transformations on mutually disjoint pairs of the 32 indices). In the type I picture these 16 $U(1)$'s are the world-volume gauge fields on each of the 16 D9-branes in the theory. A flux f on one D9-brane is accompanied by a flux $-f$ on the orientifold image of that D9-brane. (This is essentially the T-dual of the relationship between diagonal branes and their orientifold images in the intersecting brane models described in

Section 6.4.4.) We consider a configuration of the $SO(32)$ theory where we have "stacks" of N_a D9-branes with flux f_a, $a = 1, \ldots, K$

$$F = \begin{pmatrix} \left.\begin{matrix} f_1 & & \\ & \ddots & \\ & & f_1 \end{matrix}\right\} N_1 & & 0 \\ & \left.\begin{matrix} f_2 & & \\ & \ddots & \\ & & f_2 \end{matrix}\right\} N_2 & \\ 0 & & \ddots \end{pmatrix}. \tag{6.165}$$

The constraint that the total instanton number is 24 becomes

$$\frac{1}{8\pi^2}\int_s \mathrm{Tr} F \wedge F = 24 \;\Rightarrow\; \sum_a N_a f_a \cdot f_a = -24\,. \tag{6.166}$$

The constraints from supersymmetry (analogous to (6.150)) give

$$\int f^a \wedge \Omega = 0 \quad \int f^a \wedge \bar{\Omega} = 0 \quad \int f^a \wedge J = 0 \tag{6.167}$$

$$\Rightarrow\; f^a \cdot x = f^a \cdot y = f^a \cdot J = 0\,. \tag{6.168}$$

The set of abelian instanton configurations on K3 is thus parameterized by the set of fluxes $\{f_a\}$. These fluxes generate an even lattice

$$L \subset \Gamma^{3,19}\,, \tag{6.169}$$

which characterizes the theory. In fact, the structure of the gauge group and matter content of any 6D theory constructed in this way are dependent only on the integers N_a and the matrix of inner products $m_{ab} = f_a \cdot f_b$. (Note that the matrix m may be degenerate, in which case the dimensionality of L is smaller than that of m.) As long as the lattice L is negative-definite, there always exists a perpendicular positive-definite 3-plane in $\Gamma^{3,19}$, so that there are some moduli for which (6.168) holds and the theory is supersymmetric. The gauge group of the 6D theory is given by

$$G = U(N_1) \times U(N_2) \times \cdots \times U(N_K) \times SO(M), \qquad M = 32 - 2\sum_a N_a\,. \tag{6.170}$$

The matter content can be determined from a simple index theory calculation and is given in the following table

Rep. (+ c.c.)	# hypermultiplets
(N_a, N_b)	$(-2-(f_a+f_b)^2)$
$(N_a, \bar{N}_b)$	$(-2-(f_a-f_b)^2)$
Antisym. $U(N_a)$	$(-2-4f_a^2)$
$(N_a, 2M)$	$(-2-f_a^2)$
Neutral	20

It is easy to verify that anomalies cancel for any theory with this spectrum; for example, a small calculation shows that $H - V + 29T = 273$.

There are a few technical subtleties that we have glossed over here. Some of the $U(1)$ factors may be anomalous and removed by a Stückelberg mechanism, as in the intersecting brane models discussed above.[80] Furthermore, there is a possibility of further massless states that may enhance the gauge spectrum when $J \cdot f = 0, f^2 = -2$ corresponding to a rational curve on the K3 shrinking to a point.[50]

Given that the spectrum and symmetry group of a 6D model constructed in this fashion is determined uniquely by the integers N_a and inner products of vectors f_a in the lattice L, it is natural to ask for what N_a, L models can be constructed. Given N_a, m_{ab} satisfying $\sum_a N_a m_{aa} = -24$, the criterion for existence of a model of this type is that there exist a lattice embedding

$$L(m) \hookrightarrow \Gamma^{3,19} . \tag{6.171}$$

It can be shown[50] using Nikulin's lattice embedding theorems that such an embedding is always possible for stack sizes and matrix m compatible with the 24-instanton condition (6.166). Furthermore, such an embedding is often unique, with degeneracies sometimes associated with possible overlattice embeddings that would give rise to additional discrete structure in the 6D theory.

Example:

As a simple example, consider a model with gauge group

$$G = U(4) \times U(4) \times SO(16) . \tag{6.172}$$

We have $N_1 = N_2 = 4$, so (6.166) becomes

$$4m_{11} + 4m_{22} = -24 , \tag{6.173}$$

with m_{aa} an even integer for each a. The only solution (up to exchange of stacks) is

$$f_1 \cdot f_1 = -2, \; f_2 \cdot f_2 = -4 . \tag{6.174}$$

There are 5 distinct choices for m_{12} giving different theories; we consider the case where $m_{12} = 0$. From the table above we can compute the matter spectrum

$$\text{matter} = 4\times(\mathbf{4},\bar{\mathbf{4}},\mathbf{1}) + 4\times(\mathbf{4},\mathbf{4},\mathbf{1}) + 6\times(\mathbf{6},\mathbf{1},\mathbf{1}) + 14\times(\mathbf{1},\mathbf{6},\mathbf{1}) \\ +2\times(\mathbf{1},\mathbf{4},\mathbf{16}) + 20\times(\mathbf{1},\mathbf{1},\mathbf{1})\,. \tag{6.175}$$

It is straightforward to check that this ($T = 1$) model satisfies the anomaly equations and has anomaly lattice

$$\Lambda = \begin{pmatrix} 8 & 6 & 14 & -2 \\ 6 & 4 & 8 & 0 \\ 14 & 8 & 12 & 4 \\ -2 & 0 & 4 & -4 \end{pmatrix}. \tag{6.176}$$

Note that the two $U(1)$ factors must be included in the number of vector multiplets for the condition (6.117) to be satisfied, and that both kinds of bifundamental hypermultiplets $(\mathbf{4},\mathbf{4})$ and $(\mathbf{4},\bar{\mathbf{4}})$ contribute to $b_1 \cdot b_2$. The matrix defining the anomaly lattice (6.176) is degenerate, which is expected since the dimension of this lattice must be $T + 1 = 2$. Indeed, the anomaly vectors can be chosen in the coordinate system (6.115) to be

$$-a = (2,2) \qquad b_1 = (1,2) \tag{6.177}$$

$$b_2 = (1,6) \qquad b_3 = (1,-2)\,. \tag{6.178}$$

These vectors correctly reproduce the matrix of inner products in (6.176). We will see in Section 6.4.8 how this structure can be used to easily map this model to the topological data for an F-theory compactification.

The general approach described here for compactification of the $SO(32)$ theory on K3 can be repeated for the gauge group $E_8 \times E_8$ with analogous results.[108]

The models described here represent only a small slice of the full space of type I/heterotic compactifications on K3. More generally, the instanton structure can be nonabelian; the moduli space of vacua in this class is given by the moduli space of instantons on K3 with instanton number 24. Tools for understanding instanton moduli spaces of this type were developed by Friedman, Morgan, and Witten.[109] By embedding nonabelian instantons with total instanton number 24 in a subgroup of the heterotic group, many other theories can be constructed.[110–112] For example, as pointed out by Kachru and Vafa, putting all the instantons in an $SU(2)$ subgroup of one of the factors of $E_8 \times E_8$ gives a theory with gauge group

$$G = E_7 \times E_8 \tag{6.179}$$

and charged matter content

$$\text{matter} = 10 \times (\mathbf{56}, 1)\,. \tag{6.180}$$

Using the group theory coefficients

$$E_7 : \; A_{56} = 1, C_{56} = 1/24, A_{\text{adj}} = 3, C_{\text{adj}} = 1/6 \tag{6.181}$$

$$E_8 : \; A_{\text{adj}} = 1, C_{\text{adj}} = 1/100 \tag{6.182}$$

the anomaly lattice for this theory is

$$\Lambda = \begin{pmatrix} 8 & -14 & 10 \\ -14 & 12 & 0 \\ 10 & 0 & -12 \end{pmatrix}. \tag{6.183}$$

In the $T = 1$ basis (6.115) we have

$$b_7 = (1, 6), \;\; b_8 = (1, -6)\,. \tag{6.184}$$

The continuous moduli space of theories including nonabelian instantons connects many of the models that appear with distinct gauge groups in the preceding analysis. Generically, the bundle structure will break the $SO(32)$ gauge group to $SO(8)$. The different models described above are connected by Higgsing to a generic model, moving in moduli space, and un-Higgsing to restore a different gauge group. This connects part of the moduli space of 6D theories; we describe a more extensive connectivity of this moduli space in the context of F-theory in Section 6.4.7.

Note that while the magnetized brane models just described are morally "T-dual" to the intersecting models described in the previous section, the specific models arising from intersecting brane models of IIB on K3 do not appear in the magnetized brane picture. For example, the model from (6.151) with gauge group $SU(8)$ does not arise as a magnetized brane model. This is because the orbifold quotient needed to realize K3 breaks the naive T-duality associated with T^4. Further structure is needed to make the connection between these classes of models more precise, some of which we describe briefly in the following section.

6.4.6. *Other constructions of 6D type I/heterotic vacua*

A wide range of other 6D models have been constructed from string theory, using a variety of techniques related to the heterotic/type I constructions described in the previous section. We mention briefly some of these approaches and models to give a sense of the range of possibilities. A more comprehensive review of the range of models known can be found in the review by Ibáñez and Uranga.[113] As we discuss in the following section, F-theory provides a powerful unifying framework that seems to encompass all of the vacua constructed from these other approaches. The F-theory description of each of these models can be determined directly from the anomaly structure, as we discuss in Section 6.4.8. There are other types of 6D models related to the type I compactification on K3 that are constructed by generalizing the geometry of K3 to a "non-geometric space".[114] These models also can be associated with F-theory constructions.

One class of models was constructed by Bianchi and Sagnotti based on combining the $\mathbb{Z}_2$ orientifold action in the IIB theory with a space-time orbifold action giving $T^4/\mathbb{Z}_2$.[115–118] One model of this type that was considered in more detail by Gimon and Polchinski gives a gauge group

$$G = SU(16) \times SU(16)\,, \tag{6.185}$$

with matter content

$$\text{matter} = 2 \times (\mathbf{120}, \mathbf{1}) + 2 \times (\mathbf{1}, \mathbf{120}) + 1 \times (\mathbf{1}, \mathbf{1})\,. \tag{6.186}$$

For each gauge group factor the anomaly cancellation in this model follows as for the example with matter (6.126). This kind of type I model on K3 orbifolds can also give rise to theories with additional tensor multiplets.[119–121]

A variety of heterotic compactifications on various toroidal orbifold limits of K3 of the form $T^4/\mathbb{Z}_k$ were constructed by Erler and others.[79,122] Tables of some of these models can be found in the work of Aldazabal *et al.*,[123] and include models such as one with

$$G = SU(16), \qquad \text{matter} = 2 \times \mathbf{120} + 16 \times \mathbf{16}\,. \tag{6.187}$$

Note that this is the model considered in (6.128) with the maximum possible value of N. Many heterotic orbifold models were considered by Honecker and Trapletti[80,108] and connected to smooth K3 constructions.

From the heterotic point of view, more exotic theories with additional tensors can arise when instantons shrink to a point. This corresponds to a limit where the 5-brane charge encoded in the instanton congeals into a localized 5-brane and can separate from the ambient gauge theory.[100] The 5-brane world-volume theory carries a self-dual two-form field that corresponds to an additional tensor in the 6D theory. By shrinking k coincident instantons, a group $Sp(k)$ arises on the 5-brane world-volume. The gauge groups can become even more exotic when the instantons shrink to a point with an orbifold singularity. A variety of models of this type have been constructed.[124–126] An extreme case of the kind of large gauge group that can arise was identified by Aspinwall, Gross and Morrison[127] in a theory with 192 vector multiplets and

$$G = E_8^{17} \times F_4^{16} \times G_2^{32} \times SU(2)^{32} \,. \qquad (6.188)$$

A further complication that can arise in these heterotic constructions is the appearance of bundles without vector structure.[128,129] This refers to a situation where an instanton attached to an orbifold singularity describes a $\mathrm{Spin}(32)/\mathbb{Z}_2$ bundle but not an $SO(32)$ bundle.

From the heterotic/type I point of view, the wide range of possible models and underlying physical mechanisms, including all possible numbers of tensor fields, seems difficult to connect into a single systematic framework. We now turn to the point of view of F-theory, which provides a unifying perspective on the complicated network of possible 6D theories.

6.4.7. *Six-dimensional vacua of F-theory*

We now consider F-theory compactifications to six dimensions. The story is very similar in spirit to the description of eight-dimensional F-theory vacua in Section 6.3.4, though the details are more complicated. We begin by developing the structure of F-theory further to incorporate compactification on elliptically fibered Calabi-Yau threefolds. We then show how the anomaly data from any 6D theory can be used to identify the topological data for an F-theory realization of the theory, if one exists. This brings us to a point where we can discuss the global structure of the space of 6D $\mathcal{N} = 1$ supergravity theories.

For a compactification of F-theory to six dimensions, we consider a Calabi-Yau threefold X that is elliptically fibered over a complex surface B. The canonical class of the base is $K = c_1(T^*) \in H^2(B)$, with a description in local complex coordinates $\sim dz_1 \wedge dz_2$. As in the case of a surface elliptically fibered over $\mathbb{P}^1$, there is a Weierstrass description of the elliptic fibration in terms of local functions f, g on the base

$$-y^2 + x^3 + f(s,t)x + g(s,t) = 0 \,, \qquad (6.189)$$

where s, t are local coordinates on the base. The functions f, g are given globally by sections of $\mathcal{O}(-4K)$ and $\mathcal{O}(-6K)$, as in the 8D story. Again, codimension one singularities in the fibration give rise to the gauge group of the 6D theory. These are understood through the Kodaira classification as in 8D. In 6D theories, however, there are also codimension two singularities.

These give rise to matter in the 6D theory. While some of the simplest kinds of codimension two singularities, giving rise to the simplest types of matter, are well understood, the mathematical description of codimension two singularities is as yet incomplete.

One approach to systematically constructing many elliptically-fibered Calabi-Yau geometries for F-theory compactifications is through the use of toric geometry.[130–133] We do not discuss this approach in detail here, but it provides a concrete set of tools for explicitly constructing and analyzing a broad subset of the space of elliptically-fibered spaces used in F-theory constructions.

6.4.7.1. *Codimension one singularities and gauge groups*

The discriminant locus where the fibration is singular is given by the set of points where $\Delta = 4f^3 + 27g^2 = 0$. This is generically a codimension one locus on the base, which can be characterized as a divisor

$$\sum_i n_i H_i, \qquad n_i \in \mathbb{Z} \tag{6.190}$$

where H_i are irreducible algebraic hypersurfaces on the base B. The divisor class of the discriminant locus is again given through the Kodaira condition for the total space to be Calabi-Yau

$$-12K = \Delta\,. \tag{6.191}$$

This divisor is *effective*, meaning that the coefficients n_i in an expansion of the form (6.190) satisfy $n_i \geq 0$. The irreducible components of Δ can give rise to nonabelian gauge group factors, through the Kodaira classification. We denote by ξ_i the irreducible components associated with nonabelian factors and by ν_i the corresponding multiplicity (*i.e.* the degree of Δ along ξ_i). For example, an A_{N-1} singularity on a divisor class ξ has multiplicity $\nu = N$, while an E_8 singularity has multiplicity $\nu = 10$. Each ξ_i is associated with an algebraic curve on the surface B. In the IIB picture, multiple 7-branes are wrapped on the curves ξ_i. There is a residual part of the discriminant locus, which we denote by Y, that is not sufficiently singular to give rise to nonabelian gauge factors. The total discriminant locus can thus be written

$$-12K = \Delta = \sum_i \nu_i \xi_i + Y\,, \tag{6.192}$$

where ξ_i gives rise to the nonabelian gauge group factor

$$\xi_i \to G_i\,. \tag{6.193}$$

The semisimple part of the gauge group for the 6D theory is then given by

$$G = G_1 \times \cdots \times G_k/\Gamma \tag{6.194}$$

where Γ is a discrete group. The abelian part of the gauge group arises in a more subtle fashion from the global structure of the elliptic fibration, as in 8D. More precisely, the rank of the abelian part of the gauge group is given by the dimension of the *Mordell-Weil* group of rational sections of the fibration.[55,134] The structure of the discrete group Γ is determined by the torsion part of the Mordell-Weil group.[135] In terms of the total space X of the Calabi-Yau threefold, the total rank r of the gauge group, including both abelian and nonabelian factors, is given by[46]

$$r = h_{1,1}(X) - h_{1,1}(B) - 1\,. \tag{6.195}$$

While this is easy to compute when the geometry of the threefold is known, it is less straightforward to compute the abelian part of the gauge group given only the local characterization of the fibration, and requires either the explicit Weierstrass model or a detailed characterization of the codimension one and two components of the discriminant locus.

A further complication that can arise in 6D F-theory compactifications is the appearance of non-simply-laced groups outside the A-D-E classification.[136–138] This can occur when the set of singular curves associated with a codimension one singularity undergoes a monodromy around a nontrivial cycle in the base, so that the associated Dynkin diagram is mapped into itself in a nontrivial fashion (*i.e.*, through an outer automorphism). For example, an A_7 singularity that undergoes a monodromy reflecting the Dynkin diagram from one end to the other contracts the Dynkin diagram through the reflection, giving rise to a group $Sp(4)$ associated with a C_4 singularity. Treating such groups in F-theory requires a slightly more careful analysis of the singularity structure on the dimension one discriminant locus, for which a systematic procedure known as the "Tate algorithm" is helpful.[139]

6.4.7.2. *Codimension two singularities and matter*

Matter in the 6D theories is produced from codimension two singularities in the elliptic fibration. In general, at codimension two points in the base B, the singularity structure of the fibration can worsen. This can lead to additional points in the base where the total space of the fibration is singular and must be blown up into two-cycles to give a smooth Calabi-Yau threefold. These two-cycles represent additional matter fields in the theory. In general, the codimension two singularity at a point x in the base gives an enhancement of the Kodaira type of the singularity along one or more codimension one curves ξ_i passing through the point x. In the simplest kind of codimension two singularity, as described by Katz and Vafa,[140] the singularity at x is resolved by blowing up a Kodaira type associated with a group of one rank higher than the codimension one singularity/singularities intersecting that point. In this case the matter content is found by decomposing the adjoint of the larger group into the gauge factor(s) associated with the codimension one locus. In other cases, the codimension two singularity can give rise to matter in a more complicated fashion.[141,142] One general class of codimension two singularities has been analyzed by Miranda,[143] though more general classes of singularities are possible that have not yet been completely classified.

Example:

Consider a local singularity of type A_{n-1} along a curve ξ_1, which we can take to be on the line $t = 0$, with $\Delta \sim t^n$. If there is also a codimension one singularity of type A_{m-1} along the curve ξ_2 given by the line $s = 0$, with $\Delta \sim s^m$, then at the origin where the curves intersect, the total singularity type is A_{n+m-1}, since ord $(\Delta) = n + m$. The two components of the codimension one singularity locus along ξ_1, ξ_2 give a group $SU(n) \times SU(m)$. The enhanced singularity at the origin is associated with the adjoint of $SU(n + m)$ (although this is not a part of the gauge symmetry). Decomposing the adjoint representation of $SU(n + m)$ gives a bifundamental field in the $(n, \bar{m}) + (\bar{n}, m)$ representation of the two gauge group factors. This is the F-theory realization of the bifundamental strings arising from intersecting branes.

The mechanism by which matter arises in F-theory is thus related to the standard mechanism realizing matter from strings connecting intersecting branes. But in F-theory there is a much wider range of possible matter structures.[140,142,144] For example, a trifundamental matter field can be produced by a local enhancement of an $A_1 \times A_2 \times A_4$ singularity to E_8

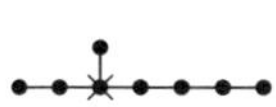 $\Rightarrow (\square_3, \square_2, \square_4)$

In addition to matter from localized singularities, there are global contributions to matter in F-theory. In particular, a codimension one singularity locus associated with a gauge group factor G_i that is wrapped on a curve of genus g in the base gives rise to g adjoint representations of G_i.

6.4.7.3. *Bases for elliptically fibered Calabi-Yau threefolds*

We now consider the question of which bases B can be used for an F-theory compactification on an elliptically fibered Calabi-Yau threefold. A full discussion of the mathematics underlying the answer to this question is beyond the scope of these lectures, but we summarize the main points here.

The only bases that can support an elliptically fibered Calabi-Yau threefold are given by the following set of spaces:[46,145,146]

T^4, K3: These spaces both have $K = 0$. So there is no discriminant locus and the fibration is trivial. Thus, these are simple IIB compactifications, and the general structure of F-theory is not needed. These models give theories with enhanced supersymmetry. Models with enhanced supersymmetry can also arise from compactification on hyperelliptic surfaces or surfaces of the form $(T^2 \times \mathbb{P}^1)/G$ with G a discrete group; we do not consider any of these models further here.

Enriques surface: This is an orbifold of K3 with no fixed points. The space has $-12K = 0$, with K a torsion class. So there are no nonabelian gauge group factors on the discriminant locus. F-theory can nonetheless be compactified on the Enriques surface.[147] The dyon charge lattice for the theory is[148] $\Gamma^{1,9}$.

$\mathbb{P}^2$: The two-dimensional complex projective space $\mathbb{P}^2$ has $H_2(\mathbb{P}^2) = \mathbb{Z}$, with $K = -3H$, where H is the hyperplane divisor generating $H_2(\mathbb{P}^2)$.

$\mathbb{F}_m$: Hirzebruch surfaces with $m \leq 12$ (described in more detail below).

Blow-ups of $\mathbb{P}^2, \mathbb{F}_m$ at one or more points: Blowing up one or more points on these spaces gives surfaces with increasingly large $H_2(B, \mathbb{Z})$.

It is also possible that there may be consistent F-theory compactifications on bases with orbifold singularities, whose resolution gives one of the spaces above;[46] understanding such compactifications is an interesting open problem.

The fact that no other surfaces besides those listed above can be used as bases for an elliptically fibered Calabi-Yau manifold follows from results in *minimal surface theory*. Basically, the idea of minimal surface theory is that on any surface which contains a divisor class describing a curve C satisfying $C \cdot K = C \cdot C = -1$, the curve C can be blown down to give a simpler smooth surface.[149] Minimal surfaces are those which admit no further curves which can be blown down in this fashion. The only minimal surfaces which can be bases for elliptically fibered Calabi-Yau manifolds are those listed above.

It will be helpful to have on hand some more details of the structure of the Hirzebruch surfaces $\mathbb{F}_m$. These surfaces can be described as $\mathbb{P}^1$ bundles over $\mathbb{P}^1$; the surface $\mathbb{F}_m$ is essentially equivalent to the line bundle over $\mathbb{P}^1$ with first Chern class $c_1 = -m$ with all fibers compactified to $\mathbb{P}^1$. For example,

$$\mathbb{F}_0 = \mathbb{P}^1 \times \mathbb{P}^1 . \tag{6.196}$$

The surface $\mathbb{F}_1$ is also described by blowing up $\mathbb{P}^2$ at a point. The surface $\mathbb{F}_2$ is given by compactifying $T^*\mathbb{P}^1$ on each fiber.

A basis for the linear space of divisors of $\mathbb{F}_m$ is given by the divisors D_v, D_s, where D_v is a section of the $\mathbb{P}^1$ fibration with $D_v^2 = -m$, and D_s is a fiber of the fibration. The intersection matrix in this basis is given by

$$\begin{pmatrix} D_v \cdot D_v & D_v \cdot D_s \\ D_s \cdot D_v & D_s \cdot D_s \end{pmatrix} = \begin{pmatrix} -m & 1 \\ 1 & 0 \end{pmatrix} . \tag{6.197}$$

Geometrically we can picture D_v as a horizontal line depicting the section and D_s as a vertical line depicting the fiber

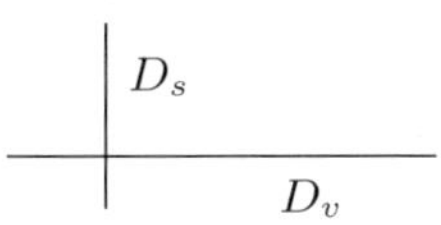

The irreducible effective divisor classes in $\mathbb{F}_m$ are given by

$$D_v, \; aD_v + bD_s, \qquad b \geq ma . \tag{6.198}$$

The canonical class on $\mathbb{F}_m$ is

$$-K = 2D_v + (2+m)D_s . \tag{6.199}$$

It is easy to check that for any m, $K^2 = 8$ on $\mathbb{F}_m$.

From the divisor structure of $\mathbb{F}_m$ we can see that $\mathbb{F}_m$ can only be the base for an elliptically fibered Calabi-Yau threefold with singularities that are in the Kodaira table when $m \leq 12$. From $f = -4K$, we see that for $m > 2$, f must have an irreducible component proportional to D_v, since from eq. (6.198) any irreducible component other than D_v must have a coefficient of D_s at least m times the coefficient of D_v. Writing

$$f = x_f D_v + y = x_f D_v + (AD_v + BD_s) = 8D_v + (8+4m)D_s \tag{6.200}$$

where y is a sum of irreducible components other than D_v, we have $B = 8 + 4m \geq mA$, so $A \leq 4 + 8/m$. It follows that

$$x_f \geq 4 - 8/m\,. \tag{6.201}$$

A similar analysis for $g = -6K$ and $\Delta = -12K$ gives

$$x_g \geq 6 - 12/m, \qquad x_\Delta \geq 12 - 24/m\,. \tag{6.202}$$

For $m > 12$, the divisor class D_v thus carries a singularity of degrees $\deg(f) \geq 4$, $\deg(g) > 5$, $\deg(\Delta) > 10$. Since D_v^2 is negative, there are no deformations of this divisor class, so there is a singularity worse than the Kodaira classification allows on this locus at $m > 12$.

It is easy to verify that for $\mathbb{F}_m$, the lattice $\Gamma = H_2(B, \mathbb{Z})$ spanned by D_v, D_s can be written in an appropriate basis as

$$\Gamma = \Gamma_0 = \begin{pmatrix} 0 & 1 \\ 1 & 0 \end{pmatrix}, m \text{ even}, \qquad \Gamma = \Gamma_1 = \begin{pmatrix} 1 & 0 \\ 0 & -1 \end{pmatrix}, m \text{ odd}\,. \tag{6.203}$$

In fact, all the even Hirzebruch surfaces $\mathbb{F}_{2m}$ are topologically equivalent to one another, as are all the odd surfaces $\mathbb{F}_{2m+1}$. The difference between two even or odd Hirzebruch surfaces with distinct m lies in the complex structure, which is associated with a different set of effective irreducible divisors for each m.

Blowing up a generic point on a surface B adds an additional divisor class E (the exceptional divisor), giving a new surface $\tilde{B}$ with

$$E^2 = -1, \qquad \tilde{K} \cdot E = -1\,. \tag{6.204}$$

In the new surface, $\tilde{K} = K + E$. The dimension of the second homology group increases by 1 with the blow-up so that $h_{1,1}(\tilde{B}) = h_{1,1}(B) + 1$. For example, if we blow up a generic point on $\mathbb{F}_1$ the resulting surface is the *del Pezzo* surface dP_2. The intersection form on dP_2 is given by adding a new dimension corresponding to the orthogonal generator E to Γ_1 from eq. (6.203), and taking $K_{dP_2} = K_{\mathbb{F}_1} + E$,

$$\Gamma_{dP_2} = \begin{pmatrix} 1 & 0 & 0 \\ 0 & -1 & 0 \\ 0 & 0 & -1 \end{pmatrix}, \tag{6.205}$$

in a basis where $K_{dP_2} = (-3, 1, 1)$. Blowing up a generic point on dP_k gives dP_{k+1}. Blowing up non-generic points gives a more complicated tree of resulting surfaces. Each blow-up, however, has the effect of reducing K^2

by 1 through the addition of the new exceptional divisor. In general, for any base we have

$$K^2 = 10 - h_{1,1}(B)\,. \tag{6.206}$$

The number of tensor multiplets in the 6D supergravity theory associated with F-theory on an elliptic fibration over the base B is given by

$$T = h_{1,1}(B) - 1\,. \tag{6.207}$$

The scalar fields in the T tensor multiplets correspond to the relative Kähler (volume) moduli for the $h_{1,1}$ two-cycles in the base, leaving out the overall volume modulus of B, which is controlled by a scalar hypermultiplet. Thus, fibrations over $\mathbb{P}^2$ correspond to theories with no tensor multiplets, fibrations over $\mathbb{F}_m$ give theories with one tensor multiplet, and each additional blow-up of the base gives an additional tensor multiplet. Combining (6.206) and (6.207) gives

$$K^2 = 9 - T = 10 - h_{1,1}(B)\,. \tag{6.208}$$

The mathematical connection between a Calabi-Yau elliptically fibered over a base B and a Calabi-Yau fibered over a base $\tilde{B}$ realized by blowing up B at a point is mirrored in the physics of the associated supergravity theories. In the supergravity theory, such a transition is realized by a phase transition in the theory in which a single tensor multiplet is exchanged for 29 scalar multiplets.[46,150,151] Starting from the theory on $\tilde{B}$ with the larger value of T, this transition can be described by a limit in which a string becomes tensionless as $j \cdot x \to 0$ for some $x \in \Gamma$. From the F-theory point of view, the transition from B to $\tilde{B}$ can be seen as arising from a tuning of the Weierstrass parameters for an elliptic fibration over B so that a singularity arises that is worse than any singularity in the Kodaira Table. For example, at a codimension two singularity where deg f = 4, deg $g = 6$, and deg $\Delta = 12$, the singularity must be resolved by blowing up a point in the base. These transitions connect F-theory models with different bases B and different numbers of tensor multiplets T. We discuss this connectivity in the space of supergravity theories further below in Section 6.4.9.2.

We now consider a simple example of an F-theory compactification on an elliptically fibered threefold, focusing on the topological data. As additional examples, many of the other theories constructed through other methods are mapped into F-theory in Section 6.4.8.

Example: $SU(N)$ on $\mathbb{F}_2$

This is a 6D supergravity theory with $T = 1$. If the $SU(N)$ is realized by N D7-branes on the divisor class

$$\xi = D_v \tag{6.209}$$

then we have

$$-12K = 24D_v + 48D_s = ND_v + Y\,, \tag{6.210}$$

where $Y = (24 - N)D_v + 48D_s$. Matter in the fundamental representation of the $SU(N)$ arises at intersections between $\xi = D_v$ and Y. The number of such intersections is

$$\xi \cdot Y = D_v \cdot [(24 - N)D_v + 48D_s] = 48 - 2(24 - N) = 2N\,, \tag{6.211}$$

so the theory has $2N$ matter fields in the fundamental representation of $SU(N)$. It is easy to check that this is an anomaly-free spectrum. An explicit Weierstrass model for this F-theory compactification for $N \leq 14$ has been identified.[152]

To fit the set of models described through other constructions into the context of F-theory, it is useful to exploit the close correspondence between the anomaly structure of 6D supergravity and the structure of F-theory compactifications, to which we now turn.

6.4.8. *Mapping 6D supergravities to F-theory*

As discussed in Section 6.4.2, every 6D supergravity theory contains a signature $(1, T)$ lattice Γ of dyonic string charges that couple to the (anti-)self-dual fields $B^{\pm}$ in the theory. From the F-theory point of view these dyonic strings are realized by D3-branes wrapping cycles in $H_2(B, \mathbb{Z})$ of the base. Thus,

$$\Gamma = H_2(B, \mathbb{Z}) \tag{6.212}$$

with inner product given by the intersection form on B. In F-theory the self-duality of the lattice Γ follows immediately from Poincaré duality.

We now explicitly describe how the anomaly lattice Λ of a 6D supergravity theory is mapped into the charge lattice Γ for theories with an F-theory realization.[78,152] The divisor classes associated with nonabelian gauge group factors in a 6D F-theory compactification live in Γ since

$$\xi_i \in H_2(B, \mathbb{Z})\,. \tag{6.213}$$

The nonabelian gauge group factor G_i comes from a stack of 7-branes wrapped on the cycle ξ_i. The corresponding element b_i in the anomaly lattice comes from a gauge dyonic string in space-time associated with an instanton in G_i. In F-theory this instanton gives a 3-brane wrapped on ξ_i within the 7-brane world-volume. Thus, we can associate the element b_i of the anomaly lattice with ξ_i

$$b_i \to \xi_i \,. \tag{6.214}$$

From the anomaly condition (6.119) and the condition (6.208) we have $a^2 = K^2 = 9 - T$. This suggests that

$$a \to K \,. \tag{6.215}$$

Indeed, both (6.214) and (6.215) can be checked by computing the intersections in the F-theory picture and the corresponding anomaly lattice[77,141]

$$-a \cdot b = -K \cdot \xi_i \tag{6.216}$$

$$b_i \cdot b_j = \xi_i \cdot \xi_j . \tag{6.217}$$

This gives a clear picture of how the topological data needed for an F-theory compactification can be identified from the structure of the 6D supergravity spectrum and anomaly lattice. Given a 6D theory, the number of tensor multiplets T determines $h_{1,1}(B)$ through (6.207). There must then exist a lattice embedding

$$\Lambda \hookrightarrow \Gamma \tag{6.218}$$

that maps $a, b_i \to K, \xi_i$ as in (6.215), (6.214). This map takes the $(1, T)$ vector j of scalars in the tensor multiplets to the Kähler moduli of the F-theory compactification

$$j \to J \,. \tag{6.219}$$

The mapping (6.218) is not necessarily uniquely defined given only the nonabelian gauge group and matter content of the 6D theory. There may be multiple possible surfaces B giving compatible F-theory compactifications, and for a given surface the lattice map may not be uniquely defined, although in many cases it can be shown that the embedding is unique up to automorphisms of Γ using theorems of Nikulin. To uniquely determine the F-theory model corresponding to a given 6D supergravity, further information may be needed. Knowing the dyon charge lattice of the low-energy theory and having an explicit description of the space of possible Kähler moduli encoded in j (*i.e.*, the Kähler cone), for example, is sufficient to uniquely determine the intersection form on B, along with the set of effective divisors. This information uniquely determines the F-theory realization. Knowing the $U(1)$ content of the theory can also in principle help in determining the topology of the F-theory geometry. In many simple cases, however, as we see in examples below, the nonabelian gauge group and matter content are already sufficient to uniquely determine the corresponding F-theory realization.

In the case $T = 1$, the map (6.218) can be written explicitly in terms of the divisors D_v, D_s on $\mathbb{F}_m$ satisfying (6.197). The F-theory divisor corresponding to an anomaly vector b is given by[152]

$$b = \frac{1}{2}(\alpha, \tilde{\alpha}) \;\rightarrow\; \xi = \frac{\alpha}{2}(D_v + \frac{m}{2} D_s) + \frac{\tilde{\alpha}}{2} D_s \tag{6.220}$$

where we have expressed b in terms of $\alpha, \tilde{\alpha}$ as in (6.115). This map is also compatible with the corresponding expression for a

$$-a = (2, 2) \rightarrow -K = 2D_v + (m + 2)D_s \,. \tag{6.221}$$

Note that the anomaly formalism is in this case invariant under exchange of $\alpha, \tilde{\alpha}$. This can in some cases give distinct realizations of a given 6D model (which are often related through duality, such as for models on $\mathbb{F}_0 = \mathbb{P}^1 \times \mathbb{P}^1$, where exchange of $\alpha, \tilde{\alpha}$ corresponds to an exchange of D_v and D_s.)

Dualities between heterotic and F-theory models have been worked out in many special cases and classes of examples in various parts of the literature; this connection is explored in many of the papers referenced in the discussion of heterotic orbifold models, in particular in the work of Friedman, Morgan and Witten,[109] as well as in much of the F-theory literature. The map (6.218) gives a simple unified way of identifying such dualities in terms of the discrete data of the 6D supergravity theory and the corresponding F-theory topology. We now consider a number of explicit examples of this map.

Example:

Consider again the theory with gauge group $G = SU(N)$ and $2N$ fundamental matter fields, with $T = 1$. A straightforward calculation gives

$$b = \frac{1}{2}(\alpha, \tilde{\alpha}) = (1, -1) \rightarrow \xi = D_v + (m/2 - 1)D_s \,. \tag{6.222}$$

The map will only give an integral effective divisor for $m = 2$. This gives $\xi = D_v$ on $\mathbb{F}_2$, as described in (6.209).

Example:

Consider the intersecting brane model example (6.151) with gauge group $U(8)$, 8 matter fields in the antisymmetric two-index representation and one adjoint matter field. We saw from (6.154) that the anomaly lattice is generated by a single vector $b = -a$. Using (6.218), we see that this model maps to $\mathbb{F}_0$ with

$$b = (2,2) \to \xi = 2D_v + 2D_s\,. \tag{6.223}$$

The genus of this curve is given by

$$(K + \xi) \cdot \xi = (a + b) \cdot b = 0 = 2g - 2\,. \tag{6.224}$$

So the $SU(8)$ group lives on a discriminant locus curve of genus one, as expected given the one adjoint representation.

Example:

Now consider the example of a type I/heterotic K3 compactification with gauge group (6.172) and matter content (6.175). From the form of the anomaly vector $b_3 = (1, -2)$ associated with the gauge group factor $SO(16)$, we see that the only possible value for m that gives an effective divisor for the image of b_3 under (6.220) is $m = 4$. The divisor classes associated with the different gauge group factors are then

$$\begin{aligned} b_1 &= (1,2) \to D_u = D_v + 4D_s \\ b_2 &= (1,6) \to D_v + 8D_s \\ b_3 &= (1,-2) \to D_v\,. \end{aligned} \tag{6.225}$$

So we expect a D_8 singularity on the divisor D_v, and A_3 singularities on $D_v + 4D_s$ and $D_v + 8D_s$. This general structure is typical for smooth heterotic compactifications of the $SO(32)$ theory on K3; the F-theory realization of such models is always associated with the base $\mathbb{F}_4$, and the residual $SO(M)$ factor resides on D_v.[46]

Example: $E_7 \times E_8$ on $\mathbb{F}_{12}$

We discussed above the model (6.179) with gauge group $E_7 \times E_8$. From the vector $b_8 = (1, -6)$ in (6.184), we see that (6.218) only gives an effective divisor for $m = 12$. The E_8 and E_7 loci are then $D_v, D_u = D_v + 12D_s$ respectively. There is no bifundamental matter, as expected, since $D_v \cdot D_u = 0$. This is an example of a heterotic $E_8 \times E_8$ vacuum with all instantons in one E_8. For general heterotic $E_8 \times E_8$ compactifications, the resulting F-theory model has a base $\mathbb{F}_m$ where $12 \pm m$ of the instantons are embedded in each E_8 factor.[46]

Example: Gimon-Polchinski model

Consider now the Gimon-Polchinski model (6.185), with gauge group $G - SU(16) \times SU(16)$, two matter fields transforming under the antisymmetric **120** for each gauge group factor, and one bifundamental matter field. It is easy to compute that the anomaly matrix for this model is

$$\Lambda = \begin{pmatrix} 8 & -2 & -2 \\ -2 & 0 & 1 \\ -2 & 1 & 0 \end{pmatrix} . \tag{6.226}$$

It follows that in the canonical $T = 1$ basis (6.115),

$$b_1 = (1, 0), \qquad b_2 = (0, 1) . \tag{6.227}$$

This theory must therefore be realized through an F-theory compactification on $\mathbb{F}_0 = \mathbb{P}^1 \times \mathbb{P}^1$, with divisor classes $b_1 \to D_v, b_2 \to D_s$. This F-theory realization of the Gimon-Polchinski model has been explored in some detail as a useful example for studying the heterotic-F-theory duality, in particular clarifying the role of instantons with Spin(32)/$\mathbb{Z}_2$ structure but not $SO(32)$ vector structure.[128,129,153]

The reader may find it illuminating to work out a few further examples. As far as the author is aware all known 6D supergravity theories that arise from any kind of string compactification admit an embedding into F-theory through a map of the form (6.218), at least at the level of the discrete data characterizing the gauge group and matter content of the theory.

6.4.9. *Global structure of the space of 6D $\mathcal{N} = 1$ supergravities*

6.4.9.1. *Consistency conditions, and matching $\mathcal{G}$ to $\mathcal{V}$*

In Section 6.4.2 we summarized the known constraints on the space $\mathcal{G}^{6D,\mathcal{N}=1}$ (which we refer to simply as $\mathcal{G}$ for the remainder of this section). For $T < 9$ there are a finite set of possible combinations of nonabelian gauge groups and matter content compatible with anomaly cancellation conditions and proper-sign gauge kinetic terms. For $T \geq 9$ there are infinite families of models that satisfy the known consistency conditions. Consistent supergravity theories must furthermore satisfy the condition that the dyonic string charge lattice Γ is unimodular.

Given the finite constraint on the set of possible groups and matter for $T < 9$, it is possible in principle to enumerate all models consistent with the known consistency constraints. This can be done in practice more easily for smaller values of T. For $T = 0$ the constraints are strongest, as the vectors in the anomaly lattice are just integers $b \in \mathbb{Z}$, and a complete classification of models is tractable.[83] For $T = 1$ the space of possible combinations of groups and representations becomes larger; an exploration of part of this space of theories with some restriction on matter representations was initiated by Avramis and Kehagias.[154] Using the same methods that lead to the finiteness bounds, a complete classification of models has been carried out for gauge groups with $SU(N)$ and certain matter representations,[152] such a classification can be done for arbitrary gauge groups and matter. Incorporating the unimodularity constraint on the dyon charge lattice, such a systematic classification of models can in principle be continued up to $T = 8$.

Six-dimensional F-theory constructions, as described in Section 6.4.7, give a wide range of string vacua, and define a set $\mathcal{V} = \mathcal{V}^{6D,\mathcal{N}=1}$ of known string vacuum constructions in 6D. For any 6D supergravity theory admitting an F-theory realization, the anomaly lattice of the 6D theory is mapped into the topological structure of the F-theory construction through the map (6.218). This provides a framework for identifying general features of models lying in the set $\mathcal{G} \setminus \mathcal{V}$, which satisfy all known consistency conditions and yet have no known string theory construction. Basically, we just need to determine which models in $\mathcal{G}$ do not lead to acceptable data for an F-theory model. There are a number of specific ways in which the map (6.218) can fail to take the anomaly lattice of an apparently-consistent theory to valid topological data for an F-theory compactification. We list a number of these possible failure modes and comment on each.

• Unimodular embedding.

A necessary condition for the existence of an embedding in F-theory through (6.218) is the existence of an embedding of Λ in a unimodular lattice. As discussed at the end of Section 6.4.2, however, it has been shown[88] that every consistent 6D theory has a unimodular charge lattice, of which Λ is a sublattice. Thus, every consistent theory has an anomaly lattice that can be embedded in some unimodular lattice Γ. Γ can then be taken to be the homology lattice $H_2(B, \mathbb{Z})$ of an F-theory base.

• Exotic matter fields.

There are a variety of 6D supergravity models that satisfy the known consistency conditions and that have exotic matter representations not realized through known codimension two singularity types in F-theory.[78,83,152] As examples, consider the following theories (listing only charged matter for each theory)

$$T = 1, G = SU(8), \ \text{matter} = \ydiagram{1} + 3\,\ydiagram{2} + 2\,\ydiagram{1,1,1} + \ydiagram{1,1,1,1} \tag{6.228}$$

$$T = 0, G = SU(4), \ \text{matter} = 1 \times \ydiagram{2,2} + 64 \times \ydiagram{1} \tag{6.229}$$

$$T = 0, G = SU(8), \ \text{matter} = 1 \times \ydiagram{2,2} \tag{6.230}$$

Each of these theories contains a matter representation that cannot be realized using a standard codimension two F-theory singularity. The range of possible matter fields that may be realized using more exotic codimension two singularities is, however, quite large.[142] It may be possible to implement at least the first two of these models using more exotic types of codimension two singularity. One clue to this connection is given by a relationship between the group theory of matter representations and the corresponding F-theory geometry.[83] For any representation we can define the quantity

$$g_R = \frac{1}{12}\left(2C_R + B_R - A_R\right) . \tag{6.231}$$

From the anomaly relations it is possible to show that for models with $T = 0$ the set of representations transforming under a gauge factor G satisfies

$$\sum_R x_R\, g_R = \frac{(b-1)(b-2)}{2} . \tag{6.232}$$

In a $T = 0$ F-theory realization, b is the degree of the polynomial defining the curve ξ on $\mathbb{P}^2$ carrying the singularity associated with the gauge factor G. The quantity (6.232) is precisely the arithmetic genus of such a curve.[51] The arithmetic genus carries a contribution from the topological (geometric) genus as well as a contribution from singular points on the curve. So it is natural to expect that g_R encodes the arithmetic genus contribution of a singularity on the curve ξ carrying the group factor G. This correspondence works, for example, for the two-index symmetric representation of $SU(N)$, which has $g_{\ydiagram{2}} = 1$, and is produced by an ordinary double point singularity contributing 1 to the arithmetic genus.[77,142] A complete analysis of codimension two singularities in F-theory is a challenge for future research. Note, however, that some representations, such as the "box" representation of $SU(8)$ appearing in (6.230) cannot appear in any known kind of F-theory model for other reasons; in the case of the $SU(8)$ box representation, the theory violates the bound associated with the Kodaira condition (which is discussed further below). There are also constraints on matter representations that can be realized from the heterotic point of view.[155,156] Understanding how the heterotic, F-theory, and supergravity constraints on allowed representations may be related is an interesting open problem.

- Weierstrass formulation

In the discussion here we have focused on the topological structure of the F-theory realization, in particular on the divisor classes supporting the singularities associated with a given nonabelian gauge group. For a complete F-theory realization, an explicit Weierstrass model is needed, where f, g are described explicitly as sections of the appropriate line bundles $\mathcal{O}(-4K), \mathcal{O}(-6K)$. In a variety of cases that have been studied, Weierstrass models can be found whenever a map of the form (6.218) can be found whose image gives divisor classes satisfying all the topological constraints from F-theory (including the positivity and Kodaira constraints formulated below). In cases where the topological conditions are satisfied, the number of degrees of freedom in the Weierstrass model that remain unfixed when the coefficients are tuned to achieve the desired gauge group precisely matches the number of uncharged scalar fields in the low-energy theory.[152] This led Kumar, Morrison, and the author to conjecture that the degrees of freedom will match in all cases, and that a Weierstrass model will exist whenever the topological constraints are satisfied. Finding a proof of this conjecture is left as an open problem for further research.

- Positivity conditions and effective divisors

There is a sign condition on the divisors appearing in the image of the map (6.218). The divisors ξ_i (associated with the curves where the 7-branes are wrapped) must be effective divisors; this is equivalent to the statement that there exists a moduli vector J such that

$$J \cdot \xi_i > 0 \; \forall i \,. \tag{6.233}$$

We understand this condition physically in the 6D supergravity theory as the constraint that the gauge kinetic term for G_i must have the proper sign, *i.e.* $j \cdot b_i > 0$.

It is also the case for all F-theory models that $-K$ is an effective divisor, so that $J \cdot (-K) > 0$. This condition in the supergravity theory states that

$$j \cdot a < 0 \,. \tag{6.234}$$

This fixes the sign of the curvature-squared term in the 6D theory of the form $(a \cdot j)R^2$. It is not clear whether this sign constraint is necessary for consistency of the low-energy theory. It is possible that the sign of this term is fixed through a causality constraint analogous to those studied by Adams *et al.*[157] A complete explanation of this constraint from the supergravity point of view is left as another open question for future work.

- Kodaira constraint

  Related to the positivity conditions just mentioned, there is also a condition on the residual divisor locus $Y = -12K - \sum_i \nu_i \xi_i$, which states that this divisor is effective. This condition arises from the Kodaira constraint (6.192). In the 6D supergravity theory this constraint states that

$$j \cdot (-12a - \sum_i \nu_i b_i) > 0\,. \tag{6.235}$$

All of the known infinite families of models that satisfy the other macroscopic 6D constraints violate this "Kodaira bound". Thus, proving this bound would potentially reduce the space of possibly consistent combinations of T, G, and $\mathcal{M}$ to a finite set $\mathcal{G}$. One possible route to proving (6.235) would be to follow an approach along the lines described in Section 6.3.5. It may be that a careful analysis of world-sheet anomalies on the dyonic strings of a 6D supergravity theory will constrain the set of allowed models in such a way that (6.235) is necessary for consistency of the theory. Further study is needed, however, to see whether this speculation is borne out in practice.

6.4.9.2. *Connectivity and finiteness of the space of theories*

Many, if not all, 6D supergravity theories with different gauge groups and matter content are connected through continuous deformations in the scalar moduli space. Moving in the moduli space can lead to a familiar Higgs type transition where a gauge group is broken, removing scalar fields and vector fields from the theory in such a way that the difference $H - V$ is unchanged and the anomaly relation (6.117) remains valid. From the F-theory point of view this kind of transition arises from changing the coefficients in the Weierstrass functions f, g so that the structure of the discriminant locus changes. It is clear from the F-theory point of view that any two models associated with the same base B for the elliptic fibration are continuously connected through a deformation of the Weierstrass equation.

There are, however, further connections in the space of supergravity theories. As discussed in Section 6.4.7.3, supergravity theories with different numbers of tensor multiplets T can be connected through exotic phase transitions where a tensor multiplet T is exchanged for 29 hypermultiplets (with a possible additional change in the vector multiplet structure) in such a way that the anomaly relation (6.117) is preserved. Such a transition occurs in a limit of the moduli space where $j \cdot x \to 0$ for some $x \in \Gamma$ in the lattice of dyonic string charges. Thus, this transition is associated with a string becoming tensionless.[46,150,151] From the F-theory point of view this corresponds to a cycle in the base B shrinking to a point, *i.e.* to a blow-down (the opposite of a blow-up) of a rational curve in the surface. From the heterotic point of view, the transition from the theory with fewer tensor multiplets occurs when an instanton shrinks to a point, as mentioned above, producing a 5-brane carrying the additional tensor field. Since all smooth F-theory bases corresponding to theories with one 6D supersymmetry can be connected by a sequence of blow-up and blow-down operations, this suggests that the full space of 6D $\mathcal{N} = 1$ supergravities is a single connected moduli space[i]. This would mean that there is really a single unified quantum theory of 6D supergravity since the entire moduli space is in principle visible in the structure of fluctuations of massless scalar fields around any given vacuum. This gives a simple and appealing picture of supergravity in six dimensions.

From the point of view of F-theory, the number of distinct bases B, as well as the number of combinations of T, G, and matter representations that can be realized must be finite. We summarize briefly the argument for finiteness;[78] the related fact that the number of topological classes of elliptically fibered Calabi-Yau threefolds (up to birational equivalence) is finite was proven by Gross.[159] From the minimal surface point of view, any F-theory model can be viewed as a Weierstrass model over either $\mathbb{P}^2$ or one of the Hirzebruch surfaces $\mathbb{F}_m$. If the number of tensor multiplets is greater than 1, then the Weierstrass model can be thought of as a Weierstrass model on $\mathbb{F}_m$ with singularities tuned to require one or more blow-ups of points on B, as discussed in Section 6.4.7.3. Each different combination of base, gauge group, and matter content, then, corresponds to a locally closed set in the finite-dimensional space of Weierstrass models. (A locally closed set is an intersection of a closed set with an open set. Each condition on the discriminant locus gives a closed set, with each further condition giving an enhanced gauge group or further blow-up giving a smaller closed set within the original set.) A fundamental theorem from commutative algebra and algebraic geometry, the Hilbert basis theorem, guarantees that there are only a finite number of distinct such locally closed sets. Heuristically this

[i]Extremal transitions associated with theories having the Enriques surface as base are identified in Ref. 158.

follows from the observation that each gauge group enhancement or blow-up requires tuning a finite number of parameters, and the total number of parameters is finite and equal to the number of neutral scalar fields in the generic model, 273 on $\mathbb{P}^2$ and 244 on $\mathbb{F}_m$. Since the known constraints on $\mathcal{G}$ give an infinite set of distinct models, at present, the "apparent swampland" of 6D $\mathcal{N} = 1$ theories is infinite. Finding an argument for the Kodaira constraint as a consistency condition for any supergravity theory could eliminate the infinite number of apparently consistent models that are not realized in F-theory.

While the set of allowed F-theory models is provably finite, the precise extent of the set of F-theory compactifications has not been determined. For example, the maximum number of tensor multiplets T which can appear has not been definitively identified. The largest number known is $T = 192$, for the theory with gauge group (6.188).[127,132] The structure of the anomaly lattice provides insight into the the extremal transitions where T changes; this may help in charting the connectivity and global structure of the space of F-theory models and definitively identifying the upper bound on T. Ideally such an understanding would lead to a clear picture of how the space of models arising from string/F-theory is bounded using considerations based only on the macroscopic structure of the supergravity theory.

6.5. Comments on Four Dimensions and Other Concluding Remarks

In these lectures we have described some global aspects of the space of supergravity theories in dimensions 10, 8, and 6. We have used constraints on these theories to determine a set $\mathcal{G}$ of "apparently consistent" theories in each dimension. We have also described the set $\mathcal{V}$ of known string theory compactifications in each dimension. We have focused on the discrete data characterizing each supergravity theory: the spectrum of fields and the gauge symmetry group of the theory. In 10D, $\mathcal{G} = \mathcal{V}$, so that we have "string universality"; any theory of supergravity in ten dimensions that cannot be realized in string theory is inconsistent (at least at the level of the field content and symmetry of the theory). In 8D and 6D, $\mathcal{G} \supset \mathcal{V}$, and the set of possible models in the "apparent swampland" $\mathcal{G} \setminus \mathcal{V}$ is infinite. In each case, however, we have a set of fairly well-understood criteria for determining which models in $\mathcal{G}$ admit a string realization (and hence are in $\mathcal{V}$). It is possible that both in 8D and in 6D these criteria will eventually be understood in terms of additional consistency constraints on the low-energy theory; we have outlined possible approaches that may lead to such an understanding. It is also possible that new approaches to string compactification may be found that will expand the space $\mathcal{V}$. Whether or not it can be proven that string universality holds for supergravity theories in eight and/or six dimensions, focusing on theories in the set $\mathcal{G} \setminus \mathcal{V}$ is a promising direction for further progress. By searching for new quantum consistency conditions that rule out such theories, or novel realizations of such theories in string theory, our understanding of the global space of quantum gravity theories is improved.

An important direction on which we have not focused in these lectures is the investigation of the relationship between macroscopic and stringy consistency conditions for more detailed aspects of supergravity theories, beyond the field content and symmetry structure. For example, for 6D $\mathcal{N} = 1$ supergravity theories, the hypermultiplets live in a quaternionic Kähler manifold. To go beyond the discrete version of string universality, it would be necessary to demonstrate that the set of such manifolds, along with their metrics, is determined uniquely from the low-energy theory to match the set of possibilities arising from string compactifications. Compact quaternionic Kähler manifolds have been classified.[160–162] But a full analysis of which of these manifolds are compatible with string theory remains as a challenge for the future. In most supergravity theories there are also terms that might be added to the action of the theory, such as higher-derivative terms. For string realizations of these theories, the coefficients of these additional terms in the low-energy theory are fixed by the quantum physics of string theory. For string universality to hold in a complete fashion, there need to be consistency conditions that fix these coefficients to the values determined by string theory. It has been shown that in some cases, supersymmetry places very strong constraints on such coefficients.[26] But a systematic understanding of the extent to which such constraints uniquely constrain low-energy theories to those realized through string theory is another direction open for future progress.

Extending the kind of global analysis done here to four dimensions would obviously be an interesting and important endeavor. In principle this could be a very promising direction for future research. String theory constructions do impose strong constraints on which low-energy 4D supergravity theories can be realized, just as in higher dimensions. In many constructions these consistency conditions come from tadpole conditions closely analogous to the Kodaira constraint and related tadpole conditions we have described here. Anomalies have provided a clear window on these constraints in six and ten dimensions. Just as in eight dimensions, however, for four-dimensional theories some principles beyond space-time anomalies seem to be required to identify the string theoretic constraints from the structure of the low-energy theory. A number of new consistency conditions on 4D low-energy theories have been identified in recent years,[1,157,163] though not directly related to string constructions. One natural approach that we have discussed here for deriving further constraints on theories both in 8D and in 6D is the analysis of anomalies in the world-volume of solitonic strings in supergravity theories. The same approach could be used for 4D theories. 4D supersymmetric gravity theories generally contain axions that play the same role as B fields in higher dimensions. (Axions are dual to two-form fields in 4D.) Most if not all supersymmetric string vacuum constructions give rise to strings in 4D that couple to these axions just as strings in higher dimensions couple to two-form fields $B_{\mu\nu}$. Thus, a careful analysis of the world-volume theories of axionic strings in supersymmetric 4D gravity theories may give new constraints on what structures are allowed in these theories.

One complication in approaching 4D theories with less than $\mathcal{N} = 4$ supersymmetry is that the set of string vacua is vast, and still poorly understood. Even for $\mathcal{N} = 2$ theories there are fundamental mathematics problems that must be solved to get a global picture of the space of string vacua. Compactifications of type II string theory on Calabi-Yau threefolds (not necessarily elliptically fibered) give rise to such $\mathcal{N} = 2$ supergravity theories. There is at this time no proof that the set of topological classes of Calabi-Yau threefolds is finite, and no systematic classification. While we were able to use the mathematical structure of lattices to great efficacy in the description of higher-dimensional supergravity theories, the analogous structure in 4D is more complex. For example, the 2-homology and intersection form on a complex surface is described by a lattice that must be self-dual due to Poincaré duality. For a general Calabi-Yau threefold, the intersection structure is given by a trilinear form $C : (H_2(B, \mathbb{Z}))^3 \to \mathbb{Z}$. The space of such trilinear forms, and the constraints on such forms, are much less well understood than the space of unimodular lattices. Thus, even for $\mathcal{N} = 2$ theories in 4D, new mathematics probably must be developed to attain a true global classification of theories.

The situation becomes even more complicated for $\mathcal{N} = 1$ theories. Studying different string approaches to $\mathcal{N} = 1$ 4D supergravity theories has been a major industry for decades. The close correspondence identified in six dimensions between the anomaly structure in supergravity theories and the topology of F-theory compactifications, along with the observation that in 6D F-theory provides a fairly universal framework for describing string vacua, suggest that F-theory is a natural place to look for a global characterization of the space of string vacua. From the F-theory point of view, however, the set of possible compactifications to 4D depends upon elliptically fibered Calabi-Yau fourfolds. Such fourfolds are fibered over a three-complex dimensional base manifold B. The classification of three-folds is in a much more limited state of development than that of surfaces. For surfaces, the minimal model approach leads to a systematic characterization of possible F-theory bases, and provides a geometric understanding of the connection between these bases. For threefolds, the program of Mori[164] aspires to a similar classification. But the mathematics is much more complicated, and seems to need significantly more development to be usable by physicists as an approach to a global characterization of the set of bases that can be used for 4D F-theory constructions. While in recent years 4D F-theory models have provided a rich new perspective on string phenomenology,[165–171] the full range of possible 4D vacua is still poorly understood. Some geometric constraints on a class of 4D F-theory models have been found by Cordova[172] based on the assumption that gravity can decouple from the low-energy physics. Inclusion of fluxes ("G-flux") in F-theory should significantly expand the range of possible 4D vacua, but the tools for understanding this from a global/nonperturbative point of view are just being developed.[173–175] As for the 6D models discussed here, U(1) factors also have some subtle features in global constructions, and are a subject of current interest in 4D F-theory models,[176] where anomaly constraints also play a role.[177,178]

Other very general approaches to building new string vacua suggest that we may have only seen the tip of the iceberg of 4D string constructions. Non-Kähler compactifications,[179–184] non-geometric fluxes,[114,185–189] G_2 compactifications of M-theory,[190–192] asymmetric orbifolds,[40] and other constructions all suggest that much work must be done to begin to get some global grasp of the range of possibilities.

Finally, the discrete nature of the set of 4D $\mathcal{N} = 1$ vacua makes it clear that attaining a global classification of models, as well as determining macroscopic constraints, will require qualitatively different insights from those used in the simpler classes of theories in higher dimensions and/or with more supersymmetry. For the 8D and 6D $\mathcal{N} = 1$ supergravities we have studied here, the theories are connected in a single continuous moduli space of Minkowski vacua. In 4D, the inclusion of fluxes and other discrete structure can stabilize many moduli, giving a "discretium" of isolated regions of (generally AdS) supersymmetric vacua.[193] The number of discrete families of supersymmetric 4D vacua is infinite in some families of string compactifications.[194] Showing that certain apparently continuous parameters in the description of the low-energy theory can only take the specific values associated with string vacua presents an added challenge for the program of identifying constraints that rule out all models other than those realized in string theory. It may be that statistical methods[195,196] are needed to make sense of this large set of discrete solutions. Even in 4D, however, in certain cases such as intersecting brane models on a toroidal orbifold it has been possible to place finite bounds on the set of possible vacuum solutions,[197] and to systematically enumerate these models.[198,199] Generalizing such results to more generic backgrounds such as smooth Calabi-Yau manifolds, which may be possible using the trilinear intersection form, would be a promising step forward in attaining a more global picture of large classes of $\mathcal{N} = 1$ 4D vacua.

Despite the preceding cautionary remarks regarding the challenges of developing a global analysis of 4D $\mathcal{N} = 1$ vacua, a number of the lessons learned from higher-dimensional supergravity theories should be relevant for 4D physics. Consideration of 6D theories satisfying anomaly cancellation and other constraints has have suggested some new structures that may arise in string theory, such as novel types of matter from codimension two singularities in F-theory. Understanding the stringy derivation of such phenomena in six dimensions will also give new tools for describing 4D physics. In 6D, the close correspondence between the structures underlying supergravity theories and F-theory has provided a "bottom-up" map which can uniquely identify the F-theory geometry of any string vacuum realizing a supergravity model with given symmetries and spectra. Progress towards explicitly realizing such a correspondence in four dimensions would help narrow the range of string constructions which might correspond with observed physics. This systematic characterization of string vacua has also helped to identify new supergravity constraints in six dimensions, such as the self-dual nature of the dyonic charge lattice, which suggests that new constraints on 4D gravity theories remain to be discovered. Insights into the duality relationships between different string theories in higher dimensions has played an important role in our developing understanding of string theory over the last 15 years. Incorporating these relationships into a global picture of the space of 6D string vacua will help provide insight into how the different string vacuum constructions are related in 4D.

Acknowledgments

The perspective presented in these lectures incorporates the contributions of many people. I would particularly like to thank Vijay Kumar and David Morrison, with whom many of the ideas developed in these lectures have been explored. Thanks also to my other collaborators in the part of the work presented here in which I was involved in the original research: Allan Adams, Oliver DeWolfe, Michael Douglas, Satoshi Nagaoka, Daniel Park, Vladimir Rosenhaus, and Nathan Seiberg. Thanks to Oliver DeWolfe, Vijay Kumar, David Morrison and Daniel Park for comments on a preliminary version of these notes. I am grateful for helpful discussions and correspondence with Massimo Bianchi, Ralph Blumenhagen, Tom Banks, Mirjam Cvetic, Frederik Denef, Noam Elkies, Dan Freedman, Michael Green, Vladimir Ivashchuk, Jim Halverson, Gabriele Honecker, Shamit Kachru, John McGreevy, Greg Moore, Joe Polchinski, Bert Schellekens, John Schwarz, Ashoke Sen, Isadore Singer, Cumrun Vafa, Edward Witten, and Barton Zwiebach. I would like to thank the Simons Center for Geometry and Physics for hospitality during the completion of these notes. And of course, many thanks to the organizers of the TASI 2010 summer school for making these lectures possible and to the students at the school for many insightful and provocative questions and discussions. This research was supported by the DOE under contract #DE-FC02-94ER40818.

References

1. C. Vafa (2005), hep-th/0509212.
2. H. Ooguri and C. Vafa, *Nucl.Phys.* **B766**, 21 (2007), hep-th/0605264.
3. V. Kumar and W. Taylor (2009), arXiv:0906.0987.
4. J. de Boer, R. Dijkgraaf, K. Hori, A. Keurentjes, J. Morgan *et al.*, *Adv.Theor.Math.Phys.* **4**, 995 (2002), hep-th/0103170.
5. M. B. Green, J. H. Schwarz and E. Witten, *Superstring theory. Vol. 1: Introduction* Cambridge, Uk: Univ. Pr. (1987) 469 P. (Cambridge Monographs On Mathematical Physics).
6. M. B. Green, J. H. Schwarz and E. Witten, *Superstring theory. Vol. 2: Loop amplitudes, anomalies and phenomenology* Cambridge, Uk: Univ. Pr. (1987) 596 P. (Cambridge Monographs On Mathematical Physics).
7. J. Polchinski, *String theory. Vol. 1: An introduction to the bosonic string* Cambridge, UK: Univ. Pr. (1998) 402 p.
8. J. Polchinski, *String theory. Vol. 2: Superstring theory and beyond* Cambridge, UK: Univ. Pr. (1998) 531 p.
9. E. Salam, A. and E. Sezgin, E., *Supergravities in Diverse Dimensions. Vol.*

1, 2 Amsterdam, Netherlands: North-Holland/World Scientific/Elsevier (1989) 1499 p.
10. T. Banks (2010), 1007.4001.
11. E. Witten, *Nucl.Phys.* **B443**, 85 (1995), hep-th/9503124.
12. E. Bergshoeff, M. de Roo, B. de Wit and P. van Nieuwenhuizen, *Nucl.Phys.* **B195**, 97 (1982).
13. G. Chapline and N. Manton, *Phys.Lett.* **B120**, 105 (1983).
14. J. A. Harvey (2005), hep-th/0509097.
15. L. Alvarez-Gaume and E. Witten, *Nucl.Phys.* **B234**, p. 269 (1984).
16. M. B. Green and J. H. Schwarz, *Phys.Lett.* **B149**, 117 (1984).
17. A. Adams, O. DeWolfe and W. Taylor, *Phys.Rev.Lett.* **105**, p. 071601 (2010), arXiv:1006.1352.
18. P. Townsend (1995), hep-th/9507048.
19. N. Berkovits, *JHEP* **0004**, p. 018 (2000), hep-th/0001035.
20. O. A. Bedoya and N. Berkovits (2009), arXiv:0910.2254.
21. B. de Wit, J. Hoppe and H. Nicolai, *Nucl.Phys.* **B305**, p. 545 (1988).
22. T. Banks, W. Fischler, S. Shenker and L. Susskind, *Phys.Rev.* **D55**, 5112 (1997), hep-th/9610043.
23. W. Taylor, *Rev.Mod.Phys.* **73**, 419 (2001), hep-th/0101126.
24. J. M. Maldacena, *Adv.Theor.Math.Phys.* **2**, 231 (1998), hep-th/9711200.
25. O. Aharony, S. S. Gubser, J. M. Maldacena, H. Ooguri and Y. Oz, *Phys.Rept.* **323**, 183 (2000), hep-th/9905111.
26. M. B. Green and S. Sethi, *Phys.Rev.* **D59**, p. 046006 (1999), hep-th/9808061.
27. D. J. Gross, J. A. Harvey, E. J. Martinec and R. Rohm, *Phys.Rev.Lett.* **54**, 502 (1985).
28. J. Polchinski and E. Witten, *Nucl.Phys.* **B460**, 525 (1996), hep-th/9510169.
29. C. M. Hull and P. K. Townsend, *Nucl. Phys.* **B438**, 109 (1995), hep-th/9410167.
30. A. Salam and E. Sezgin, *Phys.Lett.* **B154**, p. 37 (1985).
31. M. Awada and P. Townsend, *Phys.Lett.* **B156**, p. 51 (1985).
32. J. Wess and J. Bagger, *Supersymmetry and supergravity* Princeton, NJ, USA: Princeton University Press (1992) 259 p.
33. D. Joyce, *Compact manifolds with special holonomy* Oxford, Oxford University Press (2000).
34. K. Becker, M. Becker, D. R. Morrison, H. Ooguri, Y. Oz *et al.*, *Nucl.Phys.* **B480**, 225 (1996), hep-th/9608116.
35. D. R. Morrison, 653 (1999), hep-th/0411120.
36. J. P. Gauntlett, G. Gibbons, G. Papadopoulos and P. Townsend, *Nucl.Phys.* **B500**, 133 (1997), hep-th/9702202.
37. M. Gross, D. Huybrechts and D. Joyce, *Calabi-Yau Manifolds and Related Geometries* Berlin: Springer-Verlag (2003).
38. J. H. Conway and N. J. A. Sloane, *Sphere packings, lattices, and groups* New York, NY, USA: Springer-Verlag (1999) 703 p.
39. K. Narain, *Phys.Lett.* **B169**, p. 41 (1986).
40. K. Narain, M. Sarmadi and E. Witten, *Nucl.Phys.* **B279**, p. 369 (1987).

41. E. Witten, *JHEP* **9802**, p. 006 (1998), hep-th/9712028.
42. O. J. Ganor, D. R. Morrison and N. Seiberg, *Nucl.Phys.* **B487**, 93 (1997), hep-th/9610251.
43. V. V. Nikulin, *Mathematics of the USSR-Izvestiya* **14 (1)**, 103 (1980).
44. C. Vafa, *Nucl.Phys.* **B469**, 403 (1996), hep-th/9602022.
45. D. R. Morrison and C. Vafa, *Nucl.Phys.* **B476**, 437 (1996), hep-th/9603161.
46. D. R. Morrison and C. Vafa, *Nucl.Phys.* **B473**, 74 (1996), hep-th/9602114.
47. F. Denef (2008), arXiv:0803.1194.
48. P. S. Aspinwall, 421 (1996), hep-th/9611137.
49. R. Blumenhagen, V. Braun, B. Kors and D. Lust, *JHEP* **0207**, p. 026 (2002), hep-th/0206038.
50. V. Kumar and W. Taylor, *JHEP* **0905**, p. 066 (2009), arXiv:0903.0386.
51. D. Perrin, *Algebraic Geometry: an Introduction* London: Springer-Verlag (2008) 258 p.
52. W. P. Barth, K. Hulek, C. A. M. Peters and A. Van de Ven, *Compact Complex Surfaces* Berlin: Springer-Verlag (2004) 436 p.
53. O. DeWolfe and B. Zwiebach, *Nucl.Phys.* **B541**, 509 (1999), hep-th/9804210.
54. O. DeWolfe, T. Hauer, A. Iqbal and B. Zwiebach, *Adv.Theor.Math.Phys.* **3**, 1785 (1999), hep-th/9812028.
55. M. Fukae, Y. Yamada and S.-K. Yang, *Nucl.Phys.* **B572**, 71 (2000), hep-th/9909122.
56. K. Kodaira, *Ann.Math.* **77**, p. 563 (1963).
57. K. Kodaira, *Ann.Math.* **78**, p. 1 (1963).
58. M. Daflon Barrozo, *Nucl.Phys.* **B574**, 189 (2000), hep-th/9909178.
59. A. Sen, *Nucl.Phys.* **B475**, 562 (1996), hep-th/9605150.
60. C. Bachas and E. Kiritsis, *Nucl.Phys.Proc.Suppl.* **55B**, 194 (1997), hep-th/9611205.
61. C. Bachas, C. Fabre, E. Kiritsis, N. Obers and P. Vanhove, *Nucl.Phys.* **B509**, 33 (1998), hep-th/9707126.
62. W. Lerche and S. Stieberger, *Adv.Theor.Math.Phys.* **2**, 1105 (1998), hep-th/9804176.
63. W. Lerche, 15 (1999), hep-th/9910207.
64. T. Banks and N. Seiberg (2010), arXiv:1011.5120.
65. T. Banks, N. Seiberg and E. Silverstein, *Phys.Lett.* **B401**, 30 (1997), hep-th/9703052.
66. A. M. Uranga, *Nucl.Phys.* **B598**, 225 (2001), hep-th/0011048.
67. S. Chaudhuri, G. Hockney and J. D. Lykken, *Phys.Rev.Lett.* **75**, 2264 (1995), hep-th/9505054.
68. W. Lerche, C. Schweigert, R. Minasian and S. Theisen, *Phys.Lett.* **B424**, 53 (1998), hep-th/9711104.
69. N. Seiberg, *Phys.Lett.* **B390**, 169 (1997), hep-th/9609161.
70. W. Nahm, *Nucl.Phys.* **B135**, p. 149 (1978).
71. H. Nishino and E. Sezgin, *Phys.Lett.* **B144**, p. 187 (1984), Revised version.
72. H. Nishino and E. Sezgin, *Nucl.Phys.* **B278**, 353 (1986).
73. L. Romans, *Nucl.Phys.* **B276**, p. 71 (1986).

74. M. B. Green, J. H. Schwarz and P. C. West, *Nucl.Phys.* **B254**, 327 (1985).
75. H. Nishino and J. Gates, S.James, *Nucl.Phys.* **B268**, p. 532 (1986).
76. A. Sagnotti, *Phys.Lett.* **B294**, 196 (1992), hep-th/9210127.
77. V. Sadov, *Phys.Lett.* **B388**, 45 (1996), hep-th/9606008.
78. V. Kumar, D. R. Morrison and W. Taylor, *JHEP* **1011**, p. 118 (2010), arXiv:1008.1062.
79. J. Erler, *J.Math.Phys.* **35**, 1819 (1994), hep-th/9304104.
80. G. Honecker, *Nucl.Phys.* **B748**, 126 (2006), hep-th/0602101.
81. D. S. Park and W. Taylor (To appear).
82. V. Kumar and W. Taylor, *JHEP* **0912**, p. 050 (2009), arXiv:0910.1586.
83. V. Kumar, D. Park and W. Taylor (2010), arXiv:1011.0726.
84. E. Witten, *Phys.Lett.* **B117**, 324 (1982).
85. M. Bershadsky and C. Vafa (1997), hep-th/9703167.
86. J. H. Schwarz, *Phys.Lett.* **B371**, 223 (1996), hep-th/9512053.
87. S. Deser, A. Gomberoff, M. Henneaux and C. Teitelboim, *Nucl.Phys.* **B520**, 179 (1998), hep-th/9712189.
88. N. Seiberg and W. Taylor (2011), arXiv:1103.0019.
89. M. Hopkins and I. Singer, *J.Diff.Geom.* **70**, 329 (2005), math/0211216.
90. D.-E. Diaconescu, G. W. Moore and E. Witten, *Adv.Theor.Math.Phys.* **6**, 1031 (2003), hep-th/0005090.
91. G. W. Moore and E. Witten, *JHEP* **0005**, p. 032 (2000), hep-th/9912279.
92. D. S. Freed, G. W. Moore and G. Segal, *Commun.Math.Phys.* **271**, 247 (2007), hep-th/0605198.
93. M. Duff, H. Lu and C. Pope, *Phys.Lett.* **B378**, 101 (1996), hep-th/9603037.
94. M. Cvetic and J. Halverson (2011), arXiv:1101.2907.
95. M. Berkooz, M. R. Douglas and R. G. Leigh, *Nucl.Phys.* **B480**, 265 (1996), hep-th/9606139.
96. W. Taylor, *Phys. Lett.* **B394**, 283 (1997), hep-th/9611042.
97. R. Blumenhagen, M. Cvetic, P. Langacker and G. Shiu, *Ann.Rev.Nucl.Part.Sci.* **55**, 71 (2005), hep-th/0502005.
98. S. Nagaoka and W. Taylor (To appear).
99. M. R. Douglas (1995), hep-th/9512077.
100. E. Witten, *Nucl.Phys.* **B460**, 541 (1996), hep-th/9511030.
101. W. Taylor (1997), hep-th/9801182.
102. W. Taylor and M. Van Raamsdonk, *Nucl.Phys.* **B558**, 63 (1999), hep-th/9904095.
103. W. Taylor and M. Van Raamsdonk, *Nucl.Phys.* **B573**, 703 (2000), hep-th/9910052.
104. R. C. Myers, *JHEP* **9912**, p. 022 (1999), hep-th/9910053.
105. M. Bershadsky, C. Vafa and V. Sadov, *Nucl.Phys.* **B463**, 420 (1996), hep-th/9511222.
106. J. F. Morales, C. A. Scrucca and M. Serone, *Nucl.Phys.* **B552**, 291 (1999), hep-th/9812071.
107. J. Stefanski, Bogdan, *Nucl.Phys.* **B548**, 275 (1999), hep-th/9812088.
108. G. Honecker and M. Trapletti, *JHEP* **0701**, p. 051 (2007), hep-th/0612030.

109. R. Friedman, J. Morgan and E. Witten, *Commun.Math.Phys.* **187**, 679 (1997), hep-th/9701162.
110. S. Kachru and C. Vafa, *Nucl.Phys.* **B450**, 69 (1995), hep-th/9505105.
111. M. Duff, R. Minasian and E. Witten, *Nucl.Phys.* **B465**, 413 (1996), hep-th/9601036.
112. G. Aldazabal, A. Font, L. E. Ibanez and A. Uranga, *Nucl.Phys.* **B492**, 119 (1997), hep-th/9607121.
113. L. E. Ibanez and A. Uranga (1997), hep-th/9707075.
114. S. Hellerman, J. McGreevy and B. Williams, *JHEP* **0401**, p. 024 (2004), hep-th/0208174.
115. M. Bianchi and A. Sagnotti, *Phys.Lett.* **B247**, 517 (1990).
116. M. Bianchi and A. Sagnotti, *Nucl.Phys.* **B361**, 519 (1991).
117. E. G. Gimon and J. Polchinski, *Phys.Rev.* **D54**, 1667 (1996), hep-th/9601038.
118. A. Dabholkar and J. Park, *Nucl.Phys.* **B472**, 207 (1996), hep-th/9602030.
119. E. G. Gimon and C. V. Johnson, *Nucl.Phys.* **B477**, 715 (1996), hep-th/9604129.
120. A. Dabholkar and J. Park, *Nucl.Phys.* **B477**, 701 (1996), hep-th/9604178.
121. J. Polchinski, *Phys.Rev.* **D55**, 6423 (1997), hep-th/9606165.
122. G. Aldazabal, A. Font, L. E. Ibanez and F. Quevedo, *Nucl.Phys.* **B461**, 85 (1996), hep-th/9510093.
123. G. Aldazabal, A. Font, L. E. Ibanez, A. Uranga and G. Violero, *Nucl.Phys.* **B519**, 239 (1998), hep-th/9706158.
124. A. Dabholkar and J. Park, *Phys.Lett.* **B394**, 302 (1997), hep-th/9607041.
125. R. Gopakumar and S. Mukhi, *Nucl.Phys.* **B479**, 260 (1996), hep-th/9607057.
126. P. S. Aspinwall and M. Gross, *Phys.Lett.* **B387**, 735 (1996), hep-th/9605131.
127. P. S. Aspinwall and D. R. Morrison, *Nucl.Phys.* **B503**, 533 (1997), hep-th/9705104.
128. M. Berkooz, R. G. Leigh, J. Polchinski, J. H. Schwarz, N. Seiberg *et al.*, *Nucl.Phys.* **B475**, 115 (1996), hep-th/9605184.
129. P. S. Aspinwall, *Nucl.Phys.* **B496**, 149 (1997), hep-th/9612108.
130. P. Candelas and A. Font, *Nucl.Phys.* **B511**, 295 (1998), hep-th/9603170.
131. P. Candelas, E. Perevalov and G. Rajesh, *Nucl.Phys.* **B502**, 613 (1997), hep-th/9606133.
132. P. Candelas, E. Perevalov and G. Rajesh, *Nucl.Phys.* **B507**, 445 (1997), hep-th/9704097.
133. P. Candelas, E. Perevalov and G. Rajesh, *Nucl.Phys.* **B519**, 225 (1998), hep-th/9707049.
134. K. Hulek and R. Kloosterman (2008), arXiv:0806.2025.
135. P. S. Aspinwall and D. R. Morrison, *JHEP* **9807**, p. 012 (1998), hep-th/9805206.
136. M. Bershadsky, K. A. Intriligator, S. Kachru, D. R. Morrison, V. Sadov *et al.*, *Nucl.Phys.* **B481**, 215 (1996), hep-th/9605200.
137. P. Berglund, A. Klemm, P. Mayr and S. Theisen, *Nucl.Phys.* **B558**, 178 (1999), hep-th/9805189.

138. P. S. Aspinwall, S. H. Katz and D. R. Morrison, *Adv.Theor.Math.Phys.* **4**, 95 (2000), hep-th/0002012.
139. J. Tate, *Modular functions of one variable IV, Lecture Notes in Math, vol. 476* Berlin: Springer-Verlag (1975).
140. S. H. Katz and C. Vafa, *Nucl.Phys.* **B497**, 146 (1997), hep-th/9606086.
141. A. Grassi and D. R. Morrison (2000), math/0005196.
142. D. R. Morrison and W. Taylor (To appear).
143. R. Miranda in: The birational geometry of degenerations, editors Friedman, R. and Morrison, D. R., Birkhauser (1983).
144. A. Grassi and D. R. Morrison (To appear).
145. A. Grassi, *Math.Ann.* **290**, p. 287 (1991).
146. A. Grassi, *Internat.J.Math.* **4**, p. 203 (1993).
147. J. D. Blum and A. Zaffaroni, *Phys.Lett.* **B387**, 71 (1996), hep-th/9607019.
148. S. Ferrara, J. A. Harvey, A. Strominger and C. Vafa, *Phys.Lett.* **B361**, 59 (1995), hep-th/9505162.
149. R. Friedman, *Algebraic Surfaces and Holomorphic Vector Bundles* New York: Springer-Verlag (1998).
150. N. Seiberg and E. Witten, *Nucl.Phys.* **B471**, 121 (1996), hep-th/9603003.
151. E. Witten, *Nucl.Phys.* **B471**, 195 (1996), hep-th/9603150.
152. V. Kumar, D. R. Morrison and W. Taylor, *JHEP* **1002**, p. 099 (2010), arXiv:0911.3393.
153. A. Sen, *Nucl.Phys.* **B498**, 135 (1997), hep-th/9702061.
154. S. D. Avramis and A. Kehagias, *JHEP* **0510**, p. 052 (2005), hep-th/0508172.
155. K. R. Dienes and J. March-Russell, *Nucl.Phys.* **B479**, 113 (1996), hep-th/9604112.
156. K. R. Dienes, *Nucl.Phys.* **B488**, 141 (1997), hep-ph/9606467.
157. A. Adams, N. Arkani-Hamed, S. Dubovsky, A. Nicolis and R. Rattazzi, *JHEP* **0610**, p. 014 (2006), hep-th/0602178.
158. P. S. Aspinwall, *Nucl.Phys.* **B460**, 57 (1996), hep-th/9510142.
159. M. Gross, *Duke Math.Jour.* **74**, p. 271 (1994).
160. Alekseevskii, *Math.USSR.Izvestija.* **9**, p. 297 (1975).
161. B. de Wit and A. Van Proeyen, *Commun.Math.Phys.* **149**, 307 (1992), hep-th/9112027.
162. K. Dasgupta, V. Hussin and A. Wissanji, *Nucl.Phys.* **B793**, 34 (2008), arXiv:0708.1023.
163. N. Arkani-Hamed, L. Motl, A. Nicolis and C. Vafa, *JHEP* **0706**, p. 060 (2007), hep-th/0601001.
164. K. Matsuki, *Introduction to the Mori Program* Berlin: Springer-Verlag (2002).
165. R. Donagi and M. Wijnholt (2008), arXiv:0802.2969.
166. C. Beasley, J. J. Heckman and C. Vafa, *JHEP* **0901**, p. 058 (2009), arXiv:0802.3391.
167. C. Beasley, J. J. Heckman and C. Vafa, *JHEP* **0901**, p. 059 (2009), arXiv:0806.0102.
168. J. Marsano, N. Saulina and S. Schafer-Nameki, *JHEP* **0908**, p. 030 (2009), arXiv:0904.3932.

169. R. Blumenhagen, T. W. Grimm, B. Jurke and T. Weigand, *Nucl.Phys.* **B829**, 325 (2010), arXiv:0908.1784.
170. J. J. Heckman, *Ann.Rev.Nucl.Part.Sci.* (2010), arXiv:1001.0577.
171. T. Weigand, *Class.Quant.Grav.* **27**, p. 214004 (2010), arXiv:1009.3497.
172. C. Cordova (2009), arXiv:0910.2955.
173. G. Curio and R. Y. Donagi, *Nucl.Phys.* **B518**, 603 (1998), hep-th/9801057.
174. H. Hayashi, R. Tatar, Y. Toda, T. Watari and M. Yamazaki, *Nucl.Phys.* **B806**, 224 (2009), arXiv:0805.1057.
175. J. Marsano, N. Saulina and S. Schafer-Nameki, *JHEP* **1011**, p. 088 (2010), arXiv:1006.0483.
176. T. W. Grimm and T. Weigand, *Phys.Rev.* **D82**, p. 086009 (2010), 1006.0226.
177. J. Marsano, *Phys.Rev.Lett.* **106**, p. 081601 (2011), arXiv:1011.2212.
178. M. J. Dolan, J. Marsano, N. Saulina and S. Schafer-Nameki (2011), arXiv:1102.0290.
179. A. Strominger, *Nucl.Phys.* **B274**, p. 253 (1986).
180. J.-X. Fu and S.-T. Yau, *J.Diff.Geom.* **78**, 369 (2009), hep-th/0604063.
181. K. Becker, M. Becker, J.-X. Fu, L.-S. Tseng and S.-T. Yau, *Nucl.Phys.* **B751**, 108 (2006), hep-th/0604137.
182. A. Adams, M. Ernebjerg and J. M. Lapan, *Adv.Theor.Math.Phys.* **12**, 817 (2008), hep-th/0611084.
183. M. Becker, L.-S. Tseng and S.-T. Yau, *Nucl.Phys.* **B786**, 119 (2007), hep-th/0612290.
184. K. Becker and S. Sethi, *Nucl.Phys.* **B820**, 1 (2009), arXiv:0903.3769.
185. A. Dabholkar and C. Hull, *JHEP* **0309**, p. 054 (2003), hep-th/0210209.
186. S. Kachru, M. B. Schulz, P. K. Tripathy and S. P. Trivedi, *JHEP* **0303**, p. 061 (2003), hep-th/0211182.
187. C. Hull, *JHEP* **0510**, p. 065 (2005), hep-th/0406102.
188. J. Shelton, W. Taylor and B. Wecht, *JHEP* **0510**, p. 085 (2005), hep-th/0508133.
189. B. Wecht, *Class.Quant.Grav.* **24**, S773 (2007), arXiv:0708.3984.
190. M. Duff, B. Nilsson and C. Pope, *Phys.Rept.* **130**, 1 (1986).
191. G. Papadopoulos and P. Townsend, *Phys.Lett.* **B357**, 300 (1995), hep-th/9506150.
192. B. S. Acharya and E. Witten (2001), hep-th/0109152.
193. M. R. Douglas and S. Kachru, *Rev.Mod.Phys.* **79**, 733 (2007), hep-th/0610102.
194. O. DeWolfe, A. Giryavets, S. Kachru and W. Taylor, *JHEP* **0507**, p. 066 (2005), hep-th/0505160.
195. M. R. Douglas, *JHEP* **0305**, p. 046 (2003), hep-th/0303194.
196. F. Denef and M. R. Douglas, *JHEP* **0405**, p. 072 (2004), hep-th/0404116.
197. M. R. Douglas and W. Taylor, *JHEP* **0701**, p. 031 (2007), hep-th/0606109.
198. F. Gmeiner, R. Blumenhagen, G. Honecker, D. Lust and T. Weigand, *JHEP* **0601**, p. 004 (2006), hep-th/0510170.
199. V. Rosenhaus and W. Taylor, *JHEP* **0906**, p. 073 (2009), arXiv:0905.1951.

Chapter 7

TASI Lectures on Complex Structures

Frederik Denef

Center for the Fundamental Laws of Nature, Harvard University,
17 Oxford Street, Cambridge, MA 02138, USA
Simons Center for Geometry and Physics,
Stony Brook, NY 11794-3636, USA
Institute for Theoretical Physics, University of Leuven,
Celestijnenlaan 200D, B-3001 Heverlee, Belgium
denef@physics.harvard.edu

These lecture notes give an introduction to a number of ideas and methods that have been useful in the study of complex systems ranging from spin glasses to D-branes on Calabi-Yau manifolds. Topics include the replica formalism, Parisi's solution of the Sherrington-Kirkpatrick model, overlap order parameters, supersymmetric quantum mechanics, D-brane landscapes and their black hole duals.

7.1. Introduction

7.1.1. *The case for complexity*

If you are a string theorist, chances are that you think of a complex structure as something squaring to minus one, rather than as something consisting of many intricately interacting degrees of freedom. These lecture notes, however, are concerned with the latter.

There is a deeply rooted belief in the natural sciences that the more fundamental a theory gets, the less important complexity becomes. In particular there has traditionally been an almost unquestioned assumption that physics at subatomic scales must be intrinsically simple, and that complexity is only relevant in the macroscopic or living world. Certainly there is a decrease in complexity when going from cells to proteins and from there to amino acids, atoms and finally elementary particles. It would seem logical that any step further down this reductive tree — for example explaining the origin of elementary particles and their properties, accounting for the entropy of black holes or understanding the initial conditions of the universe — should involve only structures and concepts of the utmost simplicity and elegance.

By now considerable theoretical and experimental evidence has accumulated suggesting this is wrong. In particular, whenever quantum mechanics and gravity pair up — as in cosmological dynamics on the largest space and time scales, the determination of effective low energy parameters in theories with compact extra dimensions, and the stringy microphysics of black holes — complexity appears not only unavoidable, but seems to some extent essential. On the theoretical side I am referring here more specifically to the highly complex, fractal-like iterated structures arising in eternal inflation, the perplexing complexity of the string theory landscape, and the closely related complexity of wrapped D-brane systems that, among many other applications, have given us the first quantitative explanation of the entropy of black holes.[a] On the experimental side, the most influential development has been the gathering of abundant cosmological precision data: detailed cosmic microwave spectroscopy providing convincing evidence for the slow roll inflation hypothesis, and the measurement of an impossibly tiny yet nonzero vacuum energy density, of just the right magnitude to allow self-reproducing resonances to crawl out of gravitationally collapsed dust at precisely the right time to see the onset of cosmic acceleration. Whether we like it or not, these developments, and the absence of plausible alternative explanations for these and other fine-tuning conundrums, have added significant credibility to the idea that some fine-tunings may have no other explanation than being tautological post-selection effects in a vast, eternally inflating multiverse, scanning a huge ensemble of low energy effective field theories, only a tiny fraction of which will allow structures to evolve of sufficient complexity to be capable of organizing summer schools. This idea requires no fundamentally new concepts of any kind, works by exactly the same semiclassical mechanisms as ordinary inflation, eliminates some of the worst fine-tuning problems in physics, provides in principle a precise, quantitative definition of naturalness, and appears to fit naturally into a UV complete, unifying framework incorporating all known principles of physics, that has no external parameters yet spawns all the necessary

[a] Except for very specific items, references will be given at the end of this section and in subsequent sections.

complexity, out of simple equations following unambiguously from an even simpler idea: the quantization of strings.

The price to pay for this remarkable feat is that it forces us to consider structures and concepts far more diverse and complex than those it seeks to explain, without immediate prospects of directly observing them. The Standard Models of elementary particles and cosmology is arguably a lot simpler than any of the string compactifications that could produce it; for example it could be F-theory compactified on the Calabi-Yau fourfold hypersurface in weighted projective space $\mathbb{CP}^{1,1,84,516,1204,1806}$, which boasts about three hundred thousand deformation moduli, up to seventy-five thousand D3-branes, and a flux-induced superpotential specified by a choice of almost two million integers multiplying an equal amount of independent period integrals [1, 2]. According to simple estimates, this compactification, assuming all moduli can be stabilized, gives rise to more than 10^{10^5} intricately interconnected flux compactifications. Similarly, a macroscopic extremal Reissner-Nordstrom black hole looks considerably simpler than the corresponding weakly coupled D-brane systems one is led to consider in microscopic computations of the black hole entropy. For example a wrapped D4 producing a modest extremal black hole of say the mass of the sun easily involves D4-branes with 10^{100} moduli subject to superpotentials specified by 10^{100} flux quanta. Finally, even in simple toy model landscapes, the large scale structure generated by eternal inflation is infinitely more complex than what we observe in our universe.

Although the importance and urgency of a better understanding of these matters is clear to many, there has been widespread reluctance to face this kind of stringy complexity directly. Instead, the dominant approach has been to try to isolate particular phenomena of interest, for example by studying simplified compactification models like local Calabi-Yau spaces, effective field theory models, non-disordered toy models of the landscape, D-brane models for black holes restricted to charge regimes where the dominant contributions to the entropy are highly structured by symmetry (Cardy regime), or coarse grained asymptotic parameter distributions in ensembles of compactifications allowing similar complexity-minimizing limits. There are of course obvious excellent reasons to follow this reductionist approach. However, there are also excellent reasons to try to probe complexity itself:

(1) In contrast to supersymmetric AdS compactifications, a specific choice of compactification data does in general *not* correspond to a superselec-

tion sector in cosmologies with a positive vacuum energy, as quantum and thermal fluctuations will force changes in geometric moduli, fluxes and topology. Thus, any complete, nonperturbative description of a sector of string theory rich enough to describe our own universe should be able to encode not just one compactification, but the full space of internal space configurations that can be dynamically reached in one way or another.[b] Since various topological transitions between for example Calabi-Yau manifolds are known to be perfectly sensible physical processes in string theory, the space of such interconnected configurations is likely to be huge. If we imagine for a moment that there exists a complete holographic description of eternally inflating cosmologies in string theory, say in the form of a field theory living at future infinity, then this field theory would somehow have to encode the full complexity of those googols of geometries — it would literally be a theory of everything, and it couldn't be anything less.

(2) There is often striking organization, universality and elegance emerging in disordered systems, as has become clear over the past decades in studies of spin and structural glasses, neural networks and other complex systems. Conventional notions of symmetry are largely irrelevant in such systems, but other, equally powerful structures and their associated order parameters appear in their place. This includes hierarchical cluster organization of the state space, replica symmetry breaking and overlap order parameters. Uncovering these led to highly nontrivial exact solutions of various models of complex, disordered systems. Much like symmetries in ordered systems, these structures also determine to a large extent the dynamics and other physical properties. They have a wide range of applicability in fields as diverse as condensed matter physics, neuroscience, biology and computer science, and have led to practical applications such as new efficient algorithms for optimization, data mining and artificial intelligence. Thus, rather than an annoyance to be avoided, complexity can be the essence, and the key feature to focus on. The lessons learned from these investigations are bound to be useful in the context of complex systems in string theory and cosmology.

(3) Conversely, string dualities such as holography may provide a new and useful approach to the general theory of complex systems. What makes this particularly promising compared to other instances of

[b]It is not necessary that such configurations also support metastable "vacua".

applied holography is that the natural large N limit is also the limit one is a priori interested in here. Concrete examples are given by supersymmetric branes wrapped on compact cycles in Calabi-Yau manifolds, which exhibit many of the typical characteristics of mean field models of glasses.

7.1.2. *Contents and goals of these notes*

In these lecture notes, I will give an introduction to some of the concepts and techniques which have been useful in studies of complex systems of many degrees of freedom, with intricate, disordered interactions. Although I will discuss ideas and techniques developed in the theory of glasses as well as ideas and techniques developed in string theory, there will be little or no discussion of applications of one to the other. For this I refer to the work that will appear in [3–5], and in which any new idea that might be present in these notes originated. The focus will be on the basics, meaning material usually assumed to be known in most of the recent specialized literature. Rather than to give a comprehensive review, I will treat a number of specific topics in a more or less self-contained and hopefully pedagogical way, to avoid variants of step 7 of [6], and to allow the reader to learn how to actually compute a number of things rather than to just get a flavor of the ideas. This comes at a price of having to leave out many interesting and important topics (including some that were discussed to a certain extent in my actual lectures at TASI). In particular, despite its importance in the motivation given above, I will say almost nothing directly here about the landscape of string compactifications, and focus instead on space localized D-brane systems arising as the weak coupling description of charged black holes. This has three additional reasons: (*i*) I have already written extensive lecture notes on string vacua [7], and wanted to avoid overlap altogether. (*ii*) The internal space geometry of the space-localized D-brane systems describing black holes is identical to that of space-filling D-brane systems describing compactifications, and the techniques used to analyze the former can directly be transported to the latter. The advantages of looking first at space-localized branes are furthermore numerous: They are conceptually and technically easier, allow arbitrarily large charges and hence a proper thermodynamic limit, and have holographic dual descriptions as black holes, providing effective "experimental measurements" of thermodynamic quantities like the entropy. (*iii*) It prevents, for the time being, potential diffusion of confusion from partially unresolved conceptual problems of quantum gravity in a cosmological setting (as reviewed by Tom Banks at this school [8]) into what should be an exposition of well-understood and diversely applicable methods.

More specifically, I will cover the following topics:

(1) An introduction to the theory of spin glasses, in particular Parisi's solution of the Sherrington-Kirkpatrick model, using the replica formalism. This was the first nontrivial energy landscape to be studied and understood in detail in physics. Special attention is given to Parisi's overlap order parameters, which allow to detect a nontrivial equilibrium state space structure without having to be able to explicitly know those equilibrium states. A generalization of this order parameter for arbitrary quantum systems is proposed at the end of the section. Some other complex systems such as the Hopfield model for memory and learning are briefly discussed. Other, non-replica approaches such as the cavity method and Langevin dynamics are important but not treated in any detail in these notes.

(2) An introduction to supersymmetric quantum mechanics, where simple but powerful concepts such as the Witten index make it possible to compute exact quantum ground state degeneracies of highly complex systems. The computation of nonperturbative lifting effects due to landscape barrier tunneling as well as relations to Morse theory are discussed in some detail. Finally, the map between nonsupersymmetric Langevin dynamics and supersymmetric quantum mechanics is introduced, and its appearance in glass theory is outlined.

(3) An introduction to the low energy quantum mechanics description of D-branes wrapped on compact cycles in Calabi-Yau manifolds. These are the prime examples of complex systems in string theory, and excellent models to introduce techniques ubiquitous in the field of string compactification. Again I start from scratch, assuming only some background knowledge in elementary differential geometry and the basics of D-branes in string theory. I describe in detail the low energy reduction of the D4-brane wrapped on a high degree 4-cycle. This results in a supersymmetric quantum mechanics of high complexity, which nevertheless is manageable thanks to the underlying supersymmetric geometry. If the D4 charge is of order N, the brane allows of order N^3 worldvolume magnetic fluxes, inducing an extremely complicated superpotential for its order N^3 moduli, and leading to an enormous

energy landscape with exponentially many minima. Nevertheless it is possible to explicitly construct vast numbers of exact supersymmetric critical points, by giving the critical point condition the interpretation of "capturing" holomorphic 2-cycles. At large D0-charge, this method becomes inadequate, but at the same time it becomes easier to give good estimates for the number of ground states. A formula for the index is derived by directly applying the machinery of Lagrangian supersymmetric quantum mechanics, treating the worldvolume fluxes as momenta canonically conjugate to angular variables. This reproduces the well known continuum approximation at leading order, but also provides all corrections in terms of differential euler characteristics of critical point loci. Finally, bound states with mobile D0-branes are introduced and counted in a rudimentary fashion, a microscopic formula for the D4-D0 is derived, and some open ends and generalizations are summed up to conclude.

(4) An overview of the construction of the zoo of multicentered supersymmetric black hole bound state solutions that are the holographic duals of the D-brane energy landscapes discussed in the previous section.

7.1.3. *Some general references*

For the topics covered in these lectures, I will give references along the way. For the topics mentioned above but which will not be treated in any detail in what follows, here are a few general referencesthat may be useful: [7, 9–22, 22–42].

7.2. Spin Glasses and Beyond

7.2.1. *Introduction*

A spin glass [43–45] is a system of localized spins with disordered interactions. An example is copper sprinkled with manganese atoms at random positions. The sign of the spin-spin interaction potential between the manganese atoms oscillates as a function of distance and is relatively long range, so we get effectively randomized mixed ferromagnetic and antiferromagnetic interactions. Some spin pairs will want to align, while others will wish to anti-align. If among three given spins, two pairs want to align and one pair wants to anti-align, or if all three pairs want to anti-align, one pair will not get what it wants and the triangle is called frustrated. The presence of many frustrated triangles typically leads to exponentially many local minima of the energy, and thus to a landscape.

This landscape is highly complex; the problem of finding the ground state in even the simplest spin glass models is effectively intractable. Nevertheless, as we shall see, some models of spin glasses have exactly solvable thermodynamics, thanks to a remarkable underlying mathematical structure which greatly simplifies the analysis but nevertheless leads to a very rich phase space structure. Its most striking consequence is *ultrametricity* of the state space. An ultrametric space is a metric space with a distance function d that satisfies something stronger than the standard triangle inequality, namely

$$d(a,b) \leq \max\{d(b,c), d(c,a)\} . \tag{7.1}$$

What this means is that all triangles are isosceles, with the unequal side the shortest of the three. Ultrametric spaces appear in various branches of mathematics; for example the p-adic distance[c] between rational numbers is ultrametric. For nice introductions to the mathematics of ultrametric spaces, see chapter 9 of [46], and [47]. Ultrametric spaces are also familiar in biology: If we define the distance between two current species as how far back in time one has to go to find a common ancestor, this distance is trivially ultrametric. Indeed, picking a turkey, a monkey and a donkey for example, the distance turkey-donkey and turkey-monkey is the same, while the distance monkey-donkey is the shorter one of the three. Alternatively, one can define a species distance as the degree of difference of DNA. Since this will be roughly proportional to how far back a common ancestor is found, this distance will again be essentially ultrametric. The same is true for any representative set of proteins. This is used with great success to reconstruct evolutionary trees (for a recent account see e.g. [48]). The lesson here for us is that ultrametricity is equivalent to the possibility of clustering objects and organizing them in a *hierarchical tree*, similar to the evolution tree of species.

In the case of spin glasses, the points in the metric space are the different equilibrium states, that is the different ergodic components or superselection sectors in which the standard Gibbs measure breaks up in the glass phase. We will describe this in detail in the subsequent sections, here we give a qualitative picture. The distance between different equilibrium states is given by some measure of microscopic dissimilarity, for example the sum

[c]The p-adic norm of a rational number is $|q|_p = p^{-k}$ when q can be written as $q = p^k m/n$ where m and n contain no powers of p. The p-adic distance is then $d_p(a,b) \equiv |a-b|_p$.

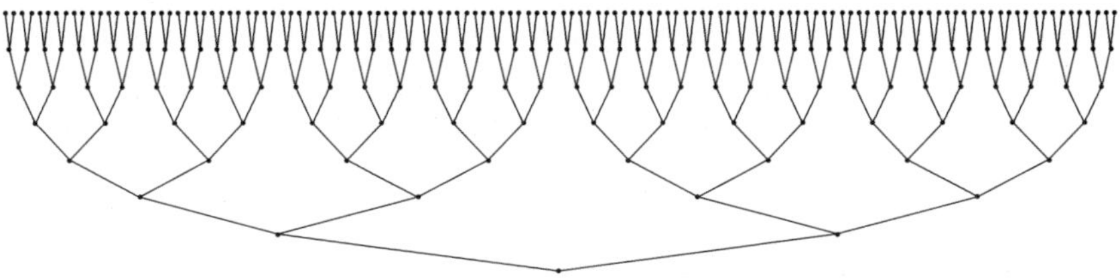

Fig. 7.1. An ultrametric tree. Distances between the points in the top layer are set by the vertical distance to the first common ancestor node.

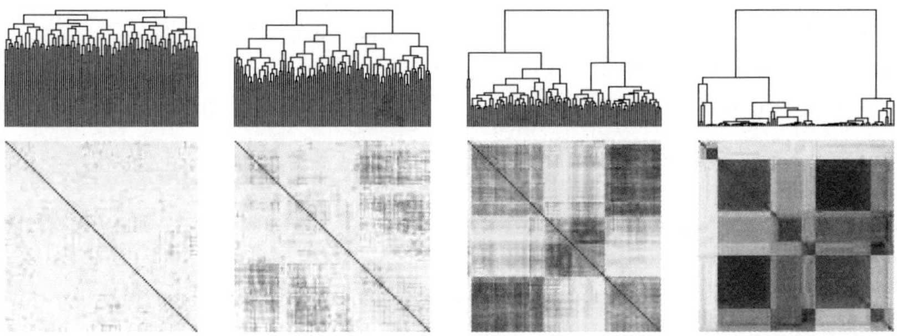

Fig. 7.2. [5] Dendrogram plots and overlap matrices for the SK model with $N = 800$ spins. We used parallel tempering Monte Carlo [53, 54] to reach thermal equilibrium, with 50 replicas at equally spaced temperatures between $T = 0.1$ and $T = 1.2$. Based on overlaps of 100 configurations sampled with separation of 100 sweeps, clustered using Mathematica. Results are shown for $T/T_c = 1.2, 0.86, 0.55, 0.12$. Red = maximal positive overlap $q = +1$ (i.e. minimal distance), white = overlap $q = 0$, dark blue = maximal negative overlap $q = -1$. Although the data is too limited to draw firm conclusions, the plots are suggestive of hierarchical clustering.

of local magnetization differences squared. Remarkably, in mean-field models such as the Sherrington-Kirkpatrick model [49], this space turns out to be ultrametric [50–52]: In the thermodynamic limit, with probability 1, triples of magnetization distances are isosceles. Equilibrium states get hierarchically organized in clusters. The cluster division is independent of the distance measure used, as long as it is statistically representative, meaning zero distance implies equal states. Figure 7.2 shows the results of a simulation illustrating this phenomenon.

What is remarkable is that such a nontrivial tree structure appears in a purely static setting, without any apparent underlying evolution process creating it.[d] Instead, the role of evolutionary time is played by temperature or more generally energy scale: At high temperatures there is only one ergodic component of phase space. When the temperature is lowered below the spin glass phase transition, the phase space starts breaking up into distinct parts that no longer talk to each other, separated by effectively unsurpassable free energy barriers. Initially these valleys in the free energy landscape will all still look very similar, but upon further cooling, their mutual magnetization distance grows, and eventually they in turn will start breaking up in components separated by free energy barriers. This branching process proceeds all the way down to zero temperature, generating an ultrametric tree of states. Any pair of ground states defines a "time" on this tree, proportional to their ultrametric distance.

As we will see, ultrametricity plays an important role in the exact solution of mean field spin glass models, and determines many of their static and dynamic properties. It plays in this sense a role similar to that of symmetries and symmetry breaking in physics. In fact, its emergence is closely related to the breaking of an auxiliary symmetry, called replica symmetry, which naturally arises in the description of disordered systems. The order parameter (or rather the order *function*) capturing the symmetry breaking pattern exhibits ultrametricity as one of its characteristic features. We will discuss this extensively in what follows.

For more realistic, local (short range) spin glass models in three dimensions, for which the mean field approximation is poor, there is no consensus on the physical presence or relevance of ultrametricity, and different schools exist with different favorite models and results, the most widespread being the droplet model of [55]. On the other hand, in systems where little or no locality is present, such as neural networks or combinatorial optimization problems, mean field methods and ultrametricity are ubiquitous [43]. A recent numerical study of ultrametricity in simple Ising type models of different degrees of locality can be found in [56].

For a review of ultrametricity in physics, see [57]. General criteria for the appearance ultrametricity were formulated in [58]. Introductory texts on spin glass theory include [43, 59–61]

[d] To be precise, the remarkable thing is the existence of a *nontrivial* ultrametric structure, with multiple branchings. Trivial ultrametricity, meaning all triangles are equilateral, is very easy to realize in infinite dimensional spaces. For example all pairs of randomly chosen points on an infinite dimensional sphere are with probability 1 at the same distance from each other.

In the following we will make the above qualitative description more precise, introduce the new order parameters and explain the mysterious but powerful replica trick and the phenomenon of replica symmetry breaking associated with ultrametricity.

7.2.2. *The Parisi order parameter*

7.2.2.1. *Spin glass models*

The most basic models for spin glasses are formally similar to the classical Ising model, with N spins $s_i = \pm 1$ interacting according to a Hamiltonian of the form[e]

$$H = \sum_{ij} J_{ij} s_i s_j \,. \tag{7.2}$$

In the case of the Ising model, we have $J_{ij} = -J$ for nearest neighbors (ij) and zero otherwise. At high temperatures the magnetization $M \equiv \frac{1}{N}\sum_i m_i$ is zero (m_i is the average value of s_i). In dimension 2 or higher, below a critical temperature T_c, the magnetization acquires a nonzero value and the $\mathbb{Z}_2$ spin flip symmetry is spontaneously broken. In the mean field approximation, $T_c = J$ and M is a solution of the mean field consistency equation $M = \tanh(MJ/T)$.

In 1975, to model spin glasses, Edwards and Anderson [62] proposed to study the above Hamiltonian, still with nearest neighbor couplings J_{ij}, but now drawn randomly and independently from a Gaussian distribution with zero mean and standard deviation J. Although at finite N everything in this model will depend on the actual values of J_{ij}, in the thermodynamic limit $N \to \infty$, intensive quantities such as the (spatially averaged) magnetization or free energy density are self-averaging: they become equal to their J_{ij}-averaged value with unit probability. This makes computations possible. Due to the disorder, the magnetization M vanishes at all temperatures in this model. Nevertheless, there exists a critical temperature below which the spins freeze, in the sense that they acquire locally preferred directions. This disordered frozen phase is the spin glass phase. The magnetization is clearly not a good order parameter to detect this situation. Instead

---

[e] In the statistical mechanics literature, the common convention is to put a minus sign in front of the sum over the spins, so ferromagnetic couplings are positive. We will not do this in these lecture notes.

Edwards and Anderson introduced a new order parameter:

$$q_{EA} \equiv \frac{1}{N} \sum_i m_i^2 \,. \tag{7.3}$$

Here the m_i are the local magnetizations in the equilibrium state the system finds itself in, which in an actual physical setting can be thought of as the time averaged value of the spin s_i. For the Ising model m_i is independent of i, so $q_{EA} = M^2$. The Edwards-Anderson parameter is related to the experimentally measured magnetic susceptibility: $\chi = (1 - q_{EA})/T$.

Though the meaning of time averaging is physically clear, we have not specified any dynamics for this model, so we need a purely static definition for m_i. This is subtle. It is *not* the canonical ensemble average $\langle s_i \rangle$ obtained from the Gibbs probability measure $p(s) \propto e^{-\beta H(s)}$, as this average is trivially zero due to the $\mathbb{Z}_2$ symmetry of H. The same issue arises already for the Ising model, but there it is clear what is going on: below T_c the Gibbs probability measure splits in two "superselection" sectors, characterized by opposite values of the magnetization. The two sectors can be separated by switching on a small background magnetic field h, which shifts $H \to H + h \sum_i s_i$, lifting the $\mathbb{Z}_2$ degeneracy and eliminating one of the two sectors, depending on the sign of h. We can then define the magnetization for each sector as $m_{i\pm} \equiv \lim_{h \to 0\pm} \lim_{N \to \infty} \langle s_i \rangle_h$.

For spin glasses this does not work; since $M = 0$, switching on a constant h will not lift the degeneracy. Switching on an inhomogeneous field h_i tailored to the m_i of the frozen equilibrium state we want to single out would of course do the job, but it is impossible to know in advance what the required profile is going to be; it could be anything, and it will be different for different values of J_{ij}. To make things worse, there may be many distinct equilibrium states, all with different values of the m_i, and again there is no way of telling without actually solving the system. There is no standard symmetry and no order apparent in the system at any temperature. What "order" parameter could possibly distinguish between those phases?

These questions were considerably sharpened and then answered through the study of an even more simplified spin glass model, proposed by Sherrington and Kirkpatrick [49] in 1975. The Hamiltonian is as above except that now the couplings J_{ij} are nonzero for *all* pairs of spins, not just nearest neighbors; they are all independent Gaussian random variables with zero mean and variance $J^2/2N$,[f]

$$\overline{J_{ij}J_{kl}} = \tfrac{J^2}{2N}\delta_{ik}\delta_{jl}\,. \tag{7.4}$$

Overlines will denote averages over the J_{ij} throughout these notes. The model is completely nonlocal; spatial geometry has no meaning in this setup. It is exactly solvable, but the solution turned out much more interesting than Sherrington and Kirkpatrick originally imagined. The simple solution they originally proposed was correct at high temperatures, but manifestly wrong near $T = 0$, where it predicted for instance negative entropy densities. After considerable effort, the correct solution was finally found in 1979, in a seminal breakthrough by Parisi [50, 51]. (It took another 25 years before the solution was rigorously proven to be correct [63, 64].) In the course of the process, Parisi uncovered the proper order parameters to fully describe the spin glass phase, as well as a hidden "statistical" symmetry group whose breaking they parametrize. We will describe the solution in section 7.2.3, but first we describe the physical meaning of the order parameter.[g] To do this we must introduce the notion of thermodynamic pure states.

7.2.2.2. *Pure states*

The probability measure characterizing the Gibbs state is

$$p_G(s) = \frac{1}{Z}\,e^{-\beta H(s)}\,, \quad Z = \sum_s e^{-\beta H(s)}\,. \tag{7.5}$$

For a given probability measure p, we denote the expectation value of an observable $A(s)$ by $\langle A\rangle_p$. For the Gibbs state we usually just write $\langle A\rangle_G = \langle A\rangle$. We say p satisfies cluster decomposition if correlation functions of local observables factorize in the thermodynamic limit for almost all points. That is, for any finite r, in the limit $N \to \infty$,

$$\langle A_{i_1}B_{i_2}\cdots C_{i_r}\rangle_p = \langle A_{i_1}\rangle_p\langle B_{i_2}\rangle_p\cdots\langle C_{i_r}\rangle_p + R_{i_1 i_2\cdots i_r}\,, \tag{7.6}$$

where the remainder R is negligible on average: $\lim_{N\to\infty}\frac{1}{N^r}\sum_{i_1\cdots i_r}\times |R_{i_1\cdots i_r}| = 0$. Here A_i, B_i, ..., C_i are local observables, like for example $A_i = s_i$ or $B_i = s_{i-1}s_is_{i+3}$. In particular the clustering property

[f]The variance is scaled with N such that the typical size of the interaction potential at each lattice site remains finite in the thermodynamic limit: $\overline{(\sum_j J_{ij}s^j)^2} = J^2/2$. The factor of 2 is added for consistency with standard conventions, in which usually $J_{ij} = J_{ji}$ is imposed, which we will not do.

[g]Historically it was first obtained as a formal mathematical object in the replica formalism. The physical interpretation was given later in [65].

implies that intensive quantities like the free energy density have definite, non-fluctuating values. In local theories (7.6) is equivalent to the property that correlation functions factorize in the limit of infinite spatial separation.

The Gibbs measure for the Ising model satisfies cluster decomposition above the critical temperature, but not below: at high temperatures we have $\langle s_i s_j \rangle = 0 = \langle s_i \rangle \langle s_j \rangle$ when $|i - j| \to \infty$, but at low temperatures we have instead $\langle s_i s_j \rangle = M^2 \neq \langle s_i \rangle \langle s_j \rangle = 0 \cdot 0$. However we can canonically decompose the Gibbs measure into a direct sum of probability measures which do satisfy cluster decomposition. Thus, for the Ising example, we have below the critical temperature

$$p_G(s) = \tfrac{1}{2} p_+(s) + \tfrac{1}{2} p_-(s) \,, \tag{7.7}$$

where $p_\pm(s) = \lim_{h \to \pm 0} \lim_{N \to \infty} \frac{1}{Z} e^{-\beta(H(s) + h \sum_i s_i)}$. Sets of finite measure according to p_+ have zero measure according to p_- and vice versa. The measures $p_\pm$ do satisfy cluster decomposition: $\langle s_i s_j \rangle_+ = M^2 = \langle s_i \rangle_+ \langle s_j \rangle_+$. The superselection sectors described by $p_\pm$ are called "pure states" in the statistical mechanics literature.

Rigorously defining pure states for general systems is subtle. For a discussion in the context of spin glasses we refer to page 89 (appendix 1) of [66]. What is known rigorously is that any probability measure p_G defined for a system of *infinite* size, which locally behaves like the Gibbs measure (i.e. it gives relative probabilities proportional to $e^{-\beta \Delta H}$ for finite size fluctuations), can always be uniquely decomposed into pure states p_α as

$$p_G(s) = \sum w_\alpha \, p_\alpha(s) \,. \tag{7.8}$$

Here $w_\alpha > 0$ is the Gibbs probability of being in the pure state α. The pure states satisfy the clustering property and cannot be further decomposed. This gives them an implicit definition.[h]

7.2.2.3. *Overlap distributions*

As mentioned earlier, for disordered systems, it is in general impossible to explicitly find the actual decomposition into pure states. But granting the decomposition exists, one can formally define the *overlap* $q_{\alpha\beta}$ between pure

[h] More intuitively, in physical systems, pure states correspond to distinct ergodic components, individually invariant under time evolution and not further decomposable into smaller time invariant components. In ergodic components the ergodic theorem implies that the time average of any observable equals its ensemble (phase space) average.

states as

$$q_{\alpha\beta} \equiv \frac{1}{N}\sum_i \langle s_i\rangle_\alpha \langle s_i\rangle_\beta \,. \tag{7.9}$$

As a special case, notice that $q_{\alpha\alpha}$ is nothing but the Edwards-Andersen order parameter q_{EA} defined in (7.3). Although not at all obvious, it turns out [67] that $q_{\alpha\alpha}$ is independent of the pure state α (and of the disorder realization J_{ij}), so q_{EA} is actually an invariant of the system, depending only on T/J.[i] Using this, we see that the overlap is closely related to the Euclidean distance between states:

$$d(\alpha,\beta)^2 = \frac{1}{N}\sum_i (m_{i\alpha} - m_{i\beta})^2 = 2(q_{EA} - q_{\alpha\beta})\,. \tag{7.10}$$

At this point the overlap matrix may still seem like an abstract, incomputable quantity. However, consider the *overlap probability distribution*:

$$P(q) = \sum_{\alpha\beta} w_\alpha w_\beta \,\delta(q - q_{\alpha\beta})\,. \tag{7.11}$$

This is the probability of finding an overlap $q_{\alpha\beta} = q$ when one samples the Gibbs state. The wonderful thing is that this quantity is actually computable without any knowledge of the actual decomposition into pure states. It can be rewritten purely in terms of the Gibbs state as the overlap distribution for two identical replicas of the system, with spins $s^{(1)}$ and $s^{(2)}$:

$$P(q) = \left\langle \delta\left(q - \tfrac{1}{N}\textstyle\sum_i s_i^{(1)} s_i^{(2)}\right)\right\rangle_{n=2}\,, \tag{7.12}$$

where $n = 2$ means we are considering two replicas:

$$\left\langle A(s^{(1)}, s^{(2)})\right\rangle_{n=2} \equiv \frac{1}{Z^2}\sum_s e^{-\beta H(s^{(1)}) - \beta H(s^{(2)})}\, A(s^{(1)}, s^{(2)})\,. \tag{7.13}$$

---

[i]Since the SK model is completely nonlocal and does not a priori distinguish any site j relative to a given site i, we can think of the permutation symmetry acting on the site indices i as a gauge symmetry, similar to the diffeomorphism group in gravity (where J_{ij} is the analog of the spacetime metric). Then the local magnetization distribution $\mathcal{P}(m) \equiv \frac{1}{N}\sum_i \delta(m - m_{i\alpha}) = \overline{\delta(m - m_{i\alpha})}$ encodes all gauge invariant information based on the "vevs" $m_{i\alpha}$. It is shown in [67] that again this quantity is independent of the state α. So we can say that there is no gauge invariant distinction between different pure states based on just the magnetizations; all equilibrium states look the same as far as non-fluctuating quantities are concerned.

To prove this, one shows the moments are equal. Consider for example its second moment:

$$
\begin{aligned}
\langle q^2 \rangle &= \int dq\, P(q)\, q^2 \\
&= \sum_{\alpha\beta} w_\alpha w_\beta \, \frac{1}{N} \sum_i \langle s_i \rangle_\alpha \langle s_i \rangle_\beta \, \frac{1}{N} \sum_j \langle s_j \rangle_\alpha \langle s_j \rangle_\beta \\
&= \frac{1}{N^2} \sum_{\alpha\beta} w_\alpha w_\beta \sum_{ij} \langle s_i s_j \rangle_\alpha \langle s_i s_j \rangle_\beta \\
&= \frac{1}{N^2} \sum_{ij} \langle s_i s_j \rangle \langle s_i s_j \rangle \\
&= \frac{1}{N^2} \left\langle \left(\sum_i s_i^{(1)} s_i^{(2)} \right)^2 \right\rangle_{n=2} .
\end{aligned}
$$

To get to the third line we used the clustering property (7.6), and for the fourth line we used (7.8) in reverse to express everything in terms of Gibbs state expectation values. The higher moments are treated analogously.

Equation (7.12) in principle allows us to compute or approximate $P(q)$ by standard methods, for example by using Monte Carlo simulations sampling the Boltzmann-Gibbs distribution. It captures the presence and properties of pure states very well and is therefore a good order parameter for the spin glass phase. If only one pure state is present, namely the Gibbs state itself, as is the case at high temperatures, then $q_{11} = q_{EA} = 0$ due to the $\mathbb{Z}_2$ symmetry, and $P(q) = \delta(q)$. If the system freezes and splits into one $\mathbb{Z}_2$ pair of pure states, with local magnetizations m_i and $-m_i$, then for both states the self-overlap is $q_{EA} = \frac{1}{N}\sum_i m_i^2$ and $P(q) = \frac{1}{2}\delta(q - q_{EA}) + \frac{1}{2}\delta(q + q_{EA})$. If there are many pure states, there will be many delta-functions, and if there is an infinite number of them, $P(q)$ becomes continuous.

Computing $P(q)$ analytically for a given coupling matrix J_{ij} is not possible. It is however possible to compute its average $\overline{P(q)}$, by a variety of methods that can also be used to compute more basic thermodynamic quantities; the most prominent ones are the replica, cavity (TAP), Langevin dynamics and supersymmetry methods [43, 59, 69–72]. In these lecture notes we will focus on the replica method. To already get an idea of where all this is heading, a result of such computations is shown in Fig. 7.3, displaying the disorder-averaged $\overline{P(q)}$ for various temperatures below T_c. The function is smooth except for a delta-function peak at the highest value of q, which is the self-overlap $q_{\alpha\alpha} = q_{EA}$. Its presence is a consequence of the state

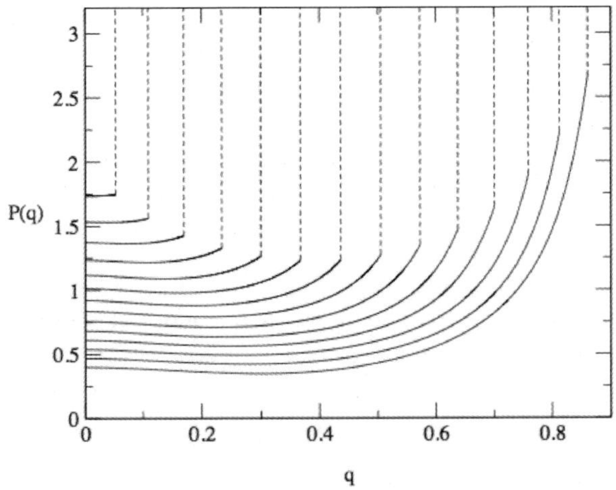

Fig. 7.3. From [68]. Disorder-averaged overlap distribution $\overline{P(q)}$ for the SK model at various temperatures, from $T = 0.95$ on right to $T = 0.30$ on left in steps of 0.05 ($T_c = 1$). The dotted lines represent delta-functions, localized at $q_{\alpha\alpha} = q_{EA}$.

and disorder independence of q_{EA}, leading to a term $(\sum_\alpha w_\alpha^2)\delta(q - q_{EA})$ in (7.11). The nonvanishing of $\overline{\sum_\alpha w_\alpha^2}$ in turn indicates the presence of states of finite weight w_α. The continuity of $\overline{P(q)}$ shows that there are in general multiple pure states, but it does not imply $P_J(q)$ is continuous at fixed J; in fact because some w_α are finite, $P_J(q)$ must have multiple delta peaks, but with J-dependent locations, so integrating over J smooths them out. Indeed, unlike the magnetization or the free energy, $P_J(q)$ is not a self-averaging quantity; it fluctuates between different realizations of the disorder.

7.2.2.4. *Ultrametricity*

One can also define more refined overlap distributions, such as the overlap triangle distribution

$$P(q_1, q_2, q_3) \equiv \sum_{\alpha\beta\gamma} w_\alpha w_\beta w_\gamma \, \delta(q_1 - q_{\beta\gamma})\delta(q_2 - q_{\gamma\alpha}) \, \delta(q_3 - q_{\alpha\beta}) \,, \qquad (7.14)$$

which similar to (7.12) can be written in terms of the overlap distribution of three independent replicas in the Gibbs state. Restricting to triples with positive overlaps,[j] one finds [52]

$$P(q_1, q_2, q_3) = \frac{1}{2}\int_0^{q_1} dq' P(q') P(q_1)\,\delta(q_1 - q_2)\,\delta(q_2 - q_3) \\ + \frac{1}{2}\Big(P(q_1)P(q_2)\,\theta(q_1 - q_2)\,\delta(q_2 - q_3) + \text{permutations}\Big)\,, \tag{7.15}$$

where θ is the step function and $P(q)$ is as in (7.11). This manifestly exhibits ultrametricity, with the first term encoding equilateral and the last three terms more general isosceles triangles.

7.2.3. *Replica solution of the SK model*

7.2.3.1. *The replica trick*

We now turn to the actual computations in the SK model, first using the replica formalism [49–51, 62]. The Hamiltonian is, as stated earlier:

$$H_J[s] = \sum_{i \neq j} J_{ij} s_i s_j\,, \tag{7.16}$$

where $i, j = 1, \ldots, N$, $s_i = \pm 1$ and J_{ij} is drawn randomly out of a Gaussian distribution with mean 0 and variance $\frac{J^2}{2N}$, i.e.

$$p(J_{ij}) \propto e^{-\frac{N}{J^2} J_{ij}^2}\,. \tag{7.17}$$

Other distributions would give equivalent results; only the first two moments matter. It is impossible to solve this system exactly for arbitrary given couplings J_{ij} — just finding the absolute minimum energy configuration is already an NP-hard problem [73]. Fortunately the usual intensive thermodynamic quantities of interest are *self-averaging*, that is they are independent of the random matrix J_{ij} with probability going to 1 in the thermodynamic limit. Thus we can compute them by computing averages. For example the free energy density is

$$F = \overline{F_J} \equiv \int \prod_{ij} dJ_{ij}\, p(J_{ij})\, F_J\,, \quad F_J \equiv -\frac{1}{\beta N} \log Z_J\,, \quad Z_J \equiv \sum_{s \in \{\pm 1\}^N} e^{-\beta H_J[s]}\,. \tag{7.18}$$

[j] This is necessary because without this restriction, the $\mathbb{Z}_2$ symmetry trivially destroys the ultrametric structure: A sign flip of the spins in state α will flip the sign of two out of three overlaps of an $(\alpha\beta\gamma)$ triangle, and clearly this does not preserve isoscelesness. This also explains the structure in the blue regions of Fig. 7.2.

Crucially, the average $\overline{\log Z_J}$ has to be computed *after* taking the logarithm. Such an average is called *quenched*, the disorder represented by the J_{ij} is called quenched disorder, etc. Computing the average first, i.e. on the partition function itself, is called *annealed* averaging. Physically, this corresponds to a situation in which the couplings themselves are fluctuating variables. The annealed average is much easier to compute than the quenched average, but it is not the situation we are in; in a real spin glass for example the couplings are determined by the positions of the impurity atoms in the host crystal, which vary randomly over space but do not fluctuate in time, or if they do, on much longer time scales than the spins fluctuate.

To deal with the quenched average, one can use the replica trick. This is based on the observation that

$$\log Z = \lim_{n\to 0} \frac{Z^n - 1}{n} = \frac{\partial}{\partial n} Z^n|_{n=0}\,, \tag{7.19}$$

together with the fact that for positive integers n, the average of Z^n is just an annealed average of the disorder coupled to n replicas of the original system, and therefore easy to compute. Of course, the subtle part is the "analytic continuation" to $n = 0$, which is hard to make rigorous.[k] However, its effective power in a large range of applications is undeniable, so the lack of a rigorous framework is no doubt more a failure of our arsenal of rigorous frameworks than of the idea itself.

Introducing replica indices $a, b = 1, \ldots, n$, the coupling-averaged n-fold replicated partition function is, putting $J = 1$,

$$\overline{Z^n} = \int dJ_{ij}\, e^{-NJ_{ij}^2} \sum_{s\in\{\pm 1\}^{nN}} e^{-\beta J_{ij} s_a^i s_a^j}\,. \tag{7.20}$$

Repeated indices are summed over and we absorb the Gaussian normalization factors in the measure dJ_{ij}. At this point, the replicas do not interact with each other. However, because they all couple to the same J_{ij}, inte-

[k] A perhaps somewhat more precise version starts from the observation that Z^n is an entire function of n, with Taylor expansion $Z^n = e^{n\log Z} = \sum_{k=0}^{\infty} \frac{n^k}{k!} (\log Z)^k$ valid for all n, so $\overline{Z^n} = 1 + n\,\overline{\log Z} + \cdots$ for all n. The idea is then that the term linear in n can be extracted by considering arbitrary integers n and computing $\overline{Z^n}$ as an expansion in n. However this is in principle ambiguous: for example $1 + \sin(\pi n)$ is entire and evaluates to 1 for all integers, yet it has a nonvanishing first order term in its Taylor expansion. For a deeper analysis see e.g. [74, 75].

grating out the disorder induces effective interactions between the replicas:

$$\overline{Z^n} = \sum_{s \in \{\pm 1\}^{nN}} e^{\frac{\beta^2}{4N} s_a^i s_a^j s_b^i s_b^j} . \tag{7.21}$$

The positive sign in the exponent means the interaction is attractive. Now, reverse-mimicking how replicas got coupled by integrating out the site coupling matrix J_{ij}, we *de*couple the lattice sites by integrating *in* a "replica coupling matrix" Q_{ab}:

$$\begin{aligned} \overline{Z^n} &= \sum_{s \in \{\pm 1\}^{nN}} \int dQ \, e^{-N \frac{\beta^2}{4} (Q_{ab})^2} \, e^{\frac{\beta^2}{2} Q_{ab} s_a^i s_b^i} \\ &= \int dQ \, e^{-N \frac{\beta^2}{4} (Q_{ab})^2} \Bigg(\sum_{S \in \{\pm 1\}^n} e^{\frac{\beta^2}{2} Q_{ab} S_a S_b} \Bigg)^N . \end{aligned} \tag{7.22}$$

The S-sum here runs over the n replica copies only; lattice indices no longer occur. The sum over the lattice sites has simply produced an overall power of N. This is awesome because it means that in the thermodynamic limit $N \to \infty$, we can evaluate the integral over Q in a saddle point approximation. It all boils down now to finding the critical points of

$$\mathcal{F}(Q) \equiv \frac{\beta}{4} Q_{ab}^2 - \frac{1}{\beta} \log \mathcal{Z}(Q) , \qquad \mathcal{Z}(Q) \equiv \sum_{S \in \{\pm 1\}^n} e^{\frac{\beta^2}{2} Q_{ab} S_a S_b} . \tag{7.23}$$

Denoting the dominant critical point(s) of $\mathcal{F}(Q)$ by $Q_\star$, we thus get $\overline{Z^n} = e^{-\beta N \mathcal{F}(Q_\star)}$. If we can find the saddle points for general n, we are done. Using the trick (7.19) and noting that consistency with $Z^0 = 1$ requires $\mathcal{F}(Q_\star)|_{n=0} = 0$, we obtain the free energy density:

$$F = \frac{\partial}{\partial n} \mathcal{F}(Q_\star)|_{n=0} . \tag{7.24}$$

To summarize, what we have done is trade summing over the lattice for summing over replicas, which allows us to do a saddle point computation of the quenched average. Before proceeding to find the solution, we pause to ponder the meaning of the matrix Q_{ab} that apparently captures the large N behavior of the model.

7.2.3.2. *Meaning of Q_{ab}*

So far the physical meaning of the matrix Q_{ab} is obscure, although the conspicuous notational similarity with the overlap matrix $q_{\alpha\beta}$ introduced in (7.9) suggests the two are related. We will now show that this is indeed the case, in that its structure in the limit $n \to 0$ will produce for us the Parisi order parameter $P(q)$ defined in (7.11), or more precisely $\overline{P(q)}$. To see this, first observe that the saddle point equations $\mathcal{F}'(Q_\star) = 0$ can be written as the self-consistency equation

$$Q_{\star ab} = \langle S_a S_b \rangle_{Q_\star} \equiv \frac{1}{\mathcal{Z}(Q_\star)} \sum_{S \in \{\pm 1\}^n} S_a S_b \, e^{\frac{\beta^2}{2} Q_{\star cd} S_c S_d} \,. \tag{7.25}$$

Hence Q_{ab} equals the overlap *in replica space*. Now consider the expression (7.12) for $P(q)$:

$$P(q) = \sum_{s_1, s_2 \in \{\pm 1\}^N} \frac{e^{-\beta(H[s_1]+H[s_2])}}{Z^2} \, \delta(q - \frac{1}{N} \sum_{i=1}^N s_1^i s_2^i) \,. \tag{7.26}$$

This depends on the random couplings J_{ij}. We want to average it but are facing a problem similar to the problem we had when we wanted to compute the average of $\log Z$: now the problem is the explicit appearance of J in the factors $1/Z$. We deploy again the replica trick, this time based on

$$P(q) = \lim_{n \to 0} \sum_{s \in \{\pm 1\}^{nN}} e^{-\beta \sum_{a=1}^n H[s_a]} \, \delta(q - \frac{1}{N} \sum_{i=1}^N s_1^i s_2^i) \,, \tag{7.27}$$

where we have singled out the first two replicas to appear in the delta function. Indeed, the right hand side can for $n \geq 2$ be written as

$$Z^{n-2} \sum_{s_1, s_2 \in \{\pm 1\}^N} e^{-\beta(H[s_1]+H[s_2])} \, \delta(q - \frac{1}{N} \sum_{i=1}^N s_1^i s_2^i) \,, \tag{7.28}$$

formally reducing to (7.26) when continued to $n \to 0$. Since in (7.27) there are no longer any denominators, we can compute the disorder average by simple Gaussian integration.

Using (7.27) and manipulations similar to those leading up to (7.22), one computes the finite moments $\overline{\langle q^k \rangle} \equiv \int dq \, \overline{P(q)} \, q^k$. Dropping $1/N$ suppressed terms:

$$\begin{aligned}
\overline{\langle q^k \rangle} &= \lim_{n \to 0} \int dQ \, e^{-\frac{\beta^2 N}{4} (Q_{ab})^2} \Big(\sum_{S \in \{\pm 1\}^n} e^{\frac{\beta^2}{2} Q_{ab} S_a S_b} \Big)^{N-k} \\
&\quad \times \Big(\sum_{S \in \{\pm 1\}^n} e^{\frac{\beta^2}{2} Q_{ab} S_a S_b} \, S_1 S_2 \Big)^k \\
&= \lim_{n \to 0} \int dQ \, e^{-\beta N \mathcal{F}(Q)} \Big(\sum_{S \in \{\pm 1\}^n} \frac{e^{\frac{\beta^2}{2} Q_{ab} S_a S_b}}{\mathcal{Z}(Q)} S_1 S_2 \Big)^k \,,
\end{aligned} \tag{7.29}$$

where we have absorbed in the measure dQ a normalization factor $1/\int dQ\, e^{-\beta N \mathcal{F}(Q)}$, which ensures $\overline{\langle q^0 \rangle} = 1$. The functions $\mathcal{Z}(Q)$ and $\mathcal{F}(Q)$ were defined in (7.23). We may replace $S_1 S_2$ by $S_a S_b$ with $a \neq b$ and average over replicas, since there is no distinction between replicas. The resulting expression has the advantage of being the same at distinct saddle points related by the permutation symmetry of the replicas. In view of (7.25), the saddle point evaluation then gives

$$\overline{\langle q^k \rangle} = \lim_{n\to 0} \sum_{[Q_\star]} W_{Q_\star} \frac{1}{n(n-1)} \sum_{a\neq b} Q_{\star ab}^k = -\lim_{n\to 0} \sum_{[Q_\star]} W_{Q_\star} \sum_{b\neq 1} Q_{\star 1b}^k \,. \quad (7.30)$$

The sum over $Q_\star$ is a sum over *distinct* replica permutation symmetry orbits, i.e. it is nontrivial only if there are different saddle points *not* related by the permutation symmetry of the replicas, as may be the case when there are additional symmetries beyond the replica permutation symmetry. For each such orbit $Q_\star$, $W_{Q_\star}$ denotes its relative weight. In the case at hand, we do have an additional $\mathbb{Z}_2$ spin flip symmetry in the system. If this is the only other source of saddle degeneracy, we have $W_{Q_\star} = \frac{1}{2}$. In general we conclude:

$$\overline{P(q)} = -\lim_{n\to 0} \sum_{[Q_\star]} W_{Q_\star} \sum_{b\neq 1} \delta(q - Q_{\star 1b})\,, \quad (7.31)$$

and if there is no degeneracy besides the $\mathbb{Z}_2$ one (as one would expect generically), this becomes

$$\overline{P(q)} = -\lim_{n\to 0} \sum_{b\neq 1} \tfrac{1}{2}\delta(q - Q_{\star 1b}) + \tfrac{1}{2}\delta(q + Q_{\star 1b})\,. \quad (7.32)$$

Thus the pure state overlap distribution will be entirely determined by the saddle point solution $Q_\star$ continued to $n \to 0$. To obtain the probability to find an overlap q, all we need to do is compute the fraction of entries $Q_{\star 1b}$ that are equal to q. The weird looking minus sign is not a typo: it must be there because $\sum_{b\neq 1} 1 = (n-1) \to -1$ when $n \to 0$. Still, counting the number of entries Q_{1b} in a 0×0 matrix may cause a feeling of unease in the reader. In the following section we will try to make sense of this, and conclude that, paradoxically, the properly continued 0×0 matrix Q_{ab} in fact has infinitely many degrees of freedom!

7.2.3.3. *Replica symmetric solution*

Finding the critical points of $\mathcal{F}(Q)$ for general n is still a nontrivial task. Some obvious consequences of (7.25) are $Q_{\star aa} = 1$ and the fact that it is a positive definite symmetric matrix, but to make further progress, one must make an ansatz for the form of the solution. The simplest one, which was the one used by Sherrington and Kirkpatrick [49], is the replica symmetric (RS) ansatz, which is the unique ansatz leaving the permutation symmetry unbroken (we will drop the explicit $\star$ subscripts from here on):

$$Q_{ab} = u\delta_{ab} + q(1-\delta_{ab})\,. \tag{7.33}$$

So $Q_{ab} = u$ if $a = b$ and $Q_{ab} = q$ if $a \neq b$. The relation (7.32) then gives

$$\overline{P(q')} = \frac{1}{2}\big(\delta(q'-q) + \delta(q'+q)\big)\,. \tag{7.34}$$

Thus the RS ansatz is equivalent to assuming not more than one $\mathbb{Z}_2$ pair of pure states. If $q = 0$, there is just one pure state and the system is in the paramagnetic phase.

We wish to extremize $\mathcal{F}(Q) = \frac{\beta}{4}Q_{ab}^2 - \frac{1}{\beta}\log\sum e^{\frac{\beta^2}{2}Q_{ab}S_aS_b}$. From the ansatz (7.33) we obtain

$$Q_{ab}^2 = nu^2 + n(n-1)q^2\,, \qquad e^{\frac{\beta^2}{2}Q_{ab}S_aS_b} = e^{\frac{\beta^2}{2}\left(n(u-q)+q(\sum_a S_a)^2\right)}\,. \tag{7.35}$$

Assuming without loss of generality $q \geq 0$, we linearize the term involving S_a in the exponent with the transformation

$$\sum_{S\in\{\pm1\}^n} e^{\frac{\beta^2}{2}q(\sum_a S_a)^2} = \frac{1}{\sqrt{2\pi q}}\int dz\, e^{-\frac{z^2}{2q}} \sum_{S\in\{\pm1\}^n} e^{z\beta\sum_a S_a} \tag{7.36}$$

$$= \frac{1}{\sqrt{2\pi q}}\int dz\, e^{-\frac{z^2}{2q}}\,[2\cosh(\beta z)]^n \tag{7.37}$$

$$= 1 + n\,\frac{1}{\sqrt{2\pi q}}\int dz\, e^{-\frac{z^2}{2q}}\log[2\cosh(\beta z)] + \mathcal{O}(n^2)\,, \tag{7.38}$$

whence

$$\partial_n\mathcal{F}|_{n=0} = \frac{\beta}{4}(u^2-q^2) - \frac{\beta}{2}(u-q) - \frac{1}{\beta}\frac{1}{\sqrt{2\pi q}}\int dz\, e^{-\frac{z^2}{2q}}\log[2\cosh(\beta z)]\,. \tag{7.39}$$

This is to be extremized with respect to u and q. Extremizing u is trivial:

$$u = 1\,, \tag{7.40}$$

reproducing the result $Q_{aa} = 1$ we arrived at earlier directly from (7.25). We set $u = 1$ in what follows. There is no simple closed form solution for the saddle point value of q, so let us consider limiting cases. In the high temperature limit $\beta \to 0$, we expand in powers of β:

$$\partial_n \mathcal{F}|_{n=0} = -\frac{\beta}{4}\left(1+q^2\right) - \frac{\log 2}{\beta} + \mathcal{O}(\beta^2)\,. \tag{7.41}$$

The extremum is at $q=0$: As expected for high temperatures the system is in its paramagnetic phase. The free energy and entropy densities at $q=0$ are, from (7.24):

$$F = -\frac{\log 2}{\beta} - \frac{\beta}{4}\,, \qquad S = \beta^2 \partial_\beta F = \log 2 - \frac{\beta^2}{4}\,. \tag{7.42}$$

The $\log 2$ corresponds the two-fold spin degeneracy at each lattice site.

Below a critical temperature T_c, $q=0$ ceases to be the thermodynamically stable saddle point. To find this transition point, we expand (7.39) for small q

$$\partial_n \mathcal{F}|_{n=0} = -\frac{\log 2}{\beta} - \frac{\beta}{4} + \frac{\beta\left(\beta^2-1\right)}{4}\,q^2 - \frac{\beta^5}{3}\,q^3 + \mathcal{O}(q^4)\,. \tag{7.43}$$

The coefficient of q^2 changes sign when $\beta = 1$, and a new saddle point with $q \approx \beta - 1 > 0$ becomes the thermodynamically stable one.[1] This signals the spin glass phase transition. Recalling we chose units such that the coupling variance parameter $J=1$, we conclude

$$T_c = J\,. \tag{7.44}$$

This seems all fine, and is consistent with Monte Carlo simulations. However, in the low temperature limit $\beta \to \infty$ something awkward happens. We have

$$\partial_n \mathcal{F}|_{n=0} = -\frac{(1-q)^2}{4}\,\beta - \sqrt{\frac{2q}{\pi}} + \mathcal{O}(\beta^{-2})\,. \tag{7.45}$$

The relevant saddle point is at $q = 1 - \sqrt{\frac{2}{\pi}\frac{1}{\beta}}$. This means the self-overlap q_{EA} of the pure states approaches 1, as expected. But the free energy and

[1]What "stable" means in the limit $n \to 0$ is actually rather subtle [76]. The reader may notice for example that in the case at hand, the stable saddle points are local *maxima* in q-space, whereas ordinarily in physical parameter spaces, stable equilibria are those that *minimize* the free energy. This "inverted" rule finds its origin in the formally negative dimension of the fluctuation modes of Q_{ab} in the limit $n=0$. For example, the expression $U = \frac{1}{n}\sum_{a,b} Q_{ab}^2$ is manifestly positive definite for positive integers n. It has a positive definite Hessian (the $n^2 \times n^2$ unit matrix) for all n and the $Q=0$ extremum of U is stable. Nevertheless, inserting the RS ansatz (7.33) gives $U(q) = 1 - q^2$. It is in general nontrivial to do a full stability analysis in the replica formalism. For more discussion see [59, 76].

entropy density are, to leading order

$$F = -\sqrt{\frac{2}{\pi}} + \frac{1}{2\pi\beta} \approx -0.798\,, \qquad S = -\frac{1}{2\pi} \approx -0.159\,. \tag{7.46}$$

The free energy disagrees with numerical simulations, which indicate $F \approx -0.76$ at $T = 0$. More dramatically, the negative entropy clearly does not make any sense. Therefore the solution must be wrong at low temperatures. The replica symmetric ansatz (7.33) is apparently too restrictive. The replica symmetry must be broken.

7.2.3.4. *Replica symmetry breaking and Parisi matrices*

We want to relax the replica symmetric ansatz (7.33), which was $Q_{ab} = \delta_{ab} + q(1 - \delta_{ab})$.[m]

The effective replica Hamiltonian $H_R = -\frac{\beta N}{4} \sum_{ab} \left(\frac{1}{N} \sum_i s_a^i s_b^i \right)^2$ appearing in (7.21) describes an attractive interaction between the replicas: It is energetically favorable if the replicas line up. Competing with this is the fact that lining up means less phase space. Below T_c, the energy gain is more important than the entropy loss and we get a nonzero overlap expectation value between the replicas. In the RS ansatz the degree of correlation is assumed to be the same for all pairs of replicas. Relaxing this assumption means considering the possibility that groups of replicas form clusters that have larger overlaps amongst themselves than with other clusters. The maximally symmetric situation in such a scenario corresponds to the case in which the clusters are indistinguishable from each other, in particular equal in size and mutual overlaps. In this case there is also no absolute distinction between the replicas, only relative to some fixed other replica.

To find the thermodynamically relevant critical points of $\mathcal{F}(Q)$ (defined in equation (7.23)), it is therefore natural to consider [50, 51] a minimal permutation symmetry breaking scheme $\mathcal{S}_n \to \mathcal{S}_{n/m} \times \mathcal{S}_m$, by splitting up[n] the n replicas in clusters of size m, and assuming the overlap between distinct replicas within one cluster to be q_0 and the overlap between replicas in different clusters to be $q_1 < q_0$. Refining the replica labeling by $a = a_1 a_0$

[m]We put the diagonal part of Q_{ab} equal to δ_{ab} here and it what follows. This is justified by observing that the diagonal part of Q_{ab} decouples from the off-diagonal part in (7.25), and that for any saddle point $Q_{aa} = 1$.

[n]When m does not divide n, it is of course not possible to split the replicas in equal blocks of size m. The idea is to consider values of n and m for which it is possible, and then to analytically continue the resulting expressions, treating m, like n, as a continuous parameter. In this section we only consider the integer n case. The continuation will be discussed in section 7.2.3.5.

where $a_1 = 1, \ldots, n/m$ labels the clusters and $a_0 = 1, \ldots, m$ the replicas inside each cluster, this translates to

$$Q_{a_1 a_0, b_1 b_0} = \delta_{a_1 b_1} \delta_{a_0 b_0} + q_0 \, \delta_{a_1 b_1} \epsilon_{a_0 b_0} + q_1 \, \epsilon_{a_1 b_1} \, , \qquad \epsilon_{ab} \equiv 1 - \delta_{ab} \tag{7.47}$$

(so $\epsilon_{ab} = 1$ if $a \neq b$ and 0 if $a = b$). The idea is then to substitute this new ansatz in $\mathcal{F}(Q)$, extract the $\mathcal{O}(n)$ term, and look for saddle points by varying q_0, q_1 and m.

The clustering process can be iterated, by breaking up the clusters into smaller clusters and then those in turn into even smaller ones. When there are K distinct nontrivial cluster sizes, this is referred to as replica symmetry breaking at level K (K-RSB).

To write things out explicitly, put $m_0 \equiv 1$ and denote the size of the smallest nontrivial cluster by m_1, the next larger one by m_2 and so on, up to $m_{K+1} = n$. Let q_i be the overlap between replicas within a cluster of size m_{i+1} (excluding those contained in an even smaller cluster). Labeling replicas by $a = a_K a_{K-1} \cdots a_1 a_0$ where $a_i = 1, \ldots, \frac{m_{i+1}}{m_i}$ labels the clusters of size m_i, the ansatz can be written as

$$Q_{a_K \cdots a_0, b_K \cdots b_0} = \delta_{ab} + \sum_{i=0}^{K} q_i \, \delta_{a_K b_K} \cdots \delta_{a_{i+1} b_{i+1}} \, \epsilon_{a_i b_i} \tag{7.48}$$

$$= \sum_i \Delta_i \, \delta_{a_K b_K} \cdots \delta_{a_i b_i} \, , \qquad \Delta_i \equiv q_{i-1} - q_i > 0 \, , \tag{7.49}$$

where $q_i \equiv 1$ if $i < 0$ and $q_i \equiv 0$ if $i > K$. For example for $K = 2$ this becomes

$$\begin{aligned} Q_{a_2 a_1 a_0, b_2 b_1 b_0} &= \delta_{a_2 b_2} \delta_{a_1 b_1} \delta_{a_0 b_0} + q_0 \, \delta_{a_2 b_2} \delta_{a_1 b_1} \epsilon_{a_0 b_0} + q_1 \, \delta_{a_2 b_2} \epsilon_{a_1 b_1} + q_2 \, \epsilon_{a_2 b_2} \\ &= (1 - q_0) \delta_{a_2 b_2} \delta_{a_1 b_1} \delta_{a_0 b_0} + (q_0 - q_1) \delta_{a_2 b_2} \delta_{a_1 b_1} \\ &\quad + (q_1 - q_2) \delta_{a_2 b_2} + q_2 \, . \end{aligned}$$

Reading this from right to left corresponds to zooming in to finer structures of the overlap matrix. An example with $K = 3$ is shown in Fig. 7.4.

The hierarchical block structure is equivalent to having an ultrametric structure in the overlap matrix. In other words, it can be organized as a tree, as illustrated in Fig. 7.4. Defining an ultrametric reference distance r_{ab} between two replicas to be, say, the size m of the smallest cluster to which they both belong, we can reformulate the Parisi ansatz simply as the statement that the overlap only depends on the distance:

$$Q_{ab} = q(r_{ab}) \, , \tag{7.50}$$

where $q(r)$ is an arbitrary function (encoding the q_i).

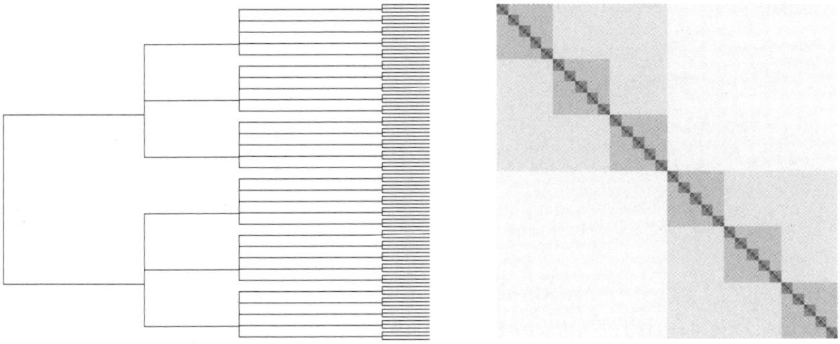

Fig. 7.4. Tree and matrix representation of a Parisi matrix Q_{ab}. Different colors and different tree connection heights correspond to different values of Q_{ab}, with darker red representing larger values. In this example, $n = 90$, $K = 3$, $\{m_0, m_1, m_2, m_3, m_4\} = \{1, 3, 15, 45, 90\}$, $\{u, q_0, q_1, q_2, q_3, q_4\} = \{1, 0.9, 0.6, 0.4, 0.1, 0\}$.

We want to evaluate the replica free energy $\mathcal{F}(Q) = \frac{\beta}{4} Q_{ab}^2 - \frac{1}{\beta} \log \sum_S e^{\frac{\beta^2}{2} Q_{ab} S_a S_b}$ defined in (7.23). The first term is straightforward:

$$\sum_{ab} Q_{ab}^2 = n \left(1 + \sum_{i=0}^{K} q_i^2 \left(m_{i+1} - m_i \right) \right). \tag{7.51}$$

The interpretation of the coefficient $(m_{i+1} - m_i)$ multiplying each q_i^2 is clear: it is the number of replicas having overlap q_i with some fixed reference replica. The second term in $\mathcal{F}(Q)$, proportional to $\log \mathcal{Z}(Q)$, where $\mathcal{Z}(Q) \equiv \sum_S e^{\frac{\beta^2}{2} Q_{ab} S^a S^b}$, is more interesting. Using (7.49), we have

$$\mathcal{Z} = \sum_S \exp\left[\frac{\beta^2}{2} \sum_i \Delta_i \sum_{a_K \cdots a_i} \left(\sum_{a_{i-1} \cdots a_0} S_{a_K \cdots a_0} \right)^2 \right]. \tag{7.52}$$

The squares in the exponential can be linearized by Gaussian transforms similar to (7.36). The resulting expression initially involves Gaussian integrals over many variables $z^{(i)}_{a_K \cdots a_i}$, but they can be evaluated iteratively in clusters starting from the smallest cluster and integrating up to larger and larger distance scales (i.e. larger i / larger m_i / smaller q_i). The result of this little exercise is conveniently and suggestively expressed in terms of convolutions with the Green's function of the heat equation,

$$G_q(z) \equiv \frac{1}{\sqrt{2\pi q}} \exp\left[-\frac{z^2}{2q} \right], \tag{7.53}$$

producing $\mathcal{Z}$ as the outcome of the following recursion:

$$\mathcal{Z}_0(z) \equiv \sum_s e^{-\beta z s} = 2\cosh(\beta z)\,, \tag{7.54}$$

$$\mathcal{Z}_{i+1}(z) \equiv \left(\int dz'\, G_{|q_i - q_{i-1}|}(z - z')\, \mathcal{Z}_i(z') \right)^{k_i}, \quad k_i = \frac{m_{i+1}}{m_i}\,, \tag{7.55}$$

$$\mathcal{Z} = \int dz\, G_{q_K}(z)\, \mathcal{Z}_{K+1}(z)\,. \tag{7.56}$$

The powers k_i arise from identical copies of Gaussian integrals, the copies corresponding to different values of the sub-index $a_i = 1, \ldots, k_i$. The first step in the recursion is easy enough: $\mathcal{Z}_1(z) = \left(e^{\frac{\beta^2}{2}(1-q_0)}(2\cosh(\beta z))\right)^{m_1}$, but past this point one has to resort to expansions or numerical evaluation. As a simple check of the above result, notice that when we remove the distinction between the clusters, i.e. $q_K = \cdots = q_1 = q_0$ (but keeping $q_{-1} \equiv 1$, $q_{K+1} \equiv 0$), we have $G_{|q_i - q_{i-1}|}(z - z') = \delta(z - z')$ for $1 \leq i \leq K$, so $\mathcal{Z} = \int dz\, G_{q_0}(z) \left(e^{\frac{\beta^2}{2}(1-q_0)}(2\cosh(\beta z))\right)^{m_{K+1}}$. Recalling $m_{K+1} = n$, we see this correctly reproduces the replica symmetric formulae of section 7.2.3.3.

The recursion (7.55) can be thought of as an exact Wilsonian renormalization group action, evolving from the UV (small clusters) to the IR (large clusters), rescaling the number of degrees of freedom in jumps by factors $k_i = m_{i+1}/m_i$. We can make the equations more familiar looking by defining a function $\mathcal{Z}(q, z)$ for all $q \in [0, 1]$ such that $\mathcal{Z}(q_i - \epsilon, z) = \mathcal{Z}_{i+1}(z)$. The equations (7.55) and the identification of G as the Green's function of the heat equation show that we can take this function $\mathcal{Z}(q, z)$ to be the single spin sum $2\cosh(\beta z)$ at $q = 1$, and then evolve it down in q according to the heat equation $\partial_q \mathcal{Z} = -\frac{1}{2}\partial_z^2 \mathcal{Z}$, with a jump $\mathcal{Z} \to \mathcal{Z}^{k_i}$ whenever q crosses a q_i. As the latter transformation is nonlinear, it is more convenient to work with $g(q, z) = \log \mathcal{Z}(q, z)$, since on g it acts linearly: $g \to k_i g$. (The price to pay for this is that the heat equation becomes nonlinear.) To implement this without reference to the indices i, we define a function $m(q)$ (illustrated in Fig. 7.5) as

$$m(q) = \text{size of clusters with overlap} > q\,. \tag{7.57}$$

So in particular $m(q_i) = m_i$ and $\partial_q m(q) = \sum_i (m_i - m_{i+1})\, \delta(q - q_i)$. Then $g(q, z) = \log \mathcal{Z}(q, z)$ is the solution to

$$\mathcal{D}_q g = -\frac{1}{2}\left[\left(\frac{\partial g}{\partial z}\right)^2 + \frac{\partial^2 g}{\partial z^2} \right], \qquad \mathcal{D}_q \equiv \frac{\partial}{\partial q} - \frac{1}{m}\frac{\partial m}{\partial q}\,, \tag{7.58}$$

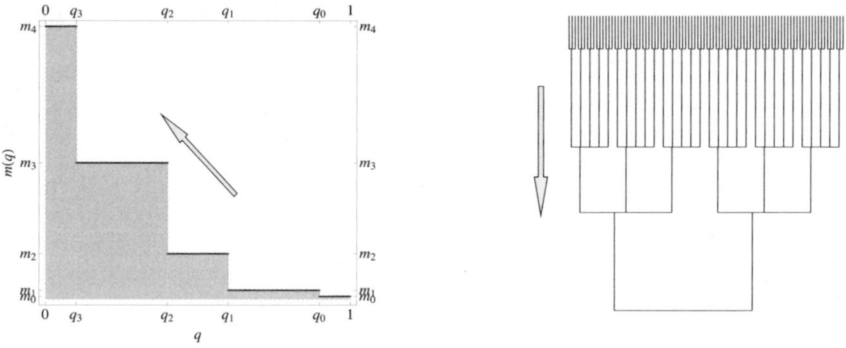

Fig. 7.5. The cluster size function $m(q)$ for the $K = 3$ example of Fig. 7.4, and the corresponding tree. The arrow indicates the direction of the RG flow.

evolving down from $q = 1$ to $q = 0$ with initial condition

$$\lim_{q \nearrow 1} g(q, z) = \log\left(2\cosh(\beta z)\right) , \tag{7.59}$$

and final identification

$$\log \mathcal{Z} = \lim_{q \searrow 0} g(q, 0) . \tag{7.60}$$

The "gauge connection" $A_q = (\partial_q m)/m$ keeps track of the increase in the number of degrees of freedom, while the nonlinear equation in g is equivalent to the heat equation. The connection may be gauged away by redefining $f \equiv \frac{1}{m} g$, so $\partial_q f = \frac{1}{m}\mathcal{D}_q g$, for which

$$\frac{\partial f}{\partial q} = -\frac{1}{2}\left[m(q)\left(\frac{\partial f}{\partial z}\right)^2 + \frac{\partial^2 f}{\partial z^2}\right] . \tag{7.61}$$

We thus see that the iterative system is nothing but a particularly simple example of the exact renormalization group equations, with q the analog of the scale and $f(q, z)$ the analog of the Wilsonian action.

For some purposes it is more convenient to take the scale variable to be some arbitrary auxiliary parameter λ, parametrizing the (q, m) staircase (Fig. 7.5) by *continuous* functions $q(\lambda)$ and $m(\lambda)$. Letting λ evolve from $\lambda = 0$ to $\lambda = 1$, with

$$(q, m)|_{\lambda=0} = (1, 1) , \qquad (q, m)_{\lambda=1} = (0, n) , \tag{7.62}$$

and denoting λ-derivatives by a dot, the RG equation becomes

$$\dot{f} = -\frac{\dot{q}}{2}\left[m(\lambda)\left(\frac{\partial f}{\partial z}\right)^2 + \frac{\partial^2 f}{\partial z^2}\right] , \qquad f(z)|_{\lambda=0} = \log\bigl(2\cosh(\beta z)\bigr) . \tag{7.63}$$

Obviously this introduces a reparametrization gauge symmetry. Putting everything together, we get

$$\frac{1}{n}\mathcal{F}[q,m] = \frac{\beta}{4}\left(1 + \int_{\lambda=0}^{1} dm\, q^2\right) - \frac{1}{\beta} f(z=0)|_{\lambda=1} \,. \tag{7.64}$$

The remaining tasks are (i) make sense of the analytic continuation to $n = 0$ and (ii) extremize the functional $\mathcal{F}[q,m]$.

7.2.3.5. $n \to 0$ *with RSB*

Given that $1 = m_0 < m_1 < m_2 < \cdots < m_K < m_{K+1} = n$ in the Parisi matrix construction, it is not obvious how to properly continue to $n = 0 < 1$, to say the least. To get an idea of what could constitute a sensible continuation, consider the overlap distribution (7.32), which in K-RSB becomes, taking into account the crucial overall sign of (7.32),[o]

$$\overline{P(q)} = \lim_{n\to 0} \sum_{i=0}^{K} (m_i - m_{i+1})\, \delta(q - q_i) = \lim_{n\to 0} \partial_q m(q) \,. \tag{7.65}$$

For this to make sense as a probability density we need $1 = m_0 > m_1 > m_2 > \cdots > m_{K+1} = n = 0$, i.e. the above cluster size inequalities must be inverted, and the m_i obviously can no longer be integers; we will allow them to be arbitrary real numbers. Equivalently, $m(q)$ must become an *increasing* function from the unit interval to the unit interval, to make $\partial_q m$ a proper probability density. This also inverts the interpretation of $m(q)$ from (7.57) to

$$\lim_{n\to 0} m(q) = \text{probability of finding an overlap} \leq q \,. \tag{7.66}$$

In general the function $m(q)$ may have continuously increasing parts, which can be thought of as the $K \to \infty$ limit of the discretized construction. This is illustrated in Fig. 7.6. Recalling the discussion of section 7.2.2.3, getting a smooth function should indeed not surprise us: even if individual realizations of the disorder J_{ij} produce a discrete overlap structure like the one depicted on the right of the figure, if this structure itself is sensitive to the disorder, averaging over the J_{ij} will smooth out the discreteness.

This suggests what we should aim for. Looking back at the equations (7.62)-(7.64), we see that in fact, the required structure is very naturally

[o]To avoid having to duplicate everything, we restrict again to positive overlaps here, i.e. we condition the probabilities on $q \geq 0$.

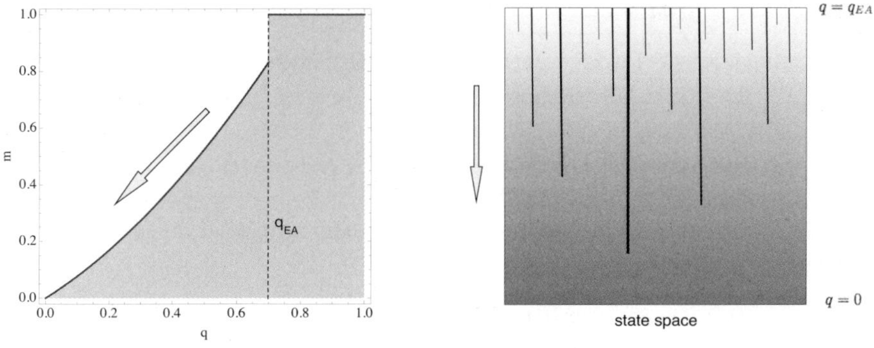

Fig. 7.6. (Fictitious) example of (m, q)-line after continuation to $n = 0$. The arrow again indicates the direction of the RG flow. The diagram on the right is a representation of the ultrametric state space tree for some choice of the disorder J_{ij}. The vertical lines are separating the branches of the tree. Different compartments at the top (the leaves) correspond to different pure states α, and their width represents their probability weight w_α. The value of q where two branches separate sets their mutual overlap (so the taller the wall, the smaller the overlap). Thus, going up along the q-axis, $m(q)$ increases in discrete jumps. But different choices of J_{ij} lead to different barrier structures, and averaging produces the smooth overlap distribution on the left. (The residual step up is due to the disorder independence of the self-overlap q_{EA}.)

obtained by simply taking $n \to 0$ in (7.62) while keeping (7.63) and (7.64) the same. Indeed, $m(\lambda)$ becomes now a decreasing function while $q(\lambda)$ remains decreasing. Everything else stays the same. In particular the intrinsic ultrametric structure is preserved. Notice that now the quadratic terms $\int_{\lambda=0}^{1} dm\, q^2$ in (7.64) become negative definite, with $q = 0$ being the maximum. This is not an indication of thermodynamic instability, but rather the converse, as mentioned already in footnote l. In the limit $n \to 0$, thermodynamically stable states correspond to maxima of the free energy $\mathcal{F}(Q)$ instead of minima.

Admittedly, the above is not a particularly solid justification for the proposed continuation to $n = 0$. One can probably do better, but I will not try to do this here.[p]

[p]The continuation procedure is somewhat reminiscent of continuation in the p-adic numbers: A p-adic integral over a region with p-adic norm between 1 and n is a finite sum with p^n terms, and it naturally continues to a p-adic integral over a region with p-adic norm between $n = 0$ and 1, which is an infinite series. Similarities with p-adic and adelic cumbers were explored in chapter 9 of [46] and [77, 78]. However to reproduce the above one still needs to do a formal continuation $p \to 1$.

7.2.3.6. *The Parisi solution*

To summarize, the Parisi solution[q] of the SK model is (going back to the gauge $\lambda = 1 - q$), in all its glory:

$$F(\beta) = \max_{m(q)} \left[\frac{\beta}{4} \left(1 - \int_{q=0}^{1} dm \, q^2 \right) - \frac{1}{\beta} f_m(z=0)|_{q=0} \right], \qquad (7.67)$$

where $m(q)$ is an non-decreasing function on the unit interval satisfying the boundary conditions

$$m(0) = 0, \qquad m(1) = 1, \qquad (7.68)$$

and $f_m(q, z)$ is the solution to the following flow equation running from $q = 1$ to $q = 0$:

$$\frac{\partial f}{\partial q} = -\frac{1}{2} \left[m(q) \left(\frac{\partial f}{\partial z} \right)^2 + \frac{\partial^2 f}{\partial z^2} \right], \qquad f(z)|_{q=1} = \log\big(2\cosh(\beta z)\big). \quad (7.69)$$

Although this may still look somewhat unwieldy, in practice a simple trial family of $m(q)$ functions (for example stepwise constant functions with just a few steps K) can be treated easily and already lead to excellent approximations of the exact result. For example [51], for $K = 1$ one gets for the ground state free energy $F = -0.765$ and entropy $S = -0.01$, already significantly better than the RS result (7.46). For $K = 2$ this improves further to $F = -0.764$, $S = -0.004$. In particular we see that the entropy converges rapidly to the exact $S = 0$. Detailed computations at various temperatures can be found in [68]. A recent stability analysis of the low T solution is given in [79].

7.2.4. *Beyond spin glasses*

This is as far as we will go in describing the explicit solution of the Sherrington-Kirkpatrick model. We have left many topics untouched, in particular the various other ways in which the model can be solved (for which we refer to [43, 59] as a starting point). We hope we have made it clear that the hierarchical organization of the state space was the central structure in the story. This appears to be typical for disordered mean field (i.e. nonlocal) models. Other popular examples of such models include the

[q]To compare to [51]: x, h there equals m resp. z here, and the apparent extra terms there are generated by the initial flow from $q = 1$ to $q = q_{EA}$ during which $m(q) = 1$, which is included here but not there. This part of the flow is easy to solve because it is basically exactly the heat equation: $f(q, z) = \frac{\beta^2}{2}(1 - q) + \log(2\cosh(\beta z))$, as is easily checked.

Ising and spherical p-spin models [61, 80, 81]. The latter has continuous variables s_i constrained to lie on an N-dimensional sphere $\sum_i s_i^2 = N$, with a Hamiltonian $H = \sum_{i_1 \cdots i_p} J_{i_1 \cdots i_p} s_{i_1} \cdots s_{i_p}$, where the J are random couplings. It is known [61] that when $p = 2$, the replica symmetric solution is exact, while for $p \geq 3$, the 1RSB solution is exact. In the limit $p \to \infty$, the model becomes equivalent to the "random energy model" [80] and can be solved exactly and explicitly [81]. One can interpolate between spherical spin models and Ising spin models such as SK by switching on an additional non-random potential $V = \sum_i \lambda (s_i^2 - 1)^2$, where $p = 2$, $\lambda \to \infty$ reproduces the SK model. Various generalizations to higher spins, $SU(N)$ spins, rotors, quantum spins etc have been studied too, for example in [44, 82–84].

As mentioned in section 7.2.2.1, such mean field models are not expected to be a good approximations for physical, short range spin glasses in spaces with a moderate number of dimensions (including three), and it is not clear to what extent the hierarchical organization of the mean field models is a useful zeroth order expansion point to understand the physics of such systems. But there are other systems of interest, in physics and beyond, where nonlocal interactions occur naturally, and where the conceptual framework developed for the mean field theory of spin glasses has turned out to be very useful. This includes the theory of neural networks, either as models for the brain or for various forms of machine learning, like the way Google figures out all the time what you really mean. One of the oldest and most famous such models is the Hopfield model [85], which is effectively described by a Hamiltonian $H = \sum_{ij} J_{ij} s_i s_j$, where the couplings J_{ij} encode synaptic strengths and $s_i = \pm 1$ encodes the on/off firing state of the i-th neuron. Stable firing states of the neurons correspond to minima of H. N-component patterns ξ_i with $\xi_i = \pm 1$ are stored according to a learning rule $J_{ij} \to J_{ij} - \xi_i \xi_j$ (or variants thereof), which energetically favors the stability of neuron states $s_i = \xi_i$. After M patterns are learned, the couplings are thus $J_{ij} = J_{ij}^{(0)} - \sum_{\mu=1}^{M} \xi_i^\mu \xi_j^\mu$. However, as is well known, learning does not equal remembering. Memory retrieval dynamics in this context is modeled by a standard Monte Carlo type relaxation process, which tries to minimize the energy. Remembering a learned pattern ξ^μ then means that the change in coupling effectively has given rise to a stable minimum $s_\star$ at or very near the pattern ξ^μ. Clearly this model shares many characteristics with mean field spin glass models, and mean field spin glass theory has indeed been very useful in the analysis of these models. For example in [86] it was shown that with the above learning rule, the memory undergoes a phase transition when it tries to learn too much: for

$M < M_c$ with $M_c \approx 1.4\,N$ it remembers practically everything, while for $M > M_c$ it gets completely confused and remembers practically nothing.[r] In this framework, the phase in which learned patterns are remembered is similar to the spin glass phase (many highly metastable states), hierarchical clustering in state space gets interpreted as categorization and organization of memories in concepts, and so on. A collection of some of the original papers along these lines can be found in [43].

Another important[s] branch of human activity where the ideas of spin glass theory have found fruitful applications is optimization algorithms for NP-hard problems. Canonical examples of such problems are the satisfiability problem and the traveling salesman problem, which find incarnations in practical problems going from optimal chip design to the problem of finding string vacua satisfying certain conditions [40]. For an early review we refer again to [43]. Recent developments include for example the relation of clustering phase transitions to effective hardness of optimization problem instances and ways to exploit these insights in devising new efficient algorithms [89, 90].

7.2.5. *Beyond spins*

The statistical mechanics community is fond of Ising spins, and for a good reason: there is a lot in the real world that can be naturally modeled by large collections of interacting binary degrees of freedom. String theorists on the other hand face the sad reality that the complex systems that are natural in their world are rarely as simple. On the other hand, Parisi's idea of using the overlap distribution to characterize the emergence of different pure states / ergodic components / superselection sectors in an intrinsic, quantitative and computable way is an attractive one. Thus, when pondering for example the state space structure of complex D-brane systems [5] or de Sitter space [4], one is led to the question what the analog is of the pure state spin overlap $q_{\alpha\beta} \equiv \frac{1}{N}\sum_i \langle s_i \rangle_\alpha \langle s_i \rangle_\beta$ defined in (7.9) for general quantum systems. For the more general systems of interest the degrees of freedom could be fluxes, scalars, fermions, KK modes, or whatever. Since typically there is no canonical scalar product between such degrees of freedom, a naive direct generalization of the overlap will not work.

A simple construction that is fully quantum and makes sense in general,

[r]It turns out that the key to retain the ability to remember new things is forgetting old things [87].

[s]As measured by the 2×10^4 citations of e.g. [88].

and that is equivalent to the standard spin overlap in the Ising spin case, is the following. Since we want to consider systems for which we can at least formally take a thermodynamic limit, we assume we can randomly sample[t] N independent "representative" degrees of freedom x_i, $i = 1, \ldots, N$ of the system, with the thermodynamic limit of infinite system size corresponding to $N \to \infty$. For example consider a scalar field theory living in a large box of volume V. We can subdivide the large box into smaller identical boxes of some fixed volume v, and then uniformly sample at some fixed sampling density a subset of N of the smaller boxes, which we label by $i = 1, \ldots, N$. Finally, in each of the sampled boxes i, we compute the average value of the scalar field, and we call this x_i. Another example would be a $U(M)$ matrix quantum mechanics describing a wrapped D-brane system. Here we may sample a uniform subset of N of the M^2 matrix degrees of freedom. A final example is again an Ising spin glass, for which we sample N spins.

The state of the complete system is assumed to be described by some density matrix ρ. Expectation values are computed as usual as $\langle A \rangle_\rho \equiv \mathrm{Tr}\,(\rho A)$. For each degree of freedom i there is a corresponding reduced density matrix ρ^i obtained by tracing out all other degrees of freedom, so

$$\mathrm{Tr}_{\mathcal{H}_i}\left(\rho^i A_i\right) = \mathrm{Tr}\,(\rho A_i)\ . \tag{7.70}$$

Here $\mathcal{H}_i$ is the Hilbert space associated to the degree of freedom x_i (for the scalar field example this would be $\mathcal{H}_i = \mathcal{L}^2(\mathbb{R})$, whereas for (quantum) Ising spins it is $\mathcal{H}_i = \mathbb{C}^2$), and A_i is an observable referring to $\mathcal{H}_i$ alone (so for the scalar this could for example be $A_i = x_i^2$, or $A_i = -i\partial_{x^i}$).

A natural definition of overlap between two states ρ_α, ρ_β is then

$$Q_{\alpha\beta} \equiv \frac{1}{N} \sum_{i=1}^{N} \mathrm{Tr}_{\mathcal{H}_i}\, \rho_\alpha^i \rho_\beta^i\,. \tag{7.71}$$

When the system consists of Ising spins, this reduces to the standard magnetization (or spin) overlap, up to a rescaling and a shift. For indeed, the general density matrix for a spin 1/2 degree of freedom can be written as $\rho = \frac{1}{2}\,(\mathbf{1} + \vec{m}\cdot\vec{s})$, where $\vec{m} = \langle \vec{s} \rangle_\rho$ is the magnetization, whence

[t]Sampling is not necessary for the construction, but it spares us the trouble of having to know and work with the complete set of degrees of freedom, which may lead to UV divergences or may not even be known. Moreover the overlap $q_{\alpha\beta}$ and corresponding spin glass order parameter $P(q)$ could equally well have been defined by sampling say every other spin instead of all spins; ultrametricity implies that the results would have been identical. In general this need not be true of course, so the notion of overlap may depend on the choice of degrees of freedom that are sampled.

$Q_{\alpha\beta} = \frac{1}{N}\sum_i \frac{1}{2}(1 + \vec{m}^i_\alpha \cdot \vec{m}^i_\beta)$.[u] For continuous degrees of freedom this can also be written in terms of the associated Wigner densities on phase space:

$$Q_{\alpha\beta} = \frac{1}{N}\sum_{i=1}^{N} 2\pi\hbar \int drdp\, W^i_\alpha(r,p)\, W^i_\beta(r,p)\,, \tag{7.72}$$

where $W^i_\alpha(r,p) \equiv \frac{1}{2\pi\hbar}\int ds\, e^{ips/\hbar}\langle r - s/2|\rho^i_\alpha|r + s/2\rangle$. In the classical limit this becomes (after suitable coarse graining to kill off highly oscillatory modes) the overlap of ordinary phase space probability densities, and we can think of $Q_{\alpha\beta}$ as the fraction of degrees of freedom (x^i, p_i) found in the same elementary phase space cell (of size $2\pi\hbar$) when randomly sampling from two states α and β.

We can mimic (7.9) even more closely, by writing the reduced density matrix ρ^i as the expectation value $\langle E^i\rangle$ of the operator E^i defined by

$$E^i_{rs} \equiv \mathcal{P}^\dagger_{x_i=r}\mathcal{P}_{x_i=s} = |x_i = r\rangle\langle x_i = s|\,, \tag{7.73}$$

where $\mathcal{P}_{x_i=r}$ is the projection operator on the eigenspace $x_i = r$. Then we get

$$Q_{\alpha\beta} = \frac{1}{N}\sum_{i=1}^{N}\sum_{rs}\langle E^i_{rs}\rangle_\alpha \langle E^i_{sr}\rangle_\beta\,. \tag{7.74}$$

Finally, for some states ρ, an analog of the clustering property (7.6) will hold, for others not. We define this analog to be the property that correlation functions of "local" observables factorize in the thermodynamic limit for almost all evaluation points. That is, for any finite r

$$\lim_{N\to\infty}\frac{1}{N^r}\sum_{i_1,\cdots,i_r}\Big|\langle A_{i_1}B_{i_2}\cdots C_{i_r}\rangle_\rho - \langle A_{i_1}\rangle_\rho\langle B_{i_2}\rangle_\rho\cdots\langle C_{i_r}\rangle_\rho\Big| = 0\,. \tag{7.75}$$

By "local" we mean here observables A_i referring to $\mathcal{H}_i$ alone (we could relax this definition by allowing dependence on other, "nearby" degrees of freedom, but this will be good enough for our purposes).

With these definitions set up, we can proceed in complete parallel with the Ising spin glass case, and define "pure states" ρ_α, decompositions $\rho = w_\alpha\rho_\alpha$ of non-pure states (such as $\rho = e^{-\beta H}$ for glassy systems at low temperatures), and their overlap distributions $P(q) = \sum_{\alpha\beta} w_\alpha w_\beta \delta(q - Q_{\alpha\beta})$. Whenever ρ fails to satisfy cluster decomposition (signaling a nontrivial phase structure), $P(q)$ becomes nontrivial. For example for an isotropic

[u] This is valid for quantum Ising spins. Classical Ising spins are trivially obtained from this by restricting to the up/down eigenstates of s_z, yielding $Q_{\alpha\beta} = \frac{1}{2}\left(1 + q_{\alpha\beta}\right)$ with $q_{\alpha\beta}$ as defined in (7.9).

quantum ferromagnet, the low temperature pure states will be labeled by a vector $\vec{m}$ such that for all i, $\rho^i_{\vec{m}} = \frac{1}{2}(1 + \vec{m} \cdot \vec{s})$, and the overlap distribution $P(q)$ is uniform on the interval $[\frac{1-m^2}{2}, \frac{1+m^2}{2}]$.

Moreover $P(q)$ will again be computable in principle for any known state ρ, even if we do not know how to describe its decomposition into pure states. Explicitly, the analog of (7.12) that allows this is

$$P(q) = \left\langle \delta \left(q - \tfrac{1}{N} \textstyle\sum_{irs} E^i_{1rs} E^i_{2sr} \right) \right\rangle_{n=2}, \qquad (7.76)$$

where again the subscript '$n = 2$' means we consider two replicas of the system, with density matrix $\rho \otimes \rho$, and E_1 depends on the first replica's degrees of freedom while E_2 depends on the second replica's degrees of freedom. In the classical limit, $P(q)$ becomes the probability distribution for finding a fraction q of the degrees of freedom (x^i, p_i) of the two replicas in the same elementary phase space cell.

The operator inside the delta-function has a more concrete interpretation. For a given i, let

$$X^i_{12} \equiv \sum_{rs} E^i_{1rs} E^i_{2sr} = \sum_{rs} |x^i_1 = r, x^i_2 = s\rangle\langle x^i_1 = s, x^i_2 = r| \,. \qquad (7.77)$$

Notice the $r \leftrightarrow s$ flip on the right hand side, which we can equivalently think of as a $x^i_1 \leftrightarrow x^i_2$ flip. Thus, what this operator X^i_{12} does for a given i (when acting on a wave function say) is exchange the replica variables x^i_1, x^i_2. The first moment of $P(q)$ is then

$$\langle q \rangle = \frac{1}{N} \sum_i \mathrm{Tr}(\rho_1 X^i_{12}\, \rho_2) = \frac{1}{N} \sum_i \mathrm{Tr}_{\mathcal{H}_i} (\rho^i)^2 = \frac{1}{N} \sum_i e^{-S^i(2)}, \qquad (7.78)$$

where $S^i(2)$ is the second Renyi entropy [91] of ρ^i. If $\rho = |\Psi\rangle\langle\Psi|$, this gives information on how entangled the x_i are in the wave function Ψ; if Ψ factorizes (no entanglement), then $\langle q \rangle = 1$. For the isotropic ferromagnet (strong entanglement), we have $\langle q \rangle = \frac{1}{2}$. The second moment is

$$\begin{aligned} \langle q^2 \rangle &= \frac{1}{N^2} \sum_{ij} \mathrm{Tr}(\rho_1 X^i_{12} X^j_{12}\, \rho_2) = \frac{1}{N^2} \sum_{i \neq j} \mathrm{Tr}_{\mathcal{H}_{ij}} (\rho^{ij})^2 + \frac{1}{N} \\ &= \frac{1}{N^2} \sum_{i \neq j} e^{-S^{ij}(2)}, \end{aligned} \qquad (7.79)$$

where ρ^{ij} is the density matrix obtained by tracing out all degrees of freedom except x_i, x_j, and in the last expression we dropped the $1/N$ term since it is understood that the thermodynamic limit $N \to \infty$ is taken. The ratio $\langle q^2 \rangle / \langle q \rangle^2$ gives information about the degree of long range correlation (and failure of cluster decomposition) in the system. For a thermodynamic pure state it equals 1, for the isotropic ferromagnet it equals $1 + \frac{m^4}{3}$. Moments of order k are obtained analogously by fixing k degrees of freedom.

Thus, for rather general systems, we have arrived at an infinite set of in principle computable order parameters that detect the emergence of a nontrivial state space landscape. Applications are discussed in [4, 5].

7.3. Supersymmetric Quantum Mechanics

7.3.1. *Introduction*

Complex systems are complex. As a result, it is hard in general to quantitatively determine even their basic features. When it can be done, such as for the Sherrington-Kirkpatrick model, the results often have a striking richness. Complex systems arising in string theory, such as the geometrically highly complex wrapped D-branes reviewed in section 7.4, are often significantly more intricate than simple spin models. This makes straightforward analysis more daunting. Although in principle the probability overlaps defined in section 7.2.5 define an intrinsic order parameter that could be used to analyze those stringy systems along the lines of the SK model, in practice this may still be difficult. This is even true for much simpler quantities like the entropy.

On the other hand, the existence of dualities in string theory and the presence of supersymmetry often tremendously simplify things. In this section we will focus on supersymmetry. The main simplification that arises due to this is that many quantities of interest can be computed exactly in a semiclassical limit. The primary example of this is the Witten index [92], which counts ground state degeneracies in supersymmetric quantum mechanics, and is generically invariant under continuous deformations of the theory. Famously this allowed Strominger and Vafa [93] to microscopically compute the entropy of certain classes of extremal black holes arising in string theory, by tuning the string coupling constant from the strongly self-gravitating black hole regime to a weakly coupled, non-self-gravitating regime. In this regime the system has a well-understood D-brane description, and the Witten index can be computed in a relatively straightforward and precise way.

The ground state degeneracy is not the only thing supersymmetry allows us to control. Through its relation with Morse theory, the structure of supersymmetric quantum mechanics gives useful insights into the saddle point structure of energy landscapes and the way they are dynamically connected, and makes it possible to systematically compute nonperturbative level splittings and their associated exponentially slow relaxation rates. Clearly this is of potential interest in the study of complex systems. In fact, through a formal map between nonsupersymmetric classical statistical mechanics and supersymmetric quantum mechanics, this has ramifications beyond the realm of supersymmetric systems (see section 7.3.8).

We will review these fundamental aspects of supersymmetric quantum mechanics in what follows. I will not assume any familiarity with the topic, and start from scratch, building up the topic along the lines of chapter 10 of [94], an excellent introduction to supersymmetric quantum mechanics and its interplay with geometry. A review focusing more on relations to exactly solvable systems is [95], and finally Witten's original [92] is a classic.

Specific applications to string theory are left to section 7.4.

7.3.2. *Definition*

By definition, the Hamiltonian of a supersymmetric quantum mechanical system can be written as

$$H = \frac{1}{2}\{Q, Q^\dagger\}\,, \tag{7.80}$$

where Q, the supercharge, is an operator satisfying

$$Q^2 = \left(Q^\dagger\right)^2 = 0\,. \tag{7.81}$$

The Hilbert space is $\mathbb{Z}_2$ graded, splitting up in a even or "bosonic" part, and an odd or "fermionic" part:[v]

$$\mathcal{H} = \mathcal{H}_B \oplus \mathcal{H}_F\,. \tag{7.82}$$

The operator $(-1)^F$ is defined as $+1$ on $\mathcal{H}_B$ and -1 on $\mathcal{H}_F$. The operator Q is odd/fermionic, i.e. it maps from $\mathcal{H}_B$ into $\mathcal{H}_F$ and from $\mathcal{H}_F$ into $\mathcal{H}_B$:

$$[(-1)^F, Q] = -Q\,. \tag{7.83}$$

Some immediate consequences are

- $[(-1)^F, H] = 0$, $[Q, H] = 0$, i.e. $(-1)^F$ and Q are symmetries.
- $H \geq 0$, all energies are positive or zero.

[v]At this level "bosonic" and "fermionic" are just conventional names. They do not necessarily mean that the quantum mechanical states represent bosonic or fermionic particles in the usual sense of the word, although in specific situations the notion may be correlated.

- $H|\alpha\rangle = 0 \Leftrightarrow Q|\alpha\rangle = 0 = Q^\dagger|\alpha\rangle$, zero energy states are supersymmetric.
- For each *positive* energy level $\mathcal{H}_E = \mathcal{H}_{E,B} \oplus \mathcal{H}_{E,F}$, $E > 0$, the fermionic and bosonic subspaces are *isomorphic*. The isomorphism is given by the Hermitian supercharge

$$Q_1 \equiv Q + Q^\dagger \,, \tag{7.84}$$

which satisfies $Q_1^2 = 2E$ on $\mathcal{H}_E$ and is therefore invertible when $E \neq 0$.

The latter property is particularly important. It means that all positive energy eigenstates come in boson-fermion pairs. This is not necessarily true for zero energy ground states. The Witten index Ω quantifies the degree to which this is not true:

$$\Omega \equiv \dim \mathcal{H}_{0,B} - \dim \mathcal{H}_{0,F} \,. \tag{7.85}$$

Since all positive energy states come in even-odd pairs, this can equally well be written in the following alternative ways:

$$\Omega = \mathrm{Tr}\,(-1)^F \tag{7.86}$$

$$= \mathrm{Tr}\,(-1)^F e^{-\beta H} \tag{7.87}$$

$$= \int \mathcal{D}(\cdots)\, e^{-S(\cdots)} \,. \tag{7.88}$$

The last line represents the usual Euclidean path integral representation of the partition function, except that now *both* fermionic and bosonic variables satisfy periodic boundary conditions, instead of the usual antiperiodic boundary conditions on the fermions. This is due to the insertion of $(-1)^F$, which flips the sign of the fermionic variables. (The example below may make this more clear.)

The most important property of the Witten index is that it is invariant under generic continuous deformations. This is clear from its definition: although the *total* number of supersymmetric ground states may vary when we vary the parameters of the model, as states move in and out of the $E = 0$ level, all such arrivals and departures must come in bose-fermi pairs, and therefore the index will not be affected.[w]

[w]Although generically robust, there are important exceptions to this argument. In particular, when the gap we have implicitly assumed between the zero and first excited energy levels vanishes in the course of the deformation, the Witten index may jump. This can happen for example when the deformation passes through a degeneration allowing a ground state wave function to spread out all the way to infinite distance, becoming non-normalizable and therefore no longer a proper state in the Hilbert space, but rather part of the continuum of scattering states.

7.3.3. *One dimensional example*

A simple example is obtained by considering a bosonic (i.e. commuting) variable x and a fermionic (i.e. anticommuting) variable $\bar{\psi}$, wave functions

$$\Phi(x,\bar{\psi}) = \Phi_B(x) + \Phi_F(x)\bar{\psi}\,, \tag{7.89}$$

and canonical commutation relations $[x,p]=i$, $\{\bar{\psi},\psi\}=1$, where $\psi \equiv \bar{\psi}^\dagger$. So acting on wave functions we have

$$p = -i\partial_x\,, \qquad \psi = \partial_{\bar{\psi}}\,. \tag{7.90}$$

Then we consider the supercharge

$$Q \equiv \bar{\psi}\,(ip + h'(x))\,, \qquad Q^\dagger = \psi\,(-ip + h'(x))\,. \tag{7.91}$$

Here $h(x)$ is an arbitrary function, sometimes called the superpotential. Expressed in components $\Phi = \binom{\Phi_B}{\Phi_F}$ this becomes

$$Q = \begin{pmatrix} 0 & 0 \\ \partial_x + h'(x) & 0 \end{pmatrix}, \tag{7.92}$$

and the Hamiltonian

$$\begin{aligned} H &= \frac{1}{2}\{Q,Q^\dagger\} \\ &= \frac{1}{2}\left(-\partial_x^2 + h'(x)^2 + h''(x)[\bar{\psi},\psi]\right) \\ &= \frac{1}{2}\begin{pmatrix} -\partial_x^2 + h'(x)^2 - h''(x) & 0 \\ 0 & \partial_x^2 + h'(x)^2 + h''(x) \end{pmatrix}. \end{aligned} \tag{7.93}$$

We can think of this as the Hamiltonian of a spin 1/2 particle of unit mass in a potential $V(x) = \frac{1}{2}h'(x)^2$ and a magnetic field $\frac{1}{2}h''(x)$. In the case $h(x) = \frac{\omega}{2}x^2$, this becomes the harmonic oscillator potential $V(x) = \frac{1}{2}\omega^2 x^2$ plus a constant magnetic field ω. Assuming $\omega > 0$, the bosonic (i.e. $\Phi_F = 0$) energy levels are then $E_B = 0, \omega, 2\omega, 3\omega, \ldots$ and the fermionic energy levels $E_F = \omega, 2\omega, 3\omega, \ldots$. We explicitly see the pairing of $E > 0$ states and the mismatch at $E = 0$, captured by the Witten index $\Omega = +1$. When on the other hand $\omega < 0$, we get $E_B = |\omega|, 2|\omega|, \ldots$ and $E_F = 0, |\omega|, 2|\omega|, \ldots$, so $\Omega = -1$.

More generally, it is easy to explicitly construct solutions to $Q\Phi = Q^\dagger\Phi = 0$:

$$\Phi = A_B\, e^{-h(x)} + A_F\, e^{+h(x)}\,\bar{\psi}\,.^{\mathrm{x}} \tag{7.94}$$

[x]Here $\bar{\psi}$ is, as in (7.89), a Grassman variable. Equivalently we could write $\Phi = A_B\, e^{-h(x)}|0\rangle + A_F\, e^{+h(x)}\,\bar{\psi}|0\rangle$, with $|0\rangle$ the "fermionic vacuum state", i.e. $\psi|0\rangle = 0$.

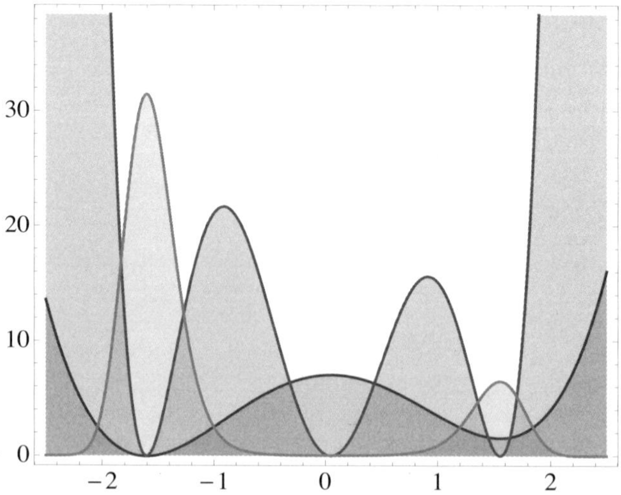

Fig. 7.7. 1d susy QM with $h(x) = x^4 - 5\,x^2 + 7$. Blue: $h(x)$, red: $V(x) = h'(x)^2/2$, green: $\Phi_0(x) \propto e^{-h(x)}$.

For these to be true supersymmetric ground states, Φ has to be normalizable. When $\lim_{|x|\to\infty} h(x) = +\infty$, this requires $A_F = 0$, and thus $\Omega = +1$. When $\lim_{|x|\to\infty} h(x) = -\infty$, we get instead $A_B = 0$ and $\Omega = -1$. In all other cases, none of the solutions are normalizable, and $\Omega = 0$. An example is shown in Fig. 7.7.

7.3.4. *More degrees of freedom*

We can easily generalize this to multiple degrees of freedom x^I and $\bar{\psi}^I$, $I = 1, \ldots, N$, leading to 2^N-component wave functions

$$\Phi(\bar{\psi}, x) = \sum_{b_1,\cdots,b_N=0,1} \Phi_{b_1\cdots b_N}(x)\,(\bar{\psi}^1)^{b_1}\cdots(\bar{\psi}^N)^{b_N}\,, \tag{7.95}$$

and a fermion number quantum number F counting the number of $\bar{\psi}^I$. We take the supercharge to be

$$Q = \sum_I \bar{\psi}^I\,(ip_I + \partial_I h(x))\,, \tag{7.96}$$

which satisfies $Q^2 = 0$. The Hamiltonian is then

$$H = \sum_I \frac{1}{2}p_I^2 + \frac{1}{2}(\partial_I h)^2 + \frac{1}{2}\sum_{IJ}[\bar{\psi}^I, \psi^J]\partial_I\partial_J h\,. \tag{7.97}$$

It is now no longer possible in general to find explicit solutions to $Q\Phi = Q^\dagger\Phi = 0$, but when $h(x)$ is a nondegenerate quadratic function, say

$$h(x) = \frac{1}{2}\sum_{I=1}^{N} \omega_I (x^I)^2 \,, \tag{7.98}$$

then the problem just factorizes into N copies of the 1d harmonic oscillator situation we solved above, and thus the unique normalizable supersymmetric ground state is

$$\Phi_0(x) = e^{-\frac{1}{2}\sum_I |\omega_I|(x^I)^2} \prod_{I:\,\omega_I<0} \bar{\psi}^I \,. \tag{7.99}$$

This has fermion number F equal to the number of negative eigenvalues of the Hessian of h at the critical point $x = 0$. This number μ is called the Morse index of the critical point. The Witten index is $\Omega = (-1)^\mu$.

Using this, we can compute the Witten index for any $h(x)$ that has only isolated, nondegenerate critical points.[y] Indeed, since the Witten index is invariant under continuous deformations that do not change the asymptotics of $h(x)$, we can consider the deformation $h(x) \to \lambda h(x)$ and compute Ω in the limit $\lambda \to \infty$. This limit corresponds to a semiclassical limit. The classical supersymmetric ground states are given by the critical points of $h(x)$. The corresponding perturbative[z] semiclassical wave functions are to lowest order Gaussians centered at the critical points, of the form (7.99). All of these remain supersymmetric to all orders in perturbation theory. (Roughly this is because perturbation theory around one critical point does not know about the presence of other critical points, and if there is only one critical point, the Witten index is 1.) However, nonperturbative tunneling effects may lift this degeneracy, just like tunneling splits up the doubly degenerate perturbative ground state energy level of a double well potential in ordinary quantum mechanics. But because all lifting must occur in bose-fermi pairs, it will not affect the total Witten index, so we can compute the index simply by adding the contributions of all critical points $x_\star$ of h:

$$\Omega = \sum_{x_\star:\, h'(x_\star)=0} (-1)^{\mu(x_\star)} \,. \tag{7.100}$$

[y] Such a function is called a Morse function, and it corresponds to the generic case. We will always assume h is a Morse functions in this section.

[z] More precisely, the perturbation theory for each critical point $x_\star$ of h is a series in inverse powers of λ and is obtained by changing variables from x to y defined by $x^I = x^I_\star + y^I/\sqrt{\lambda}$ so $H = \lambda(H_0 + \lambda^{-1/2}H_1 + \cdots)$ where H_0 is the harmonic oscillator part of the Hamiltonian, obtained by truncating h at quadratic order in y, and H_1 the part corresponding to the higher order terms.

Checking this for the 1d example shown in Fig. 7.7, we see that indeed there h has two minima and one maximum, so $\Omega = +1+1-1 = +1$, in agreement with our earlier result.

When h is a harmonic function, as is the case when it is for example equal to the real part of a holomorphic function $W(z)$ of variables $z^a = x^{2a-1} + ix^{2a}$, $a = 1, \cdots, n = N/2$, all critical points have Morse index n, and the Witten index is equal to $(-1)^n$ times the absolute number of critical points of W. This situation occurs naturally in the description of wrapped D-branes in Calabi-Yau compactifications of type II string theory.

7.3.5. *Curved spaces*

On a compact N-dimensional Riemannian manifold M, we can naturally define a supersymmetric quantum mechanics with Hilbert space given by the space of complex valued differential forms $\Phi = \Phi_0 + \Phi_I dx^I + \Phi_{IJ} dx^I \wedge dx^J + \cdots$ on M. The elementary 1-forms dx^I get identified with the Grassmann coordinates $\bar{\psi}^I$. The inner product of two differential forms $\Phi^{(a)} = \sum_{F=0}^N \Phi_{I_1 \cdots I_F} \, dx^{I_1} \wedge \cdots \wedge dx^{I_F}$, $a = 1, 2$ is

$$\langle \Phi^{(1)} | \Phi^{(2)} \rangle = \int \overline{\Phi^{(1)}} \wedge \star \Phi^{(2)} = \sum_{F=0}^{N} \int d^N x \, \sqrt{g} \, g^{I_1 J_1} \cdots g^{I_F J_F} \, \overline{\Phi^{(1)}_{I_1 \cdots I_F}} \, \Phi^{(2)}_{J_1 \cdots J_F} \, , \tag{7.101}$$

where g_{IJ} is the metric on M. The fermion number F is the form degree and the supercharge in the absence of a potential is the exterior derivative

$$Q = d = dx^I \nabla_I \, , \tag{7.102}$$

where ∇_I is the covariant derivative. The Hamiltonian is the Laplace-Beltrami operator

$$\Delta = \frac{1}{2} \left(dd^\dagger + d^\dagger d \right) \, . \tag{7.103}$$

Comparing with our previous notation, we have, at the level of operators, the identifications

$$\bar{\psi}^I = dx^I \wedge \, , \qquad \psi^I = g^{IJ} \frac{\partial}{\partial \bar{\psi}^J} = g^{IJ} \iota_{\partial_I} \, , \tag{7.104}$$

implying in particular the anticommutation relations $\{\psi^I, \bar{\psi}^J\} = g^{IJ}$. More explicitly, using the relation between curvature and covariant derivative commutators, the Hamiltonian can be written as

$$H = -\frac{1}{2} g^{IJ} \nabla_I \nabla_J + \frac{1}{2} R_{IJKL} \psi^I \bar{\psi}^J \psi^K \bar{\psi}^L \, . \tag{7.105}$$

The supersymmetric ground states are the harmonic differential forms, and the Witten index is the alternating sum of the number of harmonic forms at each form degree, which is nothing but the Euler characteristic of M:

$$\Omega = \sum_{F=0}^{N} (-1)^p \, b^p(M) = \chi(M) \,. \tag{7.106}$$

The Betti number b^p is also equal to the number of homologically independent $(N-p)-$*cycles*. On a compact manifold we have moreover $b^p = b^{N-p}$. An immediate consequence of this is that $\Omega = 0$ when N is odd. For a Riemann surface of genus g, we have $b^0 = b^2 = 1$ and $b^1 = 2g$, so $\chi = 2 - 2g$.

Adding a potential can be implemented by putting

$$Q_h = e^{-h} d e^h = d + dh \,. \tag{7.107}$$

This gives a potential term $V = g^{IJ}\partial_I h \partial_J h$ in the Hamiltonian. Switching on a potential will not change the Witten index, since the spectrum is guaranteed to be discrete on a compact space. In fact, it will not even change the absolute number of supersymmetric ground states in each fermion number. To prove this one uses the very useful fact[aa] that the space of supersymmetric ground states is isomorphic to the Q-cohomology:

$$\mathcal{H}_{\rm susy} \simeq \frac{\ker Q}{\operatorname{im} Q} \,. \tag{7.108}$$

This is combined with the observation that there is an isomorphism between the cohomologies of $Q_0 = d$ and $Q_h = e^{-h} d e^h$ provided by the map $|\alpha\rangle \to e^{-h}|\alpha\rangle$.

An example of a Riemann surface, embedded in $\mathbb{R}^3$ is shown in Fig. 7.8. This Riemann surface is known as a *pretzel* and acquires a natural metric from its embedding in $\mathbb{R}^3$. For the function h we take the height along the vertical axis. There are 3 minima, 9 saddle points and 2 maxima, each of which gives rise in the large pretzel limit to a *perturbative* susy ground state, here of fermion number 0, 1 and 2 respectively. We know we can compute the Witten index in perturbation theory, yielding $\Omega = 3 - 9 + 2 = -4$, so the Riemann surface is of genus 3, which is confirmed by visual inspection. We know also that $b_0 = 1 = b_2$, and thus $b_1 = 6$. Therefore, since the Betti numbers count the number of supersymmetric ground states at given

[aa]This is not hard to show. First, for positive energy states, we have that if $Q|\alpha\rangle = 0$, then $|\alpha\rangle = \frac{1}{2E}(Q^\dagger Q + QQ^\dagger)|\alpha\rangle = Q\left(\frac{Q^\dagger}{2E}|\alpha\rangle\right)$, so all positive energy states are trivial in cohomology. For zero energy states on the other hand we have $Q|\alpha\rangle = 0 = Q^\dagger|\alpha\rangle$, so if $|\alpha\rangle = Q|\beta\rangle$, then $\langle\alpha|\alpha\rangle = \langle\beta|Q^\dagger|\alpha\rangle = 0$, i.e. $|\alpha\rangle = 0$. So each zero energy state is nontrivial in cohomology.

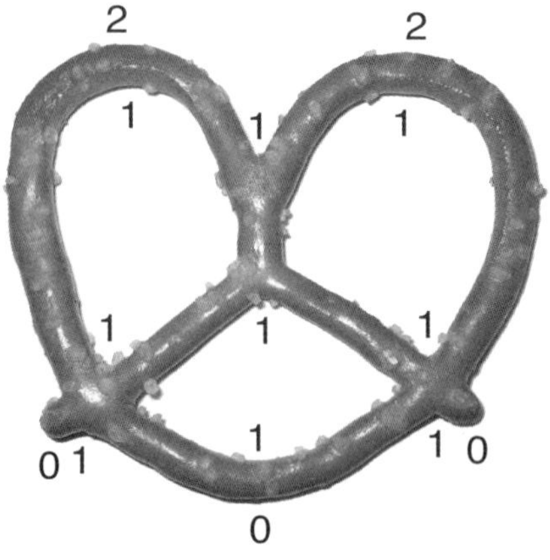

Fig. 7.8. Critical points of the vertical height function on a pretzel, with fermion number (Morse index) indicated. The Witten index is $\Omega = 3 - 9 + 2 = -4$.

fermion number, several of the perturbative ground states must be lifted by tunneling effects: only one linear combination of the $F = 0$ and the $F = 2$ states can remain supersymmetric, and six linear combinations of the $F = 1$. The remaining three $F = 1$ states pair up with two $F = 0$ and one $F = 2$ state and get lifted. The lifting will be exponentially small in the pretzel size (which sets the scale of h).

7.3.6. *Lagrangian formalism*

So far we have described everything in a Hamiltonian framework. Semi-classical tunneling amplitudes responsible for lifting perturbative ground states are however most easily computed in a Lagrangian framework. The Lagrangian framework and its associated path integral are also most convenient to efficiently obtain a differential geometric expression for the Witten index, as the integral over M of a differential form made out of the curvature, namely the Euler density.

The Lagrangian is obtained from the Hamiltonian in the usual way. We are primarily interested in the Euclidean Lagrangian. Starting from (7.105) with the addition of the potential term, this is

$$L = \frac{1}{2} g_{IJ} \dot{x}^I \dot{x}^J + \frac{1}{2} g^{IJ} \partial_I h \partial_J h$$
$$+ g_{IJ} \bar{\psi}^I D_\tau \psi^J + \frac{1}{2} R_{IJKL} \psi^I \bar{\psi}^J \psi^K \bar{\psi}^L + \nabla_I \partial_J h \, \bar{\psi}^I \psi^J \,, \quad (7.109)$$

where τ is the Euclidean time conjugate to H and $\dot{x} = \frac{dx}{d\tau}$, $D_\tau \psi^I = \partial_t \psi^I + \Gamma^I_{JK} \partial_t x^J \psi^K$, $\nabla_I V_J = \partial_I V_J - \Gamma^K_{IJ} V_K$. The supersymmetries act on the fields as $\delta(\cdots) = [\epsilon Q + \bar{\epsilon} Q^\dagger, \cdots]$:

$$\delta x^I = \epsilon \bar{\psi}^I - \bar{\epsilon} \psi^I \tag{7.110}$$

$$\delta \psi^I = \epsilon(-\dot{x}^I - \Gamma^I_{JK} \bar{\psi}^J \psi^K + g^{IJ} \partial_J h) \tag{7.111}$$

$$\delta \bar{\psi}^I = \bar{\epsilon}(\dot{x}^I - \Gamma^I_{JK} \bar{\psi}^J \psi^K + g^{IJ} \partial_J h)\,, \tag{7.112}$$

leaving the action invariant.

As an application we derive a differential geometric expression for the Witten index, which is essentially the generalized Gauss-Bonnet formula [96]. The path integral representation of the Witten index is

$$\Omega = \mathrm{Tr}\,(-1)^F e^{-\beta H} = \int \mathcal{D}x\, \mathcal{D}\bar{\psi}\, \mathcal{D}\psi\, e^{-\int_0^\beta L\, dt}\,, \tag{7.113}$$

with periodic boundary conditions on all fields: $\phi(\beta) = \phi(0)$ for $\phi = x, \psi, \bar{\psi}$. In the limit $\beta \to 0$, only the constant trajectories contribute to the path integral. Thus, taking $h = 0$ and working in an orthonormal frame, the path integral reduces to the finite dimensional integral

$$\Omega = \frac{1}{(2\pi\beta)^{N/2}} \int d^N x\, d^N \bar{\psi}\, d^N \psi\, e^{-\frac{\beta}{2} R_{IJKL} \psi^I \bar{\psi}^J \psi^K \bar{\psi}^L}\,, \tag{7.114}$$

where the normalization can be fixed for example by comparing path integral and canonical expressions for the free propagator in flat space $\langle x | e^{-\beta H_0} | x \rangle$. Using the symmetries of the Riemann tensor and the Grassmann integral representation of the Pfaffian of a matrix M, $\mathrm{Pf}\, M = \int d\psi^1 \cdots d\psi^N\, e^{-\frac{1}{2}\psi^i M_{ij} \psi^j}$ (with $(\mathrm{Pf}\, M)^2 = \det M$), this becomes

$$\Omega = \int_M \mathrm{Pf}\, \mathcal{R}\,, \qquad \mathcal{R}^I{}_J \equiv \frac{1}{4\pi} R^I{}_{JKL}\, dx^K \wedge dx^L\,. \tag{7.115}$$

This expression, which gives the Euler characteristic of M in terms of the integrated Euler density, is known as the generalized Gauss-Bonnet theorem.

7.3.7. *Instantons and lifting*

By completing the squares, the bosonic part of the action (7.109) can be brought into the form

$$S_B = s\big(h(x_f) - h(x_i)\big) + \int_{-T/2}^{T/2} d\tau\, \frac{1}{2} g_{IJ} \left(\dot{x}^I - s g^{IK} \partial_K h\right) \left(\dot{x}^J - s g^{JL} \partial_L h\right)\,, \tag{7.116}$$

where $s = \pm 1$. The first term comes from integrating a total derivative, and $x(-T/2) = x_i$ and $x(T/2) = x_f$ are the start and end points of the trajectory. This implies the lower bound $S_B \geq s\big(h(x_f) - h(x_i)\big)$ for $s = \pm 1$, hence $S_B \geq |h(x_f) - h(x_i)|$, which is saturated for steepest descent or ascent trajectories:

$$\dot{x}^I = s g^{IJ} \partial_J h(x) , \qquad s = \text{sign}\big(h(x_f) - h(x_i)\big) . \tag{7.117}$$

These gradient flow trajectories (with the fermions put to zero) are automatically solutions to the Euclidean equations of motion, and moreover from the supersymmetry variations (7.111)-(7.112) it can be seen that ascending ($s = +1$) trajectories are annihilated by the Q-supersymmetry, while descending trajectories are annihilated by the $Q^\dagger$-supersymmetry. Such minimal action trajectories, in the limit $T \to \infty$, are called instantons.

Instantons play a key role in the computation of nonperturbative lifting of ground states. For concreteness consider the example shown in Fig. 7.9, a two dimensional flat unit torus with superpotential $h(x, y) = \lambda \sin(2\pi x) \sin(2\pi y)$. There are 8 critical points: 2 minima, 4 saddle points and 2 maxima, giving $\Omega = 2 - 4 + 2 = 0 = \chi(T^2)$. The Betti numbers for T^2 are $b_0 = 1$, $b_1 = 2$, $b_2 = 1$, so again some of the perturbative ground states must be lifted by tunneling effects. The relevant tunneling trajectories responsible for this are precisely the instantons introduced above.

The general story goes as follows. Denoting the perturbative ground states by $|i\rangle$, lifting means that the matrix elements of the supercharge reduced to the perturbative zero energy level,

$$\hat{Q}_{ij} \equiv \langle i|Q|j\rangle , \tag{7.118}$$

although identically zero in perturbation theory, is not identically zero when nonperturbative effects are taken into account. Notice that since Q has fermion number 1, Q_{ij} can only be nonzero if the fermion numbers of $|i\rangle$ and $|j\rangle$ statisfy the selection rule $F_i = F_j + 1$. The matrix elements Q_{ij} can be given a path integral representation. It can be shown that thanks to supersymmetry, this path integral (generically) *only* gets contributions from instanton trajectories annihilated by Q, i.e. paths of steepest ascent, interpolating between critical points $j \to i$ with $F_i = F_j + 1$. Moreover the Gaussian approximation to this contribution is exact, resulting in the very simple expression

$$\hat{Q}_{ij} = \sum_{\gamma : j \to i} n_\gamma \, e^{-h_{ij}} , \tag{7.119}$$

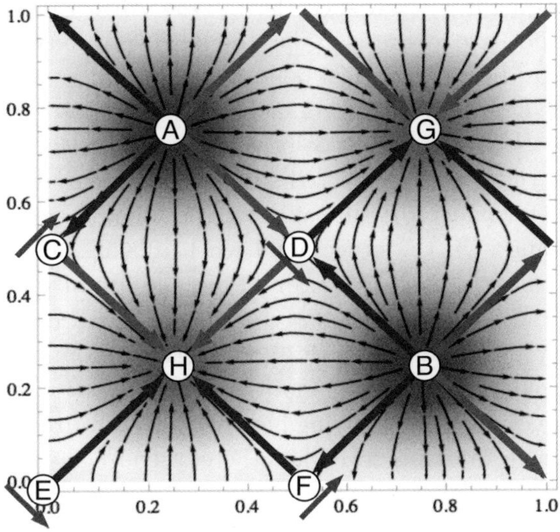

Fig. 7.9. Steepest ascent gradient flow lines for the function $h(x, y) = \lambda \sin(2\pi x) \sin(2\pi y)$ on the 2-torus defined by the identifications $x \simeq x + 1$, $y \simeq y + 1$. The critical points A, B are minima, G, H are maxima, C, D, E, F are saddle points. Big green arrows are paths which give a positive contribution to Q, big purple arrows give negative contributions. Little red arrows indicate wave function orientations.

where the sum is over steepest ascent flows from j to i satisfying $F_i - F_j = 1$. The prefactor $n_\gamma = \pm 1$ and h_{ij} is the instanton action, $h_{ij} = |h(i) - h(j)|$. This simplification in the path integral is referred to as localization. We will not derive this here; see [94] chapter 10 for a detailed pedagogical account.

The sign of n_γ is determined by orientation considerations. This is best explained by going back to the torus example. First we need to choose an orientation for the perturbative ground states $|i\rangle$. In the figure the critical points are denoted by the letters A to G, and we will denote the perturbative ground states by these letters too. Then $|A\rangle$ and $|B\rangle$ are 0-forms, $|C\rangle, |D\rangle, |E\rangle, |F\rangle$ are 1-forms, and $|G\rangle, |H\rangle$ are 2-forms. For example, to lowest order in perturbation theory,

$$|A\rangle = \alpha\, e^{-\frac{\lambda}{2}(2\pi)^2(u^2+v^2)}, \quad \alpha = \sqrt{2\pi\lambda}, \quad u = x - \tfrac{1}{4}, \quad v = y - \tfrac{3}{4}, \tag{7.120}$$

$$|C\rangle = \alpha\, e^{-\frac{\lambda}{2}(2\pi)^2(u^2+v^2)}\, du\,, \quad u = x + y - \tfrac{1}{2}, \quad v = x - y + \tfrac{1}{2}, \tag{7.121}$$

$$|G\rangle = \alpha\, e^{-\frac{\lambda}{2}(2\pi)^2(u^2+v^2)}\, du \wedge dv\,, \quad u = x - \tfrac{3}{4}, \quad v = y - \tfrac{3}{4}. \tag{7.122}$$

Note that whereas the top and bottom forms can be given a natural orientation inherited from the orientation of the torus itself, this is not so for the 1-forms. We choose them such that at the critical point, the form is positive in the positive x-direction, i.e. the local coordinate u that is such that $h \sim -u^2 + v^2$ near the critical point is taken to increase when going to the right. In the figure, the orientation of the local u-coordinate is indicated by the red arrows. The arrow is a local frame for the unstable direction. More generally, the choice of ordered standard local coordinates near each critical point determines a local frame for the unstable directions of h at each critical point. On the other hand when moving along an instanton trajectory, we can transport this frame from the initial point to the final point, and then add to it a new vector pointing in the direction of the path, to obtain a new frame for the unstable directions around the final point (which indeed generically has one more unstable direction than the initial point, in the direction of the arriving path). If this new frame has the same orientation as the frame we already had there, then $n_\gamma = +1$, and otherwise $n_\gamma = -1$.

Using these rules, and noticing that in this example $\Delta h = \lambda$ for successive critical points, we get for example $\hat{Q}|A\rangle = e^{-\lambda}(|D\rangle + |F\rangle - |C\rangle - |E\rangle)$, $\hat{Q}|C\rangle = e^{-\lambda}(|H\rangle - |G\rangle)$, $\hat{Q}|G\rangle = 0$. In matrix form, we thus compute $\hat{Q}$, $\hat{Q}^\dagger$ and $\hat{H} = \frac{1}{2}\{\hat{Q}, \hat{Q}^\dagger\}$ (writing zeros as blanks):

$$\hat{Q} = e^{-\lambda} \begin{pmatrix} & & -1 & 1 & -1 & 1 & & \\ & & 1 & -1 & 1 & -1 & & \\ & & & & & & -1 & 1 \\ & & & & & & -1 & 1 \\ & & & & & & 1 & -1 \\ & & & & & & 1 & -1 \\ & & & & & & & \\ & & & & & & & \end{pmatrix}, \quad \hat{H} = 2\,e^{-2\lambda} \begin{pmatrix} 1 & -1 & & & & & & \\ -1 & 1 & & & & & & \\ & & 1 & & & -1 & & \\ & & & 1 & -1 & & & \\ & & & -1 & 1 & & & \\ & & -1 & & & 1 & & \\ & & & & & & 1 & -1 \\ & & & & & & -1 & 1 \end{pmatrix}.$$

The space of true supersymmetric ground states is isomorphic to the cohomology $\ker \hat{Q} / \operatorname{im} \hat{Q}$ of the matrix $\hat{Q}$, which is relatively easily computed. This is called the Morse-Witten cohomology. Alternatively, we can find the actual supersymmetric ground states (within the approximation of working within the finite dimensional space of perturbative supersymmetric ground states, i.e. to leading order in $e^{-\lambda}$), as the joint null space of $\hat{Q}$ and $\hat{Q}^\dagger$, or equivalently the null space of $\hat{H}$. The reduced Hamiltonian $\hat{H}$ furthermore gives the dynamical transition rates between different perturbative ground states, and diagonalizing it gives the energy level splittings.

The conclusion in our example is that the supersymmetric ground states in this approximation are given by the symmetric combinations $|A\rangle + |B\rangle$, $|C\rangle + |F\rangle$, $|D\rangle + |E\rangle$, $|G\rangle + |H\rangle$, consistent with the Betti numbers $(1, 2, 1)$. The four antisymmetric combinations orthogonal to those all have lifted energies equal to $4e^{-2\lambda}$.

Of course in more general examples, the splitting pattern and linear combinations will be less simple. But in principle at least there is a general algorithm to compute it in terms of simple, generically isolated, "atomic" ($\Delta F = 1$) gradient flows, which is quite remarkable; in nonsupersymmetric quantum mechanics in dimensions greater than 1, doing something similar is typically very hard if not impossible.

For complex energy landscapes $h(x)$ in high dimensional spaces, it will typically be more appropriate to compute statistical, thermodynamic quantities, but again the structure of supersymmetry will lead to significant simplifications.

As a check on the above results, and also to get further insight into the nature of the approximation, observe that it is actually easy to write down the exact bottom and top fermion number ground state wave functions; they are:

$$|F = 0\rangle = e^{-h}\,, \qquad |F = \mathrm{dim}\rangle = e^{+h} \prod_I \bar{\psi}^I\,. \tag{7.123}$$

The approximation we made replaces this to lowest order in perturbation theory by an even superposition of Gaussians centered at the critical points. This is indeed an excellent approximation at large λ.

Note that in this limit perturbative ground states are exponentially close to being real ground states, and physically the number of perturbative ground states would therefore be a better measure for the ground state degeneracy than the Witten index. However, at generic, order 1 couplings, this is no longer the case, and one may reasonably expect all states that are not protected to be lifted to finite energies, making the Witten index an accurate, physically meaningful measure for the number of true ground states. On the other hand, as many examples in string theory have taught us, one must of course be careful that in the apparent strong coupling regime there isn't a dual weak coupling description hiding, which would again potentially produce many new near-exact ground states, not captured by the Witten index.

This is the case in particular in thermodynamic, "large N" limits. For example the SK model and other glassy systems develop many pseudo ground states in the thermodynamic limit. Although not supersymmetric, there is an interesting map from classical statistical mechanics systems to supersymmetric quantum mechanics systems, showing this is not just an analogy, but really the same thing. To this we turn next.

7.3.8. *Supersymmetry, stochastic systems and glasses*

Supersymmetric quantum mechanics can be interpreted as a BRST description of a classical stochastic system [70–72, 97, 98]. This goes as follows. Consider a stochastic dynamical Langevin equation for a particle with position $x(t)$ in a potential $h(x)$:

$$\frac{dx}{dt} = -\partial_x h(x) + \eta(t) \,, \tag{7.124}$$

where η is a Gaussian random noise function with variance α, $P(\eta) \propto e^{-\eta^2/2\alpha}$ (the noise can be thought of as thermal fluctuations with temperature proportional to α). Starting from some initial point x_i at time $t = 0$, the probability density of finding the configuration at a point x_f at time t is given by the path integral

$$P(x_f, t|x_i, 0) = \int \mathcal{D}x\, \mathcal{D}\eta\, e^{-\int_0^t \frac{1}{2\alpha}\eta^2}\, \delta\big(\frac{dx}{dt} + h'(x) - \eta\big)\, |\det\big(\frac{d}{dt} + h''(x)\big)| \,, \tag{7.125}$$

appropriately normalized and with boundary conditions $x(0) = x_i$, $x(t) = x_f$. The combination of delta function and determinant ensures that the path integration over x simply picks out with unit weight the unique solution to the first order linear differential equation (7.124) for given η. It is the functional analog of the finite dimensional formula $\int dx\, \delta(A{\cdot}x{+}B)\, |\det A| = 1$ where A is a constant matrix and B is a constant vector. Ignoring for the time being the absolute value signs, we can write the determinant as a fermionic path integral, so that amusingly, after also integrating out η, we obtain exactly the Euclidean propagator of supersymmetric quantum mechanics with superpotential $h(x)$:

$$P(x_f, t|x_i, 0) = \int \mathcal{D}x\, \mathcal{D}\bar\psi\, \mathcal{D}\psi\, e^{-\int_0^t \frac{1}{2\alpha}(\dot x + h'(x))^2 + \bar\psi\dot\psi + \bar\psi h''(x)\psi} \tag{7.126}$$

$$= e^{(h(x_i)-h(x_f))/\alpha}$$
$$\times \int \mathcal{D}x\, \mathcal{D}\bar\psi\, \mathcal{D}\psi\, e^{-\int_0^t \frac{1}{2\alpha}(\dot x^2 + h'(x)^2) + \bar\psi\dot\psi + \bar\psi h''(x)\psi} \tag{7.127}$$

$$= e^{(h(x_i)-h(x_f))/\alpha}\, \langle x_f| e^{-tH_{\rm susy}/\alpha} |x_i\rangle \,. \tag{7.128}$$

The last line is the propagator between states in the zero fermion number sector, with elapsed Euclidean time t and Planck's constant $\hbar = \alpha$ (before we were setting $\hbar \equiv 1$). Inserting a complete set of ($F = 0$) energy eigenstates $|n\rangle$, we have

$$\langle x_f| e^{-tH_{\rm susy}/\alpha} |x_i\rangle = \sum_n \langle x_f|n\rangle\langle n|x_i\rangle e^{-tE_n/\alpha} \,. \tag{7.129}$$

In the long time limit $t \to \infty$, only the ground state survives. As we have seen, when $Z \equiv \int dx\, e^{-2h(x)/\alpha} < \infty$, the Witten index is 1 and the unique supersymmetric ground state is $\langle x|0\rangle = \frac{1}{\sqrt{Z}} e^{-h(x)/\alpha}$. We conclude

$$P(x_f, \infty | x_i, 0) = \frac{1}{Z} e^{-\beta\, h(x_f)}\,, \qquad \beta \equiv \frac{2}{\alpha}\,, \tag{7.130}$$

with $Z = \int dx\, e^{-\beta h(x)}$. This is indeed the expected Boltzmann equilibrium distribution with inverse temperature β, which alternatively can be derived from the Fokker-Planck equation associated to the Langevin equation. In fact the Fokker-Planck equation can easily be seen to be equivalent to the Schrödinger equation in the zero fermion number sector (identifying $P(x) \propto e^{h(x)}\psi(x)$), providing an alternative derivation of the above identifications.

Note that when the Witten index is -1, i.e. $e^{+h(x)}$ is normalizable, there is no future equilibrium state, but there is a past equilibrium state. When it is zero, there is no future and no past equilibrium. If there are multiple *perturbative* supersymmetric ground states with fermion number zero (i.e. local minima of $h(x)$), then in the small α (i.e. large β) limit, there will be highly metastable states. According to (7.129), the time scale for decay of these metastable states is set by the nonperturbatively lifted energy, which as we have seen in the previous section can be computed from simple instantons and the Morse-Witten complex.

All this generalizes in a straightforward way to higher dimensional and curved spaces. If $h(x)$ is the Hamiltonian of some classical statistical mechanical system, then in the thermodynamic limit $N \to \infty$ there may be degenerations that change the number of exact supersymmetric ground states at a given fermion number and even the Witten index. This is because degeneracy lifting instanton effects may vanish altogether, thus turning perturbative ground states into exact ground states. When this happens at fermion number zero, ergodicity is broken and we get multiple exact equilibrium states. Whether or not this happens will depend on the distribution of energy barriers, i.e. the distribution of supersymmetric ground states with low fermion number, and how they are connected by instantons, i.e. the Morse-Witten complex. This should capture, if present, the ultrametric structure of the state space.

Some discussion of spin glasses in the language of supersymmetric quantum mechanics can be found in e.g. [59, 71, 72, 99]. Recent considerations in string theory involving the above connection between stochastic equations and supersymmetry include [100, 101]. A nice example of the relation between transition rate distributions and universal properties of relaxation dynamics can be found in [102], where the typical $\log(1 + t_w/t)$ behavior of return to equilibrium after a perturbation of duration t_w of a glass was derived from general arguments based on the form of a transition matrix with exponentially suppressed off-diagonal matrix elements. Such a transition matrix and Hamiltonian are typical for systems with a wide distribution of relevant instanton actions.

7.4. D-Brane Landscapes

7.4.1. *Introduction*

We now move on to the description of complex systems that appear in the context of string theory, and review some of the tools that have been used to describe their basic features, such as ground state degeneracies and distributions. One of the interesting things string theory adds to conventional analysis is the power of holographic duality [103], which in favorable circumstances can reduce a quantitative understanding of the strong coupling physics of the model under consideration to simple computations in gravity. This works particularly well for thermodynamics, as black holes are the holographic duals to thermodynamic states, capturing the strong coupling thermodynamics of the model with often rather astonishing simplicity.

More specifically, we will have a look at the complex configuration spaces that arise when wrapping D-branes around various cycles of compactification manifolds, with various worldvolume fields turned on. Such brane configurations arise in type II constructions of string vacua, where they are among other things responsible for the particle physics content of the compactification. They also arise in the description of charged black hole microstates, extrapolated to weak coupling. In the weak coupling limit (large volumes, small string coupling), the only difference between the two as far as their perturbative description is concerned is the fact that in the former case, branes are filling the observable, noncompact space, while in the latter, they are localized at a point. But in particular the geometric internal space description of the configuration is mathematically identical in the two cases. Of course the actual dynamics and other physics of the two systems will be very different, one describing the universe we may live in, the other an object we may look at, but the fact that they are related at the level of classical configuration spaces is one of the beautiful incarnations of the unifying nature of string theory. Thus, in particular, the largeness of the landscape and the largeness of the entropy of a black hole are intimately connected.

Despite this map, the fact that branes fill all of space has one dramatic consequence, and that is that the charges (brane wrapping numbers etc) are tightly constrained by so-called tadpole cancellation conditions. What this means is simply that whenever we have charged objects (such as a branes) with compact transversal dimensions (as is the case for space-filling branes but not space-localized ones), the sum of all charges must necessarily be zero. This is a consequence of Gauss' law, or more colloquially, the fact that flux lines otherwise have no place to go to; when there's a source, there must be a sink. Combined with the requirement of stability (typically realized by staying close to a supersymmetric configuration), this imposes strict bounds on the allowed charges.[bb] On the other hand, for space localized branes, no such charge restrictions arise, and in particular we can take the charges to infinity and consider thermodynamic limits, which can be expected to have universal properties. This motivates us to consider localized brane systems.

At finite coupling another difference between the two creeps in: the presence of gravity. This has a very different effect in the two settings. In the space-filling case, "vacuum" solutions with slightly different internal brane configurations can give rise to very different asymptotic geometries. As emphasized by Tom Banks at this school [8], for various reasons including the fact that our most successful descriptions of quantum gravity use fixed spacetime asymptotics in an essential way, this makes it unclear if we can even think of these different solutions as being part of the same theory, or, as he would put it, it makes it clear that these solutions are not part of the same theory. This does not mean that the presence of a space of brane configurations is of no physical relevance here of course; they do all *exist* after all, and it is certainly of interest to classify the set of all possibilities. Furthermore local fluctuations as well as larger fluctuations in the form of bubble nucleations can certainly probe — albeit in a limited sense — the configuration space. But it does mean that the situation is conceptually much more subtle than, say, a collection of manganese atoms in copper. In

[bb]These bounds are set by the background curvature or orientifold contributions to the charge, which in turn are set by topological invariants such as the Euler number of the compactification manifold. Although these numbers can get fairly high, for example there are known [2] elliptic CY 4-folds suitable for F-theory compactifications allowing up to $\mathcal{O}(10^5)$ transversal D3-branes, constructions get increasingly sparse at the higher end, leading many to suspect the set is finite. For reasons we do not understand any deeper than through case by case examination of various conspiracies, the string theory landscape is uncannily "boxed in", making it for example very hard, and plausibly impossible [38], to find infinite families of parametrically controlled solutions with positive cosmological constant.

contrast, for space-localized branes, these issues do not arise. Gravitational backreaction just turns the weakly coupled brane system into a black hole, leaving the asymptotics unaffected. Even better, thanks to the effects of gravity, we get a host of thermodynamic parameters for free, courtesy of the Bekenstein-Hawking entropy formula. This gives another motivation to study brane systems associated to black holes.

Besides their intrinsic interest as models of complex, glassy systems with intricate landscapes and known holographic dual descriptions, the study of such localized brane systems can also teach us things about the space-filling systems. Given their close physical and mathematical relation, it is clear that they contain information about each other. For example, their Witten indices, suitably defined, are the same. Indeed, Witten indices of supersymmetric quantum field theories are traditionally computed by reduction to a quantum mechanical model. In the present context one can go further, and use the computational power of gravity (manifested through for example the Bekenstein-Hawking formula or its refinements [104]) to extract information about ensembles of possible string vacua. This idea was made precise in [105]; I will not have much to say about it in these lectures, except that it serves as yet another motivation to study space-localized branes.

The analysis of D-brane systems as mean-field glassy systems, or in other words connecting this section and section 7.2, will be addressed in [5].

7.4.2. *Low energy physics of the weakly coupled wrapped D4*

In this section we derive in detail the quantum mechanics describing the low energy dynamics of a D4-brane wrapped on a 4-cycle, in the limit of small string coupling and large volume. Besides being an interesting and accessible example of a complex system in string theory with applications to black hole physics, mathematics and string model building, it also provides a clean example illustrating various useful Kaluza-Klein reduction techniques, in particular the relation between fluxes and superpotentials. (The idea to consider such an open string landscape was proposed in [106].)

7.4.2.1. *Actions*

Consider type IIA string theory on $\mathbb{R}^4 \times X$, where X is some compact 6-dimensional manifold. The part of the IIA bulk action relevant for our purposes is, in units with the string scale $\ell_s = 2\pi\sqrt{\alpha'} \equiv 1$:

$$S_{\mathrm{IIA}} = S_{\mathrm{IIA,NSNS}} + S_{\mathrm{IIA,RR}} \tag{7.131}$$

$$S_{\mathrm{IIA,NSNS}} = 2\pi \int_{\mathbb{R}^4\times X} d^{10}x\,\sqrt{-g}\left(\frac{1}{g_s^2}R - \frac{1}{g_s^2}H_{MNP}H^{MNP}\right) \tag{7.132}$$

$$S_{\mathrm{IIA,RR}} = -2\pi \int_{\mathbb{R}^4\times X} d^{10}x\,\sqrt{-g} \times \left(\frac{1}{2}F^{(2)}_{MN}F^{(2)MN} + \frac{1}{2}F^{(4)}_{MNPQ}F^{(4)MNPQ}\right), \tag{7.133}$$

where $H = dB$ is the NSNS 3-form field strength, $F^{(p+2)} = dC^{(p+1)}$ are the RR field strengths and g_s is the string coupling constant. The theory contains Dp-branes with p even. In particular a single D4 wrapped around a 4-cycle Σ in the compact manifold X has, in the limit of large volume and weak string coupling, an effective worldvolume action

$$S_{D4} = S_{\mathrm{D4,NSNS}} + S_{\mathrm{D4,RR}} \tag{7.134}$$

$$S_{\mathrm{D4,NSNS}} = -\frac{2\pi}{g_s}\int_{\mathbb{R}\times\Sigma} d^5x\,\sqrt{-h}\left(1 + \frac{1}{2}F_{IJ}F^{IJ}\right) \tag{7.135}$$

$$S_{\mathrm{D4,RR}} = 2\pi \int_{\mathbb{R}\times\Sigma} F\wedge C^{(3)} + \left(\frac{1}{2}F\wedge F + \frac{1}{24}e(R)\right)\wedge C^{(1)}, \tag{7.136}$$

where h_{IJ} is the induced brane metric, $F = dA + B$, A is the $U(1)$ gauge connection living on the brane and $e(R)$ is a quadratic polynomial in the tangent and normal bundle curvature forms. When X is a Calabi-Yau manifold, this is the Euler density $e(R)$:

$$e(R) = \mathrm{Pf}\,\mathcal{R}, \qquad \mathcal{R}^I{}_J \equiv \frac{1}{4\pi}R^I{}_{JKL}\,dx^K\wedge dx^L, \tag{7.137}$$

which we encountered earlier in (7.115).

The couplings to the RR fields given by $S_{\mathrm{D4,RR}}$ are topological in the sense that they are metric-independent. They specify the gauge flux and curvature induced D2- and D0-charges on the brane.

We have ignored curvature corrections to $S_{\mathrm{D4,NSNS}}$, but in the end we will fix this by matching the energy required by supersymmetry for a given charge. A slightly more precise version of this action and references to derivations in the literature can be found in section 2 of [107].

7.4.2.2. *Bulk Kaluza-Klein reduction*

The low energy dynamics of the wrapped D4 system is obtained by Kaluza-Klein reduction. This is done most easily, and leads to the richest structure, when there is residual supersymmetry. An $\mathcal{N} = 1$ theory in four dimensions does not have pointlike objects that preserve part of the supersymmetry, so one has to consider at least $\mathcal{N} = 2$, i.e. compactifications with eight preserved supercharges, with the wrapped D-brane breaking half of those. When no bulk magnetic fluxes are turned on, this requires the compactification manifold X to be Calabi-Yau, i.e. Ricci-flat and Kähler. Concretely this means that there exist complex coordinates y^m, $m = 1, 2, 3$ and a holomorphic complex orthonormal frame $\theta^m = \theta^m{}_n dy^n$, such that the metric is of the form

$$ds^2 = |\theta^1|^2 + |\theta^2|^2 + |\theta^3|^2 \,, \tag{7.138}$$

with associated Kähler form J and covariantly constant holomorphic 3-form Ω given by

$$J = \frac{i}{2}\left(\theta^1 \wedge \bar\theta^1 + \theta^2 \wedge \bar\theta^2 + \theta^3 \wedge \bar\theta^3\right) \,, \qquad \Omega = \theta^1 \wedge \theta^2 \wedge \theta^3 \,. \tag{7.139}$$

The Calabi-Yau condition means that J and Ω are globally well-defined. The Kähler form has one holomorphic and one anti-holomorphic index and is therefore called a $(1,1)$-form. Similarly Ω is a $(3,0)$-form. With the above definitions we have

$$i \int_X \Omega \wedge \overline{\Omega} = 8 \int_X \frac{J^3}{6} = 8 V_X \,, \tag{7.140}$$

where V_X is the volume of X.

As a concrete example we can take X to be the quintic Calabi-Yau, described by a homogeneous degree 5 polynomial equation in $\mathbb{CP}^4$. The four complex dimensional projective space $\mathbb{CP}^4$ is the set of all nonzero $(x_1, x_2, x_3, x_4, x_5) \in \mathbb{C}^5$, modulo the equivalence relation $x \simeq \lambda x$, $\lambda \in \mathbb{C}^*$. The simplest choice of polynomial is of Fermat form:

$$X : Q_5(x) \equiv x_1^5 + x_2^5 + x_3^5 + x_4^5 + x_5^5 = 0 \,. \tag{7.141}$$

Deformations of the metric preserving Calabi-Yauness are called (geometric) moduli. The four dimensional low energy effective field theory will contain these moduli as massless scalars. In general it is not possible to write down explicit expressions for the Ricci flat Calabi-Yau metric, but Yau's theorem states that for a given Kähler class (i.e. the cohomology class of the Kähler form) and complex structure (i.e. a choice of complex coordinates), there is a unique Ricci flat Kähler metric. Thus, there are two kinds of moduli, those specifying the cohomology class of the Kähler form and those specifying the complex structure. In the case at hand there is just one Kähler modulus, which parametrizes the overall scale of the metric, i.e. the volume of X — the second cohomology of X is one-dimensional. The different complex structures are simply parametrized by the choice of defining polynomial $Q_5(x)$, modulo linear coordinate transformations of the x_i. This space is 101-dimensional.

More generally, the dimension of the second cohomology of X will be $b^2(X)$-dimensional, i.e. there are $b^2(X)$ independent harmonic 2-forms on X. Equivalently, because of Poincaré duality,[cc] there are $b^2(X)$ independent homology 4-cycles in X. Let D_A, $A = 1, \ldots, B_2 \equiv b^2(X)$ be an integral basis of harmonic 2-forms. Then

$$J = J^A D_A \,, \tag{7.142}$$

and the J^A are the Kähler moduli. We choose the signs of the 2-forms D_A such that $J^A > 0$.

The reduction of the RR potentials produces, among other fields, electrostatic potentials in four dimensions coupling to D2 and D0 charges:

$$C^{(3)} = \phi^A \, dt \wedge D_A + \cdots \,, \qquad C^{(1)} = \phi^0 \, dt + \cdots \,. \tag{7.143}$$

7.4.2.3. *Brane Kaluza-Klein reduction*

In the weak coupling limit $g_s \to 0$, the backreaction of the wrapped D4 on the bulk geometry is negligible, due to the lower power in $1/g_s$ appearing in the D4-action. We can therefore consider the D-brane to be a probe in a background geometry specified by arbitrary, fixed bulk moduli. For the wrapped brane to be supersymmetric, it must wrap a minimal volume 4-cycle, or somewhat stronger even, it must be holomorphic [108]. In the case of the quintic this is equivalent to saying it can be described by a degree N homogeneous polynomial equation

$$\Sigma \,:\, P_N(x_1, x_2, x_3, x_4, x_5) = 0 \,. \tag{7.144}$$

The degree N can be identified with the multiplicity of the D4-charge. For example for $N = 1$, the most general polynomial is $P_1(x) = a_1 x_1 +$

[cc] Poincaré duality is a natural isomorphism between p-cycle homology classes and $(n-p)$-form cohomology classes in an n-dimensional space. For a p-cycle C locally given by $n-p$ equations $f_i(x) = 0$, the associated dual cohomology class can be represented by the closed delta-function $(n-p)$-form $\hat{C} = \delta(f_1)\, df_1 \wedge \cdots \wedge \delta(f_{n-p})\, df_{n-p}$. Note that for two cycles C and C' with dimension adding up to n, we can thus write $\int \hat{C} \wedge \hat{C}' = \int_C \hat{C}' = \#(C \cap C')$, where the last expression counts intersection points for generic representatives, with signs according to orientations. Usually we will drop the hat in the dual to avoid cluttering.

$\cdots + a_5 x_5$. The complex coefficients c are deformation moduli of Σ. Since overall rescaling of the coefficients does not change the zero set, we have a four complex dimensional deformation moduli space, topologically $\mathbb{CP}^4$. An example of a degree N polynomial is $(P_1(x))^N$, which corresponds to N coincident branes wrapping $P_1(x) = 0$, with gauge group enhanced to $U(N)$. The most general degree N homogeneous polynomial on the other hand gives a smooth singly wrapped brane with gauge group $U(1)$. In the case at hand it has $d(N) = \frac{5N^3}{6} + \frac{50N}{12} - 1$ independent deformation moduli z^a, as can be verified by direct monomial counting, taking into account that when $N \geq 5$, adding a term $q_{N-5}(x)Q_5(x)$ to $P_N(x)$, with Q_5 as in (7.141), does not change the zero set. More generally one can deduce the number of deformations using standard algebraic geometry techniques [109]. For a D4-brane wrapping a homology class[dd] $[\Sigma] = N^A D_A$ with $N^A > 0$ this gives

$$d = \frac{1}{6} D_{ABC} N^A N^B N^C + \frac{1}{12} c_{2,A} N^A - 1 \,. \tag{7.145}$$

Here the "triple intersection numbers" D_{ABC} are defined as

$$D_{ABC} \equiv \int_X D_A \wedge D_B \wedge D_C = \#(D_A \cap D_B \cap D_C) \tag{7.146}$$

and the $c_{2,A}$ are topological numbers (the second Chern class of X) depending on X only. For the quintic we see by comparing that $D_{111} = 5$ and $c_{2,1} = 50$. The same techniques allow straightforward computation of the generic Euler characteristic of Σ, namely

$$\chi = D_{ABC} N^A N^B N^C + c_{2,A} N^A \,, \tag{7.147}$$

which for the quintic leads to $\chi = 5\,N^3 + 50\,N$. The second Betti number, i.e. the number of independent 2-cycles or 2-forms on Σ is obtained from this as $b_2(\Sigma) = \chi - 2 - b_1 - b_3$. When $N^A > 0$ the Betti numbers $b_1 = b_3$ of Σ are inherited from X, so they vanish except in the higher susy cases when X is T^6 ($b_1 = 6$) or $T^2 \times K3$ ($b_1 = 2$). In any case, b_2 grows as N^3, and therefore, crucially, we will get a huge magnetic flux degeneracy on these branes in the large N limit. Indeed we can turn on harmonic gauge field strengths $F = dA$ on Σ characterized by b_2 integers S_i, the flux quanta:

$$F = \sum_{i=1}^{b_2(\Sigma)} S_i \, \sigma^i \,, \tag{7.148}$$

[dd]Here the D_A are the Poincaré duals to the D_A introduced in (7.142), for which as mentioned in footnote cc we use the same notation.

where the σ^i form a basis of integrally quantized harmonic 2-forms on Σ. In terms of Poincaré dual 2-cycles, we can equivalently write

$$S_i = \eta_{ij} \int F \wedge \sigma^j = \eta_{ij} \int_{\sigma^j} F \,, \qquad \eta^{ij} \equiv \int_\Sigma \sigma^i \wedge \sigma^j = \#(\sigma^i \cap \sigma^j) \,, \quad (7.149)$$

with η_{ij} the inverse of η^{ij}. The matrix η^{ij} is called the intersection form of Σ. It is integral and unimodular, but in general not positive definite — in fact as we will see later at large N it has signature $(b_2^+, b_2^-) \propto (\frac{1}{3}, \frac{2}{3}) \times N^3$.

The flux and curvature induced D2 and D0 brane charge can be read off from (7.136) with the reduction (7.143):

$$S_{\text{D4,RR}} = 2\pi \int dt \, \left(-q_0\, \phi^0 + q_A\, \phi^A\right) \quad (7.150)$$

$$q_0 = -\frac{1}{24}\, \chi(\Sigma) - \frac{1}{2} \int_\Sigma F \wedge F = -\frac{\chi}{24} - \frac{1}{2} \eta^{ij}\, S_i S_j \quad (7.151)$$

$$q_A = \int_\Sigma F \wedge D_A = D_A^i S_i \;, \quad (7.152)$$

where the integers D_A^i are given by

$$D_{A,i} = \int_\Sigma \sigma^i \wedge D_A = \#(\sigma^i \cap D_A) \,. \quad (7.153)$$

Kaluza-Klein reduction of the D4 produces a supersymmetric quantum mechanics with $d(\Sigma)$ complex continuous degrees of freedom z^a and $b_2(\Sigma)$ discrete flux degrees of freedom S_i. The moduli z^a parametrize a moduli space $\mathcal{M}$. The discrete fluxes can be thought of as quantized momenta dual to periodic coordinates that do not explicitly appear in the Hamiltonian. The wrapped D-brane with $F = 0$ preserves 4 supercharges; it is the dimensional reduction from 4 to 1 of a four dimensional $\mathcal{N} = 1$ theory. In a sector with fixed flux we therefore expect the low energy effective action for the z^a to be of the general chiral multiplet form [110][ee]

$$S = \frac{2\pi}{g_s} \int dt\, \frac{1}{2} \left(g_{a\bar{b}} \dot{z}^a \dot{\bar{z}}^b - g^{a\bar{b}} \partial_a W(z) \bar{\partial}_{\bar{b}} \bar{W}(\bar{z}) \right) + \cdots \,, \quad (7.154)$$

where the $+\cdots$ consists of terms independent of z, $W(z)$ is a holomorphic superpotential (absent when $F = 0$ but generically nonzero when $F \neq 0$), and $g_{a\bar{b}}$ a Kähler metric:

$$g_{a\bar{b}} = \partial_a \bar{\partial}_{\bar{b}}\, K(z, \bar{z}) \,. \quad (7.155)$$

[ee] Note that this is of the form (7.109) (in its Lorentzian version) with $h = \frac{1}{2}(W + \bar{W})$. The special form of h is due to the presence of four supersymmetries Q_α, $Q_\alpha^\dagger$, $\alpha = 1, 2$, rather than the generic minimal amount of 2 supercharges.

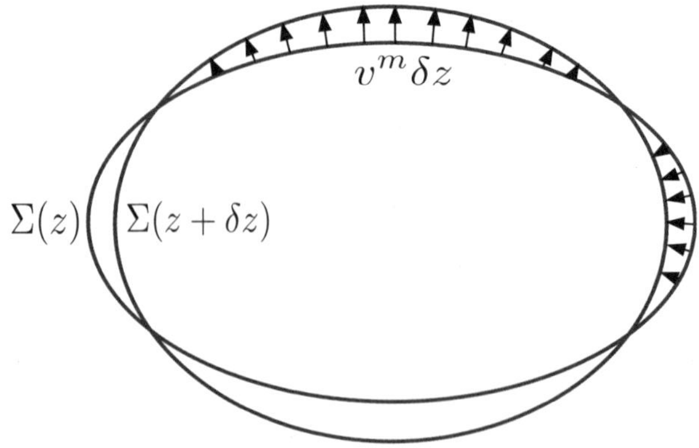

Fig. 7.10. Infinitesimal variation of Σ along its moduli space.

We have explicitly retained the overall prefactor $\frac{2\pi}{g_s}$ instead of absorbing it in the metric and superpotential. In this way g_s can be thought of as Planck's constant in the supersymmetric quantum mechanics.

To identify the metric, we should take the brane to be slowly moving along its moduli space and expand the first term in (7.135) to second order in the velocities. An infinitesimal displacement δz^a along the moduli space causes an infinitesimal normal displacement $\delta y^m = v_a^m \delta z^a$ of Σ inside X, where the v_a^m depend holomorphically on the coordinates of Σ. This is illustrated in Fig. 7.10. Thus, when slowly moving along the moduli space with velocity $\dot{z}^a$, the first term in (7.135) becomes

$$-\int_{\mathbb{R}\times\Sigma} d^5x\,\sqrt{-h} = \int dt \int_\Sigma dV \left(-1 + \frac{1}{2} g_{m\bar{n}} v_a^m \bar{v}_{\bar{b}}^{\bar{n}}\, \dot{z}^a \dot{\bar{z}}^{\bar{b}}\right), \qquad (7.156)$$

where dV is the volume element on Σ. Choosing the orthonormal frame θ appearing in (7.139) such that θ^1 and θ^2 lie along Σ (i.e. they span $T^*\Sigma$) while θ^3 is normal to it, we can write $dV = \left(\frac{i}{2}\theta^1 \wedge \bar{\theta}^1\right) \wedge \left(\frac{i}{2}\theta^2 \wedge \bar{\theta}^2\right)$ so the kinetic term becomes

$$\left(\frac{1}{8}\int_\Sigma \theta^1 \wedge \theta^2 v_a^3 \wedge \bar{\theta}^1 \wedge \bar{\theta}^2 \bar{v}_{\bar{b}}^{\bar{3}}\right) \dot{z}^a \dot{\bar{z}}^{\bar{b}} = \left(\frac{1}{8}\int_\Sigma \omega_a \wedge \bar{\omega}_{\bar{b}}\right) \dot{z}^a \dot{\bar{z}}^{\bar{b}}, \qquad (7.157)$$

where

$$\omega_a = \Omega \cdot v_a = \theta^1 \wedge \theta^2\, v_a^3 \qquad (7.158)$$

is the contraction of Ω with the vector field v_a. It is a well-defined holomorphic (2,0) form on Σ. This map is an isomorphism between deformations

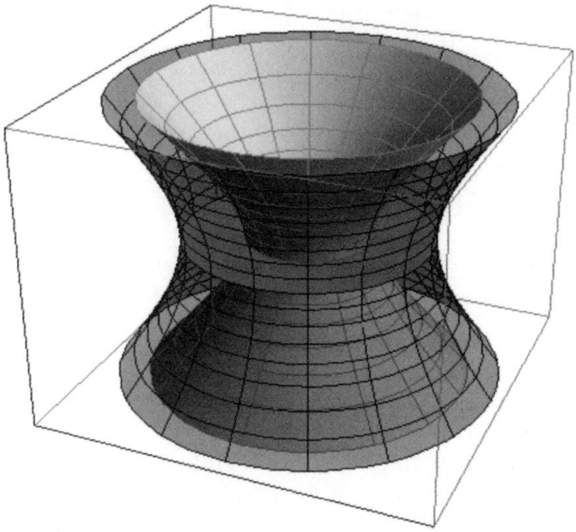

Fig. 7.11. The inner (yellow) hyperboloid represents $\Sigma(0)$, the outer (blue) one is $\Sigma(z)$, the green horizontal disk stretched between the two represents the 3-chain $\Gamma(z)$, and its boundary in $\Sigma(z)$ is $\sigma(z)$.

of Σ and harmonic (2, 0)-forms on Σ. By comparing to (7.154) we thus identify

$$g_{a\bar{b}} = \frac{1}{4} \int_{\Sigma} \omega_a \wedge \bar{\omega}_{\bar{b}} \,. \tag{7.159}$$

To show that this is Kähler, we first expand the ω_a in the integral basis introduced in (7.148):

$$\omega_a = \omega_{ai}\sigma^i \,, \qquad \omega_{ai} = \eta_{ij} \int_{\sigma^j} \omega_a \,. \tag{7.160}$$

Now fix an arbitrary reference point $z^a \equiv 0$ in the moduli space $\mathcal{M}$ parametrizing the holomorphic cycles $\Sigma(z)$. If we move away from this point, the 2-cycles σ^i will sweep out 3-chains $\Gamma^i(z)$. This is illustrated in Fig. 7.11. We can integrate the holomorphic 3-form over these 3-chains, producing the holomorphic 3-chain periods

$$\Pi^i(z) = \int_{\Gamma^i(z)} \Omega \,. \tag{7.161}$$

Because Σ is holomorphic this is independent of the choice of representative of σ^i inside $\Sigma(z)$. By the definition of the ω_a, we then have

$$\partial_a \Pi^i(z) = \int_{\sigma^i(z)} \Omega \cdot v_a(z) = \int_{\sigma^i} \omega_a \,, \tag{7.162}$$

and combining this with (7.160) and (7.159) gives us

$$g_{a\bar{b}} = \frac{1}{4}\,\partial_a \Pi^i\,\eta_{ij}\,\bar{\partial}_{\bar{b}}\bar{\Pi}^j = \partial_a \bar{\partial}_{\bar{b}} K(z,\bar{z})\,, \qquad K = \frac{1}{4}\,\Pi^i\,\eta_{ij}\,\bar{\Pi}^j\,. \tag{7.163}$$

This makes it manifest that the metric is Kähler.

To identify the superpotential, we consider the second term in (7.135), which for a given magnetic flux F gives a potential energy

$$\int_\Sigma \frac{1}{2} F \wedge \star F = \int_\Sigma \frac{1}{4}(F + \star F) \wedge (F + \star F) - \int_\Sigma \frac{1}{2} F \wedge F\,, \tag{7.164}$$

where $\star$ is the Hodge star operator on Σ and we used $\star^2 = 1$. The second term is topological — it does not depend on the z^a, and (7.151) shows it is nothing but the D0-charge induced by the fluxes. Thus, the magnetic potential energy is bounded below by the flux induced D0-charge, with the bound saturated by anti-self-dual flux $F = -\star F$. In the following we will show that the first term is (almost) the $|\partial W|^2$ term of (7.154).

The $\star$ operator commutes with the decomposition of 2-forms according to their number of (holomorphic, antiholomorphic) indices.[ff] It acts as $+1$ on $(2,0)$ forms: $\star\omega^{(2,0)} = \omega^{(2,0)}$. This is easily checked by considering a holomorphic orthonormal frame of Σ: $\star(\theta^1 \wedge \theta^2) = +\theta^1 \wedge \theta^2$. The complex conjugate $(0,2)$-forms are self-dual as well. For $(1,1)$-forms there are two cases to consider: $(1,1)$-forms proportional to the Kähler form are again self-dual, while all $(1,1)$ forms orthogonal to that are anti-selfdual, i.e. $\star = -1$. This is checked by considering $\theta^1 \wedge \bar{\theta}^1 + \theta^2 \wedge \bar{\theta}^2$ resp. $\theta^1 \wedge \bar{\theta}^1 - \theta^2 \wedge \bar{\theta}^2$.

Accordingly the first term in (7.164) is

$$\int_\Sigma \frac{1}{4}(F + \star F) \wedge (F + \star F) = \int_\Sigma 2\,F^{(2,0)} \wedge F^{(0,2)} + F_+^{(1,1)} \wedge F_+^{(1,1)}\,. \tag{7.165}$$

The component $F_+^{(1,1)}$ is obtained by projection onto the Kähler form $J = J^A D_A$ (pulled back to Σ):

$$F_+^{(1,1)} = f_+ J\,, \quad f_+ = \frac{1}{\int_\Sigma J^2} \int_\Sigma F \wedge J = \frac{1}{\int_\Sigma J^2}\, q_A J^A\,,$$
$$\int_\Sigma J^2 = D_{ABC} N^A J^B J^C\,, \tag{7.166}$$

[ff]This is true because Σ is a Kähler manifold, being a complex submanifold of the Kähler manifold X.

where we used (7.152). This expression is again independent of the D4 moduli z^a. The second term in (7.165) thus equals $\frac{1}{\int_\Sigma J^2}(q_A J^A)^2$. Note this is independent of the overall scale of J.

The remaining term does depend on the moduli z^a and we will see it can be identified with $|\partial W|^2$. Expand $F^{(2,0)}$ in the basis of $(2,0)$-forms defined in (7.158):

$$F^{(2,0)} = f^a \omega_a \,, \qquad f^a = \frac{1}{4}\, g^{a\bar{b}} \int_\Sigma F \wedge \bar{\omega}_{\bar{b}} \,, \tag{7.167}$$

with $g_{a\bar{b}}$ defined in (7.159) and $g^{a\bar{b}}$ its inverse. Using the decomposition (7.148) and the relation to chain period derivatives (7.162), we write

$$\int_\Sigma F \wedge \bar{\omega}_{\bar{b}} = S_i \int_{\sigma^i} \bar{\omega}_{\bar{b}} = \bar{\partial}_{\bar{b}} \left(S_i \bar{\Pi}^i(\bar{z}) \right) \,, \tag{7.168}$$

and the first term in (7.165) becomes $\frac{1}{2} g^{a\bar{b}} \partial_a (S_i \Pi^i) \bar{\partial}_{\bar{b}} (S_j \bar{\Pi}^j)$. Comparing to (7.154), we identify the flux induced superpotential

$$W(z) = S_i \Pi^i(z) \,. \tag{7.169}$$

We put everything together below.

7.4.3. *D4 supersymmetric quantum mechanics*

7.4.3.1. *Bosonic action*

To summarize, the low energy degrees of freedom at weak coupling and large volume of a generic[gg] D4-brane wrapped on a 4-cycle $\Sigma = N^A D_A$ in a generic Calabi-Yau 3-fold with triple intersection numbers D_{ABC} and Kähler form $J = J^A D_A$ ($J^A, N^A > 0$) are $d = b_{2+} - 1$ complex deformation moduli z^a and $b_2 = \chi - 2$ integers S_i parametrizing the $U(1)$ flux $F = S_i \sigma^i$, with $b_{2+} = \frac{1}{6} D_{ABC} N^A N^B N^C + \frac{1}{12} c_{2,A} N^A$ and $\chi = D_{ABC} N^A N^B N^C + c_{2,A} N^A$. Up to a constant energy term the bosonic part of the action is

$$S = S_{\rm dyn} + S_{\rm top} \tag{7.170}$$

$$S_{\rm dyn} = \frac{2\pi}{g_s} \int dt \left[\frac{1}{2} g_{a\bar{b}} \dot{z}^a \dot{\bar{z}}^b - \frac{1}{2} g^{a\bar{b}} \partial_a W \bar{\partial}_{\bar{b}} \bar{W} \right] \tag{7.171}$$

$$S_{\rm top} = -\frac{2\pi}{g_s} \int dt \left[\left(q_0 + (q_A \tilde{J}^A)^2 \right) + g_s \left(q_0\, \phi^0 - q_A\, \phi^A \right) \right] \,, \tag{7.172}$$

[gg]Generic means here that the brane worldvolume is smooth and the gauge group is $U(1)$. Coincident branes will give rise to an enhanced nonabelian gauge symmetry and intersecting branes will give rise to additional light bifundamentals. We also do not include pointlike D0-branes yet. We will return to this in section 7.4.4.

with superpotential and Kähler metric given by

$$W(z) = S_i \Pi^i(z), \qquad \Pi^i(z) = \int_{\Gamma^i(z)} \Omega \tag{7.173}$$

$$g_{a\bar{b}} = \partial_a \bar{\partial}_{\bar{b}} \big(\frac{1}{4} \Pi^i \, \eta_{ij} \, \overline{\Pi^j} \big) \,, \tag{7.174}$$

flux-dependent D0- and D2-charges

$$q_0 = -\frac{\chi}{24} - \frac{1}{2} \eta^{ij} S_i S_j = -\frac{\chi}{24} - \frac{1}{2} \int_\Sigma F \wedge F \tag{7.175}$$

$$q_A = D_A^i S_i = \int_\Sigma D_A \wedge F \,, \tag{7.176}$$

and unit Kähler form

$$\tilde{J} = \frac{1}{(D_{ABC} N^A J^B J^C)^{1/2}} \, J = \frac{1}{(\int_\Sigma J^2)^{1/2}} \, J \,. \tag{7.177}$$

The matrix $\eta^{ij} = \#(\sigma^i \cap \sigma^j)$ is the intersection form of Σ and $D_A^i = \#(D_A \cap \sigma^i)$. The 3-chains $\Gamma^i(z)$ are 3-chains swept out by the 2-cycles σ^i on Σ when moving from a fixed reference point to z in the moduli space. We recall that for the quintic $D_{111} = 5$, $\tilde{J}^1 = \frac{1}{\sqrt{5N}}$, $c_{2,1} = 50$.

This is a reliable description at weak coupling and low energies. What exactly does weak coupling mean? We recall from our general discussion of supersymmetric quantum mechanics in section 7.3 that weak coupling corresponds to a limit in which the superpotential h is scaled up as $h \to \lambda h$ with $\lambda \to \infty$, while keeping the metric constant. To see how this relates to other scalings, consider the simplest case of a single variable with bosonic Lagrangian $L = A\dot{x}^2 + B(dh/dx)^2$. Then we can redefine $x = y/\sqrt{A}$ such that $L = \dot{y}^2 + AB(dh/dy)^2$. So weak coupling means $AB \to \infty$. Applying this to the case at hand and remembering that because of (7.140) the metric and periods scale as $g_{a\bar{b}} \propto V_X$, $\Pi^i \propto \sqrt{V_X}$ with the volume of X, we have effectively $A \propto V_X/g_s$ and $B \propto 1/g_s$. Thus weak coupling means

$$\frac{V_X}{g_s^2} \to \infty \,. \tag{7.178}$$

The left hand side is essentially the ratio of the string length squared over the 4d Planck length squared, ℓ_s^2/ℓ_4^2. Hence the weak coupling regime is the regime in which the string length is much larger than the 4d Planck length, and the low energy supergravity description breaks down.

We see from (7.171) that switching on the flux quanta S_i generates a highly complex potential energy function for the z^a, leading at fixed q_0 and q_A (and large q_0) to a vast energy landscape with exponentially many minima, parametrized by the values of S_i and z^a for which $\partial W_S(z) = 0$. This is kinematically very similar to the landscape of flux vacua in string theory, except of course that here we have a quantum mechanical system rather than a universe.

The potential is in general not a single valued function on the moduli space $\mathcal{M}$, due to possible monodromies $\sigma^i \to M^i{}_j \sigma^j$ of the 2-cycle basis when going around noncontractible loops in $\mathcal{M}$. It does become single valued on the covering space $\widetilde{\mathcal{M}}$, also known as Teichmüller space, but then one has to quotient the theory by the covering group, which acts as a discrete gauge symmetry on z^a and S_i. The metric $g_{a\bar{b}}$ and the charges q_0 and q_A on the other hand are single valued, since they can be expressed without reference to the basis σ^i.

The different zero energy classical minima of the potential can mix quantum mechanically through tunneling instantons. By completing the squares in the Euclidean version of the action one finds a bound $S^E_{\rm dyn} \geq \frac{2\pi}{g_s} \operatorname{Re}(e^{-i\alpha} \Delta W)$ for any real α, saturated when

$$\dot{z}^a = -e^{-i\alpha} g^{a\bar{b}} \bar{\partial}_{\bar{b}} \bar{W}(\bar{z}) \,. \tag{7.179}$$

The strongest bound is obtained with $\alpha = \arg(\Delta W)$, so this is what we should take α to be in order to be able to find instanton solutions. The instanton action is then

$$S^E_{\rm dyn} = \frac{2\pi}{g_s} |\Delta W| \,, \tag{7.180}$$

and the trajectories are straight lines when projected to the W-plane.

7.4.3.2. *Geometry of moduli space*

Before turning to the supersymmetric completion of the model, it may be useful to quickly review some basic formulae in Kähler geometry as well as more specialized expressions applicable to the model under consideration. When a metric is Kähler, i.e. $g_{a\bar{b}} = \partial_a \bar{\partial}_{\bar{b}} K$, all Christoffel symbols $\Gamma^K_{IJ} = \frac{1}{2} g^{KL} (\partial_I g_{LJ} + \partial_J g_{IL} - \partial_L g_{IJ})$ vanish except when all indices are holomorphic or all indices are anti-holomorphic. When all indices are holomorphic, we have

$$\Gamma^a_{bc} = g^{a\bar{d}} \partial_b g_{c\bar{d}} = \tfrac{1}{4} \bar{\partial}^a \bar{\Pi}^i \, \partial_b \partial_c \Pi_i = \tfrac{1}{4} \int_X \bar{\omega}^a \wedge \partial_b \omega_c \,. \tag{7.181}$$

We are lowering and raising indices with $g_{a\bar{b}}$ and η_{ij}. The first expression is true for general Kähler manifolds, the remainder specializes to

(7.174). Equivalently, $\int \bar{\omega}_{\bar{a}} \wedge \nabla_b \omega_c = 0$. The Riemann curvature, defined by $[\nabla_I, \nabla_J] X^K = R_{IJL}{}^K X^L$, also simplifies:

$$R_{a\bar{b}c\bar{d}} = g_{e\bar{d}} \bar{\partial}_{\bar{b}} \Gamma^e_{ac} = \tfrac{1}{4} \bar{\nabla}_{\bar{b}} \bar{\partial}_{\bar{d}} \bar{\Pi}^i \, \nabla_a \partial_c \Pi_i = \tfrac{1}{4} \int_X \bar{\nabla}_{\bar{b}} \bar{\omega}_{\bar{d}} \wedge \nabla_a \omega_c \,. \tag{7.182}$$

Again, the first equation is valid for general Kähler manifolds. Components of the curvature tensor not related by symmetries to the above ones (such as $R_{abc\bar{d}}$) all vanish. Finally, in addition to the usual curvature symmetries $R_{IJKL} = -R_{JIKL} = -R_{IJLK} = R_{KLIJ}$ and $R_{I[JKL]} = 0$, a Kähler manifold also satisfies $R_{a\bar{b}c\bar{d}} = R_{c\bar{b}a\bar{d}}$.

Geometrically, the fact that $\int \bar{\omega}_{\bar{a}} \wedge \nabla_b \omega_c = 0$ means that $\nabla_b \omega_c$ is of type $(1,1)$. Indeed, taking a derivative of a $(2,0)$-form turns it into a $(2,0)$-form plus a $(1,1)$-form (because the pullback of a $(2,0)$ form from $\Sigma(z)$ to $\Sigma(z+\delta z)$ will at most produce one extra antiholomorphic leg to first order in δz). The same is then true for $\nabla_b \omega_c$, and $\int \bar{\omega}_{\bar{a}} \wedge \nabla_b \omega_c = 0$ further implies that the $(2,0)$ part is actually zero, leaving only a $(1,1)$ part.

Keeping in mind the (p,q)-type of various forms, the following orthogonality properties hold:

$$\partial_a \Pi^i \, \partial_b \Pi_i = 0 \,, \qquad \partial_a \Pi^i \, \nabla_b \partial_c \Pi_i = 0 \,, \qquad \partial_a \Pi^i \, D_{Ai} = 0 \,. \tag{7.183}$$

Further orthogonality properties can be derived from these by differentiation, for example $\nabla_b \partial_a \Pi^i \, D_{Ai} = 0$.

7.4.3.3. *Hamiltonian and supersymmetric completion*

To find the supersymmetric completion of this model, we could either reduce the fermionic part of the full D4-action, or we can infer it from the structure of the bosonic part. We will go for the latter route. To do so, we first write down the bosonic Hamiltonian derived from the action given earlier. This is $H^{(0)} = H^{(0)}_{\rm dyn} + H^{(0)}_{\rm top}$ with

$$H^{(0)}_{\rm dyn} = \tfrac{g_s}{\pi} \, g^{a\bar{b}} p_a \bar{p}_{\bar{b}} + \tfrac{\pi}{g_s} \, g^{a\bar{b}} \, \partial_a \Pi^i S_i \, \bar{\partial}_{\bar{b}} \bar{\Pi}^j S_j \,, \tag{7.184}$$

$$H^{(0)}_{\rm top} = \tfrac{2\pi}{g_s} \left(q_0 + (\tilde{J}^A q_A)^2 \right) . \tag{7.185}$$

Here $p_a = \frac{\pi}{g_s} g_{a\bar{b}} \dot{\bar{z}}^a$ and we dropped the "chemical potential terms" proportional to ϕ^0 and ϕ^A here. If we consider the S_i as constants, this is of the form of a 4d $\mathcal{N} = 1$ theory of chiral superfields reduced to 1d, and the supersymmetric completion would be immediate. However the S_i are *not* dynamically conserved in this system, since motion from a point z in the moduli space $\mathcal{M}$ to itself along a noncontractible loop will in general change $S_i \to M_i{}^j S_j$ due to monodromy. On the other hand q_0, q_A and $H^{(0)}_{\rm top}$ are monodromy invariant, and hence they are conserved. In particular for an instanton, the change ΔW that appears in (7.180) is due to a change Δz with constant S_i on *Teichmuller space*, but it will in general lead to a change ΔS when reduced back to the moduli space.[hh] Therefore we should consider the S_i to be dynamical. We can think of the S_i as quantized momenta conjugate to angular coordinates ϑ^i that do not appear explicitly in the Hamiltonian, i.e. $S_i = -i\partial_{\vartheta^i}$. The ϑ^i can then be viewed as the potentials obtained by KK reduction of the 2-form potential on the D4 that is dual to the $U(1)$ gauge field. The appropriate metric for these coordinates can be read off from the above expression for the Hamiltonian, which can be cast in the form

$$H^{(0)} = \tfrac{g_s}{\pi}\, g^{a\bar{b}} p_a \bar{p}_{\bar{b}} + \tfrac{\pi}{g_s} g^{ij} S_i S_j\,, \tag{7.186}$$

where we have defined a metric on flux space

$$g^{ij} \equiv 2\left(\Pi^i_a \bar{\Pi}^{ja} + \bar{\Pi}^i_{\bar{a}} \Pi^{j\bar{a}} + \tilde{J}^i \tilde{J}^j\right) - \eta^{ij}\,, \qquad \Pi^i_a \equiv \frac{1}{2}\partial_a \Pi^i\,, \qquad \tilde{J}^i \equiv \tilde{J}^A D^i_A\,. \tag{7.187}$$

In this expression we use η_{ij} and $g_{a\bar{b}}$ for index raising and lowering. The three terms inside the brackets are projectors onto $(2,0)$-forms, $(0,2)$-forms and the $(1,1)$ direction parallel to the Kähler form; together they project to the space of self-dual forms. That these terms are properly normalized projectors follows from the relation (7.159) and from the definition (7.177).

In fact this metric is nothing but the the Hodge $\star$-product, and the decomposition into projections is just (7.164) with (7.165) again. Consistent with this, we have that $g^i{}_j g^j{}_k = \delta^i_k$, which is the property $\star^2 = 1$. This also shows that the inverse of g_{ij} is $g^{ij} = \eta^{ik}\eta^{jl} g_{kl}$.

The above Hamiltonian is just that of a free supersymmetric quantum mechanics in a curved space. Accordingly we could construct the usual supercharges $Q = d$ and $Q^\dagger = d^\dagger$ and take $H = \frac{1}{2}\{Q, Q^\dagger\}$, i.e. the Laplacian. However then the $Q = 0$ supersymmetric ground states of the system would have $H_{\rm dyn} = H_{\rm top} = 0$, which corresponds to D-brane states without any flux. This is too restrictive. We only need $H_{\rm dyn} = 0$ for a state to be supersymmetric, so we need to find supercharges that square to $H_{\rm dyn}$ only. These are basically the dimensionally reduced 4d $\mathcal{N} = 1$ supersymmetry

[hh] Incidentally, an open D2-instanton wrapping a special Lagrangian 3-chain Γ in X with boundary $\partial\Gamma = \sigma_+ - \sigma_-$ on Σ has an action $S_{\rm D2} = \frac{2\pi}{g_s} \,|\int_\Gamma \Omega| = \frac{2\pi}{g_s}|\Delta W|$, where ΔW is the change in superpotential when $F \to F + \sigma_+ - \sigma_-$. This is exactly the same as for monodromy instantons, suggesting a possible identification of the two.

generators, except that we interpret the S_i now as dynamical momenta: $S_i = -i\partial_i$. Introducing the fermionic operators $\bar{\psi}^a = dz^a\wedge$, $\bar{\psi}^{\bar{a}} = d\bar{z}^{\bar{a}}\wedge$ and their conjugates $\bar{\psi}^{\bar{a}} = (\psi^a)^\dagger$, $\bar{\psi}^a = (\psi^{\bar{a}})^\dagger$, we define the supercharges

$$Q_- = \sqrt{\tfrac{g_s}{\pi}}\,\bar{\psi}^a\nabla_a + \sqrt{\tfrac{\pi}{g_s}}\,\psi^{\bar{a}}\,\partial_{\bar{a}}\bar{\Pi}^i\,\partial_i\,, \quad \bar{Q}_+ = \sqrt{\tfrac{g_s}{\pi}}\,\psi^{\bar{a}}\bar{\nabla}_{\bar{a}} + \sqrt{\tfrac{\pi}{g_s}}\,\bar{\psi}^a\,\partial_a\Pi^i\,\partial_i\,. \tag{7.188}$$

and their conjugates $\bar{Q}_- = (Q_-)^\dagger$, $Q_+ = (\bar{Q}_+)^\dagger$. Equivalently

$$Q_- = \sqrt{\tfrac{g_s}{\pi}}\,\partial + \sqrt{\tfrac{\pi}{g_s}}\,\bar{\partial}\bar{\Pi}^i\,\partial_i\,, \qquad \bar{Q}_+ = \sqrt{\tfrac{g_s}{\pi}}\,\bar{\partial} + \sqrt{\tfrac{\pi}{g_s}}\,\partial\Pi^i\,\partial_i\,. \tag{7.189}$$

The supercharges satisfy the extended supersymmetry algebra[ii]

$$\{Q_\alpha, \bar{Q}_\beta\} = \delta_{\alpha\beta}H_{\rm dyn}, \qquad \{Q_\alpha, Q_\beta\} = 0 = \{\bar{Q}_\alpha, \bar{Q}_\beta\}\,, \tag{7.190}$$

where $H_{\rm dyn}$ reduces to (7.184) in the zero fermion number sector. This defines the summetric completion of $H^{(0)}_{\rm dyn}$. Working out e.g. $\{Q_-, Q_-^\dagger\}$ we get explicitly, acting on wave functions $\Phi(\bar{\psi}, x)$ (i.e. differential forms):

$$\begin{aligned} H_{\rm dyn} = &-\tfrac{g_s}{\pi}\bar{\nabla}^a\nabla_a + \tfrac{g_s}{\pi}R_{a\bar{b}c\bar{d}}\,\bar{\psi}^a\bar{\psi}^{\bar{b}}\psi^c\psi^{\bar{d}} - \tfrac{\pi}{g_s}\,\partial_a\Pi^i\bar{\partial}^a\bar{\Pi}^j\partial_i\partial_j \\ &- \bar{\psi}^a\psi^b\nabla_a\partial_b\Pi^i\,\partial_i + \bar{\psi}^{\bar{b}}\psi^{\bar{a}}\bar{\nabla}_{\bar{a}}\partial_{\bar{b}}\bar{\Pi}^i\partial_i\,. \end{aligned}$$

Here, again acting on differential forms, $\nabla_a = \partial_a + \Gamma^b_{ac}\bar{\psi}^c\psi_b$, and $p_a = -i\partial_a$, $S_i = -i\partial_i$. Evidently neither the ϑ^i nor their fermionic superpartners (i.e. something like $d\vartheta^i$) appear explicitly in the Hamiltonian. Nevertheless they do not completely decouple from the z-degrees of freedom, again because of global monodromies — the ϑ-torus fibration over the moduli space $\mathcal{M}$ has nontrivial identifications when going around nontrivial loops.

To summarize, the full supersymmetric Hamiltonian including the RR potential terms is, in terms of the conserved charges:

$$H = H_{\rm dyn} + \tfrac{2\pi}{g_s}\big(q_0 + (\tilde{J}^A q_A)^2\big) + 2\pi(\phi^0 q_0 + \phi^A q_A)\,. \tag{7.191}$$

Making the ϑ momenta S_i explicit in the topological part of the Hamiltonian, this becomes

$$H = H_{\rm dyn} - 2\pi\big(\tfrac{1}{g_s} + \phi^0\big)\big(\tfrac{\chi}{24} + \tfrac{1}{2}\eta^{ij}S_iS_j\big) + \tfrac{2\pi}{g_s}(\tilde{J}^iS_i)^2 + 2\pi\phi^iS_i\,. \tag{7.192}$$

[ii]Because a 4d $\mathcal{N} = 1$ theory has four supersymmetries, the system has twice as much supersymmetry as the minimal susy QM case studied in section 7.3. As "the" Q-operator (satisfying $Q^2 = 0$, $\{Q, Q^\dagger\} = H_{\rm dyn}$), we can therefore choose from a family of linear combinations of the supercharges. The phase α appearing in the instanton flow equation (7.179) determines which supersymmetry is preserved by the instanton, and therefore which supersymmetry gets corrected.

Unpacking the whole thing gives us

$$H = -\tfrac{g_s}{\pi}\bar{\nabla}^a\nabla_a + \tfrac{g_s}{\pi}R_{a\bar{b}c\bar{d}}\bar{\psi}^a\bar{\psi}^{\bar{b}}\psi^c\psi^{\bar{d}} - 2\pi(\tfrac{1}{g_s}+\phi^0)\tfrac{\chi}{24}$$
$$+\pi\big(\tfrac{1}{g_s}g^{ij} - \phi^0\eta^{ij}\big)S_iS_j + 2\pi L^iS_i\,. \quad (7.193)$$

Here $\phi^i = \phi^A D^i_A$, g^{ij} is the metric (7.187), and

$$L^i = \phi^i - \frac{i}{2\pi}\bar{\psi}^a\psi^b\nabla_a\partial_b\Pi^i + \frac{i}{2\pi}\bar{\psi}^{\bar{b}}\psi^{\bar{a}}\bar{\nabla}_{\bar{a}}\bar{\partial}_{\bar{b}}\bar{\Pi}^i\,. \quad (7.194)$$

It is often convenient to split quantities like S_i and L_i into self-dual and antiself-dual parts, as this leads to more transparent "left-moving" and "right-moving" expressions. Thus, for instance

$$\big(\tfrac{1}{g_s}g^{ij} - \phi^0\eta^{ij}\big)S_iS_j = (\tfrac{1}{g_s} - \phi^0)S_+^2 - (\tfrac{1}{g_s}+\phi^0)S_-^2\,, \quad (7.195)$$

where squares denote contraction with η, i.e. $X^2 = X_iX^i = \eta^{ij}X_iX_j$ (so $X_+^2 > 0$, $X_-^2 < 0$). The projectors introduced in (7.187) can of course be used to give an explicit expression for X_+ and hence also for $X_- = X - X_+$.

7.4.3.4. *Supersymmetric Lagrangian*

Legendre transforming the Hamiltonian, with $\dot{z}^a = \frac{\partial H}{\partial p_a}$, $\dot{\vartheta}^i = \frac{\partial H}{\partial S_i}$, we get the corresponding Lagrangian

$$\mathcal{L} = \frac{\pi}{g_s}g_{a\bar{b}}\dot{z}^a\dot{\bar{z}}^{\bar{b}} + i\bar{\psi}_a D_t\psi^a + i\bar{\psi}_{\bar{b}}D_t\psi^{\bar{b}} - \frac{g_s}{\pi}R_{a\bar{b}c\bar{d}}\bar{\psi}^a\bar{\psi}^{\bar{b}}\psi^c\psi^{\bar{d}}$$
$$+ 2\pi(\tfrac{1}{g_s}+\phi^0)\frac{\chi}{24} + \frac{1}{4\pi(\frac{1}{g_s}-\phi^0)}\big(\dot{\vartheta}_+ - 2\pi L_+\big)^2$$
$$- \frac{1}{4\pi(\frac{1}{g_s}+\phi^0)}\big(\dot{\vartheta}_- - 2\pi L_-\big)^2\,, \quad (7.196)$$

where $D_t\psi^a = \dot{\psi}^a + \dot{z}^b\Gamma^a_{bc}\psi^c$. Note that $L_+ = \phi_+$ as $(\nabla_a\partial_b\Pi)_+ = 0$.

7.4.4. *Constructing supersymmetric flux configurations*

We have constructed the supersymmetric quantum mechanics describing the low energy physics of smooth D4-branes with $U(1)$ gauge symmetry and abelian fluxes F turned on. In the thermodynamic limit $N \to \infty$ we get $\mathcal{O}(N^3)$ discrete flux degrees of freedom S_i interacting with each other in a complicated, nonlocal way through coupling to $\mathcal{O}(N^3)$ continuous degrees of freedom z^a. The effective Hamiltonian (7.186) is somewhat reminiscent of models for spin glasses or the Hopfield model for neural networks, but more intricate, and a priori without quenched disorder, making it more like a structural glass than a spin glass.[jj] In any case this suggests the presence of many effectively stable states. We will now make this quantitative, ignoring for the time being the possibility of brane degenerations and nonabelian gauge group enhancement.

Classical supersymmetric brane configurations are those for which $\Pi^i_a S_i = 0$, i.e. $\partial W_F(z) = 0$. As the 3-chain periods are very hard to compute explicitly, it might seem hopeless to try to construct any of those configurations explicitly. But a little thought shows otherwise. The condition $\partial W_F = 0$ is equivalent to

$$F^{(0,2)} = F^{(2,0)} = 0 \,. \tag{7.197}$$

In other words if we switch on a flux $F = S_i \omega^i$, the brane moduli must adapt themselves to make the flux of type $(1,1)$. A geometric interpretation of this is that the flux becomes Poincaré dual to a linear combination of holomorphic 2-cycles. Indeed the Poincaré dual of a holomorphic 2-cycle σ is clearly of type $(1,1)$ since for every $(2,0)$-form ω we have $\int \sigma \wedge \omega = \int_\sigma \omega = 0$. The last equation is true because σ is the image of some map $y^m(u)$ with u a single complex coordinate, and so the pullback of a $(2,0)$ form will be proportional to $du \wedge du = 0$. Thus, any linear combination (with arbitrary signs) of fluxes Poincaré dual to holomorphic 2-cycle will be of type $(1,1)$. The converse is also true as a special (proven) case of the Hodge conjecture. The subspace of moduli space where extra integral harmonic forms become of type $(1,1)$ is sometimes called the (generalized) Noether-Lefschetz locus in algebraic geometry.

This interpretation gives a relatively simple way of explicitly constructing classical supersymmetric ground states. Let us illustrate this with an example. Consider again the quintic Calabi-Yau $Q_5(x) = 0$ and let us assume that we have chosen the polynomial Q_5 such that it contains the following holomorphic 2-cycle, parametrized by $(u,v) \in \mathbb{CP}^1$:

$$\gamma : (x_1, x_2, x_3, x_4, x_5) = (u, -u, v, -v, 0) \,. \tag{7.198}$$

The Fermat quintic $\sum_i x_i^5 = 0$ for example contains γ. The topology of γ is that of $\mathbb{CP}^1$, i.e. it is a 2-sphere. Such "degree 1 rational curves" are generically isolated in a Calabi-Yau. This can be checked in the case at hand by simple counting: the most general linear map $(u,v) \to (x_1, \cdots, x_5)$, has

[jj] A spin glass has quenched disorder built into the Hamiltonian. A structural glass has a simple Hamiltonian, and spontaneously generates its own disorder. This includes the material normal people refer to when they use the word glass.

10 coefficients, 4 of those can be eliminated by linear coordinate transformations of (u, v), and requiring $Q_5(x(u,v)) = 0$ for all (u, v) produces 6 equations on the remaining coefficients (namely the coefficients of u^5, $u^4 v$, $\ldots$, v^5 after expanding out), so we generically expect a discrete solution set. Indeed in the generic quintic there are 2875 isolated degree 1 rational curves.

Now consider our single wrapped D4 ($N = 1$), described by a linear equation

$$\Sigma : \sum_i a_i x_i = 0 \,. \tag{7.199}$$

Requiring Σ to contain γ produces the equations $a_1 = a_2$, $a_3 = a_4$. Therefore, if we switch on a flux Poincaré dual to γ on Σ, rigidity of the rational curve means that in order to preserve supersymmetry, these equations must remain satisfied when deforming Σ. Hence the D4 moduli space is reduced from $\mathbb{CP}^4$ to $\mathbb{CP}^2$. If we also require say the rational curve $\gamma' : x = (0, u, -u, v, -v)$ to lie on Σ, enforced by switching on $F = \gamma'$, we end up with $a_1 = a_2 = a_3 = a_4 = a_5$, i.e. all moduli have been lifted. A simple 2d analog of this is illustrated in Fig. 7.12. By combining different pairs of the 2875 degree 1 rational curves we can thus explicitly construct $\frac{2875 \times 2874}{2} = 4131375$ different isolated supersymmetric configurations (assuming there are no duplicates or degenerates). In fact for each choice of pair (γ, γ') there is an infinite number of different flux states, obtained by switching on a more general linear combination of the harmonic forms γ and γ':[kk]

$$F = k\gamma + k'\gamma' \,, \tag{7.200}$$

where $k, k' \in \mathbb{Z}$. Of course these are not all degenerate in total energy. They all have $H_{\rm dyn} = 0$ but they will have different $H_{\rm top}$. To compute this we should find the D2- and D0-charges. They are

$$q_1(F) = \int_\Sigma D_1 \wedge F = k + k' \,, \quad q_0(F) = -\frac{\chi}{24} - \frac{1}{2}\int_\Sigma F \wedge F = -\frac{\chi}{24} + \frac{3}{2}(k^2 + k'^2) \,, \tag{7.201}$$

with $\chi = 55$. Here we used the fact that the self-intersection number of γ inside Σ equals -3. This can be computed either by explicitly deforming γ inside Σ (necessarily nonholomorphically) and counting intersections, or by using some algebraic geometry (as in section 4.3 of [107]). The total energy

[kk] Actually we should shift F by the "half-flux" $\frac{1}{2}D_1$, because for $N = 1$, Σ is not spin [111, 112]. This ensures proper charge quantization. We will ignore this here.

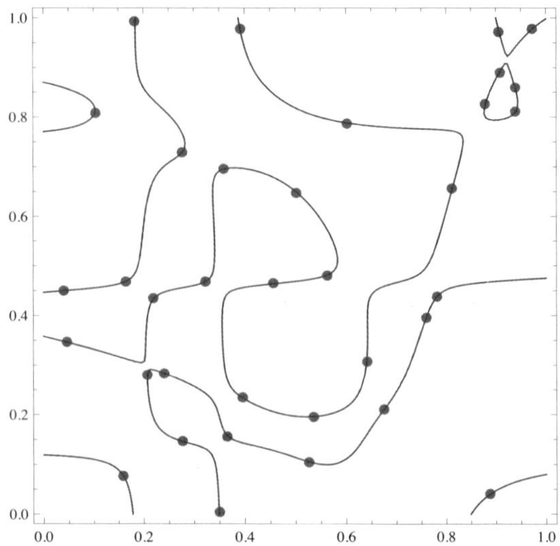

Fig. 7.12. Analog in 2d of the problem of fitting a holomorphic 4-cycle to contain a given collection of rigid holomorphic 2-spheres in a 6-dimensional space: here we are fitting a 1-cycle given by a degree 5 polynomial in x and y to contain 36 randomly generated points, thus fixing all moduli of the 1-cycle.

$H = H_{\text{top}}$ is given by

$$\frac{g_s}{2\pi} H = q_0 + \frac{q_1^2}{5} = -\frac{\chi}{24} + \frac{3}{2}(k^2 + k'^2) + \frac{1}{5}(k + k')^2 \,. \tag{7.202}$$

Thus we get a nondegenerate lattice of vacua for each generic pair of degree 1 rational curves.

This can be generalized in various ways. Instead of degree 1 rational curves, we can consider degree d rational curves. The same simple counting argument as before indicates again that they are generically isolated: there are $5(d+1) - 4$ coefficients modulo reparametrizations, and there are $5d + 1$ equations, which happens to be the same number. The number of rational curves grows exponentially with the degree. There are for example 609250 degree 2 curves and 704288164978454686113488249750 degree 10 curves. Furthermore, we can consider arbitrary D4-charge N. Requiring such a degree N 4-cycle to contain a degree d 2-cycle freezes $Nd+1$ moduli, out of $\frac{5N^3+25N}{6}$. So at large N the 4-cycle can "store" up to $\frac{5N^2}{6d}$ degree d rational curves by switching on the appropriate fluxes. If d is sufficiently large so that the number of degree d curves $N_d \sim e^{\kappa d}$ is much larger than this number, then this leads, naively at least, to up to $\sim e^{\frac{5\kappa}{6}N^2}$ susy configurations. Switching on one degree d flux quantum increases q_0 by $\frac{dN}{2}+1$, and q_2 by $\pm d$ depending on the sign of the flux (see section 4.3 of [107] for a derivation). So at large N the energy in the maximal amount of $\frac{5N^2}{6d}$ stored curves is $\frac{g_s}{2\pi}H \approx q_0 \approx -\frac{\chi}{24} + \frac{5N^2}{6d} \times \frac{dN}{2} \approx -\frac{5N^3}{24} + \frac{5N^3}{12} = \frac{5N^3}{24}$.[11]

Although the above discussion is for weak coupling, there is strong evidence that some of this structure carries over the strong coupling black hole regime, in the form of the structure of multicentered black hole bound states (cf section 7.5). We refer to [107] for more discussion on this.

7.4.5. *Counting supersymmetric states*

Although the above construction is explicit, and the number of ground state configurations that can be built in this way is huge, it is still only a small subset, especially at large q_0. In this regime the generic ground state configuration is isolated, and their total number and distribution over the moduli space can be computed using the statistical methods to count flux vacua developed in [36, 113, 114] and reviewed in section 6 of [7]. Those methods map the classical critical point counting problem, after making a continuum approximation, to the computation a supersymmetric finite dimensional integral, essentially a finite dimensional version of what we did is section 7.3.8. In the present case we already have a supersymmetric quantum mechanics to start from. Not surprisingly, it is closely related to the effective supersymmetric quantum mechanics used in the statistical approach.

In what follows we will consider the problem by applying the general machinery of supersymmetric quantum mechanics reviewed is section 7.3. This clarifies and complements the results obtained by the methods reviewed in [7] (and applied to the case at hand in appendix G of [107]). In particular it allows us to go beyond the continuum approximation. The computations will get a little technical, for which I apologize. I chose to include them here because it may be useful for some readers to see a nontrivial example worked out, and because as far as I know this has not been done in the literature (the treatment has some overlap with [107], but we

[11]Notice in particular that the energy becomes positive before all moduli are frozen. If what we were studying were not D4-branes but D7-branes in IIB orientifolds, this would basically mean that we cannot fix all moduli in this way while respecting tadpole cancelation, going against common genericity arguments. This is not conclusive though since this construction certainly does not exhaust all possibilities.

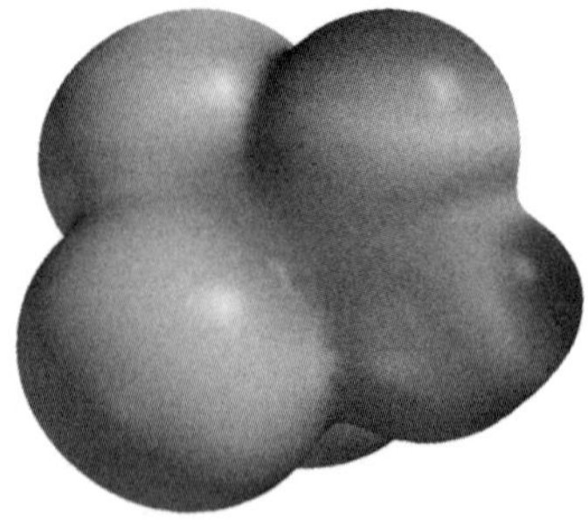

Fig. 7.13. Artist impression of a smooth abelian D4 configuration with $U(1)$ magnetic fluxes turned on (flux density represented by colors).

will not use S-duality as an input, but rather derive it directly from the susy quantum mechanics).

From our general consideration of supersymmetric quantum mechanics in the previous section, it is not hard to come up with a formula for the Witten index $\Omega(q)$ as a weighted sum of Euler characteristics of critical point loci $\mathcal{M}_S$ of the superpotential W_S, summed over fluxes/momenta S with the given total charge q:

$$\Omega(q) = \sum_S (-1)^{d-d_S} \chi(\mathcal{M}_S) \,. \tag{7.203}$$

The sign $(-1)^{d-d_S}$ appears because m transversal dimensions contribute m (the Morse index) to the fermion number. This is a useful formula for supersymmetric configurations with small D0-charge, but not for large D0-charge. We will try to extract the large charge asymptotics from the partition function.

The supersymmetric partition function generating the Witten indices $\Omega(q)$ in each conserved charge sector is, with H as in (7.191)-(7.193):

$$\begin{aligned} Z(\beta, g_s, \phi) &= \mathrm{Tr}\,(-1)^F\, e^{-\beta H(g_s,\phi)} \\ &= \sum_q \Omega(q)\, e^{-\frac{2\pi\beta}{g_s}\left(q_0+(\tilde{J}^A q_A)^2\right)-2\pi\beta(\phi^0 q_0+\phi^A q_A)} \,. \end{aligned} \tag{7.204}$$

Introducing rescaled RR potentials

$$\varphi \equiv \beta\phi \,, \tag{7.205}$$

we obtain the following expression for the Witten index $\Omega(q)$:

$$\Omega(q) = \lim_{\beta\to 0} \int d\varphi\, e^{2\pi\varphi\cdot q}\, Z(\beta,\varphi) \,, \tag{7.206}$$

where the φ integrals in principle are over a unit imaginary interval, e.g. from $-i/2$ to $i/2$. For computational purposes it is more convenient to work with integrals over the entire imaginary axis, because it allows easy implementation of contour integration techniques. Extending the integration ranges in the integrals above only produces trivial additional delta-function factors, which in the end can be stripped off manually.

We will now try to compute the Witten indices from the Euclidean path integral representation of Z in the limit $\beta \to 0$, analogous to the derivation leading to (7.115):

$$Z = \int \mathcal{D}z\, \mathcal{D}\psi\, \mathcal{D}\vartheta\, e^{-S[z,\psi,\vartheta]}\,, \tag{7.207}$$

with periodic boundary conditions on all fields: $z^a(\tau+\beta) = z^a(\tau)$, $\psi^a(\tau+\beta) = \psi^a(\tau)$, $\vartheta^i(\tau+\beta) = \vartheta^i(\tau) + 2\pi n^i$, with $n^i \in \mathbb{Z}$. By Wick rotating (7.196) to rescaled Euclidean time

$$t = -i\beta\tau\,, \tag{7.208}$$

we obtain the Euclidean action

$$\begin{aligned} S = \int_0^1 d\tau \Big(& \frac{\pi}{g_s\beta}\,|\dot{z}|^2 + \bar{\psi}D_t\psi + \frac{g_s\beta}{\pi}\, R\,\bar{\psi}\bar{\psi}\psi\psi - 2\pi u\,\frac{\chi}{24} \\ & + \frac{1}{4\pi v}\big(\dot{\vartheta} + 2\pi i \ell\big)_+^2 - \frac{1}{4\pi u}\big(\dot{\vartheta} + 2\pi i \ell\big)_-^2 \Big), \end{aligned} \tag{7.209}$$

where we suppressed indices for clarity, denoted $\ell \equiv L/\beta$, i.e.

$$\ell^i \equiv \varphi^i - i\beta\, \tfrac{1}{2\pi}\bar{\psi}\psi\nabla\partial\Pi^i + i\beta\, \tfrac{1}{2\pi}\bar{\psi}\psi\bar{\nabla}\bar{\partial}\bar{\Pi}^i\,, \tag{7.210}$$

and introduced "light cone" coordinates

$$u \equiv \frac{\beta}{g_s} + \varphi^0\,, \qquad v \equiv \frac{\beta}{g_s} - \varphi^0\,. \tag{7.211}$$

As in (7.114), in the limit $\beta \to 0$, the path integral localizes to constant paths (z,ψ), with the fluctuation determinants for z and ψ canceling each other. The path integral over ϑ does not localize to constant paths, since the ϑ kinetic terms remain finite when $\beta \to 0$ (for $\varphi^0 \neq 0$). The ϑ path integral is essentially the partition function of a free particle on a b_2-dimensional torus. Making the sum over different winding sectors $\vartheta^i(\tau+1) = \vartheta^i(\tau) + 2\pi n^i$ explicit, this is

$$\sum_n e^{-\frac{\pi}{v}(n+i\ell)_+^2 + \frac{\pi}{u}(n+i\ell)_-^2} \int \mathcal{D}\vartheta\, e^{-\frac{1}{4\pi}\int_0^1 d\tau\,(\frac{1}{v}\dot{\vartheta}_+^2 - \frac{1}{u}\dot{\vartheta}_-^2)}\,. \tag{7.212}$$

The first factor corresponds to the action weights for the classical straight line trajectories in each sector, and the remaining path integral corresponds to the fluctuations from the classical paths and thus has strictly periodic boundary conditions $\vartheta^i(\tau+1) = \vartheta^i(\tau)$. This is just the standard Euclidean free particle propagator integrated over the ϑ-torus:

$$\int \mathcal{D}\vartheta \, e^{-\frac{1}{4\pi}\int_0^1 d\tau \, [\frac{1}{v}\dot{\vartheta}_+^2 - \frac{1}{u}\dot{\vartheta}_-^2]} = u^{-b_2^-/2} \, v^{-b_2^+/2} \, . \qquad (7.213)$$

By making use of (7.182) and the various orthogonality and (anti-)self-duality properties discussed there, as well as the projectors introduced in (7.187) to write explicit expressions for n_+ and $n_- = n - n_+$, the sum in (7.212) together with the 4-fermion curvature term in (7.209) reduces after some tender and care to

$$\sum_n e^{-\frac{\pi}{v}(n+i\varphi)_J^2 + \frac{\pi}{u}(n+i\varphi)_{\perp J}^2} \, e^{-\beta\left(\frac{\pi}{g_s uv}|n\Pi'|^2 - \frac{1}{u}(\bar{\psi}\psi \, n\Pi'' + \mathrm{cc}) - \frac{g_s v}{\pi u} R\bar{\psi}\bar{\psi}\psi\psi\right)} \, , \qquad (7.214)$$

where we separated out a factor in the summand that is independent of the dynamical variables z, ψ, in which we also introduced the notation $X_J^i \equiv \tilde{J}^i X_i$ for the projection of X in the direction of J, and $X_{\perp J}$ for its orthogonal complement (w.r.t. η_{ij}). Now consider the z, ψ integral of the other factor:

$$\int d^{2d}z \, d^d\psi \, d^d\bar{\psi} \, e^{-\beta\left(\frac{\pi}{g_s uv}|n\Pi'|^2 - \frac{1}{u}(\bar{\psi}\psi \, n\Pi'' + \mathrm{cc}) - \frac{g_s v}{\pi u} R\bar{\psi}\bar{\psi}\psi\psi\right)} \, . \qquad (7.215)$$

This integral has the typical zero dimensional supersymmetric form with superpotential $W_n(z) = n^i\Pi_i(z)$. It localizes on the critical point locus $\mathcal{M}_n$ of W_n, meaning the Gaussian approximation to the integral is exact.[mm] The Gaussian integral for quadratic fluctuations normal to $\mathcal{M}_n$ produces a factor $\frac{g_s uv}{\beta}$ for each complex normal direction, while the corresponding normal fermions produce a factor $\frac{\beta^2}{u^2}$. Together this gives a factor $\frac{\beta v}{u}$ for each normal direction. The tangential directions to $\mathcal{M}_n$ remain integrated over, and the corresponding tangent fermion integral gives for each complex tangent dimension a factor $-\frac{g_s \beta v}{\pi u} R$. Finally, there will be an overall path integral normalization factor $(2\pi g_s \beta)^{-d}$ as in (7.114). Altogether the above integral thus reduces to $(-1)^{d_n} e(\mathcal{M}_n) \, v^d \, u^{-d}$, where $d_n \equiv \dim \mathcal{M}_n$ and

[mm] Actually, because n can transform by a monodromy when going around a loop in moduli space, this only holds up to boundary terms for a given n, but these cancel between different values of n. To see localization more directly, note that since the Witten index is invariant under a change of g_s, we can make the integrand as sharply peaked on $\mathcal{M}_n$ as we want, with vanishing boundary terms when $g_s \to 0$.

$e(\mathcal{M}_n)$ is the Euler density on $\mathcal{M}_n$, as in (7.115). The sign is physically meaningful; we expand on this below.

Putting everything together, we find the generating function:

$$Z = (u/v)^{1/2}\, u^{-b_2/2}\, e^{2\pi u \frac{\chi}{24}} \sum_n (-1)^{d_n} \chi_n\, e^{-\frac{\pi}{v}(n+i\varphi)_J^2 + \frac{\pi}{u}(n+i\varphi)_{\perp J}^2}\,, \quad (7.216)$$

where we defined the differential geometric Euler characteristics

$$\chi_n \equiv \int_{\mathcal{M}_n} e(R)\,, \quad (7.217)$$

and we used $d = \frac{b_{2+}-1}{2}$. Note that since by definition Ω_n is nonzero only if $(n_{\perp J})_+$ can be made to vanish, all terms in this series have $n_{\perp J}^2 \leq 0$.

Defining the dual charges $\tilde{q}_A = D_{Ai} n^i$, $\tilde{q}_0 = -\frac{\chi}{24} - \frac{1}{2}n^2$, we can also express this as

$$Z = (\tfrac{u}{v})^{1/2}\, u^{-b_2/2}\, e^{2\pi \frac{\chi}{24}(u-\frac{1}{u}) - \frac{\pi}{u}\varphi_-^2 + \frac{\pi}{v}\varphi_+^2}(-1)^d \\ \times \sum_{\tilde{q}} \Omega(\tilde{q})\, e^{-\frac{2\pi}{u}\tilde{q}_0 - \pi(\frac{1}{u}+\frac{1}{v})\tilde{q}_+^2 + \frac{2\pi i}{u}\tilde{q}_-\tilde{\varphi}_- - \frac{2\pi i}{v}\tilde{q}_+\tilde{\varphi}_+}\,, \quad (7.218)$$

with as in (7.203) $\Omega(\tilde{q}) = \sum_n (-1)^{d_n - d}\chi_n$, summing over the values of n with the given charges $\tilde{q}$. Comparing this to the original (7.204), which can be written as

$$Z = \sum_q \Omega(q)\, e^{-2\pi u q_0 - \pi(u+v)q_+^2 - 2\pi q\varphi}\,, \quad (7.219)$$

we see that what we have shown is roughly that Z is a modular (Jacobi) form under the modular transformation $u \to 1/u$, $v \to 1/v$, transforming similar to an ordinary theta function (although what we have here is much more nontrivial than a theta function, due to the nontrivial moduli dynamics). The duality exchanges winding and momentum modes, and there are many ways of understanding it: electromagnetic duality of the D-brane theory, S-duality of the parent string theory, modularity of the parent CFT in M-theory, T-duality of the torus, a version of Poisson resummation, etc. The modular transformation relates the large and small flux / D0-charge regimes.

In any case, we can now extract the large q_0 asymptotics of $\Omega(q)$ from Z. For simplicity we will consider here the case $q_A = 0$, but this is easily extended. Performing the Gaussian integrals over the φ^A in (7.206) is easy. We get

$$\int_{-i\infty}^{i\infty} d\varphi^1 \cdots d\varphi^{B_2}\, Z(u,v,\varphi) = c\, u^{-k} e^{2\pi u \frac{\chi}{24}} \sum_n (-1)^{d_n} \chi_n\, e^{\frac{\pi}{u} n_\perp^2}\,. \quad (7.220)$$

Here $k \equiv \frac{b_2 - B_2}{2}$ with $b_2 = \dim H^{(}\Sigma)$, $B_2 = \dim H^2(X)$, and $n_\perp$ is the component of n orthogonal to all $\varphi = \varphi^A D_A$, i.e. the component of n orthogonal to the pullback of $H^2(X)$ in $H^2(\Sigma)$. Explicitly $n_\perp^i = n^i - (n^j D_j^A) D_A^i$, where $D^A = D^{AB} D_B$, D^{AB} being the inverse of $D_{AB} \equiv D_{ABC} N^C$, with D_{ABC} defined in (7.146). The constant c is the Gaussian determinant factor: $c = (\det D_{AB})^{-1/2}$. For the example of the quintic we have $k = (5N^3 + 50N - 3)/2$, $D_{11} = 5N$, $c = \frac{1}{\sqrt{5N}}$. The sum over n is trivially divergent because adding an arbitrary $m = m^A D_A \in H^2(X)$ to n does not alter $n_\perp$. From the point of view of the integral over φ^A this is due to the redundancy $\phi^A \to \phi^A - im^A$, $n \to n + m^A D_A$, which we introduced ourselves a little earlier when we extended the integration domain from an interval to the full imaginary axis. The upshot is that we can take this into account simply by restricting the sum over $n \in H^2(\Sigma, \mathbb{Z})$ to a sum over the quotient $n \in H^2(\Sigma, \mathbb{Z})/H^2(\Sigma, \mathbb{Z})$.

The $q_0 \to \infty$ asymptotics of Ω will be captured by the $u \to 0$ behavior of Z. In this regime all $n \neq 0$ exponential corrections can be dropped. The φ^0 integral can then be evaluated exactly by closing the contour. There is a pole of order k at $u = 0$ i.e. $\varphi^0 = -\beta < 0$. If $q_0' \equiv q_0 + \frac{\chi}{24} < 0$ we must close the contour to the right and the integral vanishes. If $q_0' > 0$, the contour is closed to the left, and the residue theorem gives

$$\Omega(q_0, 0) \approx 2\pi |c| \, (-1)^d \chi(\mathcal{M}) \, \frac{(2\pi q_0')^{k-1}}{(k-1)!} \sim \left(\frac{2\pi e q_0'}{k} \right)^k . \qquad (7.221)$$

The last approximation is valid for $k \gg 1$. It is obtained as a Stirling approximation, or equivalently as the saddle point approximation of the integral. The saddle point lies at $\varphi^0 = 2\pi q_0'/k$. This implies we need at least $q_0' \gg k$ to justify dropping the $n \neq 0$ terms, or equivalently $q_0 \gg \chi \sim N^3$.

This expresses the number of supersymmetric flux configurations as the volume of a $2k$-dimensional shell of radius squared $2q_0'$, which loosely (but not literally because η_{ij} is indefinite) can be thought of as the shell in flux space for which the D0-charge $-\frac{\chi}{24} - \frac{1}{2}\eta^{ij} S_i S_j$ equals q_0 and the D2-charge is constrained to be zero. This reproduces the large q_0 asymptotics for the number of flux vacua found in various contexts in [7, 36, 107, 113, 114], based on an approximation in which fluxes were taken to be continuous.

Note that $\chi(\mathcal{M}) = \int_\mathcal{M} e(R)$ is the Witten index for the pure D4 without the flux degrees of freedom. Since this space is topologically $\mathbb{CP}^d$, one may expect $\chi(\mathcal{M}) = \chi(\mathbb{CP}^d) = d + 1 \sim N^3/6$. This is not obviously correct because of possible singularities in $\mathcal{M}$, but arguments were given in [107] that this is nevertheless the correct identification. The sign factor $(-1)^d$

has a physical meaning: In the large q_0 limit, almost all supersymmetric configurations will be isolated critical points of the superpotential $W_S(z)$. Isolated critical points lead to susy states with fermion number equal to their Morse index, which here always equals d because the superpotential is holomorphic. This explains the $(-1)^d$. The sign $(-1)^{d_n}$ in the terms with $n \neq 0$ further suggests (but does not prove) that these correspond to the contributions from non-isolated configurations, with $d - d_n$ residual moduli.

It is of course easy to get exponentially large numbers of flux configurations out of these estimates, even for modest charges. For example for the quintic with, say, $N = 5$, $q_0 = 365$, we get $\Omega \sim 10^{500}$, widely considered to be a very large number. (Similar estimates in the context of mathematically very similar constructions of type IIB string vacua with D7-branes form the basis for suspicions that there exists a staggeringly huge landscape of string vacua sweeping out a for all practical purposes dense set in parameter space, giving a simple solution to the cosmological constant problem but obliterating hopes of top-down predictivity [7, 31–34, 36, 39, 113, 114].)

On the other hand note that when $2\pi e q_0' < k$, the above estimates give exponentially *small* estimates. Clearly then, in this regime, it must be that the $n \neq 0$ terms dominate the degeneracy. If it is indeed true, as suggested above, that $n \neq 0$ contributions can be identified with non-isolated configurations, this means that in this regime, non-isolated configurations become entropically dominant.

For $q_0' \ll k$ we could try to build up the spectrum along the lines of section 7.4.4, and this could in turn be used to infer the corrections in the regime $q_0' \gg k$. However in the intermediate regime $q_0' \sim k$, the system is extremely complex.[nn] Trying to count degeneracies in this regime may be like trying to compute the boiling point of water from first principles. In the case of water we can just measure the boiling point and be done with it. We let nature do the computation for us. In the case of D-branes, this would not appear to be an option. However, there is something analogous, provided we enlarge our task to the problem of counting *all* D4-D2-D0 bound states, not just those corresponding to abelian flux configurations of the D4. In that case, we can construct the corresponding BPS black hole solutions,

[nn]This happens to be the relevant regime for applications to type IIB string theory vacua [7, 105]. In this context space-localized D4-branes are replaced with space-filling D7-branes, carrying $U(1)$ fluxes inducing D5- and D3-charges. The D3-tadpole cancelation condition dictates that the total D3-charge vanishes, hence $q_0' = \frac{\chi}{24} \approx \frac{k}{12}$ and $\frac{2\pi e q_0'}{k} \approx \frac{\pi e}{6} \approx 1.42$, barely above the threshold.

and simply read off their Bekenstein-Hawking entropy $S = \log \Omega(q)$. We let gravity do the computation for us! To include all entropically relevant D4-D2-D0 bound states we need in particular consider bound states with localized D0-branes. We turn to this next.

7.4.6. *Bound states with D0-branes*

So far we have only considered D4-brane and $U(1)$ worldvolume flux degrees of freedom. A simple but entropically important extension is to include bound states with pointlike D0-branes. For a D4 wrapped on a fixed smooth Σ, the number of such bound states is easy to compute [115]. We can think of the D0-branes as a gas of noninteracting particles. D0-branes form bound states among themselves of arbitrary D0-charge k (the easiest way to see this is to consider their uplift to M-theory, where they are KK modes along the M-theory circle, k being the KK momentum), so each particle in the D4-D0 gas is characterized by a charge quantum number $k > 0$, as well as by the supersymmetric 1-particle state $|p\alpha\rangle$ it occupies. As we have seen in section 7.3.5, these 1-particle states are given by the harmonic differential forms on Σ. The label $p = 0, 1, 2, 3, 4$ in $|p\alpha\rangle$ indicates the fermion number (form degree) and $\alpha = 1, \ldots, b_p$. By the usual Fock space construction, an arbitrary multi-particle state can be represented by specifying occupation numbers $n_{k,p,\alpha}$, which take values in $\mathbb{Z}^+$ if p is even and in $\{0,1\}$ is p is odd. In the absence of fluxes, the total D0-charge of the D4-D0 state is $Q = -\frac{\chi}{24} + \sum_{k,p,\alpha} k\, n_{kp\alpha}$ and the fermion number is $F = \sum_{k,p,\alpha} p\, n_{kp\alpha}$. Here we are still ignoring the moduli degrees of freedom of Σ, considering those frozen for the time being. The generating function for the degeneracies d_{QF} of supersymmetric ground states of total D0-charge Q and fermion number F is then

$$\begin{aligned}
\sum_{QF} d_{QF}\, y^F t^Q &= t^{-\frac{\chi}{24}} \sum_{\{n_{k\alpha}\}} y^{\sum_{kp\alpha} p\, n_{kp\alpha}}\, t^{\sum_{kp\alpha} k\, n_{kp\alpha}} \\
&= t^{-\frac{\chi}{24}} \prod_{kp\alpha} \sum_{n} y^{np} t^{nk} \\
&= t^{-\frac{\chi}{24}} \prod_{k} \frac{\prod_{p\,\text{odd}} (1 + y^p t^k)^{b_p}}{\prod_{p\,\text{even}} (1 - y^p t^k)^{b_p}} \,.
\end{aligned} \tag{7.222}$$

A generating function for the Witten indices $\Omega_{\text{D0}}(Q) = \sum_F (-1)^F d_{QF}$ is obtained by setting $y = -1$ in the above:

$$\sum_{Q} \Omega_{\text{D0}}(Q)\, t^Q = t^{-\frac{\chi}{24}} \prod_{k} (1 - t^k)^{-\chi} = \eta(t)^{-\chi}, \tag{7.223}$$

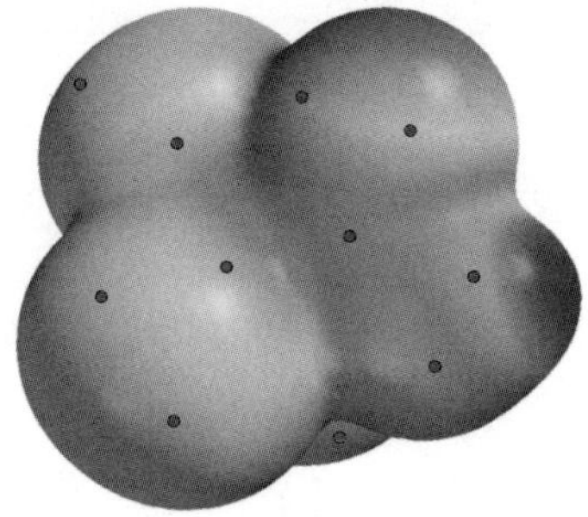

Fig. 7.14. Artist impression of a smooth abelian D4 configuration with $U(1)$ magnetic fluxes and mobile D0-branes bound to it.

where $\chi = \sum_p (-1)^p b_p$ is the Euler characteristic of Σ, and $\eta(t)$ is the Dedekind eta-function.

If we ignore the fact that there can be nontrivial interplay between the D4 and D0 moduli,[oo] adding the D0-branes to the D4 simply has the effect of multiplying the partition function Z in (7.206) by $e^{-2\pi u \frac{\chi}{24}} \eta(e^{-2\pi u})^{-\chi}$. Using the modularity property of the eta function, we furthermore have

$$\eta(e^{-2\pi u})^{-\chi} = u^{\chi/2}\, \eta(e^{-\frac{2\pi}{u}})^{-\chi}\,. \tag{7.224}$$

Combining this with the D4 partition function (7.216), using $\chi = b_2 + 2$, and for notational simplicity at the cost of slight formality setting $v = -u = -\varphi^0$ in (7.216), we get, within our no D4-D0 interplay approximation:

$$Z_{\rm D4+D0} = \eta(e^{-2\pi\varphi^0})^{-\frac{\chi}{24}} \sum_n (-1)^{d-d_n} \chi_n \, e^{2\pi\varphi^0 \frac{n^2}{2} - 2\pi\varphi^0 n} \tag{7.225}$$

$$= \varphi^0 \, e^{\frac{2\pi}{\varphi^0}(\frac{\chi}{24} - \frac{\varphi^2}{2})} \prod_k (1 - e^{-\frac{2\pi k}{\varphi^0}})^{-\chi} \times \sum_n (-1)^{d_n} \chi_n \, e^{\frac{2\pi}{\varphi^0}(\frac{n^2}{2} + in\varphi)}\,. \tag{7.226}$$

Again we see that Z transforms as a modular form under $\varphi^0 \to 1/\varphi^0$. Notice in particular that the large positive and negative weights of the D4 resp. the D0-part have canceled.

[oo]Ignoring this is actually not a good idea at small q_0, but at large q_0 it is a good approximation since most D4 flux configurations will be isolated anyway in this regime.

To extract the large q_0 asymptotics, we retaining only the leading terms at small positive φ^0. After integrating out φ:

$$\Omega(q) \approx \int d\varphi^0 \, \varphi_0^{\frac{B_2}{2}+1} \, e^{2\pi \hat{q}_0 \varphi^0 + \frac{2\pi\chi}{24\varphi^0}} \, (-1)^d \chi(\mathcal{M}) \,, \tag{7.227}$$

where $\hat{q}_0 \equiv q_0 + \frac{D^{AB} q_A q_B}{2}$. The saddle point of this integral is at $\varphi^0 = \sqrt{\frac{\chi}{24\,\hat{q}_0}}$, with value

$$\Omega(q) \approx e^{2\pi\sqrt{\frac{\hat{q}_0 \chi}{6}}} \,. \tag{7.228}$$

We recall that $\chi = D_{ABC} N^A N^B N^C + c_{2,A} N^A$, so this is an explicit formula for the index as a function of the charges. The logarithm of this expression can therefore directly be compared to the corresponding black hole entropy.

This is reliable when the saddle point value of φ^0 is small, i.e. $\hat{q}_0 \gg \chi$. When the latter is not the case, the saddle point value of φ^0 is not small, and so, as in the case without D0-branes, even within the model limitations we have made (such as ignoring singular or nonabelian D-branes), the above approximation for $\Omega(q)$ is not necessarily reliable. We will see in the next section that the agreement with the black hole entropy is excellent when $\hat{q}_0 \gg \chi$. In fact for the leading order matching in this regime, we do not even need to include the fluxes and moduli degrees of freedom, and a much simpler derivation is possible [93, 109, 115]. The match does get better at subleading order when the D4 degrees of freedom are included. But when $\hat{q}_0 \lesssim \chi$, the correct entropy is not anywhere near (7.228). In this regime, the flux and D4-moduli degrees of freedom dominate the entropy.

7.4.7. *Some extensions*

With the goal of introducing in detail a complex, glassy system that occurs naturally in string theory and has a precise geometric description and well-controlled holographic counterparts, we have given a fully explicit construction of the supersymmetric quantum mechanics describing the ground state sector of wrapped D4-branes bound to D0-branes, and discussed explicit constructions of supersymmetric ground states and their counting in some detail. Some of this was a review of parts of [107] and the ideas and results used there, with some simplifications and some points worked out more explicitly. Our discussion was necessarily incomplete in scope. Rather than try to give an overview of the huge related literature, let me mention a few immediate extensions that could have directly followed this part, with a small, non-representative, biased sample of possible starting points for further reading:

(1) The lift to M-theory as an M5 wrapped on $S^1 \times \Sigma$, reducing to a $(0,4)$ CFT. This is in some ways a more natural framework as it unifies D4 and D0 degrees of freedom and allows for a holographic dual description beyond the ground state sector [109, 116, 117].
(2) Quiver quantum mechanics: these are 1d gauged linear sigma models, providing a simple but very rich class of models describing complex D-brane systems. The field content is represented by quiver diagrams, nodes being partonic branes and arrows open strings between them [118]. Quantization allows explicit interpolation between the weakly coupled geometric regime and the strongly coupled black hole bound state regimes [119] (see also below).
(3) A more refined description of D4-D0 bound states that takes into account the interplay between D4 and D0 moduli, nonabelian degrees of freedom, and so on. One conceptually simple but efficient approach is the brane-anti-brane tachyon condensation picture [107, 120–125].
(4) Counting of ground states away from the regime $\hat{q}_0 \gg \chi$. Relation to Gromov-Witten, Gopakumar-Vafa and Donaldson-Thomas invariants and the OSV conjecture [104, 107, 126–131].
(5) Wall crossing phenomena: e.g. the possibility for a D4 to split into a D6 and an anti-D6 away from the strict large volume limit [107, 119, 121, 132–135].
(6) Applications to global aspects of IIB model building [125, 136]

It would certainly be useful to have a fully explicit example in which in particular the period vectors can be computed exactly. This may be possible along the lines of [137].

We end with some references to key papers in the history of the subject. The idea that branes can be wrapped on nontrivial compact cycles to obtain charged particles in four dimensions appeared first in [138] (and shown to be necessary for the consistency of string theory in [139]). The discovery of the perturbative string description of D-branes [140] made it possible to quantize and count states at weak coupling, leading to the computation of the Bekenstein-Hawking entropy in [93]. Many of the ideas in the more systematic development of the physics and mathematics of wrapped D-branes in Calabi-Yau manifolds, including stability and wall crossing, originated in [121, 141], with precursors in [132, 142].

7.5. Black Hole Duals

7.5.1. *Introduction*

In the appropriate regime, D-branes wrapped on compact cycles manifest themselves as black holes in the low energy effective field theory. It turns out that an intricate zoo of stationary supersymmetric bound states of such black holes exists, like giant molecules, all of which collapse to a D-brane localized at a single point in space in the limit $g_s \to 0$.

In this final section we summarize the explicit solutions of such black hole bound states, and briefly discuss their entropy compared to the microscopic picture developed in the previous section. The complexity of the black hole solutions is correlated with the complexity of the D-brane landscapes that arise at weak coupling. We will focus again on type IIA Calabi-Yau compactifications.

7.5.2. *BPS solutions of $\mathcal{N} = 2$ supergravity*

The four dimensional effective theory of type IIA string theory compactified on a Calabi-Yau manifold X is $\mathcal{N} = 2$ supergravity coupled to $B_2 = b_2(X)$ abelian $\mathcal{N} = 2$ vector multiplets. There are $B_2 + 1$ gauge fields $\mathcal{A}^\Lambda$, $\Lambda = 0, 1, \cdots, B_2$, obtained as in (7.143) by KK reduction of RR potentials: $C^{(1)} = \mathcal{A}^0$, $C^{(3)} = \mathcal{A}^A \wedge D_A$. The B_2 vector multiplets furthermore each contain a complex scalar, obtained by KK reduction of the complexified Kähler form: $B + iJ = t^A D_A$. They also contain spin 1/2 fermions, but we don't need those. Finally, the theory contains massless hypermultiplets too, but they do not affect the solutions of interest to us and can be consistently put to constant values. We will work in this section in units with the 4d Newton constant $G_N \equiv 1$.

Wrapped D-branes manifest themselves as charged point particles in four dimensions, sourcing the vector multiplet fields. We have both electric (D2/D0) and magnetic (D4/D6) monopole charges. In general, the lattice L of magnetic-electric charges Γ carries a fundamental symplectic product, which in a symplectic basis has the canonical form

$$\langle \Gamma, \tilde{\Gamma} \rangle \equiv \Gamma^\Lambda \tilde{\Gamma}_\Lambda - \Gamma_\Lambda \tilde{\Gamma}^\Lambda. \qquad (7.229)$$

Upper indices denote magnetic and lower indices electric components. In the IIA case at hand, the electric charges are the D0 and D2 charges Q_0 and Q_A, and the dual magnetic charges are the D6 and D4 charges N^0

and N^A.[pp] In terms of these, the symplectic product is $\langle (N,Q),(\tilde{N},\tilde{Q})\rangle = N^0\tilde{Q}_0 + N^A\tilde{Q}_A - Q_0\tilde{N}^0 - Q_A\tilde{N}^A$. Integrality of this product is equivalent to Dirac quantization.

In the weak string coupling regime (7.178), the wrapped D-branes are well described as point particles moving in flat $\mathbb{R}^3$, interacting with each other primarily through the lightest stretched open string modes. The coupling $g_4 \equiv g_s/\sqrt{V_X} \propto \ell_p/\ell_s$ is in a hypermultiplet and can therefore be tuned at will. When it gets larger, excited open string modes become important, until eventually the interactions are better described by massless closed string exchange, i.e. graviton, photon and scalar exchange. This is the regime in which the supergravity description becomes valid. When the charges are large, the wrapped D-brane states manifest themselves as weakly curved black hole solutions.

A single centered BPS (i.e. supersymmetric) solution to the equations of motion is necessarily static and spherically symmetric, with a metric of the form $ds^2 = -e^{2U}dt^2 + e^{-2U}d\vec{x}^2$, and all fields functions of $r = |\vec{x}|$ only. The BPS equations of motion take the first order gradient flow form typical for supersymmetric solutions [143, 144]:

$$\dot{U} = -e^U|Z|, \tag{7.230}$$

$$\dot{t}^A = -2e^U g^{A\bar{B}}\partial_{\bar{B}}|Z|, \tag{7.231}$$

where $g_{A\bar{B}}$ is the metric on the vector multiplet moduli space, the dot denotes derivation with respect to $\tau \equiv 1/r$, and $Z(\Gamma,t)$ is the *central charge* of the magnetic-electric charge Γ in a background with moduli t^A. It is a complex function on the vector multiplet moduli space, holomorphic up to a normalization factor, and linear in the charge vector Γ:

$$Z(\Gamma;t) = \frac{1}{|\langle V,\bar{V}\rangle|^{1/2}}\langle \Gamma, V\rangle\,, \tag{7.232}$$

where

$$V^\Lambda = X^\Lambda\,, \qquad V_\Lambda = \frac{\partial F}{\partial X^\Lambda}\,, \qquad X^A = t^A X^0\,, \qquad A = 1,\ldots,B_2\,. \tag{7.233}$$

Here $F(X)$ is the *prepotential* of the $\mathcal{N}=2$ theory, which determines all couplings and metrics in the 4d action. In general it is a locally defined holomorphic function, homogeneous of degree 2. For our type IIA theory,

[pp] Usually these are denoted by P^0 and P^A, but to be consistent with the previous sections we use N^0 and N^A instead.

it takes the form

$$F(X) = -\frac{1}{6} D_{ABC} X^A X^B X^C + \cdots \tag{7.234}$$

where D_{ABC} was defined in (7.146) and the ellipsis denote string worldsheet instanton corrections, which are exponentially suppressed in the Kähler moduli $J^A = \mathrm{Im}\, t^A$ and therefore negligible when the Calabi-Yau is large. Dropping those, (7.232) boils down to

$$Z = \frac{1}{\left(\frac{8\, D_{ABC} J^A J^B J^C}{6}\right)^{1/2}} \times \left(\frac{1}{6} N^0 D_{ABC} t^A t^B t^C - \frac{1}{2} N^A D_{ABC} t^B t^C + Q_A t^A + Q_0\right). \tag{7.235}$$

The central charge gets its name because it appears as a charge commuting with everything in the $\mathcal{N} = 2$ supersymmetry algebra. Its absolute value equals the lowest mass a particle of charge Γ can possibly have in a background specified by t. This bound is saturated for BPS states. Its phase determines the supercharges preserved by the BPS state. Two BPS objects are mutually supersymmetric if their central charge phases line up.

The gradient flow equations (7.230) drive $|Z(\Gamma, t)|$ to its minimal value $|Z_\star|$.[qq] If this minimal value is zero at a nonsingular point in moduli space, no solution exists. If it is nonzero, we get a black hole with near horizon solution $t = \text{const.} = t_\star$, $e^{-2U(r)} = \frac{|Z_\star|^2}{r^2}$, as can be checked directly from the above BPS flow equations. This describes $\mathrm{AdS}_2 \times S^2$ with S^2 horizon area $A = 4\pi |Z_\star|^2$ and therefore Bekenstein-Hawking entropy

$$S(\Gamma) = \pi |Z_\star(\Gamma)|^2 \,. \tag{7.236}$$

When the D6-charge N^0 is zero, the minimization of $|Z|$ is straightforward and there is a simple closed form expression for $S(\Gamma)$ (still in the large volume approximation) [146]:

$$S(N,Q)|_{N^0=0} = 2\pi \sqrt{\frac{\widehat{Q}_0 D_{ABC} N^A N^B N^C}{6}}\,, \qquad \widehat{Q}_0 = Q_0 + \frac{1}{2} D^{AB} Q_A Q_B \,, \tag{7.237}$$

where $D_{AB} = D_{ABC} N^C$ and D^{AB} is its inverse. Note that to leading order in the large charge limit, this is exactly the same as the microscopic result

[qq] Although all regular critical points of $|Z|$ are isolated local minima [145], in the presence of singularities at finite distance in the moduli space, there can be multiple basins of attraction, but we will ignore this here, as it does not occur in the large volume approximation.

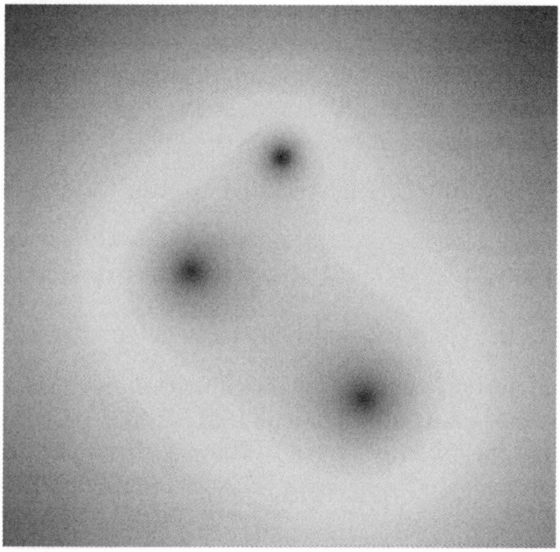

Fig. 7.15. Warp factor e^U for some three centered $D6 - \overline{D6} - D4$ bound state, for a choice of charges that doesn't really matter because this figure is primarily ornamental.

(7.227)! The subleading term can be reproduced macroscopically as well, from the Wald entropy in the presence of an R^2 term [147]. When N^0 is not zero there is no general closed form solution for $S(\Gamma)$, except when Q_A is chosen to be proportional to $D_{ABC}N^B N^C$. In that case we define for some fixed $K^A > 0$ the charge parameters (n^0, n, q, q_0) by $(N^0, N^A, Q_A, Q_0) = (n^0, n\, K^A, q\,(K^2)_A, q_0\, K^3)$, where K^3 is a short for $D_{ABC}K^A K^B K^C$ and $(K^2)_A$ for $D_{ABC}K^B K^C$. Then we can write

$$S = \frac{\pi}{3}\, K^3 \sqrt{3\, n^2\, q^2 - 8\, n^0\, q^3 + 6\, n^3\, q^0 - 18\, n\, n^0\, q\, q^0 - 9\, n^{0\,2}\, q^{0\,2}}\,. \quad (7.238)$$

As we will see below, once the entropy function $S(\Gamma)$ is known, it is easy to write down fully explicit expressions for all fields at all points in space.

Quite remarkably, these theories also have multicentered, supersymmetric, stationary *black hole bound states*, like giant molecules. These are genuine bound states in the sense that the centers are constrained by nontrivial potentials, generated by scalar, electomagnetic and gravitational forces [107, 117, 133, 148–150]. They have a metric of the form

$$ds^2 = -e^{2U}(dt + \omega_i dx^i)^2 + e^{-2U} d\vec{x}^2, \quad (7.239)$$

where U and ω depend on $\vec{x}$, and they are fully characterized by harmonic functions H^Λ, H_Λ with sources at the positions $\vec{x}_i$ of the charges Γ_i:

$$H = \sum_i \frac{\Gamma_i}{|\vec{x} - \vec{x}_i|} + h\,, \tag{7.240}$$

where the constant term h is determined by the total charge Γ and the asymptotic moduli:

$$h = -2\,\mathrm{Im}\big(e^{-i\alpha}\tilde{V}\big)|_{r=\infty}, \qquad \alpha = \arg Z(\Gamma)\,, \qquad \tilde{V} = V/|\langle V, \bar{V}\rangle|^{1/2}\,, \tag{7.241}$$

and V is as in (7.233).

As mentioned earlier, once the entropy function $S(\Gamma)$ is known on charge space, the complete solution is known [150], simply by substituting $H(\vec{x})$ for Γ in S and its derivatives:

$$e^{-2U} = \frac{1}{\pi}S(H)\,, \qquad \mathcal{A}^\Lambda = \frac{1}{\pi}\frac{\partial \log S(H)}{\partial H_\Lambda}\,(dt + \omega) + \mathcal{A}^\Lambda_{\mathrm{mon}}, \qquad t^A = \frac{H^A - \frac{i}{\pi}\frac{\partial S}{\partial H_A}}{H^0 - \frac{i}{\pi}\frac{\partial S}{\partial H_0}}\,. \tag{7.242}$$

The one-form $\mathcal{A}^\Lambda_{\mathrm{mon}}$ is the vector potential for a system of Dirac magnetic monopoles of charge N_i^Λ located at the positions $\vec{x}_i$. The off-diagonal components ω of the metric are the solutions to

$$\nabla \times \omega = \langle \nabla H, H \rangle\,, \tag{7.243}$$

where ∇ is the flat space gradient. This equation implies an important integrability condition: $\nabla \cdot (\nabla \times \omega) = 0 \Rightarrow \langle \nabla^2 H, H\rangle = 0$, from which, using $\nabla^2 \frac{1}{|\vec{x}|} = -4\pi\delta^3(x)$, we get for every center i a condition:

$$\sum_j \frac{\langle \Gamma_i, \Gamma_j\rangle}{|\vec{x}_i - \vec{x}_j|} = -\langle \Gamma_i, h\rangle = 2\,\mathrm{Im}\big(e^{-i\alpha} Z(\Gamma_i, t)\big)\big|_{r=\infty}\,. \tag{7.244}$$

This imposes constraints on the positions $\vec{x}_i$; solutions are the BPS equilibrium positions of the black holes in their mutual force fields. For the 2-centered case we have in particular

$$|\vec{x}_1 - \vec{x}_2| = \frac{\langle \Gamma_1, \Gamma_2\rangle}{2}\frac{|Z_1 + Z_2|}{\mathrm{Im}(Z_1 \bar{Z}_2)}\,, \tag{7.245}$$

where $Z_i = Z(\Gamma_i, t)|_{r=\infty}$. When the right hand side goes from positive to negative through a wall where Z_1 and Z_2 line up, the equilibrium separation diverges and the bound state decays. This is the supergravity incarnation of the wall crossing phenomenon.

For completeness we also give the solution to (7.243) [150]. For a 2-centered bound state with centers at $(0, 0, L)$ and $(0, 0, -L)$, up to residual diffeomorphism gauge transformations $t \to t + f(\vec{x})$, $\omega \to \omega - df$:

$$\omega = \frac{\langle \Gamma_1, \Gamma_2 \rangle}{2L} \left(\frac{L^2 - r^2}{(L^4 + r^4 - 2L^2 r^2 \cos 2\theta)^{1/2}} + 1 - \cos\theta_1 + \cos\theta_2 \right) d\phi \,. \tag{7.246}$$

Here (r, ϕ, θ) are standard spherical coordinates centered at the origin and θ_1, θ_2 are angles of spherical coordinates centered at the two particle positions. The integrability condition is equivalent to the absence of physical singularities along the z-axis. Because (7.243) is linear, the solution for more centers is obtained by superposition.

The solutions are generically stationary but not static: they have intrinsic angular momentum, given by

$$\vec{J} = \frac{1}{2} \sum_{i<j} \langle \Gamma_i, \Gamma_j \rangle \, \hat{x}_{ij} \,, \qquad \hat{x}_{ij} = \frac{\vec{x}_i - \vec{x}_j}{|\vec{x}_i - \vec{x}_j|} \,. \tag{7.247}$$

This generalizes the well-known electromagnetic field angular momentum sourced by monopole-electron pairs (generated by the $\vec{E} \times \vec{B}$ Poynting vector circling around the electron-monopole axis).

The integrability conditions are a necessary but not sufficient condition for the existence of solutions. One also needs to check if the entropy function $S(H)$ is positive everywhere. This is in general difficult to do. In [107, 133] the conjecture was formulated that the existence of solutions is equivalent to the existence of the (much easier to analyze) *attractor flow trees*. An attractor flow tree is built out of single attractor flows. The tree starts at the background value of the moduli and terminates at the attractor points of the constituent charges. Each edge E of an attractor flow tree is given by a single charge attractor flow for some charge Γ_E. Charge and energy are conserved at the vertices, i.e. for each vertex splitting $E \to (E_1, E_2)$, $\Gamma_E = \Gamma_{E_1} + \Gamma_{E_2}$ and $|Z(\Gamma_E)| = |Z(\Gamma_{E_1})| + |Z(\Gamma_{E_2})|$. The last condition is equivalent to requiring the vertices to lie on a line of marginal stability: $\arg Z(\Gamma_{E_1}) = \arg Z(\Gamma_{E_2})$. Flow trees can be thought of as "skeletons" for the supergravity solutions, or as giving a recipe for adiabatically assembling or disassembling multicentered bound states. This gives a strong physical argument for the conjecture. We refer to section 3.2.2 of [107] for a more detailed discussion.

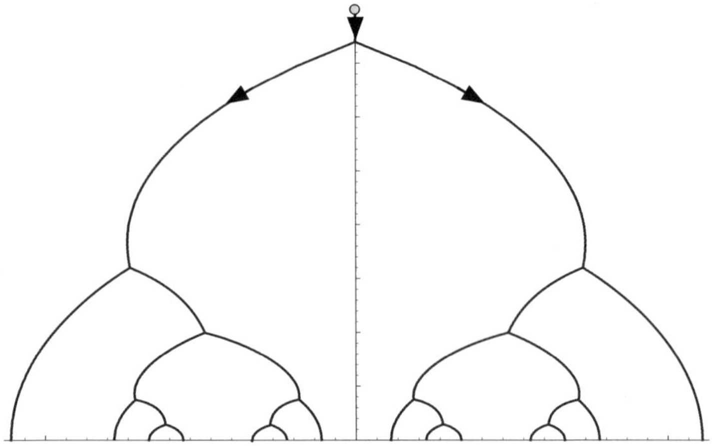

Fig. 7.16. (from [107]) Attractor flow tree in the t-plane for a particular bound state of 7 flux-carrying D6-branes and 7 flux-carrying anti-D6-branes. The flow starts at the yellow dot at the top and goes down. The iterated structure can be thought of as representing clutering of individual centers. In the case at hand all attractor points lie on the boundary of moduli space, which is due the special choice of charges. The symmetric form is also special to our particular choice. The main motivation to pick those charges was esthetics and dramatic effect. More generic examples can be found in [107].

7.5.3. *Entropy*

Comparing the macroscopic and microscopic D4-D0 entropies (7.237) and (7.227), we find perfect agreement to leading order when $\hat{Q}_0 \gg N^3$. The subleading correction can be reproduced from the Wald entropy in the presence of an R^2 correction term in the 4d effective gravity theory [147]. In fact, in the same regime, an infinite series of corrections matches [104, 107, 128–130]

However, as was clear from the discussion in section 4, microcopically things get much more complicated when $\hat{Q}_0$ becomes of the order of $\chi \sim N^3$: the D4-degrees of freedom start getting important and eventually dominate the entropy. On the black hole side things get much more complicated too in exactly the same regime: Multicentered solutions appear, and at some point they start dominating the entropy [107].

In particular one clearly exits the easy regime where $\hat{Q}_0 \gg N^3$ whenever *all* charges are uniformly scaled up sufficiently. Thus, more precisely, when charges Γ are scaled up as $\Gamma \to \Lambda\Gamma$, there may exist for example two-centered solutions with horizon entropy scaling as Λ^3, while the single centered entropy scales as Λ^2.

Let us give a concrete example of this phenomenon, taken from [151]. Consider the Calabi-Yau $X = T_1^2 \times T_2^2 \times T_3^2$, a product of three two-tori. Let Σ be the 4-cycle $(T_1^2 \times T_2^2) + (T_2^2 \times T_3^2) + (T_3^2 \times T_1^2)$ and let σ be the 2-cycle $T_1^2 + T_2^2 + T_3^2$. Then the entropy function of a charge Γ corresponding to wrapping N^0 D6-branes on X, N D4-branes on Σ, Q D2-branes on σ and Q_0 D0-branes is given by $S(\Gamma) = \pi\sqrt{4N^3Q_0 + 3N^2Q^2 + 6N^0NQQ_0 - 4N^0Q^3 - (N^0Q_0)^2}$. Now consider a total charge

$$\Gamma = (N^0, N, Q, Q_0) = \Lambda(0, 6, 0, -12), \tag{7.248}$$

in a background in which the area of each T^2 equals v. Then for any v, there exists a single centered solution with horizon entropy

$$S_1 = 72\sqrt{2}\pi\,\Lambda^2. \tag{7.249}$$

However, there also exists D6-anti-D6 two-centered black hole bound states, for instance with charges

$$\Gamma_1 = (1, 3\Lambda, 6\Lambda^2, -6\Lambda), \qquad \Gamma_2 = (-1, 3\Lambda, -6\Lambda^2, -6\Lambda), \tag{7.250}$$

provided $v > \sqrt{18}\Lambda$. The constant terms in the harmonic functions (7.240) are $h = (0, \frac{1}{\sqrt{2v}}, 0, -\sqrt{\frac{v^3}{2}})$, and the equilibrium separation (7.245) at large Λ is $|\vec{x}_1 - \vec{x}_2| \approx \frac{108\sqrt{2}\sqrt{v}\,\Lambda}{(v/\Lambda)^2 - 18}$. The resulting metric is well defined, with $S(H)$ real and positive everywhere. The two centers have equal horizon entropy, summing up to a total entropy

$$S_2 \approx 12\pi\sqrt{3}\,\Lambda^3\,, \tag{7.251}$$

which is indeed parametrically larger than the single centered entropy (7.249).

When v is kept fixed while sending $\Lambda \to \infty$, the equilibrium separation diverges and then formally goes negative, meaning these 2-centered solutions disappear from the spectrum. But the low energy effective quantum mechanics of section 7.4.3 is supposed to become increasingly reliable in the limit of large volume $v \to \infty$ (and weak string coupling), so its Witten index for given charges must be the total index of all black hole configurations in the limit $v \to \infty$. Thus we conclude that at large Λ, the usual weakly coupled, weakly curved wrapped D4-branes are not computing the entropy of a single D4-D2-D0 black hole, but predominantly that of multi-centered configurations. The single center D4 black hole entropy can only be microscopically reproduced from a weak coupling computation in the regime $\hat{Q}_0 \gg N^3$.

This occurs in particular also in the lift to M-theory [117], where a precise holographic correspondence exists between a 1+1 dimensional (0,4) CFT [109] and quantum gravity in $AdS_3 \times S^2$. It was pointed out there that the solution with the highest entropy in the $\Lambda \to \infty$ limit appears to be a configuration in which all entropy carrying charges has been concentrated in one of the D6-charged black holes.

7.5.4. *Landscape structure*

In the minimal charge case, namely $\hat{Q}_0 = -\frac{\chi}{24} \sim -\frac{N^3}{24}$, the 4d supergravity solution is a pure D6-anti-D6 bound state, where the D6 branes carry flux but no entropic degrees of freedom [107]. Quantization of the centers leads to a spin $j = d/2$ multiplet of supersymmetric ground states with degeneracy $2j + 1 = d + 1$, in agreement with the microscopic picture where this arises as the Euler characteristic of the moduli space $\mathcal{M} = \mathbb{CP}^d$. The lifted M-theory solution becomes pure "spinning" $AdS_3 \times S^2$ in this case.

Some identifications between the microscopic structure of the D-brane landscape and the macroscopic black hole solutions have been made (where it should be kept in mind of course that the map from D-brane microstates to black hole macroscopic states must be many-to-one, due to the existence of black hole entropy). For example switching on fluxes represented by holomorphic 2-cycles as in section 7.4.4 corresponds to "dressing up" the D6 or anti-D6 by halos of D2-D0 particles [107], or equivalently in M-theory by adding M2-branes to the north or south poles of the S^2 [129, 130]. But the full map at higher D0-charge is not well understood.

Dynamics near zero energy should proceed through tunneling of charged particles in and out of the black hole centers in this picture. This is under study in [3].

Acknowledgments

I would like to thank the organizers of TASI 2010, Tom Banks, Michael Dine, K.T. Mahanthappa and Subir Sachdev, for giving me the opportunity to teach on these topics. I am very much indebted to Dionysios Anninos, Tarek Anous, Jacob Barandes, Marcus Benna, Hyeyoun Chung, Mike Douglas, Bram Gaasbeek, Hajar Ebrahim, Greg Moore and Andy Strominger, whose insights, collaborations and discussions were crucial in the genesis of these notes. Many of the ideas motivating the theme of these lectures were developed over the past year together with Dionysios Anninos, and we would like to thank Charlies Kitchen for hospitality. I'm grateful to A.P. Young and Subir Sachdev for clarifying conversations about spin glasses. Finally, special thanks to Dionysios Anninos, Tarek Anous, Jacob Barandes and Marcus Benna for a careful reading of parts of these notes and their useful suggestions for improvements. This work was supported in part by DOE grant DE-FG02-91ER40654.

References

[1] M. Lynker, Landau-Ginzburg vacua of string, M- and F-theory at c=12, *Nuclear Physics B.* **550**(1-2), 123–150 (June, 1999). ISSN 05503213. doi: 10.1016/S0550-3213(99)00204-7. URL `http://arxiv.org/abs/hep-th/9812195`.

[2] M. Kreuzer. Calabi-Yau data. URL `http://tph16.tuwien.ac.at/%7ekreuzer/CY`.

[3] J. Barandes, H. Chung, F. Denef, H. Ebrahim, and P. Petrov, Tunneling transitions between black hole bound states, *to appear.*

[4] D. Anninos and F. Denef, Replica order parameter for de Sitter dynamics, *to appear.*

[5] D. Anninos, T. Anous, J. Barandes, and F. Denef, String Glasses, *to appear.*

[6] A. Goose. Prerequisites. URL `http://abstrusegoose.com/272`.

[7] F. Denef, Les Houches lectures on constructing string vacua, *arXiv 0803.1194.* (2008). URL `http://arxiv.org/pdf/0803.1194`.

[8] T. Banks, TASI Lectures on Holographic Space-Time, SUSY and Gravitational Effective Field Theory, *arXiv:1007.4001* (July. 2010). URL `http://arxiv.org/abs/arXiv:1007.4001`.

[9] A. Starobinsky, Dynamics of phase transition in the new inflationary universe scenario and generation of perturbations, *Physics Letters B.* **117**(3-4), 175–178 (Nov., 1982). ISSN 03702693. doi: 10.1016/0370-2693(82)90541-X. URL `http://dx.doi.org/10.1016/0370-2693(82)90541-X`.

[10] A. Linde, Chaotic inflation, *Physics Letters B.* **129**(3-4), 177–181 (Sept., 1983). ISSN 03702693. doi: 10.1016/0370-2693(83)90837-7. URL `http://dx.doi.org/10.1016/0370-2693(83)90837-7`.

[11] A. Linde, D. Linde, and A. Mezhlumian, From the Big Bang theory to the theory of a stationary universe, *Physical Review D.* **49**(4), 1783, (1994). URL `http://arxiv.org/abs/gr-qc/9306035`.

[12] S. Winitzki, Eternal fractal in the universe, *Physical Review D.* **65** (8), 083506, (2002). URL `http://prd.aps.org/abstract/PRD/v65/i8/e083506`.

[13] A. Strominger, The dS/CFT correspondence, *Journal of High Energy Physics.* **2001**(10), 034–034 (Oct., 2001). ISSN 1029-8479. doi: 10.1088/1126-6708/2001/10/034. URL `http://arxiv.org/abs/hep-th/0106113`.

[14] A. Strominger, Inflation and the dS/CFT Correspondence, *Journal of High Energy Physics.* **2001**(11), 049–049 (Nov., 2001). ISSN 1029-

8479. doi: 10.1088/1126-6708/2001/11/049. URL `http://arxiv.org/abs/hep-th/0110087`.

[15] M. Spradlin and A. Strominger, Les Houches lectures on de Sitter space, *Arxiv preprint hep-th/0110007* (Sept. 2001). URL `http://arxiv.org/abs/hep-th/0110007`.

[16] E. Witten, Quantum Gravity In De Sitter Space, *arXiv: hep-th/0106109* (June. 2001). URL `http://arxiv.org/abs/hep-th/0106109`.

[17] J. Maldacena, Non-gaussian features of primordial fluctuations in single field inflationary models, *Journal of High Energy Physics.* **2003**(05), 013–013 (May, 2003). ISSN 1029-8479. doi: 10.1088/1126-6708/2003/05/013. URL `http://arxiv.org/abs/astro-ph/0210603`.

[18] D. Anninos and G. Ng, Asymptotic Symmetries and Charges in De Sitter Space, *Arxiv preprint arXiv:1009.4730* (Sept. 2010). URL `http://arxiv.org/abs/1009.4730`.

[19] R. Bousso, B. Freivogel, and S. Leichenauer, Geometric origin of coincidences and hierarchies in the landscape, *Arxiv preprint arXiv:1012.2869* (Dec. 2010). URL `http://arxiv.org/abs/1012.2869`.

[20] R. Bousso, B. Freivogel, S. Leichenauer, and V. Rosenhaus, Boundary definition of a multiverse measure, *Physical Review D.* **82**(12), 39 (Dec., 2010). ISSN 1550-7998. doi: 10.1103/PhysRevD.82.125032. URL `http://arxiv.org/abs/1005.2783`.

[21] R. Bousso, B. Freivogel, Y. Sekino, S. Shenker, L. Susskind, I.-S. Yang, and C.-P. Yeh, Future foam: Nontrivial topology from bubble collisions in eternal inflation, *Physical Review D.* **78**(6), 23 (Sept., 2008). ISSN 1550-7998. doi: 10.1103/PhysRevD.78.063538. URL `http://arxiv.org/abs/0807.1947`.

[22] R. Bousso, The cosmological constant, *General Relativity and Gravitation.* **40**(2-3), 607–637 (Dec., 2007). ISSN 0001-7701. doi: 10.1007/s10714-007-0557-5. URL `http://arxiv.org/abs/0708.4231`.

[23] J. Garriga and A. Vilenkin, Holographic multiverse, *Journal of Cosmology and Astroparticle Physics.* **2009**, 021, (2009). URL `http://arxiv.org/abs/0809.4257`.

[24] A. Vilenkin, Holographic multiverse and the measure problem, *Arxiv preprint arXiv:1103.1132* (Mar. 2011). URL `http://arxiv.org/abs/1103.1132`.

[25] Wikipedia. Multiverse. URL `http://en.wikipedia.org/wiki/Multiverse`.

[26] H. Everett, "Relative State" Formulation of Quantum Mechanics, *Reviews of Modern Physics.* **29**(3), 454–462 (July, 1957). ISSN 0034-6861. doi: 10.1103/RevModPhys.29.454. URL `http://rmp.aps.org/abstract/RMP/v29/i3/p454_1`.

[27] S. Giddings, Axion-induced topology change in quantum gravity and string theory, *Nuclear Physics B.* **306**(4), 890–907 (Sept., 1988). ISSN 05503213. doi: 10.1016/0550-3213(88)90446-4. URL `http://dx.doi.org/10.1016/0550-3213(88)90446-4`.

[28] S. Coleman, Black holes as red herrings: topological fluctuations and the loss of quantum coherence, *Nuclear Physics B.* **307**(4), 867–882, (1988).

ISSN 0550-3213. URL http://linkinghub.elsevier.com/retrieve/pii/0550321388901101.

[29] N. Arkani-Hamed, J. Orgera, and J. Polchinski, Euclidean wormholes in string theory, *Journal of High Energy Physics*. **2007**, 018, (2007). URL http://iopscience.iop.org/1126-6708/2007/12/018.

[30] S. Weinberg, Anthropic Bound on the Cosmological Constant, *Physical Review Letters*. **59**(22), 2607–2610 (Nov., 1987). ISSN 0031-9007. doi: 10.1103/PhysRevLett.59.2607. URL http://www.adsabs.harvard.edu/abs/1987PhRvL..59.2607W.

[31] A. Strominger, Superstrings with torsion, *Nuclear Physics B*. **274**(2), 253–284 (Sept., 1986). ISSN 05503213. doi: 10.1016/0550-3213(86)90286-5. URL http://dx.doi.org/10.1016/0550-3213(86)90286-5.

[32] R. Bousso and J. Polchinski, Quantization of Four-form Fluxes and Dynamical Neutralization of the Cosmological Constant, *JHEP*. **006**, 0006, (2000). URL http://arxiv.org/abs/hep-th/0004134.

[33] S. Kachru, R. Kallosh, A. Linde, and S. Trivedi, de Sitter vacua in string theory, *Physical Review D*. **68**(4) (Aug., 2003). ISSN 0556-2821. doi: 10.1103/PhysRevD.68.046005. URL http://arxiv.org/abs/hep-th/0301240.

[34] L. Susskind, The anthropic landscape of string theory, *Universe or multiverse*. pp. 247–66 (Feb., 2007). URL http://arxiv.org/abs/hep-th/0302219.

[35] A. N. Schellekens, The Landscape "avant la lettre", *ArXiv: physics/0604134* (Apr. 2006). URL http://arxiv.org/abs/physics/0604134.

[36] M. R. Douglas, The statistics of string/M theory vacua, *Journal of High Energy Physics*. **2003**(05), 046–046 (May, 2003). ISSN 1029-8479. doi: 10.1088/1126-6708/2003/05/046. URL http://arxiv.org/abs/hep-th/0303194.

[37] M. Grana, Flux compactifications in string theory: A comprehensive review, *Physics Reports*. **423**(3), 91–158 (Jan., 2006). ISSN 03701573. doi: 10.1016/j.physrep.2005.10.008. URL http://arxiv.org/abs/hep-th/0509003.

[38] B. S. Acharya and M. R. Douglas, A Finite Landscape?, *hep-th/0606212* (June. 2006). URL http://arxiv.org/abs/hep-th/0606212.

[39] M. Douglas and S. Kachru, Flux compactification, *Reviews of Modern Physics*. **79**(2), 733–796 (May, 2007). ISSN 0034-6861. doi: 10.1103/RevModPhys.79.733. URL http://rmp.aps.org/abstract/RMP/v79/i2/p733_1.

[40] F. Denef and M. R. Douglas, Computational complexity of the landscape: Part I, *Annals of Physics*. **322**(5), 1096–1142 (May, 2007). ISSN 00034916. doi: 10.1016/j.aop.2006.07.013. URL http://arxiv.org/abs/hep-th/0602072.

[41] M. R. Douglas, Effective potential and warp factor dynamics, *Journal of High Energy Physics*. **2010**(3), 1–32–32 (Mar., 2010). ISSN 1029-8479. doi: 10.1007/JHEP03(2010)071. URL http://arxiv.org/abs/0911.3378.

[42] A. Nabutovsky and S. Weinberger, The fractal nature of Riem/Diff I, *Geometriae Dedicata.* **101**(1), 1–54, (2003). ISSN 0046-5755. URL http://www.math.toronto.edu/nabutovsky/.
[43] M. Mézard, G. Parisi, and M. A. Virasoro, *Spin Glass Theory and Beyond.* vol. 9, *Lecture Notes in Physics*, (World Scientific, 1987). ISBN 9971501155. URL http://books.google.com/books?id=ZIF9QgAACAAJ.
[44] K. Binder and A. Young, Spin glasses: Experimental facts, theoretical concepts, and open questions, *Reviews of Modern physics.* **58**(4), 801–976, (1986). ISSN 1539-0756. URL http://link.aps.org/doi/10.1103/RevModPhys.58.801.
[45] K. H. Fisher and J. A. Hertz, Spin Glasses, *American Journal of Physics.* **60**(5), 478, (1992). ISSN 00029505. doi: 10.1119/1.16856. URL http://books.google.com/books?id=zXqel8dS-rIC.
[46] P. Cartier, J. B. Bost, H. Cohen, D. Zagier, R. Gergondey, H. M. Stark, E. Reyssat, F. Beukers, G. Christol, M. Senechal, A. Katz, J. Bellissard, P. Cvitanovic, and J. C. Yoccoz, *From Number Theory to Physics.* (Springer, 1992). ISBN 3540533427. URL http://books.google.com/books?id=UUFIQgAACAAJ.
[47] L. Brekke, P. Freund, M. Olson, and E. Witten, Non-archimedean string dynamics, *Nuclear Physics B.* **302**(3), 365–402 (June, 1988). ISSN 0550-3213. doi: 10.1016/0550-3213(88)90207-6. URL http://dx.doi.org/10.1016/0550-3213(88)90207-6.
[48] S. Roch, Toward extracting all phylogenetic information from matrices of evolutionary distances., *Science (New York, N.Y.).* **327**(5971), 1376–9 (Mar., 2010). ISSN 1095-9203. doi: 10.1126/science.1182300. URL http://www.ncbi.nlm.nih.gov/pubmed/20223986.
[49] D. Sherrington and S. Kirkpatrick, Solvable Model of a Spin-Glass, *Physical Review Letters.* **35**(26), 1792–1796 (Dec., 1975). ISSN 0031-9007. doi: 10.1103/PhysRevLett.35.1792. URL http://prl.aps.org/abstract/PRL/v35/i26/p1792_1.
[50] G. Parisi, Infinite number of order parameters for spin-glasses, *Physical Review Letters.* **43**(23), 1754–1756, (1979). ISSN 1079-7114. URL http://link.aps.org/doi/10.1103/PhysRevLett.43.1754.
[51] G. Parisi, A sequence of approximated solutions to the SK model for spin glasses, *Journal of Physics A: Mathematical and General.* **13**, L115, (1980). URL http://iopscience.iop.org/0305-4470/13/4/009.
[52] M. Mézard, G. Parisi, N. Sourlas, G. Toulouse, and M. Virasoro, Nature of the spin-glass phase, *Physical Review Letters.* **52**(13), 1156–1159, (1984). ISSN 1079-7114. URL http://link.aps.org/doi/10.1103/PhysRevLett.52.1156.
[53] E. Marinari, Optimized monte carlo methods, *Advances in Computer Simulation.* pp. 50–81, (1998). URL http://dx.doi.org/10.1007/BFb0105459.
[54] D. J. Earl and M. W. Deem, Parallel tempering: Theory, applications, and new perspectives, *Physical Chemistry Chemical Physics.* **7** (23), 3910 (Sept., 2005). ISSN 1463-9076. doi: 10.1039/b509983h. URL http://pubs.rsc.org/en/Content/ArticleHTML/2005/CP/B509983H.

[55] D. Fisher and D. Huse, Ordered phase of short-range Ising spin-glasses, *Physical Review Letters.* **56**(15), 1601–1604, (1986). ISSN 1079-7114. URL http://link.aps.org/doi/10.1103/PhysRevLett.56.1601.

[56] H. G. Katzgraber and A. K. Hartmann, Ultrametricity and clustering of States in spin glasses: a one-dimensional view., *Physical review letters.* **102** (3), 037207 (Jan., 2009). ISSN 0031-9007. URL http://dx.doi.org/10.1103/PhysRevLett.102.037207.

[57] R. Rammal, G. Toulouse, and M. Virasoro, Ultrametricity for physicists, *Reviews of Modern Physics.* **58**(3), 765–788, (1986). ISSN 1539-0756. URL http://link.aps.org/doi/10.1103/RevModPhys.58.765.

[58] G. Parisi and F. Ricci-Tersenghi, On the origin of ultrametricity, *Journal of Physics A: Mathematical and General.* **33**, 113, (2000). URL http://iopscience.iop.org/0305-4470/33/1/307.

[59] C. De Dominicis and I. Giardina, *Random fields and spin glasses: a field theory approach.* (Cambridge Univ Press, 2006). ISBN 0521847834. URL http://books.google.com/books?id=58iM6EFKx9cC.

[60] T. Castellani and A. Cavagna, Spin-glass theory for pedestrians, *Journal of Statistical Mechanics: Theory and Experiment.* **2005**, P05012, (2005). URL http://arxiv.org/abs/cond-mat/0505032.

[61] A. Crisanti and H. J. Sommers, The spherical p-spin interaction spin glass model: the statics, *Zeitschrift fur Physik B Condensed Matter.* **87**(3), 341–354 (Oct., 1992). ISSN 0722-3277. doi: 10.1007/BF01309287. URL http://www.springerlink.com/content/x720420477168352/.

[62] S. Edwards and P. Anderson, Theory of spin glasses, *Journal of Physics F: Metal Physics.* **5**, 965, (1975). URL http://iopscience.iop.org/0305-4608/5/5/017.

[63] F. Guerra, Broken replica symmetry bounds in the mean field spin glass model, *Communications in Mathematical Physics.* **233**(1), 1–12, (2003). ISSN 0010-3616. URL http://dx.doi.org/10.1007/s00220-002-0773-5.

[64] M. Talagrand, The Parisi formula, *Annals of Mathematics-Second Series.* **163**(1), 221–264, (2006). ISSN 0003-486X. URL http://annals.princeton.edu/annals/2006/163-1/annals-v163-n1-p04-s.pdf.

[65] G. Parisi, Order parameter for spin-glasses, *Physical Review Letters.* **50** (24), 1946–1948, (1983). ISSN 1079-7114. URL http://link.aps.org/doi/10.1103/PhysRevLett.50.1946.

[66] A. P. Young, Ed., *Spin glasses and random fields.* (World Scientific Pub Co Inc, 1998). ISBN 9810232403. URL http://books.google.com/books?id=9hsCr-i6b5wC.

[67] M. Mézard and M. Virasoro, The microstructure of ultrametricity, *Journal de Physique.* **46**(8), 1293–1307, (1985). ISSN 0302-0738. URL http://dx.doi.org/10.1051/jphys:019850046080129300.

[68] A. Crisanti and T. Rizzo, Analysis of the-replica symmetry breaking solution of the Sherrington-Kirkpatrick model, *Physical Review E.* **65**(4), 46137, (2002). ISSN 1550-2376. URL http://link.aps.org/doi/10.1103/PhysRevE.65.046137.

[69] D. J. Thouless, P. W. Anderson, and R. G. Palmer, Solution of 'Solvable

model of a spin glass', *Philosophical Magazine.* **35**(3), 593–601 (Mar., 1977). ISSN 1478-6435. doi: 10.1080/14786437708235992. URL `http://adsabs.harvard.edu/abs/1977PMag...35..593T`.

[70] G. Parisi and N. Sourlas, Random Magnetic Fields, Supersymmetry, and Negative Dimensions, *Physical Review Letters.* **43**(11), 744–745, (1979). ISSN 1079-7114. URL `http://link.aps.org/doi/10.1103/PhysRevLett.43.744`.

[71] J. Kurchan, Supersymmetry in spin glass dynamics, *Journal de Physique I.* **2**(7), 1333–1352, (1992). URL `http://hal.archives-ouvertes.fr/jpa-00246625/`.

[72] J. Kurchan, Supersymmetry, replica and dynamic treatments of disordered systems: a parallel presentation, *Arxiv preprint cond-mat/0209399.* pp. 1–19, (2002). URL `http://arxiv.org/pdf/cond-mat/0209399`.

[73] F. Barahona, On the computational complexity of Ising spin glass models, *Journal of Physics A: Mathematical and General.* **15**, 3241, (1982). URL `http://iopscience.iop.org/0305-4470/15/10/028`.

[74] I. Kondor, Parisi's mean-field solution for spin glasses as an analytic continuation in the replica number, *Journal of Physics A: Mathematical and General.* **16**, L127, (1983). URL `http://iopscience.iop.org/0305-4470/16/4/006`.

[75] J. Verbaarschot and M. Zirnbauer, Critique of the replica trick, *Journal of Physics A: Mathematical and General.* **18**, 1093, (1985). URL `http://iopscience.iop.org/0305-4470/18/7/018`.

[76] J. Almeida and D. Thouless, Stability of the Sherrington-Kirkpatrick solution of a spin glass model, *Journal of Physics A: Mathematical and General.* **11**, 983, (1978). URL `http://iopscience.iop.org/0305-4470/11/5/028`.

[77] V. Avetisov, A. Bikulov, and S. Kozyrev, Application of p-adic analysis to models of breaking of replica symmetry, *Journal of Physics A: Mathematical and General.* **32**(50), 8785, (1999). ISSN 0305-4470. doi: http://dx.doi.org/10.1088/0305-4470/32/50/301. URL `http://iopscience.iop.org/0305-4470/32/50/301`.

[78] G. Parisi and N. Sourlas, P-adic numbers and replica symmetry breaking, *The European Physical Journal B.* **14**(3), 535–542 (Mar., 2000). ISSN 1434-6028. doi: 10.1007/s100510051063. URL `http://dx.doi.org/10.1007/s100510051063`.

[79] A. Crisanti and C. De Dominicis, Stability of the Parisi Solution for the Sherrington-Kirkpatrick model near T= 0, *Arxiv preprint arXiv:1101.5233.* (2011). URL `http://arxiv.org/abs/1101.5233`.

[80] B. Derrida, Random-energy model: Limit of a family of disordered models, *Physical Review Letters.* **45**(2), 79–82, (1980). ISSN 1079-7114. URL `http://link.aps.org/doi/10.1103/PhysRevLett.45.79`.

[81] D. Gross and M. Mézard, The simplest spin glass, *Nuclear Physics B.* **240**(4), 431–452, (1984). ISSN 0550-3213. URL `http://dx.doi.org/10.1209/0295-5075/20/3/002`.

[82] A. Bray and M. Moore, Replica theory of quantum spin glasses, *Journal of Physics C: Solid State Physics.* **13**, L655, (1980). URL `http://iopscience.iop.org/0022-3719/13/24/005`.

[83] J. Ye, S. Sachdev, and N. Read, Solvable spin glass of quantum rotors, *Physical review letters*. **70**(25), 4011–4014, (1993). ISSN 1079-7114. URL http://link.aps.org/doi/10.1103/PhysRevLett.70.4011.

[84] L. Cugliandolo, D. Grempel, and C. da Silva Santos, Imaginary-time replica formalism study of a quantum spherical p-spin-glass model, *Physical Review B*. **64**(1), 14403, (2001). ISSN 1550-235X. URL http://link.aps.org/doi/10.1103/PhysRevB.64.014403.

[85] J. Hopfield, Neural networks and physical systems with emergent collective computational abilities Biophysics, *Proceedings of the National Academy of Sciences of the United States of America*. **79**(8), 2554, (1982). URL http://www.pnas.org/content/79/8/2554.short.

[86] D. Amit, H. Gutfreund, and H. Sompolinsky, Spin-glass models of neural networks, *Physical Review A*. **32**(2), 1007–1018, (1985). ISSN 1094-1622. URL http://link.aps.org/doi/10.1103/PhysRevA.32.1007.

[87] G. Parisi, A memory which forgets, *Journal of Physics A: Mathematical and General*. **19**(M), L617, (1986). URL http://iopscience.iop.org/0305-4470/19/10/011.

[88] S. Kirkpatrick, Optimization by simulated annealing: Quantitative studies, *Journal of Statistical Physics*. **34**(5-6), 975–986 (Mar., 1984). ISSN 0022-4715. doi: 10.1007/BF01009452. URL http://dx.doi.org/10.1007/BF01009452.

[89] M. Mézard, T. Mora, and R. Zecchina, Clustering of Solutions in the Random Satisfiability Problem, *Physical Review Letters*. **94**(19) (May, 2005). ISSN 0031-9007. doi: 10.1103/PhysRevLett.94.197205. URL http://prl.aps.org/abstract/PRL/v94/i19/e197205.

[90] M. Mézard and R. Zecchina, Random K-satisfiability problem: From an analytic solution to an efficient algorithm, *Physical Review E*. **66**(5) (Nov., 2002). ISSN 1063-651X. doi: 10.1103/PhysRevE.66.056126. URL http://pre.aps.org/abstract/PRE/v66/i5/e056126.

[91] A. Renyi. On measures of entropy and information. In *Fourth Berkeley Symposium on Mathematical Statistics and Probability*, vol. 547, pp. 547–561, (1961). URL http://l.academicdirect.org/Horticulture/GAs/Refs/Renyi_1961.pdf.

[92] E. Witten, Constraints on supersymmetry breaking, *Nuclear Physics B*. **202**(2), 253–316 (July, 1982). ISSN 05503213. doi: 10.1016/0550-3213(82)90071-2. URL http://dx.doi.org/10.1016/0550-3213(82)90071-2.

[93] A. Strominger and C. Vafa, Microscopic origin of the Bekenstein-Hawking entropy, *Physics Letters B*. **379**(1-4), 99–104 (June, 1996). ISSN 03702693. doi: 10.1016/0370-2693(96)00345-0. URL http://arxiv.org/abs/hep-th/9601029.

[94] K. Hori, S. Katz, A. Klemm, R. Pandharipande, R. Thomas, C. Vafa, R. Vakil, and E. Zaslow, *Mirror symmetry*. (Clay mathematics monographs, 2003). ISBN 0821829556. URL http://books.google.com/books?id=uGYRaAhFGxOC&pgis=1.

[95] F. Cooper, A. Khare, and U. Sukhatme, Supersymmetry and quantum mechanics, *Physics Reports*. **251**(5-6), 267–385 (Jan., 1995). ISSN 03701573.

doi: 10.1016/0370-1573(94)00080-M. URL http://dx.doi.org/10.1016/0370-1573(94)00080-M.

[96] L. Alvarez-Gaume, Supersymmetry and the Atiyah-Singer index theorem, *Communications in Mathematical Physics*. **90**(2), 161–173 (June, 1983). ISSN 0010-3616. doi: 10.1007/BF01205500. URL http://dx.doi.org/10.1007/BF01205500.

[97] S. Cecotti and L. Girardello, Stochastic and Parastochastic Aspects of Supersymmetric Functional Measures : A New Approach to Supersymmetry, *Annals of Physics*. **99**, 81–99, (1983). URL http://dx.doi.org/10.1016/0003-4916(83)90172-0.

[98] P. Damgaard and H. Huffel, Stochastic quantization, *Physics Reports*. **152** (5-6), 227–398 (Aug., 1987). ISSN 03701573. doi: 10.1016/0370-1573(87)90144-X. URL http://dx.doi.org/10.1016/0370-1573(87)90144-X.

[99] S. Franz and J. Kurchan, On the Computation of Static Expectation Values from Dynamics in Spin Glasses, *Europhysics Letters (EPL)*. **20**(3), 197–203 (Oct., 1992). ISSN 0295-5075. doi: 10.1209/0295-5075/20/3/002. URL http://dx.doi.org/10.1209/0295-5075/20/3/002.

[100] S. Cecotti and C. Vafa, 2d Wall-Crossing, R-twisting, and a Supersymmetric Index, *Arxiv preprint arXiv:1002.3638* (Feb. 2010). URL http://arxiv.org/abs/1002.3638.

[101] R. Dijkgraaf, D. Orlando, and S. Reffert, Relating field theories via stochastic quantization, *Nuclear Physics B*. **824**(3), 365–386 (Jan., 2010). ISSN 05503213. doi: 10.1016/j.nuclphysb.2009.07.018. URL http://arxiv.org/abs/0903.0732.

[102] A. Amir, Y. Oreg, and Y. Imry, Mean-field model for electron-glass dynamics, *Physical Review B*. **77**(16), 1–8 (Apr., 2008). ISSN 1098-0121. doi: 10.1103/PhysRevB.77.165207. URL http://link.aps.org/doi/10.1103/PhysRevB.77.165207.

[103] O. Aharony, S. Gubser, J. Maldacena, H. Ooguri, and Y. Oz, Large N field theories, string theory and gravity, *Physics Reports*. **323**(3-4), 183–386 (Jan., 2000). ISSN 0370-1573. doi: 10.1016/S0370-1573(99)00083-6. URL http://arxiv.org/abs/hep-th/9905111.

[104] H. Ooguri, A. Strominger, and C. Vafa, Black hole attractors and the topological string, *Physical Review D*. **70**(10), 106007, (2004). ISSN 1550-2368. URL http://link.aps.org/doi/10.1103/PhysRevD.70.106007.

[105] F. Denef, M. Esole, and M. Padi, Orientiholes, *Journal of High Energy Physics*. **03**, 045 (Mar., 2010). ISSN 1029-8479. doi: 10.1007/JHEP03(2010)045. URL http://arxiv.org/abs/0901.2540.

[106] J. Gomis, D. Mateos, and F. Marchesano, An open string landscape, *Journal of High Energy Physics*. **2005**(11), 021–021 (Nov., 2005). ISSN 1029-8479. doi: 10.1088/1126-6708/2005/11/021. URL http://arxiv.org/abs/hep-th/0506179.

[107] F. Denef and G. W. Moore, Split States, Entropy Enigmas, Holes and Halos, *Arxiv preprint hep-th/0702146*. (2007). URL http://arxiv.org/abs/hep-th/0702146.

[108] M. Mariño, R. Minasian, G. W. Moore, and A. Strominger, Nonlinear in-

stantons from supersymmetric p -branes, *Journal of High Energy Physics.* **2000**(01), 005–005 (Jan., 2000). ISSN 1029-8479. doi: 10.1088/1126-6708/2000/01/005. URL `http://arxiv.org/abs/hep-th/9911206`.

[109] J. Maldacena, A. Strominger, and E. Witten, Black hole entropy in M-Theory, *Journal of High Energy Physics.* **1997**(12), 002–002 (Dec., 1997). ISSN 1029-8479. doi: 10.1088/1126-6708/1997/12/002. URL `http://arxiv.org/abs/hep-th/9711053`.

[110] J. Wess and J. Bagger, *Supersymmetry and supergravity.* (Princeton University Press, 1992). ISBN 0691025304. URL `http://books.google.com/books?id=4QrQZ_Rjq4UC&pgis=1`.

[111] D. Freed, Anomalies in string theory with D-branes, *Arxiv preprint hep-th/9907189* (July. 1999). URL `http://arxiv.org/abs/hep-th/9907189`.

[112] R. Minasian and G. W. Moore, K-theory and Ramond-Ramond charge, *Journal of High Energy Physics.* **1997**(11), 002–002 (Nov., 1997). ISSN 1029-8479. doi: 10.1088/1126-6708/1997/11/002. URL `http://arxiv.org/abs/hep-th/9710230`.

[113] S. K. Ashok and M. R. Douglas, Counting Flux Vacua, *Journal of High Energy Physics.* **2004**(01), 060–060 (Jan., 2004). ISSN 1029-8479. doi: 10.1088/1126-6708/2004/01/060. URL `http://arxiv.org/abs/hep-th/0307049`.

[114] F. Denef and M. R. Douglas, Distributions of flux vacua, *Journal of High Energy Physics.* **2004**(05), 072–072 (May, 2004). ISSN 1029-8479. doi: 10.1088/1126-6708/2004/05/072. URL `http://arxiv.org/abs/hep-th/0404116`.

[115] C. Vafa, Black Holes and Calabi-Yau Threefolds, *Adv.Theor.Math.Phys.* **2**, 207 (Nov., 1998). URL `http://arxiv.org/abs/hep-th/9711067`.

[116] R. Minasian, G. Moore, and D. Tsimpis, Calabi-Yau black holes and (0, 4) sigma models, *Communications in Mathematical Physics.* **209**(2), 325–352 (Apr., 2000). ISSN 0010-3616. URL `http://arxiv.org/abs/hep-th/9904217`.

[117] J. de Boer, F. Denef, S. El-Showk, I. Messamah, and D. V. den Bleeken, Black hole bound states in AdS 3 S 2, *Journal of High Energy Physics.* **2008**(11), 050–050 (Nov., 2008). ISSN 1029-8479. doi: 10.1088/1126-6708/2008/11/050. URL `http://stacks.iop.org/1126-6708/2008/i=11/a=050`.

[118] M. Douglas and G. Moore, D-branes, Quivers, and ALE Instantons, *Arxiv preprint hep-th/9603167* (Mar. 1996). URL `http://arxiv.org/abs/hep-th/9603167`.

[119] F. Denef, Quantum quivers and Hall/hole halos, *Journal of High Energy Physics.* **2002**, 023, (2002). URL `http://arxiv.org/abs/hep-th/0206072`.

[120] E. Witten, D-branes and K-theory, *Journal of High Energy Physics.* **1998**(12), 019–019 (Dec., 1998). ISSN 1029-8479. doi: 10.1088/1126-6708/1998/12/019. URL `http://arxiv.org/abs/hep-th/9810188`.

[121] I. Brunner, M. R. Douglas, A. Lawrence, and C. Römelsberger, D-branes on the quintic, *Journal of High Energy Physics.* **2000**(08), 015–015 (Aug., 2000). ISSN 1029-8479. doi: 10.1088/1126-6708/2000/08/015. URL `http://arxiv.org/abs/hep-th/9906200`.

[122] Y. Oz, D. Waldram, and T. Pantev, Brane-antibrane systems on Calabi-Yau spaces, *Journal of High Energy Physics.* **2001**(02), 045–045 (Feb., 2001). ISSN 1029-8479. doi: 10.1088/1126-6708/2001/02/045. URL `http://arxiv.org/abs/hep-th/0009112`.

[123] D. Gaiotto, M. Guica, L. Huang, A. Simons, A. Strominger, and X. Yin, D4-D0 branes on the quintic, *Journal of High Energy Physics.* **03**, 19 (Sept., 2006). URL `http://arxiv.org/abs/hep-th/0509168`.

[124] D. Gaiotto and L. Huang, D4-branes on complete intersection in toric variety, *hep-th/0612295* (Dec. 2006). URL `http://arxiv.org/abs/hep-th/0612295`.

[125] A. Collinucci, F. Denef, and M. Esole, D-brane deconstructions in IIB orientifolds, *Journal of High Energy Physics.* **2009**(02), 005–005 (Feb., 2009). ISSN 1029-8479. doi: 10.1088/1126-6708/2009/02/005. URL `http://arxiv.org/abs/0805.1573`.

[126] M. Guica and A. Strominger, Cargèse Lectures on String Theory with Eight Supercharges, *Nuclear Physics B - Proceedings Supplements.* **171**, 39–68 (Sept., 2007). ISSN 09205632. doi: 10.1016/j.nuclphysbps.2007.06.007. URL `http://arxiv.org/abs/0704.3295`.

[127] D. Gaiotto, A. Strominger, and X. Yin, The M5-brane elliptic genus: modularity and BPS states, *Journal of High Energy Physics.* **2007**(08), 070–070 (Aug., 2007). ISSN 1029-8479. doi: 10.1088/1126-6708/2007/08/070. URL `http://arxiv.org/abs/hep-th/0607010`.

[128] C. Beasley, D. Gaiotto, M. Guica, L. Huang, A. Strominger, and X. Yin, Why ZBH = Ztop^2, *hep-th/0608021* (Aug. 2006). URL `http://arxiv.org/abs/hep-th/0608021`.

[129] D. Gaiotto, A. Strominger, and X. Yin, From AdS 3 /CFT 2 to black holes/topological strings, *Journal of High Energy Physics.* **2007**(09), 050–050 (Sept., 2007). ISSN 1029-8479. doi: 10.1088/1126-6708/2007/09/050. URL `http://arxiv.org/abs/hep-th/0602046`.

[130] J. de Boer, M. C. Cheng, R. Dijkgraaf, J. Manschot, and E. Verlinde, A Farey tail for attractor black holes, *Journal of High Energy Physics.* **2006**(11), 024–024 (Nov., 2006). ISSN 1029-8479. doi: 10.1088/1126-6708/2006/11/024. URL `http://arxiv.org/abs/hep-th/0608059`.

[131] A. Collinucci and T. Wyder, The elliptic genus from split flows and Donaldson-Thomas invariants, *Journal of High Energy Physics.* **2010**(5), 1–37 (Oct., 2010). URL `http://arxiv.org/abs/0810.4301`.

[132] S. Kachru and J. McGreevy, Supersymmetric three-cycles and (super)symmetry breaking, *Physical Review D.* **61**(2), 10 (Dec., 1999). ISSN 0556-2821. doi: 10.1103/PhysRevD.61.026001. URL `http://arxiv.org/abs/hep-th/9908135`.

[133] F. Denef, Supergravity flows and D-brane stability, *Journal of High Energy Physics.* **2000**, 050, (2000). URL `http://arxiv.org/abs/hep-th/0005049`.

[134] B. Pioline, Four ways across the wall, *Arxiv preprint arXiv:1103.0261* (Mar. 2011). URL `http://arxiv.org/abs/1103.0261`.

[135] J. Manschot, B. Pioline, and A. Sen, Wall-Crossing from Boltzmann Black Hole Halos, *arXiv 1011.1258* (Nov. 2010). URL `http://arxiv.org/abs/1011.1258`.

[136] S. Cecotti, C. Cordova, J. J. Heckman, and C. Vafa, T-Branes and Monodromy, *arXiv:1010.5780* (Oct. 2010). URL `http://arxiv.org/abs/1010.5780`.

[137] M. Aganagic and C. Beem, The geometry of D-brane superpotentials, *Arxiv preprint arXiv:0909.2245* (Sept. 2009). URL `http://arxiv.org/abs/0909.2245`.

[138] A. Strominger, Heterotic solitons, *Nuclear Physics B.* **343**(1), 167–184 (Oct., 1990). ISSN 05503213. doi: 10.1016/0550-3213(90)90599-9. URL `http://dx.doi.org/10.1016/0550-3213(90)90599-9`.

[139] A. Strominger, Massless black holes and conifolds in string theory, *Nuclear Physics B.* **451**(1-2), 96–108 (Sept., 1995). ISSN 05503213. doi: 10.1016/0550-3213(95)00287-3. URL `http://arxiv.org/abs/hep-th/9504090`.

[140] J. Polchinski, Dirichlet Branes and Ramond-Ramond Charges, *Physical Review Letters.* **75**(26), 4724–4727 (Dec., 1995). ISSN 0031-9007. doi: 10.1103/PhysRevLett.75.4724. URL `http://arxiv.org/abs/hep-th/9510017`.

[141] M. R. Douglas, D-branes, categories and N=1 supersymmetry, *Journal of Mathematical Physics.* **42**(7), 2818 (Nov., 2001). ISSN 00222488. doi: 10.1063/1.1374448. URL `http://arxiv.org/abs/hep-th/0011017`.

[142] M. Berkooz, M. Douglas, and R. Leigh, Branes intersecting at angles, *Nuclear Physics B.* **480**(1-2), 265–278 (Nov., 1996). ISSN 0550-3213. doi: 10.1016/S0550-3213(96)00452-X. URL `http://arxiv.org/abs/hep-th/9606139`.

[143] S. Ferrara, R. Kallosh, and A. Strominger, N=2 extremal black holes, *Physical Review D.* **52**(10), R5412–R5416 (Nov., 1995). ISSN 0556-2821. doi: 10.1103/PhysRevD.52.R5412. URL `http://arxiv.org/abs/hep-th/9508072`.

[144] S. Ferrara, Black holes and critical points in moduli space, *Nuclear Physics B.* **500**(1-3), 75–93 (Sept., 1997). ISSN 05503213. doi: 10.1016/S0550-3213(97)00324-6. URL `http://arxiv.org/abs/hep-th/9702103`.

[145] G. Moore, Arithmetic and Attractors, *hep-th/9807087* (July. 1998). URL `http://arxiv.org/abs/hep-th/9807087`.

[146] M. Shmakova, Calabi-Yau Black Holes, *Phys. Rev.* **D56**, 540–544 (Dec., 1997). doi: 10.1103/PhysRevD.56.540. URL `http://arxiv.org/abs/hep-th/9612076`.

[147] G. L. Cardoso, B. de Wit, and T. Mohaupt, Corrections to macroscopic supersymmetric black-hole entropy, *Phys. Lett.* **B451**, 309–316 (Dec., 1999). URL `http://arxiv.org/abs/hep-th/9812082`.

[148] G. L. Cardoso, B. de Wit, J. Käppeli, and T. Mohaupt, Stationary BPS Solutions in N=2 Supergravity with R^2-Interactions, *Journal of High Energy Physics.* **2000**(12), 19 (Sept., 2000). URL `http://arxiv.org/abs/hep-th/0009234`.

[149] K. Behrndt, D. Lust, and W. Sabra, Stationary solutions of N= 2 supergravity, *Nuclear Physics B.* **510**(1-2), 264–288 (Jan., 1998). ISSN 0550-3213. doi: 10.1016/S0550-3213(97)00633-0. URL `http://linkinghub.elsevier.com/retrieve/pii/S0550321398810146`.

[150] B. Bates and F. Denef, Exact solutions for supersymmetric stationary black hole composites, *hep-th/0304094* (Apr. 2003). URL http://arxiv.org/abs/hep-th/0304094.

[151] F. Denef and G. Moore, How many black holes fit on the head of a pin?, *General Relativity and Gravitation.* **39**(10), 1539–1544 (May, 2007). ISSN 0001-7701. doi: 10.1007/s10714-007-0469-4 10.1142/S0218271808012437. URL http://arxiv.org/abs/arXiv:0705.2564.

Chapter 8

Supersymmetry from the Top Down

Michael Dine
Santa Cruz Institute for Particle Physics and Department of Physics, University of California, Santa Cruz, CA 95064, USA
dine@scipp.ucsc.edu

If supersymmetry turns out to be a symmetry of nature at low energies, the first order of business to measure the soft breaking parameters. But one will also want to understand the symmetry, and its breaking, more microscopically. Two aspects of this problem constitute the focus of these lectures. First, what sorts of dynamics might account for supersymmetry breaking, and its manifestation at low energies. Second, how might these features fit into string theory (or whatever might be the underlying theory in the ultraviolet). The last few years have seen a much improved understanding of the first set of questions, and at least a possible pathway to address the second.

8.1. String Theory at the Dawn of the LHC Era

We are on the brink of a new era in particle physics: the LHC program is finally underway. The missing element of the Standard Model, the physics responsible for electroweak symmetry breaking, is about to be discovered. This may be a simple Higgs field, as predicted in the minimal version of the theory. But there are good reasons to think that much more dramatic discoveries may be in store. The hierarchy problem provides the most compelling argument that a much richer set of phenomena should be revealed at the TeV scale. Theorists have explored a range of possibilities. Considerations of four dimensional effective field theory point towards supersymmetry or technicolor, and make a single light Higgs seem highly unlikely. String theory, understood broadly as some underlying theory of quantum gravity, may incorporate either or both of these, and points to additional possibilities, such as large extra dimensions and warping. On the other hand, string theory, and in particular the notion of a *string theory landscape* [2, 41, 51, 53, 62, 72], suggests a solution to the hierarchy problem which would lead to a single lonely Higgs.

In these lectures, we will take a "top down" view of possible physics at the LHC. We will focus mainly on supersymmetry, for reasons of time and because this is the case for which we know how to make the most concrete statements and models. Supersymmetry has well-known virtues:

(1) It offers a solution to the hierarchy problem.
(2) It leads to unification of the gauge couplings.
(3) It often provides a suitable candidate for the dark matter.
(4) It is often present in string theory.

Our enthusiasm for supersymmetry, however, should be tempered by the realization that from existing data – including early LHC data – there are, as we will discuss, reasons for skepticism.

Our approach will be "top down" in the sense that we will ask whether microscopic considerations – physics at distance scales much shorter than those which will be probed by the LHC – point towards some particular set of phenomena at TeV scales. We will consider two arenas for this problem: the dynamics responsible for supersymmetry breaking, and string theory.

8.1.1. *Two Aspects of the Hierarchy Problem*

Discussions of supersymmetry often begin with the observation that supersymmetry readily insures the cancelation of the quadratic divergences, which are the most striking manifestation of the hierarchy problem within the Standard Model (SM). But at a more primitive level, the hierarchy problem is the question: why is the scale of the weak interactions so far below that of gravity or grand unification? Supersymmetry can explain [77], not only the absence of quadratic divergences, but the appearance of a small scale. Essential to this are the so-called non-renormalization theorems, which were described in Seiberg's lectures (see also [23]). These follow from the holomorphy of the gauge couplings and the superpotential. As a result of these theorems, if supersymmetry is unbroken classically, it is unbroken to all orders of perturbation theory. These same theorems, however, indicate that it can sometimes be broken beyond perturbation theory, by effects which are exponentially small in some weak coupling. The search for such effects will be a principal theme of these lectures.

8.1.2. *Reasons for Skepticism*

While supersymmetry has many attractive features, and has established a large and devoted following (it is almost ubiquitous in the string theory lectures at this school), there are reasons to question whether it is responsible for the physics of electroweak symmetry breaking, and whether it should make its appearance at the LHC. Among these are:

(1) The "Little hierarchy" problem. The simplest approach to implementing supersymmetry at low energies, the so-called "Minimal supersymmetric standard model" (MSSM), predicts that there should be a neutral Higgs with mass lighter than M_Z, up to radiative corrections. These corrections can be substantial, and alternative models can modify the simplest prediction, but the current limit on the Higgs mass is uncomfortably large. Many implementations of supersymmetry require, for example, that the stop quark mass should be greater than 800 GeV. Tevatron limits from direct searches place (model-dependent) limits of several hundred GeV on squarks and gluinos, and the LHC is already setting much more stringent limits (or make a discovery).
(2) Unification: while unification is successful at the level of effective field theory, it is unclear why it should be generic in string theory. The string constructions described at this school, for example, do not predict unification of couplings, for typical values of the moduli. So this more microscopic viewpoint is troubling for one of the seeming successes of supersymmetry.
(3) There is another hierarchy, for which supersymmetry, or any comparable type of new physics, fails to offer any solution: the cosmological constant (c.c.). Here string theory provides some guidance, and some cause for concern. String theory, as a theory of gravity[a], must explain the exceedingly small value of the cosmological constant (dark energy). The dark energy represents a far more striking failure of dimensional analysis than M_W/M_p. To date, the only plausible explanation is provided by the notion of a *landscape* [7, 15, 75], the possibility that the theory possesses a vast array of states, of which states with small c.c. are picked out by anthropic considerations. While hardly established, this possibility – and the problem of the c.c. itself – raises the worry

[a]Defining precisely what we mean by string theory is problematic; in these lectures, we will use the term loosely to refer to whatever may be the underlying theory of quantum gravity, with the understanding that theories of strings provide examples of consistent quantum gravity theories.

that there are solutions to problems of hierarchy which cannot be understood in the framework of low energy effective field theory. In the case of a landscape, one would expect a distribution of possible values of the Higgs mass, from which what we (will) observe might be selected by some mechanism [2, 41, 51, 53, 62, 72].

Despite these cautionary notes, there are several reasons for (renewed) optimism, which we will touch upon in these lectures:

(1) The study of metastable susy breaking, initiated by work of Intriligator, Shih, and Seiberg (ISS) [52], has opened rich possibilities for model building with dynamical breaking of supersymmetry (earlier models were implausibly complex).
(2) Supersymmetry, even in a landscape, can account for hierarchies, as in traditional thinking about naturalness ($e^{-\frac{8\pi^2}{g^2}}$) [25].
(3) Supersymmetry, in a landscape, accounts for stability – i.e. the very existence of (metastable) states [31].

These notes represent the content of three lectures on low energy supersymmetry. The first two focus on issues in field theory, the third on the question of supersymmetry in string theory.

- Lecture 1: Low Energy Supersymmetry. This lecture reviews the basics of supersymmetry and its (metastable) breaking, the Minimal Supersymmetric Standard Model (MSSM) and its extensions.
- Lecture 2 is devoted to microscopic models of supersymmetry breaking and its Mediation.
- Lecture 3 takes up the question: "What might we mean by a string phenomenology?" This is a question both broad and difficult to formulate precisely. We will limit our considerations to asking: does low energy supersymmetry emerge as a prediction of string theory? While hardly settling the issue, we offer arguments, and ask if they lead to more specific predictions about the low energy spectrum.

8.2. Lecture 1. Low Scale Supersymmetry and Its Breaking

In this lecture, we consider:

(1) Some features of $N = 1$ Supersymmetry
(2) Metastable vs. stable supersymmetry breaking in the framework of simple models.

(3) The MSSM.
(4) Gauge Mediated models.

8.2.1. *A Brief Supersymmetry Review*

It is worth reviewing some basic features of $N = 1$ theories. This discussion is necessarily brief; more detail can be found in many texts and review articles, for example [3, 6, 13, 23, 42, 44, 73, 74, 76]. First, already in global supersymmetry, the supersymmetry algebra already connects an internal, fermionic, symmetry with space-time. Denoting the supersymmetry generators, Q_α, they obey the algebra:

$$\{Q_\alpha, Q^*_{\dot\beta}\} = 2\sigma^\mu_{\alpha\dot\beta} P_\mu. \tag{8.1}$$

Tracing in the Dirac indices one has:

$$Q_alpha^* Q_\alpha + Q_\alpha Q^*_{\dot\alpha} = P^0. \tag{8.2}$$

I will assume familiarity with superspace, and follow the notation of [76]. Without gravity, the effective theory should consist of fields with spin at most one. This permits only two types of supermultiplets (superfields):

(1) Chiral fields: these consist of a complex scalar and a Weyl fermion

$$\Phi(x, \theta) = \phi(y) + \sqrt{2}\theta\psi(y) + \theta^2 F, \tag{8.3}$$

with $y^\mu = x^\mu + i\theta\sigma^\mu\bar\theta$.

(2) Vector fields: These describe fields of spin 1/2 and spin one. In superspace, the field V satisfies the condition $V = V^\dagger$ (it is a *real* superfield). In the case of a $U(1)$ gauge symmetry, the superspace form of the gauge transformation is:

$$V \to V + \Lambda + \Lambda^\dagger \tag{8.4}$$

where Λ is a chiral field; this has a straightforward generalization for non-Abelian gauge symmetries. In a general gauge, V contains a number of unphysical degrees of freedom, but in the *Wess-Zumino* gauge it describes a gauge boson, and a Weyl fermion, and an auxiliary field D:

$$V = -\theta\sigma^\mu\theta^* A_\mu + i\theta^2\bar\theta\bar\lambda - i\bar\theta^2\theta\lambda + \frac{1}{2}\theta^2\bar\theta^2 D. \tag{8.5}$$

For our purposes, it is enough to consider the gauge covariant, spin-1/2 chiral field, W_α

$$W_\alpha = \lambda_\alpha + \theta_\beta\left[\delta^\beta_\alpha D + (\sigma^{\mu\nu})^\beta_\alpha F_{\mu\nu}\right] + \cdots . \tag{8.6}$$

F and D are auxiliary fields, i.e. they appear in the lagrangian without derivatives, and so their equations of motion are simply algebraic; they play an important role as they are order parameters for supersymmetry breaking. These expressions generalize immediately to non-Abelian theories, thinking of V, $F_{\mu\nu}$, λ, etc. as matrix valued fields.

The remarkable properties of supersymmetric field theories arise from the highly restricted form of any would-be supersymmetric lagrangian. At the level of terms with two derivatives, $\mathcal{L}$ is specified by three functions:

(1) The superpotential, $W(\Phi_i)$, a *holomorphic* function of the chiral fields.
(2) The Kahler potential, $K(\Phi_i, \Phi_i^\dagger)$
(3) The gauge coupling functions, $f_a(\Phi_i)$, again holomorphic functions of the fields (one such function for each gauge group).

The lagrangian density has the form, in superspace:

$$\int d^4\theta K(\Phi, \Phi^\dagger) + \int d^2\theta \left(f_a(\Phi) W_{\alpha a}^2 + W(\Phi)\right). \tag{8.7}$$

We will discuss the component lagrangians further below, but the important point is that they are fully determined by these functions and their derivatives.

8.2.1.1. *Renormalizable Interactions*

In thinking about effective theories, either as the low energy limits of theories which break supersymmetry, or of string theories, we will often be interested in general functions K, W and f. But it is is instructive to begin by first restricting our attention to the case of renormalizable theories. In this case, $K = \sum \Phi_i \Phi_i^\dagger$, $f_a = -\frac{1}{4g_a^2}$ and W is at most cubic. I will leave the details of the component lagrangian for textbooks and focus here on the scalar potential:

$$V = \sum |F_i|^2 + \frac{1}{2} \sum D_a^2, \tag{8.8}$$

where

$$F_i = \frac{\partial W}{\partial \Phi_i} \qquad D^a = \phi_i^* T^a \phi_i. \tag{8.9}$$

From the basic commutation relations, eqn. 8.1, we see that supersymmetry is spontaneously broken if and only if the vacuum energy is non-zero. Classically, supersymmetry is *unbroken* if $\langle F_i \rangle = \langle D_a \rangle = 0 \; \forall \; i, a$; conversely, it is broken if not. If it is broken, there is a Goldstone fermion ("goldstino")[b],

$$G \propto \langle F_i \rangle \psi_i + \langle D_a \rangle \lambda_a. \tag{8.10}$$

8.2.1.2. *R Symmetries*

In supersymmetry, a class of symmetries known as R-symmetries play a prominent role. Such symmetries can be continuous or discrete. Their defining property is that they transform the supercurrents,

$$Q_\alpha \to e^{i\alpha} Q_\alpha \quad Q^*_\alpha \to e^{-i\alpha} Q^*_\alpha. \tag{8.11}$$

Necessarily, the superpotential transforms as $W \to e^{2i\alpha}$ under this symmetry.

For the question of supersymmetry breaking, the importance of these symmetries is embodied in a theorem of Nelson and Seiberg [61]: In order that a generic lagrangian (one with all terms allowed by symmetries) break supersymmetry, the theory must possess an R symmetry (and in a theory with a spontaneously broken R symmetry, supersymmetry is necessarily broken). This theorem is easily proven by examining the equations $\frac{\partial W}{\partial \Phi_i} = 0$, and recalling that they are holomorphic; the proof is reviewed in Seiberg's lectures. I'll consider, instead, some examples, illustrating a variety of R symmetric lagrangians.

In general, W has R charge 2, if Q_α has charge one. Consider a theory with fields X_i, $i = 1, \ldots N$ with $R = 2$, and ϕ_a, $a = 1, \ldots M$, with R charge 0. Then the superpotential has the form:

$$W = \sum_{i=1}^{N} X_i f_i(\phi_a). \tag{8.12}$$

Suppose, first, that $N = M$. The equations $\frac{\partial W}{\partial \Phi_i} = 0$ are solved if:

$$f_i = 0; \quad X_i = 0. \tag{8.13}$$

The first set are N holomorphic equations for N unknowns, and generically have solutions. Supersymmetry is unbroken; there are a discrete set of supersymmetric ground states. Typically there will be no massless states in these vacua. The R symmetry is also unbroken, $\langle W \rangle = 0$.

Next suppose that $N < M$. Then the equations $f_i = 0$ contain more unknowns than equations; they generally have an $M - N$ (complex) dimensional space of solutions, known as a moduli space. In perturbation theory,

[b] The existence of the Goldstino follows from features of the supersymmetry current, just as for Goldstone bosons; see, for example [23, 74].

as a consequence of non-renormalization theorems, this degeneracy is not lifted. There are massless particles associated with these moduli (it costs no energy to change the values of certain fields).

If $N > M$, the equations $F_i = 0$, in general, do not have solutions; supersymmetry is broken. These are the O'Raifeartaigh models [64]. Now the equations $\frac{\partial W}{\partial \phi_i} = 0$ do not determine the X_i's, and classically, there are, again, moduli. Quantum mechanically, however, this degeneracy is lifted.

8.2.1.3. *Aside 1: The Non-renormalization Theorems*

Quite generally, supersymmetric theories have the property that, if supersymmetry is not broken at tree level, then to all orders of perturbation theory, there are no corrections to the superpotential and to the gauge coupling functions. These theorems were originally proven by examining detailed properties of Feynman diagrams, but they can be understood far more simply [68, 69].

To illustrate, consider a theory with two chiral fields, ϕ and Φ, the first light and the second heavy:

$$W = \frac{m}{2}\Phi^2 + \frac{\lambda}{3}\phi^2\Phi + \frac{\lambda'}{3}\phi^3. \tag{8.14}$$

First set $\lambda = 0$. Then the theory has an R symmetry under which Φ has unit R charge, and ϕ has R charge 2/3. The introduction of λ breaks the R symmetry, but we can take advantage of the fact that the superpotential is a holomorphic function of (the complex parameter) λ, and think of λ as itself the expectation value of a (non-dynamical) chiral field. Then we can assign λ R charge $-1/3$. Now consider the effective theory at low energies for ϕ. Necessarily, any correction to the superpotential behaves as $\lambda^n\phi^m$, with $-1/3n + 2/3m = 2$. So, for example, the ϕ^4/m term has coefficient λ^2, corresponding to the leading tree diagram. ϕ^6 would be a one loop diagram, but the only such diagram is not holomorphic in λ. Similarly, higher order polynomials in ϕ would necessarily appear at only one order in coupling, and can be shown to be non-holomorphic.

Exercise: To this theory, add $\lambda\Phi^3$ and show, by arguments as above, that only tree diagrams contribute to the low energy superpotential for ϕ.

8.2.2. *Metastable Supersymmetry Breaking*

For many years, it was taken for granted that the ultimate goal of supersymmetry model building was to find theories with stable, dynamical supersymmetry breaking, and that, suitably coupled to the fields of the MSSM, one would find acceptable low energy soft breaking. In 2006, Intriligator, Shih and Seiberg [52], demonstrated a surprising result: in vectorlike, supersymmetric QCD, for a range of colors and flavors, not only are there supersymmetric vacua, as expected from the Witten index, but there are *metastable* states with broken supersymmetry. While the particular example is remarkable and surprising, more generally this work brought the realization that such metastable supersymmetry breaking is a generic phenomenon. Indeed, this should have been anticipated from the Nelson-Seiberg theorem, which asserts that, to be generic, supersymmetry breaking requires a global, continuous R symmetry. We expect that such symmetries are, at best, accidental low energy consequences of other features of some more microscopic theory. In such a case, they will be violated by higher dimension operators, and typical theories will exhibit supersymmetry-preserving ground states.

8.2.2.1. *O'Raifeartaigh Models*

Let's consider the simplest O'Raifeartaigh model in more detail. This is a model with two fields, X, Y, with R charge 2, and a field, A, with R charge 0. Imposing a Z_2 symmetry restricts the model to:

$$W = \lambda X(A^2 - f) + mYA. \tag{8.15}$$

In this model, SUSY is broken; the equations:

$$\frac{\partial W}{\partial X} = \frac{\partial W}{\partial Y} = 0 \tag{8.16}$$

are not compatible.

If $f > \mu^2$, the vacuum has $\langle A \rangle = 0 = \langle Y \rangle$; X undetermined. It is easy to work out the spectrum. For $\langle X \rangle = 0$, the fermionic components of A combine with those of Y to form a Dirac fermion of mass m, while the scalar components of A have mass-squared $m^2 \pm \lambda F_X$ (the scalar components of Y are degenerate with the fermion). More generally, one should work out the spectrum as a function of $\langle X \rangle$.

Exercise: Work out the spectrum as a function of $\langle X \rangle$, first for $X = 0$, and then at least for small X.

Quantum effects generate a potential for X. At one loop, this is known as the Coleman-Weinberg. As explained below (section 8.2.2.2), one finds that the minimum of the potential lies at $\langle X \rangle = 0$. X is lighter than other fields (by a loop factor). The scalar components of X are a "light pseudomodulus." The spinor is massless; it is the Goldstino of supersymmetry breaking;

$$\langle F_X \rangle = f \tag{8.17}$$

is the decay constant of the Goldstino.

8.2.2.2. *Aside 2: The Coleman-Weinberg Potential*

The basic idea of Coleman Weinberg calculations for the pseudomoduli potentials is simple. First calculate masses of particles as functions of the pseudomodulus. Then compute the vacuum energy as a function of $\langle X \rangle$. At lowest order, this receives a contribution of $\frac{1}{2}\hbar\omega$ from each bosonic mode, and minus $\frac{1}{2}\hbar\omega$ from each fermion (due to filling the Fermi sea). As a result:

$$V(X) = \sum (-1)^F \int \frac{d^3k}{(2\pi)^3} \frac{1}{2} \sqrt{k^2 + m_i(X)^2}. \tag{8.18}$$

Term by term, this expression is very divergent in the ultraviolet; expanding the integrand in powers of k yields terms which are quartically and quadratically divergent. Introducing a momentum-space cutoff, Λ, yields:

$$= \sum (-1)^f \left(\Lambda^4 + m_i^2 \Lambda^2 + \frac{1}{(16\pi^2)} m_i^4 \ln(\Lambda^2/m_i^2) + \dots \right).$$

The quartically divergent term vanishes because there are equal numbers of fermions and bosons. The quadratically divergent term vanishes because of the tree level sum rule:

$$\sum (-1)^F m_i^2 = 0. \tag{8.19}$$

This sum rule holds in any theory with quadratic Kahler potential. The last, logarithmically divergent term must be evaluated, when supersymmetry is broken. The cutoff dependence of this term is associated with the renormalization of the couplings of the theory (in the O'Raifeartaigh case, the coupling λ).

One finds in the O'Raifeartaigh model that the potential grows quadratically near $X = 0$, and logarithmically for large X. As a result, the R symmetry is unbroken. Shih has shown that this is quite general [70]; if all fields have R charge 0 or 2, then the R symmetry is unbroken. Shih constructed models including fields with other R-charges, and showed that in these R symmetry is typically broken for a range of parameters. One of the simplest such theories is:

$$W = X_2(\phi_1\phi_{-1} - \mu^2) + m_1\phi_1\phi_1 + m_2\phi_3\phi_{-1}. \tag{8.20}$$

We will make use of this in model building shortly.

8.2.2.3. *Continuous Symmetry from a Discrete Symmetry*

The requirement of a continuous R symmetry in order to obtain supersymmetry breaking is, at first sight, disturbing. It is generally believed that a consistent theory of quantum gravity cannot exhibit global continuous symmetries (for a recent discussion of this issue, see [11]). Discrete symmetries, however, are different; these can be gauge symmetries (for example, they can be discrete subgroups of a broken continuous gauge symmetry, or discrete remnants of higher dimensional space-times symmetries). Such exact symmetries have the potential to give rise to *approximate* continuous global symmetries.

As an example, the continuous symmetry of the OR model might arise as an accidental consequence of a discrete, Z_N R symmetry. This could simply be a subgroup of the R symmetry of the "naive" model. For example:

$$X \to e^{\frac{4\pi i}{N}} X;\ Y \to e^{\frac{4\pi i}{N}} Y \tag{8.21}$$

corresponding to $\alpha = \frac{2\pi}{N}$ in eqn. 8.21 above.

For general N, this symmetry is enough to ensure that, keeping only renormalizable terms, the lagrangian is that of equation 8.15. But higher dimension terms can break the continuous R symmetry. Suppose, for example, $N = 5$. The discrete symmetry now allows couplings such as

$$\delta W = \frac{1}{M^3}\left(aX^6 + bY^6 + cX^4Y^2 + dX^2Y^4 + \dots\right). \tag{8.22}$$

Note that W transforms, as it must, under the discrete R symmetry, $W \to e^{\frac{4\pi i}{N}} W$.

The theory now has N supersymmetric minima, with

$$X \sim \left(\mu^2 M^3\right)^{1/5} \alpha^k \tag{8.23}$$

where $\alpha = e^{\frac{2\pi i}{5}}$, $k = 1, \dots, 5$. Classically, the original point near the origin is no longer stationary.

For large M, these vacua are "far away" from the origin. Near the origin, the higher dimension (irrelevant) operator has negligible effect, so the Coleman-Weinberg calculation, even though suppressed by a loop factor, gives the dominant contribution to the potential. The potential still exhibits a local minimum, however its global minima are the supersymmetric ones.

8.2.2.4. *Metastability*

The broken supersymmetry state near the origin, at least in the limit of global supersymmetry, will eventually decay to one of the supersymmetric minima far away. We can ask how quickly this decay occurs. We would need a separate lecture to discuss tunneling in quantum field theory (some remarks on this subject appear in Banks' lectures in this volume). Suffice it to say that in models such as those introduced above, the metastable supersymmetric state can be extremely long lived. In particular, the system has to tunnel a "long way" (compared with characteristic energy scales) to reach the "true" vacuum. Thinking (correctly) by analogy to WKB, the amplitude is exponentially suppressed by a power of the ratio of these scales. An elementary discussion appears in [35].

8.2.3. *Macroscopic Supersymmetry: The MSSM and Soft Supersymmetry Breaking*

If one simply writes a supersymmetric version of the Standard Model, it is not hard to show that supersymmetry cannot be spontaneously broken in a realistic fashion. So it is generally assumed that the dynamics responsible for supersymmetry breaking operates at a scale well above the weak scale, and in particular above the mass scale of the superpartners of ordinary fields. At lower energies, one has a supersymmetric theory, consisting of the SM fields and their superpartners, and perhaps some limited number of additional fields, described by a supersymmetric effective field theory with explicit soft breaking of supersymmetry. As a result, we can divide our considerations into "macroscopic" supersymmetry – the phenomenological description of this effective theory – and microscopic supersymmetry, the detailed mechanism by which supersymmetry is broken and this breaking is communicated to the partners of the SM fields. This section is devoted to this macroscopic picture; then, in section 8.3 we will turn to the more microscopic questions.

The MSSM is a supersymmetric generalization of the Standard Model(SM). Its field content and lagrangian are characterized by:

(1) Gauge group $SU(3) \times SU(2) \times U(1)$; correspondingly there are (twelve vector multiplets.

(2) Chiral field for each fermion of the SM: $Q_f, \bar{U}_f, \bar{D}_f, L_f, \bar{E}_f$.
(3) Two Higgs doublets, H_U, H_D.
(4) The superpotential of the MSSM contains a generalization of the Standard Model Yukawa couplings:

$$W_y = y_U H_U Q\bar{U} + y_D H_D Q\bar{D} + y_L H_D\bar{E}. \tag{8.24}$$

y_U and y_D are 3×3 matrices in the space of flavors.

8.2.3.1. *Soft Breaking Parameters*

Needless to say, it is important that supersymmetry be broken. For this purpose, one can try to construct a complete model of spontaneous supersymmetry breaking, or one can settle for an effective theory which is supersymmetric up to *explicit soft breakings.* The term "soft" refers to the fact that these breakings only have mild effects at short distances; in particular, they do not appreciably affect the renormalizable (marginal) operators, while they are themselves at most corrected logarithmically. It is easy to list the possible soft terms [46]:

(1) Mass terms for squarks, sleptons, and Higgs fields:

$$\begin{aligned} \mathcal{L}_{scalars} &= Q^* m_Q^2 Q + \bar{U}^* m_U^2 \bar{U} + \bar{D}^* m_D^2 \bar{D} \\ &\quad + L^* m_L^2 L + \bar{E}^* m_E \bar{E} \\ &\quad + m_{H_U}^2 |H_U|^2 + m_{H_U}^2 |H_U|^2 + B_\mu H_U H_D + \text{c.c.} \end{aligned} \tag{8.25}$$

m_Q^2, m_U^2, etc., are hermitian matrices in the space of flavors. Each has 9 real parameters.
(2) Cubic couplings of the scalars:

$$\begin{aligned} \mathcal{L}_A &= H_U Q\ A_U\ \bar{U} + H_D Q\ A_D\ \bar{D} \\ &\quad + H_D L\ A_E \bar{E} + \text{c.c.} \end{aligned} \tag{8.26}$$

The matrices A_U, A_D, A_E are complex matrices, each with 18 real entries.
(3) Mass terms for the U(1) (b), $SU(2)$ (w), and $SU(3)$ (λ) gauginos:

$$m_1 bb + m_2 ww + m_3 \lambda\lambda. \tag{8.27}$$

These represent 6 additional parameters.

(4) μ term for the Higgs field,

$$W_\mu = \mu H_U H_D \tag{8.28}$$

representing two additional parameters.

So we have the following counting of parameters beyond those of the SM:

(1) $\phi\phi^*$ mass matrices are 3×3 Hermitian (45 parameters).
(2) Cubic terms are described by 3 complex matrices (54 parameters).
(3) The soft Higgs mass terms add an additional 4 parameters.
(4) The μ term adds two.
(5) The gaugino masses add 6.

So there appear to be 111 new parameters.

But the Higgs sector of the SM has two parameters. In addition, the supersymmetric part of the MSSM lagrangian has symmetries which are broken by the general soft breaking terms (including μ among the soft breakings):

(1) Two of three separate lepton numbers.
(2) A "Peccei-Quinn" symmetry, under which H_U and H_D rotate by the same phase, and the quarks and leptons transform suitably.
(3) A continuous "R" symmetry, which we will explain in more detail below.

Redefining fields using these four transformations reduces the number of parameters to 105.

If supersymmetry is discovered, determining these parameters, and hopefully understanding them more microscopically, will be the main business of particle physics for some time. The phenomenology of these parameters has been the subject of extensive study; we will focus here on a limited set of issues.

8.2.3.2. *Constraints on the Soft Breaking Parameters*

Over the years, there have been extensive direct searches (LEP, Fermilab) for superpartners of ordinary particles, and these severely constrain the spectrum. For example, squark and gluino masses must be greater than 100's of GeV, while chargino masses of order 100 GeV; early LHC running has already substantially strengthened the gluino limit. But beyond these direct searches, the spectrum must have special features to explain

(1) absence of Flavor Changing Neutral Currents (suppression of $K \leftrightarrow \bar{K}$, $D \leftrightarrow \bar{D}$ mixing; $B \to s + \gamma$, $\mu \to e + \gamma$, ...) [59]
(2) suppression of CP violation (d_n; phases in $K\bar{K}$ mixing).

Both would be accounted for if the spectrum is highly degenerate, and CP violating phases in the soft breaking lagrangian are suppressed. This happens in many gauge mediated models, as we will discuss shortly, and in special regions of some superstring moduli spaces [55]. Other possible explanations include flavor symmetries [10].

8.2.3.3. *The Little Hierarchy: Perhaps the Greatest Challenge for Supersymmetry*

While with low energy supersymmetry, radiative corrections to the Higgs mass are *far* smaller than within the Standard Model, current experimental constraints still render them uncomfortably large. The largest contribution to the Higgs mass arises from top quark loops. There are two graphs, one with an intermediate top squark, one with a top quark; they cancel if supersymmetry is unbroken. The result of a simple computation is

$$\delta m_{H_U}^2 = -6\frac{y_t^2}{16\pi^2}\tilde{m}_t^2 \ln(\Lambda^2/\tilde{m}_t^2). \tag{8.29}$$

Even for modest values of the coupling, given the limits on squark masses, this can be substantial. The fine tuning seems to be order 1%.

Exercise Derive eqn. 8.29.

But the experimental limit, $m_H > 114$ GeV, poses another problem. At tree level, in the MSSM, $m_H \leq m_Z$. This traces to the fact that the quartic couplings of the Higgs, in the MSSM, are determined by the gauge couplings. Fortunately (for the viability of the model) loop corrections involving the top quark can substantially correct the Higgs quartic coupling, and increase the Higgs mass mass [43, 49, 63]. The leading contribution is proportional to $\log(\tilde{m}_t/m_t)$, and is readily calculated:

$$\delta\lambda \approx 3\frac{y_t^4}{16\pi^2}\log(\tilde{m}_t^2/m_t^2). \tag{8.30}$$

This is to be compared with the tree level term, of order $\frac{g^2+g'^2}{8}$, which is not terribly large. Still, evading the LEP bound typically requires $\tilde{m}_t > 800$ GeV. This exacerbates the problem of tuning, which now appears, over much of the parameter space, to be worse than 1 %.

Exercise Derive eqn. 8.30.

A variety of solutions have been proposed to this problem, and there is not space to review them all here. I will mention one, which will be tied to ideas we will develop subsequently. Suppose that there is some additional physics at a scale somewhat above the scale of the various superparticles. Then the Higgs coupling can be corrected by dimension five term in the superpotential or dimension six in the Kahler potential [39]

$$\delta W = \frac{1}{M} H_U H_D H_U H_D \quad \delta K = Z^\dagger Z H_U^\dagger H_U H_U^\dagger H_U. \tag{8.31}$$

For plausible values of M, and including radiative corrections as well, these couplings can lift the Higgs mass somewhat above the LEP bound.

A possible origin for this operator might be an extra, massive singlet, coupled to the Higgs:

$$W_S = \frac{M}{2} S^2 + \lambda S H_U H_D. \tag{8.32}$$

Models with an additional singlet beyond the MSSM fields are known collectively as the "Next to Minimal Supersymmetric Standard Model", NMSSM. Model builders make different assumptions about this theory; most forbid the mass term of eqn. 8.32 as unnatural. We discuss this and related issues later.

8.3. Microscopic Supersymmetry

Having explored the MSSM and its generalizations, we turn now to more microscopic considerations. First, we will simply assume that some dynamics is responsible for supersymmetry breaking, and ask how this breaking might be communicated. There are many approaches which have been considered, but we will focus on two, which have captured the most attention: Gravity mediation and gauge mediation.

8.3.1. *Supergravity*

In both cases, we need to know something about supergravity. We expect supersymmetry to be a local symmetry. Supergravity is a non-renormalizable theory; it is necessarily applicable only over a limited range of energies, and cannot be used for computation of quantum effects; some "ultraviolet completion" is required. But, as we will see, if supersymmetry is broken at a scale well below M_p, Planck scale effects can potentially control important aspects of low energy physics, and these can be described in terms of a lagrangian with local supersymmetry. The most general supergravity lagrangian with terms up to two derivatives appears in [18, 76]; a good introduction is also provided by [74]. Much like the global case, the general lagrangian is specified by a Kahler potential, superpotential, and gauge coupling functions, Here, we will content ourselves with describing some features which will be important for model building and certain more general theoretical issues.

Perhaps most important for us will be the form of the scalar potential: In units with $M_p = 1$ (here M_p is the reduced Planck mass, approximately 2×10^{18} GeV):

$$V = e^K \left[D_i W g^{i\bar{i}} D_{\bar{i}} W^* - 3|W|^2 \right]. \tag{8.33}$$

$D_i W \equiv F_i$ is the order parameter for susy breaking:

$$D_i W = \frac{\partial W}{\partial \phi_i} + \frac{\partial K}{\partial \phi_i} W. \tag{8.34}$$

If supersymmetry is *unbroken*, space time is Minkowski (if $W = 0$), It is AdS if ($W \neq 0$). If supersymmetry is broken and space is approximately flat space ($\langle V \rangle = 0$), then

$$m_{3/2} =\approx \langle e^{K/2} W \rangle. \tag{8.35}$$

8.3.2. *Mediating Supersymmetry Breaking*

The classes of models called "gauge mediated" and "gravity mediated" are distinguished principally by the scale at which supersymmetry is broken. If the F_i's are large enough, terms in the supergravity lagrangian (more generally, higher dimension operators) suppressed by M_p are important at the weak (TeV) scale. This requires:

$$F_i = D_i W \approx (TeV) M_p \equiv M_{int}^2. \tag{8.36}$$

For such F_i, we will speak of the supersymmetry breaking as "gravity mediated". We will refer to $M_{int} \approx 10^{11}$ GeV as the "intermediate scale", as it is the geometric mean of the scale of weak interactions and the Planck scale. If the scale is lower, we will use the term "gauge mediated". More precisely, gauge mediated models are models where supersymmetry breaking is transmitted principally through gauge interactions. In practice, as we will explain shortly, it is difficult to construct low scale models which are *not* gauge mediated; this is the rationale for our terminology.

8.3.3. *Intermediate Scale Supersymmetry Breaking ("Gravity Mediation")*

In the case of intermediate scale breaking, non-renormalizable couplings are responsible for the essential features of the physics at low energies. Such couplings are inherently sensitive to high scale physics. Lacking an ultraviolet completion of the theory, such as a (fully understood) string theory, one can only speculate about the origin and nature of these couplings; in general, they must be viewed as free parameters. As a result, there is enormous freedom in building models; one can readily fill out the full set of parameters of the MSSM. As a result, one must make strong assumptions about the microscopic physics in order to be consistent with existing low energy constraints. If, for example, we have a field, Z, responsible for supersymmetry breaking,

$$e^{K/2}F_Z = m_{3/2}M_p \tag{8.37}$$

then if K is a polynomial in Z and the other fields in the theory, all terms up to at least those quartic in fields are important in determining the low energy features of the theory. Suppose, for example, an O'Raifeartaigh-like model breaks supersymmetry. Choosing the constant in the superpotential, W_0 is chosen so that the cosmological constant is very small.

$$F_Z \equiv D_Z W = \frac{\partial W}{\partial Z} + \frac{\partial K}{\partial Z} W \neq 0 \tag{8.38}$$

along with

$$W_0 = \frac{1}{\sqrt{3}}\langle F\rangle \tag{8.39}$$

leads to soft masses for squarks, sleptons. In particular, for the MSSM fields, ϕ_i, the terms in the potential:

$$V(\phi) \approx \frac{\partial K}{\partial \phi_i}\frac{\partial K}{\partial \phi_{\bar{j}}}^* g^{i\bar{j}}|W|^2 \tag{8.40}$$

contribute to the mass-squared of all fields an amount of order $m_{3/2}^2 = e^K|W|^2$. Couplings such as $\int d^2\theta Z W_\alpha^2$ c an give mass to gauginos.

If the Kahler potential terms for the ϕ_i fields are simply

$$K_\phi = \sum \phi_i^\dagger \phi_i \tag{8.41}$$

then all of the ϕ_i fields acquire a mass-squared equal to $m_{3/2}^2$. However, terms in the Kahler potential:

$$\delta K = \frac{\gamma^{\bar{i}j}}{M^2} Z^\dagger Z \phi_{\bar{i}}^\dagger \phi_j \tag{8.42}$$

yields flavor-dependent masses for squarks and sleptons. No symmetry forbids such terms (approximate flavor symmetries might constrain them, however).

Exercise: Verify that by choice of the γ_{ij}'s one can explore the full parameter space of the MSSM.

Other potential difficulties with intermediate scale models include cosmological problems, such as the gravitino overproduction and moduli problems [9, 20]. We will not elaborate these here; cosmology also constrains gauge mediated models.

8.3.3.1. *Low Scale Supersymmetry Breaking: Gauge Mediation*

In the low scale case, the soft breaking effects at low energies should be calculable, without requiring further ultraviolet completion; this is the arena for field theory "model building." It is not hard to show that within the MSSM, there is no mechanism which can break supersymmetry suitably. So additional degrees of freedom are certainly required. One can contemplate many possibilities, both for the number and gauge transformation properties of the fields, and their couplings to MSSM fields. One faces several challenges:

(1) Obtaining positive mass-squared for partners of squarks and sleptons. This turns out to be achieved simply if the gauge couplings of the MSSM (i.e. the supersymmetric version of the Standard Model gauge coupling) mediate the breaking of supersymmetry. Yukawa couplings to new fields associated with supersymmetry breaking tend to be problematic.
(2) Suppressing flavor changing neutral currents. This tends to require some sort of flavor symmetry. If gauge interactions are the dominant source of squark and slepton masses, one immediately has an approximate flavor symmetry. With Yukawa couplings to new fields, the challenges are more serious.
(3) Other model building issues include: obtaining suitable gaugino masses and a μ term for Higgs fields.

In the rest of this lecture, we will focus exclusively on gauge mediation [48]. First we describe the simplest model of gauge mediation, "Minimal Gauge Mediation", which is remarkably predictive [32, 36–38]. Then we turn to the general case [60].

8.3.3.2. *Minimal Gauge Mediation*

The main premiss underlying gauge mediation can be simply described: in the limit that the gauge couplings vanish, the hidden and visible sectors decouple.[c] Perhaps the simplest model of gauge mediation, known as Minimal Gauge Mediation, involves a chiral field, X, whose vacuum expectation value is assumed to take the form:

$$\langle X \rangle = x + \theta^2 F. \tag{8.43}$$

X is coupled to a vector-like set of fields, transforming as 5 and $\bar{5}$ of $SU(5)$:

$$W = X(\lambda_\ell \bar{\ell}\ell + \lambda_q \bar{q}q). \tag{8.44}$$

For $F < X$, $\ell, \bar{\ell}, q, \bar{q}$ are massive, with supersymmetry breaking splittings of order F. The fermion masses are given by:

$$m_q = \lambda_q x \quad m_\ell = \lambda_\ell x \tag{8.45}$$

while the scalar splittings are

$$\Delta m_q^2 = \lambda_q F \qquad \Delta m_\ell^2 = \lambda_\ell F. \tag{8.46}$$

In such a model, masses for gauginos are generated at one loop; for scalars at two loops. The gaugino mass computation is quite simple. The two loop scalar masses are not very difficult, as one is working at zero momentum. The latter calculation can be done quite efficiently using supergraph techniques; an elegant alternative uses background field arguments [4, 47]. The result for the gaugino masses is:

$$m_{\lambda_i} = \frac{\alpha_i}{\pi}\Lambda, \tag{8.47}$$

For the squark and slepton masses, one finds

$$\widetilde{m}^2 = 2\Lambda^2 [C_3 \left(\frac{\alpha_3}{4\pi}\right)^2 + C_2 \left(\frac{\alpha_2}{4\pi}\right)^2 + \frac{5}{3}\left(\frac{Y}{2}\right)^2 \left(\frac{\alpha_1}{4\pi}\right)^2], \tag{8.48}$$

where $\Lambda = F_x/x$. $C_3 = 4/3$ for color triplets and zero for singlets, $C_2 = 3/4$ for weak doublets and zero for singlets.

Exercise: Derive eqn. 8.47.

Examining eqns. 8.47, 8.48 one can infer the following remarkable features of MGM:

[c] This definition was most clearly stated in [60], but some care is required, since, as we will see, additional features are needed for a realistic model.

(1) One parameter describes the masses of the three gauginos and the squarks and sleptons
(2) Flavor-changing neutral currents are automatically suppressed; each of the matrices m_Q^2, etc., is automatically proportional to the unit matrix. The corrections are tiny, and the A terms are highly suppressed (they receive no one contributions before three loop order).
(3) CP conservation is automatic
(4) This model cannot generate a μ term; the term is protected by symmetries. Some further structure is necessary.

8.3.3.3. *General Gauge Mediation*

Much work has been devoted to understanding the properties of this simple model, but it is natural to ask: just how general are these features? It turns out that they are peculiar to our assumption of a single set of messengers and just one singlet responsible for supersymmetry breaking and R symmetry breaking. Meade, Seiberg and Shih have formulated the problem of gauge mediation in a general way, and dubbed this formulation *General Gauge Mediation* (GGM). They study the problem in terms of correlation functions of (gauge) supercurrents. Analyzing the restrictions imposed by Lorentz invariance and supersymmetry on these correlation functions, they find that the general gauge-mediated spectrum is described by three complex parameters and three real parameters. The spectrum can be significantly different than that of the MGM, but the masses are still only functions of gauge quantum numbers and flavor problems are still mitigated.

The basic structure of the spectrum is readily described. In the formulas for fermion masses we introduce a separate complex parameter m_i, $i = 1, \ldots 3$ for each Majorana gaugino. Similarly, for the scalars, we introduce a real parameter Λ_c^2 for the contributions from $SU(3)$ gauge fields, Λ_w^2 for those from $SU(2)$ gauge fields, and Λ_Y^2 for those from hypercharge gauge fields:

$$\widetilde{m}^2 = 2\left[C_3\left(\frac{\alpha_3}{4\pi}\right)^2 \Lambda_c^2 + C_2\left(\frac{\alpha_2}{4\pi}\right)^2 \Lambda_w^2 + \frac{5}{3}\left(\frac{Y}{2}\right)^2\left(\frac{\alpha_1}{4\pi}\right)^2 \Lambda_Y^2\right]. \quad (8.49)$$

One can construct models which exhibit the full set of parameters [16, 17]. In MGM, the messengers of each set of quantum numbers each have a supersymmetric contribution to their masses, λM, while the supersymmetry breaking contribution to the scalar masses goes as λM^2, so in the ratio the coupling cancels out. In GGM model building, additional fields and couplings lead to more complicated relations.

One feature which of MGM which is not immediately inherited by GGM is the suppression of new sources of CP violation. Because the gaugino masses are independent parameters, in particular, they introduce additional phases which are inherently CP violating. Providing a natural explanation of the suppression of these phases is one of the main challenges of GGM model building.

8.4. Lecture 2: Microscopic Models of Supersymmetry Breaking

In this second lecture, we will continue our consideration of more microscopic models of supersymmetry and its breaking. This lecture covers:

(1) Low Energy, Dynamical Supersymmetry Breaking: A connection to the Cosmological Constant.
(2) The importance of Discrete R Symmetries.
(3) Gaugino condensation and its generalizations.
(4) Building models of Low Energy Dynamical Supersymmetry Breaking.
(5) Assessment.
(6) A theorem about the superpotential.

8.4.1. *Low Energy Supersymmetry Breaking and the Cosmological Constant*

In this lecture, we will focus on low energy supersymmetry breaking. While we won't consider string constructions per se, we will consider an important connection with gravity: the cosmological constant. We will not be attempting to provide a new explanation, but rather simply asking about the features of the low energy lagrangian in a world with approximate SUSY and small Λ. We will argue that this may be a guide to the microscopic mechanism of supersymmetry breaking.

With supersymmetry, there is an inevitable connection of *low* energy physics and gravity:

$$\langle |W|^2 \rangle = 3\langle |F|^2 \rangle M_p^2 + \text{ tiny}. \tag{8.50}$$

So not only do we require that F be small, but also W. Why? A few possible answers have been offered:

(1) Some sort of accident? For example, in the KKLT scenario [54], one assumes tuning of W relative to F (presumably anthropically).
(2) R symmetries can account for small W (Banks). We we will see, $< W >$ can be correlated naturally with the scale of supersymmetry breaking.

These remarks suggest a possible role for R symmetries. In string theory (gravity theory) such symmetries are necessarily discrete. and they are, at least at the level of textbook models, ubiquitous. They can arise, for example, as discrete subgroups of a higher dimensional Lorentz group, preserved by compactification. As such, they are necessarily discrete gauge symmetries, expected to survive in the quantum theory. Discrete R symmetries are interesting from several points of view:

(1) They can account for the small W_0 needed to understand the cosmological constant.
(2) They can give rise to approximate continuous R symmetries at low energies which can account for supersymmetry breaking (Nelson-Seiberg).
(3) They can account for small, dimensionful parameters.
(4) They can provide needed suppression of proton decay and other rare processes.

8.4.1.1. *Continuous R Symmetries from Discrete Symmetries*

Recall that the basic OR model possesses a continuous R symmetry::

$$W = X_2(A_0^2 - f) + mA_0Y_2 \tag{8.51}$$

(subscripts denote R charges). If, e.g., $|m^2| > |f|$, $F_X = f$. We have seen that this model can arise as the low energy limit of a model with a discrete R symmetry:

$$X_2 \to e^{\frac{2\pi i}{N}} X_2;\ Y_2 \to e^{\frac{2\pi i}{N}} Y_2;\ A_0 \to A_0. \tag{8.52}$$

This symmetry allows higher dimension (non-renormalizable) terms such as

$$\delta W = \frac{X^{N-n}Y^{n+1}}{M_p^{N-2}}. \tag{8.53}$$

The model has N supersymmetric vacua, far away from the supersymmetry-breaking vacuum near the origin.. Physics in this vacuum exhibits an approximate, accidental R symmetry. The state with broken supersymmetry is *highly* metastable.

One can treat Shih's model, eqn. 8.20, in a similar fashion. Coupling the field of that model to messengers, as in the MGM (eqn. 8.44), one can build a realistic model of gauge mediation. Additional tunneling instabilities arise as there are now additional supersymmetric vacua, in which some of the messenger fields are non-vanishing. Again, however, the desired metastable state can readily be *highly* metastable. We will discuss these models further later.

8.4.2. *Retrofitting the O'Raifeartaigh Models*

Up to this point, in these lectures, we have distinguished a notion of "macroscopic physics", phenomena occurring at the TeV scale, and "microscopic physics", associated with supersymmetry breaking. The notion of "microscopic", however, already requires some refinement in light of our discussion of gauge mediation. Here we have a mass scale associated with the messengers, and a potentially very different scale associated with the fields which break supersymmetry (the fields X of the various O'Raifeartaigh models, for example). Even these scales may arise from dynamics at still higher scales. While such a structure may seem arcane, we will see that it can be quite natural, and even (automatically) compatible with the order of magnitude of W required to obtain small cosmological constant.

In this section, we will describe a simple strategy for building models with metastable, dynamical supersymmetry breaking, known as "retrofitting" [27]. This breaking will be induced by dynamics at a higher scale which dynamically breaks a discrete R symmetry, without breaking supersymmetry. The prototype for such theories are pure gauge theories, in which gaugino condensation breaks a Z_N symmetry, in the case of $SU(N)$. In the following subsection, we first generalize gaugino condensation to theories which include order parameters of dimension one [34, 56]. We will then be in a position to build models in which all dimensionful parameters arise through dimensional transmutation, including the μ term of the MSSM, and possible parameters of the NMSSM. We will see that the expectation value of the superpotential plays an important role. Since W transforms under any R symmetry, $\langle W \rangle$ itself is an order parameter for R breaking. In the context of supergravity theories, this is particularly important. We will see that the relations among scales implied by the retrofitting hypothesis are of the correct order of magnitude to account for the smallness of the cosmological constant; this is not true of many other schemes for supersymmetry breaking, where additional scales must be introduced by hand.

8.4.2.1. *Generalizing Gaugino Condensation*

In this section I will assume some familiarity with basic aspects of supersymmetric dynamics. One can skip this section, and simply accept the basic result, that one can construct models in which R symmetry is dynamically broken, with order parameters of dimension one as well as dimension three. Introductory references on supersymmetry dynamics include Seiberg's lectures in this volume, and [23, 35, 65].

There is a huge literature on gaugino condensation, but one should ask: what is the essence of this phenomenon. Almost all discussion focuses on the fact that an $SU(N)$ gauge theory without matter has a Z_N discrete symmetry, broken by gaugino condensation, a non-zero value of the dimension three order parameter,

$$\langle\lambda\lambda\rangle = \Lambda^3 e^{\frac{2\pi i k}{N}}. \tag{8.54}$$

More generally, any non-abelian gauge theory without matter exhibits such a phenomenon.

But, if we forget the details of the models, we might extract three features:

(1) Breaking of a discrete R symmetry.
(2) All scales arise through dimensional transmutation.
(3) Order parameter of dimension 3.

If we relax the second item, then supersymmetric QCD with massive quarks already breaks a discrete R symmetry, and possesses gauge-invariant order parameters of dimension two. But for thinking about supersymmetric models and dynamical supersymmetry breaking, it is more interesting to relax the third item, i.e. we define gaugino condensation as [34]:

(1) Breaking of a discrete R symmetry.
(2) All scales arise through dimensional transmutation.

A simple class of generalizations with gauge invariant order parameters of dimension one is provided by supersymmetric QCD with N colors and N_f flavors, $N_f < N$, and with N_f^2 gauge singlet chiral fields, $S_{f,f'}$. For the superpotential, take:

$$W = yS_{ff'}\bar{Q}_f Q'_f + \lambda \mathrm{Tr} S^3. \tag{8.55}$$

To simplify the writing, we have assumed an $SU(N_f)$ flavor symmetry; this is not necessary to any of our considerations here. This theory possesses a

$Z_{2(3N-N_f)}$ R symmetry. This can be seen by noting that an instanton produces $2N$ gaugino zero modes, and $2N_F$ fermionic (Q and $\bar{Q}$) zero modes. This symmetry is spontaneously broken by $\langle S\rangle$; $\langle\bar{Q}Q\rangle$; $\langle W_\alpha^2\rangle$, $\langle W\rangle$.

The dynamics responsible for this breaking can be understood along the lines developed in Seiberg's lectures. Suppose, for example, that $\lambda \ll y$. Then we might guess that S will acquire a large vev, giving large masses to the quarks,

$$m_q = ys. \tag{8.56}$$

In this case, one can integrate out the quarks, leaving a pure $SU(N)$ gauge theory, and the singlets $S_{f,f'}$. The singlet superpotential follows by noting that the scale, Λ, of the low energy gauge theory depends on the masses of the quarks, which in turn depend on S. So

$$W(S) = \lambda S^3 + \langle\lambda\lambda\rangle_S. \tag{8.57}$$

$$\langle\lambda\lambda\rangle = \mu^3 e^{-3\frac{8\pi^2}{b_{LE}g^2(\mu)}} \tag{8.58}$$

$$= \mu^3 e^{-3\frac{8\pi^2}{g_{LE}g^2(M)}+3\frac{b_0}{b_{LE}}\ln(\mu/M)},$$

where μ is the scale at which we match the couplings of the high and low energy theories, $\mu = m_q$, and

$$b_0 = 3N - N_F;\ b_{LE} = 3N. \tag{8.59}$$

So

$$\langle\lambda\lambda\rangle = M^{\frac{3N-N_f}{N}} e^{-\frac{8\pi^2}{Ng^2(M)}} \mu^{\frac{N_f}{N}}. \tag{8.60}$$

In our case, $\mu = yS$, so the effective superpotential has the form

$$W(S) = \lambda S^3 + (yS)^{N_f/N}\Lambda^{3-N_f/N}. \tag{8.61}$$

This has roots

$$S = \Lambda\left(\frac{y^{N_f/N}}{\lambda}\right)^{\frac{N}{3N-N_F}} \tag{8.62}$$

times a Z_{3N-N_F} phase.

Note that this analysis is self-consistent; S is indeed large for small λ. The dynamics in other ranges of couplings has alternative descriptions, but the result that the discrete symmetry is spontaneously broken, while supersymmetry is unbroken, always holds.

8.4.2.2. *Gauge Mediation/Retrofitting*

Given our models of gaugino condensation, it is a simple matter to generate the various dimensionful couplings of O'Raifeartaigh models dynamically. In the model of 8.15, for example, we can make the replacements:

$$X(A^2 - \mu^2) + mAY \tag{8.63}$$

$$\rightarrow \frac{XW_\alpha^2}{M_p} + \gamma SAY.$$

Note that $\langle W \rangle \approx \Lambda^3$, $\langle S \rangle \sim \Lambda$, and $m^2 \gg f$. SUSY breaking is metastable, as in our earlier perturbed O'Raifeartaigh models (again, the supersymmetric vacua are far away).

Exercise: Verify that eqns. 8.63,8.55 respect a suitable discrete R symmetry.

8.4.2.3. *Gauge Mediation and the Cosmological Constant*

A traditional objection to gauge mediated models[d] is that the smallness of the c.c. requires a large constant in W, unrelated to anything else.

But we have just seen that in retrofitted models, one naturally expects $\langle W \rangle \approx FM_p^2$, i.e. of the correct order of magnitude to (almost) cancel the susy-breaking contributions to the c.c. This makes retrofitting, or something like it, almost inevitable in gauge mediation.

Other small mass parameters, such as the μ-term, arise readily from dynamical breaking of the discrete R symmetry. For example

$$W_\mu = \frac{S^2}{M_p} H_U H_D \tag{8.64}$$

gives rise to a μ term due to the expectation value of S.

In traditional approaches to gauge mediation, the μ term is problematic, not so much because it is hard to generate the term itself, but because B_μ tends to be too large. If, for example, μ, like the soft breaking masses is generated at, say, two loop order, B_μ is also typically generated at two loop order, meaning that $B_\mu \gg \mu^2$. This tends to lead to problematic hierarchies of mass scales. But in the present case, because S has only a tiny F-component, the corresponding B term is extremely small. A larger contribution arises from renormalization group running from the messenger scale to the TeV scale. This is suppressed by a loop factor but enhanced by

[d] I first heard this from Tom Banks.

a logarithm. As a result, the expectation value of H_D is suppressed relative to that of H_U, leading to a prediction of a large value of $\tan\beta$,

$$\tan(\beta) = \frac{\langle H_U \rangle}{\langle H_D \rangle}. \tag{8.65}$$

With these ingredients, it is almost *too* easy to build realistic models of gauge mediation/dynamical supersymmetry breaking with all scales dynamical, no μ problem, and prediction of a large $\tan\beta$.

8.4.2.4. *R Symmetry Breaking in Supergravity*

Even in supergravity theories, the scale of the superpotential is small compared to the Planck scale. As we have already mentioned, in the KKLT construction this is not particularly natural; one has to assume a selection of small W_0. One might hope to account for this phenomenon through R symmetries.

In supergravity (superstring) theories, there are natural candidates for Goldstino fields. These are the fermionic partners of the (pseudo) moduli. Classically, by definition, these fields have vanishing superpotential. They might acquire a superpotential through non-perturbative effects:

$$W = f\ M_p\ g(X/M_p). \tag{8.66}$$

For $X \ll M_p$, there might be an approximate R, along the lines required by Nelson and Seiberg, perhaps due to discrete R symmetries. But it is unclear how one can get a *large* enough W under these circumstances to cancel the c.c.; $\langle W \rangle$ would be suppressed by both R breaking and susy breaking. Alternatively, one could, again, retrofit scales, as in the low scale models.

These sorts of questions motivate study of W itself as an order parameter for R-symmetry breaking. In the next section, we will prove a theorem about the size of $\langle W \rangle$ in models with continuous R symmetries.

8.4.2.5. *A Theorem About the Superpotential*

While not critical to our subsequent discussion, it is interesting that there is a quite general statement that one can make about $\langle W \rangle$ in any globally supersymmetric theory with a continuous R symmetry.

Theorem [29]: In any theory with spontaneous breaking of a continuous R-symmetry and SUSY:

$$|\langle W \rangle| \leq \frac{1}{2}|F| f_a$$

where F is the Goldstino decay constant and f_a is the R-axion decay constant.

We will content ourselves with demonstrating the result for O'Raifeartaigh models with quadratic Kahler potentials and arbitrary superpotentials; the bound can be shown to hold quite generally, even in strongly coupled theories.

Consider a generic renormalizable O'Raifeartaigh model with an R-symmetry $\Phi_i \to e^{iq_i\xi}\Phi_i$.

$$K = \sum_i \Phi_i \bar{\Phi}_i, \quad W(\Phi_i) = f_i \Phi_i + m_{ij}\Phi_i\Phi_j + \lambda_{ijk}\Phi_i\Phi_j\Phi_k.$$

If the theory breaks SUSY at $\phi_i^{(0)}$ then classically it has a pseudomoduli space parameterized by the goldstino superpartner [57, 66].

$$G = \sum_i \left(\frac{\partial W}{\partial \phi_i}\right)\psi_i, \qquad \phi = \sum_i \left(\frac{\partial W}{\partial \phi_i}\right)\delta\phi_i,$$

Exercise: Verify that ϕ is massless.

Wherever the R-symmetry is broken there is also a flat direction corresponding to the R-axion. Define two complex vectors $w_i = q_i\phi_i$ and $v_i^\dagger = \frac{\partial W}{\partial \phi_i}$. Since the superpotential has R-charge 2,

$$2\langle W\rangle = \sum_j q_j\phi_j \frac{\partial W(\phi_i)}{\partial \phi_j} = \langle v, w\rangle.$$

On the pseudomoduli space we can write

$$|F|^2 = \sum_i \left(\frac{\partial W}{\partial \phi_i}\right)\left(\frac{\partial W}{\partial \phi_i}\right)^* = \langle v, v\rangle.$$

Parameterizing $\phi_i(x) = \langle\phi_i(x)\rangle e^{iq_i a(x)}$ we obtain for the R-axion kinetic term:

$$\left(\sum_i |\phi_i(x)|^2 q_i^2\right)(\partial a)^2 \Rightarrow f_a^2 = \langle w, w\rangle.$$

Then by the Cauchy-Schwarz inequality:

$$4|\langle W\rangle|^2 = |\langle v, w\rangle|^2 \leq \langle v, v\rangle\langle w, w\rangle = |F|^2 f_a^2$$

which is the bound to be established.

It is worth noting that:

- The bound is saturated if $v \propto w$, in which case the R-axion is the Goldstino superpartner.
- Adding gauge interactions strengthens the bound because the D terms contribution to the potential makes $|F|^2$ larger.

The proof can readily be generalized to arbitrary Kahler potential. Further input is needed to prove the result in full generality, i.e. for strongly coupled field theories as well.

8.5. Lecture 3. Supersymmetry in String Theory

At this school, and in most string papers, it is taken as a given that low energy supersymmetry is a consequence of string theory. But, as we will see, this is by no means self-evident. In this lecture, we will outline some of the issues. We will argue that if one could assert that low energy supersymmetry *is* an outcome of string theory, this is a dramatic prediction. Even more exciting would be to make some statement about the form of supersymmetry breaking. But it is quite possible that string theory predicts no such thing.

In this lecture we will consider:

(1) What might it mean for string theory to make contact with nature. We will argue that the *landscape* is the only plausible setting we have contemplated to date.
(2) The elephant in the room: The cosmological constant.
(3) The Banks/Weinberg proposal [7, 75].
(4) The Bousso-Polchinski model [15] (string theory fluxes) as an implementation (details for Denef).
(5) KKLT as a model. What serves as small parameter? Why is a small parameter important?
(6) Distributions of theories.
(7) Supersymmetry in string theory and the landscape.
(8) KKLT as a realization of intermediate scale susy breaking.
(9) A new look at susy breaking in the KKLT framework.
(10) The landscape perspective on intermediate scale breaking.
(11) Assessment.
(12) Discrete symmetries in string theory and the landscape.
(13) Strong CP and axions in string theory and the landscape (KKLT).
(14) Assessment: string theory predictions(?)

8.5.1. *String Theory and Nature*

In Miriam Cvetic's lectures, you heard how string theory can come close to reproducing many features of the Standard Model: the gauge group, the number of generations, and at least some features of Yukawa couplings. The constructions she described typically come with less desirable features, especially extra massless particles.

But these constructions raise an obvious question: there seem to be myriad possibilities. What principle governs which nature chooses, if any? Why, say, intersecting branes, and not heterotic constructions, non-geometric models, F theory constructions, or something perhaps not yet known. What sets a particular value of the allowed fluxes? One possible response is that eventually we will find *the* model which describes everything, work out its consequences, and make other predictions. Another is that we might find some *principle* which provides the answer, pointing to a unique string vacuum state. Finally, there is the point of view advocated by Banks at this school, that the different string "vacua" are actually different theories of quantum gravity; indeed, there are simply many different theories of quantum gravity, just as there are many possible different field theories.

But there is at least one fact which points to a different possibility; this is is the cosmological constant.

8.5.2. *The Cosmological Constant Problem*

At one level the cosmological constant problem is simply one of dimensional analysis: one would naturally expect that the c.c. would be of order some microscopic scale to the fourth power. M_p^4 would give a result 120 orders of magnitude larger than the observed cosmological constant, but even M_Z^4 would miss the observed value by more than 55 orders of magnitude. Were we to suppose that, for some reason, the cosmological constant is zero classically, quantum corrections would seem, inevitably, to be huge. The problem is illustrated by our earlier expression for the vacuum energy, 8.2.2.2, which is quartically divergent. Typically, in string models, if there is no supersymmetry, one obtains a result of the predicted order of magnitude, with a suitable cutoff (e.g. the string scale; such a calculation was first carried out by Rohm [67]). (In practice, since there are typically moduli, and certainly moduli in any case where weak coupling computations make sense, one is actually calculating a potential for the moduli.) So there is no evidence that string theory performs some magic with regards to this problem.

8.5.2.1. *Banks, Weinberg: A proposal*

Now I will use words that Tom told you one should never use, they are at the heart of a brilliant idea which he put forward [7] and Steven Weinberg turned into a dramatic prediction [75]. In order to understand the smallness of Λ, suppose that the underlying theory has many, many vacuum states, with a more or less uniform distribution of c.c.'s (Bousso and Polchinski dubbed such a distribution a "discretuum" [15]). Suppose that the system makes transitions between these states, or in some other way samples all of the different states (e.g. they all exist more or less simultaneously). Now imagine a star trek type figure, traveling around this vast universe. (This requires all sorts of superluminal phenomenon, but we won't worry about this, which is to say that we don't understand the issues too well, not that they are unimportant!). Most universes will not be habitable; the c.c. will be of order, say, the Planck scale. But sometimes the c.c. will be smaller, and the universe will be comparatively flat. Under what conditions might this star trek character find intelligent life? That's a tough question. As a proxy, Weinberg asked: under what circumstances, assuming all of the other laws of nature are the same, will this observer find galaxies. Weinberg noted, first, negative c.c. is unacceptable; if the resulting Hubble parameter is not much smaller than (one over) a billion years or so, the universe will undergo a big crunch long before galaxies, much less life, form. Similarly, for large, positive c.c., the would be structures will be ripped apart before galaxies can form. This sort of argument gives a c.c. about 100 times as large as observed. . (This was actually a prediction). This is remarkably good (Weinberg was originally rather negative about the result). On a log scale, it's excellent. More refined versions of the argument [58] reduce this discrepancy by an order of magnitude.

8.5.3. *Bousso-Polchinski and KKLT*

Bousso and Polchinski (BP) proposed that such a discretuum might arise in string theory as a result of fluxes. The idea is simply that if there are many possible fluxes (say χ), each of which can take N different values, one has of order N^χ states. If, say, $\chi \sim 100$ and $N \sim 10$, this number is huge. As a model, BP assumed that there is one (meta-)stable *vacuum* or *state* for each flux choice, but it is not clear this is reasonable. It is certainly not clear, for example, that moduli are fixed for every choice of fluxes. KKLT put forward a plausible implementation of the BP proposal in string theory which included a dynamical mechanism to fix all of the moduli, and a candidate for a small parameter which would allow exploration of some states. Central to their scenario are the 3-form fluxes. Again, if there are χ types of fluxes, each taking L values, of order L^χ states. For known Calabi-Yau manifolds, this number can easily be enormous, large enough to give the sort of discretuum required.

The elements of the KKLT proposal are readily enumerated; more detailed aspects of the construction are described in Denef's lectures at this school.

(1) The first ingredient is a IIB Orientifold of CY, or F-theory on CY Four-Fold. with $h_{2,1}$ complex structure moduli, $h_{1,1}$ Kahler moduli and the dilaton (and their superpartners).
(2) Three form fluxes fix the complex structure moduli and dilaton, along lines discussed in [12, 19, 45].
(3) The fixing of the complex structure moduli leaves a low energy theory which is approximately supersymmetric, with a small parameter, $W_0 = \langle W \rangle$.
(4) Non-perturbative dynamics fix the remaining Kahler moduli in terms of the small number, W_0.
(5) Additional branes are required in order to obtain matter, and perhaps for susy breaking.

In the spirit of these lectures, we will focus on the low energy effective theory for the Kahler moduli and other light fields. The details of steps 1-2 are discussed in Denef's lectures.

8.5.3.1. *Fixing the Kahler moduli*

For simplicity, we'll suppose that there is only one Kahler modulus (new issues which arise in the presence of multiple Kahler moduli are discussed in [1, 14, 28]; these will be discussed in section 8.5.5.1). The Kahler modulus sits in a chiral multiplet, ρ; the other scalar in the multiplet is an axion, which, from the microscopic point of view, respects a discrete shift symmetry. In other words,

$$\rho = \rho + ia \tag{8.67}$$

and the theory is symmetric under $a \to a + 2\pi$ (in a suitable normalization). The low energy theory, after integrating out the complex structure moduli is described by a *supersymmetric* effective action with

$$W = W_0 \qquad K = -3\ln(\rho + \rho^\dagger) \tag{8.68}$$

as well as some additional fields (gauge fields and charged matter) localized on branes. As it stands, this low energy theory breaks supersymmetry, with a potential which, at the classical level, is independent of ρ (a so-called "no-scale" model).

In general, the parameter W_0 is of order one. But KKLT and Douglas [21] argued that the large number of possible flux choices would give rise to a distribution of the parameters of this low energy lagrangian. W_0, in particular, would be distributed uniformly as a complex variable. If there are many states, then, in some, W_0 will be small. W_0 serves as the small parameter of the KKLT scenario (note that if the number of states is finite, W_0 cannot be arbitrarily small, so there is not really a systematic expansion).

In order to account for the stabilization of ρ, it is assumed that either stringy instanton effects or effects in the low energy theory, such as gaugino condensation, give rise to an additional term in the superpotential

$$W = W_0 - Ae^{-\rho/b}. \tag{8.69}$$

For W_0 small, the superpotential has a supersymmetric stationary point:

$$D_\rho W \approx aA/be^{-\rho b} - \frac{3}{\rho + \rho^\dagger} W_0 = 0, \tag{8.70}$$

$$\rho = \rho_0 \approx -b \ln(|W_0|). \tag{8.71}$$

This nominally justifies a large ρ expansion, i.e. the α' expansion (ref. [45] proposed a mechanism to achieve weak string coupling).

8.5.3.2. *Supersymmetry Breaking in KKLT*

So far in this discussion we have not accounted for supersymmetry breaking. KKLT proposed that $\overline{D3}$ branes could break susy. Indeed, such branes, if present, would appear to break supersymmetry *explicitly* in the low energy theory. This is somewhat confusing, since it is not clear how this could be described in an effective low energy theory. After all, one is supposing that a system with a light gravitino is not supersymmetric, which would appear to be inconsistent. The resolution may, perhaps, be that for such (warped) branes, the low energy theory is not a four dimensional theory with a finite number of fields. One can avoid this issue by simply hypothesizing there is another (four dimensional) field theory sector, again perhaps localized on a brane, which spontaneously breaks susy, perhaps along lines of models we discussed in the previous lecture. What is important is that this extra sector gives a positive contribution to the cosmological constant (you can fill in Banks objections at this point).

So KKLT provided a plausible (but hardly rigorous) scenario to understand:

(1) The existence of a large number of (metastable) states.
(2) Fixing of moduli.
(3) Breaking of supersymmetry.
(4) Distribution of parameters of low energy physics, including the c.c. and the scale of supersymmetry breaking.

Based on this analysis, the likely existence of a "landscape" of vacua in string theory has become widely (though certainly not universally) accepted. Adopting this viewpoint, Douglas and Denef embarked on a program of studying the statistics of KKLT-like vacua. The most primitive question they investigated was counting, but they also studied distributions of various quantities in addition to W_0, such as the supersymmetry-breaking scale [21, 22].

If we adopt this view, we still would like to know which features of the KKLT construction might be generic. Are typical states in the landscape approximately supersymmetric? Or is supersymmetry, perhaps, exponentially rare? Within the supersymmetric states, are features of the spectrum similar to those of KKLT? For example, the modulus, ρ, is parameterically heavy (by a power of ρ) compared to $m_{3/2}$. Is this typical? Are moduli often heavy in this sense?

8.5.4. *What Might We Extract from the Landscape?*

If the landscape viewpoint is correct, we can't hope to find "the state" which describes the world around us. I like to describe this problem as follows. Imagine an army of graduate students. Each is given a state to study, specified by a bar code. Graduate student A calculates the c.c. to fourth order and finds it's very small. She becomes exited. She goes on to third order, which takes her five years. She is even more excited, as she is even closer to the observed value. She keeps going. After 40 years she completes eighth order. But, oh well, she is not within errors. She returns to her adviser, and is given another bar code. Wati Taylor has described, at this school, a more plausible program involving searches for correlations among various quantities, such as numbers of generations and features of Yukawa couplings.

But I would suggest that the most promising questions are those connected with questions of naturalness. This is precisely because the phenomena we are trying to explain seem at first sight unlikely. These are

(1) The c.c. (already discussed).
(2) The strong CP problem.
(3) Fine tuning of the electroweak scale.
(4) Cosmological issues such as inflation.

There has been much work on the last of these, but it is very unclear what might be generic. I'll focus on the second and third. The reader should be warned that we are entering, here, a zone of (possibly biased) speculation.

8.5.5. *Supersymmetry in the Landscape*

At first sight, supersymmetry would seem special. Even if it is easier to explain hierarchies, we are talking about such large numbers of states that the number of non-susy states exhibiting huge hierarchies, could well overwhelm those in which the scale arises naturally [41, 72]. To consider this question, we can divide the landscape into three branches:

(1) States with no supersymmetry.
(2) States with approximate supersymmetry.
(3) States with approximate supersymmetry and discrete R symmetries.

One possible argument against the first branch and in favor of branches two and three invokes stability (now I will invoke a simple-minded version of Banks' concerns about "states"). It is usually said that tunneling amplitudes are naturally small, but this requires the existence of some small parameter (a small coupling, a small ratio of energy splitting to barrier height, a small ratio of energy scales to the distance over which one tunnels...). In the landscape, a typical state with small cc would be surrounded by vast number of states with large, negative cc. What prevents decays to big crunch space-times? In a typical state, one expects no parameter which would account for the smallness of decay rates. Note that *every* decay channel must be suppressed and that there are potentially an exponentially large number of such channels [31] . One can imagine various features which might be typical of many landscape states and which might suppress decays:

(1) Weak string coupling.
(2) Large compactification volume.

(3) Warping.
(4) Supersymmetry.

Items 1 and 3 appear not to be lead to metastability in generic situations [31]. Large compactification volume, to be effective, requires that the volume scale as a power of the typical flux. No non-supersymmetric landscape model studied to date has this feature, but perhaps it is possible. Supersymmetry *does* generically lead to stability. It is, indeed, a theorem that, with unbroken supersymmetry, Minkowski space is stable. With small supersymmetry breaking, one finds that the decay amplitudes vanish, or are exponentially small [30]:

$$\Gamma \approx e^{-M_p^2/m_{3/2}^2}. \tag{8.72}$$

So, even though the conditions for a supersymmetric vacuum (e.g. conditions for a supersymmetric stationary state of some complicated action) may be special, requiring some degree of metastability – that, say, some stationary point of the effective action be in any sense a "state" – might favor supersymmetry. One can study this question in toy landscapes; whether one make a definite statement requires a deeper understanding of the landscape.

Even if one accepts that some degree of supersymmetry is generally needed to account for stability, by itself this would not explain why very low scale supersymmetry should emerge. After all, examining equation 8.72, it is clear that if $m_{3/2}/M_p \sim 10^{-3}$, the decay amplitude is already unimaginably small. In terms of our classification of states, in other words, stability perhaps accounts for why we are not on the first (non-supersymmetric) branch, but not why *very* low energy supersymmetry should be favored. But from facts which are understood about supersymmetric landscapes, it would appear that the explanation could simply be conventional naturalness. For example, studying IIB landscapes, Douglas and Denef found instances with a uniform distribution of gauge couplings [21]. It would seem plausible that, in a significant fraction of states, supersymmetry is broken dynamically, i.e.

$$m_{3/2} \propto e^{-\frac{8\pi^2}{bg^2}}, \tag{8.73}$$

The uniform distribution of g^2 then implies a logarithmic distribution of $m_{3/2}^2$ (as in the original arguments for hierarchies, e.g. [71]). Coupling this with the assumption of a uniform distribution of W_0, and requiring small c.c., gives a distribution of supersymmetry breaking scales which is uniform on a log scale. If W_0 is itself fixed dynamically, as might be the case in models with discrete R symmetries, then low scale supersymmetry breaking is favored even more strongly [33].

There are further questions one can sensibly ask, even given our limited understanding of these issues. For example, one might argue that the third branch, with R symmetries, while favoring low energy supersymmetry breaking, is itself disfavored. After all, in order that the low energy theory exhibit an R symmetry, in the case of flux vacua, it is necessary to set to zero all fluxes which transform non-trivially under the symmetry. Given the assumption that the vast number of states arises because there are many types of fluxes, which take many values, this has the effect, typically, of reducing the number of flux types by an order one fraction [26], and thus reducing the number of available states by an exponential amount (say $e^{600} \rightarrow e^{200}$). On the other hand, one can advance rather primitive cosmological arguments that such states might be favored [24].

8.5.5.1. *Axions in the Landscape*

The landscape suggests the possibility that all quantities relevant to low energy physics are randomly distributed, unless selected anthropically. But not all of the quantities in our low energy effective theory look random. This applies to the quark and lepton mass matrices, which exhibit curious patterns. Most dramatically, though, it applies to the θ parameter of QCD, a pure number, which, if random, one might expect to be of order one. Instead, we know $\theta < 10^{-9}$. No known anthropic argument would seem to select for such a small value, which would seem otherwise highly improbable [8, 40].

A natural response is that in string theory, axions seem ubiquitous, so perhaps there is an automatic solution here. But in the case of KKLT, we have just seen that, while there is an axion candidate, it is fixed at the same time that ρ is fixed (it is fixed by the phase of W_0 in eqn. 8.71). As a result, it is very massive, and can play no role in solving the strong CP problem. This has lead to significant pessimism about the strong CP problem in string theory.

But perhaps KKLT is a little too simple as a model. In particular, typical string compactifications have multiple Kahler moduli. This leads to more interesting possibilities [1, 14, 28]. Consider the case of two Kahler moduli, ρ_1, ρ_2. It is now possible that one linear combination of these is fixed in an approximately supersymmetric fashion, as in KKLT, while the other is fixed by supersymmetry-violating dynamics. The "axionic" component of this multiplet need not be fixed by these dynamics. The requirements can be understood by examining a simple model:

$$W = W_0 + Ae^{-\frac{(n_1\rho_1+n_2\rho_2+n_3\rho_3)}{b}} + Ce^{-\rho_1} + De^{-\rho_2} + \cdots . \qquad (8.74)$$

The second term might be generated by gaugino condensation, say, on some brane, with $b > 1$, while the third and fourth could be generated by high scale, "stringy" instantons. The first term would fix, in a supersymmetric fashion one linear combination of moduli. In other words, the real and imaginary parts of one field would gain mass approximately supersymmetrically. Supersymmetry breaking effects could fix the real part of the other linear combination, leaving a light axion, which would obtain mass only from QCD and the (exponentially suppressed) third and fourth terms in W.

More precisely, in this model one has a "heavy", complex field,

$$\Phi = n_1\rho_1 + n_2\rho_2 + n_3\rho_3 \qquad (8.75)$$

and two light fields. As in the KKLT model,

$$\Phi \sim b\log(|W_0|). \qquad (8.76)$$

The Φ field has mass of order $\Phi m_{3/2}^2$. At lower energies, one has one light modulus, ϕ (as well as any additional matter fields on branes, etc.); the effective action consists simply of the Kahler potential for these fields and a constant superpotential. Supersymmetry breaking in this theory (which can be due to ϕ itself) can fix the real part of ϕ, leaving a light axion.

Apart from possibly accounting for the QCD axion, the fact that there might be many ρ_i-type fields could give rise to the "axiverse" of ref. [5]. The idea is that if the masses of axions arise from $e^{-\rho_i}$ type factors, their masses might be roughly uniformly distributed in energy scale, leading to interesting possible cosmological and astrophysical phenomena.

8.5.6. *A Top Down View of TeV Physics*

This combination of considerations provides a coherent, principled (though not necessarily true!) picture in which perhaps low energy supersymmetry is about to be found at LHC. But it could well be a house of cards. It stands on many shaky assumptions, most crucially the existence of a landscape, and secondarily a landscape which resembles the limited sets of configurations which have been studied by string theorists. Perhaps even within a landscape framework some other phenomena (warping, large extra dimensions, technicolor?) is responsible for electroweak symmetry breaking. Or worse, the landscape idea points to the alternative possibility that the electroweak scale is determined anthropically and we are about to find a single light Higgs. Arguments have been offered that this may not be the case [50], but the stress should be on *may*. It is one thing to dislike anthropic arguments; another to rule them out. I hope I have outlined some questions and some possible approaches, but perhaps your young, fresh minds will come up with better ways of thinking about these questions. And perhaps, within a few years (maybe even only one!), we will have experimental verification of some of these ideas, or unexpected clues as to what physics lies Beyond the Standard Model.

References

[1] Bobby Samir Acharya, Konstantin Bobkov, and Piyush Kumar. An M Theory Solution to the Strong CP Problem and Constraints on the Axiverse. 2010.

[2] V. Agrawal, Stephen M. Barr, John F. Donoghue, and D. Seckel. The anthropic principle and the mass scale of the standard model. *Phys. Rev.*, D57:5480–5492, 1998.

[3] I. J. R. Aitchison. Supersymmetry in particle physics: An elementary introduction. SLAC-R-865.

[4] Nima Arkani-Hamed, Gian F. Giudice, Markus A. Luty, and Riccardo Rattazzi. Supersymmetry-breaking loops from analytic continuation into superspace. *Phys. Rev.*, D58:115005, 1998.

[5] Asimina Arvanitaki, Savas Dimopoulos, Sergei Dubovsky, Nemanja Kaloper, and John March-Russell. String Axiverse. *Phys. Rev.*, D81:123530, 2010.

[6] H. Baer and X. Tata. Weak scale supersymmetry: From superfields to scattering events. Cambridge, UK: Univ. Pr. (2006) 537 p.

[7] Tom Banks. T C P, Quantum Gravity, the Cosmological Constant and All That.. *Nucl. Phys.*, B249:332, 1985.

[8] Tom Banks, Michael Dine, and Elie Gorbatov. Is there a string theory landscape? *JHEP*, 0408:058, 2004.

[9] Tom Banks, David B. Kaplan, and Ann E. Nelson. Cosmological Implications of Dynamical Supersymmetry Breaking. *Phys. Rev.*, D49:779–787, 1994.

[10] Tom Banks, Yosef Nir, and Nathan Seiberg. Missing (up) mass, accidental anomalous symmetries and the strong CP problem. 1994.

[11] Tom Banks and Nathan Seiberg. Symmetries and Strings in Field Theory and Gravity. 2010.

[12] Katrin Becker and Melanie Becker. M-Theory on Eight-Manifolds. *Nucl. Phys.*, B477:155–167, 1996.

[13] P. Binetruy. Supersymmetry: Theory, experiment and cosmology. Oxford, UK: Oxford Univ. Pr. (2006) 520 p.
[14] Konstantin Bobkov, Volker Braun, Piyush Kumar, and Stuart Raby. Stabilizing All Kahler Moduli in Type IIB Orientifolds. 2010.
[15] Raphael Bousso and Joseph Polchinski. Quantization of four form fluxes and dynamical neutralization of the cosmological constant. *JHEP*, 0006:006, 2000.
[16] Matthew Buican, Patrick Meade, Nathan Seiberg, and David Shih. Exploring General Gauge Mediation. *JHEP*, 03:016, 2009.
[17] Linda M. Carpenter, Michael Dine, Guido Festuccia, and John D. Mason. Implementing General Gauge Mediation. *Phys. Rev.*, D79:035002, 2009.
[18] E. Cremmer et al. Spontaneous Symmetry Breaking and Higgs Effect in Supergravity Without Cosmological Constant. *Nucl. Phys.*, B147:105, 1979.
[19] Jan de Boer et al. Triples, fluxes, and strings. *Adv. Theor. Math. Phys.*, 4:995–1186, 2002.
[20] B. de Carlos, J. A. Casas, F. Quevedo, and E. Roulet. Model independent properties and cosmological implications of the dilaton and moduli sectors of 4-d strings. *Phys. Lett.*, B318:447–456, 1993.
[21] Frederik Denef and Michael R. Douglas. Distributions of flux vacua. *JHEP*, 05:072, 2004.
[22] Frederik Denef and Michael R. Douglas. Distributions of nonsupersymmetric flux vacua. *JHEP*, 03:061, 2005.
[23] M. Dine. Supersymmetry and string theory: Beyond the standard model. Cambridge, UK: Cambridge Univ. Pr. (2007) 515 p.
[24] M. Dine, G. Festuccia, and A. Morisse. Symmetric Points in the Landscape as Cosmological Attractors. *JHEP*, 12:013, 2009.
[25] M. Dine, D. O'Neil, and Zheng Sun. Branches of the landscape. *JHEP*, 07:014, 2005.
[26] M. Dine and Zheng Sun. R symmetries in the landscape. *JHEP*, 01:129, 2006.
[27] Michael Dine, Jonathan L. Feng, and Eva Silverstein. Retrofitting O'Raifeartaigh models with dynamical scales. *Phys. Rev.*, D74:095012, 2006.
[28] Michael Dine, Guido Festuccia, John Kehayias, and Weitao Wu. Axions in the Landscape and String Theory. 2010.
[29] Michael Dine, Guido Festuccia, and Zohar Komargodski. A Bound on the Superpotential. *JHEP*, 03:011, 2010.
[30] Michael Dine, Guido Festuccia, and Alexander Morisse. The Fate of Nearly Supersymmetric Vacua. *JHEP*, 09:013, 2009.
[31] Michael Dine, Guido Festuccia, Alexander Morisse, and Korneel van den Broek. Metastable Domains of the Landscape. *JHEP*, 06:014, 2008.
[32] Michael Dine and Willy Fischler. A Phenomenological Model of Particle Physics Based on Supersymmetry. *Phys. Lett.*, B110:227, 1982.
[33] Michael Dine, Elie Gorbatov, and Scott D. Thomas. Low energy supersymmetry from the landscape. *JHEP*, 08:098, 2008.
[34] Michael Dine and John Kehayias. Discrete R Symmetries and Low Energy Supersymmetry. *Phys. Rev.*, D82:055014, 2010.

[35] Michael Dine and John D. Mason. Supersymmetry and Its Dynamical Breaking. 2010.
[36] Michael Dine and Ann E. Nelson. Dynamical supersymmetry breaking at low-energies. *Phys. Rev.*, D48:1277–1287, 1993.
[37] Michael Dine, Ann E. Nelson, Yosef Nir, and Yuri Shirman. New tools for low-energy dynamical supersymmetry breaking. *Phys. Rev.*, D53:2658–2669, 1996.
[38] Michael Dine, Ann E. Nelson, and Yuri Shirman. Low-energy dynamical supersymmetry breaking simplified. *Phys. Rev.*, D51:1362–1370, 1995.
[39] Michael Dine, Nathan Seiberg, and Scott Thomas. Higgs Physics as a Window Beyond the MSSM (BMSSM). *Phys. Rev.*, D76:095004, 2007.
[40] John F. Donoghue. Dynamics of M theory vacua. *Phys.Rev.*, D69:106012, 2004.
[41] Michael R. Douglas. Statistical analysis of the supersymmetry breaking scale. 2004.
[42] M. Drees, R. Godbole, and P. Roy. Theory and phenomenology of sparticles: An account of four-dimensional N=1 supersymmetry in high energy physics. Hackensack, USA: World Scientific (2004) 555 p.
[43] John R. Ellis, Giovanni Ridolfi, and Fabio Zwirner. On radiative corrections to supersymmetric Higgs boson masses and their implications for LEP searches. *Phys. Lett.*, B262:477–484, 1991.
[44] S.J. Gates, Marcus T. Grisaru, M. Rocek, and W. Siegel. Superspace Or One Thousand and One Lessons in Supersymmetry. *Front.Phys.*, 58:1–548, 1983.
[45] Steven B. Giddings, Shamit Kachru, and Joseph Polchinski. Hierarchies from fluxes in string compactifications. *Phys. Rev.*, D66:106006, 2002.
[46] L. Girardello and Marcus T. Grisaru. Soft Breaking of Supersymmetry. *Nucl. Phys.*, B194:65, 1982.
[47] G. F. Giudice and R. Rattazzi. Extracting Supersymmetry-Breaking Effects from Wave- Function Renormalization. *Nucl. Phys.*, B511:25–44, 1998.
[48] G. F. Giudice and R. Rattazzi. Theories with gauge-mediated supersymmetry breaking. *Phys. Rept.*, 322:419–499, 1999.
[49] Howard E. Haber and Ralf Hempfling. Can the mass of the lightest Higgs boson of the minimal supersymmetric model be larger than m(Z)? *Phys. Rev. Lett.*, 66:1815–1818, 1991.
[50] Roni Harnik, Graham D. Kribs, and Gilad Perez. A Universe without weak interactions. *Phys.Rev.*, D74:035006, 2006.
[51] Craig J. Hogan. Nuclear astrophysics of worlds in the string landscape. *Phys. Rev.*, D74:123514, 2006.
[52] Kenneth A. Intriligator, Nathan Seiberg, and David Shih. Dynamical SUSY breaking in meta-stable vacua. *JHEP*, 04:021, 2006.
[53] Tesla E. Jeltema and Marc Sher. The triple-alpha process and the anthropically allowed values of the weak scale. *Phys. Rev.*, D61:017301, 2000.
[54] Shamit Kachru, Renata Kallosh, Andrei D. Linde, and Sandip P. Trivedi. De Sitter vacua in string theory. *Phys. Rev.*, D68:046005, 2003.
[55] Vadim S. Kaplunovsky and Jan Louis. Model independent analysis of soft

terms in effective supergravity and in string theory. *Phys. Lett.*, B306:269–275, 1993.

[56] John Kehayias. Generalized Gaugino Condensation in Super Yang-Mills Theories: Discrete R-Symmetries and Vacua. *Phys. Rev.*, D82:125041, 2010.

[57] Zohar Komargodski and David Shih. Notes on SUSY and R-Symmetry Breaking in Wess-Zumino Models. *JHEP*, 04:093, 2009.

[58] Hugo Martel, Paul R. Shapiro, and Steven Weinberg. Likely Values of the Cosmological Constant.

[59] A. Masiero and O. Vives. Flavour and CP violation in supersymmetry. *New J. Phys.*, 4:4, 2002.

[60] Patrick Meade, Nathan Seiberg, and David Shih. General Gauge Mediation. *Prog. Theor. Phys. Suppl.*, 177:143–158, 2009.

[61] Ann E. Nelson and Nathan Seiberg. R symmetry breaking versus supersymmetry breaking. *Nucl. Phys.*, B416:46–62, 1994.

[62] Heinz Oberhummer, Attila Csoto, and Helmut Schlattl. Bridging the mass gaps at A = 5 and A = 8 in nucleosynthesis. *Nucl. Phys.*, A689:269–279, 2001.

[63] Y. Okada, Masahiro Yamaguchi, and T. Yanagida. Renormalization group analysis on the Higgs mass in the softly broken supersymmetric standard model. *Phys. Lett.*, B262:54–58, 1991.

[64] L. O'Raifeartaigh. Spontaneous Symmetry Breaking for Chiral Scalar Superfields. *Nucl.Phys.*, B96:331, 1975.

[65] Michael Edward Peskin. Duality in supersymmetric Yang-Mills theory. 1997.

[66] Sebastien Ray. Some properties of meta-stable supersymmetry-breaking vacua in Wess-Zumino models. *Phys. Lett.*, B642:137–141, 2006.

[67] Ryan Rohm. Spontaneous Supersymmetry Breaking in Supersymmetric String Theories. *Nucl. Phys.*, B237:553, 1984.

[68] Nathan Seiberg. Exact results on the space of vacua of four-dimensional SUSY gauge theories. *Phys. Rev.*, D49:6857–6863, 1994.

[69] Nathan Seiberg. The Power of holomorphy: Exact results in 4-D SUSY field theories. 1994.

[70] David Shih. Spontaneous R-symmetry breaking in O'Raifeartaigh models. *JHEP*, 02:091, 2008.

[71] Leonard Susskind. Dynamics of Spontaneous Symmetry Breaking in the Weinberg- Salam Theory. *Phys. Rev.*, D20:2619–2625, 1979.

[72] Leonard Susskind. Supersymmetry breaking in the anthropic landscape. 2004.

[73] J. Terning. Modern supersymmetry: Dynamics and duality. Oxford, UK: Clarendon (2006) 324 p.

[74] Steven Weinberg. The quantum theory of fields. Vol. 3: Supersymmetry. Cambridge, UK: Univ. Pr. (2000) 419 p.

[75] Steven Weinberg. Anthropic Bound on the Cosmological Constant. *Phys. Rev. Lett.*, 59:2607, 1987.

[76] J. Wess and J. Bagger. Supersymmetry and supergravity. Princeton, USA: Univ. Pr. (1992) 259 p.

[77] Edward Witten. Dynamical Breaking of Supersymmetry. *Nucl. Phys.*, B188:513, 1981.

AdS/CFT Applications

Chapter 9

The Landscape of the Hubbard Model

Subir Sachdev
Department of Physics, Harvard University,
Cambridge MA 02138, USA
sachdev@physics.harvard.edu

I present a pedagogical survey of a variety of quantum phases of the Hubbard model. The honeycomb lattice model has a conformal field theory connecting the semi-metal to the insulator with Néel order. States with fractionalized excitations are linked to the deconfined phases of gauge theories. I also consider the confining phases of such gauge theories, and show how Berry phases of monopoles induce valence bond solid order. The triangular lattice model can display a metal-insulator transition from a Fermi liquid to a deconfined spin liquid, and I describe the theory of this transition. The bilayer triangular lattice is used to illustrate another compressible metallic phase, the 'fractionalized Fermi liquid'. I make numerous connections of these phases and critical points to the AdS/CFT correspondence. In particular, I argue that two recent holographic constructions connect respectively to the Fermi liquid and fractionalized Fermi liquid phases.

9.1. Introduction

The Hubbard model is the simplest of a class of models describing electrons moving on a lattice with repulsive electron-electron interactions. Despite its apparent simplicity, it has become clear in the past two decades that it can display a very rich phase diagram, with a plethora of interesting phases. The most common phase is, of course, the Fermi liquid (FL), which is adiabatically connected to the metallic phase of non-interacting electrons. However, electron-electron interactions can break one or more symmetries of the Hamiltonian, and this leads to phases such as antiferromagnets, charge or spin density waves, or superconductors. Also of great interest are quantum phases which do not break any symmetries, but are nevertheless qualitatively distinct from the non-interacting electron states: such states are characterized by emergent gauge excitations, fractionalization of quasiparticle excitations, and non-trivial ground state degeneracies which depend upon the global topology of the lattice—it is often stated that such states have 'topological' order. Finally, there are interesting quantum phase transitions between such phases, and such quantum critical points are often described by strongly-coupled quantum field theories. In some cases, the quantum critical points can broaden into gapless quantum critical phases.

This article will present a pedagogical review of a small sample of this landscape of phases and critical points. My aim is to describe the appearance of a variety of non-trivial phases in the simplest possible context. For the honeycomb lattice with a density of one electron per site, such phases naturally have low energy excitations which have a relativistic form at low energies. Consequently, in the vicinity of quantum phase transitions, such phases and their critical points are amenable to a description by relativistic quantum field theories. In some cases, the critical points are also conformally invariant, and so are described by conformal field theories (CFTs). In these cases, the AdS/CFT correspondence can be directly applied, and I will describe the insights that have been gained from such an approach.

However, once we move away from commensurate electron densities, the quantum phases and critical points of electron lattice models rarely have any relativistic invariance in their low energy theory. I will describe here the simplest examples of 'topologically ordered' phases at generic electron densities. It is important to note that the lack of relativistic invariance does not rule out application of the AdS/CFT correspondence. We can begin from a relativistically invariant gravity dual theory and dope it with charge carriers by turning on a chemical potential: then even the gravity theory is not relativistically invariant at low energies, and we can hope to match its low energy physics to a condensed matter system. There has been a large effort to apply the AdS/CFT correspondence along this direction in the past few years. I have discussed some of this work in another recent review article,[1] which should be viewed as a companion to the present article. I will also discuss the generic density phases here, with an emphasis on the general low energy structure of the 'fractionalized Fermi liquid' (FL*) phase,[2,3] which I believe is closely related to the generic density phases that have appeared using the AdS/CFT correspondence.[4] Further details on the connection of these phases of Hubbard-like models and the AdS/CFT correspondence appear in a recent paper.[5]

I begin by introducing the Hubbard model. It is defined by the Hamiltonian

$$H = -\sum_{i,j} t_{ij} c_{i\alpha}^\dagger c_{j\alpha} + \sum_i \left[-\mu \left(n_{i\uparrow} + n_{i\downarrow} \right) + U_i \left(n_{i\uparrow} - \frac{1}{2} \right) \left(n_{i\downarrow} - \frac{1}{2} \right) \right]. \tag{9.1}$$

Here $c_{i\alpha}$, $\alpha = \uparrow, \downarrow$ are annihilation operators on the site i of a regular lattice, and t_{ij} is a Hermitian, short-range matrix containing the 'hopping matrix elements' which move the electrons between different lattice sites. The density of electrons is controlled by the chemical potential μ which couples to the total electron density, with

$$n_{i\uparrow} \equiv c_{i\uparrow}^\dagger c_{i\uparrow} \quad , \quad n_{i\downarrow} \equiv c_{i\downarrow}^\dagger c_{i\downarrow}. \tag{9.2}$$

The electrons repel each other with an on-site interaction U_i; in most cases we will take $U_i = U$ site-independent, but it will also be useful later to allow for a site-dependent U_i. For completeness, we also note the algebra of the fermion operators:

$$\begin{aligned} c_{i\alpha} c_{j\beta}^\dagger + c_{j\beta}^\dagger c_{i\alpha} &= \delta_{ij} \delta_{\alpha\beta} \\ c_{i\alpha} c_{j\beta} + c_{j\beta} c_{i\alpha} &= 0. \end{aligned} \tag{9.3}$$

The equations (9.1), (9.2), and (9.3) constitute a self-contained and complete mathematical statement of the problem of the landscape of the Hubbard model. It is remarkable that a problem that is so simple to state has such a rich phase structure as a function of the lattice choice, the fermion density, and the spatial forms of t_{ij} and U_i.

Sections 9.2, 9.4, and 9.5 will deal exclusively with the honeycomb lattice at a density of one electron per site ("half-filling"), so that $\langle n_{i\uparrow} \rangle = \langle n_{i\downarrow} \rangle = 1/2$. The emphasis on the honeycomb lattice is not motivated by its particular physical importance (although, it is the lattice of graphene), but by its simplicity as a context for introducing various technical methods, quantum phases and critical points. In Section 9.2, we will consider the semi-metal and the insulating antiferromagnet, and show that a phase transition between them is described by a relativistic field theory, which is a version of the Gross-Neveu-Yukawa model; Section 9.3 will use this field theory to present a general discussion of the physics at non-zero temperatures in the vicinity of a quantum critical point. We will focus on the transport of conserved charges, and describe insights gained from the AdS/CFT correspondence. Section 9.4 will consider the problem of restoring the spin rotation symmetry from the antiferromagnet, while remaining in an insulating phase: this will lead to a description in terms of a U(1) gauge theory, and the appearance of an insulating phase with valence bond solid (VBS) order. Finally, Section 9.5 will combine all the phases of the half-filled honeycomb lattice discussed so far in a single phase diagram: this will require introduction of a SU(2) gauge theory. We will find an interesting multi-critical point in Section 9.5, which has many features in common with the supersymmetric CFTs studied using the AdS/CFT correspondence.

Sections 9.4 and 9.5 can be skipped in a first reading, without significant loss of continuity.

We will move away from half-filling in Sections 9.6 and 9.7, where we will turn our attention to metallic phases with Fermi surfaces. Section 9.6 considers the Hubbard model on the triangular lattice, and describes a phase diagram which includes Fermi liquid (FL) and spin liquid phases. Section 9.7 extends our discussion to the the Hubbard model on a bilayer triangular lattice, which has been realized in experiments[6] on ^{3}He. Here we will present a gauge theory of another metallic phase, the fractionalized Fermi liquid (FL*). We will also discuss the connections to compressible metallic phases obtained using the AdS/CFT correspondence, complementing the recent discussion in Ref. 5.

9.2. Semi-metal and Antiferromagnetism on the Honeycomb Lattice

9.2.1. *Preliminaries*

We will consider the Hubbard model (9.1) with the sites i on locations $\boldsymbol{r}_i$ on the honeycomb lattice shown in Fig. 9.1. Here, we set up some notation allowing us to analyze the geometry of this lattice.

We work with a lattice with unit nearest neighbor spacing. We define unit length vectors which connect nearest-neighbor sites

$$\boldsymbol{e}_1 = (1, 0) \quad , \quad \boldsymbol{e}_2 = (-1/2, \sqrt{3}/2) \quad , \quad \boldsymbol{e}_3 = (-1/2, -\sqrt{3}/2). \tag{9.4}$$

Note that $\boldsymbol{e}_i \cdot \boldsymbol{e}_j = -1/2$ for $i \neq j$, and $\boldsymbol{e}_1 + \boldsymbol{e}_2 + \boldsymbol{e}_3 = 0$. The lattice can be divided into the A and B sublattices, as shown in Fig. 9.1. We take the origin of co-ordinates of the lattice at the center of an *empty hexagon*. The A sublattice sites closest to the origin are at $\boldsymbol{e}_1$, $\boldsymbol{e}_2$, and $\boldsymbol{e}_3$, while the B sublattice sites closest to the origin are at $-\boldsymbol{e}_1$, $-\boldsymbol{e}_2$, and $-\boldsymbol{e}_3$.

The unit cell of the hexagonal lattice contains 2 sites, one each from the A and B sublattices. These unit cells form a triangular Bravais lattice consisting of the centers of the hexagons. The triangular lattice points closest

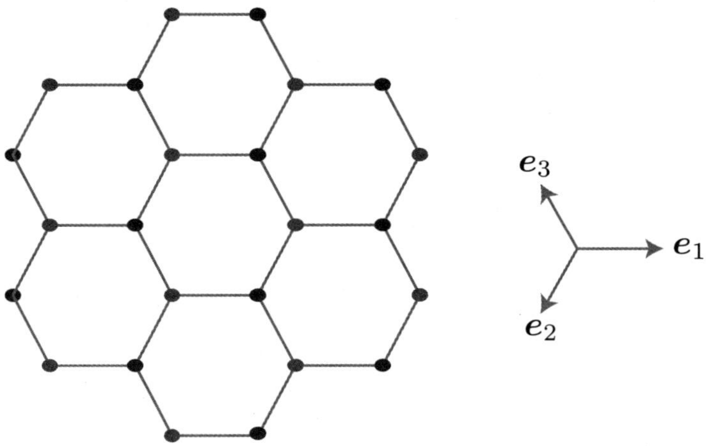

Fig. 9.1. The honeycomb lattice with its A (green) and B (red) sublattices.

to the origin are $\pm(\boldsymbol{e}_1 - \boldsymbol{e}_2)$, $\pm(\boldsymbol{e}_2 - \boldsymbol{e}_3)$, and $\pm(\boldsymbol{e}_3 - \boldsymbol{e}_1)$. The reciprocal lattice is a set of wavevectors $\boldsymbol{G}$ such that $\boldsymbol{G} \cdot \boldsymbol{r} = 2\pi \times$ integer, where $\boldsymbol{r}$ is the center of any hexagon of the honeycomb lattice. The reciprocal lattice is also a triangular lattice, and it consists of the points $\sum_i n_i \boldsymbol{G}_i$, where n_i are integers and

$$\boldsymbol{G}_1 = \frac{4\pi}{3}\boldsymbol{e}_1 \quad , \quad \boldsymbol{G}_2 = \frac{4\pi}{3}\boldsymbol{e}_2 \quad , \quad \boldsymbol{G}_3 = \frac{4\pi}{3}\boldsymbol{e}_3. \tag{9.5}$$

The unit cell of the reciprocal lattice is called the first Brillouin zone. This is a hexagon whose vertices are given by

$$\boldsymbol{Q}_1 = \frac{1}{3}(\boldsymbol{G}_2 - \boldsymbol{G}_3) \quad , \quad \boldsymbol{Q}_2 = \frac{1}{3}(\boldsymbol{G}_3 - \boldsymbol{G}_1) \quad , \quad \boldsymbol{Q}_3 = \frac{1}{3}(\boldsymbol{G}_1 - \boldsymbol{G}_2), \tag{9.6}$$

and $-\boldsymbol{Q}_1$, $-\boldsymbol{Q}_2$, and $-\boldsymbol{Q}_3$; see Fig. 9.2. Integrals and sums over momentum space will implicitly extend only over the first Brillouin zone. This is the 'ultraviolet cutoff' imposed by the underlying lattice structure.

We define the Fourier transform of the electrons on the A sublattice by

$$c_{A\alpha}(\boldsymbol{k}) = \frac{1}{\sqrt{\mathcal{N}}} \sum_{i \in A} c_{i\alpha} e^{-i\boldsymbol{k}\cdot\boldsymbol{r}_i}, \tag{9.7}$$

where $\mathcal{N}$ is the number of sites on one sublattice; similarly for $c_{B\alpha}$. Note that $c_{A\alpha}(\boldsymbol{k} + \boldsymbol{G}) = c_{A\alpha}(\boldsymbol{k})$: consequently, sums over momentum have to be restricted to the first Brillouin zone to avoid double counting. Thus the inverse of Eq. (9.7) sums over $\boldsymbol{k}$ in the first Brillouin zone.

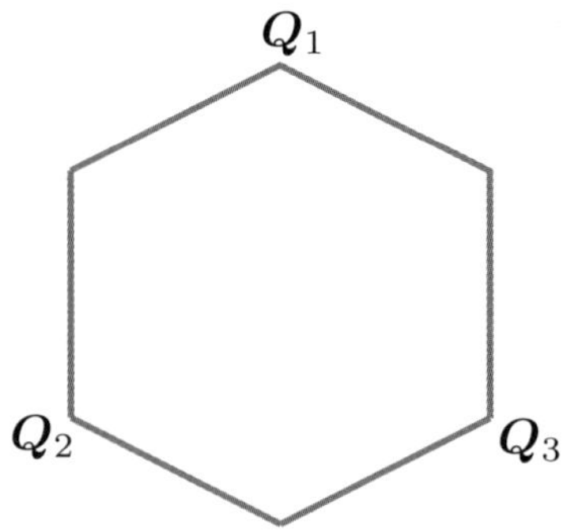

Fig. 9.2. The first Brillouin zone of the honeycomb lattice.

9.2.2. *Semi-metal*

We begin with free electrons in the honeycomb lattice, $U = 0$, with only nearest-neighbor electron hopping $t_{ij} = t$. Using Eq. (9.7), we can write the hopping Hamiltonian as

$$H_0 = -t\sum_{\boldsymbol{k}} \left(e^{i\boldsymbol{k}\cdot\boldsymbol{e}_1} + e^{i\boldsymbol{k}\cdot\boldsymbol{e}_2} + e^{i\boldsymbol{k}\cdot\boldsymbol{e}_3}\right) c^\dagger_{A\alpha}(\boldsymbol{k})c_{B\alpha}(\boldsymbol{k}) + \text{H.c.}$$
$$-\mu\sum_{\boldsymbol{k}} \left(c^\dagger_{A\alpha}(\boldsymbol{k})c_{A\alpha}(\boldsymbol{k}) + c^\dagger_{B\alpha}(\boldsymbol{k})c_{B\alpha}(\boldsymbol{k})\right). \tag{9.8}$$

We introduce Pauli matrices τ^a $(a = x, y, z)$ which act on the A, B sublattice space; then this Hamiltonian can be written as

$$H_0 = \sum_{\boldsymbol{k}} c^\dagger(\boldsymbol{k})\Big[-\mu - t\Big(\cos(\boldsymbol{k}\cdot\boldsymbol{e}_1) + \cos(\boldsymbol{k}\cdot\boldsymbol{e}_2) + \cos(\boldsymbol{k}\cdot\boldsymbol{e}_3)\Big)\tau^x$$
$$+ t\Big(\sin(\boldsymbol{k}\cdot\boldsymbol{e}_1) + \sin(\boldsymbol{k}\cdot\boldsymbol{e}_2) + \sin(\boldsymbol{k}\cdot\boldsymbol{e}_3)\Big)\tau^y\Big]c(\boldsymbol{k}), \tag{9.9}$$

where the sublattice and spin indices on the electrons are now implicit: the $c(\boldsymbol{k})$ are 4-component fermion operators.

The energy eigenvalues are easily determined to be

$$-\mu \pm \left|e^{i\boldsymbol{k}\cdot\boldsymbol{e}_1} + e^{i\boldsymbol{k}\cdot\boldsymbol{e}_2} + e^{i\boldsymbol{k}\cdot\boldsymbol{e}_3}\right| \tag{9.10}$$

and these are plotted in Fig. 9.3. At half-filling, exactly half the states should be occupied in the ground state, and for the spectrum in Eq. (9.10) this is achieved at $\mu = 0$.

A crucial feature of any metallic state is the Fermi surface: this is boundary between the occupied and empty states in momentum space. In two spatial dimensions, this boundary is generically a line in momentum space,

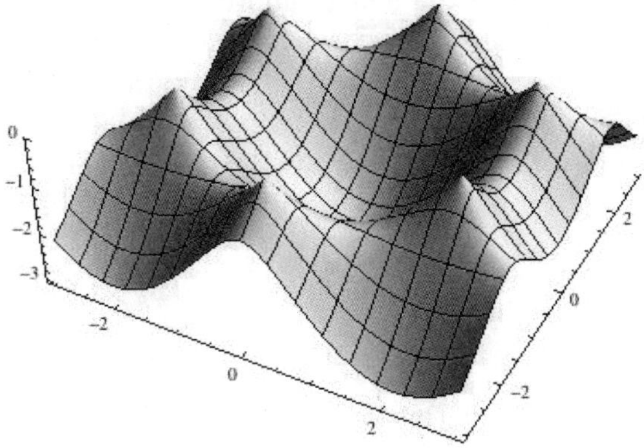

Fig. 9.3. The lower band of the dispersion in Eq. (9.10) for $\mu = 0$.

and this is the case for the dispersion in Eq. (9.10) for $\mu \neq 0$. However, for the $\mu = 0$, the honeycomb lattice has the special property that the occupied and empty states meet only at a discrete set of single points in momentum space: this should be clear from the dispersion plotted in Fig. 9.3. Only 2 of these points are distinct, in that they are not separated by a reciprocal lattice vector $\boldsymbol{G}$. So the half-filled honeycomb lattice has 2 'Fermi points', and realizes a 'semi-metal' phase. The low energy excitations of the semi-metal consist of particles and holes across the Fermi point, and these have a lower density of states than in a metallic phase with a Fermi line. We also note that the Fermi-point structure is protected by a sublattice exchange symmetry: it is not special to the nearest-neighbor hopping model, and it also survives the inclusion of electron-electron interactions.

We obtain a very useful, and universal, theory for the low energy excitations of the semi-metal by expanding (9.9) in the vicinity of the Fermi points. The distinct Fermi points are present at $\boldsymbol{Q}_1$ and $-\boldsymbol{Q}_1$; all other Fermi points are separated from these two points by a reciprocal lattice vector $\boldsymbol{G}$. So we define continuum Fermi field which reside in 'valleys' in the vicinity of these points by

$$\begin{aligned} C_{A1\alpha}(\boldsymbol{k}) &= \sqrt{A}\, c_{A\alpha}(\boldsymbol{Q}_1 + \boldsymbol{k}) \\ C_{A2\alpha}(\boldsymbol{k}) &= \sqrt{A}\, c_{A\alpha}(-\boldsymbol{Q}_1 + \boldsymbol{k}) \\ C_{B1\alpha}(\boldsymbol{k}) &= \sqrt{A}\, c_{B\alpha}(\boldsymbol{Q}_1 + \boldsymbol{k}) \\ C_{B2\alpha}(\boldsymbol{k}) &= \sqrt{A}\, c_{B\alpha}(-\boldsymbol{Q}_1 + \boldsymbol{k}), \end{aligned} \tag{9.11}$$

where A is the total area of the honeycomb lattice, and the momentum $\boldsymbol{k}$ is small. The field C is a 8-component continuum canonical Fermi field: the components correspond to spin ($\uparrow$, $\downarrow$), sublattice (A, B), and valley $(1, 2)$ indices. We will also use Pauli matrices which act on the spin (σ^a), sublattice (τ^a), and valley (ρ^a) space.

Inserting Eq. (9.11) into Eq. (9.9), we obtain the continuum Hamiltonian

$$H_0 = \int \frac{d^2k}{4\pi^2} C^\dagger(\boldsymbol{k}) \Big(v\tau^y k_x + v\tau^x \rho^z k_y \Big) C(\boldsymbol{k}), \tag{9.12}$$

where $v = 3t/2$. From now on we rescale time to set $v = 1$. Diagonalizing Eq. (9.12), we obtain the relativistic spectrum

$$\pm\sqrt{k_x^2 + k_y^2}, \tag{9.13}$$

which corresponds to the values of Eq. (9.10) near the Fermi points.

The relativistic structure of H_0 can be made explicit by rewriting it as the Lagrangian of massless Dirac fermions. Define $\overline{C} = C^\dagger \rho^z \tau^z$. Then we can write the Euclidean time (τ) Lagrangian density of the semi-metal phase as

$$\mathcal{L}_0 = \overline{C} \left(\partial_\tau \gamma_0 + \partial_x \gamma_1 + \partial_y \gamma_2 \right) C \tag{9.14}$$

where ω is the frequency associated with imaginary time, and the Dirac γ matrices are

$$\gamma_0 = -\rho^z \tau^z \quad \gamma_1 = \rho^z \tau^x \quad \gamma_2 = -\tau^y. \tag{9.15}$$

In addition to relativistic invariance, this form makes it clear the free-fermion Lagrangian has a large group of 'flavor' symmetries that acts on the 8×8 fermion space and commute with the γ matrices. Most of these symmetries are not obeyed by higher-order gradients, or by fermion interaction terms which descend from the Hubbard model.

Let us now turn on a small repulsion, U, between the fermions in the semi-metal. Because of the point-like nature of the Fermi surface, it is easier to determine the consequences of this interaction here than in a metallic phase with a Fermi line of gapless excitations. We can use traditional renormalization group (RG) methods to conclude that a weak U is irrelevant in the infrared: the computation is left as an exercise below. Consequently, the semi-metal state is a stable phase which is present over a finite range of parameters.

Exercise: Observe that $\mathcal{L}_0$ is invariant under the scaling transformation $x' = xe^{-\ell}$ and $\tau' = \tau e^{-\ell}$. Write the Hubbard interaction U in terms of the Dirac fermions, and show that it has the tree-level scaling transformation $U' = Ue^{-\ell}$. So argue that all short-range interactions are *irrelevant* in the Dirac semi-metal phase.

9.2.3. *Antiferromagnet*

Although a small U is irrelevant, new phases can and do appear at large U. To see this, let us return to the lattice Hubbard model in Eq. (9.1), and consider the limit of large $U_i = U$. We will assume $\mu = 0$ and half-filling in the remainder of this section.

At $U = \infty$, the eigenstates are simple products over the states on each site. Each site has 4 states:

$$|0\rangle \quad , \quad c_{i\uparrow}^{\dagger}|0\rangle \quad , \quad c_{i\downarrow}^{\dagger}|0\rangle \quad , \quad c_{i\uparrow}^{\dagger}c_{i\downarrow}^{\dagger}|0\rangle, \tag{9.16}$$

where $|0\rangle$ is the empty state. The energies of these states are $U/4$, $-U/4$, $-U/4$, and $U/4$ respectively. Thus the ground state on each site is doubly-degenerate, corresponding to the spin-up and spin-down states of a single electron. The lattice model has a degeneracy of $2^{2\mathcal{N}}$, and so a non-zero entropy density (recall that $\mathcal{N}$ is the number of sites on one sublattice).

Any small perturbation away from the $U = \infty$ limit is likely to lift this exponential large degeneracy. So we need to account for the electron hopping t. At first order, electron hopping moves an electron from one singly-occupied site to another, yielding a final state with one empty and one doubly occupied site. This final state has an energy U higher than the initial state, and so is not part of the low energy manifold. So by the rules of degenerate perturbation theory, there is no correction to the energy of all the $2^{2\mathcal{N}}$ ground states at first order in t.

At second order in t, we have to use the effective Hamiltonian method. This performs a canonical transformation to eliminate the couplings from the ground states to all the states excited by energy U, while obtaining a modified Hamiltonian which acts on the $2^{2\mathcal{N}}$ ground states. This method is described in text books on quantum mechanics, and we leave its application here as an exercise. The resulting effective Hamiltonian is the Heisenberg antiferromagnet:

$$H_J = \sum_{i<j} J_{ij} S_i^a S_j^a \quad , \quad J_{ij} = \frac{4t_{ij}^2}{U}, \tag{9.17}$$

where J_{ij} is the exchange interaction and S_i^a are the spin operators on site i

$$S_i^a = \frac{1}{2} c_{i\alpha}^\dagger \sigma_{\alpha\beta}^a c_{i\beta}. \tag{9.18}$$

Note that these spin operators preserve the electron occupation number on every site, and so act within the subspace of the $2^{2\mathcal{N}}$ low energy states. The Hamiltonian H_J lifts the macroscopic degeneracy, and the entropy density of the new ground state will be zero.

Exercise: Use the effective Hamiltonian method described in Ref. 7 to obtain Eq. (9.17). At second order in U, it is sufficient to consider the 2-site Hubbard model. This has a total of 16 states, and 4 ground states at $U = \infty$. Derive the effective Hamiltonian which acts on these 4 states at order t^2.

Although we cannot compute the exact ground state of H_J on the honeycomb lattice with nearest-neighbor exchange, numerical studies[8] leave little doubt to its basic structure. The ground state is adiabatically connected to that obtained by treating the S_i^a as classical vectors in spin space: it has antiferromagnetic (or Néel) order which breaks the global SU(2) spin rotation symmetry, by a spontaneous polarization of the spins on opposite orientations on the two sublattices

$$\eta_i \left\langle S_i^a \right\rangle = N^a, \tag{9.19}$$

where $\eta_i = 1$ ($\eta_i = -1$) on sublattice A (B), and N^a is the vector Néel order parameter; see Fig. 9.4. Classically this state minimizes the exchange coupling in Eq. (9.17) because $J_{ij} > 0$. Quantum fluctuations for spin $S = 1/2$ reduce the spontaneous moment from its classical value, but a non-zero moment remains on the honeycomb lattice.

What is the electronic excitation spectrum in the antiferromagnet? To determine this, it is useful to write the Néel order parameter in terms of the continuum Dirac fields introduced in Section 9.2.2. We observe

$$\sum_i \eta_i S_i^a = \sum_{\boldsymbol{k}} \left(c_{A\alpha}^\dagger \sigma_{\alpha\beta}^a c_{A\beta} - c_{B\alpha}^\dagger \sigma_{\alpha\beta}^a c_{B\beta} \right) = \int \frac{d^2k}{4\pi^2} C^\dagger \tau^z \sigma^a C. \tag{9.20}$$

Thus the Néel order parameter N^a is given by the fermion bilinear

$$N^a = \left\langle C^\dagger \tau^z \sigma^a C \right\rangle = \left\langle \overline{C} \rho^z \sigma^a C \right\rangle, \tag{9.21}$$

and the vacuum expectation value (VEV) is non-zero in the antiferromagnet. We can expect that electron-electron interactions will induce a coupling

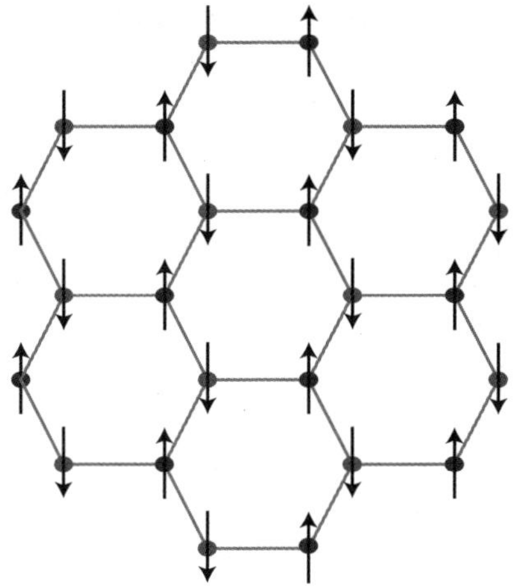

Fig. 9.4. The large U state with antiferromagnetic (Néel) order.

between the fermion excitations and this VEV in the low energy Hamiltonian for the Néel phase. Choosing Néel ordering in the z direction

$$N^a = N_0 \delta_{az}, \tag{9.22}$$

we anticipate that H_0 in Eq. (9.12) is modified in the Néel phase to

$$H_N = \int \frac{d^2k}{4\pi^2} C^\dagger(\boldsymbol{k}) \Big(\tau^y k_x + \tau^x \rho^z k_y - \lambda N_0 \tau^z \sigma^z \Big) C(\boldsymbol{k}), \tag{9.23}$$

where λ is a coupling determined by the electron interactions, and we have assumed Néel order polarized in the z direction. This effective Hamiltonian will be explicitly derived in the next subsection. We can now easily diagonalize H_N to deduce that the electronic excitations have energy

$$\pm \sqrt{k_x^2 + k_y^2 + \lambda^2 N_0^2}. \tag{9.24}$$

This is the spectrum of massive Dirac fermions. So the Fermi point has disappeared, and an energy gap has opened in the fermion excitation spectrum. In condensed matter language, the phase with antiferromagnetic order is an insulator, and not a semi-metal: transmission of electronic charge will require creation of gapped particle and hole excitations.

9.2.4. *Quantum phase transition*

We have now described a semi-metal phase for small U, and an antiferromagnetic insulator for large U. Both are robust phases, whose existence has been reliably established. We now consider connecting these two phases at intermediate values of U. This is a complex subject, and careful numerical studies are only just emerging for the model with nearest-neighbor hopping.[8] It is already clear, however, that by varying the form of the microscopic coupling we can obtain a rich variety of intermediate phases.[9–14] In the present subsection we consider the simplest possibility: there are no new intermediate phases, and only a direct quantum phase transition between the semi-metal and the antiferromagnetic insulator.[15–17]

We can derive the field theory for this direct transition either by symmetry considerations, or by an explicit derivation from the Hubbard model. Let us initially follow the second route. We start with the Hubbard Hamiltonian in Eq. (9.1), use the operator identity (valid on each site i):

$$U\left(n_\uparrow - \frac{1}{2}\right)\left(n_\downarrow - \frac{1}{2}\right) = -\frac{2U}{3}S^{a2} + \frac{U}{4}. \tag{9.25}$$

Then, in the fermion coherent state path integral for the Hubbard model, we apply a 'Hubbard-Stratonovich' transformation to the interaction term; this amounts to using the identity

$$\begin{aligned} &\exp\left(\frac{2U}{3}\sum_i \int d\tau S_i^{a2}\right) \\ &\quad = \int \mathcal{D}X_i^a(\tau)\exp\left(-\sum_i \int d\tau\left[\frac{3}{8}X_i^{a2} - \sqrt{U}X_i^a S_i^a\right]\right). \end{aligned} \tag{9.26}$$

The fermion path integral is now a bilinear in the fermions, and we can, at least formally, integrate out the fermions in the form of a functional determinant. We imagine doing this temporarily, and then look for the saddle point of the resulting effective action for the X_i^a. At the saddle-point we find that the lowest energy is achieved when the vector has opposite orientations on the A and B sublattices. Anticipating this, we look for a continuum limit in terms of a field φ^a where

$$X_i^a = \eta_i \varphi^a. \tag{9.27}$$

Using Eq. (9.20), the continuum limit of the coupling between the field φ^a and the fermions in Eq. (9.26) is given by

$$X_i^a c_{i\alpha}^\dagger \sigma_{\alpha\beta}^a c_{i\beta} = \varphi^a C^\dagger \tau^z \sigma^a C = \varphi^a \overline{C} \rho^z \sigma^a C. \tag{9.28}$$

From this it is clear that φ^a is a dynamical quantum field which represents the fluctuations of the local Néel order, and

$$\langle \varphi^a \rangle \propto N^a. \tag{9.29}$$

Now we can take the continuum limit of all the terms in the coherent state path integral for the lattice Hubbard model and obtain the following continuum Lagrangian density

$$\mathcal{L} = \overline{C}\gamma_\mu \partial_\mu C + \frac{1}{2}\left[(\partial_\mu \varphi^a)^2 + s\varphi^{a2}\right] + \frac{u}{24}\left(\varphi^{a2}\right)^2 - \lambda \varphi^a \overline{C} \rho^z \sigma^a C. \quad (9.30)$$

This is a relativistic quantum field theory for the 8-component fermion field C and the 3-component real scalar φ^a, related to the Gross-Neveu-Yukawa model. We have included gradient terms and quartic in the Lagrangian for φ^a: these are not present in the derivation outlined above from the lattice Hubbard model, but are clearly induced by higher energy fermions are integrated out. The Lagrangian includes various phenomenological couplings constants (s, u, λ); as these constants are varied, $\mathcal{L}$ can describe *both* the semi-metal and insulating antiferromagnet phases, and also the quantum critical point between them.

Note that the matrix $\rho^z\sigma^a$ commutes with all the γ_μ; hence $\rho^z\sigma^a$ is a matrix in "flavor" space. So if we consider C as 2-component Dirac fermions, then these Dirac fermions carry an additional 4-component flavor index.

The semi-metal phase is the one where φ^a has vanishing VEV. In mean-field theory, this appears for $s > 0$. The φ^a excitations are then massive, and these constitute a triplet of gapped 'spin-excitons' associated with fluctuations of the local antiferromagnetic order. The Dirac fermions are massless, and represent the Fermi point excitations of the semi-metal.

The Néel phase has a non-zero VEV, $\langle \varphi^a \rangle \neq 0$, and appears in mean-field theory for $s < 0$. Here the Dirac fermions acquire a gap, indicating that the Fermi point has vanished, and we are now in an insulating phase. The fluctuations of φ are a doublet of Goldstone modes ('spin waves') and a longitudinal massive Higgs boson.

Finally, we are ready to address the quantum critical point between these phases. In mean-field theory, this transition occurs at $s = 0$. As is customary in condensed matter physics, it is useful to carry out an RG analysis near this point. The tree-level analysis is carried out in the following exercise.

Exercise: Perform a tree-level RG transformation on $\mathcal{L}$. The quadratic gradient terms are invariant under $C' = Ce^{\ell}$ and $\varphi' = \varphi e^{\ell/2}$. Show that this leads to $s' = se^{2\ell}$. Thus s is a relevant perturbation which drives the system into either the semi-metal or antiferromagnetic insulator. The quantum critical point is reached by tuning s to its critical value ($= 0$ at tree level). Show that the couplings u and λ are both relevant perturbations at this critical point. Thus, while interactions are irrelevant in the Dirac semi-metal (and in the insulator), they are strongly relevant at the quantum-critical point.

Further study of this quantum critical point requires a RG analysis which goes beyond tree-level. Such an analysis can be controlled in an expansion in $1/N$ (where N is the number of fermion flavors) or $(3-d)$ (where d is the spatial dimensionality. For reviews see Ref. 18 or Chapter 17 of Ref. 19. The main conclusion of such analyses is that there is an RG fixed point at which the φ^{a2} is the only relevant perturbation. Non-linearities such as λ and u all reach stable fixed point values of order unity. This non-trivial fixed point implies that the physics of the quantum critical point is highly non-trivial and strongly coupled. The RG fixed point is scale- and relativistic-invariant, and this implies that it is also conformally invariant. Thus the quantum critical point is described by a CFT in 2+1 spacetime dimensions: a CFT3.

We will not describe the critical theory in any detail here. However, we will note some important characteristics of correlation functions at the quantum critical point. The electron Green's function has the following structure

$$\left\langle C(k,\omega); C^\dagger(k,\omega)\right\rangle \sim \frac{i\omega + k_x\tau^y + k_y\tau^x\rho^z}{(\omega^2 + k_x^2 + k_y^2)^{1-\eta/2}} \qquad (9.31)$$

where $\eta > 0$ is the *anomalous dimension* of the fermion. This leads to a fermion spectral density which has no quasiparticle pole: thus the quantum critical point has no well-defined quasiparticle excitations. This distinguishes it from both the semi-metal and insulating antiferromagnetic phases that flank it on either side: both had excitations with infinitely-sharp quasiparticle peaks. Similar anomalous dimensions appear in the correlations of the bosonic order parameter φ^a.

9.2.5. *Quantum impurities*

We briefly note the physics of quantum impurities in the honeycomb lattice, which were discussed more completely in the companion review.[1] The translational invariance of the honeycomb lattice will be broken *only* in this subsection.

Imagine removing a single atom from the honeycomb lattice, as shown in Fig. 9.5 At $U = 0$, the electronic spectrum of such a lattice was described in

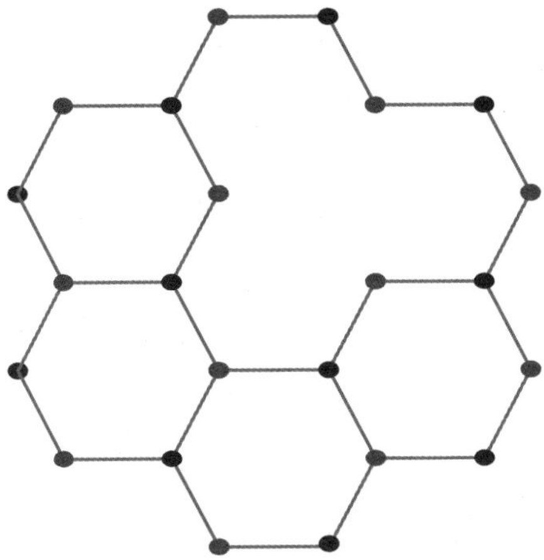

Fig. 9.5. The honeycomb lattice with a vacancy.

Ref. 20. They showed that there was a quasi-localized state in the vicinity of the impurity exactly at zero energy, which decayed as $1/r$ at a distance r from the impurity. Upon including U, we expect this state to capture a single electron whose $S = 1/2$ spin interacts with the bulk excitations.

Let us represent the impurity by a localized fermion operator $\chi_\alpha(\tau)$. Note that this fermion has no dependence upon the spatial co-ordinate $\boldsymbol{r}$, and is a function only of the time τ. Now we can couple this fermion to the bulk excitations which were described by $\mathcal{L}$ in Eq. (9.30) in the vicinity of the semi-metal to antiferromagnetic insulator transition. The full system is described by the Lagrangian $\mathcal{L} + \mathcal{L}_{\text{imp}}$ where

$$\mathcal{L}_{\text{imp}} = \chi_\alpha^\dagger \frac{\partial \chi_\alpha}{\partial \tau} - h\, \chi_\alpha^\dagger \sigma^a_{\alpha\beta} \chi_\beta\, \varphi^a(\boldsymbol{r} = 0, \tau); \tag{9.32}$$

note that whereas the Lagrangian $\mathcal{L}$ is integrated over spacetime, the Lagrangian $\mathcal{L}_{\text{imp}}$ is only integrated over time. There are many possible couplings between the impurity and bulk degrees of freedom which are allowed by the symmetry of the problem, but we have only included a single one. This is easily seen to be the only term which is relevant under the RG which applies in the vicinity of the bulk quantum phase transition.

The RG flow of the bulk-impurity coupling h was described in Refs. 21–23. It was found that h approached a fixed-point coupling in the infrared, just like the couplings u and λ in $\mathcal{L}$. Thus no new couplings are necessary to describe the low energy physics of the impurity provided we are not too far from the semi-metal to antiferromagnetic insulator quantum critical point.

Further details of the impurity dynamics may be found in the companion review,[1] where it was described for a closely related bulk quantum critical point. A close analogy was also drawn between these impurity problems and defects in super Yang-Mills theories; the latter can be solved[24] by intersecting brane models in string theory, and led to a description of the impurity criticality using a AdS_2 geometry.

9.2.6. *Electrical transport*

We now revert to the honeycomb lattice without impurities.

An important set of observables which do *not* acquire anomalous dimensions at the quantum critical point are the currents associated with global conservation laws. As the simplest example here, let us consider correlations of the conserved electric charge of the electrons, and the associated electrical conductivity σ. At zero temperature ($T = 0$), we have $\sigma = 0$ in the insulator, while the semi-metal and the quantum critical point have finite non-zero values of σ, as we will now see.

The conserved electrical current is

$$J_\mu = -i\overline{C}\gamma_\mu C. \tag{9.33}$$

Let us compute its two-point correlator, $K_{\mu\nu}(k)$ at a spacetime momentum k_μ. At leading order, this is given by a one fermion loop diagram which evaluates to

$$\begin{aligned} K_{\mu\nu}(k) &= \int \frac{d^3p}{8\pi^3} \frac{\mathrm{Tr}\left[\gamma_\mu(i\gamma_\lambda p_\lambda + m\rho^z\sigma^z)\gamma_\nu(i\gamma_\delta(k_\delta+p_\delta) + m\rho^z\sigma^z)\right]}{(p^2+m^2)((p+k)^2+m^2)} \\ &= -\frac{2}{\pi}\left(\delta_{\mu\nu} - \frac{k_\mu k_\nu}{k^2}\right)\int_0^1 dx \frac{k^2 x(1-x)}{\sqrt{m^2 + k^2 x(1-x)}}, \end{aligned} \tag{9.34}$$

where the mass $m = 0$ in the semi-metal and at the quantum critical point, while $m = |\lambda N_0|$ in the insulator. Note that the current correlation is purely transverse, and this follows from the requirement of current conservation

$$k_\mu K_{\mu\nu} = 0. \tag{9.35}$$

Of particular interest to us is the K_{00} component, after analytic continuation to Minkowski space where the spacetime momentum k_μ is replaced by (ω, k). The conductivity is obtained from this correlator via the Kubo formula

$$\sigma(\omega) = \lim_{k\to 0} \frac{-i\omega}{k^2} K_{00}(\omega, k). \tag{9.36}$$

In the insulator, where $m > 0$, analysis of the integrand in Eq. (9.34) shows that that the spectral weight of the density correlator has a gap of $2m$ at $k = 0$, and the conductivity in Eq. (9.36) vanishes. These properties are as expected in any insulator.

In the metal, and at the critical point, where $m = 0$, the fermionic spectrum is gapless, and so is that of the charge correlator. The density correlator in Eq. (9.34) and the conductivity in Eq. (9.36) evaluate to the simple universal results

$$\begin{aligned} K_{00}(\omega, k) &= \frac{1}{4}\frac{k^2}{\sqrt{k^2-\omega^2}} \\ \sigma(\omega) &= \frac{1}{4}. \end{aligned} \tag{9.37}$$

How about beyond the one-loop results? The insulator maintains a gap to charged excitations, and so the conductivity remains at zero. In the semi-metal, the fermions are gapless, but they couple only to the gapped fluctuations of the Néel order φ^a. Examination of the perturbation theory shows that these have no effect on the current correlators at small momenta and frequency, and so the results in Eq. (9.37) are *exact* in the limit of small ω and k in the semi-metal.

At the quantum critical point, we have to consider the strong critical fluctuations associated with fixed-point values of the Yukawa coupling λ and the quartic bosonic interaction u. These can be examined in the $(3 - d)$ or the $1/N$ expansion, and require evaulation of multi-loop diagrams. We will not describe the computations here, but note a remarkable feature: all divergences associated with the critical fluctuations cancel, and the final result is universal. The values of none of the couplings of the Lagrangian in Eq. (9.30) matters because these are all pinned by the RG fixed point. There are no anomalous dimensions, and the results in Eq. (9.37) generalize to

$$\begin{aligned} K_{00}(\omega, k) &= \mathcal{K}\frac{k^2}{\sqrt{k^2-\omega^2}} \\ \sigma(\omega) &= \mathcal{K}, \end{aligned} \tag{9.38}$$

where $\mathcal{K}$ is a universal number dependent only upon the universality class of the quantum critical point. The value of the $\mathcal{K}$ for the Gross-Neveu-Yukawa model in Eq. (9.30) is not known exactly, but can be estimated by computations in the $(3 - d)$ or $1/N$ expansions.

9.3. Non-zero Temperatures and the AdS/CFT Correspondence

We begin by some general remarks on the influence of a non-zero temperature, T, on a generic, strongly-coupled quantum critical point. Let us consider a quantum-critical point which has only a single relevant perturbation, s, as is the case for the model in Eq. (9.30) (the generalization to several relevant perturbations is immediate). So near the quantum critical point, the RG flow is described by

$$\frac{ds}{d\ell} = \frac{1}{\nu} s. \tag{9.39}$$

In standard condensed matter notation, the eigenvalue of the relevant flow is written in terms of ν, the correlation length exponent. Now in the quantum field theory in Euclidean time, a non-zero T corresponds to placing the theory on a cylinder of circumference $1/T$. Such a finite size is clearly relevant in the infrared, and also indicates that $1/T$ transforms just like the temporal length under the RG. We write this as

$$\frac{dT}{d\ell} = zT, \tag{9.40}$$

where z is the dynamic critical exponent. All the theories for the honeycomb lattice at half filling have $z = 1$, but we allow z to be arbitrary here.

Eqs. (9.39) and (9.40) are of course trivial to integrate

$$s(\ell) = s e^{\ell/\nu} \quad , \quad T(\ell) = T e^{z\ell}, \tag{9.41}$$

but the results teach us an important lesson which is summarized in the canonical quantum-critical phase diagram shown in Fig. 9.6. We ask the question: which of the relevant perturbations, s or T, is more important? To answer this question, we integrate the RG equations to a scale $\ell = \ell^*$ until the winner reaches a value of unity. This allows us to conclude that for $T > |s|^{z\nu}$, thermal effects are more important than any deviation of the coupling from the RG fixed point. Conversely, for $T < |s|^{z\nu}$ the couplings flow far from the critical fixed point before any thermally excited states need be considered. These considerations lead to the two distinct regimes show in Fig. 9.6.

In the blue-colored regimes of Fig. 9.6, where $T < |s|^{z\nu}$, the physics of the two non-critical phases dominates. For the model of Eq. (9.30), these are the semi-metal or antiferromagnetic insulator phases, both of which have well-defined quasiparticle excitations. Consequently, the long-time

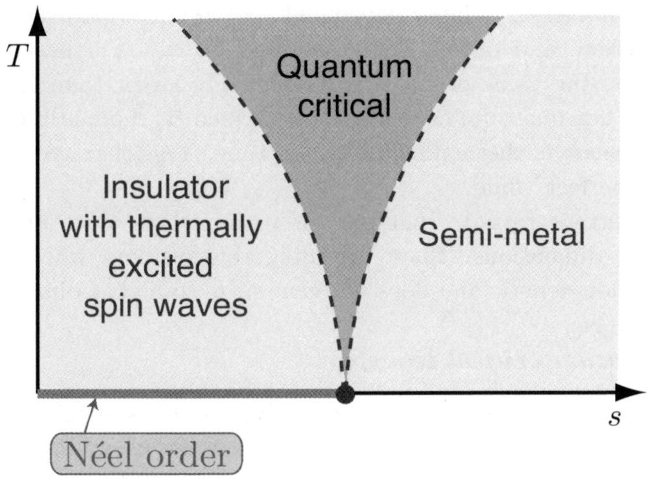

Fig. 9.6. Canonical quantum critical crossover phase diagram. The dashed lines occur for $T \sim |s|^{z\nu}$, and indicate crossovers between the orange- and blue-shaded regions which are generic from any strongly-coupled quantum critical point. Specific features of the blue-shaded regions for the theory Eq. (9.30) of the transition from the semi-metal to the Néel-ordered insulator are also indicated. The Néel order vanishes for any $T > 0$ because non-Abelian continuous symmetries cannot be broken in two spatial dimensions.

dynamics can be written using quasi-classical models of the interactions of these quasiparticles.

Our interest here is primarily in the orange-colored regime of quantum criticality, $T > |s|^{z\nu}$. Here T is the primary perturbation to the quantum critical theory. The deviation of the couplings from $T = 0$ RG fixed point is unimportant, and the system behaves as it is described by the universal quantum-critical Lagrangian in the entire regime. For the relativistic model considered here, the strongly coupled CFT describes the dynamics of the orange-colored region.

It has been argued[19] that a central general property of quantum critical dynamics is the short time over which the system relaxes back to thermal equilibrium. We imagine perturbing the system away from equilibrium, and measuring the time, τ_{eq} over which it relaxes back to local equilibrium (the adjective 'local' implies that we exclude diffusion of globally conserved charges which can take a long time to reach equilibrium across the entire system). In the regime of strongly-coupled quantum criticality this is given by

$$\tau_{\text{eq}} = \mathcal{C}\frac{\hbar}{k_B T} \tag{9.42}$$

where $\mathcal{C}$ is a universal number dependent only upon the universality class of the transition, and precise definition used for $\tau_{\rm eq}$. Furthermore, in all other regimes, the value of $\tau_{\rm eq}$ is parametrically larger than the value in Eq. (9.42). Thus quantum criticality is described by a quantum fluid with the *shortest possible* thermal equilibration time. This characteristic makes it a "nearly perfect" fluid.

It is important to note that our discussion above does *not* apply to CFTs in 1+1 dimensions. These are integrable systems, whose long-time dynamics is non-generic and does not generalize to higher dimensions.

9.3.1. *Quantum critical transport*

Let us explore the ideas above by examining the behavior of the electron conductivity of the model in Eq. (9.30) in the quantum-critical regime. At one-loop order, we can set $m = 0$, and then repeat the computation in Eq. (9.34) at $T > 0$. This only requires replacing the integral over the loop frequency by a summation over the Matsubara frequencies, which are quantized by odd multiples of πT. Such a computation, via Eq. (9.36) leads to the conductivity[25]

$$\begin{aligned}\mathrm{Re}[\sigma(\omega)] &= (2T\ln 2)\,\delta(\omega) + \frac{1}{4}\tanh\left(\frac{|\omega|}{4T}\right)\\ \mathrm{Im}[\sigma(\omega)] &= \int_{-\infty}^{\infty}\frac{d\Omega}{\pi}\mathcal{P}\left(\frac{\mathrm{Re}[\sigma(\Omega)]-1/4}{\omega-\Omega}\right),\end{aligned} \tag{9.43}$$

where $\mathcal{P}$ is the principal part. Note that this reduces to Eq. (9.37) in the limit $\omega \gg T$. However, the most important new feature of Eq. (9.43) arises for $\omega \ll T$, where we find a delta function at zero frequency in the real part. Thus the d.c. conductivity is infinite at this order, arising from the collisionless transport of thermally excited carriers.

Exercise: Evaluate K_{00} from Eq. (9.34) at $T > 0$. First perform the trace over the Dirac matrices, and then the summation over the frequency. Subtract from your answer the result of integrating over the frequency; this subtraction can be compensated by the $T = 0$ result in Eq. (9.34). The remaining expressions are explicitly convergent in the ultraviolet, and the integration over spatial momenta can be evaluated. Finally, analytically continue the answer to real frequencies to obtain Eq. (9.43).

The relaxational processes associated with Eq. (9.42) should lead to collisions between the thermally excited carriers and broaden the delta function at zero frequency. However, this relaxation does appear in a direct perturbative analysis of the critical theory in powers of $(3-d)$ or $1/N$. As has been discussed elsewhere,[25–31] an infinite order resummation is required, whose simplest realization requires solution of a quantum Boltzmann equation. Such a solution shows that the delta function acquires a width of order $(3-d)^2T$ or T/N, and so there is a large d.c. conductivity of order $(3-d)^{-2}$ or N. Thus $\sigma(\omega)$ has the form of 'Drude peak' at zero frequency, and the behavior in Eq. (9.38) for $\omega \gg T$. However, the accuracy of such a Boltzmann equation computation is untested, and it is likely that such perturbative analyses of quantum-critical dynamics are quantitatively unreliable.

We will be satisfied here by scaling arguments which generalize the $T = 0$ quantum-critical results in Eq. (9.38) to the $T > 0$ quantum critical region in Fig. 9.6. The quantum-critical relaxational processes invalidate the form in Eq. (9.38) for the density correlation function, and we instead expect the form dictated by the hydrodynamic diffusion of charge. Thus for K_{00}, Eq. (9.38) applies only for $\omega \gg T$, while

$$K_{00}(\omega, k) = \chi \frac{Dk^2}{Dk^2 - i\omega} \quad , \quad \omega \ll T. \tag{9.44}$$

Here χ is the charge susceptibility (here it is the compressibility), and D is the charge diffusion constant. Associated with Eq. (9.42), these have universal values in the quantum critical region:

$$\chi = \mathcal{C}_\chi T \quad , \quad D = \frac{\mathcal{C}_D}{T}, \tag{9.45}$$

where again $\mathcal{C}_\chi$ and $\mathcal{C}_D$ are universal numbers. For the conductivity, we expect a crossover from the collisionless critical dynamics at frequencies $\omega \gg T$, to a hydrodynamic collision-dominated form for $\omega \ll T$. This entire crossover is universal, and is described by a universal crossover function

$$\sigma(\omega) = \mathcal{K}_\sigma(\omega/T). \tag{9.46}$$

The result in Eq. (9.38) applies for $\omega \gg T$, and so

$$\mathcal{K}_\sigma(\infty) = \mathcal{K}. \tag{9.47}$$

For the hydrodynamic transport, we apply the Kubo formula in Eq. (9.36) to Eq. (9.44) and obtain

$$\mathcal{K}_\sigma(0) = \mathcal{C}_\chi \mathcal{C}_D \tag{9.48}$$

which is a version of Einstein's relation for Brownian motion.

9.3.2. *The AdS/CFT correspondence*

Portions of this section have been adapted from Chapter 15 of Ref. 19.

It turns out the AdS/CFT correspondence is 'just what the doctor ordered' to compute strongly-coupled quantum critical dynamics and transport in the orange-colored region of Fig. 9.6. This is a consequence of a crucial property: even at the level of the classical gravity approximation in the AdS theory, the system relaxes back to thermal equilibrium in a time which obeys Eq. (9.42). No other method in condensed matter physics shares this remarkable feature. We will review specific computations by this method of the universal function $\mathcal{K}_\sigma(\omega/T)$, and of the collisionless-to -hydrodynamic crossover in the density correlation function.

The CFT solvable by the AdS/CFT correspondence may be viewed as a generalization of the CFT described by Eq. (9.30). It has a closer resemblance to the SU(2) gauge theory we consider later in Eq. (9.81). We take the structure of critical matter fields coupled to a gauge field, and generalize it to a relativistically invariant model with a non-Abelian SU(N) gauge group and the maximal possible supersymmetry. The resulting supersymmetric Yang-Mills (SYM) theory has only one independent coupling constant g. Under the RG, it is believed that g flows to an attractive fixed point at a non-zero coupling $g = g^*$; the fixed point then defines a supersymmetric conformal field theory in 2+1 dimensions (a SCFT3). We are interested here in computing the transport properties of the SCFT, as a paradigm of quantum critical transport at a strongly interacting quantum critical point.

The solution proceeds by a dual formulation as a four-dimensional supergravity theory on a spacetime with uniform negative curvature: anti-de Sitter space, or AdS_4. The solution is also easily extended to non-zero temperatures, and allows direct computation of the correlators of conserved charges in real time. At $T > 0$ a black hole appears in the gravity theory, resulting in an AdS-Schwarzschild spacetime, and T is the Hawking temperature of the black hole; the real time solutions also extend to $T > 0$.

The reader is referred to the original paper,[32] and to the TASI lectures by Son for an explicit description of the method. In the AdS/CFT correspondence, every globally conserved quantity in the CFT gets mapped onto a gauge field in AdS. Moreover, in the leading classical gravity theory on AdS, different global charges commute with each other, and so can be considered separately. In the end, we have a U(1) gauge field on AdS for every global conservation law of the CFT. The low energy effective field theory on AdS_4 has the standard Einstein-Maxwell action for gravity+electromagnetism:

$$\mathcal{S}_M = \frac{1}{g_4^2} \int d^4x \sqrt{-g} \left[-\frac{1}{4} F_{ab} F^{ab} \right] . \tag{9.49}$$

Here g_{ab} is the AdS-Schwarzschild metric (g is its determinant), F_{ab} is the Maxwell flux tensor, and g_4 is a dimensionless coupling constant fixed by the value of N in the SU(N) SYM theory. This 4-dimensional Maxwell theory can be used to compute the density correlation function, $K_{00}(\omega, k)$, of the 3-dimensional SYM theory, and the results are shown in Figs. 9.7 and 9.8.

The most important feature of these results is that the expected limiting forms in the collisionless (Eq. (9.38)) and collision-dominated (Eq. (9.44)) are obeyed. Thus the results do display the collisionless to collision-dominated crossover at a frequency of order $k_B T/\hbar$, as we expected from the physical discussion in Section 9.3.1.

At this point, we describe some technical aspects of the results which turn out to have important physical implications. For this, let us generalize the constraints on $K_{\mu\nu}$ from current conservation in Eq. (9.35) to non-zero temperatures. At $T > 0$, we do not expect $K_{\mu\nu}$ to be relativistically covariant, and so can only constrain it by spatial isotropy and density

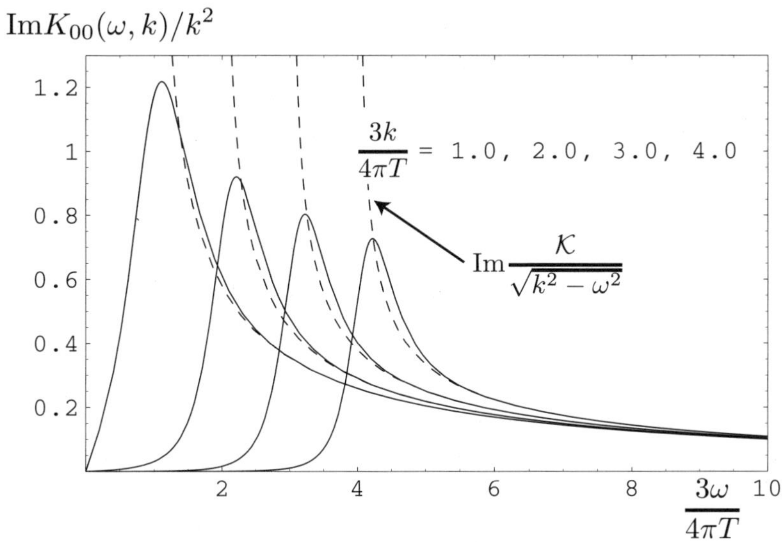

Fig. 9.7. Spectral weight of the density correlation function of the SCFT3 with $\mathcal{N} = 8$ supersymmetry in the collisionless regime.

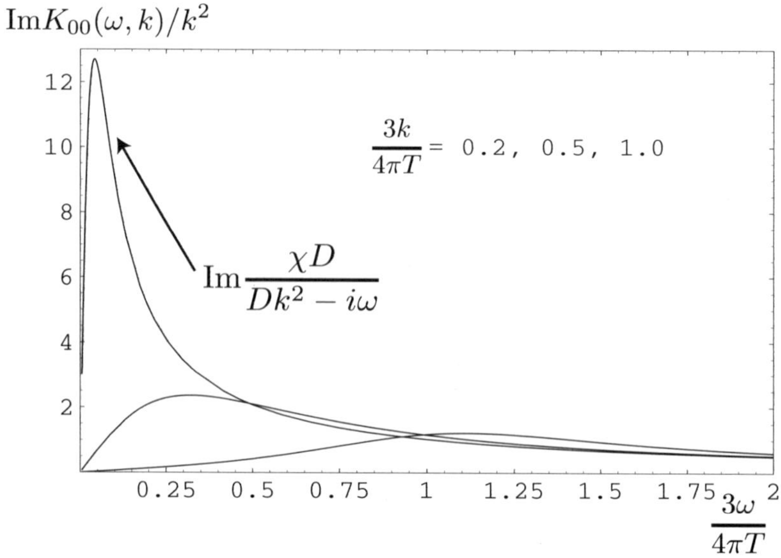

Fig. 9.8. As in Fig. 9.7, but for the collision-dominated regime.

conservation. These two constraints, along with dimensional analyses, lead to the most general form

$$K_{\mu\nu}(\omega, \boldsymbol{k}) = \sqrt{k^2 - \omega^2}\Big(P^T_{\mu\nu}\, K^T(\omega, k) + P^L_{\mu\nu}\, K^L(\omega, k)\Big), \tag{9.50}$$

where $p_\mu \equiv (-\omega, \boldsymbol{k})$ and $k = |\boldsymbol{k}|$. The $K^{L,T}$ are dimensionless functions of the arguments, and depend upon ω and the magnitude of the 2-vector $\boldsymbol{k}$. Also $P^T_{\mu\nu}$ and $P^L_{\mu\nu}$ are orthogonal projectors defined by

$$P^T_{00} = P^T_{0i} = P^T_{i0} = 0 \quad , \quad P^T_{ij} = \delta_{ij} - \frac{k_i k_j}{k^2} \quad , \quad P^L_{\mu\nu} = \left(\eta_{\mu\nu} - \frac{k_\mu k_\nu}{p^2}\right) - P^T_{\mu\nu}, \tag{9.51}$$

with $\eta_{\mu\nu} = \mathrm{diag}(-1, 1, 1)$, and the indices i, j running over the 2 spatial components. Thus, in the general case at $T > 0$, the full density and current responses are described in terms of two functions $K^{L,T}(k, \omega)$, representing current fluctuations longitudinal and transverse to the momentum. These two functions are not entirely independent. At $T > 0$, we expect all correlations to be smooth functions at $k = 0$: this is because all correlations are expected to decay exponentially to zero as a function of spatial separation. However, this is only possible from (9.50) if we have the additional relation

$$K^T(\omega, 0) = K^L(\omega, 0). \tag{9.52}$$

Finally, we note that application of the Kubo formula in Eq. (9.36) to Eq. (9.50) yields

$$\sigma(\omega) = K^L(\omega, 0). \tag{9.53}$$

The relations of the previous paragraph are completely general and apply to any theory. Specializing to the AdS-Schwarzschild solution of SYM3 as determined by the Einstein-Maxwell theory in Eq. (9.49), the results were found to obey a simple and remarkable identity:[32]

$$K^L(\omega, k) K^T(\omega, k) = \mathcal{K}^2 \tag{9.54}$$

where $\mathcal{K}$ is a known pure number, independent of ω and k. This identity is a consequence of the self-dual structure of Eq. (9.49): the Maxwell action is in 3+1 dimensions is well-known to have a self-dual structure corresponding to the exchange of electric and magnetic fields. Thus we have the important and key result that every global charge in a CFT3 maps onto a self-dual theory in the leading gravity approximation on AdS_4. The identity in (9.54) is a consequence of this emergent self-duality of CFT3s.

The combination of (9.54) and (9.52) now fully determine the response functions at zero momenta: $K^L(\omega, 0) = K^T(\omega, 0) = \mathcal{K}$. Computing the conductivity from Eq. (9.53), we then have

$$\sigma(\omega) = \mathcal{K}_\sigma(\omega/T) = \mathcal{K}; \tag{9.55}$$

i.e. the scaling function in Eq. (9.46) is *independent* of ω and equal to the value in Eq. (9.47). This result is an important surprise: the conductivity of the classical gravity theory on AdS_4 is frequency-independent. Furthermore, its value is fixed by self-duality to be the constant $\mathcal{K}$ appearing in the self-duality relation (9.54). All these remarkable results are a direct consequence of the self-duality of the U(1) Maxwell theory on AdS_4.

Given the strong consequences of self-duality relation in Eq. (9.54), it is useful to ask whether it can be valid for CFTs beyond those described by the classical Einstein-Maxwell theory on AdS_4. This question was addressed recently by Myers *et al.*[33] The examined the general structure of the higher-derivative corrections to Eq. (9.49), and argued that for the current correlations the leading terms could always be transformed into the following form which has only *one* dimensionless constant γ (L is the radius of AdS_4):

$$\mathcal{S}'_M = \frac{1}{g_4^2} \int d^4x \sqrt{-g} \left[-\frac{1}{4} F_{ab} F^{ab} + \gamma L^2 C_{abcd} F^{ab} F^{cd} \right], \tag{9.56}$$

where the extra four-derivative interaction is expressed in terms of the Weyl tensor C_{abcd}. A crucual observation of Ref. 33 was that stability and causality constraints on the effective theory restrict $|\gamma| < 1/12$. A generalized duality relation applies also to $\mathcal{S}'_M$. However this is *not* a *self*-duality. The dual CFT has current correlation functions which were characterized by functions $\widetilde{K}^{L,T}(\omega, k)$ which are distinct from those of the direct CFT $K^{L,T}(\omega, k)$, and the self-duality relation of Eq. (9.54) take the less restrictive form

$$K^L(\omega, k)\widetilde{K}^T(\omega, k) = \mathcal{K}^2 \quad , \quad K^T(\omega, k)\widetilde{K}^L(\omega, k) = \mathcal{K}^2. \qquad (9.57)$$

These duality relation determines the correlators of the dual CFT in terms of the direct CFT, but do not fix the latter. Determination of the functions $K^L(\omega, 0) = K^T(\omega, 0)$ requires explicit computation using the extended theory $\mathcal{S}_{vec}$, and the results for the conductivity are presented in Fig. 9.9. Now the conductivity does have a non-trivial universal dependence on ω/T.

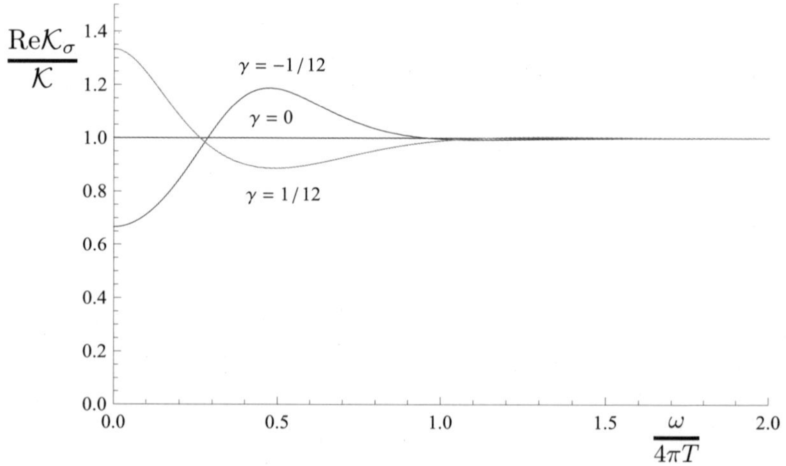

Fig. 9.9. Frequency dependent conductivity for CFT3s for which the AdS_4 theory includes the leading correction beyond the Einstein-Maxwell theory from Ref. 33. The co-efficient of this correction in the action is γ, and stability requirements impose the bound $|\gamma| < 1/12$.

However, as is clear from Fig, 9.9, stability conditions on the effective theory on AdS_4 allow only a limited range of dependence on ω/T. The smooth ω/T dependence in Fig. 9.9 should be contrasted to the very singular dependence in the free-field result in Eq. (9.43); the former is clearly more generic for a strongly-coupled CFT. It is also interesting to note that the

ω/T dependence in Fig. 9.9 for $\gamma > 0$ is very similar to the structure we discussed in Section 9.3.1 on the effect of collisional broadening of the singularities in Eq. (9.43): the AdS_4 result shows a collision-dominated Drude peak at $\omega = 0$, and a collisionless critical continuum at large ω.

Does the duality mapping of Myers *et al.*[33] have an interpretation directly in the CFT, without using the mapping to AdS_4? It has been argued[32–34] that this duality is the analog of the 'particle-vortex' duality of condensed matter physics. The latter is an exact duality of the critical theory of a complex relativistic scalar field[35] (this the theory in Eq. (9.30) without the fermions, and with φ^a having two components). In the particle-vortex duality, the world line of the complex scalars is reinterpreted as the world line of vortices in the dual theory of a dual complex scalar interacting with an emergent electromagnetic field. This particle-vortex duality also allows us to interpret the structure of the results in Fig. 9.9 for $\gamma < 0$. Note that from Eq. (9.57) the conductivity of the direct CFT maps onto the resistivity of the dual CFT. Thus for $\gamma < 0$, it is the dual CFT which will have a conductivity which has the structure discussed in Section 9.3.1, with a collisionally-broadened Drude peak at $\omega = 0$. Thus we conclude that a Boltzmann-like picture of transport applies better to the particle interpretation of the CFT for $\gamma > 0$, and to the vortex interpretation for $\gamma < 0$.

Let us summarize the lessons we have learnt from the AdS theory of quantum critical transport in strongly interacting systems in 2+1 dimensions. This theory should be view as complementary to the quasiparticle-based theory, whose implications were discussed in Section 9.3.1. The lessons are:

- There are a large class of strongly interacting 2+1 dimensional quantum liquids which relax back to thermal equilibrium in the shortest possible time of order $\hbar/(k_B T)$, as we indicated in (9.42).
- The quasiparticle transport theory[25–31] starts from the free theory with an infinite thermal equilibration time, and includes the effect of weak interactions using the Boltzmann equation. Complementary to this is the quantum-critical transport theory applicable for the shortest possible equilibration time of order $\hbar/(k_B T)$, which is the classical Einstein-Maxwell theory on AdS_4.
- The Einstein-Maxwell theory exhibits collisionless dynamics for $\omega \gg T$, and collision-dominated dynamics for $\omega \ll T$, as we displayed in Figs. 9.7 and 9.8.

- All continuous global symmetries are represented by a self-dual Einstein-Maxwell theory.
- This emergent self-duality implies that, in systems with particle-hole symmetry, $\sigma(\omega)$ is frequency-independent in the Einstein-Maxwell theory and equal to the self-dual value.
- For systems with particle-hole symmetry, a frequency dependent conductivity is obtained[33] upon considering corrections to the effective Einstein-Maxwell theory, with the forms in Fig. 9.9. Stability conditions on the effective theory allow only a limited range of frequency dependence, which depend upon the single parameter $|\gamma| < 1/12$. For $\gamma > 0$, the frequency dependence has the form expected from collision-dominated transport of particles, while for $\gamma < 0$ it is characteristic of the transport of vortices. It is remarkable that the physical pictures expected from the Boltzmann transport analysis correspond to precisely those obtained from the holographic approach.
- Such quantum-critical fluids also have universal momentum transport. By extending the scaling arguments to momentum transport we would conclude that the ratio of the shear viscosity to the entropy density η/s should equal a universal number characterizing the collision-dominated regime. This number was computed in the Einstein-Maxwell theory by Kovtun *et al.*[36,37] and found to equal $\hbar/(4\pi k_B)$. The shortest possible relaxation time implies that η is also the smallest possible and so these fluids are 'nearly perfect'.

9.4. U(1) Gauge Theory and the Valence Bond Solid on the Honeycomb Lattice

We now return to the honeycomb lattice at half-filling. In Section 9.2.4 we described a quantum phase transition in which two characteristics of the ground state changed simultaneously. In the charge sector, the one electron excitation gap opened up leading to a transition from the semi-metal to the insulator. And in the spin sector, the breaking of SU(2) spin rotation symmetry led to Néel order in the insulator. However, in the Mott picture, the insulating behavior is tied to repulsion between the electrons, which keeps them apart, rather than to any specific symmetry breaking. This would suggest that it is possible to have an insulating state while preserving spin rotation invariance. We will explore such a possibility in the present section.

Readers not interested in issues related to electron fractionalization and emergent gauge fields in insulators may skip ahead to the discussion of metallic phases in Section 9.6.

Our approach will be begin in the Néel-ordered insulator, and restore spin rotation invariance by allowing for slow angular fluctuations in the local *orientation* of the Néel order parameter φ^a. At the same time, it will also pay to transform the fermions to a 'rotating reference frame' so that their spin is measured relative to the orientation of the local Néel order.[38–40] This transformation is most conveniently done using spinor variables. So let us decompose the vector Néel order φ^a into a complex two-component bosonic spinor z_α by

$$\varphi^a = z_\alpha^* \sigma^a_{\alpha\beta} z_\beta. \tag{9.58}$$

Such a decomposition is familiar from early work of D'Adda *et al.*[41] and Witten,[42] who established the equivalence between the O(3) non-linear σ−model and the CP^1 model in 2 spacetime dimensions. A similar equivalence does not immediately apply in the 3 spacetime dimensional case of interest here because point defects in spacetime have to be treated with some care.[43] In particular, the theory for the fluctuations of the vector field φ^a must allow for point spacetime defects ('instantons') where $\varphi^a = 0$, which are known in the condensed matter literature as 'hedgehogs'. Note that these hedgehogs are present even in a 'fixed-length', non-linear σ-model in which we set $\varphi^{a2} = 1$; such models require ultraviolet regularization, and the hedgehogs are invariably permitted in the regulated theory *e.g.* with a lattice regularization. Ignoring these defects momentarily, let us proceed as in the earlier work.[41,42] The parameterization in Eq. (9.58) is invariant under the U(1) gauge transformation

$$z_\alpha \to z_\alpha e^{i\zeta} \tag{9.59}$$

and so the theory for the z_α must be a U(1) gauge theory involving an emergent U(1) gauge field A_μ. The boson only terms in Eq. (9.30) are equivalent[41,42] to a U(1) gauge theory for the complex scalars z_α

$$\mathcal{L}_z = |(\partial_\mu - iA_\mu)z_\alpha|^2 + s|z_\alpha|^2 + u(|z_\alpha|^2)^2. \tag{9.60}$$

Here the gauge field A_μ is dynamical, and will acquire a Maxwell action after high energy z_α modes have been integrated out.

Let us now discuss the point defects. Eqn (9.58) implies[43] that the hedgehogs in φ^a become Dirac monopoles in A_μ: these are tunnelling events associated with a change in the total A_μ flux by 2π. Such monopoles are permitted by the U(1) gauge theory in Eq. (9.60) only if the U(1) gauge field is *compact*. So we must account for the dynamics of such a compact U(1) gauge theory to completely account for the fluctuations of the local antiferromagnetic order. The dynamics of the matter fields can suppress the monopoles in some cases,[44–46] and this can then lead to deconfined critical points or phases with a gapless U(1) photon excitation associated with an effectively non-compact U(1) gauge field. We will find an example of this phenomenon in Section 9.7.

Let us now turn to the fermionic excitations in antiferromagnetic insulator. We transform these to the rotating reference frame by writing[38–40]

$$\begin{pmatrix} c_\uparrow \\ c_\downarrow \end{pmatrix} = \begin{pmatrix} z_\uparrow & -z_\downarrow^* \\ z_\downarrow & z_\uparrow^* \end{pmatrix} \begin{pmatrix} \psi_+ \\ \psi_- \end{pmatrix} \tag{9.61}$$

where ψ_p, $p = \pm$, are the "electrons" in the rotating reference frame. The index p measures the spin-projection along the direction of the local Néel order. However, more properly it is the "charge" under the emergent U(1) gauge field because Eq. (9.61) is invariant under Eq. (9.59) and

$$\psi_+ \to \psi_+ e^{-i\varsigma} \quad , \quad \psi_- \to \psi_- e^{i\varsigma}. \tag{9.62}$$

An important consequence of Eqs. (9.61) and (9.58) is the identity

$$\varphi^a c_\alpha^\dagger \sigma^a_{\alpha\beta} c_\beta = (|z_\alpha|^2)^2 \left(\psi_+^\dagger \psi_+ - \psi_-^\dagger \psi_-\right). \tag{9.63}$$

Thus the effective moment acting on the ψ fermions is always along the z axis, as expected by the transformation to a rotating reference frame.

Let us now take the continuum limit for the fermions in the rotating reference frame. We follow exactly the same mapping as in Eq. (9.11) to map the lattice ψ fermions to continuum 8-component Ψ fermions. Based upon Eq. (9.63), we also expect the Ψ fermions to experience a field polarized along the σ^z direction. Combined with gauge invariance, and the structure of Eq. (9.30), we are led to the following Lagrangian density for Ψ:

$$\mathcal{L}_\Psi = \overline{\Psi}\gamma_\mu(\partial_\mu + i\sigma^z A_\mu)\Psi - \lambda N_0 \overline{\Psi}\rho^z\sigma^z\Psi. \tag{9.64}$$

Note this is the theory of Dirac fermions of mass $|\lambda N_0|$ coupled to a U(1) gauge field. The coupling of the fermions to the U(1) gauge field in Eq. (9.64) can also be derived explicitly by substituting Eq. (9.61) into the last term in Eq. (9.30), and using the expression for the U(1) gauge field in the CP^1 model. We will present a more explicit derivation of an emergent gauge field for the case of the triangular lattice in Section 9.6 below.

The Lagrangian $\mathcal{L}_z + \mathcal{L}_\Psi$ is then our U(1) gauge theory for the fluctuating Néel state, complementary to the Gross-Neveu-Yukawa theory in Eq. (9.30). The remainder of this section is devoted to understanding the physical properties of $\mathcal{L}_z + \mathcal{L}_\Psi$.

Let us now discuss the phases of this U(1) gauge theory.

First, we have the Higgs phase, where $s < 0$ and z_α is condensed. Here the U(1) photon is gapped, and spin rotation invariance is broken. This is just the insulating Néel state, and its properties are identical to the Néel ordered state described by Eq. (9.30).

The other phase with $s > 0$ has z_α gapped and spin rotation invariance is preserved. However, as is clear from Eq. (9.64), the fermionic spectrum remains gapped. Thus this phase is clearly not the semi-metal of Eq. (9.30). Instead it is a new insulating phase with spin rotation invariance preserved. Thus we have achieved our objective of describing an insulator without Néel order.

However, this insulator is not a featureless state with a spin and a charge gap, as we will now show. The interesting physics arises from an interplay of the monopole events with the filled band of fermionic states. If we integrate out this filled band via $\mathcal{L}_\Psi$, we generate an effective Maxwell action for the U(1) gauge field

$$\mathcal{L}_A = \frac{1}{12\pi|\lambda N_0|}(\epsilon_{\mu\nu\lambda}\partial_\nu A_\lambda)^2. \tag{9.65}$$

We recall that the U(1) gauge field is compact, and it was shown by Polyakov[47] that such a gauge field always acquires a mass gap and is in a confining phase in 2+1 dimensions. The confinement is caused by the proliferaction of monopole tunneling events. Here we will show[13,43,46,48] that the monopole operator has non-trivial transformation properties under the symmetry group of the honeycomb lattice: consequently, the proliferation of monopoles in the confining phase breaks a lattice symmetry due to the appearance of valence bond solid (VBS) order.

To see this, it is useful to add an external source B_μ to the fermion Lagrangian in Eq. (9.64) so that $\mathcal{L}_\Psi$ becomes

$$\mathcal{L}_\Psi = \overline{\Psi}\gamma_\mu(\partial_\mu + i\sigma^z A_\mu)\Psi - \lambda N_0 \overline{\Psi}\rho^z\sigma^z\Psi - \frac{i}{2}B_\mu\overline{\Psi}\gamma_\mu\rho^z\Psi. \tag{9.66}$$

This source has been judiciously chosen so that when we integrate out the fermions, the action for the A_μ gauge field in Eq. (9.65) acquires a mutual

Chern-Simons term[13,48]

$$\mathcal{L}_A = \frac{1}{12\pi|\lambda N_0|}(\epsilon_{\mu\nu\lambda}\partial_\nu A_\lambda)^2 + \frac{i}{2\pi}B_\mu\epsilon_{\mu\nu\lambda}\partial_\nu A_\lambda. \tag{9.67}$$

Let us now proceed with Polyakov's duality mapping on Eq. (9.67) to obtain an effective theory of monopoles: the B_μ source term will allow us to deduce the connection between the monopole operator and the underlying lattice fermions. The first step corresponds to decoupling the Maxwell term by a Hubbard-Stratonovich field, Y_μ, to obtain

$$\mathcal{L}_A = \frac{3|\lambda N_0|}{4\pi}Y_\mu^2 + \frac{i}{2\pi}Y_\mu\epsilon_{\mu\nu\lambda}\partial_\nu A_\lambda + +\frac{i}{2\pi}B_\mu\epsilon_{\mu\nu\lambda}\partial_\nu A_\lambda. \tag{9.68}$$

Now, we integrate over A_μ, and this yields the constraint

$$Y_\mu = \partial_\mu\phi - B_\mu. \tag{9.69}$$

where ϕ is the scalar field which is dual to the photon. We have judiciously chosen factors of (2π) above to ensure a normalization so that $e^{i\phi}$ is the monopole operator. Finally, inserting Eq. (9.69) into (9.68) we obtain[13,48]

$$\mathcal{L}_\phi = \frac{3|\lambda N_0|}{4\pi}(\partial_\mu\phi - B_\mu)^2. \tag{9.70}$$

In the absence of the external source B_μ this is a free scalar field theory, which implies that the monopole operator $e^{i\phi}$ has long-range correlations in 2+1 dimensions. In other words, the free photon phase described by Eq. (9.65) has a non-zero VEV with $\langle e^{i\phi}\rangle \neq 0$.

The B_μ term in Eq. (9.70) will help us link the monopole operator to the underlying electrons.[48] First, we notice that the theory in Eq. (9.66) actually enjoys a gauge invariance under which

$$\Psi \to \exp\left(i\frac{\rho^z}{2}\theta\right)\Psi \quad , \quad B_\mu \to B_\mu - \partial_\mu\theta \tag{9.71}$$

where θ is a field with an arbitrary spacetime dependence. (Note that this gauge invariance is distinct from that associated with the A_μ gauge field in Eq. (9.62), under which $\Psi \to \exp(-i\sigma^z\zeta)\Psi$.) Now we observe that this gauge invariance extends also to Eq. (9.70), under which

$$e^{i\phi} \to e^{i\theta}e^{i\phi}. \tag{9.72}$$

The combination of Eqs. (9.71) and (9.72) now allows us to identify the operator $e^{i\phi}$. We look for a fermion bilinear of the form $\overline{\Psi}M\Psi$ so it transforms like Eq. (9.72) under the gauge transformation in Eq. (9.71). Moreover, the

Lorentz invariance of the theory implies that the matrix M should commute with the γ_μ matrices in Eq. (9.15). This leads us to the unique choice

$$e^{i\phi} \sim \overline{\Psi}\tau^y(\rho^x + i\rho^y)\Psi \sim \overline{C}\tau^y(\rho^x + i\rho^y)C. \tag{9.73}$$

It now remains to use the geometric definitions in Section 9.2.1 and Eq. (9.11) to deduce the physical interpretation of the fermion bilinear in Eq. (9.73). A careful analysis[48] along these lines shows that $e^{i\phi}$ is an operator associated with the valence bond solid (VBS) order in Fig. 9.10, and the VEVs of the operators in Eq. (9.73) imply long-range VBS order.

Exercise: Compute the transformations of the fermion bilinear in Eq. (9.73) under honeycomb lattice symmetries such as translations, reflections, and rotations by 60 degrees. All these transformations map the pattern in Fig. 9.10 either to itself or to 2 equivalents patterns. Assign the weights 1, $e^{2\pi i/3}$, and $e^{4\pi i/3}$ to these patterns, and show that their transformations coincide with those of Eq. (9.73).

Thus we reach our main conclusion: the insulating phase without Néel order as described by the U(1) gauge theory $\mathcal{L}_z + \mathcal{L}_\Psi$ has long-range VBS order. This order onsets with the confinement induced by the proliferation of monopoles.

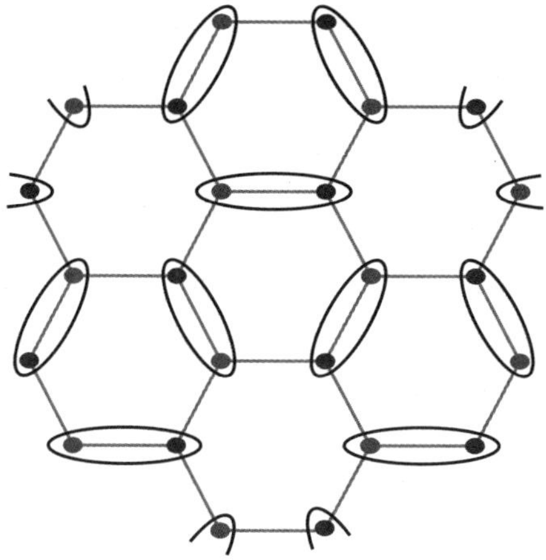

Fig. 9.10. A schematic illustration of the valence bond solid (VBS). The ellipses represent singlet valence bonds between the electrons. These reside preferentially in the pattern shown in the VBS state. Expectation values of all spin-singlet observables, such as $S_i^a S_j^a$ or $c_{i\alpha}^\dagger c_{j\alpha} + c_{j\alpha}^\dagger c_{i\alpha}$, are different on the links with the ellipses than those without.

It is interesting to note that the matrices in the fermion bilinears associated with VBS order $\sim \overline{C}\tau^y(\rho^x + i\rho^y)C$ and Néel order $\sim \overline{C}\rho^z\sigma^a C$ all anti-commute with each other, and commute with the γ_μ matrices in Eq. (9.15). This can be used to formulate the present theory without gauge fields but using a Wess-Zumino-Witten term:[48–52] we will not explore this approach here.

We can also use the methods of this section to address the nature of the transition between the Néel and VBS phases. We will not go into details here, but this transition has been proposed[45,46] to be a deconfined critical point at which monopoles are suppressed, and the critical theory is the non-compact version of the U(1) gauge theory given by $\mathcal{L}_z$.

9.5. SU(2) Gauge Theory and a Phase Diagram for the Half-filled Honeycomb Lattice

Sections 9.2 and 9.4 have so far described 3 possible phases of the honeycomb lattice at half-filling: the semi-metal, the insulator with Néel order, and the insulator with VBS order. The first two phases appear in the theory $\mathcal{L}$ in Eq. (9.30), while the latter two appear in the theory $\mathcal{L}_z + \mathcal{L}_\Psi$ in Eqs. (9.60) and (9.64). This is an unsatisfactory state of affairs: we would like to write down a single unified theory in which all 3 phases appear. One approach, implicitly mentioned above, is to extend Eq. (9.30) by including an additional two-component real scalar field representing the VBS order parameter, and couple it to the fermion bilinear $\overline{C}\tau^y(\rho^x + i\rho^y)C$ appearing in Eq. (9.73). Integrating out the fermions in the background of a spatially varying 5-component scalar representing the Néel and VBS orders yields a Wess-Zumino-Witten term for the scalar field.[48–52] The resulting theory is difficult to work with, and little is known about it in the regime where all three phases can meet.

Here we will present an alternative approach which allows for exotic phases using an emergent SU(2) gauge field. We will find that the resulting phase diagram has a fourth semi-metallic phase with an emergent topological order, and an interesting multicritical point.

Our starting point is the observation that the decomposition of the electron into spinful bosons and spinless charged fermions in Eq. (9.61) has a larger gauge invariance[40] than U(1). Rewriting eq. (9.61) using a natural matrix notation

$$c = R\psi \qquad (9.74)$$

where

$$R \equiv \begin{pmatrix} z_\uparrow & -z_\downarrow^* \\ z_\downarrow & z_\uparrow^* \end{pmatrix}, \tag{9.75}$$

we note that Eq. (9.74) is invariant under the gauge transformation generated by $\mathrm{SU}(2)_g$ matrix U under which

$$R \to R U^\dagger \quad , \quad \psi \to U \psi \quad , \quad c \to c. \tag{9.76}$$

This $\mathrm{SU}(2)_g$ gauge transformation should be distinguished from the global SU(2) spin rotation V, under which

$$R \to V R \quad , \quad \psi \to \psi \quad , \quad c \to V c. \tag{9.77}$$

Turning to the Néel order φ^a, this clearly transforms as a **3** under the global SU(2). However, the parameterization for the Néel order in Eq. (9.58) is not invariant the $\mathrm{SU}(2)_g$ gauge tranformation in Eq. (9.76). As written, Eq. (9.58) is invariant only under the U(1) gauge transformation in Eq. (9.59) which was the reason for our original choice of a U(1) gauge theory in Section 9.4. Thus we cannot use Eq. (9.58) as our definition of the Néel order in the present SU(2) gauge theory.

It is more natural to proceed here[13,40] by defining the scalar fields using bilinears of the fermions. Thus using Eq. (9.28) and extending to continuum 8-component fermions near the Fermi points, we define

$$\varphi^a = \overline{C} \rho^z \sigma^a C. \tag{9.78}$$

From this definition it is clear that φ^a transforms as a **3** under the global SU(2), while it is invariant under $\mathrm{SU}(2)_g$, just as expected.

We can also define the corresponding scalar using the ψ fermions:[40]

$$\Phi^a = \overline{\Psi} \rho^z \sigma^a \Psi. \tag{9.79}$$

Now Φ^a transforms as a **3** under the gauge $\mathrm{SU}(2)_g$, while it is invariant under SU(2).

From Eqs. (9.74), (9.78) and (9.79), we find that the scalar fields are related by

$$\begin{aligned} \varphi^a &= \frac{1}{2} \Phi^b \, \mathrm{Tr}\left(\sigma^a R \sigma^b R^\dagger\right) \\ \Phi^a &= \frac{1}{2} \varphi^b \, \mathrm{Tr}\left(\sigma^b R \sigma^a R^\dagger\right) (|z_\alpha|^2)^2. \end{aligned} \tag{9.80}$$

These relations generalize Eq. (9.58) from the U(1) gauge theory.

To summarize, the matter fields of our $\mathrm{SU}(2)_g$ gauge theory are the bosonic matrix R, the fermions Ψ, and the scalar Φ^a. As in Section 9.4, we will also need an emergent dynamic $SU(2)_g$ gauge field A^a_μ. Using symmetry and gauge invariance, we can now write down the following Lagrangian density[40] for the SU_g gauge theory; this combines and generalizes the Gross-Neveu-Yukawa model in Eq. (9.30), and the U(1) gauge theory in Eqs. (9.60) and (9.64).

$$\begin{aligned}\mathcal{L}_g &= \overline{\Psi}\gamma_\mu(\partial_\mu + i\sigma^a A^a_\mu)\Psi - \lambda\Phi^a\overline{\Psi}\rho^z\sigma^a\Psi \\ &+ \frac{1}{2}\left[\left(\partial_\mu\Phi^a - 2\epsilon_{Abc}A^b_\mu\Phi^c\right)^2 + s\,\Phi^{a2}\right] + \frac{u}{24}\left(\Phi^{a2}\right)^2 \\ &+ \mathrm{Tr}\left[(\partial_\mu R - iA^a_\mu R\sigma^a)(\partial_\mu R^\dagger + iA^a_\mu\sigma^a R^\dagger)\right] \\ &+ \widetilde{s}\,\mathrm{Tr}\left(RR^\dagger\right) + \widetilde{u}\left[\mathrm{Tr}\left(RR^\dagger\right)\right]^2 .\end{aligned} \tag{9.81}$$

This Lagrangian combines all three phases discussed so far, and forms the basis of our remaining discussion of the honeycomb lattice at half-filling.

In mean-field theory, the model $\mathcal{L}_g$ has 4 phases, depending upon whether one or both of the scalar fields Φ^a and R are condensed. These 4 phases can be identified using the methods developed in Section 9.2.4 and 9.4, and lead to the phase diagram shown in Fig. 9.11. First, we describe how $\mathcal{L}_g$ reproduces the phases and phase transitions already discussed:

- The Higgs phase where $\langle R\rangle \neq 0$ breaks $SU(2)_g$ completely. Using $SU(2)_g$ gauge invariance we may as well set $R = 1$. Then from Eq. (9.80), we have $\Phi^a \sim \varphi^a$, the Néel order parameter. Also the gauge boson A^a_μ is gapped and can be neglected. Then the theory $\mathcal{L}_g$ reduces to the Gross-Neveu-Yukawa model in Eq. (9.30). As discussed in Section 9.2.4, this theory has semi-metal and insulating Néel phases, and these are shown in Fig. 9.11.
- The Higgs phase where $\langle N^a\rangle \neq 0$ breaks $SU(2)_g$ down to U(1). Then only the A^z_μ (say) gauge boson is active, and the theory $\mathcal{L}_g$ reduces to the U(1) gauge theory $\mathcal{L}_z + \mathcal{L}_\Psi$ discussed in Section 9.4. The insulating Néel and insulating VBS phases found there are also shown in Fig. 9.11.

The possible new phase of $\mathcal{L}_g$ is the deconfined phase where both Φ^a and R are gapped. Then the low energy theory of $\mathcal{L}_g$ is simply massless QCD with the Lagrangian density

$$\mathcal{L}_{QCD} = \overline{\Psi}\gamma_\mu(\partial_\mu + i\sigma^a A^a_\mu)\Psi. \tag{9.82}$$

This QCD with a $SU(2)_g$ gauge group with massless 2-component Dirac fermions which carry $N_c = 2$ colors and $N_f = 2$ flavors. When N_f is

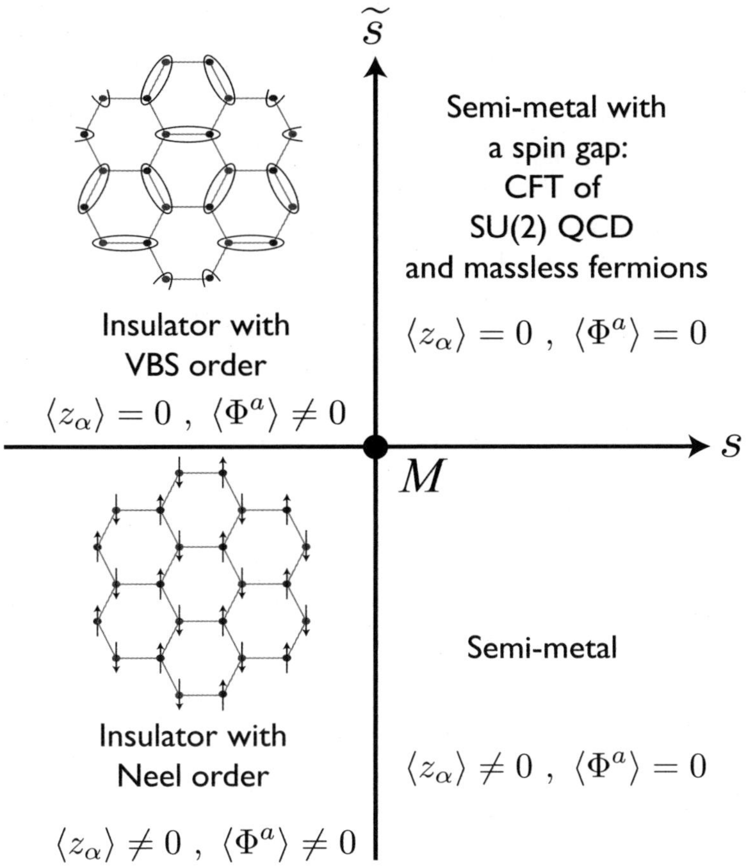

Fig. 9.11. Schematic phase diagram of the $SU(2)_g$ gauge theory $\mathcal{L}_g$ in Eq. (9.77). The two phases in the bottom are described by the Gross-Neveu-Yukawa model in Eq. (9.30), while the two phases on the left are described by the U(1) gauge theory $\mathcal{L}_z + \mathcal{L}_\Psi$ in Eqs. (9.60) and (9.64).

large enough, it can be shown from a $1/N_f$ expansion that the confining tendencies of non-Abelian gauge fields are screened, and $\mathcal{L}_{QCD}$ describes a non-trivial CFT, with anomalous dimensions for all observables which are not currents of global flavor or spacetime symmetries. It is an open question whether such a critical phase is allowed for $N_f = 2$, as we have assumed in Fig. 9.11. If not, then this phase will be unstable to confinement into one of the other phases of Fig. 9.11. If present, this deconfined phase would be a topologically ordered semi-metal with a spin gap; it is an 'algebraic charge liquid' (ACL) in the notation of Ref. 65. The gapless Ψ fermions carry electromagnetic charge, and so there is no gap to charged excitations excitations and this phase is not an insulator. However, the Ψ fermions are spinless, and SU(2) spin is only carried by the gapped bosonic excitations; hence the spin gap.

Figure 9.11 also shows an interesting multi-critical point M, where all 4 phases meet; if the massless QCD phase is confining, this would be the meeting point of 3 phases. Here the $SU(2)_g$ gauge bosons, the scalars Φ^a and R, and the fermions Ψ are all gapless and critical. Thus M realizes a non-trivial CFT which can be perturbed by the two relevant directions of the plane of Fig. 9.11. Indeed, it is not unreasonable to view this multicritical M theory as a non-supersymmetric analog of the M-theory of strings!

9.6. The Metal-Insulator Transition on the Triangular Lattice

This section will describe possible phases of the Hubbard model in Eq. (9.1) on the triangular lattice. We will now consider the case of generic density, so that unlike Sections 9.2, 9.4, and 9.5 on the honeycomb lattice we will allow $\langle n_{i\uparrow}\rangle, \langle n_{i\uparrow}\rangle \neq 1/2$, although the half-filled density will also appear in our phase diagram. Unlike the honeycomb lattice, we will ignore the possibilities of magnetically ordered phases in which the global SU(2) spin rotation symmetry is broken. The half-filled model on the triangular lattice likely does have antiferromagnetic order in the limit of large U, but we will not consider this complication here. Our purpose here is to describe the structure of possible phases without magnetic order.

The most significant difference from the honeycomb lattice is apparent in the limit of small U, when the electrons are nearly free. Then the triangular lattice ground state is a metal at all densities, unlike the semi-metal state found on the honeycomb lattice at half-filling. The semi-metal had a spectrum with a relativistic structure at low energies, a fact which we have exploited in our discussion so far. However the metal has zero energy excitations along a line in momentum space, the Fermi surface, and the fermionic excitations near the Fermi surface do not have a relativistic spectrum.

Landau's Fermi liquid (FL) theory provides a complete description of the universal properties of the metal. We will not review this theory here: the reader is referred to Chapter 18 of Ref. 19 for the author's perspective. A discussion in the context of the gauge-gravity duality appears in a recent paper.[5] For our purposes here, we need only two basic facts: (i) the fermionic excitations near the Fermi surface are essentially non-interacting electrons, and (*ii*) the area enclosed by the Fermi surface is equal to the electron density—this is Luttinger's theorem, which we state more explicitly below.

The FL metal can be described by ignoring the U interactions, and transforming Eq. (9.1) to momentum space. Unlike the honeycomb lattice, there is only one site per unit cell of the triangular lattice, and so the analog of Eq. (9.9) is now simply

$$H_0 = \sum_{\boldsymbol{k}} c_\alpha^\dagger(\boldsymbol{k}) \Big[-\mu - 2t \Big(\cos(\boldsymbol{k}\cdot\boldsymbol{e}_1) + \cos(\boldsymbol{k}\cdot\boldsymbol{e}_2) + \cos(\boldsymbol{k}\cdot\boldsymbol{e}_3) \Big) \Big] c_\alpha(\boldsymbol{k}), \quad (9.83)$$

where there are no Pauli matrices associated with sublattice index, the $\boldsymbol{e}_i$ are as in Eq. (9.4), and are shown in Fig. 9.12. The reciprocal lattice now consists of the vectors $\sum_i n_i \boldsymbol{G}_i$, where Eq. (9.5) is replaced by

$$\boldsymbol{G}_1 = \frac{4\pi}{3}(\boldsymbol{e}_1 - \boldsymbol{e}_2) \quad , \quad \boldsymbol{G}_2 = \frac{4\pi}{3}(\boldsymbol{e}_2 - \boldsymbol{e}_3) \quad , \quad \boldsymbol{G}_3 = \frac{4\pi}{3}(\boldsymbol{e}_3 - \boldsymbol{e}_1), \quad (9.84)$$

while the first Brillouin zone is a hexagon with vertices given by Eq. (9.6), as shown in Fig. 9.14. The electronic dispersion in Eq. (9.83) is plotted in Fig. 9.13: it only has simple parabolic minima at $\boldsymbol{k} = 0$, and its periodic images at $\boldsymbol{k} = \boldsymbol{G}$, and there are no Dirac points. At any chemical potential, the negative energy states are occupied, leading to a Fermi surface bounding the set of occupied states, as shown in Fig. 9.14. Luttinger theorem states

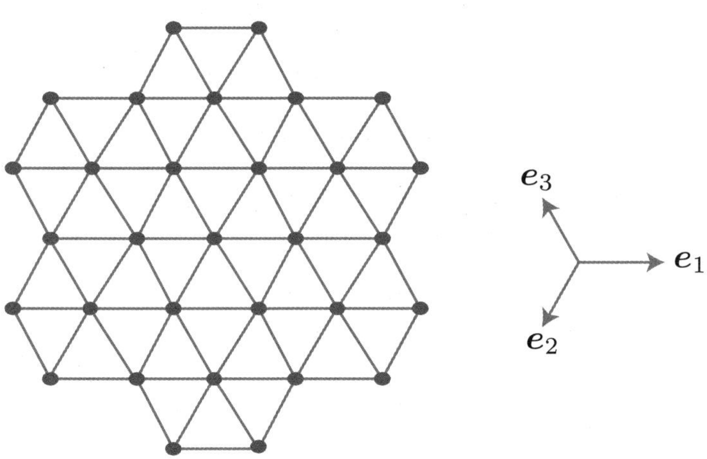

Fig. 9.12. The triangular lattice.

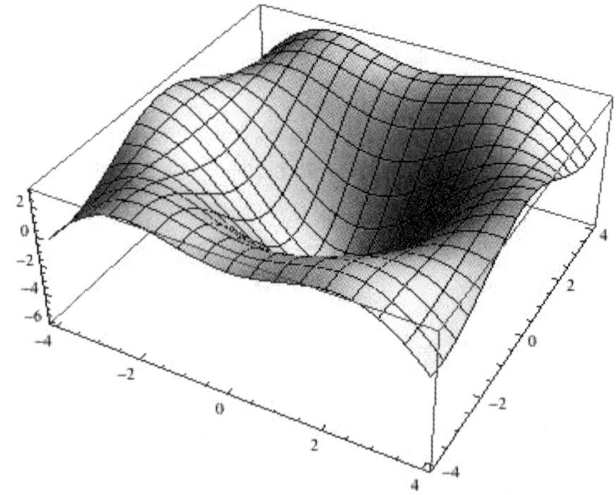

Fig. 9.13. The electronic dispersion in Eq. (9.83) for $\mu = 0$ and $t = 1$.

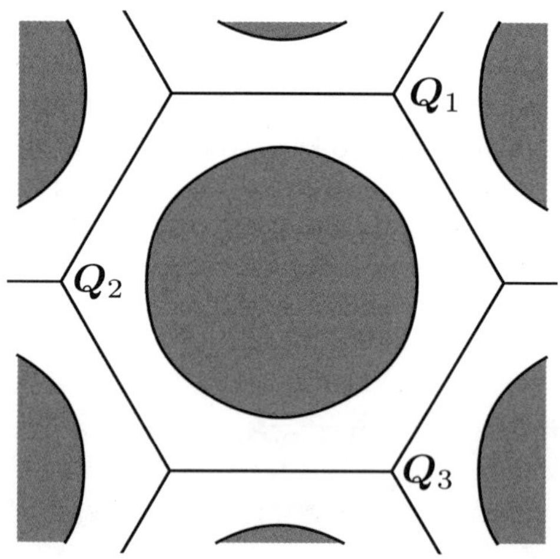

Fig. 9.14. The Fermi surface of Eq. (9.83) for $\mu = 1/2$ and $t = 1$; the occupied states are shaded. Also shown are the periodic images of the Fermi surface in their respective Brillouin zones.

that the total area of the occupied states, the shaded region of the first Brillouin zone in Fig. 9.14 occupies an area, $\mathcal{A}$, given by

$$\frac{\mathcal{A}}{2\pi^2} = \mathcal{N}, \tag{9.85}$$

where $\mathcal{N} = \sum_\alpha c_\alpha^\dagger c_\alpha$ is the total electron density. This relationship is obviously true for free electrons simply by counting occupied states, but it also remains true for interacting electrons, as reviewed recently in Ref. 5.

Now we turn up the strength of the interactions, U. For the honeycomb lattice, we presented in Sections 9.2.3 and 9.2.4 a treatment which allowed for the appearance of spontaneous magnetic moment on each site, leading to the onset of antiferromagnetic order at large U. We also found that the onset of antiferromagnetic order co-incided with appearance of the insulator *i.e.* the semi-metal to insulator transition. This co-incidence was related to the appearance of a gap in the spectrum even for an infinitesimal antiferromagnetic moment, as in Eq. (9.24). We can apply a similar treatment here to the triangular lattice. However, such an analysis finds that the onset of magnetic order does not co-incide with the metal-insulator transition. Instead, we find an intermediate metallic phase with magnetic order, in which the Fermi surface has been reconstructed into small "pockets". Such a reconstruction is observed in many correlated electron systems. However, we will not explore this route to the insulator here, and refer the reader to recent papers.[53,54]

Instead, we will explore here a distinct route to the destruction of the Fermi liquid, one which reaches directly to an insulator which is a 'spin liquid'.[55–57] The spin liquid insulator is a phase in which the spin rotation symmetry is preserved, and there is a gap to all charged excitations. In these respects, the spin liquid is similar to the insulating state discussed in Section 9.4. However, in Section 9.4 we found that such a insulator had an emergent U(1) gauge field, A_μ, and the proliferation of monopole defects in A_μ led to a confining phase in which lattice translational symmetry was broken by the appearance of VBS order. Here we will find that the triangular lattice spin liquid also has an emergent U(1) gauge field, but the presence of Fermi surfaces in the spinful excitations leads to the suppression of monopole events. Consequently, we have a deconfined phase with gapless gauge excitations, no lattice translational symmetry breaking, and the spin liquid character survives.

The key to the description of the metal insulator transition is an exact rewriting of the Hubbard model in Eq. (9.1) as a compact U(1) lattice gauge theory. We proceed with a method which parallels that in Section 9.4, of transforming to a 'rotating reference frame'. However, instead of using the frame of reference of local antiferromagnetic order, we use a quantum rotor to keep track of the charge on each lattice site. Each rotor has a periodic angular co-ordinate ϑ_i with period 2π; hence the states of the rotors are $e^{in_{ri}\vartheta_i}$ where n_{ri} is a rotor angular momentum, whose eigenvalues take all positive and negative integer values. We will use the state with all $n_{ri} = 0$ to represent the states with one electron each lattice site. The analog of the transformation to a rotating reference frame in Eq. (9.61) is now[58]

$$c_\alpha = e^{-i\vartheta} f_\alpha \tag{9.86}$$

where we have dropped the implicit site index, and f_α are neutral fermions ('spinons') which keep track of the orientation of the electron. We can now identify the 4 states on each lattice site in Eq. (9.16) with corresponding states of the rotor and spinons:

$$\begin{aligned} |0\rangle &\Leftrightarrow e^{-i\vartheta}|0\rangle \\ c_\alpha^\dagger|0\rangle &\Leftrightarrow f_\alpha^\dagger|0\rangle \\ c_\uparrow^\dagger c_\downarrow^\dagger|0\rangle &\Leftrightarrow e^{i\vartheta} f_\uparrow^\dagger f_\downarrow^\dagger|0\rangle. \end{aligned} \tag{9.87}$$

Note that these allowed states obey the constraint

$$f_\alpha^\dagger f_\alpha - n_r = 1. \tag{9.88}$$

Associated with this constraint is the U(1) gauge invariance which is the analog of Eqs. (9.59) and (9.62):

$$f_\alpha \to f_\alpha e^{i\zeta} \quad , \quad \vartheta \to \vartheta + \zeta. \tag{9.89}$$

Just as in Section 9.4, there will be an emergent gauge field A_μ in the effective theory of this model. The constraints in Eq. (9.88) will be the Gauss law of this theory.

First, let us rewrite the Hubbard model in terms of these new variables. Our degrees of freedom are the Fermi operators $f_{i\alpha}$ on each lattice site which obey the usual canonical fermion anti-commutation relations as in Eq. (9.3), and the rotor angle ϑ_i and angular momentum n_{ri} which obey

$$[\vartheta_i, n_{rj}] = i\delta_{ij}. \tag{9.90}$$

The Hubbard Hamiltonian in Eq. (9.1) is now exactly equivalent to

$$H[f, \vartheta] = -\sum_{i,j} t_{ij} f_{i\alpha}^\dagger f_{j\alpha} e^{i(\vartheta_i - \vartheta_j)} + \sum_i \left(-\mu(n_{ri} + 1) + \frac{U}{2} n_{ri}(n_{ri} + 1) \right), \tag{9.91}$$

provided our attention is restricted to the set of states which obey the constraint in Eq. (9.88) on every lattice site; note that the Hamiltonian in Eq. (9.91) commutes with constraints in (9.88), and so these can be consistently imposed. In Eq. (9.91) we have used the rotor angular momentum to measure the charge on each site, and so the dependence of the energy on μ and U can be expressed in terms of n_{ri} alone.

We can now implement the commutation relations, the Hamiltonian, and the constraint in a coherent state path integral

$$\mathcal{Z} = \int \mathcal{D}f_{i\alpha}(\tau)\mathcal{D}f_{i\alpha}^{\dagger}(\tau)\mathcal{D}\vartheta_i(\tau)\mathcal{D}n_{ri}(\tau)\mathcal{D}\lambda_i(\tau)\exp\Bigg(-\int d\tau\, H[f,\vartheta] - \int d\tau \sum_i \left[f_{i\alpha}^{\dagger}\frac{\partial f_{i\alpha}}{\partial\tau} - in_{ri}\frac{\partial\vartheta_i}{\partial\tau} + i\lambda_i(f_{i\alpha}^{\dagger}f_{i\alpha} - n_{ri} - 1)\right]\Bigg), \quad (9.92)$$

where $\vartheta_i(\tau)$ takes values on a circle with unit radius, ensuring quantization of eigenvalues of the angular momentum n_{ri} to integer values. The constraint in Eq. (9.88) is implemented using an auxilliary field $\lambda_i(\tau)$ which acts as a Lagrange multiplier.

A key observation now is that the partition function in Eq. (9.92) is invariant under a site, i, and τ-dependent U(1) gauge transformation $\zeta_i(\tau)$ where the fields transform as in Eq. (9.89), and λ transforms as

$$\lambda_i \to \lambda_i - \frac{\partial\zeta_i}{\partial\tau}. \quad (9.93)$$

In other words, λ transforms like the temporal component of a U(1) gauge field.

How do we obtain the spatial components of the gauge field? For this, we apply the Hubbard-Stratonovich transformation of Eq. (9.26) to the t_{ij} hopping term in Eq. (9.91). For this, we introduce another auxiliary complex field $Q_{ij}(\tau)$ which lives on the links of the triangular lattice and replace the hopping term by

$$\sum_{i,j}\left(\frac{|Q_{ij}(\tau)|^2}{t_{ij}} - Q_{ij}(\tau)f_{i\alpha}^{\dagger}f_{j\alpha} - Q_{ij}^{*}(\tau)e^{i(\vartheta_i-\vartheta_j)}\right). \quad (9.94)$$

We now see from Eq. (9.89), that Q_{ij} transforms under the gauge transformation in Eq. (9.89) as

$$Q_{ij} \to Q_{ij}e^{i(\zeta_i-\zeta_j)}. \quad (9.95)$$

In other words, $\arg(Q_{ij})$ is the needed spatial component of the compact U(1) gauge field.

So far, we have apparently only succeeded in making our analysis of the Hubbard model in Eq. (9.1) more complicated. Instead of the functional integral of the single complex fermion $c_{i\alpha}$, we now have a functional integral over the complex fermions $f_{i\alpha}$, the rotor ϑ_i, and the auxilliary fields λ_i and Q_{ij}. How can this be helpful? The point, of course, is that the new variables help us access new phases and critical points which were inaccessible using the electron operators, and these phases have strong correlations which are far removed from those of weakly interacting electrons.

The utility of the new representation is predicated on the assumption that the fluctuations in the auxiliary fields Q_{ij} and λ_i are small along certain directions in parameter space. So let us proceed with this assumption, and describe the structure of the phases so obtained. We parameterize

$$Q_{ij} = \overline{Q}_{ij} e^{A_{ij}} \quad , \quad \lambda_i = -i\overline{\lambda} - A_{i\tau} \tag{9.96}$$

and ignore fluctuations in the complex numbers $\overline{Q}_{ij}$, and the real number $\overline{\lambda}$. With these definitions, it is clear from Eqs. (9.93) and (9.95) that A_{ij} and A_τ form the spatial and temporal components of a U(1) gauge field, and so must enter into all physical quantities in a gauge invariant manner. The values of $\overline{Q}_{ij}$ and $\overline{\lambda}$ are determined by a suitable saddle-point analysis of the partition function, and ensure that the constraint (9.88) is obeyed. With these assumptions, the partition function separates into separate fermionic and rotor degrees of freedom interacting via their coupling to a common U(1) gauge field $(A_{i\tau}, A_{ij})$. In the continuum limit, the gauge fields become a conventional U(1) gauge field $A_\mu = (A_\tau, \mathbf{A})$. The partition function of the gauge theory is

$$\begin{aligned}
\mathcal{Z} = \int & \mathcal{D}f_{i\alpha}(\tau)\mathcal{D}f_{i\alpha}^\dagger(\tau)\mathcal{D}\vartheta_i(\tau)\mathcal{D}n_{ri}(\tau)\mathcal{D}A_{i\tau}(\tau)\mathcal{D}A_{ij}(\tau) \\
& \exp\left(-\int d\tau\left[\mathcal{L}_f + \mathcal{L}_r + i\sum_i A_{i\tau}\right]\right) \\
\mathcal{L}_f = & \sum_i f_{i\alpha}^\dagger\left(\frac{\partial}{\partial\tau} + \overline{\lambda} - iA_{i\tau}\right) f_{i\alpha} - \sum_{ij} \overline{Q}_{ij} f_{i\alpha}^\dagger e^{iA_{ij}} f_{j\alpha} \\
\mathcal{L}_r = & -i\sum_i n_{ri}\left(\frac{\partial\vartheta_i}{\partial\tau} - A_{i\tau}\right) - \sum_{ij} \overline{Q}_{ij}^* e^{i(\vartheta_i - \vartheta_j - A_{ij})} \\
& + \sum_i \left(-\overline{\lambda} n_{ri} - \mu(n_{ri}+1) + \frac{U}{2} n_{ri}(n_{ri}+1)\right).
\end{aligned} \tag{9.97}$$

Thus we have fermions $f_{i\alpha}$ moving in a band structure which is roughly the same as that of the electrons in Eq. (9.83), rotors which obey a boson Hubbard-like Hamiltonian, and both are minimally coupled to a compact U(1) gauge field.

We begin by neglecting the gauge fields, and computing the separate phase diagrams of $\mathcal{L}_f$ and $\mathcal{L}_r$.

The fermions are free, and so occupy the negative energy states determined by the chemical potential $\overline{\lambda}$.

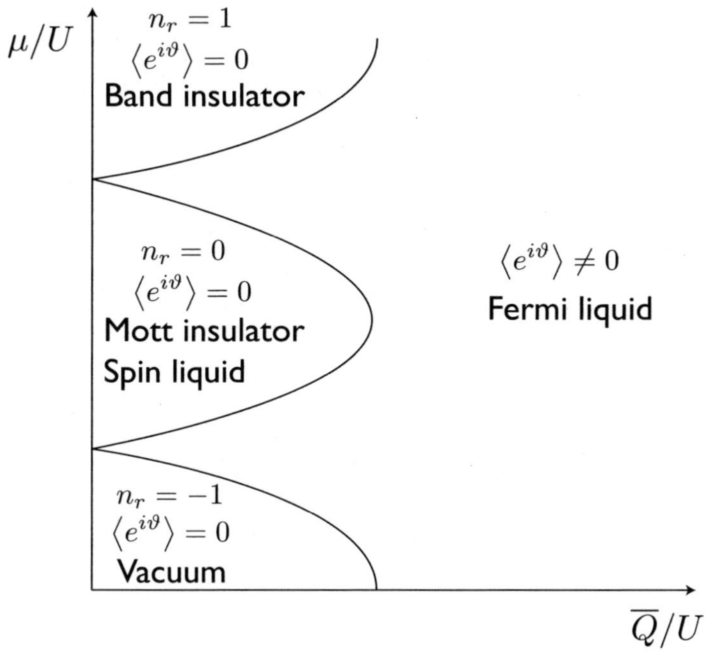

Fig. 9.15. Possible phase diagram of the electron Hubbard model in Eq. (9.1) on the triangular lattice. This phase diagram is obtained by a mean-field analysis of the theory $\mathcal{L}_r$ in Eq. (9.97), similar to that for the boson Hubbard model in Chapter 9 of Ref. 19. Only the Mott insulating lobes with $n_r = -1, 0, 1$ are compatible with the constraint in Eq. (9.88); these Mott insulating lobes have fermion density $\left\langle f_\alpha^\dagger f_\alpha \right\rangle = n_r + 1$.

The phase diagram of $\mathcal{L}_r$ is more interesting: it involves strong interactions between the rotors. It can be a analyzed in a manner similar to that of the boson Hubbard model (see Chapter 9 of Ref. 19), leading to the familiar "Mott lobe" structure shown in Fig. 9.15.

At large values of $\overline{Q}/U$ we have the analog of the superfluid states of the boson Hubbard model, in which there is a condensate of the rotor ladder operator $e^{i\vartheta}$. However, this operator is charged under the U(1) gauge field, and so this phase does not break any global symmetries. Instead it is a Higgs phase, like the Néel phase in the model of Section 9.4. In the presence of the Higgs condensate, the operator relation in Eq. (9.86) implies that $c_\alpha \sim f_\alpha$, and so the f_α fermions carry the same quantum numbers as the physical electron. Consequently, the f_α Fermi surface is simply an electron Fermi surface. Furthermore, the Higgs condensate quenches the A_μ fluctuations, and so there are no singular interactions between the Fermi surface excitations. This identifies the present phase as the familiar Fermi liquid, as identified in Fig. 9.15.

Having reproduced a previously known phase of the Hubbard model in the U(1) gauge theory, let us now examine the new phases within the 'Mott lobes' of Fig. 9.15. In these states, the rotor excitations are gapped, and the rotor angular momentum has integer expectation values. The constraint in Eq. (9.88) implies that only $n_r = -1, 0, 1$ are acceptable values, and so only these values are shown. It is clear from the representation in Eq. (9.87) that any excitation involving change in electron number must involve a rotor excitation, and so the rotor gap implies a gap in excitations carrying non-zero electron number. This identifies the present phases as insulators. Thus the phase boundary out of the lobes in Fig. 9.15 is a metal-insulator transition.

Section 9.7.2 will present an explicit demonstration of the insulating and metallic properties of the phases in Fig. 9.15 by a computation of the transport properties of a broader class of models.

The three insulators in Fig. 9.15 have very different physical characteristics.

Using the constraint in Eq. (9.88) we see that the $n_r = -1$ insulator has no f_α fermions. Consequently this is just the trivial empty state of the Hubbard model, with no electrons.

Similarly, we see that the $n_r = 1$ insulator has 2 f_α fermions on each site. This is the just the fully-filled state of the Hubbard model, with all electronic states occupied. It is a band insulator.

Finally, we turn to the most interesting insulator with $n_r = 0$. Now the electronic states are half-filled, with $\langle f_\alpha^\dagger f_\alpha \rangle = 1$. Thus there is an unpaired fermion on each site, and its spin is free to fluctuate. There is a non-trivial wavefunction in the spin sector, realizing an insulator which is a 'spin liquid'. In our present mean field theory, the spin wavefunction is specified by Fermi surface state of the f_α fermions. Going beyond mean-field theory, we have to consider the fluctuations of the A_μ gauge field, and determine if they destabilize the spin liquid, as we did in earlier Section 9.4. Here the f_α fermions carry the A_μ gauge charge, and these fermions form a Fermi surface. This is a crucial difference from Section 9.4, where the $\psi_\pm$ fermions were gapped. In Section 9.4 we found that the monopoles proliferated, leading to confinement and VBS order. Here, the gapless fermionic excitations at the Fermi surface prevent the proliferation of monopoles: the low energy fermions suppress the tunneling event associated with global change in A_μ flux.[59,60] Thus the emergent U(1) gauge field remains in a deconfined phase, and this spin liquid state is stable. These gapless gauge excitations have strong interactions with the f_α fermions, and this leads to strong critical damping of the fermions at the Fermi surface which is described by a strongly-coupled field theory.[61–63] The effect of the gauge fluctuations is also often expressed in terms of an improved trial wavefunction for the spin liquid:[55] we take the free fermion state of the f_α fermions, and apply a projection operator which removes all components which violate the constraint in Eq. (9.88). This yields the 'Gutzwiller projected' state

$$|\text{spin liquid}\rangle = \left(\prod_i \left[\frac{1-(-1)^{\sum_\alpha f_{i\alpha}^\dagger f_{i\alpha}}}{2}\right]\right)\left(\prod_{\boldsymbol{k}<k_F} f_\uparrow^\dagger(\boldsymbol{k}) f_\downarrow^\dagger(\boldsymbol{k})\right)|0\rangle, \tag{9.98}$$

where the product over $\boldsymbol{k}$ is over all points inside the Fermi surface.

Finally, we turn to an interesting quantum phase transition in Fig. 9.15. This is the transition between the spin liquid and the Fermi liquid at total electron density $\mathcal{N} = 1$, which occurs at the tip of the $n_r = 0$ Mott lobe. From the rotor sector, this looks like a Higgs transition, of the condensation of a complex scalar in the presence of a fluctuating U(1) gauge field. However, the fermionic sector is crucial in determining the nature of this transition. Indeed, in the absence of the Fermi surface, this transition would not even exist beyond mean field theory: this is because the U(1) gauge field is compact, and the scalar carries unit charge, and so the confining and Higgs phases of this gauge theory are smoothly connected. So we have to combine the Higgs theory of a complex scalar with the gapless Fermi surface excitations. Introducing a Bose field

$$b = e^{-i\vartheta}, \tag{9.99}$$

and coarse-graining to a continuum limit for the bosons, we find the field theory[57]

$$\begin{aligned}\mathcal{L} &= |(\partial_\mu + iA_\mu)b|^2 + s|b|^2 + u|b|^4 + iA_\tau\mathcal{N} \\ &\quad + f_\alpha^\dagger\left[\frac{\partial}{\partial\tau} + \epsilon_f - iA_\tau - \frac{1}{2m_f}(\boldsymbol{\nabla} - i\boldsymbol{A})^2\right]f_\alpha,\end{aligned} \tag{9.100}$$

where the energy ϵ_f is to be adjusted to yield total fermion density $\mathcal{N} = 1$. The transition is accessed by tuning s, and we move from a spin liquid for $s < s_c$, to a Fermi liquid for $s > s_c$. The critical properties of the theory in Eq. (9.100) have been studied,[57,64] and an interesting result is obtained: the Fermi surface excitations damp the gauge bosons so that they become ineffective in coupling to the critical b fluctuations. Consequently, the gauge bosons can be ignored in the b fluctuations, and the transition is in the universality class of the 2+1 dimensional XY model.

9.7. Fractionalized Fermi Liquids

In Section 9.6 we met the canonical description of a compressible metallic state, the Fermi liquid. The is the state adiabatically connected to the metallic state of non-interacting electrons. It has long-lived fermionic quasiparticle excitations along the Fermi surface, and the area enclosed by this Fermi surfaces obeys the Luttinger theorem.

Here we explore an extended model which allows for other compressible phases of strongly interacting electrons at generic densities which do not break any global symmetries, and which are not adiabatically connected to the limit of non-interacting electrons. We shall focus here on the compressible state[2,3] known as the fractionalized Fermi liquid (FL*). In principle, the FL* state can appear in a variety of models of correlated electrons, including ones with a single band, and all the sites equivalent with $U_i = U$. Such single-band FL* states have been described in recent work.[65–71] However, these single-band analyses are involved, and require intermediate steps which make them sub-optimal for a first description of the FL* state.

Instead, we will introduce the FL* state in a model with 2 types of inequivalent sites. As a simple example, consider the Hubbard model on a bilayer triangular lattice shown in Fig. 9.16. We label the two layers as A and B, and so there are 2 electron operators, $c_{Ai\alpha}$ and $c_{Bi\alpha}$. We write the Hamiltonian as

$$\begin{aligned}
H &= H_A + H_B + H_{AB} \\
H_A &= -t_A \sum_{\langle ij \rangle} c^\dagger_{Ai\alpha} c_{Aj\alpha} + \text{H.c.} + (\epsilon_A - \mu) \sum_i (n_{Ai\uparrow} + n_{Ai\downarrow}) \\
&\quad + U_A \sum_i \left(n_{Ai\uparrow} - \frac{1}{2} \right) \left(n_{Ai\downarrow} - \frac{1}{2} \right) \\
H_B &= -t_B \sum_{\langle ij \rangle} c^\dagger_{Bi\alpha} c_{Bj\alpha} + \text{H.c.} + (\epsilon_B - \mu) \sum_i (n_{Bi\uparrow} + n_{Bi\downarrow}) \\
&\quad + U_B \sum_i \left(n_{Bi\uparrow} - \frac{1}{2} \right) \left(n_{Bi\downarrow} - \frac{1}{2} \right) \\
H_{AB} &= -w \sum_i c^\dagger_{Ai\alpha} c_{Bi\alpha} + \text{H.c.}
\end{aligned} \tag{9.101}$$

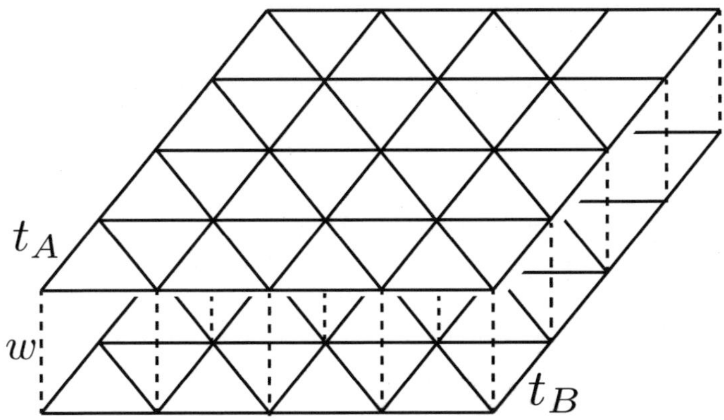

Fig. 9.16. The bilayer triangular lattice. The top layer (A) has nearest neighbor hopping t_A, the bottom layer (B) and nearest neighbor hopping t_B, and the inter-layer hopping is w. A closely related model is realized in the experiments of Ref. 6.

Here the sites i, j lie on a triangular lattice, and $\langle ij \rangle$ represents the sum over nearest-neighbor pairs. The Hubbard models on the two layers have distinct values of the hopping parameters, on-site repulsion, and on-site energies $\epsilon_{A,B}$. Finally, there is an on-site interlayer tunneling, w. Experiments[6] on bilayer films of ^{3}He adsorbed on graphite provide a remarkable realization of a closely related model.

First, let us discuss the FL state, where $U_{A,B}$ can be treated perturbatively. Diagonalizing the one-electron Hamiltonian, we find two bands corresponding to the bonding and anti-bonding states between the two layers. Let $\mathcal{N}$ be the total density of electrons for each bilayer site of the triangular lattice. So

$$\sum_{\alpha} \left(\left\langle c^{\dagger}_{A\alpha} c_{A\alpha} \right\rangle + \left\langle c^{\dagger}_{B\alpha} c_{B\alpha} \right\rangle \right) = \mathcal{N}. \tag{9.102}$$

This relation holds for every site i, and the site-index has been left implicit. Depending upon the value of $\mathcal{N}$ and interlayer tunneling w, one or both of the bands will be occupied, leading to one or two Fermi surfaces. Let the areas enclosed by the Fermi surfaces be $\mathcal{A}_1$ and $\mathcal{A}_2$; if there is only one Fermi surface, $\mathcal{A}_2 = 0$. Luttinger's theorem fixes the areas of Fermi surfaces to a value which is independent of the nature of the electron-electron interactions. There is one Luttinger theorem for each global U(1) symmetry of the Hamiltonian which is not spontaneously broken in the ground state.[5,72,73] Here, the total numbers of both up-spin and down-spin electrons are separately conserved, and so there are 2 Luttinger constraints. However, we will implicitly only consider states in which spin rotation invariance is preserved, and so there is only a single constraint. The constraint is the same as that for non-interacting electrons, which, as in Eq. (9.85), is

$$\frac{\mathcal{A}_1 + \mathcal{A}_2}{2\pi^2} = \mathcal{N}. \tag{9.103}$$

We will implicitly assume $\mathcal{N} > 1$ below.

We now wish to induce a quantum phase transition to a FL* state. This is most easily done in a model in which the bottom layer B has a density of one electron per site, while the top layer A remains dilute (as in the experiment in Ref. 6). For small hopping this is achieved for $U_A = U_B = U$ and

$$\epsilon_B < \epsilon_A < \epsilon_B + U\,. \tag{9.104}$$

Then the bottom layer B will acquire strong electronic correlations like those in Section 9.6, while the dilute gas on layer A can be treated perturbatively in the two-particle scattering amplitude. It is customary at this point to follow the analysis of Section 9.2.3, and project onto this restricted Hilbert space, while using a canonical transformation to derive an effective Hamiltonian. The restricted space has only spin degrees of freedom on B lattice sites, and as in Section 9.2.3, these spins have exchange interactions with each other. The canonical transformation also generates exchange interactions between electrons separate layers, and this is known as the Kondo exchange interaction. The resulting Hamiltonian is the Kondo-Heisenberg model. However, we will not take this step here, and continue to work with the Hubbard model in Eq. (9.101).

We will assume that layer B, with a density of one electron per site, realizes the spin liquid state discussed in Section 9.6; *i.e.* it is the $n_r = 0$ spin liquid in Fig. 9.15, with a spinon Fermi surface. We can obtain a description of this spin liquid by applying the analysis of Section 9.6 to layer. So we replace Eq. (9.86) by

$$c_{B\alpha} = e^{-i\vartheta} f_\alpha, \tag{9.105}$$

and perform the same transformations which led to Eq. (9.97). Then we take the same continuum limit as that used for Eq. (9.100), and obtain the following continuum Lagrangian $\mathcal{L}$ which captures the low energy physics of the Hubbard model in Eq. (9.101). The degrees of freedom are the A layer electrons $c_{A\alpha}$, the B layer spinons f_α, and the bosonic rotors $b = e^{-i\vartheta}$

as in Eq. (9.99). The structure of the terms also follows from general considerations of gauge invariance and the preservation of global symmetries.

$$\begin{aligned}
\mathcal{L} &= \mathcal{L}_f + \mathcal{L}_b + \mathcal{L}_c + iA_\tau \mathcal{N}_B \\
\mathcal{L}_f &= f_\alpha^\dagger \left[\frac{\partial}{\partial\tau} + \epsilon_f - iA_\tau - \frac{1}{2m_f}(\boldsymbol{\nabla} - i\boldsymbol{A})^2 \right] f_\alpha \\
\mathcal{L}_b &= \left[(\partial_\mu - (\epsilon_r - \mu)\delta_{\mu\tau} - iA_\mu + iA_{\text{ext},\mu})\, b^\dagger \right] \\
&\quad \times \left[(\partial_\mu + (\epsilon_r - \mu)\delta_{\mu\tau} + iA_\mu - iA_{\text{ext},\mu})\, b \right] + s|b|^2 + u|b|^4 \\
\mathcal{L}_c &= c_{A\alpha}^\dagger \left[\frac{\partial}{\partial\tau} - \mu - iA_{\text{ext},\tau} - \frac{1}{2m_c}(\boldsymbol{\nabla} - i\boldsymbol{A}_{\text{ext}})^2 \right] c_{A\alpha} \\
&\quad - w\left(c_{A\alpha}^\dagger b f_\alpha + b^\dagger f_\alpha^\dagger c_{A\alpha} \right). \qquad (9.106)
\end{aligned}$$

Here $A_\mu = (A_\tau, \boldsymbol{A})$ is an emergent U(1) gauge field; we have also introduced a non-fluctuating electromagnetic gauge field $A_{\text{ext},\mu}$ as a source term which couples to the current of the globally conserved electromagnetic charge; we have coarse-grained b to a complex scalar field with both amplitude and phase fluctuations; the symbol μ refers separately to the chemical potential and spacetime component, and the interpretation should be clear from the context; the final Yukawa term is allowed by the symmetries, and represents the inter-layer tunneling w; the on-site energies ϵ_f and ϵ_r are related to ϵ_A and ϵ_B and have to be tuned so that the system obeys the density constraints to be discussed below.

To review, the continuum theory in Eq. (9.106) has a $\text{U}(1)\times\text{U}(1)_{\text{ext}}$ symmetry associated with the transformations

$$\begin{aligned}
f_\alpha \to f_\alpha e^{i\zeta} \quad &, \quad b \to b^{-i\zeta} \quad , \quad c_{A\alpha} \to c_{A\alpha} \\
f_\alpha \to f_\alpha \quad &, \quad b \to b^{i\tilde{\zeta}} \quad , \quad c_{A\alpha} \to c_{A\alpha} e^{i\tilde{\zeta}}. \qquad (9.107)
\end{aligned}$$

The first U(1) symmetry is gauged by the dynamical emergent U(1) gauge field A_μ, and is the same as that in Eq. (9.89). The second U(1) symmetry remains global; the fixed external electromagnetic field $A_{\text{ext},\mu}$ couples a source term which gauges this global symmetry.

In general, there will be 2 Luttinger constraints associated with these two U(1) symmetries[5,72,73] (as before, we are ignoring spin rotation symmetries here, which is assumed to be always fully preserved). The first transformation in Eq. (9.107) leads to a Luttinger constraint on the associated conserved charge density (which is the continuum analog of Eq. (9.88))

$$\sum_\alpha \left\langle f_\alpha^\dagger f_\alpha \right\rangle - \langle \mathcal{Q}_b \rangle = \frac{\mathcal{A}_1}{2\pi^2} = \mathcal{N}_B. \qquad (9.108)$$

Here $\mathcal{N}_B$ is the density of electrons on layer B in the projected Hilbert space: our present lattice derivation was for $\mathcal{N}_B = 1$, but the continuum theory in Eq. (9.106) is sensible for any value of $\mathcal{N}_B$. The operator $\mathcal{Q}_b$ is the rotor angular momentum, given by

$$\mathcal{Q}_b = -\frac{\partial \mathcal{L}_b}{\partial \mu}. \tag{9.109}$$

Thus there must be a Fermi surface enclosing area $\mathcal{A}_1$, which counts the density of f fermions minus the bosonic rotor density.

Similarly, the second transformation of Eq. (9.88) leads to the constraint

$$\sum_\alpha \left\langle c^\dagger_{A\alpha} c_{A\alpha} \right\rangle + \langle \mathcal{Q}_b \rangle = \frac{\mathcal{A}_2}{2\pi^2} = \mathcal{N} - \mathcal{N}_B. \tag{9.110}$$

Again, there is a Fermi surface enclosing area $\mathcal{A}_2$ which counts the density of c_A fermions minus the bosonic rotor density. Thus our analysis so far appears to imply that there must be at least 2 Fermi surfaces, and their areas are constrained by the two independent relations in Eqs. (9.108) and (9.110).

This last conclusion seems rather surprising from our discussion above of the FL phase. There we found only a single constraint in Eq. (9.103) for the total areas of one or more Fermi surfaces. The only possible conclusion is that the FL phase is *not* one in which the $\mathrm{U}(1)\times\mathrm{U}(1)_{\mathrm{ext}}$ symmetry of the Lagrangian $\mathcal{L}$ in Eq. (9.106) remains unbroken. Rather the FL phase is realized as a Higgs phase in which the $\mathrm{U}(1)\times\mathrm{U}(1)_{\mathrm{ext}}$ symmetry in Eq. (9.107) is broken down to a diagonal U(1). Just as in Section 9.6, this is the Higgs phase in which the boson b condenses

$$\langle b \rangle \neq 0 \text{ in the FL phase.} \tag{9.111}$$

Once the symmetry is broken in this manner, the corresponding Luttinger constraint no longer applies.[5,72,73] Only the *sum* of the constraints in Eqs. (9.108) and (9.110) applies, and this leads immediately to the defining relation in Eq. (9.103) of the FL phase. The condensation of b also quenches the emergent U(1) gauge field, so there are no gapless gauge excitations in the FL state, again as in Section 9.6.

We now see that state of the theory $\mathcal{L}$ in which the Luttinger constraints in Eqs. (9.108) and (9.110) apply separately is a new phase: this is the advertised FL* phase, in which the boson b is uncondensed[2,3]

$$\langle b \rangle = 0 \text{ in the FL* phase.} \tag{9.112}$$

The full $U(1)\times U(1)_{\rm ext}$ symmetry is preserved, and the gauge boson A_μ becomes an emergent gapless photon. The arguments for the stability of the FL* phase towards gauge fluctuations mirror those of Section 9.6 for the stability of the spinon Fermi surface in the spin liquid.

The criteria in Eqs. (9.111) and (9.112) show that the transition between the FL and FL* states is tuned by varying the coupling s in $\mathcal{L}_b$ from negative to positive values. The transition between these phases occurs at a quantum critical point where the scalar b is also critical.

9.7.1. *Connections to holographic metals*

We now connect the above generic theory of the compressible FL and FL* phases of the Hubbard model to recent studies of compressible metallic phases via the AdS/CFT correspondence. The discussion below refers to recent work from the gravity perspective; an analysis starting the canonical supersymmetric gauge theories of gauge-gravity duality may be found in Ref. 5.

A connection was made in Ref. 4 between a mean-field solution of models like the Hubbard model in Eq. (9.101) and a particular AdS realization of a holographic metal. Specifically, the bilayer Hubbard model has been solved in a limit with infinite-range hopping matrix elements between the sites (in contrast to the nearest-neighbor hopping shown in Eq. (9.101)). A detailed correspondence was found between the low energy properties of the FL* phase of such a model and the holographic theory[24,74,75] in which the low energy limit factorized to a $AdS_2\times R^d$ geometry (d is the dimensionality of space). This work has been recently reviewed in the companion article.[1]

However, the mean-field solution of Eq. (9.101) and the $AdS_2\times R^d$ geometry share a number of artifacts: they have a non-zero ground state entropy, and the spin correlations of layer B scale with dynamic exponent $z=\infty$. These artifacts are not expected to be properties of the field theory $\mathcal{L}$ in Eq. (9.106), applicable for models with short-range interactions.

It is clearly of interest to move beyond the $AdS_2\times R^d$ factorization in the holographic theory, and derive a holographic model which has a closer correspondence with the phases of the field theory in Eq. (9.106). A number of recent theories[75–84] have examined the feedback of the finite density matter on the metric of the AdS space, and found that the AdS_2 horizon disappears at $T=0$, and is replaced by a metric with a finite value of z. Many physical properties of such holographic metals are similar to those of the field theory in Eq. (9.106), but a detailed correspondence awaits future work.

It is useful to consider these recent works in the context of a holographic RG.[74,75,85–88] In these works, the UV degrees of freedom are coupled to external sources, which in our case is $A_{\text{ext},\mu}$. Then an effective action is derived which couples the external sources to the IR degrees of freedom. Two distinct fixed-point theories have been considered in the literature: those of Nickel and Son,[85] and of Faulkner *et al.*[74] and Faulkner and Polchinski.[75] We argue here that these fixed points capture the physics of the FL and FL* phases respectively.

Let us consider, first, the theory of Nickel and Son.[85] They argued that the low energy theory had an emergent U(1) gauge field, so that the theory had a $\text{U}(1)_{\text{global}} \times \text{U}(1)_{\text{gauge}}$ symmetry. This is strikingly similar to the $\text{U}(1) \times \text{U}(1)_{\text{ext}}$ symmetry of Eq. (9.106). Indeed, we can more closely map the low energy theory of the FL phase of Eq. (9.106) to the model proposed by Nickel and Son. In the FL phase, we condense the b boson, and focus on the fluctuations of its phase $b = e^{-i\vartheta}$. Then the effective theory of the FL phase of Eq. (9.106) is

$$\mathcal{L}_{FL} = K_1 \left(\partial_\tau \vartheta - A_\tau + A_{\text{ext},\tau}\right)^2 + K_2 \left(\boldsymbol{\nabla} \vartheta - \boldsymbol{A} + \boldsymbol{A}_{\text{ext}}\right)^2 + \Pi_f(A_\mu) + \mathcal{L}_c \,, \tag{9.113}$$

where Π_f is the effective action obtain after integrating out the f spinons. The structure of Eq. (9.113) is essentially identical to Eqs. (6) and (52) of Nickel and Son.[85]

Consider, next, the corresponding low-energy theory of FL* phase. Now the b field is not condensed, and has an energy gap, Δ. So we can safely integrate it out from Eq. (9.106), and obtain an effective theory for the electrons, the spinons, and the gauge fields:

$$\begin{aligned}\mathcal{L}_{FL*} = \mathcal{L}_f &+ J_1 \left(c^\dagger_{A\alpha} \sigma^a_{\alpha\beta} c_{A\beta}\right) \left(f^\dagger_\gamma \sigma^a_{\gamma\delta} f_\delta\right) + J_2 \left(c^\dagger_{A\alpha} c_{A\alpha}\right) \left(f^\dagger_\gamma f_\gamma\right) + \mathcal{L}_c \\ &+ K_3 \left[\boldsymbol{\nabla}(A_\tau - A_{\text{ext},\tau}) - \partial_\tau(\boldsymbol{A} - \boldsymbol{A}_{\text{ext}})\right]^2 \\ &+ K_4 \left[\boldsymbol{\nabla} \times (\boldsymbol{A} - \boldsymbol{A}_{\text{ext}})\right]^2 ,\end{aligned} \tag{9.114}$$

where $J_1 \sim J_2 \sim w^2/\Delta$. The coupling J_1 is the Kondo exchange between the electrons in layer A and the spins on layer B, while J_2 couples density fluctuations of the two layers. A key property of the FL* phase is that the couplings $J_{1,2}$ can be treated perturbatively: there is no flow to strong coupling in the Kondo exchange, and the layer B spins are not screened by the conduction electrons. Let us now rewrite the matter component of Eq. (9.114) as

$$\mathcal{L}_{FL*} = \mathcal{L}_f - \frac{1}{2} \left[F^\dagger_\alpha c_{A\alpha} + c^\dagger_{A\alpha} F_\alpha\right] + \mathcal{L}_c + \ldots \tag{9.115}$$

where F_α is a IR fermion defined by[4]

$$F_\alpha \equiv -J_1 \left(\sigma^a_{\alpha\beta} f^\dagger_\gamma \sigma^a_{\gamma\delta} f_\delta\right) c_{A\beta} - J_2 \left(f^\dagger_\gamma f_\gamma\right) c_{A\alpha}. \tag{9.116}$$

Notice that both fermions in the displayed term in Eq. (9.115) are invariant under the emergent U(1); this term is a coupling between the microscopic fermion $c_{A\alpha}$ and a composite gauge-invariant fermion operator F_α representing the IR degrees of freedom. We can view F_α in Eq. (9.116) as the most general fermion operator which involves the IR fermions f_α, which is invariant under the gauge transformation associated with A_μ, and which also carries the global electron number charge associated with $A_{\text{ext},\mu}$. Then structure of Eq. (9.115) is precisely that of the semi-holographic theory of Faulkner *et al.*[74] and Faulkner and Polchinski,[75] and their IR fermion F_α is chosen by essentially identical criteria.

It would clearly be of interest to also find another fixed point of the holographic theory corresponding to the quantum-critical points between the FL and FL* phases.

9.7.2. *Transport theory*

We conclude our discussion of FL and FL* phases by presenting a general formulation of their transport properties. The arguments below are in the spirit of those of Ioffe and Larkin.[89]

We begin with a theory like $\mathcal{L}$ in Eq. (9.106), and integrate out the matter fields to obtain a Coleman-Weinberg effective action for the U(1)×U(1) gauge fields A_μ and $A_{\text{ext},\mu}$. In general, the form of this effective action is constrained only spatial isotropy and gauge invariance. Using the projectors defined in Eq. (9.51), we can write the quadratic portion of the effective action in the following form (we work here in Euclidean time, and ω_n is a Matsubara frequency)

$$\begin{aligned}\mathcal{S} = \frac{1}{2} T \sum_{\omega_n} \int \frac{d^2k}{4\pi^2} \Bigg[& A_\mu \Big(P^L_{\mu\nu} K^L_f(\omega_n, k) + P^T_{\mu\nu} K^T_f(\omega_n, k)\Big) A_\nu \\ & + (A_\mu - A_{\text{ext},\mu}) \Big(P^L_{\mu\nu} K^L_b(\omega_n, k) + P^T_{\mu\nu} K^T_b(\omega_n, k)\Big) (A_\mu - A_{\text{ext},\mu}) \\ & + A_{\text{ext},\mu} \Big(P^L_{\mu\nu} K^L_c(\omega_n, k) + P^T_{\mu\nu} K^T_c(\omega_n, k)\Big) A_{\text{ext},\mu} \Bigg].\end{aligned} \tag{9.117}$$

Here $K^{L,T}_f$ are given by the correlator of the current of the f fermions, $K^{L,T}_c$ by the correlator of the current of the c_A fermions, and $K^{L,T}_b$ by the current of the bosonic b rotors. Note that, unlike Eq. (9.50), we have not pulled out a factor of $\sqrt{k^2+\omega_n^2}$ in the definition of the $K^{L,T}$. In general, determining these functions requires a complex transport analysis of the theory in Eq. (9.106). However, in the FL and FL* phases, the simpler low energy effective theories in Eqs. (9.113) and (9.114) lead to simple forms for the bosonic correlators $K_b^{L,T}$.

In the FL phase, integrating out the phase ϑ in Eq. (9.113) we obtain

$$K_b^L(\omega_n, k) = \frac{K_1 K_2 (k^2+\omega_n^2)}{K_2 k^2 + K_1 \omega_n^2} \quad , \quad K_b^T(\omega_n, k) = K_2. \tag{9.118}$$

Thus $K_b^{L,T}$ are constants in the limits of small momenta or frequency. Indeed, had we chosen the velocity of 'light' judiciously in the definition of $P_{\mu\nu}^L$ in Eq. (9.51), we would have had $K_b^L(\omega_n) = K_1$.

In contrast, in the FL* phase, we can directly match the low energy theory in Eq. (9.114) to Eq. (9.117) and obtain

$$K_b^L(\omega_n, k) = K_3(k^2+\omega_n^2) \quad , \quad K_b^T(\omega_n, k) = K_3\omega_n^2 + K_4 k^2. \tag{9.119}$$

Now the $K_b^{L,T}$ vanish in the limit of small momentum and frequency.

We need to use the respective low energy theories of the FL and FL* phases in Eqs. (9.113) and (9.114) to determine $K_c^{L,T}$ and $K_f^{L,T}$, and then combine them with the above results for $K_b^{L,T}$ to obtain the physical conductivity. As in Nickel and Son,[85] and in Ioffe and Larkin,[89] this is obtained by implementing the equation of motion of the emergent gauge field A_μ in Eq. (9.117). This equation of motion is equivalent to the constraint that the current of the b bosons must equal the current of the f fermions, which is a consequence of the lattice constraint in Eq. (9.88). Evaluating the equation of motion $\delta\mathcal{S}/\delta A_\mu = 0$ from Eq. (9.117), and substituting the resulting value of A_μ back (after suitable gauge fixing), we obtain an effective action for the probe field $A_{\text{ext},\mu}$ alone

$$\mathcal{S}_{\text{ext}} = \frac{1}{2} T \sum_{\omega_n} \int \frac{d^2k}{4\pi^2} A_{\text{ext},\mu} \Big(P_{\mu\nu}^L K_{\text{ext}}^L(\omega_n, k) + P_{\mu\nu}^T K_{\text{ext}}^T(\omega_n, k) \Big) A_{\text{ext},\mu} \tag{9.120}$$

with

$$\begin{aligned} K_{\text{ext}}^L(\omega_n, k) &= K_c^L(\omega_n, k) + \frac{K_f^L(\omega_n, k) K_b^L(\omega_n, k)}{K_f^L(\omega_n, k) + K_b^L(\omega_n, k)} \\ K_{\text{ext}}^T(\omega_n, k) &= K_c^T(\omega_n, k) + \frac{K_f^T(\omega_n, k) K_b^T(\omega_n, k)}{K_f^T(\omega_n, k) + K_b^T(\omega_n, k)}. \end{aligned} \tag{9.121}$$

After analytic continuation to Minkowski space, these results lead to the physical conductivity via the Kubo formula in Eq. (9.36)

$$\sigma(\omega) = \frac{i}{\omega} K^L_{\text{ext}}(\omega, 0). \tag{9.122}$$

The distinction from Eq. (9.53) is a consequence of omitting in Eq. (9.117) the prefactor $\sqrt{\omega_n^2 + k^2}$ present in Eq. (9.50).

These expressions can be used along with specific computations of the dynamics of the f and c fermions: the latter can be carried out either using a Boltzmann theory of the continuum model in Eq. (9.106), or via a theory on AdS. The analysis by Nickel and Son[85] for their holographic theory is equivalent to the application of Eq. (9.121). Let us verify that the present method yields the expected FL behavior in the Higgs phase where Eq. (9.118) implies that $K^L_b(\omega, 0) = K_2$. We assume that c and f Fermi surfaces have metallic conduction with $K^L_c(\omega, 0) = -i\omega\sigma_c$ and $K^L_f(\omega, 0) = -i\omega\sigma_f$, with $\sigma_{c,f}$ the respective conductivities. Inserting these expressions in Eq. (9.122), we obtain the expected FL behavior with $\sigma = \sigma_c + \sigma_f$ in the limit $\omega \to 0$. Thus there is no superfluidity associated with the condensation of b, and the gauge fluctuations lead eventually to metallic behavior. Similarly, it is easy to show that if b and c excitations are gapped, we have insulating transport, even though the f spinons have a gapless Fermi surface.

Further theoretical work exploring the connection between the AdS and Boltzmann approaches to transport is clearly of interest.

Acknowledgments

I am very grateful to the participants of TASI 2010 in Boulder, and of the ICTS Chandrasekhar Lecture Series and Discussion Meeting on "Strongly Correlated Systems and AdS/CFT" in Bangalore, Dec 2010. Many of the ideas presented here were developed and sharpened during discussions at these meetings. This research was supported by the National Science Foundation under grant DMR-0757145 and by a MURI grant from AFOSR.

References

1. S. Sachdev, "Strange metals and the AdS/CFT correspondence," J. Stat. Mech. **1011**, P11022 (2011) [arXiv:1010.0682 [cond-mat.str-el]].
2. T. Senthil, S. Sachdev, and M. Vojta, "Fractionalized Fermi liquids," Phys. Rev. Lett. **90**, 216403 (2003) [arXiv:cond-mat/0209144].

3. T. Senthil, M. Vojta, and S. Sachdev, "Weak magnetism and non-Fermi liquids near heavy-fermion critical points," Phys. Rev. B **69**, 035111 (2004) [arXiv:cond-mat/0305193].
4. S. Sachdev, "Holographic metals and the fractionalized Fermi liquid," Phys. Rev. Lett. **105**, 151602 (2010) [arXiv:1006.3794 [hep-th]].
5. L. Huijse and S. Sachdev, "Fermi surfaces and gauge-gravity duality," arXiv:1104.5022 [hep-th].
6. M. Neumann, J. Nyéki, B. Cowan, and J. Saunders, "Bilayer ^{3}He: A Simple Two-Dimensional Heavy-Fermion System with Quantum Criticality," Science **317**, 1356 (2007).
7. C. Cohen-Tannoudji, J. Dupont-Roc, and G. Grynberg, *Atom-Photon Interactions: Basic Processes and Applications*, Complement B_1, Wiley-VCH (1998).
8. Z. Y. Meng, T. C. Lang, S. Wessel, F. F. Assaad, and A. Muramatsu, "Quantum spin-liquid emerging in two-dimensional correlated Dirac fermions," Nature **464**, 847 (2010) [arXiv:1003.5809 [cond-mat.str-el]].
9. M. Hermele, "SU(2) gauge theory of the Hubbard model and application to the honeycomb lattice," Phys. Rev. B **76**, 035125 (2007) [arXiv:cond-mat/0701134].
10. F. Wang, "Schwinger Boson Mean Field Theories of Spin Liquid States on Honeycomb Lattice: Projective Symmetry Group Analysis and Critical Field Theory," Phys. Rev. B **82**, 024419 (2010) [arXiv:1004.2693 [cond-mat.str-el]].
11. Y. -M. Lu and Y. Ran, "Spin liquids on a honeycomb lattice: Projective Symmetry Group study of Schwinger fermion mean-field theory," arXiv:1005.4229 [cond-mat.str-el].
12. Y. -M. Lu and Y. Ran, "Z_2 spin liquid and chiral antiferromagnetic phase in Hubbard model on the honeycomb lattice: Duality between Schwinger-fermion and Schwinger-boson representations," arXiv:1007.3266 [cond-mat.str-el].
13. C. Xu and S. Sachdev, "Majorana liquids: the complete fractionalization of the electron," Phys. Rev. Lett. **105**, 057201 (2010) [arXiv:1004.5431 [cond-mat.str-el]].
14. C. Xu, "Quantum Spin Hall, triplet Superconductor, and topological liquid on the honeycomb lattice," arXiv:1010.0455 [cond-mat.str-el].
15. I. F. Herbut, "Interactions and phase transitions on graphene's honeycomb lattice," Phys. Rev. Lett. **97**, 146401 (2006) [arXiv:cond-mat/0606195].
16. I. F. Herbut, V. Juricic, and B. Roy, "Theory of interacting electrons on the honeycomb lattice," Phys. Rev. B **79**, 085116 (2009) [arXiv:0811.0610 [cond-mat.str-el]].
17. I. F. Herbut, V. Juricic, and O. Vafek, "Relativistic Mott criticality in graphene," Phys. Rev. B **80**, 075432 (2009) [arXiv:0904.1019 [cond-mat.str-el]].
18. S. Sachdev, "Quantum phase transitions of correlated electrons in two dimensions," Physica A **313**, 252 (2002) [arXiv:cond-mat/0109419].
19. S. Sachdev, *Quantum Phase Transitions*, 2nd ed., Cambridge (2011).
20. V. M. Pereira, F. Guinea, J. M. B. Lopes dos Santos, N. M. R. Peres, and

A. H. Castro Neto, "Disorder Induced Localized States in Graphene," Phys. Rev. Lett. **96**, 036801 (2006) [arXiv:cond-mat/0508530].
21. S. Sachdev, C. Buragohain, and M. Vojta, "Quantum impurity in a nearly-critical two dimensional antiferromagnet," Science **286**, 2479 (1999) [arXiv:cond-mat/0004156].
22. M. Vojta, C. Buragohain, and S. Sachdev, "Quantum impurity dynamics in two-dimensional antiferromagnets and superconductors," Phys. Rev. B **61**, 15152 (2000) [arXiv:cond-mat/9912020].
23. S. Sachdev, "Static hole in a critical antiferromagnet: field-theoretic renormalization group, S. Sachdev, Physica C **357**, 78 (2001) [arXiv:cond-mat/0011233].
24. S. Kachru, A. Karch, and S. Yaida, "Holographic Lattices, Dimers, and Glasses," Phys. Rev. **D81**, 026007 (2010) [arXiv:0909.2639 [hep-th]].
25. S. Sachdev, "Non-zero temperature transport near fractional quantum Hall critical points," Phys. Rev. B **57**, 7157 (1998) [arXiv:cond-mat/9709243].
26. K. Damle and S. Sachdev, "Non-zero temperature transport near quantum critical points," Phys. Rev. B **56**, 8714 (1997) [arXiv:cond-mat/9705206].
27. L. Fritz, J. Schmalian, M. Müller, and S. Sachdev, "Quantum critical transport in clean graphene," Phys. Rev. B **78**, 085416 (2008) [arXiv:0802.4289 [cond-mat.str-el]].
28. M. Müller, L. Fritz, and S. Sachdev, "Quantum critical relativistic magnetotransport in graphene," Phys. Rev. B **78**, 115406 (2008) [arXiv:0805.1413 [cond-mat.str-el]].
29. I. F. Herbut, V. Juricic, and O. Vafek, "Coulomb interaction, ripples, and the minimal conductivity of graphene," Phys. Rev. Lett. **100** 046403 (2008) [arXiv:0707.4171 [cond-mat.mes-hall]].
30. V. Juricic, O. Vafek, I. F. Herbut, "Conductivity of interacting massless Dirac particles in graphene: Collisionless regime," arXiv:1009.3269 [cond-mat.mes-hall].
31. L. Fritz, "Quantum-Critical transport at a semimetal-to-insulator transition on the honeycomb lattice," arXiv:1012.0263 [cond-mat.str-el].
32. C. P. Herzog, P. Kovtun, S. Sachdev and D. T. Son, "Quantum critical transport, duality, and M-theory," Phys. Rev. D **75**, 085020 (2007) [arXiv:hep-th/0701036].
33. R. C. Myers, S. Sachdev, and A. Singh, "Holographic Quantum Critical Transport without Self-Duality," Phys. Rev. D **83**, 066017 (2011) [arXiv:1010.0443 [hep-th]].
34. S. Sachdev and X. Yin, "Quantum phase transitions beyond the Landau-Ginzburg paradigm and supersymmetry," Annals Phys. **325**, 2 (2010) [arXiv:0808.0191 [cond-mat.str-el]].
35. C. Dasgupta and B. I. Halperin, "Phase Transition in a Lattice Model of Superconductivity," Phys. Rev. Lett. **47**, 1556 (1981).
36. P. Kovtun, D. T. Son and A. O. Starinets, "Viscosity in strongly interacting quantum field theories from black hole physics," Phys. Rev. Lett. **94**, 111601 (2005) [arXiv:hep-th/0405231].

37. M. Müller, J. Schmalian, and L. Fritz, "Graphene - a nearly perfect fluid," Phys. Rev. Lett. **103**, 025301 (2009) [arXiv:0903.4178 [cond-mat.str-el]].
38. B. I. Shraiman and E. D. Siggia, "Mobile Vacancies in a Quantum Heisenberg Antiferromagnet," Phys. Rev. Lett. **61**, 467 (1988).
39. H. J. Schulz, "Effective action for strongly correlated fermions from functional integrals," Phys. Rev. Lett. **65**, 2462 (1990).
40. S. Sachdev, M. A. Metlitski, Y. Qi, and Cenke Xu, "Fluctuating spin density waves in metals," Phys. Rev. B **80**, 155129 (2009) [arXiv:0907.3732 [cond-mat.str-el]].
41. A. D'Adda, P. Di Vecchia, and M. Lüscher, "A $1/n$ Expandable Series of Nonlinear Sigma Models with Instantons," Nucl. Phys. **B 146**, 63 (1978).
42. E. Witten, "Instantons, the Quark Model, and the $1/n$ Expansion," Nucl Phys. **B149**, 285 (1979).
43. N. Read and S. Sachdev, "Spin-Peierls, valence bond solid, and Neel ground states of low dimensional quantum antiferromagnets," Phys. Rev. B **42**, 4568 (1990).
44. O. I. Motrunich and A. Vishwanath, "Emergent photons and new transitions in the O(3) sigma model with hedgehog suppression," Phys. Rev. B **70**, 075104 (2004) [arXiv:cond-mat/0311222].
45. T. Senthil, A. Vishwanath, L. Balents, S. Sachdev, and M. P. A. Fisher, "Deconfined quantum critical points," Science **303**, 1490 (2004) [arXiv:cond-mat/0311326].
46. T. Senthil, L. Balents, S. Sachdev, A. Vishwanath, and M. P. A. Fisher, "Quantum criticality beyond the Landau-Ginzburg-Wilson paradigm," Phys. Rev. B **70**, 144407 (2004) [arXiv:cond-mat/0312617].
47. A. M. Polyakov, "Compact gauge fields and the infrared catastrophe," Phys. Lett. B **59**, 82 (1975).
48. L. Fu, S. Sachdev, and C. Xu, "Geometric phases and competing orders in two dimensions," Phys. Rev. B **83**, 165123 (2011) [arXiv:1010.3745 [cond-mat.str-el]].
49. A. G. Abanov and P. B. Wiegmann, "Theta terms in nonlinear sigma models," Nucl. Phys. **B570**, 685-698 (2000) [arXiv:hep-th/9911025].
50. A. Tanaka and Xiao Hu, "Many-body spin Berry phases emerging from the π-flux state: antiferromagnetic/valence-bond-solid competition," Phys. Rev. Lett. **95**, 036402 (2005) [arXiv:cond-mat/0501365].
51. T. Senthil and M. P. A. Fisher, "Competing orders, non-linear sigma models, and topological terms in quantum magnets," Phys. Rev. B **74**, 064405 (2006) [arXiv:cond-mat/0510459].
52. H. Yao and D. -H. Lee, "Topological insulators and topological nonlinear sigma models," Phys. Rev. B **82**, 245117 (2010) [arXiv:1003.2230 [cond-mat.str-el]].
53. M. A. Metlitski, and S. Sachdev, "Quantum phase transitions of metals in two spatial dimensions: II. Spin density wave order," Phys. Rev. **B82**, 075128 (2010) [arXiv:1005.1288 [cond-mat.str-el]].
54. M. A. Metlitski, and S. Sachdev, "Instabilities near the onset of spin density wave order in metals," New Journal of Physics **12**, 105007 (2010) [arXiv:1007.1968 [cond-mat.str-el]].

55. O. I. Motrunich, "Variational study of triangular lattice spin-1/2 model with ring exchanges and spin liquid state in κ-$(ET)_2Cu_2(CN)_3$," Phys. Rev. B **72**, 045105 (2005) [arXiv:cond-mat/0412556].
56. S.-S. Lee and P. A. Lee, "U(1) Gauge Theory of the Hubbard Model : Spin Liquid States and Possible Application to κ-$(BEDT\text{-}TTF)_2Cu_2(CN)_3$," Phys. Rev. Lett. **95**, 036403 (2005) [arXiv:cond-mat/0502139].
57. T. Senthil, "Theory of a continuous Mott transition in two dimensions," Phys. Rev. B **78**, 045109 (2008) [arXiv:0804.1555 [cond-mat.str-el]].
58. S. Florens and A. Georges, "Slave-rotor mean field theories of strongly correlated systems and the Mott transition in finite dimensions," Phys. Rev. B **70**, 035114 (2004) [arXiv:cond-mat/0404334].
59. M. Hermele, T. Senthil, M. P. A. Fisher, P. A. Lee, N. Nagaosa, and X.-G. Wen, "On the stability of U(1) spin liquids in two dimensions," Phys. Rev. B **70**, 214437 (2004) [arXiv:cond-mat/0404751].
60. Sung-Sik Lee, "Stability of the U(1) spin liquid with spinon Fermi surface in 2+1 dimensions," Phys. Rev. B **78**, 085129 (2008) [arXiv:0804.3800 [cond-mat.str-el]].
61. Sung-Sik Lee, "Low energy effective theory of Fermi surface coupled with U(1) gauge field in 2+1 dimensions," Phys. Rev. B **80**, 165102 (2009) [arXiv:0905.4532 [cond-mat.str-el]].
62. M. A. Metlitski, and S. Sachdev, "Quantum phase transitions of metals in two spatial dimensions: I. Ising-nematic order," Phys. Rev. **B82**, 075127 (2010) [arXiv:1001.1153 [cond-mat.str-el]].
63. D. F. Mross, J. McGreevy, H. Liu, and T. Senthil, "A controlled expansion for certain non-Fermi liquid metals," Phys. Rev. B **82**, 045121 (2010) [arXiv:1003.0894 [cond-mat.str-el]].
64. R. K. Kaul, M. A. Metlitski, S. Sachdev and C. Xu, "Destruction of Neel order in the cuprates by electron-doping," Phys. Rev. B **78**, 045110 (2008) [arXiv:0804.1794 [cond-mat.str-el]].
65. R. K. Kaul, Y. B. Kim, S. Sachdev, and T. Senthil, "Algebraic Charge Liquids," Nature Physics **4**, 28 (2008) [arXiv:0706.2187 [cond-mat.str-el]].
66. R. K. Kaul, A. Kolezhuk, M. Levin, S. Sachdev, and T. Senthil, "Hole dynamics in an antiferromagnet across a deconfined quantum critical point," Phys. Rev. B **75** , 235122 (2007) [arXiv:cond-mat/0702119].
67. Y. Qi and S. Sachdev, "Effective theory of Fermi pockets in fluctuating antiferromagnets," Phys. Rev. B **81**, 115129 (2010) [arXiv:0912.0943 [cond-mat.str-el]].
68. E. G. Moon and S. Sachdev, "The underdoped cuprates as fractionalized Fermi liquids: transition to superconductivity," arXiv:1010.4567 [cond-mat.str-el].
69. T. C. Ribeiro and X.-G. Wen, "Doped carrier formulation and mean-field theory of the t-t'-t''-J model," Phys. Rev. B **74**, 155113 (2006) [arXiv:0705.2261 [cond-mat.str-el]].
70. Ying Ran and X.-G. Wen, "Dichotomy in underdoped high T_c superconductors and spinon-dopon approach to t-t'-t''-J model," arXiv:cond-mat/0611034.

71. T. C. Ribeiro and X.-G. Wen, "Electromagnetic response of high-T_c superconductors – the slave-boson and doped-carrier theories," Phys. Rev. B **77**, 144526 (2007) [arXiv:0705.2261 [cond-mat.str-el]].
72. S. Powell, S. Sachdev, and H. P. Büchler, "Depletion of the Bose-Einstein condensate in Bose-Fermi mixtures," Phys. Rev. B **72**, 024534 (2005) [arXiv:cond-mat/0502299].
73. P. Coleman, I. Paul, and J. Rech, "Sum rules and Ward identities in the Kondo lattice," Phys. Rev. B **72**, 094430 (2005) [arXiv:cond-mat/0503001].
74. T. Faulkner, H. Liu, J. McGreevy, and D. Vegh, "Emergent quantum criticality, Fermi surfaces, and AdS(2)," arXiv:0907.2694 [hep-th].
75. T. Faulkner and J. Polchinski, "Semi-Holographic Fermi Liquids," arXiv:1001.5049 [hep-th].
76. S. A. Hartnoll, J. Polchinski, E. Silverstein, and D. Tong, "Towards strange metallic holography," JHEP **1004**, 120 (2010) [arXiv:0912.1061 [hep-th]].
77. S. S. Gubser and F. D. Rocha, Phys. Rev. D **81**, 046001 (2010).
78. C. Charmousis, B. Gouteraux, B. S. Kim, E. Kiritsis, and R. Meyer, "Effective Holographic Theories for low-temperature condensed matter systems," JHEP **1011**, 151 (2010) [arXiv:1005.4690 [hep-th]].
79. S. A. Hartnoll and A. Tavanfar, "Electron stars for holographic metallic criticality," Phys. Rev. D **83**, 046003 (2011) [arXiv:1008.2828 [hep-th]].
80. S. A. Hartnoll, D. M. Hofman, and A. Tavanfar, "Holographically smeared Fermi surface: Quantum oscillations and Luttinger count in electron stars," arXiv:1011.2502 [hep-th].
81. X. Arsiwalla, J. de Boer, K. Papadodimas, and E. Verlinde, "Degenerate Stars and Gravitational Collapse in AdS/CFT," JHEP **1101**, 144 (2011) [arXiv:1010.5784 [hep-th]].
82. K. Goldstein, S. Kachru, S. Prakash, and S. P. Trivedi, "Holography of Charged Dilaton Black Holes," JHEP **1008**, 078 (2010) [arXiv:0911.3586 [hep-th]].
83. K. Goldstein, N. Iizuka, S. Kachru, S. Prakash, S. P. Trivedi, and A. Westphal, "Holography of Dyonic Dilaton Black Branes," JHEP **1010**, 027 (2010) [arXiv:1007.2490 [hep-th]].
84. S. Kachru, A. Karch, and S. Yaida, "Adventures in Holographic Dimer Models," New J. Phys. **13**, 035004 (2011) [arXiv:1009.3268 [hep-th]].
85. D. Nickel and D. T. Son, "Deconstructing holographic liquids," arXiv:1009.3094 [hep-th].
86. A. Karch, D. T. Son, and A. O. Starinets, "Holographic Quantum Liquid," Phys. Rev. Lett. **102**, 051602 (2009) [arXiv:0806.3796 [hep-th]].
87. I. Heemskerk and J. Polchinski, "Holographic and Wilsonian Renormalization Groups," arXiv:1010.1264 [hep-th].
88. T. Faulkner, H. Liu, and M. Rangamani, "Integrating out geometry: Holographic Wilsonian RG and the membrane paradigm," arXiv:1010.4036 [hep-th].
89. L. B. Ioffe and A. I. Larkin, "Gapless fermions and gauge fields in dielectrics," Phys. Rev. B **39**, 8988 (1989).

Chapter 10

Holography for Strongly Coupled Media

Dam Thanh Son

Institute for Nuclear Theory, University of Washington, Seattle, WA 98195-1550, USA

We discuss some recent attempts to apply AdS/CFT correspondence to systems with finite temperature and chemical potential, emphasizing the hydrodynamic aspects. We also discuss the system of nonrelativistic fermions at unitarity, the Schrödinger symmetry and possible directions for constructing a holographic dual of this system. This is the write-up of lectures delivered at the TASI school of 2010.

10.1. Motivation

Many problems of modern theoretical physics are related to strong coupling. One example is the problem of the hot and dense matter in QCD. The creation of hot QCD matter is the goal of relativistic heavy ion experiments, the most recent of which are RHIC and LHC. Although there are ample evidence that some form of matter with strong collective behavior is formed in ultra-relativistic heavy ion collisions, the theoretical problem of finding whether thermal equilibrium is achieved and at which temperature has still not been solved. (The problem can be made very sharp by imagining a world with very small electromagnetic fine structure constant so that nuclei can be very large. Can we make a quark gluon plasma by colliding very large nuclei at very high energy? What is the temperature of the system at thermal equilibration? We still do not have definite answer to these questions.) Assuming that system reaches equilibrium, one can ask questions about the properties of the thermal equilibrium state. While thermodynamics of the QGP at finite temperature and zero chemical potential can be studied by lattice methods, the latter becomes very inefficient in dealing with real time quantities, for example the viscosities. Current lattice methods are also incapable of treating QCD matter at finite chemical potential, a problem that hinders our understanding of the core of neutron stars.

Another example of a strong coupling problem is that of unitarity fermions (unitary Fermi gas). This system is that of nonrelativistic fermion interacting through a short-range potential fine tuned to resonance at threshold (see Section 10.6 for more discussion). The simplest version of the problem is the Bertsch problem: given a gas of spin-1/2 fermions, interacting with short-range interaction fine tuned to unitarity (defined below in the lectures), what are the properties of the ground state? This problem has became extremely important when it became possible to realize unitarity fermions in atomic trap experiments.

Various other strong coupling problems in condensed matter physics are discussed in Subir Sachdev's lectures in this school. In these lectures, we will describe some points of contact between gauge/gravity duality and the physics of the quark gluon plasma and the unitary Fermi gas.

10.2. Thermal Field Theory

There are two main formalisms used in thermal field theory. The first formalism is the Matsubara, Euclidean formalism. It is used in lattice QCD, very convenient for thermodynamic and static quantities (like correlation length), but cannot directly address dynamic, real-time quantities. The second formalism is the real-time, close time path formalism. (For more details, see Refs. 1 and 2).

In the Matsubara formalism, the theory is formulated on a Euclidean spacetime, where the time axis is compactified to an interval $0 < \tau < \beta = 1/T$. In the close-time-path formalism, one makes a detour into real time, as in Fig. 10.1.

One can turn on source on the upper and lower parts of the contour, J_1 and J_2. The partition function of field theory now is a functional of both J_1

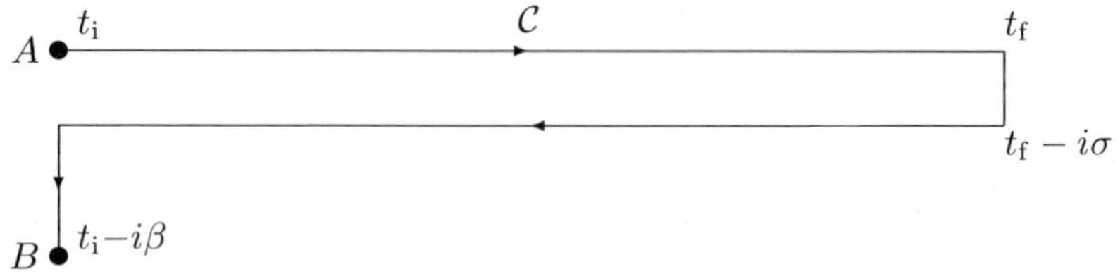

Fig. 10.1. The close time path contour.

and J_2, $Z = Z[J_1, J_2]$, and and derivatives of $\log Z$ with respect to J gives a 2×2 matrix propagators G_{ab}, where $a, b = 1, 2$. Changing σ rescales the off-diagonal elements by a trivial factor,

$$G_{12}(\omega, q) = e^{\sigma\omega} G_{12}^{\sigma=0}(\omega, q), \tag{10.1}$$

$$G_{21}(\omega, q) = e^{-\sigma\omega} G_{21}^{\sigma=0}(\omega, q). \tag{10.2}$$

For $\sigma = 0$, the propagators G_{ab} include path-ordered, reversed path-ordered, and Wightmann Green's functions. They are related by

$$G_{11} + G_{22} = G_{12} + G_{21}, \tag{10.3}$$

but the choice $\sigma = \beta/2$ leads to symmetric 2×2 propagator matrix: $G_{12} = G_{21}$. This choice of the σ is most natural for holography, as we will see.

From the point of view of the CTP formalism, putting our system in an external source J corresponds to having, in the $\sigma = 0$ choice of the contour, $J_1 = J_2 = J$. The expectation value of the operator ϕ at a point x on an upper contour is given by an integral over the whole contour, which can be written as

$$\langle \phi_1(x) \rangle = -\int dy\, (G_{11}(x - y)J(y) - G_{12}(x - y)J(y)), \qquad \sigma = 0. \tag{10.4}$$

Define the retarded propagator $G_R = G_{11} - G_{12}$ ($\sigma = 0$). The retarded propagator governs the response of a system to a small external perturbation:

$$\langle \phi(x) \rangle = -\int dy\, G_R(x - y)J(y). \tag{10.5}$$

On the other hand, for the symmetric choice $\sigma = \beta/2$, $G_R = G_{11} - e^{-\beta\omega/2} G_{12}$.

Normally, the computations of thermal Green's function rely on summing Feynman diagrams. The set of Feynman diagrams that one has to sum in order to compute, say, the viscosity, can be quite complicated.[3] In the low-momentum limit, however, the forms of many correlation functions are simple and are dictated by an effective theory—hydrodynamics.

10.3. Hydrodynamics

Consider an interacting quantum field theory at finite temperature. One can visualize such a system as a collection of particles (or quasiparticles), moving with random velocities and colliding with each other from time to time. Such a picture is too simplistic for a strongly interacting system (with no discernible quasiparticles) but it does tell us that there is an important length scale in the problem—the mean free path, which is the length which a particle travels before colliding with other particles.

Hydrodynamics can be thought of as an effective theory describing the dynamics of a finite-temperature system at distance scales much larger than the mean free path. By definition the degrees of freedom entering hydrodynamics have to have relaxation time much larger than the mean free time. Such modes include

- Density of conserved quantities. Consider, for example, the QCD plasma, and imagine a perturbation of the system where there is a net excess of charge in a volume with size $L \gg \ell$. If one waits a long time this lump of excess charge will disappear, with the charge now distributing uniformly over the whole volume. However, since charge is conserved, causality implies that the time scale for this process cannot be smaller than L/c, where c is the speed of light (in fact in many cases the time scale is much bigger than this naive estimate. For example, if the relaxation is due to diffusion, the length scale is $\sim L^2/\ell$).
- Nambu-Goldstone modes. If there is a broken continuous symmetry, Goldstone's theorem dictates that there must be a massless particle at zero temperature. If the symmetry remains broken at a finite temperature, the Nambu-Goldstone mode continuously deforms into a hydrodynamic mode. For example, in superfluid ^{4}He the phase of the condensate φ is a hydrodynamic degree of freedom (the superfluid velocity $\mathbf{v}_s$ is proportional to the gradient of φ: $\mathbf{v}_s = \nabla\varphi/m$).
- Unbroken U(1) gauge fields. At zero temperature, a U(1) gauge field which does not suffers from the Anderson-Higgs mechanism corresponds to a massless photon. At finite temperature, the electric field is screened (Debye screening) but the magnetic field is unscreened and should be included in the hydrodynamic description. An example of such a theory is magnetohydrodynamics, describing for example the interior of the Sun.

In these lectures we will consider only the simplest class of hydrodynamic theories, where the only slow degrees of freedom are the densities of conserved charges. In this case, hydrodynamics is given by the conservation equation,

$$\nabla_\mu T^{\mu\nu} = 0, \tag{10.6}$$

supplemented by the continuity equation that expresses $T^{\mu\nu}$ in terms of four variables: the local temperature T and the local fluid velocity u^μ:

$$T^{\mu\nu} = (\epsilon + P)u^\mu u^\nu + Pg^{\mu\nu} + \tau^{\mu\nu}, \tag{10.7}$$

where $\tau^{\mu\nu}$ is the correction containing terms proportional to first derivatives. It is conventional to impose the condition $u_\mu \tau^{\mu\nu} = 0$ which eliminates any ambiguity in the definition of u^μ and T. In this case one has

$$\tau^{\mu\nu} = -\eta P^{\mu\alpha}P^{\nu\beta}(\nabla_\alpha u_\beta + \nabla_\beta u_\alpha) - \zeta P^{\mu\nu}(\nabla \cdot u), \tag{10.8}$$

where η and ζ are the shear and bulk viscosities, respectively. In a conformal plasma, the stress-energy tensor is traceless, hence $\epsilon = 3P \sim T^4$ and $\zeta = 0$. In such a plasma, the shear viscosity has to scale with the temperature as $\eta \sim T^3$.

10.3.1. *Hydrodynamics and two point functions*

From the hydrodynamic equations, one can easily compute the two-point functions of between two components of the stress-energy tensor. According to the general formulas of the linear response theory, the two-point function can be computed by first turning on a weak gravitational perturbation $g_{\mu\nu} = \eta_{\mu\nu} + h_{\mu\nu}$, $h_{\mu\nu} \ll 1$, then measuring the expectation value of the stress-energy tensor $\langle T_{\mu\nu}\rangle$. The two-point function is the coefficient of proportionality between $\langle T_{\mu\nu}\rangle$ and $h_{\mu\nu}$,

$$T_{\mu\nu}(x) \sim \int dy \, \langle T^{\mu\nu}(x) T^{\alpha\beta}(y)\rangle h_{\alpha\beta}(y). \tag{10.9}$$

On the other hand, when $h_{\mu\nu}$ varies with space and time very slowly, the response of the system can be determined by hydrodynamics. One first generalizes the hydrodynamic equation to curve spacetime. Assuming the system is in thermal equilibrium in the infinite past, and $h_{\mu\nu}$ is nonzero in a finite regime in spacetime, the state of the system can be completely determined.

We can re-derive the well known Kubo's formula in this way. Let us turn on a small metric perturbation whose only nonzero component is h_{xy} which is assumed to be homogeneous in space and is time dependent, $h_{xy} = h_{xy}(t)$. Then by symmetry one can right away determine that the fluid will remain in a state with constant temperature, $T = \text{const}$, and zero spatial velocity

$u^\mu = (1, \mathbf{0})$ (a tensor perturbation cannot excite a scalar or vector mode to linear order). Nevertheless, the stress-energy tensor receives a correction

$$T^{xy} = Pg^{xy} - \eta(\nabla_x u_y + \nabla_y u_x) = -Ph_{xy} + 2\eta\Gamma^0_{xy}u_0 \tag{10.10}$$

proportional to the perturbation. Thus we find the two-point function

$$\langle T^{xy}T^{xy}\rangle = P - i\eta\omega. \tag{10.11}$$

The real part of this Green's function is a contact term, and depend on the way the two point function is defined; but one cannot get rid of the imaginary part by a redefinition of the Green function. Moreover, the imaginary part gives the value of the viscosity through the Kubo formula:

$$\eta = -\lim_{\omega\to 0}\frac{1}{\omega}\mathrm{Im}\, G_R^{xy,xy}(\omega, \mathbf{0}). \tag{10.12}$$

10.4. AdS/CFT Prescription for Correlation Function

10.4.1. *Euclidean Green's function*

Let us remind ourselves how the Euclidean Green's function is computed. For simplicity we limit ourselves to the case of an operator of dimension 4, dual to a massless scalar field ϕ. Assuming the action for the scalar field is

$$S = -\frac{K}{2}\int d^5x\,\sqrt{-g}\,g^{\mu\nu}\partial_\mu\phi\partial_\nu\phi. \tag{10.13}$$

Then the prescription tells us to solve the wave equation

$$\partial_\mu(\sqrt{-g}g^{\mu\nu}\partial_\nu\phi) = 0, \tag{10.14}$$

with boundary condition $\phi = \phi_0$ at the boundary. The solution, in momentum space, is $\phi(z,k) = \phi_0(k)f_k(z)$, where $f_k(z)$ is the solution to the field equation (at momentum k). We now rewrite S as a boundary action

$$S = \frac{K}{2}\int d^5x\frac{1}{z^3}\phi\phi'. \tag{10.15}$$

Differentiating the action with respect to the boundary value ϕ_0, we find the two point function to be

$$\langle\phi\phi\rangle_k \sim K\lim_{z\to 0}\frac{f_k'}{z^3}. \tag{10.16}$$

The boundary condition at the boundary needs to be supplemented by the boundary condition in the IR. At zero temperature, we require $\phi(z)$ to vanish as $z \to 0$. At finite temperature, spacetime is capped off at some

$z = z_0$. We require the field to be regular at the horizon; in the case of the scalar, $\phi'(z_0) = 0$. The solution to the field equation is then unique, and the AdS/CFT prescription well defined.

10.4.2. *Real-time Green's function*

In real-time, the formulation of the AdS/CFT prescription is more subtle. The AdS/CFT rules are best formulated using the whole Penrose diagram of the black hole.

In the Poincare metric the AdS black hole looks like

$$ds^2 = -\frac{r^2}{R^2}(-f dt^2 + d\mathbf{x}^2) + \frac{R^2}{r^2 f} dr^2, \tag{10.17}$$

where $f = 1 - r_0^4/r^4$. The metric can be extended pass the horizon, one recovered four quadrants in the following Penrose diagram,

Let us remind ourselves how it is done. Near the horizon, we expand $r = r_0 + \rho$. The (t, r) part of the metric can be rewritten as

$$ds^2 = 4\pi T \rho \left(-dt^2 + \frac{1}{(4\pi T)^2} \frac{d\rho^2}{\rho^2} \right), \tag{10.18}$$

where $T = r_0/\pi R^2$. This can be rewritten as $ds^2 = e^{4\pi T r_*}(-dt^2 + dr_*^2)$ where $r_* = (4\pi T)^{-1} \ln \rho$. Finally, we introduce Kruskal's coordinates

$$U = -e^{-2\pi T(t - r_*)}, \tag{10.19}$$

$$V = e^{2\pi T(t + r_*)}, \tag{10.20}$$

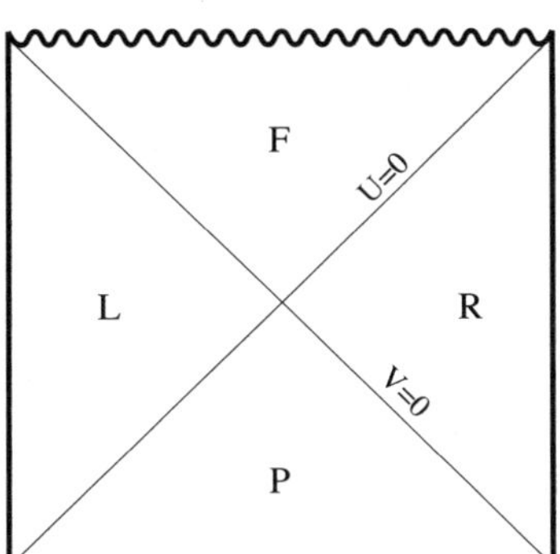

Fig. 10.2. Penrose diagram of AdS black hole.

and metric is $ds^2 = -dUdV$. The Poincare coordinates cover only $U < 0$, $V > 0$ part quadrant of the diagram. There is another copy with the same metric, corresponding to the $U > 0$, $V < 0$ part. There are two boundaries.

The extension of the AdS/CFT duality was suggested by Maldacena,[4] and then explicitly considered in Ref. 5. The idea is that the two boundaries correspond to two horizontal parts of the close time path contour. The AdS/CFT prescription is then identifies the logarithm partition function of the thermal field theory, with sources J_1 and J_2 on the two parts of the contour, with the classical action of a configuration where the bulk field ϕ reaches the values J_1 and J_2 on the right and left boundaries, respectively.

In addition, one should also put boundaries conditions near the horizon. The choice of the boundary condition should be that when the bulk field ϕ is considered as function of the complex Kruskal coordinates U and V, it is analytic in the U upper half plane and V lower half plane.

The solution to the linearized field equation can be written in terms of the mode function $f_k(r)$, defined as the radial profile of a solution to the wave equation with momentum k, and is incoming wave at the horizon. One can write the solution down separately in the right and left quadrants,

$$\begin{aligned}\phi(k,r)|_R &= ((n+1)f_k^*(r_R) - nf_k(r_R))\,\phi_1(k)\\ &\quad + \sqrt{n(n+1)}\,(f_k(r_R) - f_k^*(r_R))\,\phi_2(k),\end{aligned} \tag{10.21}$$

$$\begin{aligned}\phi(k,r)|_L &= \sqrt{n(n+1)}\,(f_k^*(r_L) - f_k(r_L))\,\phi_1(k)\\ &\quad + ((n+1)f_k(r_L) - nf_k^*(r_L))\,\phi_2(k).\end{aligned} \tag{10.22}$$

Here $n = (e^{\omega/T} - 1)^{-1}$ is the Fermi-Dirac distribution function at frequency ω. Substituting the solution into the quadratic action, using the boundary form of the on-shell action,

$$\frac{K}{2}\int\limits_R \sqrt{-g}g^{rr}\phi(-k,r)\partial_r\phi(k,r)\frac{d^4k}{(2\pi)^4} - \frac{K}{2}\int\limits_L \sqrt{-g}g^{rr}\phi(-k,r)\partial_r\phi(k,r)\frac{d^4k}{(2\pi)^4} \tag{10.23}$$

(where K is a normalization factor) and differentiate it, one obtains the CTP propagators. Taking the appropriate linear combination of G_{11} and G_{12} we then find the retarded Green function,

$$G_R(k) = -K\sqrt{-g}\,g^{rr} f_k(r)\partial_r f_k^*(r)|_{r\to\infty}. \tag{10.24}$$

This formula coincides with the prescription first proposed in Ref. 6. We now use this formula to compute the shear viscosity of the $\mathcal{N} = 4$ plasma.

10.4.3. *Viscosity*

Let us compute the a two-point function. We assume the momentum to be $q = (\omega, 0, 0, q)$, and we compute the two-point function T^{xy}, we consider gravitation perturbation with the only perturbation being $h_{xy}(t, z)$. One can show that the quadratic action of h_{xy} is that of a minimally coupled theory when written in terms of $\phi = g^{xx} h_{xy}$:

$$S = \frac{1}{2\kappa_{10}^2} \int d^{10}x \sqrt{-g}\,(R - 2\Lambda) = -\frac{V(S^5)}{4\kappa_1 0^2} \int d^5x \sqrt{-g}\, g^{\mu\nu} \partial_\mu \phi \partial_\nu \phi. \tag{10.25}$$

We now the write down the mode equation

$$\left(\frac{f(z)}{z^3} f_k'\right)' + \left(\frac{\omega^2}{z^3 f} - \frac{q^2}{z^3}\right) f_k(z) = 0. \tag{10.26}$$

The solution to this equation is

$$f_k(z) = \left(1 - \frac{z}{z_0}\right)^{-i\omega/4\pi T}. \tag{10.27}$$

Inserting the solution into the formula for G_R, we find the imaginary part of the retarded propagator,

$$\operatorname{Im} G_R(k) = -\frac{V(S^5)}{2\kappa_{10}^2} \frac{R^3}{z_0^3} i\omega. \tag{10.28}$$

To compute the real part of G_R one needs to be more careful with holographic renormalization. But we can already read out the viscosity from $\operatorname{Im} G_R$,

$$\eta = \frac{V(S^5)}{2\kappa_{10}^2} \frac{R^3}{z_0^3}. \tag{10.29}$$

This can be compared with the entropy density,

$$s = \frac{S}{V} = V(S^5) \frac{R^3}{z_0^3} \frac{2\pi}{\kappa_{10}^2}, \tag{10.30}$$

and we find that $\eta/s = 1/4\pi$. This is the common feature of all theories with gravitational dual.[7]

10.5. Fluid-Gravity Correspondence: Diffusion

There is an alternative method to compute the kinetic coefficients. This method, sometimes called fluid-gravity correspondence, allows one to see directly the emergence of the *nonlinear* hydrodynamic equations from the field equations in the bulk.[8] The approach is thus complementary to standard AdS/CFT method based on the calculations of correlation functions.

We will illustrate the technique of fluid-gravity correspondence on a very simple example where the higher-dimensional theory is an abelian gauge theory in a black hole background,

$$S = -\frac{1}{4g_{\rm YM}^2}\int d^{d+1}x\, F_{\mu\nu}F^{\mu\nu}. \tag{10.31}$$

The background is chosen to be

$$ds^2 = r^2(-f(r)dt^2 + d\mathbf{x}^2) + \frac{dr^2}{r^2 f(r)}\,, \tag{10.32}$$

where $f(r)$ is a function that vanishes at the horizon, $f(r_0) = 0$ and tends to 1 at the AdS boundary, $f(\infty) = 1$. This is the usual back hole (black brane) background.

We are interested in solution to Maxwell's equations which satisfies outgoing wave boundary conditions. To enforce the incoming-wave boundary condition, it is more convenient to use the incoming Eddington-Fikelstein coordinates,

$$ds^2 = -r^2 f dv^2 + 2dv\, dr + r^2 d\mathbf{x}^2. \tag{10.33}$$

The usefulness of the Eddington-Fikelstein coordinates is that regularity at the horizon in these coordinates correspond to incoming wave boundary conditions in the usual coordinates. We go on to construct such a solution. The Maxwell equations are

$$\partial_r(r^d F_{vr}) + r^{d+2}\partial_i F_{ir} = 0, \tag{10.34}$$

$$r^d\partial_v F_{rv} + r^{d-2}\partial_i F_{iv} + fr^d\partial_i F_{ir} = 0, \tag{10.35}$$

$$\partial_r(r^{d-2}F_{vi}) + \partial_r(fr^d F_{ri}) + r^{d-2}\partial_v F_{ri} + r^{d-4}\partial_j F_{ji} = 0. \tag{10.36}$$

We start from gauge field of a charged black hole,

$$A_0 = \frac{q}{r^{d-1}}\,. \tag{10.37}$$

This is a one-parameter family of solutions to the Maxwell equations, parameterized by the charge density q. It describes a state in the field theory with a constant charge density in complete thermal equilibrium. Note that the solution is translationally invariant in all field-theory directions, t and $\mathbf{x}$.

What happens if we make q a function of the space and time? As one can easily verify, now the configuration (10.37) is not an exact solution to the Maxwell equation. However, when q varies slowly in space and time, one should be able to still find the solution by expanding it in powers of $\partial_t q$ and $\partial_x q$, which are small parameters. This is exactly the strategy that we will follow.

First we need to settle on a power counting scheme. Anticipating the end result to be a diffusion equation $\partial_t \rho = D\nabla^2\rho$, we shall treat $\partial_i q$ as $O(\epsilon)$ and $\partial_t q$ as $O(\epsilon^2)$, where ϵ is small. We then expand the solutions, using the gauge $A_r = 0$,

$$A_v(r,x) = \frac{q(x)}{r^{d-2}} + A_v^{(1)}, \qquad A_i = A_i^{(1)}. \tag{10.38}$$

We can demand that $A_0^{(1)}$ falls off faster than $r^{-(d-2)}$ at large r. Otherwise, one can redefine $q(x)$ to absorb any $r^{-(d-2)}$ piece in $A_0^{(1)}$. Consistency requires that we treat $A_v^{(1)}$ as a quantity of order ϵ^2 and $A_i^{(1)} \sim \epsilon$

Substituting the ansatz in the the Maxwell equations, collecting terms with the same smallness in ϵ, we find

$$\partial_r(r^d \partial_r A_v^{(1)}) + r^{d+2}\partial_i \partial_r A_i^{(1)} = 0, \tag{10.39}$$

$$(d-1)\partial_v q + f r^d \partial_i \partial_r A_i^{(1)} - \frac{1}{r}\partial_i^2 q = 0, \tag{10.40}$$

$$-\partial_r\left(\frac{\partial_i q}{r}\right) + \partial_r(f r^d \partial_r A_i) = 0. \tag{10.41}$$

Integrating the last equation, we find

$$\partial_r A_i^{(1)} = \frac{C}{f r^d} + \frac{\partial_i q}{f r^{d+q}}, \tag{10.42}$$

where C is an x-dependent integration constant. Both terms in the right hand side have pole at the horizon $r = r_0$, and regularity at the horizon requires that the singularities cancel out between the two terms. Therefore we find

$$C = -r_0 \partial_i q. \tag{10.43}$$

Integrating once more, we then can find A_i. Actually for our purposes, we just need to know the asymptotic behavior of A_i at large r,

$$A_i = \frac{r_0}{d-1}\frac{\partial_i q}{r^{d-1}} + O(r^{-d}). \tag{10.44}$$

We can now substitute A_i into the second equation of (10.39), taking the large r limit and derive the following equation for q:

$$\partial_v q - \frac{r_0}{d-1}\nabla^2 q = 0, \qquad (10.45)$$

which is nothing but the diffusion equation. Thus we have found a more general solution to the Maxwell equation which is parameterized by solutions to the diffusion equation. Maxwell equation in the background of black brane metric reduces to the diffusion equation in the long-wavelength limit.

10.6. Nonrelativistic Conformal Invariance

Fermions interacting through a unitarity interaction form a simplest nonrelativistic strongly interacting system. This system is beautiful because of its simplicity and universality. It has attracted enormous attention since being realized in cold atom experiments.

Let us first define fermions at unitarity. Consider two nonrelativistic particles, interacting through a potential,

$$H = \frac{\mathbf{p}_1^2}{2} + \frac{\mathbf{p}_2^2}{2} + V(|\mathbf{x}_1 - \mathbf{x}_2|). \qquad (10.46)$$

For simplicity, we can consider V of the form of a square well potential, with size r_0 and depth $-V_0$: $V(r < r_0) = -V_0$ and $V(r > r_0) = 0$. If the potential is shallow, it does not have any bound state; but if it is deep enough it may have one, two, or more bound states. There is a critical value of $V_0 \sim r_0^{-2}$ at which the potential starts to have exactly one bound state. We tune V_0 to be exactly this value.

Then we take the range of interaction r_0 to zero, keeping V_0 always tuned to the critical value (in other words, keeping $V_0 r_0^2$ fixed). This limit is called the unitary limit, and the system of fermions interacting with this interaction the unitary Fermi gas.

The stability of such a system is not a trivial issue. It is relatively easy to see that for bosons, and for fermions of three or more different species, the finite-density system is not stable. This fact is related to the so-called Efimov effect: in the limit of zero range interaction, the Hamiltonian is unbounded from below (there is an infinite number of bound states, the lowest of which has an energy determined by the UV cutoff of the theory—the range of the potential).

10.6.1. *Quantum mechanics formulation: Boundary condition*

The quantum quantum mechanics of unitary fermions can be formulated in a way which gets rid of the interaction potential completely. Let us start with the case of two particles, one spin-up with coordinate $\mathbf{x}$, and another spin-down with coordinate $\mathbf{y}$. Neglecting the center of mass motion, the Schrödinger equation has the form

$$\frac{\partial^2}{\partial \mathbf{r}^2}\Psi(\mathbf{r}) + V(r)\Psi(\mathbf{r}) = -E\Psi(\mathbf{r}). \tag{10.47}$$

In the limit of zero range, the potential $V(r)$ is zero at any nonzero r. In the limit of $\mathbf{r} \to 0$, the right hand side can be neglected ($E \ll r^{-2}$), and we have the Laplace equation $\nabla^2_{\mathbf{r}}\Psi = 0$. Now it is known that the Laplace equation has two independent solutions, 1 and r^{-1}. The behavior of the wavefunction at small r is, in general,

$$\Psi(\mathbf{r}) = \frac{C}{r} + C_1 + O(r), \tag{10.48}$$

where C and C_1 are some numbers. In the usual problem of free particles, the wavefunction is assumed to be regular at $r = 0$, which means $C = 0$. On the other hand, from the mathematical point of view one can impose a general boundary condition

$$\Psi(\mathbf{r}) \sim \left(\frac{1}{r} - \frac{1}{a}\right), \tag{10.49}$$

with any value of a ($a = 0$ corresponding to free particles). Physically a is obtained by solving the zero-energy Schrödinger equation inside the potential $r < r_0$ and then match it to the solution to the Laplace equation outside the potential $r > r_0$; a therefore characterizes low-energy scatterings and is called the scattering length. The fine-tuning of the potential corresponds to the limit $a \to \infty$.

For the case of a general number of particles, the Hamiltonian is the sum of the kinetic terms of all particles,

$$H = \sum_i \frac{\mathbf{p}_i^2}{2m} = -\frac{1}{2m}\sum_i \frac{\partial^2}{\partial \mathbf{x}_i^2} \tag{10.50}$$

(where i numerates all particles) but the Hilbert space is nontrivial: the wave function of a system, $\Psi(\mathbf{x}_1, \ldots, \mathbf{x}_N; y_1, \ldots, \mathbf{y}_N)$ satisfies the following condition when one spin-up and one spin-down particles approach each other,

$$\Psi(\mathbf{x}_1, \ldots, \mathbf{x}_N; y_1, \ldots, \mathbf{y}_M) \to \frac{C}{|\mathbf{x}_i - y_j|} + O(1) + O(|\mathbf{x}_i - y_j|), \tag{10.51}$$

where $\mathbf{x}_i$ and $\mathbf{y}_j$ are the coordinates of the spin-up particles and spin-down particles, respectively. The Hamiltonian is trivial, but the nontriviality of the problem is in the Hilbert space.

For example, we can put a spin-up and a spin-down fermion in a harmonic potential. The Hamiltonian is now

$$H = \frac{1}{2}(\mathbf{p}_1^2 + p_2^2) + \frac{1}{2}\omega^2(\mathbf{x}_1^2 + \mathbf{x}_2^2). \tag{10.52}$$

The problem can be solved exactly even when the interaction between particles is unitary. The ground state is

$$\Psi(\mathbf{x}_1, \mathbf{x}_2) \sim \frac{e^{-\omega(|\mathbf{x}_1|^2 + |\mathbf{x}_2|^2)/2}}{|\mathbf{x}_1 - \mathbf{x}_2|}, \tag{10.53}$$

and the ground state energy is $E = 2\omega$. This is lower than the ground state energy in the case of a 3ω, in consistency with the attractiveness of the interaction.

The two-particle problem is special because it can be solved analytically. For three particles in a harmonic potentials, the energy levels are also known exactly (they are solutions to a trigonometric equation). For four particles and more (unless they have the same spin), the many-body problem cannot be solved exactly.

10.6.2. *Symmetries of unitary fermions*

A general nonrelativistic system is invariant under translation (in space and time), rotation, and Galilean boosts. In addition, the conserved mass (particle number) corresponds to a phase symmetry, $\psi \to e^{i\alpha}\psi$. These symmetries are enhanced to a new symmetry group called the Schrödinger group. The Schrödinger group contains two new symmetries

- Dilatation: $t \to \lambda^2 t$, $\mathbf{x} \to \lambda\mathbf{x}$,
- Proper conformal transformation: $t = t/(1 - \lambda t)$, $\mathbf{x} \to \mathbf{x}/(1 - \lambda t)$.

Saying that a theory theories has these symmetries means that if one has a solution to the time-dependent Schrödinger equation, $\Psi(t, \mathbf{x}_i)$, then one can generate new solutions. For example, the solution obtained by dilatation is

$$\Psi'(t, \mathbf{x}) = \lambda^{3N/2}\Psi(\lambda^2 t, \lambda x) \tag{10.54}$$

(the prefactor is to keep the normalization of Ψ). It is obvious that if Ψ solve the free Schrödinger equation, then Ψ' also does. More nontrivially, the boundary condition at short distances for unitary particle is preserved under dilatation. Similarly, the proper conformal transformation corresponds to the following family of new solutions,

$$\Psi'(t,\mathbf{x}) = C(t,\mathbf{x})\Psi\left(\frac{t}{1-\lambda t}, \frac{\mathbf{x}}{1-\lambda t}\right). \tag{10.55}$$

We leave the determination of $C(t, \mathbf{x})$ to the reader.

In the theory of unitarity fermions, the dilatation operator D and the proper conformal transformation C can be expressed in terms of the operators creating and annihilating a particle,

$$D = -\frac{i}{2}\int d\mathbf{x}, \mathbf{x}\cdot(\psi^\dagger\nabla\psi - \nabla\psi^\dagger\psi), \quad C = \frac{1}{2}\int d\mathbf{x}, \mathbf{x}^2\psi^\dagger(\mathbf{x})\psi(\mathbf{x}). \tag{10.56}$$

One can check that the operators D, C and the Hamiltonian H form a close SO(2,1) subalgebra of the Schrödinger algebra,

$$[D,\, C] = -2iC, \quad [D,\, H] = 2iH, \quad [C,\, H] = iD. \tag{10.57}$$

The full Schrödinger algebra can be found in Ref. 9.

10.6.3. *Local operators*

The local operators (for example ψ, $\psi^\dagger$, or $\psi^\dagger\psi$) depend on time t and space $\mathbf{x}$. Its commutators with time and space rotation are completely defined. The local operators can be classified by particle number by taking it commutator with the particle number operator $M = \int d^{\mathbf{x}}, \psi^\dagger\psi$. For example ψ has particle number -1 while for $\psi^\dagger$ it is $+1$. Each operator can be associated with a dimension by

$$[D,\, O(0)] = i\Delta_O O(0). \tag{10.58}$$

For example, $\Delta_\psi = 3/2$ ($d/2$ in d spatial dimensions). One example of a nontrivial composite operator is obtained when one tries to construct the product of two annihilation operators of particles with different spins, $\psi_\uparrow\psi_\downarrow$. We know that the matrix element of $\psi_\uparrow(\mathbf{x})\psi_\downarrow(\mathbf{y})$ between a two-body state and vacuum is just the wave function,

$$\langle 0|\psi_\uparrow(\mathbf{x})\psi_\downarrow(\mathbf{y})|\Psi\rangle = \Psi(\mathbf{x},\mathbf{y}). \tag{10.59}$$

The problem is that when we tries to take $\mathbf{x}\to\mathbf{y}$ to have a local operator $\psi_\uparrow\psi_\downarrow$, the matrix element diverges due to the boundary condition at $\mathbf{x}\to\mathbf{y}$. On the other hand, one can define the following operator

$$O_2(\mathbf{x}) = \lim_{\mathbf{y}\to\mathbf{x}} 4\pi|\mathbf{x}-\mathbf{y}|\psi_\uparrow(\mathbf{x})\psi_\downarrow(\mathbf{y}), \tag{10.60}$$

which has finite matrix elements between states in the Hamiltonian. Another way to write the equation above is

$$\psi_\uparrow(\mathbf{x})\psi_\downarrow(\mathbf{y}) = \frac{O_2(x)}{4\pi|\mathbf{x}-y|} + \cdots, \tag{10.61}$$

which has the form of an operator product expansion for unitarity fermions. Operator product expansions have been applied very recently for unitarity fermions; as in particle physics they are most useful at short distances. The operator $O_2^\dagger O_2$ has a special role: its expectation value is called in the literature the Tan's parameter, or the contact.

As in relativistic CFTs, one can introduce the notion of primary and descendant operators. Primary operators are called those which commute, at zero coordinates, with Galilean boosts and the proper conformal transformation: $[K_i, O(0)] = [C, O(0)] = 0$. By taking derivatives with respect to coordinates and time, descendants are obtained.

The SO(2,1) commutators are important to prove what we call the operator-state correspondence for systems with Schrödinger symmetry. Namely, a primary operator, which does not annihilate the vacuum (or its Hermitian conjugate does not annihilate the vacuum) can be put into correspondence with an eigenstate of the system of a few particles in a harmonic potential.[9] This statement can be proven by first noticing (recall the form of the operator C in Eq. (10.56) that the Hamiltonian in a harmonic potential can be written as

$$H_{\text{osc}} = H + \omega^2 C. \tag{10.62}$$

Then for an primary operator O, one can construct a state $\Psi_O\rangle$ as follows,

$$|\Psi_O\rangle = e^{-H/\omega} O^\dagger(0)|0\rangle. \tag{10.63}$$

Physically, first we use $O^\dagger$ to create a state which is localized at the origin of coordinates, and then evolve that state in imaginary time using the free-space Hamiltonian during a time $1/\omega$. The resulting state, whose wavefunction is a Gaussian-type wavepacket, can be shown, by using the SO(2,1) commutators, to be an eigenstate of the Hamiltonian H_{osc} with energy 2ω.

This operator-state correspondence can be illustrated explicitly in a few example. The operator ψ has dimension $3/2$, which matches with the ground state energy of a single particle in an isotropic harmonic potential, $3\omega/2$. The operator O_2 has dimension 2, and corresponds to the ground state of a system of one spin up and one spin down particle in a harmonic potential, whose energy was shown above to be 2ω.

10.6.4. *Schrödinger space*

If one wants to move in the direction of constructing a gravitational dual of the unitarity fermions, it seems reasonable to start by asking the question: what is the space-time that realizes the Schrödinger symmetry? (Recall that in the original Maldacena's duality, the symmetry of $AdS_5 \times S^5$ space matches with the symmetry of the quantum field theory). An example of such a space was constructed in Refs. 10 and 11. The space has *two* extra dimensions compared to one extra dimensions in standard holography. The metric is

$$ds^2 = -\frac{2(dx^+)^2}{z^4} + \frac{-2dx^+dx^- + dx^i dx^i + dz^2}{z^2}. \tag{10.64}$$

One can check that this spacetime has realizes all generators of the Schrödinger algebra as Killing vectors. In particular, the total mass and the proper conformal transformation are

$$\begin{aligned} &M : x^- \to x^- + a, \\ &C : z \to (1 - ax^+)z, \quad x^i \to (1 - ax^+)x^i, \quad x^+ \to (1 - ax^+)x^+, \\ &\qquad x^- \to x^- - \frac{a}{2}(x^i x^i + z^2). \end{aligned} \tag{10.65}$$

One can see that the translational symmetry along the direction x^- realizes the conservation of mass in the nonrelativistic theory. The simplest action which gives rise to the Schrödinger spacetime is that of Einstein gravity with negative cosmological constant, coupled to a massive gauge field with a suitably chosen mass,

Subsequently the five-dimensional Schrödinger spacetime (corresponding to two spatial directions in field theory) have been constructed in string theory. As by-product of the construction, one also found black hole solutions, which describe a medium with finite chemical potential and temperature. One might think that these solutions may be the first holographic model for the unitarity Fermi gas. Unfortunately, closer inspection reveals a serious undesirable feature: the equation of state of the black hole is $P(T, \mu) \sim \frac{T^4}{\mu^2}$, which has the correct scaling behavior but is more restrictive than required. The more general equation of state is

$$P(T, \mu) = \mu^2 F(T/\mu), \tag{10.66}$$

where the function F is not constrained. This is in contrast to the situation in relativistic holography, where fitting the equation of state of QCD is not really a problem in the bottom-up approach.

It seems that one should try to devise a more general way to realize Schrödinger symmetry. Attempts in this direction are being made. At the more general level, one should not expect the gravity dual of unitarity fermions to be a classical theory due to the lack of a large N parameter. One can generalize the unitarity fermions to a many-favor theory with $\mathrm{Sp}(2N)$ vector symmetry. This theory is trivial to solve; at large N the BCS theory becomes exact. The situation is very similar to the relativistic $\mathrm{O}(N)$ vector model. One can hope that there is a nonrelativistic high-spin theory that is dual to the $\mathrm{Sp}(2N)$ version of unitary fermions, similar to the case of the $\mathrm{O}(N)$ model.[12] As far as I know, to date no serious attempt has been made to uncover such a theory.

10.7. Summary

In these lectures we have considered some applications of gauge/gravity duality to systems with finite temperature and chemical potential. We have left out some very important applications of gauge gravity duality, most notably jet quenching and heavy quark energy loss.

Acknowledgments

I thank the organizers of the TASI school for inviting me to give this series of lectures. This work is supported, in part, by the DOE grant No. DE-FG02-00ER41132.

References

1. J. I. Kapusta and C. Gale, *Finite-temperature field theory: Principles and applications,* (Cambridge University Press, Cambridge, UK, 2006)
2. M. Le Bellac, "Thermal field theory," (Cambridge University Press, Cambridge, UK, 2000)
3. S. Jeon and L. G. Yaffe, Phys. Rev. D **53**, 5799 (1996) [arXiv:hep-ph/9512263].
4. J. M. Maldacena, JHEP **0304**, 021 (2003) [arXiv:hep-th/0106112].
5. C. P. Herzog and D. T. Son, JHEP **0303**, 046 (2003) [arXiv:hep-th/0212072].
6. D. T. Son and A. O. Starinets, JHEP **0209**, 042 (2002) [arXiv:hep-th/0205051].
7. P. Kovtun, D. T. Son and A. O. Starinets, Phys. Rev. Lett. **94**, 111601 (2005) [arXiv:hep-th/0405231].
8. S. Bhattacharyya, V. E. Hubeny, S. Minwalla and M. Rangamani, JHEP **0802**, 045 (2008) [arXiv:0712.2456 [hep-th]].

9. Y. Nishida and D. T. Son, Phys. Rev. D **76**, 086004 (2007) [arXiv:0706.3746 [hep-th]].
10. D. T. Son, Phys. Rev. D **78**, 046003 (2008) [arXiv:0804.3972 [hep-th]].
11. K. Balasubramanian and J. McGreevy, Phys. Rev. Lett. **101**, 061601 (2008) [arXiv:0804.4053 [hep-th]].
12. I. R. Klebanov and A. M. Polyakov, Phys. Lett. B **550**, 213 (2002) [arXiv:hep-th/0210114].

Chapter 11

Collisions in Anti-de Sitter Space, Conformal Symmetry, and Holographic Superconductors

Steven S. Gubser

Joseph Henry Laboratories,
Princeton, NJ 08540, USA
ssgubser@princeton.edu

In four lectures, delivered at the TASI 2010 summer school, I cover selected topics in the application of the gauge-string duality to nuclear and condensed matter physics. On the nuclear side, I focus on multiplicity estimates from trapped surfaces in AdS_5, and on the consequences of conformal symmetry for relativistic hydrodynamics. On the condensed matter side, I explain the fermion response to the zero-temperature limit of p-wave holographic superconductors.

11.1. Introduction

In my TASI 2010 lectures, I discussed two developments in applications of the gauge-string duality.[1–3] The first is aimed at nuclear physics and focuses on multiplicity estimates from trapped surfaces and $O(3)$ symmetry from colliding black holes in AdS_5. The second is aimed at condensed matter physics and treats p-wave superconductors and the response of fermions to them from holographic duals.

My lectures were among the last at the school, and they were intended as "special topics" lectures. As a result, I took more time than usual to explain how my own understanding of these subjects developed, why I worked on them, and what questions I was asking myself at the time. I did this in the hope that students would ponder whether my approach to sniffing out research problems had some relevance for them.

This writeup adheres closely to the order of presentation in my actual lectures. In section 11.2 (Lectures 1 and 2) I discuss black hole collisions in AdS_5, trapped surfaces, and $O(3)$ symmetry. These lectures are based on work done in part with S. Pufu and A. Yarom.[4–6] In sections 11.4 and 11.5

(Lectures 3 and 4), I turn to a discussion of the fermion response to p-wave holographic superconductors, based on work done with F. Rocha and A. Yarom.[7]

11.2. Lecture 1: Trapped Surfaces in AdS_5 and $O(3)$ Symmetry

11.2.1. *Overview of the main results*

When pointlike, lightlike particles collide head-on in AdS_5, a black hole forms with

$$S_{\rm BH} \geq S_{\rm trapped} \approx \pi \left(\frac{L^3}{G_5} \right)^{1/3} (2EL)^{2/3} , \tag{11.1}$$

where E is the energy of *one* of the particles, and we assume $EL \gg 1$.[4] The main meat of my first lecture was to explain how (11.1) comes about, and to draw attention to an $O(3)$ symmetry that comes up along the way. But first I will provide an overview of what the various quantities in (11.1) mean and why it seemed to me a good idea to work out the inequality (11.1).

$S_{\rm BH} = A/4G_5$ is the Bekenstein-Hawking entropy. G_5 is Newton's constant in AdS_5. L is the radius of curvature of AdS_5, whose Ricci tensor takes the form

$$R_{\mu\nu} = -\frac{4}{L^2} g_{\mu\nu} . \tag{11.2}$$

$S_{\rm trapped}$ is the area of a trapped surface: a closed, spacelike, co-dimension two surface in AdS_5 both of whose forward-directed normal vectors point inward. A rough depiction of the trapped surface is shown in figure 11.1.

Plugging numbers into (11.1) which are suitable for comparison to a top-energy gold-gold collision at RHIC ($\sqrt{s_{NN}} = 200\,{\rm GeV}$) gives $S_{\rm trapped} \approx 35{,}000$, quite close to phenomenological estimates $S_{\rm Au\,Au} \approx 38{,}000$ for central collisions. But data at RHIC energies and below favors a slower scaling with beam energy E, namely $S \propto E^\alpha$ where $\alpha \approx 1/2$ or a bit smaller: see for example the discussion by Steinberg.[8]

Now, why work on trapped surfaces in AdS_5? In no particular order, here are the reasons that I recall as motivations.

- I had read work by Eardley and Giddings[9] treating the possibility of black hole formation in pp collisions at the LHC.

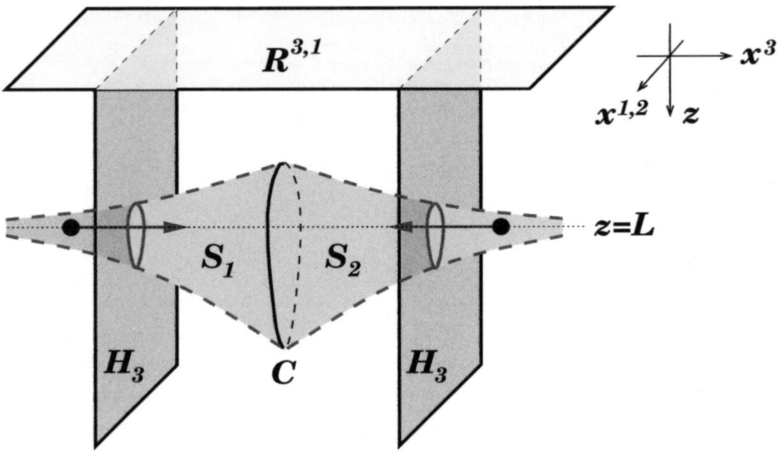

Fig. 11.1. A rough depiction of the trapped surface in AdS_5 formed in a collision of massless particles. From the original literature.[4]

- The idea of quantitatively comparing black holes in AdS_5 to the quark-gluon plasma (QGP) was well-established.
- Formation of the QGP was (and is) recognized as a hard and interesting problem.
- Black hole formation is interesting on formal grounds, and trapped surfaces provide a standard first cut at the problem.
- Penrose argued in unpublished work that trapped surfaces have to be entirely enclosed by a black hole horizon.

In explaining a derivation of (11.1), I will inevitably leave out a fair amount of detail. Readers interested in seeing the details are referred to the original literature.[4]

11.2.2. *Shock waves in AdS_5*

Before the collision of lightlike particles, the geometry we want to consider (including the back-reaction from the massless particles) is

$$ds^2 = \frac{L^2}{z^2}\left[-dudv + (dx^1)^2 + (dx^2)^2 + dz^2\right] + \delta ds^2 \tag{11.3}$$

where

$$\delta ds^2 = \frac{L}{z}\Phi(x^1, x^2, z)\left[\delta(u)du^2 + \delta(v)dv^2\right] \tag{11.4}$$

and

$$u = t - x^3 \qquad v = t + x^3 \,. \tag{11.5}$$

The scalar function Φ takes the form

$$\Phi = \frac{2G_5 E}{L} \frac{1 + 8q(1+q) - 4\sqrt{q(1+q)}(1+2q)}{\sqrt{q(1+q)}} \tag{11.6}$$

where

$$q = \frac{(x^1)^2 + (x^2)^2 + (z-L)^2}{4zL} \,. \tag{11.7}$$

As we will see in section 11.2.3, q is essentially the only combination of x^1, x^2, and z that respects an $O(3)$ symmetry which preserves the worldlines of the massless particles prior to the collision.

The stress energy tensor dual to the shock wave metric is

$$\begin{aligned} \langle T_{uu} \rangle &= \frac{2L^4 E}{\pi \left[L^2 + (x^1)^2 + (x^2)^2\right]^3} \delta(u) \\ \langle T_{vv} \rangle &= \frac{2L^4 E}{\pi \left[L^2 + (x^1)^2 + (x^2)^2\right]^3} \delta(v) \,, \end{aligned} \tag{11.8}$$

with all other components vanishing.

The shock wave metric (11.3) is an *exact* solution to

$$R_{\mu\nu} = -\frac{4}{L^2} g_{\mu\nu} + (\text{massless pointlike sources}) \tag{11.9}$$

outside the causal future of the collision plane $u = v = 0$. *Inside* this causal future, i.e. for u and v positive, it's hard to compute the metric: all the difficulties of classical black hole formation live here.

Although it would be possible to spend considerably more space explaining where the results (11.6) and (11.8) come from, let me pass on instead to trapped surfaces. Because a trapped surface $\mathbf{S}$ is spacelike and co-dimension 2, there is a "normal plane" at each point along it, spanned by one timelike and one spacelike vector. More conveniently, as shown in figure 11.2, let (n^μ, ℓ^μ) be a null basis for the normal plane, with both n^μ and ℓ^μ future-directed. Let n^μ be the more inward-pointing of the two null basis vectors. If $h_{\mu\nu}$ is the induced metric on $\mathbf{S}$, then the "expansion"

$$\Theta \equiv h^{\mu\nu} \nabla_\mu \ell_\nu \tag{11.10}$$

tells us whether deforming $\mathbf{S}$ in the ℓ^μ direction makes it bigger or smaller.

- $\Theta < 0$ everywhere on $\mathbf{S}$ means that $\mathbf{S}$ is a trapped surface.

- $\Theta = 0$ everywhere on $\mathbf{S}$ means that $\mathbf{S}$ is a marginally trapped surface.

The expansion of n^μ can be defined in a similar fashion, and, at least for simple choices of $\mathbf{S}$, it is automatically negative.

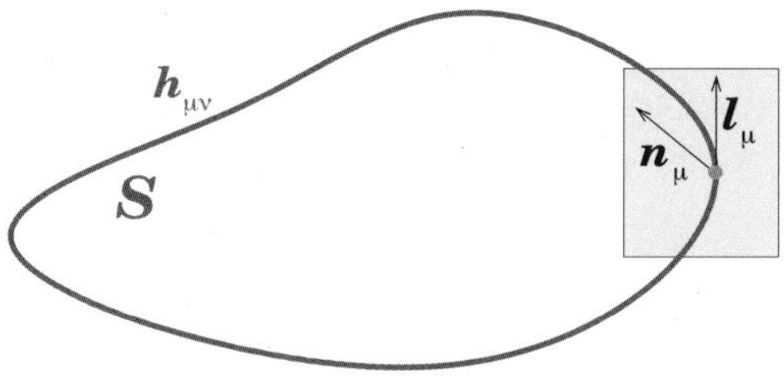

Fig. 11.2. A trapped surface $\mathbf{S}$ with induced metric $h_{\mu\nu}$ and null normal vectors (n^μ, ℓ^μ). Although the vector field ℓ^μ points more outward than n^μ, its expansion is negative.

Penrose introduced a standard choice of marginally trapped surface—understood as the outer "hull" of a family of trapped surfaces—for shocks colliding in $\mathbf{R}^{3,1}$. His choice is easily generalized to AdS_5. Figure 11.1 shows a schematic depiction of this generalized Penrose construction. The surface is the union of two halves, call them $\mathbf{S}_1$ and $\mathbf{S}_2$. $\mathbf{S}_1$ is the surface specified by the equations

$$u = 0 \qquad v = -\frac{L}{z}\Psi(x^1, x^2, z)\,, \tag{11.11}$$

while $\mathbf{S}_2$ is specified by

$$v = 0 \qquad u = -\frac{L}{z}\Psi(x^1, x^2, z)\,. \tag{11.12}$$

The function Ψ has yet to be determined, so (11.11) and (11.12) should be regarded at this point simply as ansätze. Actually, $\mathbf{S}_1$ and $\mathbf{S}_2$ should be located just slightly forward in time from the trajectories $u = 0$ and $v = 0$ of the shocks. Thus $\mathbf{S}_1$ "feels" the effects of the right-moving shock, and $\mathbf{S}_2$ feels the effects of the left-moving one.

A key point is that the cross-section of AdS_5 transverse to either of the trajectories of the massless particles is the hyperbolic space $\mathbf{H}_3$, parametrized by x^1, x^2, and z, with metric

$$ds^2_{H_3} = \frac{L^2}{z^2}\left[(dx^1)^2 + (dx^2)^2 + dz^2\right] \tag{11.13}$$

inherited from AdS_5.

With some work, and after using coordinate shifts like[a]

$$v \to v + \frac{L}{z}\Phi(x^1, x^2, z)\theta(u)\,, \tag{11.14}$$

to get rid of distributional terms in the shock-wave metric, one can check the following claims:

- The marginal trapped surface equation $\Theta = 0$ on $\mathbf{S}_1$ and $\mathbf{S}_2$ boils down to

$$\left(\square_{H_3} - \frac{3}{L^2}\right)(\Psi - \Phi) = 0\,, \tag{11.15}$$

  where $\square_{H_3}$ is the laplacian for the metric $ds^2_{H_3}$.
- Continuity of ℓ^μ as one passes from $\mathbf{S}_1$ to $\mathbf{S}_2$ across the closed curve $\mathbf{C} = \mathbf{S}_1 \cap \mathbf{S}_2$ amounts to requiring

$$\Psi\Big|_{\mathbf{C}} = 0 \qquad \text{and} \qquad (\partial\Psi)^2\Big|_{\mathbf{C}} = 4\,. \tag{11.16}$$

  Here $(\partial\Psi)^2 = g^{\mu\nu}\partial_\mu\Psi\partial_\nu\Psi$ where $g^{\mu\nu}$ is the metric on H_3.
- If we parametrize $\mathbf{S}_1$ by (x^1, x^2, z), then the induced metric on $\mathbf{S}_1$ is exactly the metric we wrote for $\mathbf{H}_3$. The same goes for $\mathbf{S}_2$.

11.2.3. *Trapped surfaces respecting the $O(3)$ symmetry*

The reason we can handle the trapped surface in an analytical fashion is that it's highly symmetrical. The purpose of this section is to explain the relevant symmetry, which we will put to greater use in the next lecture. Given a point $x^1 = x^2 = 0$, $z = L$ in $\mathbf{H}_3$—call this point P—the locus of points at a fixed geodesic distance away is a copy of S^2. The $O(3)$ we're interested in is the one that acts by ordinary rotations on this S^2. The quantity

$$q = \frac{(x^1)^2 + (x^2)^2 + (z-L)^2}{4zL} \tag{11.17}$$

which we encountered earlier is a simple function of the geodesic distance from P to (x^1, x^2, z).

$\square_{H_3}$ respects the $O(3)$ symmetry, as does $\Phi = \Phi(q)$, so it must be possible to solve the main PDE (11.15) with a function $\Psi(q)$. To find the

[a] $\theta(x)$ is the Heaviside step function, taking values 1 when $x > 0$ and 0 when $x < 0$.

area of $\mathbf{S}_1$, we don't even need to know $\Psi(q)$ in detail: it suffices to know the value q_C where $\Psi(q_C) = 0$, subject to the condition $(\partial\Psi)^2\Big|_C = 4$. These boundary conditions also respect the $O(3)$ symmetry.

Given q_C, we can compute the area A_{trapped} of the whole trapped surface $\mathbf{S}_1 \cup \mathbf{S}_2$ as twice the volume in $\mathbf{H}_3$ of the ball whose boundary is the sphere S^2 at "radius" q_C. Then S_{trapped} is computed from the formula

$$S_{\text{trapped}} = \frac{A_{\text{trapped}}}{4G_5}\,. \tag{11.18}$$

With some work, one can derive a relation between S_{trapped} and the energy E of one of the lightlike particles. (It is assumed that the energies of the two massless particles are equal. If this weren't true, an appropriate Lorentz boost would make it true.) This relation is best expressed parametrically in terms of q_C:

$$\begin{aligned}\frac{EG_5}{L^2} &= 2q_C(1+q_C)(1+2q_C)\\ \frac{S_{\text{trapped}}G_5}{L^3} &= \pi\left(x_C\sqrt{1+x_C^2} - \operatorname{arcsinh} x_C\right)\end{aligned} \tag{11.19}$$

where

$$x_C = 2\sqrt{q_C(1+q_C)}\,. \tag{11.20}$$

The main result (11.1) can be obtained by expanding the relations (11.19) at leading order in large q_C and x_C. So in a sense we're done. But I've left out a lot, both in the derivation of the result and in its significance and possible generalizations. I'll outline here a number of points which the interested reader could explore further. Many of these points are well addressed in existing literature, for instance these works[4,10,11] and references therein. My main reason for providing such a long list of questions here is that these were approximately the questions I was asking myself at the stage of understanding where I had the main result (11.1) more or less straight but was not fully confident of all the supporting details.

- How do you arrive at the shock wave metric (11.3) with the specific functional form (11.6)?
- How do you get $\langle T_{uu}\rangle = \frac{2L^4E}{\pi[L^2+(x^1)^2+(x^2)^2]^3}\delta(u)$ starting from the shock wave metric?
- How do you show that $\Theta < 0$ everywhere is the condition for a trapped surface?

- In what generality can you show that a marginally trapped surface (one with $\Theta = 0$) is the outer hull of a family of trapped surfaces?
- How do you show that trapped surfaces have to be behind an event horizon?
- How do you derive (11.15) from $\Theta = 0$ and $(\partial\Psi)^2\Big|_C = 4$ from continuity of ℓ^μ?
- How do you choose parameters G_5, L, and E for a meaningful comparison to heavy ion physics? How meaningful is this comparison?
- How do you estimate the entropy produced in a heavy ion collision starting from data?
- What other theoretical approaches are there for getting at this entropy?
- How does the main result, (11.1), generalize to other dimensions?
- Given that the dependence $S \propto E^{2/3}$ is somewhat too rapid as compared to data, are there sensible ways to change the AdS_5 calculation that would improve the match to data?
- Can one handle the case of off-center collisions, at least in some approximation?
- Can one at least approximately solve Einstein's equations in the future region, $u > 0$ and $v > u$, and extract some useful information about thermalization, stopping, and the rapidity distribution of matter produced in the collision?
- What's the significance of the $O(3)$ symmetry in field theory terms?

11.3. Lecture 2: $O(3)$ Symmetry and Bjorken Flow

Of all the questions I listed, the one that stuck with me the most was the last: Exactly what does the $O(3)$ symmetry do for you in the dual field theory? Obviously it's crucial for the entire trapped surface story: it would be very difficult to find the function Ψ if you had to deal with (11.15) as a true PDE in three variables.

Motivated by a discussion of RHIC-phenomenological hydrodynamics with U. Heinz, I decided to look into how the $O(3)$ symmetry might constrain conformal relativistic hydrodynamics. It turns out that there's a free lunch waiting to be eaten here: a rare treat! To get at it, the first step is to understand Bjorken flow.[12]

11.3.1. *Bjorken flow*

Consider a collision of highly relativistic heavy ions. The causal future of the collision plane in $\mathbf{R}^{3,1}$ can be parametrized as

$$\begin{pmatrix} t \\ x^3 \\ x^1 \\ x^2 \end{pmatrix} = \begin{pmatrix} \tau \cosh \eta \\ \tau \sinh \eta \\ x_\perp \cos \phi \\ x_\perp \sin \phi \end{pmatrix} . \tag{11.21}$$

The Minkowski metric can be expressed as

$$ds^2_{\mathbf{R}^{3,1}} = -d\tau^2 + \tau^2 d\eta^2 + dx_\perp^2 + x_\perp^2 d\phi^2 . \tag{11.22}$$

The variables $(\tau, \eta, x_\perp, \phi)$ make obvious two commuting isometries of $\mathbf{R}^{3,1}$: $SO(1,1)$ boosts, generated by $\partial/\partial\eta$, and $SO(2)$ rotations, generated by $\partial/\partial\phi$.

Part of Bjorken's setup is to assume that the post-thermalization dynamics of the QGP approximately respects both of these symmetries. For $SO(2)$, that's pretty trivial: it just means that we're focusing on head-on collisions. $SO(1,1)$ symmetry is *not at all* obvious, and it can only hold not too far from $\eta = 0$. Bjorken gave an argument[12] for $SO(1,1)$ symmetry based on how scattering of small x partons leads to a locally thermalized medium.

Bjorken also assumed symmetry under the translations generated by $\partial/\partial x^1$ and $\partial/\partial x^2$. This means that the finite size nucleus is replaced by an infinite sheet of matter filling the x^1-x^2 plane. Conformal relativistic hydrodynamics constrained by the symmetries

$$\left\{ \frac{\partial}{\partial\eta}, \frac{\partial}{\partial\phi}, \frac{\partial}{\partial x^1}, \frac{\partial}{\partial x^2} \right\} \tag{11.23}$$

can be *solved exactly.*[b] Here's how it's done.

The stress tensor takes the form

$$T_{mn} = \epsilon u_m u_n + p(g_{mn} + u_m u_n) + (\text{viscous corrections}) , \tag{11.24}$$

and conformal symmetry dictates $p = \epsilon/3$, so that $T^m{}_m = 0$. The local four-velocity u_m is constrained to have unit norm: $g^{mn} u_m u_n = -1$. The symmetries listed in (11.23) imply that u_m can only depend on τ, and that

[b]It is not necessary to make the assumption of conformal symmetry at this point. I do so for later convenience, and because the conformal equation of state $p = \epsilon/3$ is semi-realistic for highly energetic collisions.

$u_{x^1} = u_{x^2} = 0$. An additional simplification which is reasonable in the center-of-mass frame of the collision of identical nuclei is to set $u_\eta = 0$. So only u_τ is non-zero, and the unit norm constraint dictates that $u_\tau = -1$. (The sign of u_τ is fixed by requiring that u^m should be a future-directed vector, i.e. $u^\tau > 0$.)

Straightforward calculations now show that the conservation equations, $\nabla^m T_{mn} = 0$, boil down in the inviscid case to

$$\frac{\partial}{\partial\tau}\left(\tau^{4/3}\epsilon\right) = 0\,. \tag{11.25}$$

So we conclude that

$$\epsilon = \frac{\tilde{\epsilon}_0}{\tau^{4/3}}\,, \qquad u^\tau = 1\,, \quad \text{other } u^m = 0\,, \tag{11.26}$$

where $\tilde{\epsilon}_0$ is an integration constant.

11.3.2. *Massaging the symmetries*

Let's have another look at the *symmetries of Bjorken flow*, as listed in (11.23). Boost invariance commutes with the other three symmetries. We can express

$$\begin{aligned}\frac{\partial}{\partial x^1} &= \cos\phi\frac{\partial}{\partial x_\perp} - \frac{\sin\phi}{x_\perp}\frac{\partial}{\partial\phi}\\ \frac{\partial}{\partial x^2} &= -\sin\phi\frac{\partial}{\partial x_\perp} - \frac{\cos\phi}{x_\perp}\frac{\partial}{\partial\phi}\,.\end{aligned} \tag{11.27}$$

Starting from (11.27), it is easy to check the commutation relations

$$\left[\frac{\partial}{\partial\phi}, \frac{\partial}{\partial x^1}\right] = \frac{\partial}{\partial x^2} \qquad \left[\frac{\partial}{\partial\phi}, \frac{\partial}{\partial x^2}\right] = -\frac{\partial}{\partial x^1}\,. \tag{11.28}$$

These relations should remind you of the $SO(3)$ commutators

$$[J^3, J^1] = iJ^2 \qquad [J^3, J^2] = -iJ^1\,. \tag{11.29}$$

But $\left[\frac{\partial}{\partial x^1}, \frac{\partial}{\partial x^2}\right] = 0$, whereas $[J^1, J^2] = iJ^3$. So the generators $\left\{\frac{\partial}{\partial\phi}, \frac{\partial}{\partial x^1}, \frac{\partial}{\partial x^2}\right\}$ do *not* form the algebra $SO(3)$; instead they generate the group $ISO(2)$, which is a contraction of $SO(3)$, just as $\mathbf{R}^2$ is a large-radius limit of S^2.

I want to "un-contract" $ISO(2)$ back into $SO(3)$ while maintaining the property that all generators commute with $\partial/\partial\eta$. The $SO(3)$ should be part of the group $SO(4,2)$ of conformal transformations of $\mathbf{R}^{3,1}$, which is also the group of isometries of AdS_5. In fact, the $SO(3)$ I'm interested in is supposed to be the $SO(3)$ part of the $O(3)$ symmetry of $\mathbf{H}_3$ which figured prominently in section 11.2.3. If we insist that $\partial/\partial\phi$ should remain as one of the generators of $SO(3)$, then there's essentially only one possible deformation of $\partial/\partial x^1$ and $\partial/\partial x^2$ that will accomplish what we want:

$$\frac{\partial}{\partial x^1} \to \zeta \qquad\qquad \frac{\partial}{\partial x^2} \to \left[\frac{\partial}{\partial\phi}, \zeta\right], \tag{11.30}$$

where

$$\begin{aligned}\zeta \equiv 2q^2\tau x_\perp \cos\phi \frac{\partial}{\partial\tau} + (1 + q^2\tau^2 + q^2 x_\perp^2)\cos\phi \frac{\partial}{\partial x_\perp} \\ - \frac{1 + q^2\tau^2 - q^2 x_\perp^2}{x_\perp}\sin\phi \frac{\partial}{\partial\phi}\end{aligned} \tag{11.31}$$

and q is a parameter with dimensions of momentum. (Sorry, q has nothing to do with distance on $\mathbf{H}_3$.)

Recall in the shock wave story that a single shock, $\langle T_{uu}\rangle = \frac{2L^4 E}{\pi[L^2+(x^1)^2+(x^2)^2]^3}\delta(u)$, respects the $SO(3)$ symmetry in AdS_5 (but obviously, it does not respect $SO(1,1)$). A slightly subtle analysis allows one one to identify $q = 1/L$. Noting that L is, in field theory terms, the $\langle T_{uu}\rangle$-weighted root-mean-square (rms) transverse radius of the boundary shockwave, we see that $1/q$ is essentially the transverse size of the colliding object.[c]

11.3.3. *Conformal isometries and hydrodynamics*

Bjorken was able to solve *completely* for u_m just by demanding that u_m respects the symmetries $SO(1,1) \times ISO(2)$, together with setting $u_\eta = 0$, which amounts to imposing an additional $\mathbf{Z}_2$ symmetry which acts as $\eta \to -\eta$. In the previous section we saw how to replace $ISO(2)$ by $SO(3)$, which has just as many generators. So we might ask, can we use the symmetry group $SO(1,1)\times SO(3)\times\mathbf{Z}_2$ to completely determine the velocity field u_m?

[c] L is also the radius of AdS_5. The astute reader may be wondering why this radius equals the rms transverse radius of the shockwave in field theory. The answer is that I chose from the start to put the massless particles in AdS_5 at a depth $z = L$. This simplifies some formulas, but it is inessential. If I had put the massless particles at some other depth $z = z_*$, then in field theory, L would get replaced by z_*.[10]

The answer is yes, but the details are a little tricky, and to explain them I'm going to have to remind you of how Lie derivatives work. The defining relations are

$$\begin{aligned}\mathcal{L}_\xi \phi &= \xi^n \frac{\partial}{\partial x^n}\phi \\ \mathcal{L}_\xi v^m &= \xi^n \frac{\partial}{\partial x^n} v^m - v^n \frac{\partial}{\partial x^n}\xi^m \\ \mathcal{L}_\xi \omega_m &= \xi^n \frac{\partial}{\partial x^n}\omega_m + \omega_n \frac{\partial}{\partial x^m}\xi^n ,\end{aligned} \tag{11.32}$$

where ξ^m and v^m are vector fields, ω_n is a 1-form, and ϕ is a scalar, all defined on $\mathbf{R}^{3,1}$, and all assumed to have appropriate smoothness properties. $\mathcal{L}_\xi$ is linear, and it obeys Leibniz's Rule. Crucially for purposes to come,

$$\mathcal{L}_\xi g_{mn} = \nabla_m \xi_n + \nabla_n \xi_m = 0 \tag{11.33}$$

is the condition for ξ^m to be an isometry of the metric g_{mn}. What Bjorken did, in essence, to determine the four-velocity u_m, was to solve the equations

$$\mathcal{L}_\xi u_m = 0 \qquad \text{for} \qquad \xi \in \left\{ \frac{\partial}{\partial \eta}, \frac{\partial}{\partial \phi}, \frac{\partial}{\partial x^1}, \frac{\partial}{\partial x^2} \right\} , \tag{11.34}$$

with $u_\eta = 0$.

When we pass from $ISO(2)$ to $SO(3)$, the main complication is that the generators ζ and $\left[\zeta, \frac{\partial}{\partial \phi}\right]$ are not isometries of $\mathbf{R}^{3,1}$, but instead conformal isometries: for example,

$$\mathcal{L}_\zeta g_{mn} = \frac{1}{2}(\nabla_\ell \zeta^\ell) g_{mn} . \tag{11.35}$$

The result (11.35) is the infinitesimal statement of the fact that ζ^m generates conformal maps sending $g_{mn} \to \Omega^2 g_{mn}$, where the factor Ω depends on space and time.

We see from (11.24) that the projection tensor $P_{mn} = g_{mn} + u_m u_n$ plays a key role in hydrodynamics. Physically, this is the tensor which projects onto the spatial coordinates of the local rest frame of the fluid. In order for P_{mn} to transform nicely under conformal maps, we should demand that

$$\mathcal{L}_\zeta u_m = \frac{1}{4}(\nabla_\ell \zeta^\ell) u_m . \tag{11.36}$$

Together with the constraints $u_\eta = 0$, $\mathcal{L}_{\frac{\partial}{\partial \eta}} u_m = 0$, and $\mathcal{L}_{\frac{\partial}{\partial \phi}} u_m = 0$, the equation (11.36) is enough to determine u_m:

$$u^\tau = \cosh\kappa \qquad\qquad u^{x_\perp} = \sinh\kappa \tag{11.37}$$

where

$$\kappa = \operatorname{arctanh} \frac{2q^2\tau x_\perp}{1+q^2\tau^2+q^2x_\perp^2}\,, \tag{11.38}$$

and $u^\eta = u^\phi = 0$.

For Bjorken flow, the next step would be to demand $\mathcal{L}_\xi \epsilon = 0$ for all the isometries ξ (regarding ϵ as a scalar, so that $\mathcal{L}_\xi \epsilon = \xi^n \frac{\partial}{\partial x^n}\epsilon$). That would let us conclude that ϵ is a function only of τ. To obtain the explicit form for $\epsilon(\tau)$ that we listed in (11.26), it's necessary to resort to the conservation equations $\nabla^m T_{mn} = 0$.

In the $SO(3)$-symmetric case, it still make sense to require $\mathcal{L}_\xi \epsilon = 0$ for $\xi = \frac{\partial}{\partial \eta}$ and $\frac{\partial}{\partial \phi}$: this just implies $\epsilon = \epsilon(\tau, x_\perp)$. But $\mathcal{L}_\zeta \epsilon = 0$ might be the wrong equation, given that ζ is only a conformal isometry. Instead let's try

$$\mathcal{L}_\zeta \epsilon = -\frac{\alpha}{4}(\nabla_\ell \zeta^\ell)\epsilon\,, \tag{11.39}$$

where α is a constant.[d]

It's easy to show that the general solution to (11.39) is

$$\epsilon = \frac{\hat{\epsilon}(g)}{\tau^\alpha} \qquad \text{where} \qquad g = \frac{1-q^2\tau^2+q^2x_\perp^2}{2q\tau}\,, \tag{11.40}$$

and $\hat{\epsilon}(g)$ is an arbitrary function. g is essentially the only $SO(3)$-invariant combination of τ and $x_\perp$. After a bit of work, one finds that the equations $\nabla^m T_{mn} = 0$ are consistent with one another iff $\alpha = 4$, and that the general solution is

$$\hat{\epsilon}(g) = \frac{\hat{\epsilon}_0}{(1+g^2)^{4/3}}\,. \tag{11.41}$$

Thus the $SO(3)$-invariant flow takes the final form

$$\begin{aligned} \epsilon &= \frac{\hat{\epsilon}_0}{\tau^{4/3}} \frac{(2q)^{8/3}}{[1+2q^2(\tau^2+x_\perp^2)+q^4(\tau^2-x_\perp^2)^2]^{4/3}} \\ v^\perp &\equiv \frac{u^{x_\perp}}{u^\tau} = \frac{2q^2\tau x_\perp}{1+q^2\tau^2+q^2x_\perp^2}\,. \end{aligned} \tag{11.42}$$

Let me close this lecture with a few thoughts on generalizations and relevance to phenomenology and gravity duals.

[d] A considerably more systematic treat is possible here, based on the formalism introduced by Loganayagam;[13] but the way I am explaining the problem here is closer to what I actually did when I didn't know the answer.

- One can add in viscous corrections and still get exact closed form expressions for ϵ. Of course, u^m doesn't change: it is fixed by symmetry considerations alone. Assorted other generalizations are possible, as explained in work that appeared after my TASI lectures were delivered.[6]
- The transverse velocity $v^\perp$ is a quantity of considerable phenomenological interest (e.g. for single particle yields and Hanbury Brown-Twiss radii). Having a symmetry argument that determines it is interesting, even if this symmetry is somewhat broken by real heavy ion collisions.
- I should stress that I do *not* claim that this deformation of Bjorken flow will describe $\langle T_{mn} \rangle$ in the dual of a point-sourced shock wave collision in AdS_5. The $SO(3)$ symmetry is common to both situations, but in constructing a solution to hydro I have discarded all dynamical information from AdS/CFT and used instead the assumptions of that the flow has $SO(1,1)$ boost invariance and that hydrodynamics is valid.

11.4. Lecture 3: *p*-wave Holographic Superconductors

11.4.1. *Overview of the main results*

In *p*-wave holographic superconductors,[14–16] the Fermi surface degenerates to a pair of points, above each of which a Dirac cone rises, enclosing a continuum of fermion modes. There are also discrete fermion normal modes slightly outside the Dirac cones. See figure 11.3. This structure is in contrast to the normal state, where the Fermi surface is a circle (when the field theory is in $2+1$ dimensions), and a non-trivial power law governs the response at small but non-zero frequencies.[17]

The underlying lagrangian on which all the calculations on this topic will be based is

$$\mathcal{L} = R + \frac{6}{L^2} - \frac{1}{2} \operatorname{tr} F_{\mu\nu}^2 - i\Psi\Gamma^\mu D_\mu \Psi \tag{11.43}$$

in bulk spacetime dimension $D = 4$, where

$$F_{\mu\nu} = \partial_\mu A_\nu - \partial_\nu A_\mu - ig_{\text{YM}}[A_\mu, A_\nu] \tag{11.44}$$

and $A_\mu = A_\mu^a \tau^a$. The matrices $\tau^a = \frac{1}{2}\sigma^a$, with $a = 1, 2, 3$, are the generators of $SU(2)$. Ψ is a doublet of $SU(2)$, and

$$D_\mu \Psi = (\nabla_\mu - ig_{\text{YM}} A_\mu^a \tau^a)\Psi \,, \tag{11.45}$$

where ∇_μ includes the spin connection.

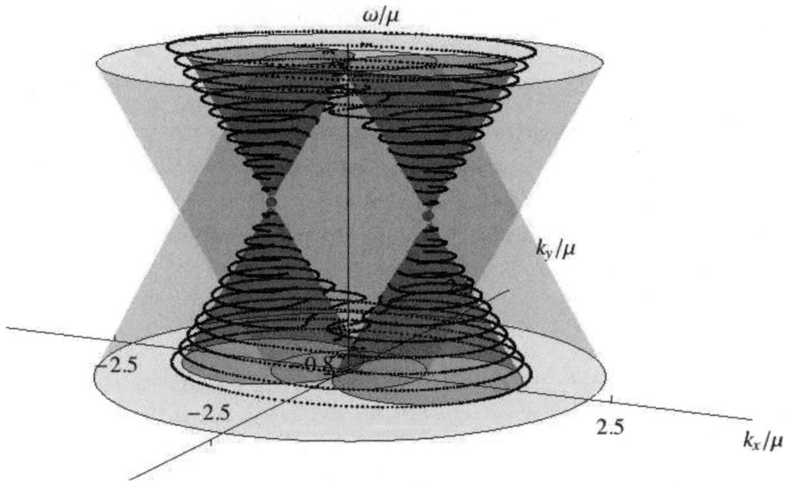

Fig. 11.3. The spectral structure of the two-point function G^Ψ of a fermion operator in a p-wave holographic superconductor at zero temperature. The dark cones are the Dirac cones whose apexes are the locations of gapless fermion excitations. The black circles around the Dirac cones are the locations of fermion normal modes. From the original literature.[7]

The simplest solution to the equations of motion is AdS_4:

$$ds^2 = \frac{L^2}{z^2}\left[-dt^2 + (dx^1)^2 + (dx^2)^2 + dz^2\right], \tag{11.46}$$

with $F_{\mu\nu} = 0$. The $T \to 0$ limit of the p-wave holographic superconductor is an AdS_4 to AdS_4 domain wall[18] (c.f. the $T \to 0$ limit of RNAdS, which interpolates between AdS_4 and $AdS_2 \times \mathbf{R}^2$). The gauge field interpolates between two flat gauge connections: A_μ^{UV} near the boundary, and A_μ^{IR} in the deep infrared.

The main results on fermion two-point functions at zero temperature[7] can be understood starting from the gauge-covariant wave-vector:

$$K_m \equiv k_m - g_{\mathrm{YM}} A_m = k_m \mathbf{1} - g_{\mathrm{YM}} A_m^a \tau^a, \tag{11.47}$$

where $\mathbf{1}$ is the 2×2 identity matrix. We also define

$$K_m^{\mathrm{IR}} = k_m - g_{\mathrm{YM}} A_m^{\mathrm{IR}} \qquad K_m^{\mathrm{UV}} = k_m - g_{\mathrm{YM}} A_m^{\mathrm{UV}}. \tag{11.48}$$

In (11.47), (11.48), and below, the index m runs over boundary directions, i.e. $m = 0, 1, 2$. Also, we require $A_z = 0$ throughout: this is a gauge choice. If $\mathcal{O}_\Psi$ is the operator in field theory dual to the fermion Ψ, then

$$\langle \mathcal{O}_\Psi^\dagger \mathcal{O}_\Psi \rangle \sim G_{\mathrm{sudden}}^\Psi(k) \equiv -i(\gamma^m q_m)^{-1}(q + \gamma^m \gamma^n q_{mn})\gamma^t, \tag{11.49}$$

where

$$
\begin{aligned}
q &= K_{\rm IR}\cosh(z_* K_{\rm UV}) \\
q_m &= -i\left[K_m^{\rm IR}\cosh(z_* K_{\rm UV}) + K_{\rm IR}K_m^{\rm UV}\frac{\sinh(z_* K_{\rm UV})}{K_{\rm UV}}\right] \\
q_{mn} &= K_m^{\rm IR}K_n^{\rm UV}\frac{\sinh(z_* K_{\rm UV})}{K_{\rm UV}}
\end{aligned}
\tag{11.50}
$$

and

$$
K_{\rm IR} = \sqrt{\eta^{mn}K_m^{\rm IR}K_n^{\rm IR}} \qquad K_{\rm UV} = \sqrt{\eta^{mn}K_m^{\rm UV}K_n^{\rm UV}}\,. \tag{11.51}
$$

As I will explain, z_* is a length scale characteristic of the bosonic background interpolating between AdS_4 and AdS_4. $G^{\Psi}_{\rm sudden}(k)$ is the expression obtained for the fermion correlator when this background is treated in a thin wall approximation. This approximation is not controlled in the sense of being approached as one dials a parameter of the lagrangian to an extreme value. However, it captures the qualitative features of the fermion two-point function, which is otherwise accessible only through numerics.

The continuous part of the spectral weight of $G^{\Psi}_{\rm sudden}(k)$ arises precisely where $K_{\rm IR}$ has a branch cut—i.e. where $K_m^{\rm IR}$ is timelike—because otherwise, $G^{\Psi}_{\rm sudden}(k)$ is a rational function of the k_m. $K_m^{\rm IR}$ is timelike inside the aforementioned Dirac cones.

The motivations for working out the fermion response to p-wave holographic superconductors were numerous:

- We knew about holographic superconductors, both s-wave and p-wave.
- We knew about fermion correlators in the normal state.
- To have some chance at successful comparison to ARPES, where Dirac cones above isolated points on the Fermi surface are observed, we knew we needed non-s-wave dynamics.
- The lagrangian we chose is almost completely determined at the two-derivative level by its symmetries: basically it's QCD with $N_c = 2$ and $N_f = 1$ (with lagrangian $-\frac{1}{2}\operatorname{tr}F^2 - i\bar\Psi \not{D}\Psi$) coupled to gravity with a negative cosmological constant (with lagrangian $R + \frac{6}{L^2}$).
- It's easy to get lagrangians similar to (11.43) out of string/M-theory low-energy effective actions.
- The AdS_4 to AdS_4 domain wall structure had recently been explained.[18]

11.4.2. *Generalities on holographic superconductors*

The main macroscopic features of superconductors are a consequence of the spontaneous breaking of $U(1)_{\rm EM}$, at finite T and finite chemical potential μ for charge carriers. In much of the theory of superconductivity (including the classic Bardeen-Cooper-Schrieffer theory of low-temperature s-wave superconductors), $U(1)_{\rm EM}$ is treated as a global symmetry for purposes of calculations of the gap, the condensate, and other properties; it later can be weakly gauged. In this spirit, consider a field theory on $\mathbf{R}^{2,1}$ with a global $U(1)$ symmetry and an AdS_4 dual. The field content of the dual gravity theory must contain an abelian gauge field, dual to the conserved current J^μ in the boundary theory. Thus, on the gravity side, we should consider

$$\mathcal{L} = R + \frac{6}{L^2} - \frac{1}{4}F_{\mu\nu}^2 + (\text{matter fields})\,, \tag{11.52}$$

where the matter fields can be charged or uncharged.

First let's consider the s-wave case.[19,20] The matter fields include

$$\mathcal{L}_\phi = -\left|(\partial_\mu - iqA_\mu)\phi\right|^2 - m^2|\phi|^2 + \dots\,, \tag{11.53}$$

where ϕ is a complex scalar. In the normal state, $\phi = 0$ because only vanishing ϕ is preserved by $U(1)$ rotations. The simplest solution with non-zero gauge field is then Reissner-Nordstrom AdS_4 (RNAdS). I will not need to consider the detailed form of RNAdS. The qualitative features shown in figure 11.4A are enough. In particular, the electric field is

$$E_z = F_{0z} = -\partial_z\Phi \tag{11.54}$$

where $A_0 = \Phi$. The chemical potential is the amount of energy it takes to push a unit of charge from the boundary into the horizon:

$$\mu = \Phi_{\rm bdy} - \Phi_{\rm horizon}\,. \tag{11.55}$$

But $\Phi_{\rm horizon}$ must be set to 0 in order for $A = \Phi dt$ to be well-defined at the horizon.

The key question is: Will the scalar condensed outside the horizon? Heuristically, the answer is YES, provided $q \neq 0$, and m is not too big, and T is sufficiently small. A naive way to reason this out is to consider the balance of forces on a test particle slightly above the horizon, as illustrated in figure 11.4B. The electrostatic force $F_{\rm up} = qE$ overcomes the gravitational force $F_{\rm down} = mg$ provided $qE > mg$. Then quanta of ϕ want to jump out of the black hole (see figure 11.4C). But because of the infinite blueshift

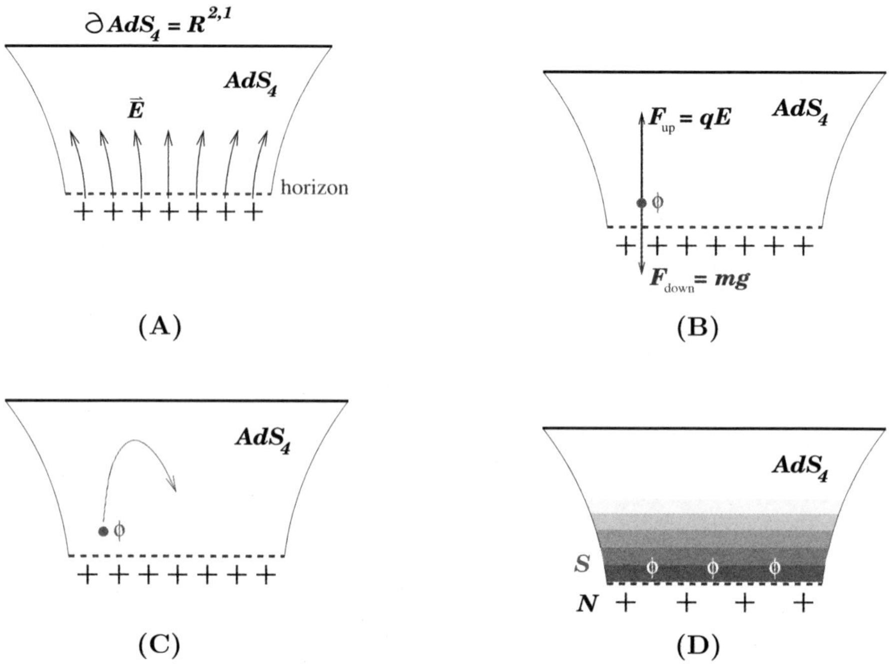

Fig. 11.4. A qualitative account of holographic superconductors, in pictures. (A) The normal state is described in terms of the RNAdS solution, with boundary $\mathbf{R}^{2,1}$. (B) The upward electrostatic force on a charged quantum of a scalar field ϕ can be greater than the downward gravitational pull. (C) The AdS_4 asymptotics prevent particles from escaping arbitrarily far from the horizon. (D) The quanta of ϕ instead condense just outside the horizon.

of the AdS_4 metric near the boundary, nothing can escape. So the simplest endpoint for the dynamics is for the charged bosonic field to condense near the horizon, as in figure 11.4D. Now, the surface gravity of the horizon is related to the Hawking temperature by $g = 2\pi T$, so we expect a condensate $\phi \neq 0$ for T less than some critical value T_c.

The p-wave case[14–16] is a variant of the s-wave story: Instead of using a complex scalar ϕ as the charged matter field, we promote $F_{\mu\nu}$ to an $SU(2)$ field strength. If the original $U(1)$ is associated with the τ^3 part of $SU(2)$, then $A_\mu^\pm \equiv A_\mu^1 \pm iA_\mu^2$ are fields with charges $q = \pm g_{\rm YM}$: this is just about how $W^\pm$ bosons arise, except that here we have no Higgs field and no $U(1)_Y$ hypercharge gauge group. As mentioned earlier, work of Basu and collaborators[18] demonstrated that in the zero-temperature limit, p-wave holographic superconductors take the form of AdS_4-to-AdS_4 domain walls.

In the next section, I will explain a simplified version of this construction in a limit where the gauge field doesn't back-react on the metric.

11.4.3. *Domain wall backgrounds in the probe limit*

In the limit $g_{\mathrm{YM}} \to \infty$, the gauge field doesn't back-react on the geometry. To see this, define

$$\hat{A}_\mu = A_\mu / g_{\mathrm{YM}} \qquad \hat{F}_{\mu\nu} = \partial_\mu \hat{A}_\nu - \partial_\nu \hat{A}_\mu - i[\hat{A}_\mu, \hat{A}_\nu] \,. \tag{11.56}$$

Then the bosonic lagrangian takes the form

$$\mathcal{L} = R + \frac{6}{L^2} - \frac{1}{2g_{\mathrm{YM}}^2} \operatorname{tr} \hat{F}_{\mu\nu}^2 \,. \tag{11.57}$$

The $1/g_{\mathrm{YM}}^2$ suppression means we can solve the equations of motion of $\mathcal{L}_{\mathrm{grav}} = R + \frac{6}{L^2}$ first to get (at zero temperature) AdS_4, as in (11.46); then we can solve the classical Yang-Mills equation in this background. AdS_4 is conformal to the $z > 0$ half of Minkowski space $\mathbf{R}^{3,1}$, and the classical Yang-Mills equations are conformally invariant. So we can solve them on $\mathbf{R}^{3,1}$ instead of on AdS_4, starting with the ansatz

$$\hat{A} = \Phi \tau^3 dt + W \tau^1 dx^1 \,. \tag{11.58}$$

The $\Phi \tau^3 dt$ term in (11.58) is needed in order to describe the $U(1)$ chemical potential. The $W\tau^1 dx^1$ term is the simplest expression that spontaneously breaks the $U(1)$ generated by τ^3. Both Φ and W are required to be functions only of z, and one can easily demonstrate that the Yang-Mills equations boil down to

$$\frac{d^2\Phi}{dz^2} = W^2 \Phi \qquad \frac{d^2 W}{dz^2} = -\Phi^2 W \,. \tag{11.59}$$

Appropriate boundary conditions are

$$\begin{aligned} &\Phi \to 0\,, \quad W \to W_{\mathrm{IR}} && \text{as } z \to \infty \text{ (the infrared)} \\ &\Phi \to \mu\,, \quad W \to 0 && \text{as } z \to 0 \text{ (the ultraviolet).} \end{aligned} \tag{11.60}$$

Requiring $W \to 0$ in the ultraviolet is the condition that the symmetry breaking must be spontaneous: we are deforming the CFT lagrangian only by the $U(1)$-symmetric term μJ_0^3, where J_0^3 is the charge density for the τ^3 part of $SU(2)$.

The solution to (11.59) with the boundary conditions (11.60) is essentially unique, and it is shown in figure 11.5. Soon we will want to make a further approximation: replace Φ and W by step functions:

$$W_{\mathrm{sudden}}(z) = W_{\mathrm{IR}}\, \theta(z_* - z) \qquad \Phi_{\mathrm{sudden}}(z) = \mu\, \theta(z - z_*)\,, \tag{11.61}$$

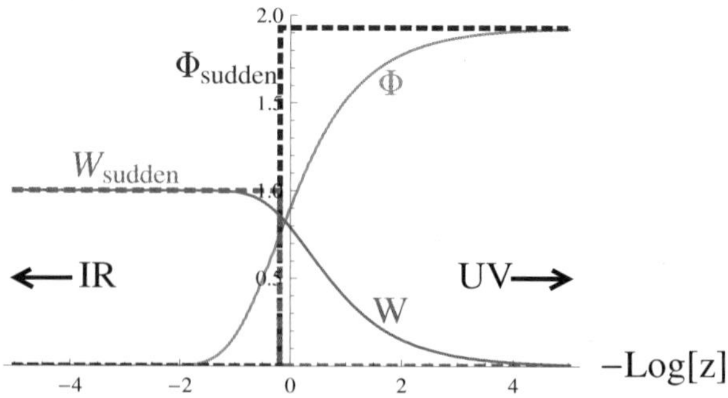

Fig. 11.5. The solution to (11.59) with boundary conditions (11.60). The step functions $\Phi_{\rm sudden}$ and $W_{\rm sudden}$, plotted as dashed lines, provide the thin wall, or sudden, approximation to this solution.

where z_* is defined so that

$$\int_0^\infty dz\, \Phi_{\rm sudden}(z) = \int_0^\infty dz\, \Phi(z)\,. \tag{11.62}$$

11.5. Lecture 4: Fermion Correlators and the Sudden Approximation

11.5.1. *Extracting the fermion two-point function*

We saw in the previous lecture that constructing a p-wave holographic superconductor at zero temperature reduces in the large $g_{\rm YM}$ limit to finding a domain wall solution to the classical Yang-Mills equations in the $z > 0$ half of flat four-dimensional Minkowski space. This is charming, because most holographic superconductors are governed by more complicated equations. What made things work is the invariance of the Yang-Mills equations under conformal transformations. The massless Dirac equation is also essentially invariant under conformal transformations: Defining

$$\psi = \left(\frac{L}{z}\right)^{3/2} \Psi\,, \tag{11.63}$$

where ψ is regarded as a fermion on $\mathbf{R}^{3,1}$ and Ψ is the original fermion on AdS_4, one finds that

$$\Gamma^\mu(\nabla_\mu - i\hat{A}_\mu)\Psi = 0 \tag{11.64}$$

on AdS_4 is equivalent to

$$\Gamma^\mu(\partial_\mu - i\hat{A}_\mu)\psi = 0 \tag{11.65}$$

on the $z > 0$ part of $\mathbf{R}^{3,1}$.

The obvious ansatz for solving (11.65) is

$$\psi(x^m, z) = e^{ik_m x^m}\hat{\psi}(z)\,. \tag{11.66}$$

Recall that we defined a covariant wave-vector as

$$K_m = k_m - \hat{A}_m\,. \tag{11.67}$$

Plugging (11.66) into (11.65) and using (11.67) leads directly to $(\Gamma^m iK_m + \Gamma^z\partial_z)\psi = 0$, and recalling that $(\Gamma^z)^2 = 1$, we see that

$$(\partial_z + i\Gamma^z\Gamma^m K_m)\hat{\psi} = 0\,. \tag{11.68}$$

Formally, the solution to this equation is

$$\hat{\psi} = P\left\{e^{-i\int_0^z dz'\,\Gamma^z\Gamma^m K_m(z')}\right\}\hat{\psi}(0)\,. \tag{11.69}$$

The solution we want for computing Green's functions has asymptotic behavior

$$\hat{\psi} \propto e^{-K_{\rm IR}z}u \qquad \text{for large } z, \tag{11.70}$$

where u is a constant spinor. The reason that the solution (11.70) is preferred over the solution proportional to $e^{+K_{\rm IR}z}$ is that only the solution (11.70) has the property that $\hat{\psi} \to 0$ as $z \to \infty$ when $K_m^{\rm IR}$ is spacelike.[e] When $K_m^{\rm IR}$ is not spacelike, then a $+i\epsilon$ prescription can be used to formulate an appropriate pole-passing description that fully defines the Green's function. For the retarded Green's function, this $+i\epsilon$ prescription can be found by demanding that (11.70) is infalling at the horizon when $K_m^{\rm IR}$ is not spacelike.

Noting the trivial identity $(\partial_z + K_{\rm IR})e^{-K_{\rm IR}z}u = 0$, we see that the condition on $\hat{\psi}(z)$ is

$$\lim_{z\to 0}(K_{\rm IR} - i\Gamma^z\Gamma^m K_m)\hat{\psi}(z) = 0\,, \tag{11.71}$$

which is equivalent to

$$(K_{\rm IR} - i\Gamma^z\Gamma^m K_m^{\rm IR})P\left\{e^{-i\int_0^\infty dz\,\Gamma^z\Gamma^m K_m(z)}\right\}\hat{\psi}(0) = 0\,. \tag{11.72}$$

[e]It is slightly subtle to say what we mean in describing $K_m^{\rm IR}$ as spacelike, because each of the $K_m^{\rm IR}$ is a matrix. Fortunately, they are commuting matrices, so one can find simultaneous eigenspaces. On any one of these simultaneous eigenspaces, the $K_m^{\rm IR}$ act as numbers, and it is these numbers which are assumed to fill out a spacelike three-vector.

Formally, (11.72) takes the form $\mathbf{P}\hat{\psi}(0) = 0$ for a matrix $\mathbf{P}$ that has four-valued Dirac and two-valued $SU(2)$ indices.

The usual basis for Γ^μ in this type of calculation is

$$\Gamma^m = \begin{pmatrix} 0 & \gamma^m \\ \gamma^m & 0 \end{pmatrix} \qquad \Gamma^z = \begin{pmatrix} -1 & 0 \\ 0 & 1 \end{pmatrix}, \tag{11.73}$$

where

$$\gamma^t = i\sigma_2 \qquad \gamma^1 = \sigma_1 \qquad \gamma^2 = \sigma_3\,, \tag{11.74}$$

so that $\{\gamma^m, \gamma^n\} = 2\eta^{mn} = 2\,\mathrm{diag}\{-1, 1, 1\}$. If in this basis we express

$$\mathbf{P} = \begin{pmatrix} P_{++} & P_{+-} \\ P_{-+} & P_{--} \end{pmatrix}, \tag{11.75}$$

then

$$G^\Psi(k) = iP_{+-}^{-1} P_{++} \gamma^t = iP_{--}^{-1} P_{-+} \gamma^t\,. \tag{11.76}$$

A justification of (11.76) can be found in the original literature.[7]

11.5.2. *Simplifications based on the sudden approximation*

Because the $K_m(z)$'s don't commute at different values of z, we need something extra to make further progress with analytical methods. So let's use the sudden approximation:

$$\hat{\psi}_{\mathrm{sudden}}(z) = \begin{cases} e^{-iz\Gamma^z\Gamma^m K_m^{\mathrm{UV}}} \hat{\psi}(0) & \text{for } 0 < z < z_* \\ e^{-i(z-z_*)\Gamma^z\Gamma^m K_m^{\mathrm{IR}}} e^{-iz_*\Gamma^z\Gamma^m K_m^{\mathrm{UV}}} \hat{\psi}(0) & \text{for } z > z_*\,, \end{cases} \tag{11.77}$$

where we take advantage of the fact that the K_m^{IR} commute with one another, as do the K_m^{UV}. Now,

$$(K_{\mathrm{IR}} - i\Gamma^z\Gamma^m K_m^{\mathrm{IR}})\hat{\psi}_{\mathrm{sudden}}(z) = 0 \qquad \text{for all } z \geq z_* \tag{11.78}$$

if and only if

$$(K_{\mathrm{IR}} - i\Gamma^z\Gamma^m K_m^{\mathrm{IR}}) e^{-iz_*\Gamma^m K_m^{\mathrm{UV}}} \hat{\psi}(0) = 0\,. \tag{11.79}$$

Thus to compute $G^\Psi_{\mathrm{sudden}}(k)$, which is the sudden approximation to the fermion two-point function $G^\Psi(k)$, we can use

$$\mathbf{P} = (K_{\mathrm{IR}} - i\Gamma^z\Gamma^m K_m^{\mathrm{IR}}) e^{-iz_*\Gamma^m K_m^{\mathrm{UV}}} = q + \Gamma^z\Gamma^m q_m + \Gamma^m\Gamma^n q_{mn}\,, \tag{11.80}$$

where q, q_m, and q_{mn} are given in (11.50). The point of the decomposition (11.80) is to decouple the spinor structure and the $SU(2)$ structure. q, q_m, and q_{mn} are all spinor singlets and $SU(2)$ adjoints.

Just a bit more work with γ^m matrices allows us to demonstrate the result I claimed earlier as (11.49) and reproduce here for convenience:

$$G^\Psi_{\text{sudden}}(k) = -i(\gamma^m q_m)^{-1}(q + \gamma^m \gamma^n q_{mn})\gamma^t \,. \tag{11.81}$$

Note that q, q_m, and q_{mn} are *analytic* functions of K_m^{UV}, because $\cosh(z_* K_{\text{UV}})$ and $\frac{\sinh(z_* K_{\text{UV}})}{K_{\text{UV}}}$ are really functions of $K_{\text{UV}}^2 \equiv \eta^{mn} K_m^{\text{UV}} K_n^{\text{UV}}$. The *spectral weight* of $G^\Psi_{\text{sudden}}(k)$ arises from poles and branch cuts of G^Ψ_{sudden} as a function of the k_m. As I already reviewed, branch cuts can only come from the square root in $K_{\text{IR}} \equiv \sqrt{\eta^{mn} K_m^{\text{IR}} K_n^{\text{IR}}}$, and therefore occur precisely when K_m^{IR} is timelike. In such a case, K_{IR} is imaginary and $e^{-K_{\text{IR}} z}$ is *oscillatory*: infalling if we're computing a retarded Green's function.

Let me now develop the analytic structure of $G^\Psi_{\text{sudden}}(k)$ in more explicit detail. The x^1 component of the gauge field is $\hat{A}_1^{\text{IR}} = W_{\text{IR}} \tau^1$, whose eigenvalues are $\pm k_*$ where

$$k_* = \frac{W_{\text{IR}}}{2} \,. \tag{11.82}$$

K_m^{IR} therefore has eigenvalues $k_m - k_{\lambda,m}$, where $\lambda = \pm 1/2$ and

$$k_{\pm\frac{1}{2},m} = (0, \pm k_*, 0) \,. \tag{11.83}$$

What really matters is if $k_m - k_{\lambda,m}$ is timelike: If it is, then there's a branch cut in the λ eigenspace of K_{IR}, and hence in G^Ψ_{sudden}. This is what leads to the Dirac cones, which rise above the two isolated points $k_{\lambda,m}$ in phase space.

Besides square root branch cuts, there is *another* way to get spectral weight: $(\gamma^m q_m)^{-1}$ might have a pole. Let's inquire when this could happen. If K_{IR}^m has a timelike part, then K_{IR} has an anti-hermitian part, and it would be non-generic for

$$\gamma^m q_m = -i \left[\gamma^m K_m^{\text{IR}} \cosh(z_* K_{\text{UV}}) + K_{\text{IR}} \gamma^m K_m^{\text{UV}} \frac{\sinh(z_* K_{\text{UV}})}{K_{\text{UV}}} \right] \tag{11.84}$$

to be non-invertible. On the other hand, if K_{IR} is hermitian (which happens when K_m^{IR} is spacelike), then adjusting one parameter (e.g. ω with k^1 and k^2 held fixed) will make $\det \gamma^m q_m$ vanish. The conclusion is that there can generically be a pole in $G^\Psi_{\text{sudden}}(k)$ outside the Dirac cones, but not inside.

I would hasten to point out that the argument of the previous paragraph is not airtight—unlike the analysis of where branch cuts appear, which is pretty obviously the complete story. In order to probe the question further,[7] we looked numerically for poles in $G^\Psi(k)$, and we indeed found just one continuous locus of zeros outside the Dirac cones. This locus eventually intersects the edge of the Dirac cones, as if the zeros were trying to get into the region where there are branch cuts. When this happens, it corresponds to a stable excitation (corresponding to a pole in $G^\Psi(k)$ at real values of k^m) becoming a sharp but finite-width resonance.

11.5.3. *Further developments*

As with the previous topics, I've left out a lot from my discussion, both in the actual computations I explained and in possible extensions, related computations, and comparisons with real-world phenomena. In particular:

- What happens when you include back-reaction of the gauge field on the geometry?
- The sudden approximation is not controlled by a small parameter (except maybe in some corners of k-space). How close is it to the true $G^\Psi(k)$?
- How does $G^\Psi(k)$ change as we go from $T = 0$ to $T = T_c$ for superconductivity?
- Little seems to depend on the choice of gauge group. How about using $SO(4)$ with a fermion in the vector **4** representation? Is there any relation to the $SO(4)$ symmetry of the Hubbard model on a bipartite lattice?
- The branch cut structure is already visible in the strict IR limit.
- Poles in $G^\Psi(k)$ correspond to normal modes where $\psi \to 0$ both for $z \to 0$ and $z \to \infty$.
- Normal modes are restricted to the "preferred region" where $K_m^{\rm IR}$ is spacelike but $K_m^{\rm UV}$ is timelike.
- There's a recent extension to a d-wave condensate.[21] This work also has the phenomenological advantage that it produces highly anisotropic Dirac cones.
- I've omitted discussion of the significant literature on
  - Fermions in s-wave holographic superconductors.
  - Conductivity at finite frequency.
  - Thermodynamic and hydrodynamic properties of p-wave superconductors.
  - Embedding holographic superconductors in string/M-theory.
- Comparison of $G^\Psi(k)$ to results of ARPES measurements is interesting: You get a peak-dip-hump structure from combination

of the normal mode and the continuum from inside the Dirac cone.

- Why not do a fermion response calculation for spin-3/2 fermions, for example the gravitini in actual supergravity theories?
- There are instabilities of holographic superconductors besides the ones that spontaneously break $U(1)$, for example the Gregory-Laflamme instability and runaways in moduli space. How do all these instabilities compete?

In short, it seems that there is still a lot to learn about holographic superconductors.

Acknowledgments

I thank my collaborators: Silviu Pufu, Fabio Rocha, and Amos Yarom. This work was supported in part by the Department of Energy under Grant No. DE-FG02-91ER40671.

References

1. J. M. Maldacena, The large N limit of superconformal field theories and supergravity, *Adv. Theor. Math. Phys.* **2**, 231–252, (1998). doi: 10.1023/A:1026654312961.
2. S. S. Gubser, I. R. Klebanov, and A. M. Polyakov, Gauge theory correlators from non-critical string theory, *Phys. Lett.* **B428**, 105–114, (1998). doi: 10.1016/S0370-2693(98)00377-3.
3. E. Witten, Anti-de Sitter space and holography, *Adv. Theor. Math. Phys.* **2**, 253–291, (1998).
4. S. S. Gubser, S. S. Pufu, and A. Yarom, Entropy production in collisions of gravitational shock waves and of heavy ions, *Phys. Rev.* **D78**, 066014, (2008). doi: 10.1103/PhysRevD.78.066014.
5. S. S. Gubser, Symmetry constraints on generalizations of Bjorken flow, *Phys. Rev.* **D82**, 085027, (2010). doi: 10.1103/PhysRevD.82.085027.
6. S. S. Gubser and A. Yarom, Conformal hydrodynamics in Minkowski and de Sitter spacetimes. (2010).
7. S. S. Gubser, F. D. Rocha, and A. Yarom, Fermion correlators in non-abelian holographic superconductors, *JHEP.* **11**, 085, (2010). doi: 10.1007/JHEP11(2010)085.
8. P. Steinberg, Landau hydrodynamics and RHIC phenomena, *Acta Phys. Hung.* **A24**, 51–57, (2005). doi: 10.1556/APH.24.2005.1-4.8.
9. D. M. Eardley and S. B. Giddings, Classical black hole production in high-energy collisions, *Phys. Rev.* **D66**, 044011, (2002). doi: 10.1103/PhysRevD.66.044011.

10. S. S. Gubser, S. S. Pufu, and A. Yarom, Off-center collisions in AdS_5 with applications to multiplicity estimates in heavy-ion collisions, *JHEP*. **11**, 050, (2009). doi: 10.1088/1126-6708/2009/11/050.
11. P. M. Chesler and L. G. Yaffe, Holography and colliding gravitational shock waves in asymptotically AdS_5 spacetime. (2010).
12. J. D. Bjorken, Highly Relativistic Nucleus-Nucleus Collisions: The Central Rapidity Region, *Phys. Rev.* **D27**, 140–151, (1983). doi: 10.1103/PhysRevD.27.140.
13. R. Loganayagam, Entropy Current in Conformal Hydrodynamics, *JHEP*. **05**, 087, (2008). doi: 10.1088/1126-6708/2008/05/087.
14. S. S. Gubser, Colorful horizons with charge in anti-de Sitter space, *Phys. Rev. Lett.* **101**, 191601, (2008). doi: 10.1103/PhysRevLett.101.191601.
15. S. S. Gubser and S. S. Pufu, The gravity dual of a p-wave superconductor, *JHEP*. **11**, 033, (2008). doi: 10.1088/1126-6708/2008/11/033.
16. M. M. Roberts and S. A. Hartnoll, Pseudogap and time reversal breaking in a holographic superconductor, *JHEP*. **08**, 035, (2008). doi: 10.1088/1126-6708/2008/08/035.
17. H. Liu, J. McGreevy, and D. Vegh, Non-Fermi liquids from holography. (2009).
18. P. Basu, J. He, A. Mukherjee, and H.-H. Shieh, Hard-gapped Holographic Superconductors. (2009).
19. S. S. Gubser, Breaking an Abelian gauge symmetry near a black hole horizon, *Phys.Rev.* **D78**, 065034, (2008). doi: 10.1103/PhysRevD.78.065034.
20. S. A. Hartnoll, C. P. Herzog, and G. T. Horowitz, Building a Holographic Superconductor, *Phys.Rev.Lett.* **101**, 031601, (2008). doi: 10.1103/PhysRevLett.101.031601.
21. F. Benini, C. P. Herzog, and A. Yarom, Holographic Fermi arcs and a d-wave gap. (2010).

Chapter 12

Emergence of Supersymmetry, Gauge Theory and String in Condensed Matter Systems

Sung-Sik Lee[1,2]

[1]*Department of Physics & Astronomy, McMaster University, 1280 Main St. W., Hamilton ON L8S4M1, Canada*

[2]*Perimeter Institute for Theoretical Physics, 31 Caroline St. N., Waterloo ON N2L2Y5, Canada*

The lecture note consists of four parts. In the first part, we review a 2+1 dimensional lattice model which realizes emergent supersymmetry at a quantum critical point. The second part is devoted to a phenomenon called fractionalization where gauge boson and fractionalized particles emerge as low energy excitations as a result of strong interactions between gauge neutral particles. In the third part, we discuss about stability and low energy effective theory of a critical spin liquid state where stringy excitations emerge in a large N limit. In the last part, we discuss about an attempt to come up with a prescription to derive holographic theory for general quantum field theory.

12.1. Introduction

A variety of phenomena in condensed matter systems ranging from metal to superconductivity can be understood based on simple Hamiltonians, such as the Hubbard model,

$$H = -\sum_{<i,j>,\sigma} t_{ij} c_{i\sigma}^{\dagger} c_{j\sigma} + U \sum_i n_{i\uparrow} n_{i\downarrow} - \mu \sum_i n_i. \tag{12.1}$$

Here t_{ij}, U and μ are hopping integral, on-site Coulomb energy and chemical potential, respectively. $c_{i\sigma}$ is the annihilation operator of electron with spin σ at site i, $n_{i\sigma} = c_{i\sigma}^{\dagger} c_{i\sigma}$ and $n_i = n_{i\uparrow} + n_{i\downarrow}$. While it is easy to write, it is impossible to solve the Hamiltonian of interacting 10^{23} electrons from the first principle calculation. The symmetry of the Hamiltonian is low and there are few kinematical constraints that simplify dynamics. Because of interactions, one can not in general disentangle a subsystem from the rest to reduce the many-body problem to one-body problem. Nonetheless, as one probes the system at low energies, various dynamical constraints emerge, which gives us a chance to understand low energy physics by focusing on fewer degrees of freedom. Presumably, the full Hilbert space has a lot of local minima, and the system 'flows' to one of the minima in the low energy limit. Around each minimum, the geometrical and topological structures of the low energy manifold determine the nature of low energy degrees of freedom. Quantum fluctuations of the low energy modes are governed by low energy effective theories which are rather insensitive to details of the microscopic Hamiltonian. The distinct set of minima and the associated low energy effective theories characterize different phases of matter (universality classes). One of the goals in condensed matter physics is to classify different phases of matter and understand universal properties of them using low energy effective theories.

Various quantum phases of matter have been identified, such as Landau Fermi liquid, band insulator, superfluid/superconductor, quantum Hall liquid, etc. The list is still growing, and new ways of characterizing different phases are being devised. However, it is likely that there are many new phases yet to be discovered.

There are some interesting aspects we need to appreciate. First, it is usually very hard to predict which phase a given microscopic Hamiltonian flows into. Often, microscopic Hamiltonians do not give much clue over what emerges in the low energy limit. Who could have predicted that the Hubbard model with impurities and defects has superconducting phase where electric current can flow without any resistivity? Second, low energy effective theories in some phases or quantum critical points in condensed matter systems are strikingly similar to what (we believe) describes the very vacuum of our universe.[1,2] For example, there exists a phase where the low energy effective theory has very high symmetry including supersymmetry even though microscopic Hamiltonian breaks almost all symmetries except for some discrete lattice symmetry and internal global symmetry. In some corner of the landscape, gauge theory emerges as a low energy effective theory of the Hubbard model. In some phases, there is no well-defined quasiparticle. Instead, some weakly coupled stringy excitations emerge as low energy excitations. How is it possible that collective fluctuations of electrons in solids have such striking similarities to the way the vacuum of our own universe fluctuates? Is the universe made of a bunch of non-relativistic 'electrons' at very short distance? Perhaps it is simply that there are not too many good ideas that are available to nature, and she has to recycle same ideas in different systems and different scales over and over. In this lecture, we will review a few examples of emergent phenomena in condensed matter systems which are potentially interesting to both condensed matter physicists and high energy physicists.

12.2. Emergent Supersymmetry

In spontaneous symmetry breaking, symmetry of microscopic model is broken at low energies as the ground state spontaneously choose a particular vacuum among degenerate vacua connected by the symmetry. In condensed matter systems, the opposite situation often arises. Namely, new symmetry which is absent in microscopic Hamiltonian can arise at low energies as the system dynamically organizes itself to show a pattern of fluctuations which obey certain symmetry in the long distance limit. Sometimes, gapless excitations (or ground state degeneracy) whose origins are not obvious from any microscopic symmetry can be protected by emergent symmetries.

12.2.1. *Emergence of (bosonic) space-time symmetry*

Consider a rotor model in one-dimensional lattice,

$$H_b = -t \sum_i (e^{i(\theta_i - \theta_{i+1})} + h.c.) + \frac{U}{2} \sum_i (n_i - \bar{n})^2, \tag{12.2}$$

where θ_i (n_i) is phase (number) of bosons at site i which satisfies $[\theta_i, n_j] = i\delta_{ij}$, and $\bar{n}$ is the average density. For integer $\bar{n}$, the long distance physics is captured by the 1+1D XY model

$$S = \kappa \int dx^2 (\partial_\mu \theta)^2, \tag{12.3}$$

where $\kappa \sim \sqrt{t/U}$. Because θ is compact, instanton is allowed where the winding number

$$\nu(t) = \int_0^L dx \partial_x \theta(t, x) \tag{12.4}$$

changes by an integer multiple of 2π. Here we consider the periodic boundary condition: $e^{i\theta(t,0)} = e^{i\theta(t,L)}$. Physically, a unit instanton describes quantum tunneling from a state with momentum k to a state with momentum $k + 2\bar{n}\pi/a$ with a the lattice spacing. This tunneling is allowed because momentum needs to be conserved only modulo $2\pi/a$ due to the underlying lattice.[a]

If $\kappa >> 1$, instantons are dynamically suppressed even though no microscopic symmetry prevents them. In this case, the absolute value of momentum is conserved, not just in modulo $2\pi/a$. Since the state with momentum $2\bar{n}\pi/a$ does not mix with the state with zero momentum, states with non-trivial windings arise as well-defined excitations which becomes gapless excitation in the thermodynamic limit. Note that this gapless excitation is not a Goldstone mode because the continuous symmetry can not be broken in 1+1D even at $T = 0$. This gapless mode is a topological excitation protected by the emergent conservation law. Within each topological sector, we can treat θ as a non-compact variable and the low energy theory becomes a free theory.

If $\kappa << 1$, the potential energy dominates and bosons are localized, exhibiting gap. In terms of the XY-model, instantons and anti-instantons proliferate, and the system can not ignore the existence of lattice anymore.

Now let's consider a simple two-dimensional lattice model where rotational symmetry emerges,

$$H_b = -t \sum_{<i,j>} (b_i^\dagger b_j + h.c.) - \mu \sum_i b_i^\dagger b_i, \tag{12.5}$$

where t is the hopping energy, μ is chemical potential, and $< i, j >$ denotes nearest neighbor sites on the square lattice. The energy spectrum is given by

$$E_k = -2t(\cos k_x + \cos k_y) - \mu. \tag{12.6}$$

The full spectrum respects only the 90 degree rotational symmetry. If the chemical potential is tuned to the bottom of the band, the energy dispersion of low energy bosons become

$$E_k = tk^2 + O(k^4). \tag{12.7}$$

In the low energy limit, higher order terms are irrelevant and the full rotational symmetry emerges.

Besides the rotational symmetry, the full Lorentz symmetry can emerge. For example, the low energy excitations of fermions at half filling on the honeycomb lattice are described by two copies of two-component Dirac fermions in 2+1 dimensions. As a result, the full Lorentz symmetry emerges

[a]In the T-dual variable $\partial_\mu \phi = \epsilon_{\mu\nu}\partial_\nu \theta$, the non-conservation of the winding number translates into the (explicitly) broken translational symmetry in the target space ϕ, and the theory is mapped into the sine-Gordon model.

in the low energy limit although the microscopic model has only six-fold rotational symmetry. When there are gapless excitations in the presence of Lorentz symmetry, usually the full conformal symmetry is realized.

12.2.2. *Emergent supersymmetry*

Supersymmetry is a symmetry which relates bosons and fermions. Since bosons and fermions have integer and half integer spins respectively in relativistic systems, supercharges that map boson into fermion (or vice versa) carry half integer spin and supersymmetry should be a part of space-time symmetry. Moreover, supersymmetry is the unique non-trivial extension of the Poincare symmetry besides the conformal symmetry. Given that all bosonic space-time symmetry can emerge in condensed matter systems, one can ask whether supersymmetry can also emerge.[b]

In 2D, it is known that supersymmetry can emerge from the dilute Ising model,[3]

$$\beta H = -J \sum_{<i,j>} \sigma_i \sigma_j - \mu \sum_i \sigma_i^2, \tag{12.8}$$

where $\sigma = \pm 1$ represents a site with spin up or down, and $\sigma = 0$ represents a vacant site. When μ is large, almost all sites are filled and the usual second order Ising transition occurs as J is tuned. As μ is lowered, the second order transition terminates at a tricritical point $\mu = \mu_c$ and the phase transition becomes first order below μ_c. The tricritical point is described by the Φ^6-theory,

$$S = \int d^2x [(\partial_\mu \Phi)^2 + \lambda_6 \Phi^6] \tag{12.9}$$

where $\Phi \sim <\sigma>$ describes the magnetic order parameter. Although there is no fermion in this action, one can construct a fermion field ψ from a string of spins through the Jordan-Wigner transformation. At the tricritical point, the scaling dimensions of Φ^2 and ψ differ exactly by 1/2. This is not an accident and these two fields form a multiplet under an emergent supersymmetry. More generally, the operators which are even (odd) under the Z_2 spin symmetry form the Neveu-Schwarz (Ramond) algebra. The dilute Ising model provides deformations of the underlying superconformal theory within $(-1)^F = 1$ sector, where F is the fermion number. For lattice realizations of other 2D superconformal field theories, see Refs. 6 and 7.

---

[b]There are large literatures on supersymmetric quantum mechanics in condensed matter systems. Here we will exclusively focus on full space-time supersymmetry.

To realize emergent supersymmetry in higher dimensions, it is desired to have an interacting theory in the IR limit.[4] Otherwise, RG flow of supersymmetry-breaking couplings would stop below certain energy scale, and supersymmetry-breaking terms generically survive in the IR limit. It is hard to have free bosons and fermions which have same velocity unless enforced by some exact microscopic symmetry. In this sense, 2+1D is a good (but not exclusive) place to look for an emergent supersymmetry. In 3+1D, it has been suggested that the $\mathcal{N} = 1$ super Yang-Mills theory can emerge from a model of gauge boson and chiral fermion in the adjoint representation.[5,c] If a subset of supersymmetry is kept as exact symmetry in lattice models, a full supersymmetry can emerge in the continuum.[8] Here, we will consider a 2+1D lattice model where supersymmetry dynamically emerge at a critical point without any lattice supersymmetry.[9]

12.2.2.1. *Model*

The Hamiltonian is composed of three parts,

$$H = H_f + H_b + H_{fb}, \tag{12.10}$$

where

$$\begin{aligned}
H_f &= -t_f \sum_{<i,j>} (f_i^\dagger f_j + h.c.), \\
H_b &= t_b \sum_{<I,J>} (e^{i(\theta_I - \theta_J)} + h.c.) + \frac{U}{2} \sum_I n_I^2, \\
H_{fb} &= h_0 \sum_I e^{i\theta_I} (f_{I+\mathbf{b}_1} f_{I-\mathbf{b}_1} + f_{I-\mathbf{b}_2} f_{I+\mathbf{b}_2} + \\
&\quad + f_{I-\mathbf{b}_1+\mathbf{b}_2} f_{I+\mathbf{b}_1-\mathbf{b}_2}) + h.c..
\end{aligned} \tag{12.11}$$

Here H_f describes spinless fermions with nearest neighbor hopping on the honeycomb lattice at half filling; H_b describes bosons with nearest neighbor hopping and an on-site repulsion on the triangular lattice which is dual to the honeycomb lattice; and H_{fb} couples the fermions and bosons. The lattice structure is shown in Fig. 12.1 (a). f_i is the fermion annihilation operator and $e^{-i\theta_I}$, the lowering operator of n_I which is conjugate to the angular variable θ_I. $< i, j >$ and $< I, J >$ denote pairs of nearest neighbor

[c] But the notion of emergent supersymmetry is less sharp in this case due to the fact that confinement sets in below an energy scale and there is no genuine IR degrees of freedom. For a given gauge coupling at the lattice scale, supersymmetry is realized as an approximate symmetry between fermionic and bosonic glue ball spectra.

sites on the honeycomb and triangular lattices, respectively. $t_f, t_b > 0$ are the hopping energies for the fermions and bosons, respectively and U is the on-site boson repulsion energy. $\mathbf{b}_1$ and $\mathbf{b}_2$ are two independent vectors which connect a site on the triangular lattice to the neighboring honeycomb lattice sites. h_0 is the pairing interaction strength associated with the process where two fermions in the f-wave channel around a hexagon are paired and become a boson at the center of the hexagon, and vice versa. In this sense, the boson can be regarded as a Cooper pair made of two spinless fermions in the f-wave wavefunction. This model has a global U(1) symmetry under which the fields transform as $f_i \to f_i e^{i\varphi}$ and $e^{-i\theta_I} \to e^{-i\theta_I} e^{i2\varphi}$.

In the low energy limit, there are two copies of Dirac fermions and two complex bosons. One set of Dirac fermion and complex boson carries momentum $\mathbf{k}_A$, and the other set carries momentum $\mathbf{k}_B$. The fermions are massless without any fine tuning, which is protected by the time reversal symmetry and the inversion symmetry. The reason why we have two complex bosons instead of one is that the boson kinetic energy is frustrated. Because $t_b > 0$, the boson kinetic energy is minimized when relative phases between neighboring bosons become π. However, not all kinetic energy terms can be minimized due to the geometrical frustration of the triangular lattice where bosons are defined. The best thing the bosons can do to minimize the kinetic energy is to form a 'spiral wave' where the phase of bosons rotates either by 120 or -120 degree around triangles. There are two such global configurations and they carry the momenta $\mathbf{k}_A$ and $\mathbf{k}_B$ respectively. Each complex boson describes condensate of bosons in each of these configurations. The effective theory for the low energy modes becomes

$$
\begin{aligned}
\mathcal{L} = & \, i \sum_{n=1}^{2} \overline{\psi}_n \left(\gamma_0 \partial_\tau + c_f \sum_{i=1}^{2} \gamma_i \partial_i \right) \psi_n \\
& + \sum_{n=1}^{2} \left[|\partial_\tau \phi_n|^2 + c_b^2 \sum_{i=1}^{2} |\partial_i \phi_n|^2 + m^2 |\phi_n|^2 \right] \\
& + \lambda_1 \sum_{n=1}^{2} |\phi_n|^4 + \lambda_2 |\phi_1|^2 |\phi_2|^2 \\
& + h \sum_{n=1}^{2} \left(\phi_n^* \psi_n^T \varepsilon \psi_n + c.c. \right) .
\end{aligned}
\tag{12.12}
$$

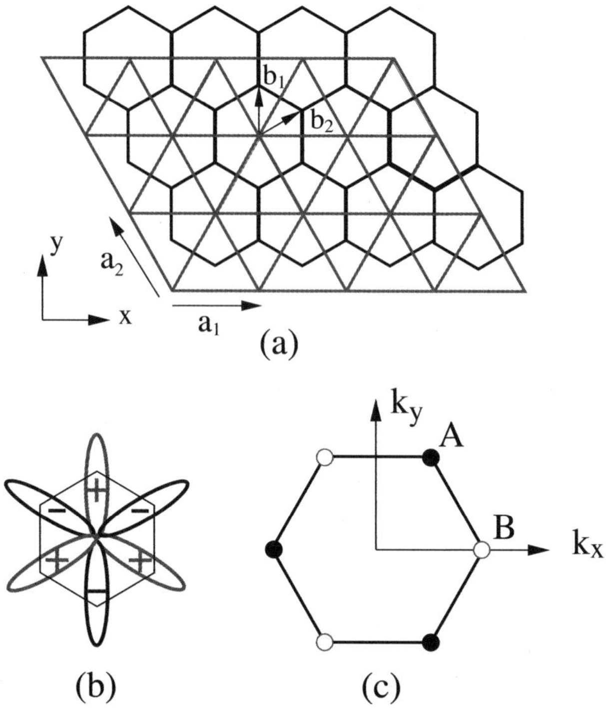

Fig. 12.1. (a) The lattice structure in the real space. Fermions are defined on the honeycomb lattice and the bosons, on the dual triangular lattice. $\mathbf{a}_1$, $\mathbf{a}_2$ are the lattice vectors with length a, and $\mathbf{b}_1$, $\mathbf{b}_2$, two independent vectors which connect a site on the triangular lattice to the nearest neighbor sites on the honeycomb lattice. (b) The phases of a fermion pair in the real space. (c) The first Brillouin zone in the momentum space. A and B indicate two inequivalent points with momenta $\mathbf{k}_A = \frac{2\pi}{a}(\frac{1}{3}, \frac{1}{\sqrt{3}})$ and $\mathbf{k}_B = \frac{2\pi}{a}(\frac{2}{3}, 0)$ where the low energy modes are located. ψ_1, ϕ_2 are located at $\mathbf{k}_A$ and ψ_2, ϕ_1, at $\mathbf{k}_B$.

A closely related field theory has been studied in the context of high T_c superconductors.[11] Although there are same number of propagating bosons and fermions[d], this Lagrangian contains four supersymmetry

[d]One complex boson is worthy of two components of complex fermions in terms of the number of initial data one has to specify to solve the equation of motion.

breaking terms,

$$
\begin{aligned}
& m \neq 0, \\
& c_b \neq c_f, \\
& \lambda_1 \neq h^2, \\
& \lambda_2 \neq 0.
\end{aligned}
\tag{12.13}
$$

One would naively expect that one has to tune at least four parameters to reach the supersymmetric point. However, it turns out that one needs to tune only the boson mass to realize supersymmetry.

12.2.2.2. *RG flow*

To control the theory, we consider the theory in $4-\epsilon$ dimensions. We use the dimensional regularization scheme where the number of fermion components and the traces of gamma matrices are fixed.

The boson mass is a relevant (supersymmetry-breaking) perturbation and we tune it (by hand) to zero. This amounts to tuning one microscopic parameter to reach the critical point which separates the normal phase and the bose condensed phase. The one-loop beta functions for the remaining couplings at the critical point is given by

$$
\begin{aligned}
\frac{dh}{dl} &= \frac{\epsilon}{2}h - \frac{1}{(4\pi c_f)^2}\left(2 + \frac{16c_f^3}{c_b(c_f+c_b)^2}\right)h^3, \\
\frac{d\lambda_1}{dl} &= \epsilon\lambda_1 - \frac{1}{(4\pi)^2}\left(\frac{20\lambda_1^2+\lambda_2^2}{c_b^2} + \frac{8h^2\lambda_1}{c_f^2} - \frac{16h^4}{c_f^2}\right), \\
\frac{d\lambda_2}{dl} &= \epsilon\lambda_2 - \frac{1}{(4\pi)^2}\left(\frac{4\lambda_2^2+16\lambda_1\lambda_2}{c_b^2} + \frac{8h^2\lambda_2}{c_f^2}\right), \\
\frac{dc_f}{dl} &= \frac{32h^2c_f(c_b-c_f)}{3(4\pi)^2c_b(c_b+c_f)^2}, \\
\frac{dc_b}{dl} &= -\frac{2h^2c_b(c_b^2-c_f^2)}{(4\pi c_bc_f)^2},
\end{aligned}
\tag{12.14}
$$

where the logarithmic scaling parameter l increases in the infrared. The RG flow is shown in Fig. 12.2.

The Gaussian (GA) fixed point, $(h^*, \lambda_1^*, \lambda_2^*) = (0,0,0)$, the Wilson-Fisher (WF) fixed point, $(h^*, \lambda_1^*, \lambda_2^*) = (0, \frac{(4\pi c_b)^2\epsilon}{20}, 0)$, and the O(4) fixed

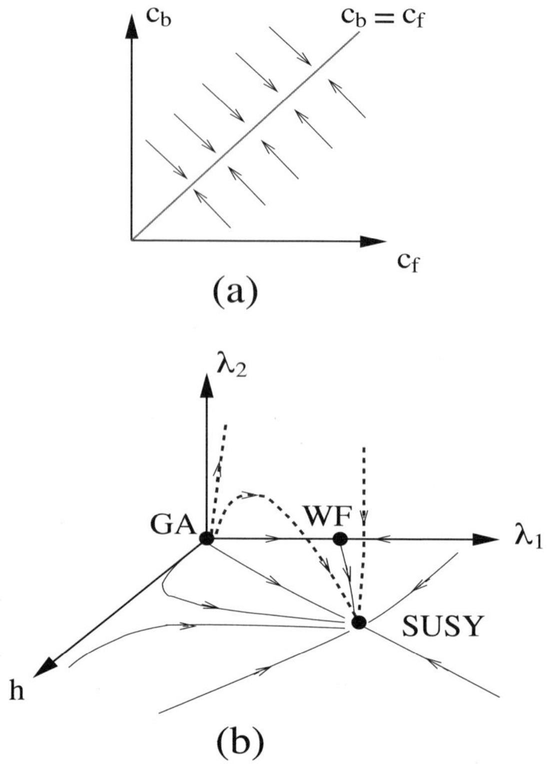

Fig. 12.2. The schematic RG flows of (a) the velocities with $h \neq 0$ and (b) λ_1, λ_2 and h in the subspace of $m = 0$. In (b), the solid lines represent the flow in the plane of (h, λ_1) and the dashed lines, the flow outside the plane.

point, $(h^*, \lambda_1^*, \lambda_2^*) = (0, \frac{(4\pi c_b)^2 \epsilon}{24}, \frac{(4\pi c_b)^2 \epsilon}{12})$ are all unstable upon turning on the pairing interaction h. If h is nonzero, the boson and fermion velocities begin to flow as can be seen from the last two equations in Eq. (12.14). Because the pairing interaction mixes the velocities of the boson and fermion, the difference of the velocities exponentially flows to zero in the low-energy limit. The converged velocity in the infrared limit is a non-universal value which we will scale to 1 in the followings. With a nonzero h, the system eventually flows to a stable fixed point,

$$(h^*, \lambda_1^*, \lambda_2^*) = (\sqrt{\frac{(4\pi)^2 \epsilon}{12}}, \frac{(4\pi)^2 \epsilon}{12}, 0). \qquad (12.15)$$

At this point, the theory becomes invariant under the supersymmetry transformation,

$$\begin{aligned} \delta_\xi \phi_n &= -\overline{\psi}_n \xi, \quad \delta_\xi \phi_n^* = \overline{\xi} \psi_n \\ \delta_\xi \psi_n &= i\partial\!\!\!/ \phi_n^* \xi - \frac{h}{2}\phi_n^2 \varepsilon \overline{\xi}^T, \quad \delta_\xi \overline{\psi}_n = i\overline{\xi}\partial\!\!\!/ \phi_n - \frac{h}{2}\phi_n^{*2} \xi^T \varepsilon, \end{aligned} \tag{12.16}$$

where ξ is a two-component complex spinor. This theory corresponds to the $\mathcal{N} = 2$ (which amounts to four supercharges) Wess-Zumino model with superpotential,

$$F = \frac{h}{3}(\Phi_1^3 + \Phi_2^3), \tag{12.17}$$

where Φ_1 and Φ_2 are two chiral multiplets. Due to the emergent superconformal symmetry, the one-loop anomalous scaling dimensions for the chiral primary fields ϕ and ψ,

$$\eta_\phi = \eta_\psi = \epsilon/3 \tag{12.18}$$

are exact.

Although the emergent supersymmetry has been obtained from a microscopic model which has both fermions and bosons, one can view bosons as composite particles (Cooper pairs) which emerge at low energies. This implies that supersymmetry can emerge from a microscopic model which contains only fermions. Therefore, the $\mathcal{N} = 2$ Wess-Zumino theory can describe the second order quantum phase transition of f-wave FFLO (Fulde-Ferrell-Larkin-Ovchinnikov) superconducting state of spinless fermions on the honeycomb lattice at half filling (see Fig. 12.3). For an alternative proposal to realize the Wess-Zumino model in condensed matter systems, see Ref. 10.

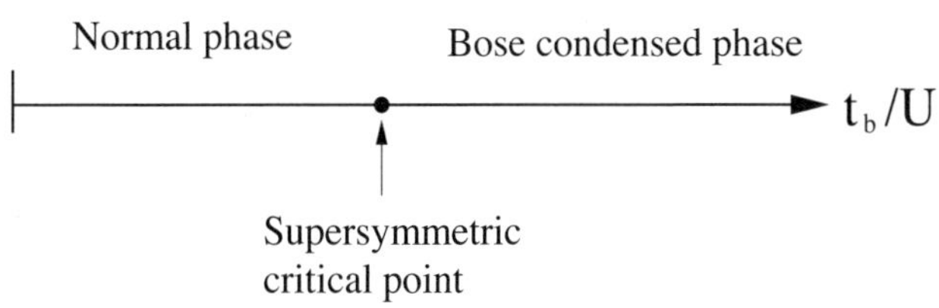

Fig. 12.3. The superconformal theory describes the quantum critical point separating the normal state and the f-wave superconducting (Bose condensed) state of spinless fermions in the honeycomb lattice.

12.3. Emergent Gauge Theory

Fractionalization is a phenomenon where a microscopic particle in many-body systems decay into multiple modes each of which carry a fractional quantum number of the original particle. In contrast to the more familiar phenomena where a composite particle breaks into its constituent particles at high energies, fractionalization is a low energy collective phenomenon where microscopic particles do not 'really' break into partons. Instead, many-body correlations make it possible for parts of microscopic particles to emerge as deconfined excitations in the low energy limit.[12] When fractionalization occurs, a gauge field emerges as a collective excitation that mediates interaction between fractionalized excitations. There exist many models which exhibit fractionalization.[13–19] In this lecture, we are going to illustrate this phenomenon using a simple model.[20]

12.3.1. *Model*

Consider a 4-dimensional Euclidean hypercubic lattice (with discretized time). At each site on the lattice, there are boson fields $e^{i\theta^{ab}}$ which carry one flavor index a and one anti-flavor index b with $a, b = 1, 2, ..., N$. We impose constraints $\theta^{ab} = -\theta^{ba}$; $e^{i\theta^{ab}}$ is anti-particle of $e^{i\theta^{ba}}$. There are $N(N-1)/2$ independent boson fields per site. In the following, we will refer to these bosons as 'mesons'. The action is

$$S = -t \sum_{<i,j>} \sum_{a,b} \cos\left(\theta_i^{ab} - \theta_j^{ab}\right) \\ -K \sum_i \sum_{a,b,c} \cos\left(\theta_i^{ab} + \theta_i^{bc} + \theta_i^{ca}\right). \qquad (12.19)$$

Here the first term is the standard Euclidean kinetic energy of bosons and the second term is a flavor conserving interactions between bosons. This model can be viewed as a low energy effective theory of excitons in a multi-band insulator.[e] But let's forget about the 'true UV theory' and take this model as our microscopic model and regard these bosons as fundamental particles.

[e] In this context, an exciton with flavor a and anti-flavor b describes a composite particle made of an electron in the a-th band and a hole in the b-th band of a multi-band insulator.

12.3.2. *Slave-particle theory*

We are interested in the strong coupling limit ($K >> 1$). In this limit, the large potential energy imposes the dynamical constraints

$$\theta_i^{ab} + \theta_i^{bc} + \theta_i^{ca} = 0 \tag{12.20}$$

for every set of a, b, c. The constraints are satisfied on a $(N-1)$-dimensional manifold in $(S^1)^{N(N-1)/2}$ on each site. The low energy manifold is parametrized by

$$\theta_i^{ab} = \phi_i^a - \phi_i^b, \tag{12.21}$$

where the new bosonic fields ϕ^a carry only one flavor quantum number contrary to the meson fields. The new bosons are called slave-particles or partons. They are 'enslaved' to each other because these partons can not escape out of mesons. Note that there is a local $U(1)$ redundancy in this parametrization and the low energy manifold is $(S^1)^N/S^1$.

Within this low energy manifold, the potential energy can be dropped and the kinetic energy becomes

$$S = -t \sum_{<i,j>} \left[\sum_a e^{i(\phi_i^a - \phi_j^a)} \right] \left[\sum_b e^{-i(\phi_i^b - \phi_j^b)} \right]. \tag{12.22}$$

Since the kinetic energy is factorized in the flavor space, one can introduce a collective dynamical field $\eta \sim \sum_b e^{-i(\phi_i^b - \phi_j^b)}$ using the Hubbard-Stratonovich transformation. Then the effective action becomes

$$S = t \sum_{<i,j>} \left[|\eta_{ij}|^2 - |\eta_{ij}| \sum_a e^{i(\phi_i^a - \phi_j^a - a_{ij})} - c.c. \right], \tag{12.23}$$

where a_{ij} is the phase of the complex field η_{ij}. Note that the full theory is invariant under the U(1) gauge transformation

$$\begin{aligned} \phi_i^a &\to \phi_i^a + \varphi_i, \\ a_{ij} &\to a_{ij} + \varphi_i - \varphi_j. \end{aligned} \tag{12.24}$$

This theory is a compact U(1) lattice gauge theory coupled with N partons. The bare gauge coupling is infinite because the gauge field does not have a bare kinetic energy term.

One may think that partons are always confined in this theory because the bare gauge coupling is infinite. However, we have to be more careful about what we mean by confinement. Since mesons themselves are fundamental particles, it is indeed impossible to literally separate one meson into two partons. In this sense, a parton is always paired with an anti-parton. However, this confinement at short distance scale does not rule out the possibility that excitations which carry the same quantum number as partons arise as low energy excitations. If such low energy excitations exist, we can say that partons are deconfined in the long distance limit.

How can we understand that partons are deconfined at low energies? To see this, let us integrate out high energy fluctuations of ϕ^a to obtain an effective theory with a cut-off $\Lambda << 1/a$ where a is the lattice spacing. The fluctuations of partons generate the Maxwell's term (and all higher order gauge invariant terms) in the low energy effective action,

$$S = \int dx^4 \left[|(\partial_\mu - ia_\mu)\Phi_a|^2 + V(\Phi_a) + \frac{1}{g^2} F_{\mu\nu}F^{\mu\nu} + ... \right], \quad (12.25)$$

where $\Phi_a = e^{i\phi^a}$. Due to screening by N charged partons, the gauge coupling is renormalized from infinity down to a finite value $g^2 \sim 1/N$. In the large N limit, the renormalized gauge coupling can be made very small and the deconfinement (Coulomb) phase can arise as a stable phase. Then gapless U(1) gauge boson (emergent photon) and fractionalized bosons (partons) arise as low energy excitations.

12.3.3. *World line picture*

Although the above simple argument is plausible, the relation between the 'slave-partons' which are confined at short distance scales and the 'liberated-partons' which are deconfined at long distance scales is rather obscure in this description. Moreover, it is not entirely clear how partons can screen the gauge field in the first place if they are always confined within neutral mesons. In other words, one can question how one can integrate out high energy fluctuations of partons reliably if they are subject to the infinitely strong gauge coupling at high energies.

For this reason, it is useful to understand fractionalization purely in terms of original mesons without introducing slave particles. Actually one can understand fractionalization intuitively in terms of world lines of mesons. The world line picture is not a quantitative description, but it is useful in that it provides a physical picture for fractionalization. In space-time, world lines of mesons can be represented as double lines with two opposite arrows, where each arrow carries a flavor current.

In the partition function, the simplest configuration is a loop of meson world line made of two overlapped single-line loops with current of flavor

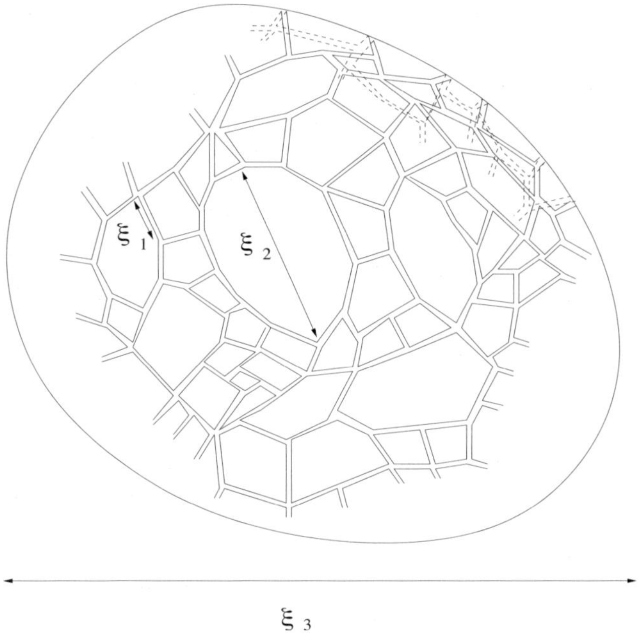

Fig. 12.4. Web of world lines.

a moving in one direction and b in the other direction. General configurations consist of a collection of connected webs made of world lines (world line web). Each world line web defines closed surfaces as depicted in Fig. 12.4. Many such closed surfaces may co-exist and interpenetrate each other. There are three different length scales in the world line web. The shortest scale is the length of double line segment (ξ_1) which corresponds to the life time of a single meson. As the interaction becomes stronger, mesons decay faster and ξ_1 becomes shorter. The next scale is the typical size of single line loops (ξ_2). This can be much larger than ξ_1 even in the strong coupling limit. This is because a single flavor a can switch as many partners as it wants during its life time before it meets with its own anti-flavor $\bar{a}$ and decays into vacuum.[f] In this case, we can view single lines as world lines of emergent long-living excitations. The largest length scale is the size of the bubble made of the web (ξ_3). By tuning three dimensionless parameters in the theory (K, N and t), it is possible to tune three scales, ξ_1/a, ξ_2/a,

[f] One's life time can be in principle much longer than the average lifetime of marriages.

ξ_3/a, where a is the lattice spacing. Here we will focus on the limit $K >> 1$ where $\xi_1 \sim a$.[g]

In the small t and small N limit (which corresponds to the limit of very massive mesons), all length scales are of order of the lattice scale. As t increases, mesons become lighter, and ξ_2 increases. As N increases, configurations with many single line loops become more important due to entropic contribution of flavors associated with each singe line loops. As a result, the tension of world line web decreases and ξ_3 increases. If $a \sim \xi_1 << \xi_2 << \xi_3$ (which can be achieved for a large N with fixed tN) the closed surface of the world line web is well-defined in the length scale $\xi_2 << x << \xi_3$. Thus the effective degrees of freedom at this scale are fluctuating membrane in space-time. The surface can be regarded as the world sheet of oriented closed string.[h] In the gauge theory picture the string corresponds to electric flux line and the string tension, the gauge coupling. The gauge group should be U(1) due to the orientedness of surfaces, and the effective theory becomes a pure U(1) gauge theory. In this scale, the world line web is very floppy and one obtains a weakly coupled pure gauge theory. The transverse fluctuations of world line web corresponds to emergent gauge bosons.

For a finite ξ_3, one eventually enters into the region where $x >> \xi_3$ as one probes longer distance scale. In this long distance limit, there are only small membranes and the vacuum is essentially empty. This corresponds to the confinement phase and ξ_3 corresponds the confining scale.

In the opposite limit, if we lower our length scale x to the scale comparable to ξ_2, one begins to notice holes on closed surfaces. One can interpret boundaries of the holes as world lines of particles. These particles should be viewed as bosons because each contribution adds up in the partition function without minus signs associated with single line loops. This is related to the fact that e^{-S} and the measure in the partition function are positive definite in the world line basis due to the absence of frustration in this model. The emergent bosons carry unit charge with respect to the gauge field. This is because world line web which corresponds to world sheet of one unit of electric flux line can end on boundaries. Moreover, these bosons carry only one flavor quantum number due to the fact that an anti-flavor b which always accompanies a flavor a is canceled by the flavor b of nearby

[g]This condition is not crucial in what follows. In general, one can tune K so that $\xi_1 >> a$.

[h]The world line web is oriented because single lines are oriented.

mesons.[i] This is analogous to the situation where one can find a nonzero charge density on a surface of a dielectric medium made of charge neutral molecules. Therefore in the scale $x \sim \xi_2$, the effective degrees of freedom are gauge bosons coupled with N fractionalized bosons which carry only one flavor quantum number.

In the shortest length scale $x << \xi_2$, loops of single lines look very large and these fractionalized bosons appear to be condensed. It corresponds to the Higgs phase of the gauge theory.

Therefore, for finite ξ_2 and ξ_3 there are crossovers from the behavior of Higgs phase to Coulomb phase and eventually to the confinement as one probes the system at progressively longer distance scales. The crossover within the confinement phase become a real phase transition to a deconfinement phase as ξ_3 is made to diverge by making N larger with fixed tN. If ξ_3 diverges while ξ_2 remains finite, the fractionalized bosons remain gapped but the gauge boson becomes gapless. This is the Coulomb phase.[j]

Although a parton is always bound with an anti-parton, it can propagate in space rather freely by switching its partners repeatedly in the Coulomb phase. As a result, partons can be effectively liberated and arise as well-defined excitations in the Coulomb phase.

One may ask why one gets a weakly coupled gauge theory rather than some string theory. One possible answer is that world line web is very dense in space-time in this case. They constantly join and split, and their contact interactions are also important. It is hard to imagine to have a well defined string in this strongly coupled soup of strings. This can be viewed as a Lagrangian (or space-time) picture for the string-net condensation.[23]

One can study different types of flavor preserving interactions. In particular, one can obtain fractionalized fermions[20,23] if one introduces a quartic interaction for mesons,

$$K_4 \sum_{a,b,c,d} \cos(\theta^{ab} + \theta^{bc} + \theta^{cd} + \theta^{da}) \tag{12.26}$$

instead of the cubic interaction. If $K_4 > 0$, the interaction is frustrated and not all terms in the potential energy can be minimized simultaneously. This frustration results in a larger low energy manifold and wavefunctions within

[i] Imagine that there is a string of mesons, $ab - bc - cd - de - ....$ If you look at this string in the scale larger than the average spacing between mesons, you will see only flavor a at the end of the string.

[j] In 2+1D, Coulomb phase can not be stable with massive partons due to non-perturbative effects.[21] This opens up the possibility that ξ_2 and ξ_3 diverge simultaneously without an additional fine tuning at a critical point.[22]

the low energy manifold acquires non-trivial sign structure, namely wave-functions describing low energy excitations are no longer positive definite. This non-trivial sign structure is responsible for the fermionic statistics of emergent excitations. The emergent fermion can be also understood from the world line picture where single line loops are endowed with minus signs in the partition function due to the frustrated interaction. Then one should interpret single line loops as world lines of fermions.

12.4. Critical Spin Liquid with Fermi Surface

12.4.1. *From spin model to gauge theory*

12.4.1.1. *Slave-particle approach to spin-liquid states*

Let us consider a system where spins are antiferromagnetically coupled in a two-dimensional lattice,

$$H = J \sum_{<i,j>} \vec{S}_i \cdot \vec{S}_j + \dots \, . \tag{12.27}$$

Here we focus on models with spin 1/2. The antiferromagnetic coupling $J > 0$ favors neighboring spins that align in anti-parallel directions. The dots denote higher order spin interactions whose specific forms are not important for the following discussion. The spin model can be viewed as a low energy effective theory of the Hubbard model at half-filled insulating phase where charges can not conduct due to a large Coulomb repulsion.

At sufficiently low temperatures $T < J$, spins usually order, spontaneously breaking the SU(2) spin rotational symmetry.[k] However, magnetic ordering can be avoided even at zero temperature if quantum fluctuations are strong enough. Two important sources for strong quantum fluctuations are proximity to metallic state and geometrical frustration. In systems which have small charge gap, the ... terms induced by charge fluctuations cause quantum fluctuations of spins (such as permutations of spins around plaquettes) which tend to disrupt magnetic ordering. Quantum fluctuations are also enhanced by geometrical frustrations. Geometrical frustrations arise when no spin configuration can minimize all interaction terms simultaneously in the classical Hamiltonian. For example, antiferromagnetic couplings on the triangular lattice can not be simultaneously minimized. Although there is no magnetic long-range order in disordered ground states,

[k] We can treat the SU(2) symmetry as an internal symmetry in the absence of the spin-orbit coupling which is typically small.

spins remain highly correlated and we refer to such correlated non-magnetic state as spin liquid[24] as opposed to spin gas. More importantly, many-body wavefunctions of spin liquids exhibit non-local entanglement[25] which can be captured only through a new quantum mechanical notion of order[26] which is beyond the scheme of conventional order parameter.

One way of understanding spin liquid states is to use slave-particle approach. For example, one decomposes a spin operator into fermion bilinear

$$\vec{S}_i = \sum_{\alpha\beta} f^\dagger_{i\alpha}\vec{\sigma}_{\alpha\beta}f_{i\beta}. \tag{12.28}$$

Here $f_{i\alpha}$ is a fermionic field which does not carry electromagnetic charge but carry only spin 1/2. For this reason, this particle is called 'spinon' This decomposition has the U(1) phase redundancy. As a result, the theory for spinon has to be in the form of gauge theory. A compact U(1) gauge theory coupled with spinons can be derived following a similar step described in the previous lecture. Although the bare gauge coupling is infinite, the coupling is renormalized to a finite value in the low energy limit. If deconfinement phase is stabilized, spinons and the gauge field arise as low energy degrees of freedom.

Since there is spin 1/2 per each site, there is one spinon per site. Although there is no bare kinetic term for spinons, they can propagate in space through the exchange interaction: simultaneous flips of neighboring spins can be viewed as two spinons exchanging their positions. As a result, spinons form a band, which then determines the low energy spectrum. In non-bipartite lattice (a lattice which can not be divided into A and B sublattices, such as the triangular lattice), the fermions generically form a Fermi surface at half filling. In this case, the low energy effective theory has a Fermi surface of spinons coupled with the emergent U(1) gauge field,[27,28]

$$\begin{aligned} S = \int d^3x \Big[&\Psi^*_j(\partial_0 - ia_0 - \mu_F)\Psi_j \\ &+ \frac{1}{2m}\Psi^*_j(-i\nabla - \mathbf{a})^2\Psi_j + \frac{1}{4g^2}f_{\mu\nu}f_{\mu\nu}\Big]. \end{aligned} \tag{12.29}$$

Here Ψ_j is the fermion field with two flavors, $j = 1, 2$ (spin up and down) and $a_\mu = (a_0, \mathbf{a})$ is the U(1) gauge field with $\mu = 0, 1, 2$. μ_F is the chemical potential and g, the gauge coupling. This is nothing but the three dimensional quantum electrodynamics (QED3) with a nonzero chemical potential.

The zeroth order question one has to ask is the stability of the deconfinement phase. Because of the presence of underlying lattice, the U(1) gauge field is compact. This allows for instanton (or monopole) which describes an event localized in time where the flux of the gauge field changes by 2π. It is known that if there is no gapless spinon, instantons always proliferate in space-time, resulting in confinement.[21,1] In this case, spin liquid states are not stable and spinons are confined. In the presence of gapless spinons, it is possible that the gauge field is screened and instanton becomes irrelevant in the low energy limit.[32] If this happens, fractionalized phase is stable and spinons arise as low energy excitations.

If there are a large number of gapless spinons which have the relativistic dispersion, instanton acquires a scaling dimension proportional to the number of flavors N[33–36] and the fractionalized phase is stable. In the presence of spinon Fermi surface, it turns out that the deconfinement phase can be stable for any nonzero fermion flavor.

12.4.1.2. *Stability of deconfinement phase in the presence of Fermi surface*

To show that instanton is irrelevant in the presence of Fermi surface, we need the following four ingredients.[37]

First, one can describe low energy particle-hole excitations near the Fermi surface in terms of an infinite copy of 1+1 dimensional fermions parametrized by the angle around the Fermi surface.[38] In this angular representation, instanton becomes a twist operator which twists the boundary condition of the 1+1D fermions by π. In the presence of an instanton, the fermions become anti-periodic around the origin in the 1+1D space-time. This can be understood in the following way. At each point on the Fermi surface, the Fermi velocity is perpendicular to the Fermi surface. A fermion (or a wave packet made of states near a point on the Fermi surface) with a given Fermi velocity explores a 1+1 dimensional subspace in real space which is perpendicular to the x-y plane in 2+1D. Since instanton is a source of 2π flux localized in space-time, the total flux of π penetrates through this plane. (Here we are assuming that there is the rotational symmetry. But this argument can be generalized to cases without the symmetry.) Therefore, a fermion moving in the plane encloses a flux close to π as it is transported around the origin at a large distance. As a result, the boundary condition of low energy fermion at each angle is twisted by π.

[1]An exception is when the time-reversal symmetry is broken. In this case, the Chern-Simons term suppresses instanton, stabilizing chiral spin liquid state.[29] Another route of stabilizing deconfinement phase is to break the U(1) gauge group into a discrete gauge group, such as Z_2.[30,31]

Second, the angle θ around the Fermi surface has a positive scaling dimension. As a result, the angle is decompactified and runs from $-\infty$ to ∞ in the low energy limit. In the low energy limit, the gauge field becomes more and more ineffective in scattering fermions from one momentum to another momentum along tangential directions to the Fermi surface. This is because the momentum of the gauge field becomes much smaller than the Fermi momentum. This amounts to saying that effective angular separation between two fixed points on the Fermi surface grows in the low energy limit.

Third, the theory is local in the space of the decompactified angle. This is due to the curvature of the Fermi surface. In the low energy limit, only those fermions very close to the Fermi surface can be excited. Since gauge boson scatters one fermion near the Fermi surface to another point near the Fermi surface, the momentum of gauge boson is almost tangential to the Fermi surface where the fermions are located. Therefore, fermions with a particular angle are coupled only with those gauge field whose momentum is tangential to the Fermi surface at the angle in the low energy limit. This means that fermions with a finite angular separation (except for those which are at the exact opposite sides of the Fermi surface) becomes essentially decoupled in the low energy limit because they are coupled only with those gauge bosons whose momenta are separated in the momentum space.

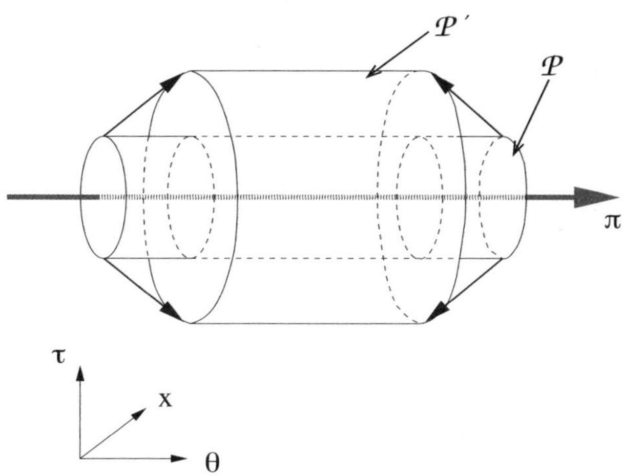

Fig. 12.5. 'Time'-evolution of a quantum state defined on the surface of a pipe extended along the angular direction, where a π-vortex is pierced through the pipe. Under the time-evolution, a point on the surface $\mathcal{P}$ is mapped to a point on the surface $\mathcal{P}'$.

Finally, instanton become a π vortex which is extended along the non-compact θ direction in the theory written in the space of θ, τ and x, where x is a real space coordinate associated with the radial momentum at each point on the Fermi surface. In analogy with the state-operator correspondence in relativistic CFT, the vortex defines a quantum state on the surface of a pipe $\mathcal{P}$ which is extended in the θ direction in the space of τ, x and θ as is shown in Fig. 12.5. The scaling dimension of instanton corresponds to the 'energy' of this quantum state associated with the radial time evolution.

Because of the locality along the decompactified angular direction, the extended π-vortex should have an infinite 'energy'. This implies that the scaling dimension of instanton diverges and instanton remains strongly irrelevant. As a result, the deconfinement phase is stable.

12.4.2. *Low energy effective theory*

Having established that deconfinement phase is stable, we can treat the theory as a non-compact U(1) gauge theory. In this lecture, we are going to analyze the low energy effective theory of Fermi surface coupled with the U(1) gauge field in 2+1D.[40–49] Although we motivated this theory in the context of spin liquid, the same theory often arises in other systems. More importantly, the theory represents a class of typical non-Fermi liquid states which arise as a result of coupling between Fermi surface and gapless bosons. Not surprisingly, many features of this theory are shared by other non-Fermi liquid states in 2+1D.[50–58]

In this system, there is no tunable parameter other than the number of fermion (vector) flavors N. Given that the physically relevant theory ($N = 2$) is a strongly coupled theory, it is natural to consider the theory with a large number of fermion flavors N. Naively one would expect that the theory becomes classical in the large N limit. However, this intuition based on relativistic field theories is incorrect in the presence of Fermi surface. Unlike those theories where gapless excitations are located only at discrete set of points in the momentum space, Fermi surface has an extended manifold of gapless points. The abundant low energy excitations subject to the strong IR quantum fluctuations in 2+1D make the theory quite nontrivial even in the large N limit.[48]

In the low energy limit, fermions whose velocities are not parallel or anti-parallel to each other are essentially decoupled. As a result, it is sufficient to consider local patches of Fermi surface in the momentum space. The

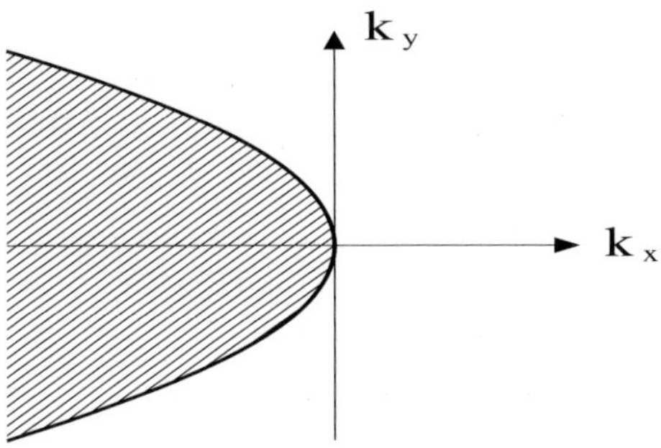

Fig. 12.6. The parabolic Fermi surface of the model in Eq. (12.30). The shaded region includes negative energy states.

minimal model in parity symmetric systems is the theory which includes an open patch of Fermi surface and one more patch whose Fermi velocity is opposite to that of the former.

Here, we will focus on one patch in the momentum space. The one-patch theory is already quite non-trivial and the full structure is yet to be understood. The two-patch theory has yet another level of complications[49] over which we have much less theoretical control. The action in the one-patch theory is

$$\begin{aligned}\mathcal{L} = &\sum_j \psi_j^*(\partial_\tau - iv_x\partial_x - v_y\partial_y^2)\psi_j \\ &+ \frac{e}{\sqrt{N}}\sum_j a\psi_j^*\psi_j \\ &+ a\left[-\partial_\tau^2 - \partial_x^2 - \partial_y^2\right]a,\end{aligned} \tag{12.30}$$

where ψ_j represent fermions with flavor $j = 1, 2, ..., N$. We have chosen our $\mathbf{k} = 0$ to be the point on the Fermi surface where the Fermi velocity is parallel to the x-direction. v_x is the Fermi velocity and $v_y \sim \frac{1}{m}$ determines the curvature of the Fermi surface. The Fermi surface is on $v_x k_x + v_y k_y^2 = 0$ as is shown in Fig. 12.6. This is a 'chiral Fermi surface' where the x-component of Fermi velocity is always positive. This chirality is what makes the one-patch theory more tractable compared to the two-patch theory. a is the transverse component of an emergent U(1) gauge boson in the Coulomb gauge $\nabla \cdot \mathbf{a} = 0$. We ignore the temporal component of the gauge field which becomes massive due to screening. The transverse gauge field remains gapless. e is the coupling between fermions and the critical boson.

Quantum fluctuations of gapless modes generate singular self energies. The one-loop quantum effective action becomes

$$\begin{aligned}\Gamma &= \sum_j \int dk \Big[i\frac{c}{N}\,\mathrm{sgn}(k_0)|k_0|^{2/3} + ik_0 + v_x k_x + v_y k_y^2 \Big] \psi_j^*(k)\psi_j(k) \\ &+ \int dk \left[\gamma \frac{|k_0|}{|k_y|} + k_0^2 + k_x^2 + k_y^2 \right] a^*(k)a(k) \\ &+ \frac{e}{\sqrt{N}} \sum_j \int dkdq \;\; a(q)\psi_j^*(k+q)\psi_j(k),\end{aligned} \tag{12.31}$$

where c and γ are constants of the order of 1. In the low energy limit, the leading terms of the quantum effective action are invariant under the scale transformation,

$$\begin{aligned} k_0 &= b^{-1}k_0^{'}, \\ k_x &= b^{-2/3}k_x^{'}, \\ k_y &= b^{-1/3}k_y^{'}, \\ \psi_a(b^{-1}k_0^{'}, b^{-2/3}k_x^{'}, b^{-1/3}k_y^{'}) &= b^{4/3}\psi_a^{'}(k_0^{'}, k_x^{'}, k_y^{'}). \end{aligned} \tag{12.32}$$

Dropping terms that are irrelevant under this scaling, we write the minimal action as

$$\begin{aligned}\mathcal{L} &= \sum_j \psi_j^*(\eta\partial_\tau - iv_x\partial_x - v_y\partial_y^2)\psi_j \\ &+ \frac{e}{\sqrt{N}} \sum_j a\psi_j^*\psi_j + a(-\partial_y^2)a.\end{aligned} \tag{12.33}$$

Note that the local time derivative term in the fermion action is also irrelevant, and η flows to zero in the low energy limit. But we can not drop this term from the beginning. Otherwise, the theory becomes completely local in time and there is no propagating mode. The role of this irrelevant η-term is to generate a non-trivial frequency dependent dynamics by maintaining the minimal causal structure of the theory before it dies off in the low energy limit. In other words, the η-term itself is irrelevant but it is crucial to generate singular self energies. But once we include the frequency dependent self energies, we can drop the η-term as far as we remember that the non-local self energies have been dynamically generated from the local Lagrangian.

The minimal action (12.33) has four marginal terms. On the other hand, there are five parameters that set the scales of energy-momentum and the fields. Out of the five parameters, only four of them can modify the coefficients of the marginal terms because the marginal terms remain invariant under the transformation (12.32). Using the remaining four parameters, one can always rescale the coefficients of the marginal terms to arbitrary values. Therefore, there is no dimensionless parameter in this theory except for the fermion flavor N. In particular, the gauge coupling e can be always scaled away. This implies that the theory with fixed N flows to a unique fixed point rather than a line of fixed point which has exact marginal deformation. In the following, we will set $v_x = v_y = e = 1$.

12.4.2.1. *Failure of a perturbative* $1/N$ *expansion*

In the naive N counting, a vertex contributes $N^{-1/2}$ and a fermion loop contributes N^1. In this counting, only the fermion RPA diagram is of the order of 1, and all other diagrams are of higher order in $1/N$. In the leading order, the propagators become

$$\begin{aligned} g_0(k) &= \frac{1}{i\eta k_0 + k_x + k_y^2}, \\ D(k) &= \frac{1}{\gamma\frac{|k_0|}{|k_y|} + k_y^2}. \end{aligned} \tag{12.34}$$

One can attempt to compute the full quantum effective action by including $1/N$ corrections perturbatively. However, it turns out vertex functions which connect fermions on the Fermi surface generically have strong IR singularity, which is cured only by loop corrections. Therefore it is crucial to include the fermion self energy in the dressed fermion propagator,

$$g(k) = \frac{1}{i\eta k_0 + i\frac{c}{N}\,\mathrm{sgn}(k_0)|k_0|^{2/3} + k_x + k_y^2}. \tag{12.35}$$

Although the self energy has an additional factor of $1/N$ compared to the bare frequency dependent term, the self energy dominates at sufficiently low energy $k_0 < 1/(\eta N)^3$ for any fixed N. Here it is important to take the low energy limit first before taking large N limit. This is the correct order of limits to probe low energy physics in any physically relevant systems with finite N.

The removal of IR singularity does not come without price. In the presence of the self energy, the IR singularity is cut-off at a scale proportional to $1/N$. As a result the IR singularity is traded with a finite piece which has an additional positive power of N. For example, the two-loop vertex

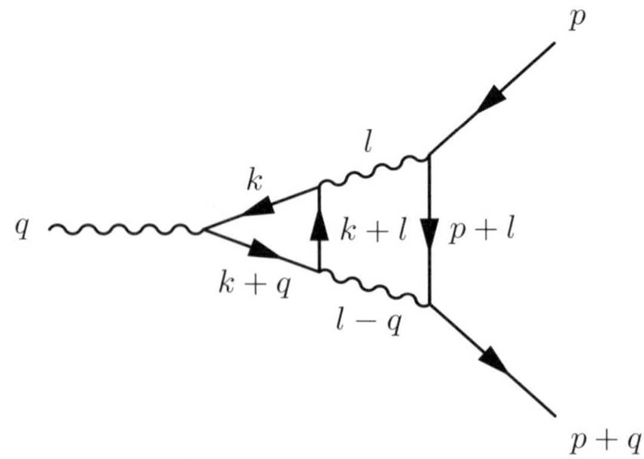

Fig. 12.7. A two-loop vertex correction.

function (Fig. 12.7) computed using the bare propagator becomes

$$\Gamma^{bare}(p = 0, q) = -\frac{N^{-3/2}}{\eta q_0^{1/3}} f_1(q_y/q_0^{1/3}) \tag{12.36}$$

when both both $p = 0$ and $p + q$ are on the Fermi surface. Here $f_1(t)$ is a non-singular universal function which is independent of N and η. Note that the vertex function diverges as q goes to zero with fixed $q_y/q_0^{1/3}$. With the dressed propagator, the IR singularity disappears but there exists an enhancement factor of N,

$$\Gamma^{dressed}(0, q) = -N^{-1/2} f_2(q_y/q_0^{1/3}), \tag{12.37}$$

where $f_2(t)$ is a non-singular universal function independent of N. Note that this two-loop vertex correction has the same order as the bare vertex, which signals a breakdown of perturbative $1/N$ expansion. Similar enhancement factors arise in other diagrams as well.

12.4.2.2. *Genus expansion*

There exists a simple way of understanding this enhancement factor systematically. The reason why the enhancement factor arises is that the fermion propagator is not always order of 1. For generic momentum away from the Fermi surface, the kinetic energy dominates and the propagator is order of 1. On the other hand, when fermions are right on the Fermi surface, the kinetic energy vanishes and there are only frequency dependent terms. At sufficiently low frequencies, the non-local self energy dominates and the propagator is enhanced to the order of N. In relativistic field theories, there are only discrete set of points in the momentum space where the kinetic energy vanishes. In contrast, there is one dimensional manifold of gapless points in the present case with Fermi surface. Whenever fermions hit the Fermi surface (there are many ways to do that), the fermion propagator gets enhanced to order of N. This is the basic reason why the naive N counting breaks down in the presence of Fermi surface.

For a given diagram, say L-loop diagram, there are $2L$ integrations of internal momenta k_x and k_y.[m] In the $2L$-dimensional space of internal momenta, in general there is a m-dimensional sub-manifold on which all internal fermions stay right on the Fermi surface as long as external fermions are on the Fermi surface. We refer to this manifold as 'singular manifold'. If we focus on the momentum integration near a generic point on the singular manifold, it generically looks like

$$I \sim \int dq_1 dq_2 ... dq_m \int dk_1 dk_2 ... dk_{2L-m} \Pi_{i=1}^{I_f} \left[\frac{1}{\alpha_j^i k_j + i\frac{1}{N} f(\omega_i)} \right]. \quad (12.38)$$

Here $q_1, q_2, ..., q_m$ are deviation of momenta from the point on the singualr manifold along the directions tangential to the singular manifold. $k_1, k_2, ..., k_{2L-m}$ are momentum components which are perpendicular to the singular manifold. I_f is the number of fermion propagators. The key point is that the fermion propagators depend only on k's but not on q's because kinetic energy of fermions stay zero within the singular manifold. The tangential momenta q's parametrize exact zero modes of Fermi surface deformations where fermions on the Fermi surface slide along the Fermi surface. If N is strictly infinite and we drop the frequency dependent term in the propagators, the fermion propagators become singular whenever fermions are on the Fermi surface. The m integrations along the tangential direction of the singular manifold do not help to remove the IR singularity. Only $2L - m$ integrations of k momenta lower the degree of IR singularity. After the integration over the all $2L$ momentum, the remaining IR singularity is order of $I_f - (2L - m)$. For a finite N, this IR singularity is cut-off at a momentum proportional to $1/N$. This means the resulting diagram has an additional factor of $N^{I_f - (2L-m)}$ as compared to the naive N counting.

What determines the dimension of the singular manifold within which fermions always remain on the Fermi surface? It turns out that the dimension of the singular manifold is given by the number of closed loops when one draws boson propagators using double lines and fermion propagators using single lines.[38,39] First, we restrict momenta of all fermions to be

[m] The frequency integrals do not play an important role as far as N counting is concerned.

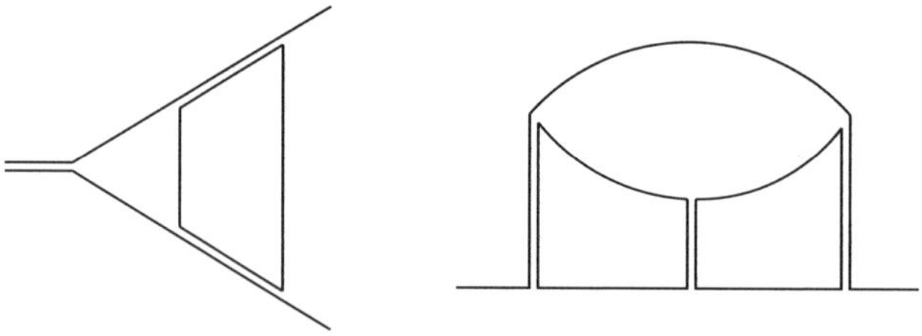

Fig. 12.8. The double line representations of (a) the 2-loop vertex correction and (b) the 3-loop fermion self energy. Double lines represent propagators of the boson, and the single lines are the propagators of the fermion. The number of single line loops (one in (a) and two in (b)) represents the dimension of the singular manifold (see the text) on which all fermions remain on the Fermi surface in the space of internal momenta.

on the Fermi surface. A momentum $\mathbf{k}_\theta$ of fermion on the Fermi surface is represented by an one-dimensional parameter θ. Then, a momentum of the boson $\mathbf{q}$ is decomposed into two momenta on the Fermi surface as $\mathbf{q} = \mathbf{k}_\theta - \mathbf{k}_{\theta'}$, where both $\mathbf{k}_\theta$ and $\mathbf{k}_{\theta'}$ are on the Fermi surface. This decomposition is unique because there is only one way of choosing such $\mathbf{k}_\theta$ and $\mathbf{k}_{\theta'}$. As far as momentum conservation is concerned, one can view the boson of momentum $\mathbf{q}$ as a composite particle made of a fermion of momentum $\mathbf{k}_\theta$ and a hole of momentum $\mathbf{k}_{\theta'}$. For example, the two-loop vertex correction and the three-loop fermion self energy correction can be drawn as Fig. 12.8 in this double line representation. In this representation, each single line represents a momentum on the Fermi surface. Momenta in the single lines that are connected to the external lines should be uniquely fixed in order for all fermions to stay on the Fermi surface. On the other hand, momenta on the single lines that form closed loops by themselves are unfixed. In other words, all fermions can stay on the Fermi surface no matter what the value of the unfixed momentum component that runs through the closed loop is. Since there is one closed loop in Fig. 12.8 (a), the dimension of the singular manifold is 1 and the enhancement factor becomes $N^{4-(4-1)} = N$ for the two-loop vertex correction. Likewise, the enhancement factor for the three-loop fermion self energy becomes $N^{5-(6-2)} = N$ which makes the three-loop fermion self energy to have the same power $1/N$ as the one-loop correction.

The enhancement factor is a direct consequence of the presence of infinitely many soft modes associated with deformations of Fermi surface. The extended Fermi surface makes it possible for virtual particle-hole excitations to maneuver on the Fermi surface without costing much energy. As a result, quantum fluctuations becomes strong when external momenta are arranged in such a way that there are sufficiently many channels for the virtual particle-hole excitations to remain on the Fermi surface. This makes higher order processes to be important even in the large N limit. We note that this effect is absent in relativistic quantum field theories where gapless modes exist only at discrete points in the momentum space.

The net power in N for general Feynman diagrams becomes

$$N^{-V/2+L_f+\left[n+\frac{E_f+2E_b}{2}-2\right]}, \tag{12.39}$$

where n is the number of single line loop, E_f (E_b) is the number of external lines for fermion (boson), and L_f is the number of fermion loop. Here $[x] = x$ for non-negative x and $[x] = 0$ for negative x. For vacuum diagram, the N-counting depends only on the topology of the Feynman diagram and it becomes

$$N^{-2g}, \tag{12.40}$$

where g is the genus of the 2d surface on which Feynman graph is drawn using the full double line representation without any crossing. In the full double line representation, we draw not only the boson propagator as a double line, but also the fermion propagator as a double line where the additional line associated with a fermion loop corresponds to the flavor degree of freedom which run from 1 to N. As a result, all planar diagrams are generically order of N^0. Some typical planar diagrams are shown in Figs. 12.9 and 12.10.

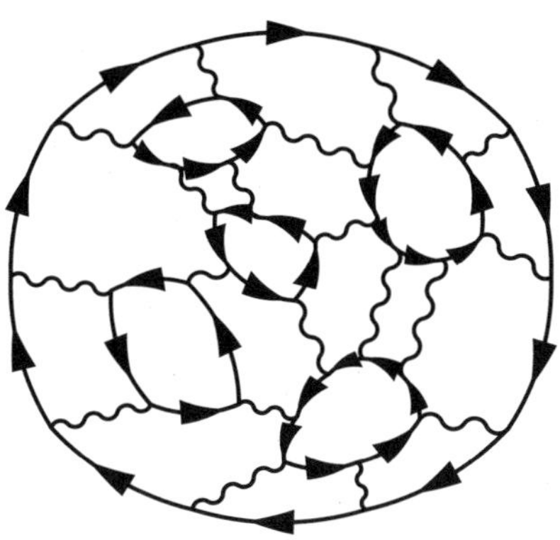

Fig. 12.9. A typical vacuum planar diagram which is of the order of N^0.

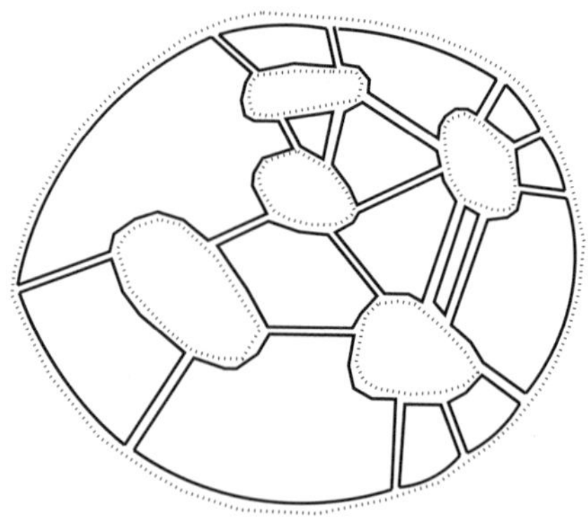

Fig. 12.10. The full double line representation of the planar diagram shown in Fig. 12.9. The solid double lines represent the boson propagator and double lines made of one solid and one dotted lines represent fermion propagators. Loops of dotted lines are added to each fermion loops. In this representation, there is a factor of N for each closed single line loop whether it is a loop made of a solid or dotted line.

The above counting is based on the local consideration on the space of internal momenta near the singular manifold. It turns out that planar diagrams precisely follow the counting obtained from this local considerations. Some non-planar diagrams may deviate from this local counting by some factor of $\log N$. The full structure of non-planar diagrams are not completely understood yet.

Power counting of diagrams with external lines can be easily obtained from the counting of vacuum diagrams. The leading contributions come from the planar diagrams where the genus of the underlying 2d surface is zero. In principle, there can be infinitely many diagrams which are order of N^0 (N^{-1}) for the boson (fermion) self energy and $N^{-1/2}$ for the three-point vertex function.

Because the one-patch theory is a chiral theory, there are strong kinematic constraints. Using this, one can prove that all planar diagrams for boson self energy vanish beyond the one loop. Moreover one can show that the beta function vanishes and fermions have no anomalous scaling dimension to the leading order of N (the contributions of planar diagrams). However, there are still infinitely many non-vanishing planar diagrams for the fermion self energy and the vertex function.

Although the present theory has fermions with vector flavors, it behaves like a matrix model in the large N limit. The reason why the genus expansion arises from this vector model is that the angle around the Fermi surface plays the role of an additional flavor. In the usual matrix model, N controls the genus expansion and the t'Hooft coupling λ controls the loop expansion. One expects that a continuum world sheet description of string emerges in a matrix model when both N and λ are large. In the present case, the effective 't Hooft coupling is order of 1. In other words, two diagrams with L-loops and $(L+1)$-loops have the same order of magnitude as far as they have same topology. This is because there is no dimensionless parameter in the theory other than N as discussed earlier. With $\lambda \sim 1$, it is unlikely that one can use a dual gravity description in a weakly curved space-time to understand this non-Fermi liquid state. Recently, it has been pointed out that the non-Fermi liquid states studied from classical gravity[60–69] show phenomenology of a system where Fermi surface is coupled with a large number of localized bosonic degrees of freedom.[70]

One may view the present theory as a weakly coupled string theory in a highly curved background. It would be interesting to find a way to stabilize the theory at a large 't Hooft coupling. However, this appears to be difficult to achieve in non-supersymmetric theories due to instabilities.

If one includes two patches of Fermi surface, even non-planar diagrams are important due to UV divergence, and the genus expansion appears to break down.[49] It is yet to be understood how Feynman diagrams are organized in this case. On the other hand, one can perturbatively control the theory if one introduces an additional control parameter besides $1/N$.[59]

12.5. Holographic Description of Quantum Field Theory

Although the AdS/CFT correspondence has been conjectured based on the superstring theory, it is possible that the underlying holographic principle is more general and a wider class of quantum field theories can be understood through holographic descriptions.[71–75] According to the holographic principle, the partition function of a D-dimensional theory can be written as a partition function of a (D+1)-dimensional theory where the information on the D-dimensional theory is encoded through boundary conditions as

$$
\begin{aligned}
Z[J(x)] &= \int D\phi(x) e^{-S^D[\phi] - \int dx J\phi} \\
&= \int D \text{ ``}J(x,z)\text{''} e^{-S^{(D+1)}[J(x,z)]}\Big|_{J(x,0)=J(x)},
\end{aligned}
\tag{12.41}
$$

where $J(x,z)$ represent (D+1)-dimensional degrees of freedom. It is most likely that the holographic description will be useful only for quantum field theories which satisfy certain conditions, such as factorization of correlation functions and a large gap in the spectrum of scaling dimensions. Nevertheless, it would be still useful to develop a general prescription for the mapping. In this lecture, we will present a prescription that may be useful in deriving holographic theory for general quantum field theory.[75]

12.5.1. *Toy-model: 0-dimensional scalar theory*

To illustrate the basic idea, we first consider the simplest field theory: 0-dimensional scalar theory. In zero dimension, the partition function is given by an ordinary integration,

$$Z[\mathcal{J}] = \int d\Phi \; e^{-S[\Phi]}. \tag{12.42}$$

We consider an action $S[\Phi] = S_M[\Phi] + S_{\mathcal{J}}[\Phi]$ with

$$\begin{aligned} S_M[\Phi] &= M^2\Phi^2, \\ S_{\mathcal{J}}[\Phi] &= \sum_{n=1}^{\infty} \mathcal{J}_n \Phi^n. \end{aligned} \tag{12.43}$$

Here S_M is the bare action with 'mass' M. $S_{\mathcal{J}}$ is a deformation with sources $\mathcal{J}_n$. For simplicity, we will consider deformations upto quartic order: $\mathcal{J}_n = 0$ for $n > 4$. Here is the prescription to construct a holographic theory.

- Step 1. Introduce an auxiliary field.[76]
  We add an auxiliary field $\tilde{\Phi}$ with mass μ,

  $$Z[\mathcal{J}] = \mu \int d\Phi d\tilde{\Phi} \; e^{-(S[\Phi]+\mu^2\tilde{\Phi}^2)}. \tag{12.44}$$

  Then, we find a new basis ϕ and $\tilde{\phi}$

  $$\begin{aligned} \Phi &= \phi + \tilde{\phi}, \\ \tilde{\Phi} &= A\phi + B\tilde{\phi}, \end{aligned} \tag{12.45}$$

  in such a way that the 'low energy field' ϕ has a mass M' which is slightly larger than the original mass M, and the 'high energy field' $\tilde{\phi}$ has a large mass m',

  $$\begin{aligned} M'^2 &= M^2 e^{2\alpha dz}, \\ m'^2 &= \frac{M^2}{2\alpha dz}, \end{aligned} \tag{12.46}$$

where dz is an infinitesimally small parameter and α is a positive constant. Quantum fluctuations for ϕ become slightly smaller than the original field Φ. The missing quantum fluctuations are compensated by the high energy field $\tilde{\phi}$. In terms of the new variables, the partition function is written as

$$Z[\mathcal{J}]=\left(\frac{Mm'}{M'}+\frac{MM'}{m'}\right)\int d\phi d\tilde{\phi}\; e^{-(S_{\mathcal{J}}[\phi+\tilde{\phi}]+M'^2\phi^2+m'^2\tilde{\phi}^2)}. \quad (12.47)$$

- Step 2. Rescale the fields.
  To maintain the same form for the quadratic action, we rescale the fields as

$$\phi \to e^{-\alpha dz}\phi, \qquad \tilde{\phi} \to e^{-\alpha dz}\tilde{\phi}. \quad (12.48)$$

  Then the couplings are rescaled as $\mathcal{J}_n \to j_n = \mathcal{J}_n e^{-n\alpha dz}$ and $m \to m' e^{-\alpha dz}$.
- Step 3. Expand the action in the power series of the low energy field.
  The new action becomes

$$\begin{aligned} S_j[\phi+\tilde{\phi}] &= S_j[\tilde{\phi}]+(j_1+2j_2\tilde{\phi}+3j_3\tilde{\phi}^2+4j_4\tilde{\phi}^3)\phi \\ &\quad +(j_2+3j_3\tilde{\phi}+6j_4\tilde{\phi}^2)\phi^2+(j_3+4j_4\tilde{\phi})\phi^3+j_4\phi^4. \end{aligned} \quad (12.49)$$

  In the standard renormalization group (RG) procedure,[77] one integrates out the high energy field to obtain an effective action for the low energy field with renormalized coupling constants. Here we take an alternative view and interpret the high energy field $\tilde{\phi}$ as fluctuating sources for the low energy field. This means that the sources for the low energy field can be regarded as dynamical fields instead of fixed coupling constants.
- Step 4. Decouple low energy field and high energy field.
  We decouple the high energy field and the low energy field by introducing Hubbard-Stratonovich fields J_n and P_n,

$$Z[\mathcal{J}] = m\int d\phi d\tilde{\phi}\Pi_{n=1}^4(dJ_n dP_n)\; e^{-(S_j'+M^2\phi^2+m^2\tilde{\phi}^2)}, \quad (12.50)$$

where

$$\begin{aligned}
S_j' &= S_j[\tilde{\phi}] \\
&+ iP_1J_1 - iP_1(j_1 + 2j_2\tilde{\phi} + 3j_3\tilde{\phi}^2 + 4j_4\tilde{\phi}^3) + J_1\phi \\
&+ iP_2J_2 - iP_2(j_2 + 3j_3\tilde{\phi} + 6j_4\tilde{\phi}^2) + J_2\phi^2 \\
&+ iP_3J_3 - iP_3(j_3 + 4j_4\tilde{\phi}) + J_3\phi^3 \\
&+ iP_4J_4 - iP_4j_4 + J_4\phi^4. \qquad (12.51)
\end{aligned}$$

- Step 5. Integrate out the high energy mode.
We integrate out $\tilde{\phi}$ to the order of dz. The auxiliary fields P and J acquire non-trivial action,

$$Z[\mathcal{J}] = \int d\phi \Pi_{n=1}^4 (dJ_n dP_n)\ e^{-(S_J[\phi] + M^2\phi^2 + S^{(1)}[J,P])}, \qquad (12.52)$$

where

$$\begin{aligned}
S^{(1)}[J,P] &= \sum_{n=1}^4 i(J_n - \mathcal{J}_n + n\alpha dz \mathcal{J}_n)P_n \\
&+ \frac{\alpha dz}{2M^2}(i\tilde{\mathcal{J}}_1 + 2P_1\tilde{\mathcal{J}}_2 + 3P_2\tilde{\mathcal{J}}_3 + 4P_3\tilde{\mathcal{J}}_4)^2, \qquad (12.53)
\end{aligned}$$

and $\tilde{\mathcal{J}}_n = (\mathcal{J}_n + J_n)/2$. One can explicitly check that the above action reproduces all renormalized coupling to the order of dz if P_n and J_n are integrated out.
- Step 6. Repeat the steps 1-5 for the low energy field
If we keep applying the same procedure to the low energy field, the partition function can be written as a functional integration over fluctuating sources and conjugate fields,

$$Z[\mathcal{J}] = \int \Pi_{n=1}^4 (DJ_n DP_n)\ e^{-S[J,P]}, \qquad (12.54)$$

where

$$\begin{aligned}
S[J,P] &= \int_0^\infty dz \Big[i(\partial_z J_n + n\alpha J_n)P_n \\
&+ \frac{\alpha}{2M^2}(iJ_1 + 2P_1J_2 + 3P_2J_3 + 4P_3J_4)^2 \Big]. \qquad (12.55)
\end{aligned}$$

Here DJ_nDP_n represent functional integrations over one dimensional fields $J_n(z), P_n(z)$ which are defined on the semi-infinite line $[0, \infty)$. The boundary value of $J_n(z)$ is fixed by the coupling constants of the original theory, $J_n(0) = \mathcal{J}_n$. $P_n(z)$ is the conjugate field of $J_n(z)$ which describes physical fluctuations of the operator ϕ^n. This can be seen from the equations of motion for J_n.

The theory given by Eqs. (12.54) and (12.55), which is exactly dual to the original theory, is an one-dimensional local quantum theory. The emergent dimension z corresponds to logarithmic energy scale.[78] The parameter α determines the rate the energy scale is changed. The partition function in Eq. (12.54) does not depend on the rate high energy modes are eliminated as far as all modes are eventually eliminated. Moreover, at each step of mode elimination, one could have chosen α differently. Therefore, α can be regarded as a function of z. If one interprets z as 'time', it is natural to identify $\alpha(z)$ as the 'lapse function', that is, $\alpha(z) = \sqrt{g_{zz}(z)}$, where $g_{zz}(z)$ is the metric. Then one can view Eqs. (12.54) and (12.55) as an one-dimensional gravitational theory with matter fields J_n. This becomes more clear if we write the Lagrangian as

$$L = P_n \partial_z J_n - \alpha H, \tag{12.56}$$

where H is the Hamiltonian (the reason why H is not Hermitian is that we started from the Euclidean field theory). However, there is one difference from the usual gravitational theory. In the Hamiltonian formalism of gravity,[79] the lapse function is a Lagrangian multiplier which imposes the constraint $H = 0$. However, in Eq. (12.54), α is not integrated over and the Hamiltonian constraint is not imposed. This is due to the presence of the boundary at $z = 0$ which explicitly breaks the reparametrization symmetry. In particular, the 'proper time' from $z = 0$ to $z = \infty$ given by

$$l = \int_0^\infty \alpha(z) dz \tag{12.57}$$

is a quantity of physical significance which measures the total warping factor. To reproduce the original partition function in Eq. (12.42) from Eq. (12.54), one has to make sure that $l = \infty$ to include all modes in the infrared limit. Therefore, l should be fixed to be infinite. As a result, one should not integrate over all possible $\alpha(z)$ some of which give different l. This is the physical reason why the Hamiltonian constraint is not imposed in the present theory. This theory can be viewed as a gravitational theory with the fixed size along the z direction.

Although there are many fields in the bulk, i.e. J_n, P_n for each n, there is only one propagating mode, and the remaining fields are non-dynamical in the sense they strictly obey constraints imposed by their conjugate fields. This is not surprising because we started with one dynamical field Φ. There is a freedom in choosing one independent field. In this case, it is convenient to choose J_3 as an independent field. If one eliminates all dependent fields, one can obtain the local bulk action for one independent field.

12.5.2. D-dimensional O(N) vector theory

The same procedure can be generalized to D-dimensional field theory. For example, the partition function for the D-dimensional O(N) vector model,

$$S[\Phi] = \int d\mathbf{x}d\mathbf{y}\ \Phi_a(\mathbf{x})G_M^{-1}(\mathbf{x}-\mathbf{y})\Phi_a(\mathbf{y}) + \int d\mathbf{x}\ \Big[\mathcal{J}_a\Phi_a + \mathcal{J}_{ab}\Phi_a\Phi_b + \mathcal{J}_{abc}\Phi_a\Phi_b\Phi_c + \mathcal{J}_{abcd}\Phi_a\Phi_b\Phi_c\Phi_d + \mathcal{J}_{ab}^{ij}\partial_i\Phi_a\partial_j\Phi_b + \mathcal{J}_{abc}^{ij}\Phi_a\partial_i\Phi_b\partial_j\Phi_c + \mathcal{J}_{abcd}^{ij}\Phi_a\Phi_b\partial_i\Phi_c\partial_j\Phi_d\Big] \quad (12.58)$$

can be written as a $(D+1)$-dimensional functional integration,

$$Z[\mathcal{J}] = \int DJDP\ e^{-S[J,P]}, \quad (12.59)$$

where the bulk action is given by

$$S[J,P] = \int d\mathbf{x}dz\Big\{iP_a(\partial J_a - \frac{2+D}{2}\alpha J_a) + iP_{ab}(\partial J_{ab} - 2\alpha J_{ab}) + iP_{ab,ij}(\partial J_{ab}^{ij}) + iP_{abc}(\partial J_{abc} - \frac{6-D}{2}\alpha J_{abc}) + iP_{abc,ij}(\partial J_{abc}^{ij} - \frac{2-d}{2}\alpha J_{abc}^{ij}) + iP_{abcd}(\partial J_{abcd} - (4-D)\alpha J_{abcd}) + iP_{abcd,ij}(\partial J_{abcd}^{ij} - (2-d)\alpha J_{abcd}^{ij})\Big\} + \frac{1}{4}\int d\mathbf{x}d\mathbf{y}dz\Big\{\alpha s_a(\mathbf{x})G'(\mathbf{x}-\mathbf{y})s_a(\mathbf{y})\Big\}, \quad (12.60)$$

with

$$s_a = \Big[iJ_a + 2P_bJ_{ab} - 2\partial_j(J_{ab}^{ij}\partial_iP_b) + 3P_{bc}J_{abc} - \partial_j(J_{abc}^{ij}\partial_iP_{bc}) + P_{bc,ij}J_{abc}^{ij} + 4P_{bcd}J_{abcd} - \frac{2}{3}\partial_j(J_{abcd}^{ij}\partial_iP_{bcd}) + 2J_{abcd}^{ij}P_{bcd,ij}\Big], \quad (12.61)$$

$G'(\mathbf{x}) \equiv M\partial_M G_M(\mathbf{x})$, and $\partial = \frac{\partial}{\partial z} - \alpha\sum_{i=1}^D x_i\frac{\partial}{\partial x_i}$. Here $\int d\mathbf{x}$ and $\int d\mathbf{y}$ are integrations on a D-dimensional manifold $\mathcal{M}^D$, and $G_M^{-1}(\mathbf{x})$ is the regularized kinetic energy with cut-off M.

The dual theory is given by the functional integrals of the source fields J and their conjugate fields P in the $(D+1)$-dimensional space $\mathcal{M}^D \times [0,\infty)$ with the boundary condition $J(\mathbf{x}, z=0) = \mathcal{J}(\mathbf{x})$. If the D-dimensional manifold $\mathcal{M}^D$ has a finite volume V, the volume at scale z is given by $Ve^{-\alpha Dz}$.

One key difference from the 0-dimensional theory is that there exist bulk fields with non-trivial spins. In Eq. (12.60), there are spin two fields which are coupled to the energy momentum tensor at the boundary. In the presence of more general deformations in the boundary theory, one needs to introduce fields with higher spins.[80,81]

One can decompose the tensor sources into singlets and traceless parts and take the large N limit where saddle point solutions become exact for singlet fields. One can integrate out all non-singlet fields and obtain an effective theory for single fields alone. However, the resulting effective action for single fields become non-local in this $O(N)$ vector model. This is because there are light non-singlet fields in the bulk and integrating over those soft modes generates non-local correlations for singlet fields. This means that we should keep light non-singlet fields as 'low energy degrees of freedom' in the bulk description if we want to use a local description.

12.5.3. *Phase Transition and Critical Behaviors*

One can understand the phase transition and the critical properties of the model in $D > 2$ using the holographic theory. As one tunes the singlet sources at the UV boundary, the shape of potential in the IR limit changes accordingly. In some parameter regime, the bulk fields are forced to spontaneously break the $O(N)$ symmetry. This is illustrated in Fig. 12.11.

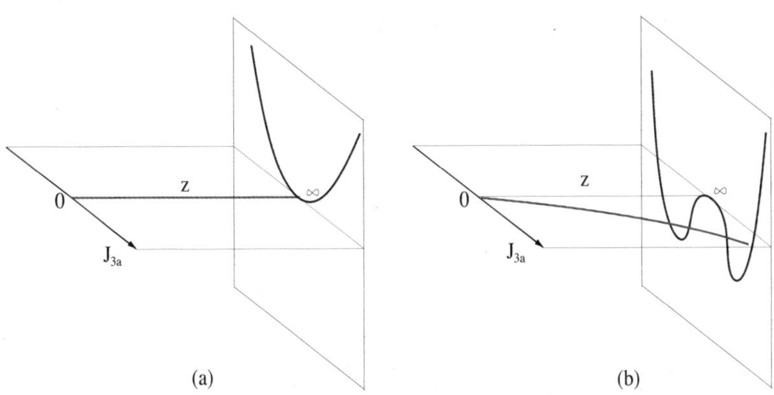

Fig. 12.11. Saddle point configuration for a non-singlet source field $J_{3a}(z)$ (a) in the disordered phase and (b) in the ordered phase. When $\mathcal{J}_2$ is sufficiently negative, a Mexican-hat potential at the IR boundary drags $J_{3a}(z)$ away from $J_{3a}(z) = 0$ in the bulk. At the critical point, J_{3a} at the IR boundary $z = \infty$ is more or less free to fluctuate, generating algebraic correlations between fields inserted at the UV boundary $z = 0$.

One can also compute correlation functions of the singlet operators at the critical point using the similar method used in the AdS/CFT correspondence. For this one integrates over all bulk fields consistent with the x-dependent UV boundary condition. The bulk action can be computed as a function of the UV sources and this gives the generating function for the boundary theory. From this one can compute the critical exponents of singlet operators, which matches with the known field theory predictions. More recently, the present prescription has been applied to large N gauge theory where a field theory of closed loops arise as a holographic dual for the $U(N)$ gauge theory.[82]

Acknowledgments

This note is based on the lectures given at TASI in June 2010. I would like to thank Thomas Banks, Michael Dine and Subir Sachdev for their kind invitation to give lectures at TASI. I am also grateful to Thomas DeGrand, K.T. Mahanthappa and Susan Spika for the hospitality during the summer school. Parts of this material were also presented at the APCTP Focus Program on Aspects of Holography and Gauge/string duality held at Pohang, Korea in August 2010. I thank Deog Ki Hong, Sang-Jin Sin and Piljin Yi for giving me the opportunity to give lectures. I would like to thank lecturers, participants and students of TASI and APCTP workshop for stimulating comments and discussions. Finally, I thank Andrey Chubukov, Guido Festuccia, Matthew Fisher, Sean Hartnoll, Michael Hermele, Yong Baek Kim, Patrick Lee, Hong Liu, Max Metlitski, Lesik Motrunich, Joe Polchinski, Subir Sachdev, T. Senthil, Mithat Unsal and Xiao-Gang Wen for many illuminating discussions in the past. This work was supported by NSERC.

References

1. X.-G. Wen, *Quantum Field Theory of Many-body Systems: From the Origin of Sound to an Origin of Light and Electrons* (Oxford Graduate Texts).
2. S. Sachdev, *Quantum Phase Transitions*, Cambridge University Press.
3. D. Friedan, Z. Qui and S. Shenkar, Phys. Lett. B **151**, 37 (1985).
4. S. Thomas, *Emergent Supersymmetry*, KITP talk (2005).
5. D. B. Kaplan, Phys. Lett. B **136**, 162 (1984).
6. P. Fendley, K. Schoutens and J. de Boer, Phys. Rev. Lett. **90**, 120402 (2003).
7. P. Fendley, B. Nienhuis and K. Schoutens, J. Phys. A **36**, 12399 (2003).
8. S. Catterall, D. B. Kaplan and Mithat Unsal, arXiv: 0903.4881.

9. S.-S. Lee, Phys. Rev. B **76**, 075103 (2007).
10. Y. Yu and K. Yang, arXiv:1005.1399.
11. L. Balents, M. P. A. Fisher and C. Nayak, Int. J. Mod. Phys. B **12**, 1033 (1998).
12. P. W. Anderson, Science **235**, 1196 (1987); P. Fazekas and P. W. Anderson, Philos. Mag. **30**, 432 (1974).
13. A. Y. Kitaev, Ann. Phys. (N.Y.) **303**, 2 (2003).
14. X.-G. Wen, Phys. Rev. Lett. **90**, 016803 (2003).
15. R. Moessner and S. L. Sondhi, Phys. Rev. Lett. **86**, 1881 (2001).
16. X.-G. Wen, Phys. Rev. Lett. **88**, 11602 (2002).
17. X.-G. Wen, Phys. Rev. B **68**, 115413 (2003).
18. O. I. Motrunich and T. Senthil, Phys. Rev. Lett. **89**, 277004 (2002).
19. O. I. Motrunich and T. Senthil, Phys. Rev. B **71**, 125102 (2005).
20. S.-S. Lee and P. A. Lee, Phys. Rev. B **72**, 235104 (2005).
21. A. M. Polyakov, Phys. Lett. B **59**, 82 (1975); Nucl.Phys. B **120**, 429 (1977).
22. T. Senthil, A. Vishwanath, L. Balents, S. Sachdev and M. P. A. Fisher, Science **303** 1490 (2004).
23. M. A. Levin and X.-G. Wen Phys. Rev. B **67**, 245316 (2003); *ibid.* **71**, 045110 (2005).
24. For more comprehensive reviews, see P. A. Lee, N. Nagaosa and X.-G. Wen, Rev. Mod. Phys. **78**, 17 (2006); S. Sachdev, arXiv:0901.4103; L. Balents, Nature **464**, 199 (2010).
25. M. B. Hastings, Phys.Rev. B **69**, 104431 (2004).
26. X.-G. Wen, Phys. Rev. B **65**, 165113 (2002).
27. O. I. Motrunich, Phys. Rev. B **72**, 045105 (2005).
28. S.-S. Lee and P. A. Lee, Phys. Rev. Lett. **95**, 036403 (2005).
29. X.-G. Wen, F. Wilczek and A. Zee, Phys. Rev. B **39**, 11413 (1989).
30. N. Read and S. Sachdev, Phys. Rev. Lett. **66**, 1773 (1991).
31. T. Senthil and M. P. A. Fisher, Phys. Rev. B **62**, 7850 (2000).
32. L. B. Ioffe and A. I. Larkin, Phys. Rev. B **39**, 8988 (1989).
33. G. Murthy and S. Sachdev, Nucl. Phys. B 344, 557 (1990).
34. V. Borokhov, A. Kapustin and X. Wu, J. High Ener. Phys. **0212** 044 (2002).
35. M. A. Metlitski, M. Hermele, T. Senthil, M. P. A. Fisher, Phys. Rev. B **78**, 214418 (2008).
36. M. Hermele, T. Senthil, M. P. A. Fisher, P. A. Lee, N. Nagaosa and X.-G. Wen, Phys. Rev. B **70**, 214437 (2004).
37. S.-S. Lee, Phys. Rev. B **78**, 085129 (2008).
38. R. Shankar, Rev. Mod. Phys. **66**, 129 (1994).
39. S.-W. Tsai, A. H. Castro Neto, R. Shankar, and D. K. Campbell, Phys. Rev. B **72**, 054531 (2005).
40. P. A. Lee, Phys. Rev. Lett. **63**, 680 (1989).
41. B. I. Halperin, P. A. Lee, and N. Read, Phys. Rev. B **47**, 7312 (1993).
42. P. A. Lee and N. Nagaosa, Phys. Rev. B **46** 5621 (1992).
43. J. Polchinski, Nucl. Phys. B **422**, 617 (1994).
44. Y. B. Kim, A. Furusaki, X.-G. Wen and P. A. Lee, Phys. Rev. B **50**, 17917 (1994).

45. C. Nayak and F. Wilczek, Nucl. Phys. B **417**, 359 (1994); **430**, 534 (1994).
46. B. L. Altshuler, L. B. Ioffe and A. J. Millis, Phys. Rev. B **50**, 14048 (1994).
47. O. I. Motrunich and M. P. A. Fisher, Phys. Rev. B **75**, 235116 (2007).
48. S.-S. Lee, Phys. Rev. B **80**, 165102 (2009).
49. M. Metlitski and S. Sachdev, arxiv:1001.1153.
50. V. Oganesyan, S. A. Kivelson, and E. Fradkin, Phys. Rev. B **64**, 195109 (2001).
51. W. Metzner, D. Rohe, and S. Andergassen, Phys. Rev. Lett. **91**, 066402 (2003).
52. V. M. Galitski, G. Refael, M. P. A. Fisher, and T. Senthil, Phys. Rev. Lett. **95**, 077002 (2005).
53. M. J. Lawler, V. Fernandez, D. G. Barci, E. Fradkin, L. Oxman, Phys. Rev. B **73**, 085101 (2006).
54. R. K. Kaul, A. Kolezhuk, M. Levin, S. Sachdev, and T. Senthil, Phys. Rev. B **75**, 235122 (2007).
55. R.K. Kaul, Y.B. Kim, S. Sachdev, T. Senthil, Nat. Phys. **4**, 28 (2008).
56. J. Rech, C. Pepin, and A. V. Chubukov, Phys. Rev. B **74**, 195126 (2006).
57. T. Senthil, Phys. Rev. B **78**, 035103 (2008).
58. D. L.Maslov and A. V. Chubukov, Phys. Rev. B **81**, 045110 (2010).
59. D. F. Mross, J. McGreevy, H. Liu and T. Senthil, arXiv:1003.0894.
60. S.-J. Rey, Prog. Theor. Phys. Supp. **177**, 128 (2009).
61. S.-S. Lee, Phys. Rev. D **79**, 086006 (2009).
62. H. Liu, J. McGreevy and D. Vegh, arXiv:0903.2477.
63. M. Cubrovic, J. Zaanen and K. Schalm, Science **325**, 439 (2009).
64. T.Faulkner, H. Liu, J. McGreevy and D. Vegh, arXiv:0907.2694.
65. F. Denef, S. A. Hartnoll, and S. Sachdev, Phys. Rev. D **80**, 126016 (2009).
66. S. A. Hartnoll, J. Polchinski, E. Silverstein and D. Tong, J. High Ener. Phys. **1004**, 120 (2010).
67. T. Faulkner, N. Iqbal, H. Liu, J. McGreevy and D. Vegh, arXiv:1003.1728.
68. T. Faulkner and J. Polchinski, arXiv:1001.5049.
69. F. Larsen, and G. van Anders, arXiv:1006.1846.
70. S. Sachdev, arXiv:1006.3794.
71. I.R. Klebanov and A.M. Polyakov, Phys. Lett. B **550**, 213 (2002).
72. S. R. Das and A. Jevicki, Phys. Rev. D **68**, 044011 (2003).
73. R. Gopakumar, Phys. Rev. D **70**, 025009 (2004); *ibid.* **70**, 025010 (2004).
74. I. Heemskerk, J. Penedones, J. Polchinski and J. Sully, J. High Energy Phys. **10**, 079 (2009).
75. S.-S. Lee, Nucl. Phys. B **832**, 567 (2010).
76. J. Polonyi, arXiv:hep-th/0110026v2.
77. J. Polchinski, Nucl. Phys. B **231**, 269 (1984).
78. J. de Boer, E. Verlinde and H. Verlinde, J. High Energy Phys. **08**, 003 (2000).
79. R. Arnowitt, S. Deser, and C. Misner, Phys. Rev. **116**, 1322 (1959).
80. M. A. Vasiliev, arXiv:hep-th/9910096.
81. S. Giombi and Xi Yin, arXiv:0912.3462.
82. S.-S. Lee, arXiv:1011.1474.

Chapter 13

Lectures on Holographic Non-Fermi Liquids and Quantum Phase Transitions*

Nabil Iqbal, Hong Liu and Mark Mezei

Center for Theoretical Physics, Massachusetts Institute of Technology, Cambridge, MA 02139, USA

In these lecture notes we review some recent attempts in searching for non-Fermi liquids and novel quantum phase transitions in holographic systems using gauge/gravity duality. We do this by studying the simplest finite density system arising from the duality, obtained by turning on a nonzero chemical potential for a $U(1)$ global symmetry of a CFT, and described on the gravity side by a charged black hole. We address the following questions of such a finite density system:

(1) Does the system have a Fermi surface? What are the properties of low energy excitations near the Fermi surface?
(2) Does the system have an instability to condensation of scalar operators? What is the critical behavior near the corresponding quantum critical point?

We find interesting parallels with those of high T_c cuprates and heavy electron systems. Playing a crucial role in our discussion is a universal intermediate-energy phase, called a "semi-local quantum liquid," which underlies the non-Fermi liquid and novel quantum critical behavior of a system. It also provides a novel mechanism for the emergence of lower energy states such as a Fermi liquid or a superconductor.

13.1. Introduction and Motivation

Understanding phases of matter for which there is no quasiparticle description presents some of the most challenging problems in physics. Examples include the quark-gluon plasma (QGP) created in heavy ion collisions at

*Lectures by HL at TASI 2010, Boulder, June 2010; KITPC workshop/school on "AdS/CM duality and other approaches," Beijing, November, 2011; School on "Holographic View of Condensed Matter Physics" at International Institute of Physics of the Federal University of Rio Grande do Norte, Natal, Brazil, July 2011.

RHIC during the past decade and now at LHC and ultracold atomic systems at unitarity. In condensed matter physics prominent examples include the quantum spin liquid phase of a magnetic insulator, the "strange metals" occurring in the normal state of the high temperature cuprate superconductors, and heavy electron systems near a quantum phase transition. There has been an accumulation of examples, but so far no satisfactory theoretical framework exists to describe them. Finding solvable examples of quantum phases with no quasiparticles should provide a valuable guide for the search for such a framework.

During the last decade, developments in string theory have revealed surprising and profound connections between gravity and many-body systems. The so-called gauge/gravity duality (or AdS/CFT, or holographic duality), relates a classical gravity theory in a weakly curved $(d+1)$-dimensional anti-de Sitter (AdS_{d+1}) spacetime to a strongly-coupled d-dimensional quantum field theory living on its boundary.[1–3] This suggests that complicated questions in strongly interacting many-body physics can be mapped to simple problems in classical gravity, i.e. to *geometry*, potentially giving rise to a new paradigm for approaching many-body dynamics. In particular, black holes appear to play a universal role in characterizing quantum phases without quasiparticles, giving important insight into dynamical behavior of such systems.

Black holes are known for their simplicity and universality; assuming rotational symmetry, the geometry of a black hole is fully specified by its mass and certain conserved charges such as electric charge, independent of other details of a gravity system (such as matter content). Moreover, for a given mass, there is a maximal allowed charge. A quark-gluon plasma has almost equal number of "quarks" and "anti-quarks" and the temperature is main characteristic energy scale. Such a state is described on the gravity side by a black hole of almost zero charge, i.e. a Schwarzschild black hole. In contrast, systems which are of interest in condensed matter physics, like strange metals, are finite density systems, with temperature much smaller than the chemical potential (a strange metal has only electrons but not positrons!). On the gravity side these finite density systems are described by black holes having close to the maximum charges. Thus, in some sense black holes provide a unified description of such seemingly distant systems like a QGP and a strange metal.

Using gravity dual to explore dynamical properties of holographic QGPs has provided important insights into the understanding of the QGP at RHIC (see e.g. Refs. 4 and 5 and references therein). The goal of these notes is to bring these holographic tools to bear on strongly correlated systems at *finite density*. In particular, we review recent progress in finding new universal classes of non-Fermi liquids and novel quantum phase transitions using this approach.

13.1.1. *Challenge: Non-Fermi liquids*

One of the cornerstones of condensed matter physics is Landau's Fermi liquid theory, which underlies our understanding of almost all metals, semiconductors, superconductors and superfluids. Since 1980's there has been an accumulation of metallic materials whose thermodynamic and transport properties differ significantly from those predicted by Fermi liquid theory. These so-called non-Fermi liquids, which include the strange metal phase of cuprate superconductors and heavy fermion materials near a quantum phase transition, present important theoretical challenges. Here we first highlight some salient aspects of the Fermi liquid theory and then point out where it breaks down.

Let us start by recalling the story for a non-interacting Fermi gas (e.g. a gas of electrons in a box) for which the many-particle states can be obtained by simply filing single-particle energy eigenstates following the Pauli exclusion principle. The ground state is then given by filling all the (single-particle) states inside a sphere[a] in momentum space with radius k_F determined by the density of fermions and with all states outside the sphere empty. The locus of points in momentum space at the boundary of this sphere, $k \equiv |\vec{k}| = k_F$, is called the Fermi surface. The low-energy excitations of the system are given by either filling a state slightly outside the Fermi surface or removing a fermion from a filled state slightly inside the Fermi surface, and are called a *particle* and *hole* respectively. These excitations are gapless and have a linear dispersion (for $k - k_F \ll k_F$):

$$\epsilon(k) = E_0(k) - \mu = \frac{k_F}{m}(k - k_F) \equiv v_F(k - k_F) \tag{13.1}$$

where $E_0(k) = \frac{k^2}{2m}$ denotes the free particle energy and we are working in the grand canonical ensemble, in which the energy of a excitation is measured from the chemical potential $\mu = E_F = \frac{k_F^2}{2m}$. Particles and holes[b] are

[a]In these lectures we will only consider rotationally invariant theories, hence we will not encounter Fermi surfaces with shapes other than the sphere.

[b]Note that while all excitations have positive energies, it is convenient to represent a hole using negative energy, i.e. in terms of a filled particle state.

distinguished by the sign of $k - k_F$. At a field theoretical level, these excitations manifest themselves as poles in the complex frequency plane of the retarded Green's function $G_R(\omega, k)$ for the electron operator in momentum space

$$G_R(\omega, \vec{k}) = \frac{1}{\omega - \epsilon(k) + i0_+} \tag{13.2}$$

where ω is again measured with respect to the chemical potential μ. Note that the retarded function is relevant as it describes the causal response of the system if we "add" an electron to the system. Fourier transforming (13.2) back in time we see that the propagator describes the propagation of a particle of energy $\epsilon(k)$:

$$G_R(t, \vec{k}) \sim \theta(t) e^{-i\epsilon(k)t} \tag{13.3}$$

which is of course as it should be for a free theory.

The situation becomes complicated once we switch on interactions between fermions, since the notion of single-particle states no longer makes sense. While one may intuitively expect that the qualitative picture for a non-interacting gas should still apply for sufficiently weak interactions, it is *a priori* not clear what should happen at finite or strong couplings. Laudau's Fermi liquid theory postulates that the above qualitative picture for a non-interacting gas in fact persists for *generic* interacting fermionic systems. In particular, it assumes that

(1) The ground state of an interacting fermionic system is characterized by a Fermi surface in momentum space at $k = k_F$.
(2) Despite (possibly strong) interactions among fundamental fermions, the *low energy* excitations near the Fermi surface nevertheless behave like weakly interacting particles and holes, which are called quasi-particles and quasi-holes. They have the same charge as fundamental fermions and satisfy Fermi statistics. The dispersion of a quasi-particle (similarly for a quasi-hole) resembles (13.1) in free theory

$$\epsilon(k) = v_F(k - k_F) + \dots \qquad v_F = \frac{k_F}{m_*}, \tag{13.4}$$

where m_* can be considered as the effective mass of the quasi-particle and is in general different from the original fermion mass m, from renormalization by many-body interactions.

Given the existence of a Fermi surface, the second postulate above is *self-consistent.* When turning on interactions, a particle (or hole) can now decay into another particle plus a number of particle-hole pairs. But it is not difficult to check that the exclusion principle constrains the phase space around a Fermi surface so much, that given *generic local interactions* among particles and holes, the decay (or scattering) rate of a particle (or hole) obeys

$$\Gamma \sim \frac{\epsilon^2}{\mu} \ll \epsilon \tag{13.5}$$

where ϵ is the energy of a particle (or hole). Thus, despite potentially strong interactions, particle or hole excitations near the Fermi surface have a long lifetime and an approximate particle picture still applies. (Below for simplicity we will refer to quasi-particles and quasi-holes collectively as quasi-particles.) Thus for a Fermi liquid, equation (13.3) should be modified to

$$G_R(t, \vec{k}) \propto e^{-i\epsilon(k)t - \frac{\Gamma}{2}t} , \qquad \Gamma \propto \epsilon^2(k) \tag{13.6}$$

which implies that near the Fermi surface the retarded function for the electron operator should have the form

$$G_R(\omega, \vec{k}) = \frac{Z}{\omega - v_F(k - k_F) + \Sigma(\omega, k)} + \dots , \tag{13.7}$$

with the self-energy $\Sigma(\omega, k)$

$$\Sigma = \frac{i\Gamma}{2} \propto i\omega^2 . \tag{13.8}$$

The residue $Z \leq 1$ of the pole, which is called the quasiparticle weight, can be interpreted as the overlap between the (approximate) one-quasiparticle state with the state generated by acting the electron operator on the vacuum.

The concept of quasi-particles is extremely powerful and makes it possible to develop a general low energy theory – *Fermi liquid theory* – independently of the precise microscopic details of a system. With some phenomenological input, the theory can then be used to predict essentially all the low energy behavior of the system. For example, the theory predicts that the specific heat is linear in temperature (see e.g. Ref. 6)

$$C_e = \gamma T + \dots \qquad \gamma \sim m_* \tag{13.9}$$

and that the low temperature resistivity increases with temperature quadratically

$$\rho_e = \rho_0 + AT^2 + \dots \,. \tag{13.10}$$

The theory has been tremendously successful in explaining *almost* all metallic states in nature.

It is important to emphasize that Fermi liquid theory does not require interactions among the fundamental constituents to be weak. For example for ^{3}He, one finds that $m_* = 2.8 m_{He}$, indicating that interactions among ^{3}He atoms are clearly not weak. There also exist so-called heavy electron compounds for which the effective mass for electron quasi-particles can be as large as $10^2 - 10^3$ times of the electron mass.

It is rather remarkable that weakly interacting quasiparticles can emerge as the low energy collective excitations of a strongly interacting many-body system, with only certain parameters (such as the effective mass) renormalized compared to the fundamental constituents of the system. The self-consistency of Fermi liquid theory can also be understood from an effective field theory perspective using the renormalization group.[7–9] Assuming the existence of quasi-particles, one can then try to write down the most general *local* effective field theory for them. One finds that due to kinematical constraints from the Fermi surface all interactions are *irrelevant* at low energies except for the forward scatterings[c] and BCS-type pairing instabilities leading to a superconductor. Note that while the renormalization group analysis shows that the Fermi liquid theory is a stable fixed point (up to superconducting instabilities), it does *not* tell us whether or why a specific microscopic theory will flow to this fixed point.

Even though the Fermi liquid theory has been tremendously successful, nature always has interesting surprises for us. Since the 1980's, there has been an accumulation of metallic materials whose thermodynamic and transport properties differ significantly from those predicted by Fermi liquid theory (see e.g. Ref. 10). One prominent class of examples of these so-called non-Fermi liquids is the strange metal phase of the cuprate superconductors, which refers to the metallic state above the superconducting transition temperature T_c near optimal doping (see Fig. 13.1). The strange metal phase exhibits thermodynamic and transport behavior significantly different from those of an ordinary metal. A particularly striking prop-

[c]The forward scatterings give rise to interactions among quasi-particles via their densities, which are incorporated in the Fermi liquid theory. Note that such interactions do not give rise to widths.

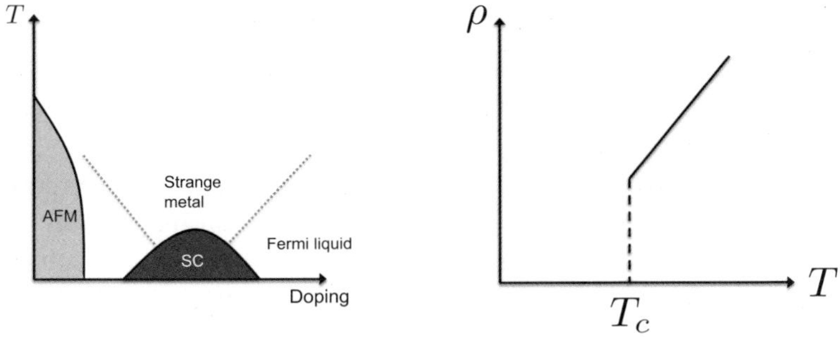

Fig. 13.1. Left: a cartoon picture of the phase diagram of cuprate superconductors; Right: a cartoon picture of linear temperature dependence of the resistivity in the strange metal phase.

erty of the strange metal phase is that the electrical resistivity increases linearly with temperature, in contrast to the quadratic temperature dependence (13.10) of an ordinary metal. This remarkably simple behavior is very robust, existing over a wide range of temperatures (sometimes until a material melts), and universal, appearing in all cuprate superconductors. Nonetheless, it has resisted a satisfactory explanation for more than twenty years.

The anomalous behavior of a strange metal implies that the Fermi liquid theory breaks down. The immediate question is whether one or both of the postulates stated earlier break down. Fortunately, this question can be answered with the help of photoemission experiments which can probe directly a Fermi surface and its low energy excitations. In an Angular Resolved Photoemission Spectropy (ARPES) experiment, incident photons knock out electrons from the sample and the intensity $I(\omega, \vec{k})$ of the electron beam is proportional to $A(\omega, \vec{k})f(\omega, \vec{k})$, where $f(\omega, \vec{k})$ is the Fermi-Dirac distribution and $A(\omega, \vec{k})$ is the electron spectral function defined by

$$A(\omega, \vec{k}) = \frac{1}{\pi} \mathrm{Im}\, G_R(\omega, k) . \tag{13.11}$$

Near the Fermi surface, the spectral function for a Fermi liquid (13.7) is given by a Lorentzian peak centered at $\epsilon(k) = v_F(k - k_F)$ with a width proportional to ϵ^2. In particular, as the Fermi surface is approached

$$A(\omega, k) \quad \to \quad Z\delta(\omega - v_F(k - k_F)), \qquad k \to k_F . \tag{13.12}$$

For a non-Fermi liquid, it is *a priori* not clear what to expect of $A(\omega, k)$. Nevertheless, as a working definition, a Fermi surface $k = k_F$ can be defined as the surface in momentum space where there are gapless excitations, which in turn should result in non-analytic behavior of $A(\omega, \vec{k})$ at $k = k_F$ and $\omega = 0$. Thus a Fermi surface and low energy excitations around it can be diagnosed by examining the non-analytic behavior of $A(\omega, k)$ obtained from ARPES experiments.

For high T_c cuprates in the strange metal region, ARPES experiments indicate that a Fermi surface still exists, but excitations exhibit a much broader peak than that for a Fermi liquid. The experimental results can be fit well to the following expression,[11] postulated as "Marginal Fermi liquid" (MFL) in Ref. 12,

$$G_R(\omega, k) = \frac{h}{\omega - v_F(k - k_F) + \Sigma(\omega, k)} \tag{13.13}$$

with the self-energy $\Sigma(\omega, k)$ given by

$$\Sigma(\omega) \approx c\omega \log \omega + d\omega, \qquad c \text{ real}, \quad d \text{ complex} . \tag{13.14}$$

We see from (13.13) that the system appears to possess gapless excitations of dispersion relation $\omega = v_F(k - k_F)$. However, the decay rate Γ of such excitations, which is given by the imaginary part of Σ, is now proportional to ω in contrast to ω^2 (13.8) for a Fermi liquid. The decay rate, which is comparable to ω, is so large, that an excitation will already have decayed before it can propagate far enough (i.e. one wavelength) to show its particle-like properties. As a result, such an excitation can no longer be treated as a quasiparticle. Also note that the residue for the pole in the complex plane scales like $Z \sim 1/\log(k - k_F)$ as the Fermi surface is approached and thus the quasiparticle weight vanishes logarithmically. Thus at the Fermi surface the overlap of an excitation with original electrons vanishes. Mathematically, the singularity of $A(\omega, k)$ at $k = k_F$ and $\omega = 0$ is much softer than that (13.12) for a Fermi liquid.

Thus the strange metal phase of cuprates has a Fermi surface (i.e. there still exist a surface in momentum space which has gapless excitations), but the quasiparticle picture breaks down. In other words, we are dealing with a *Fermi surface without quasiparticles.* From the perspective of renormalization group, it must be that the system is flowing to a nontrivial fixed point distinct from that for a Fermi liquid.

Known field theoretical examples of non-Fermi liquids which also exhibit the phenomenon of a Fermi surface without long-lived quasiparticles include Luttinger liquids in 1+1 dimensions, and a free fermion gas coupled to some gapless bosonic excitations, which can be either a transverse gauge field or certain order parameter fluctuations near a quantum critical point (see e.g. Refs. 13–16). In the latter case gapless bosonic excitations generate *nonlocal* interactions among fermionic excitations near a Fermi surface and destroy the quasiparticles. Neither of these, however, is able to explain the behavior of a strange metal phase. The former is specific to $(1+1)$-d kinematics. In the latter class of examples, the influence of gapless bosons is mostly along the forward direction, and is not enough, for example, to account for the linear temperature dependence of the resistivity. Even for this class of examples, there still exists significant technical challenges in understanding the precise nature of the IR fixed point.[14–16]

13.1.2. *Challenge: novel quantum phase transitions*

We now switch gears, turning to another challenge faced by contemporary condensed matter theory. Consider a continuous phase transition; these are generally be driven by fluctuations that can be both quantum and thermal. Thermal phase transitions are by now very well developed, based on the so-called Landau-Ginsburg-Wilson paradigm, which consists of the following postulates:

- Different orders are characterized by different symmetries. Phase transitions between orders is a consequence of symmetry breaking.
- The symmetry breaking is characterized by an order parameter and critical behavior for the phase transition is solely captured by low energy dynamics of the order parameter. For example, the free energy can be obtained from

$$Z = \int D\phi(\vec{x})\, e^{-S[\phi]} \tag{13.15}$$

where $S[\phi]$ is the effective action for the order parameter ϕ.[d]

Continuous *quantum* phase transitions, which happen at zero temperature from tuning non-thermal control parameters, have also been traditionally formulated within the Laudau paradigm.[17–23] Now the Euclidean time direction is uncompact and the quantum critical fluctuations of the order parameter live both in space and in time. As the critical point is approached in addition to a divergent correlation length ξ, one also finds

[d]Note that one can ignore the dependence of ϕ on the time direction as such dependence only give rise to gapped excitations due to compactness of the Euclidean time direction at a finite temperature.

a divergent correlation time $\xi_\tau \sim \xi^z$, with z the so-called the dynamical exponent. Thus the order parameter now lives effectively in $d+z$ dimensions, where d is the number of spatial dimensions and z is the effective time dimension. Again one expects that the critical behavior be captured by a path integral like that in (13.15) with now ϕ depending on both space and time.

This paradigm, however, has been challenged on several fronts in recent times. Firstly, there are quantum phases of matter whose order cannot not be captured by symmetries alone, with examples including topologically ordered states such as the quantum Hall phases.[24,25] Phase transitions among or out of such phases cannot be described in terms of order parameter fluctuations. Secondly, during the last decade, heavy fermion compounds have provided a variety of fascinating examples of quantum phase transitions (see e.g. Refs. 22, 26 and 27 for recent reviews), which appear not to be explainable using the standard Landau paradigm. In these examples, while the phase transition can still be characterized as symmetry breaking, new modes, which are inherently quantum and are beyond order-parameter fluctuations, appear to emerge as part of the quantum critical excitations.[e] These experimental examples also inspired new theoretical findings of quantum critical points lying outside the Landau paradigm, including "local quantum critical points"[28] and "deconfined quantum critical points".[29]

Quantum criticality is hence considerably richer and more delicate than its thermal classical counterpart. In particular, quantum critical behavior has been seen to be associated with some of the most interesting phenomena in condensed matter physics including non-Fermi liquids, novel superconductivity etc. In turn, new methods are needed to search for, study, and characterize strongly coupled quantum critical systems.

For a quantum phase transition which does have an order parameter, an important set of observables to characterize the dynamical nature of the quantum phase transition are susceptibilities of the order parameter. Suppose the order parameter is given by the expectation value of some bosonic operator $\mathcal{O}$, then the corresponding susceptibility $\chi(\omega, \vec{k})$ are given by the retarded function for $\mathcal{O}$, which captures the linear responses of the system to an infinitesimal source[f] conjugate to $\mathcal{O}$.

---

[e]In some examples, the phase transition can involve a jump in the size or shape of the Fermi surfaces.

[f]For example if $\mathcal{O}$ is the magnetization of the system, then the corresponding source is a magnetic field.

In a stable phase in which $\mathcal{O}$ is uncondensed, turning on an infinitesimal source will result in an expectation value for $\mathcal{O}$ which is proportional to the source with the proportional constant given by the susceptibility. However, if the system has an instability to the condensation of $\mathcal{O}$, turning on an infinitesimal source will lead to modes exponentially growing with time (just like a ball falling down from top of a hill). Such growing modes are reflected in the presence of singularities of $\chi(\omega, \vec{k})$ in the *upper* complex ω-plane. Thus the dynamical susceptibility provides an important diagnostic of potential instabilities of a system. Similarly, at the onset of an instability (i.e. a critical point, both thermal and quantum), the static susceptibility typically diverges, reflecting that the tendency of the system to develop an expectation value of $\mathcal{O}$ even in the absence of an external source. The divergence is characterized by a critical exponent γ (see Appendix A.4 for a review of definitions of other critical exponents)

$$\chi(k = 0, \omega = 0) \sim |g - g_c|^{-\gamma} \tag{13.16}$$

where g is the tuning parameter (which is temperature for a thermal transition) with g_c the critical point.

13.1.3. *"Local" quantum critical behavior*

A particularly interesting example which illustrates the nontrivial nature of quantum phase transitions for heavy fermions is provided by $CeCu_{6-x}Au_x$ with doping parameter x as the tuning parameter.[30] There is a quantum critical point at $x_c = 0.1$ above which the system is described by an antiferromagnetic phase. One finds near the quantum critical point the dynamical susceptibility behaves as[30]

$$\chi(\omega, \vec{k}) = \frac{Z}{(\vec{k} - \vec{Q})^2 + x_c - x + T^\alpha f\left(\frac{\omega}{T}\right)} \tag{13.17}$$

where $\vec{Q}$ is the ordering vector for the antiferromagnetic order. Equation (13.17) exhibits interesting contrast between momentum and frequency dependence:

(1) The spatial part of the susceptibility exhibits ordinary mean-field behavior.
(2) The self-energy depends only on frequency and exhibits nontrivial ω/T scaling (the exponent $\alpha = 0.75$).

The above behavior has been called "local quantum critical behavior",[28,31] and cannot be explained using the standard Laudau paradigm.

Interestingly, local quantum critical behavior are also present in the retarded response function for the Marginal Fermi Liquid (13.13). The self-energy (13.14) and its finite temperature generalization depend only on ω and exhibits ω/T scaling.[12] In contrast, the spatial momentum dependence is the same as that for a free fermion Fermi surface. Note that in,[12] the fermionic response function (13.13) results from coupling electrons to spin and charge fluctuations, whose spectral function (below χ_0 is a constant and ω_c is a UV cutoff)

$$\mathrm{Im}\,\chi(\omega) \sim \begin{cases} \chi_0 \frac{\omega}{T} & \text{for} \quad |\omega| \ll T \\ \chi_0 \mathrm{sign}\,\omega & \text{for} \quad T \ll |\omega| \ll \omega_c \end{cases} \tag{13.18}$$

is k-independent and again exhibits local quantum critical behavior.

Local quantum critical behavior described above naturally arises in theoretical models of Dynamical Mean Field Theory (DMFT),[32] based on a large spatial dimension mean field approximation. In DMFT one maps a lattice many body problem into an impurity problem and the self-energy arises from that of the impurity. Thus by construction the self-energy is momentum independent and its scaling behavior can be modeled from critical behavior of the impurity. Note that since an impurity typically has a nonzero entropy, in general such an approximation leads to a finite entropy density, indicating that it describes intermediate-energy rather than ground state physics. So it is an interesting question whether the observed local quantum critical behavior in heavy fermion materials and cuprates could be due to intermediate energy effects.[g]

13.1.4. *Scope of this review*

Holography (or gauge/gravity duality, or AdS/CFT) is the surprising statement that certain quantum field theories can be exactly equivalent to theories of quantum gravity that live in one extra spatial dimension. This equivalence is obviously conceptually startling, having far-reaching implications for our understanding of quantum gravity. Remarkably, it is also practically useful; which stems from the fact that it is a strong/weak duality: generally when the field theoretical side is strongly correlated, the gravitational description is weakly coupled and tractable. This suggests

[g]See also an argument which points to this direction at the end of Sec. 13.2.6. More on this issue in the conclusion section.

that complicated questions in strongly interacting many-body physics can be mapped to simple problems in classical gravity, i.e. to *geometry.*

In the most well-studied examples of the duality, gravity (and associated matter fields) propagating on a weakly curved anti-de Sitter spacetime in $d+1$ dimensions (AdS_{d+1}) is mapped to a strongly coupled conformally invariant quantum field theory (CFT_d) that lives at its boundary (with d dimensions). Many examples are now known in various spacetime dimensions, including $\mathcal{N}=4$ Super-Yang-Mills theory in $d=4$, and ABJM theory in $d=3$.[33–35] These theories essentially consist of elementary bosons and fermions interacting with non-Abelian gauge fields. At a heuristic level one may visualize such a theory as the continuum limit of a lattice system where the number of degrees of freedom at each lattice site is of order $O(N^2)$, with N the rank of the gauge group. The classical gravity approximation in the bulk corresponds to the strong coupling regime and the large N limit. In addition to these theories, there also exist vastly many asymptotically-AdS vacua of string theory, each of which is believed to give rise to an example of the correspondence, though an explicit description of the dual field theory is not known for most vacua.

The goal of these notes is to bring holographic tools to bear on the problem of strongly-interacting field theory states at *finite density*, with an attempt to make contact with the physical problems alluded to in the above. We do this by studying the simplest finite density system arising from the duality, obtained by turning on a nonzero chemical potential for a $U(1)$ global symmetry of a CFT_d, which is described on the gravity side by a charged black hole. We ask the following questions of such a finite density system:

(1) Does the system have a Fermi surface? What are physical properties of low energy excitations near the Fermi surface?
(2) Does the system have instabilities to condensation of scalar operators? What is the quantum critical behavior near such a critical point?

In general for a strongly coupled system, these questions are difficult to answer. Fortunately here they can be answered with the help of the gravity dual. In addition to thermodynamical quantities, the main dynamical observables that we will use to probe the system are retarded Green's function of gauge invariant operators, which could be a scalar or a spinor,

$$G_R(t,\vec{x}) \equiv i\theta(t)\,\mathrm{tr}\left(\rho[\mathcal{O}(t,\vec{x}),\mathcal{O}(0)]_{\pm}\right), \tag{13.19}$$

where ρ is the density matrix characterizing the ensemble and the notation $[,]_\pm$ refers to a commutator for scalars and an anticommutator for spinors. The retarded Green's function carries all of the real-time information about the response of the system and its imaginary part is proportional to the spectral function (or spectral weight) $A(\omega, k) = \frac{1}{\pi}\text{Im}\, G_R(\omega, k)$, which captures the density of states which can be accessed by acting $\mathcal{O}$ on the ensemble. As discussed in Sec. 13.1.1 and Sec. 13.1.2, the retarded function for a fermionic operator can be used to probe the existence of a Fermi surface and the nature of low energy excitations around the Fermi surface, while that for a bosonic order parameter provides a convenient diagnostic for possible instabilities, and encodes important dynamical information for quantum phase transitions. For real-life systems the spectral function of various operators can be directly measured; for example that for the electron operator by photo-emission experiments and that for spin polarizations by neutron scatterings.

We will see that, even in this simplest context, holographic systems offer us plenty of opportunities for studying exotic phenomena associated with Fermi surfaces without long-lived quasiparticles and novel quantum phase transitions, some of which share intriguing parallels with cuprates and heavy fermion systems discussed earlier. And gravity provides fascinating ways to characterize them. Our discussion will be mainly based on.[36–46][h] In particular, we will find a *universal intermediate-energy phase*, called "semi-local quantum liquid" in Ref. 43, which does not have a quasiparticle description, and is characterized by: (1) a finite spatial correlation length; (2) an infinite correlation time and associated nontrivial scaling behavior in the time direction; (3) a nonzero entropy density. In a semi-local quantum liquid, the self-energy for an operator depends on spatial momentum[i] k only through the ratio k/μ, where μ is the chemical potential. Thus in the kinematic regime where momentum variation is much smaller than the chemical potential (e.g. near a Fermi surface), the momentum dependence can be neglected and one finds the "local quantum critical behavior" described in Sec. 13.1.3. For example, we will find Fermi surfaces which exhibit the MFL behavior (13.13) and quantum phase transitions whose dynamical susceptibility has the behavior (13.17) or (13.18).

For a holographic system at a nonzero chemical potential, this unstable

[h]The application of holographic methods to condensed matter physics was initiated in Ref. 47 and other reviews include.[48–54]

[i]We will only consider rotationally symmetric systems in this review. k refers to the magnitude of momentum.

semi-local quantum liquid phase sets in at an energy scale of order of the chemical potential, and then orders into other phases at lower energies, including superconductors, antiferromagnetic-type states, and Fermi liquids. While the precise nature of the lower energy state depends on the specific dynamics of the individual system, the semi-local quantum liquid arises universally from these lower energy phases through deconfinement (or in condensed matter language, fractionalization).

In this review we will be mainly interested in searching for new physical scenarios and dynamical mechanisms which can arise for a class of holographic systems, rather than detailed phase structure or dynamical effects for a specific theory. Thus we will take the so-called "bottom-up" approach, i.e. we will just consider a certain type of operator spectrum without referring to a specific theory. The results should apply to any theory which contains the given spectrum. This complements the so-called "top-down" approach which starts with specific string theory embeddings and works out the precise phase structure for these theories. See Refs. 55–63 for some examples of the top-down approach.

The plan of the paper is as follows. In next section we discuss various aspects of gauge/gravity duality for a finite density system. In particular, we introduce the notation of an intermediate-energy semi-local quantum liquid phase and work out its physical properties. In Sec. 13.3 we turn to study of holographic (non)-Fermi liquids and in Sec. 13.4 possible scalar instabilities and associated quantum phase transitions. We conclude in Sec. 13.5 with a summary of main points and a discussion of possible implications for real-life systems.

13.2. Gauge/Gravity Duality at a Finite Density

In this section we discuss various aspects of gauge/gravity duality at a finite density, in preparation for our discussion of non-Fermi liquids and quantum phase transitions in subsequent sections.

13.2.1. *Some aspects of AdS/CFT*

In this subsection we highlight certain aspects of gauge/gravity duality that will be needed for our discussion. More detailed expositions on AdS/CFT are available: for example, the TASI lectures of Polchinski[64] in this volume. Other recent reviews include.[5,49,65] The reader who is already familiar with the basic ideas of holography is encouraged to skip this subsection, and the reader who only wants to compute things can probably skip to the results (13.29) and (13.30).

The gravitational action in AdS takes the form

$$S = \frac{1}{2\kappa^2}\int d^{d+1}x\,\sqrt{-g}\left(\mathcal{R} + \frac{d(d-1)}{R^2}\right) \tag{13.20}$$

where R is the AdS radius, and the Newton's constant

$$\frac{2\kappa^2}{R^{d-1}} \propto \frac{1}{N^2} \tag{13.21}$$

is small in the large N limit of the boundary theory. The simplest solution satisfying the field equations arising from (13.20) is pure anti-de Sitter space, which takes the form

$$ds^2 = R^2\left(\frac{dz^2 + \eta_{\mu\nu}dx^\mu dx^\nu}{z^2}\right) = \frac{R^2}{z^2}(dz^2 - dt^2 + d\vec{x}^2) \tag{13.22}$$

Here the $x^\mu = (t, \vec{x})$ run over the coordinates of the field theory, and z is the extra "holographic" coordinate in the gravitational description, with the AdS boundary lying at $z = 0$. As the simplest and most symmetric solution on the gravitational side this geometry represents the *vacuum* of the dual field theory. Indeed, the vacuum of the CFT should be invariant under the conformal group in d-dimensions, which is precisely the same as the isometry group of AdS_{d+1}. We draw special attention to the following isometry:

$$z \to \lambda z \qquad x^\mu \to \lambda x^\mu \tag{13.23}$$

which represents *scaling*.

One of the most important aspects of the duality is that the bulk spacetime *geometrizes the renormalization group (RG) flow* of the field theory, which can be seen as follows. Due to the warp factor $\frac{R^2}{z^2}$ in front of the Minkowski metric in (13.22), local proper energy and length scales along field theory directions in AdS are related to those in the dual field theory by a z-dependent rescaling. More explicitly, consider a physical process with *proper* energy E_{loc} at some value of z in AdS. The bulk energy E_{loc} is measured in units of the local proper time $d\tau = \frac{R}{z}dt$ and when viewed in terms of boundary time t, it corresponds to an excitation of boundary theory energy E given by

$$E = \frac{R}{z}E_{loc}\,. \tag{13.24}$$

We thus see that physical processes in the bulk with identical proper energies but occurring at different radial positions correspond to different field theory processes with energies that scale as $1/z$. In other words, given that typical local bulk energies are given by the curvature scale, i.e. $E_{loc} \sim \frac{1}{R}$, a field theory process with a characteristic energy E is associated with a bulk process at $z \sim 1/E$.[1,66,67] This relation implies that the z can be identified as renormalization group scale of the boundary theory. In particular, the high-energy (UV) limit $E \to \infty$ corresponds to $z \to 0$, i.e. to the near-boundary region, while the low-energy (IR) limit $E \to 0$ corresponds to $z \to \infty$, i.e. deep in the interior. In a conformal theory, there exist excitations of arbitrarily low energies, which is reflected in the bulk in the fact that the geometry (13.22) extends all the way to $z \to \infty$. If instead we consider a boundary theory with a mass gap m, the corresponding bulk geometry will then end smoothly at a finite value $z_0 \sim 1/m$. Similarly, at a finite temperature T, which provides an effective IR cutoff, the bulk spacetime will be cut off by an event horizon at a finite $z_0 \sim 1/T$.

Another important aspect of the duality is the field/operator mapping, *i.e.* to each (conformally primary) operator one can associate a field in the bulk. For illustration, we will consider a scalar operator $\mathcal{O}$ in the boundary theory. This is *dual* to some massive bulk scalar field ϕ with mass m, whose action can, for example, be written as

$$S_\phi = -\frac{1}{2}\int d^{d+1}x\,\sqrt{-g}\left((\nabla\phi)^2 + m^2\phi^2 + \cdots\right) . \qquad (13.25)$$

By studying the equations of motion arising from the action above one can show that near the boundary a solution to the scalar wave equation has the expansion

$$\phi(z \to 0, x^\mu) \sim A(x) z^{\Delta_-} + B(x) z^{\Delta_+} \qquad (13.26)$$

where the two exponents satisfy

$$\Delta_\pm = \frac{d}{2} \pm \nu_U \qquad \nu_U = \sqrt{\frac{d^2}{4} + m^2R^2}. \qquad (13.27)$$

Then one can show that:

- Δ_+ is the conformal dimension of the dual operator $\mathcal{O}$.
- $A(x)$, which is the coefficient of the more dominant term in (13.26) (in the $z \to 0$ limit) can be identified as the source for $\mathcal{O}$, i.e.

a nonzero $A(x)$ corresponds to adding to the boundary theory Lagrangian a source term

$$\delta S_{\text{boundary}} = \int d^d x\, A(x)\mathcal{O}(x)\ . \tag{13.28}$$

- $B(x)$, which is the coefficient of the subdominant term in (13.26) (in the $z \to 0$ limit), can be identified with the expectation value of the operator $\mathcal{O}(x)$, i.e.

$$\langle \mathcal{O}(x)\rangle = 2\nu_U B(x). \tag{13.29}$$

For example, if one finds a regular solution where $A = 0$ but $B \neq 0$, this implies that the operator $\mathcal{O}$ has spontaneously developed an expectation value, in the absence of a source.
- Given (13.28) and (13.29), one immediately sees that the linear response function (in momentum space) for $\mathcal{O}$ should be given by

$$G_R(\omega, k) = 2\nu_U \frac{B(\omega, k)}{A(\omega, k)} \tag{13.30}$$

where $B(\omega, k)$ and $A(\omega, k)$ are the Fourier transform along the boundary directions of the corresponding quantity in (13.26). In order to determine the ratio in (13.30), one needs to provide an additional boundary condition at the interior of the spacetime, typically a regularity condition. In the situation that a horizon develops in the interior, the boundary condition for computing a *retarded* correlator is that the solution for ϕ should be *infalling* at the horizon.[68]
- For $\nu_U \in (0, 1)$, both terms in (13.26) are normalizable and there are two ways to quantize ϕ by imposing Dirichlet or Neumann conditions at the AdS boundary, which are often called standard and alternative quantizations respectively, and lead to two different CFTs. Our discussion above refers to the standard quantization. In the alternative quantization the role between $A(x)$ and $B(x)$ in (13.26) are reversed, i.e. $B(x)$ corresponds to the source and $A(x)$ to the expectation value. Also the conformal dimension dimension for $\mathcal{O}$ is now given by Δ_- which lies between $(\frac{d}{2} - 1, \frac{d}{2})$ with the lower limit (corresponding to $\nu_U \to 1$) approaching that of a free particle in d spacetime dimension. In the alternative quantization, the double trace operator $\mathcal{O}^2$ is relevant, upon turning on which the system flows to the standard quantization in the IR. See Appendix A.3 for more discussion.

At a practical level, the above (and their generalizations to spinor fields) are all that we will need to perform the computations in the rest of these lectures. Note that while the factors of $2\nu_U$ in (13.29) and (13.30) can be important, we will generally omit them in our discussion below for notational simplicity. Before moving on we summarize our understanding so far:

Boundary CFT_d		Bulk AdS_{d+1}
Complicated many-body dynamics	$\leftrightsquigarrow$	Physics encoded in geometry
RG flow of the system	$\leftrightsquigarrow$	Bulk "holographic" coordinate z
Bulk scalar field ϕ	$\leftrightsquigarrow$	CFT scalar operator $\mathcal{O}$
$N \to \infty$	$\leftrightsquigarrow$	classical GR

The last entry in the table deserves further explanation; generally the specific field theories we are considering are gauge theories with gauge group $SU(N)$. It is a well-known fact[69] that such theories are in some sense classical (but not trivial!) in the large-N limit, by which we mean that correlators factorize and that the physics should be determined by some classical equations of motion, which are not those of the naive $\hbar \to 0$ classical limit. AdS/CFT now tells us that the relevant classical equations of motion are actually those of higher-dimensional Einstein gravity. One must always keep in mind the important caveat when comparing to real-life physical systems that the observables computed using classical AdS/CFT are generally exact only in the $N \to \infty$ limit. Indeed there exist interesting effects that are visible only at finite N; these are dual to quantum effects in the bulk, and we will discuss some of them in later sections (see e.g. Sec. 13.3.4 for perturbative corrections and Sec. 13.3.5 for non-perturbative effects).

13.2.2. *Finite density states and AdS_2*

Our primary interest is to study systems at a finite density (and temperature), the simplest way of realizing which in AdS/CFT is to consider a conformally invariant theory with a $U(1)$ global symmetry and turn on a nonzero chemical potential μ for the $U(1)$. The corresponding conserved current $J^\mu(x)$ is dual to a bulk gauge field $A_M(z,x)$, and the global $U(1)$ symmetry in the field theory is represented by a $U(1)$ *gauge* symmetry in the gravitational description.

Turning on a finite chemical potential μ for the current corresponds to perturbing the boundary theory by the operator

$$\delta S_{\text{boundary}} = \mu \int d^d x \; J^t. \tag{13.31}$$

Being conserved, J^t has scaling dimension $d-1$ and is thus a relevant operator. As a result, under (13.31), the system should flow nontrivially in the IR. Note that since (13.31) also breaks Lorentz symmetry, one expects the flow to be *different* in the time and spatial directions. For a strongly interacting many-body system, understanding the IR physics under relevant deformation (13.31) is very difficult, and in general is not possible using conventional techniques. Fortunately, for a theory with a gravity dual, we can readily extract answers from the gravity side, to which we now turn.

Analogously to the scalar case discussed in the previous subsection, the boundary value of the gauge field A_μ is equal to the value of the field theory source for J^μ: thus the perturbation (13.31) means that we should study a classical gravity solution where the gauge field $A_t(z)$ at the boundary takes some nonzero value $A_t(z \to 0) = \mu$. The relevant action is now (13.20) but with the addition of the Maxwell term for the gauge field,

$$S = \frac{1}{2\kappa^2} \int d^{d+1}x \, \sqrt{-g} \left[\mathcal{R} + \frac{d(d-1)}{R^2} + \frac{R^2}{g_F^2} F_{MN} F^{MN} \right] . \tag{13.32}$$

g_F is a bulk gauge coupling; the factors of R in its definition combine with the dimensional Newton's constant κ^2 out front to make it dimensionless. The solution that satisfies the relevant boundary condition is the Reissner-Nordstrom charged black hole,[70,71 j] whose metric and background gauge field are given by

$$ds^2 = \frac{R^2}{z^2}(-f dt^2 + d\vec{x}^2) + \frac{R^2}{z^2}\frac{dz^2}{f} \tag{13.33}$$

with

$$f = 1 + Q^2 z^{2d-2} - M z^d, \qquad A_t = \mu \left(1 - \frac{z^{d-2}}{z_0^{d-2}} \right) , \tag{13.34}$$

where Q, M, z_0 are constants. Compare this to (13.22): the nontrivial function $f(z)$ indicates that the physics is changing with scale. The horizon lies at $z = z_0$ where

$$f(z_0) = 0, \qquad \to \qquad M = z_0^{-d} + Q^2 z_0^{d-2} . \tag{13.35}$$

[j] Here we say "black hole" even though the horizon is planar with topology $\mathbb{R}^{d-1}$; it would perhaps be more accurate to say "black brane."

Q can be expressed in terms of the chemical potential μ as

$$Q = \sqrt{\frac{2(d-2)}{d-1}} \frac{\mu}{g_F z_0^{d-2}} . \tag{13.36}$$

The temperature of the system can be identified with the Hawking temperature of the black hole, which can be written as

$$T = \frac{d}{4\pi z_0} \left(1 - \frac{d-2}{d} Q^2 z_0^{2d-2} \right) . \tag{13.37}$$

The solution (13.33)–(13.34) has two independent parameters μ and z_0, with μ setting the unit of scale and the solution is characterized by the dimensionless number μz_0 (which can in turn be expressed in terms of the dimensionless ratio T/μ).[k]

The field theory quantities dual to this black hole are given by the following:

$$\text{Charge density:} \qquad \rho = \sqrt{2(d-1)(d-2)} \frac{R^{d-1}}{\kappa^2} \frac{Q}{g_F} \tag{13.38}$$

$$\text{Entropy density:} \qquad s = 2\pi \frac{R^{d-1}}{\kappa^2} \frac{1}{z_0^{d-1}} \tag{13.39}$$

$$\text{Energy density:} \qquad \epsilon = \frac{d-1}{2} \frac{R^{d-1}}{\kappa^2} M . \tag{13.40}$$

From (13.21), all these quantities are of order $O(N^2)$. It can be readily checked from the above equations that the first law of thermodynamics is satisfied

$$d\epsilon = T ds + \mu d\rho . \tag{13.41}$$

Up to some constant factors Q is essentially the charge density of the boundary theory. It is convenient to parameterize it as

$$Q \equiv \sqrt{\frac{d}{d-2}} \frac{1}{z_*^{d-1}} \tag{13.42}$$

by introducing a length scale z_* (which is fixed by boundary charge density). In terms of z_*, various quantities can be written as

$$\rho = \frac{R^{d-1}}{\kappa^2} \frac{1}{z_*^{d-1}} \frac{1}{e_d}, \quad \mu = \frac{d(d-1)}{d-2} \frac{1}{z_*} \left(\frac{z_0}{z_*} \right)^{d-2} e_d, \quad T = \frac{d}{4\pi z_0} \left(1 - \frac{z_0^{2d-2}}{z_*^{2d-2}} \right) \tag{13.43}$$

[k]Note that since the system is scale invariant in the vacuum, at finite density the system is characterized by the dimensionless ratio $\frac{T}{\mu}$.

where we have introduced

$$e_d \equiv \frac{g_F}{\sqrt{2d(d-1)}} \,. \tag{13.44}$$

Note that $z_* \geq z_0$ as otherwise the metric will have a naked singularity (rather than a black hole).

From (13.43), at zero temperature, we have $z_0 = z_*$ and the black hole metric can be written as

$$f = 1 + \frac{d}{d-2}\frac{z^{2d-2}}{z_*^{2d-2}} - \frac{2(d-1)}{d-2}\frac{z^d}{z_*^d}, \qquad A_t = \mu\left(1 - \frac{z^{d-2}}{z_*^{d-2}}\right) \tag{13.45}$$

where μ and z_* are related by

$$\mu_* \equiv \frac{1}{z_*} = 2(d-2)e_d\frac{\mu}{g_F^2} \,. \tag{13.46}$$

For convenience we introduce the appropriately rescaled μ_*, which will be used often below as it avoids having various numerical factors and g_F flying around. The corresponding field theory quantities can now be written as

$$\rho = \frac{R^{d-1}}{\kappa^2}\frac{1}{e_d}\frac{1}{z_*^{d-1}}, \qquad s = \frac{R^{d-1}}{\kappa^2}\frac{2\pi}{z_*^{d-1}}, \qquad \epsilon = \frac{R^{d-1}}{\kappa^2}\frac{(d-1)^2}{d-2}\frac{1}{z_*^d} \,. \tag{13.47}$$

Note that the entropy density is nonzero even at $T = 0$. This finite ground state entropy density appears to violate the Third Law of Thermodynamics; we will discuss the implications of this in Sec. 13.2.6.

For small $z \ll z_*$ we have $f \sim 1$, and the metric resembles that of pure AdS (13.22); this is telling us that physics at energy scales much larger than the chemical potential μ is simply that of the conformal vacuum, as expected. However at $z \sim z_*$ the geometry is very different: f has a double zero at the horizon $z = z_*$, with

$$f(z) \approx d(d-1)\frac{(z_* - z)^2}{z_*^2} + \cdots, \; z \to z_* \tag{13.48}$$

This implies that the horizon is actually an infinite proper distance away, as one can readily check from (13.33). In fact, the near-horizon geometry factorizes into $\mathrm{AdS}_2 \times \mathbb{R}^{d-1}$:

$$ds^2 = \frac{R_2^2}{\zeta^2}(-dt^2 + d\zeta^2) + \mu_*^2 R^2 d\vec{x}^2, \qquad A = \frac{e_d}{\zeta}dt. \tag{13.49}$$

Here we have defined a new radial coordinate ζ and R_2 is the curvature radius of AdS_2,

$$\zeta \equiv \frac{z_*^2}{d(d-1)(z_* - z)}, \qquad R_2 \equiv \frac{R}{\sqrt{d(d-1)}} \,. \tag{13.50}$$

The metric (13.49) applies to the region $\frac{z_*-z}{z_*} \ll 1$ which translates into $\mu\zeta \gg 1$. From (13.46) and (13.50), a relation which will useful later is that

$$e_d = \frac{g_F R_2}{\sqrt{2}R} \,. \tag{13.51}$$

See Fig. 13.2 for a cartoon picture of the geometry of an extremal charged black hole.

At finite temperature, in the regime $\frac{z_*-z_0}{z_*} \ll 1$, i.e. $T/\mu \ll 1$, the near horizon metric is obtained from replacing the AdS_2 factor in (13.49) by an AdS_2 black hole,[39]

$$ds^2 = \frac{R_2^2}{\zeta^2}\left(-\left(1-\frac{\zeta^2}{\zeta_0^2}\right)dt^2 + \frac{d\zeta^2}{1-\frac{\zeta^2}{\zeta_0^2}}\right) + \mu_*^2 R^2 d\vec{x}^2 \tag{13.52}$$

with

$$A_t = \frac{e_d}{\zeta}\left(1-\frac{\zeta}{\zeta_0}\right), \qquad \zeta_0 \equiv \frac{z_*^2}{d(d-1)(z_*-z_0)} \tag{13.53}$$

and the temperature becomes

$$T = \frac{1}{2\pi\zeta_0} \,. \tag{13.54}$$

At a temperature comparable to μ, the AdS_2 structure will be lost.

As discussed in the last subsection, the radial direction z can be identified as the RG scale of the boundary theory. The emergence of the near-horizon AdS_2 region thus holds the key to understanding the low energy physics of the boundary theory at finite density. The fact that at $T = 0$ the horizon ($\zeta \to \infty$) lies at infinite proper distance away and the warp factor in (13.49) imply that the system should again possess gapless degrees of freedom of arbitrarily low energies. In particular, the metric (13.49) has a scaling isometry

$$t \to \lambda t, \qquad \zeta \to \lambda\zeta, \qquad \vec{x} \to \vec{x} \tag{13.55}$$

under which only the time coordinate scales (compare this to the relevant scaling symmetry for Lorentz-invariant AdS_{d+1} (13.23) in which the spatial and time coordinates scale in the same way.) Note that the AdS_2 region is in fact invariant under a full $SL(2,\mathbb{R})$ symmetry group, and likely a Virasoro symmetry.

Thus gravity tells us that at low energies the boundary system flows to a fixed point which is dual to $AdS_2 \times \mathbb{R}^{d-1}$.[39] See Figs. 13.3 and 13.4 for illustration. This fixed point has previously been called the "IR CFT_1"

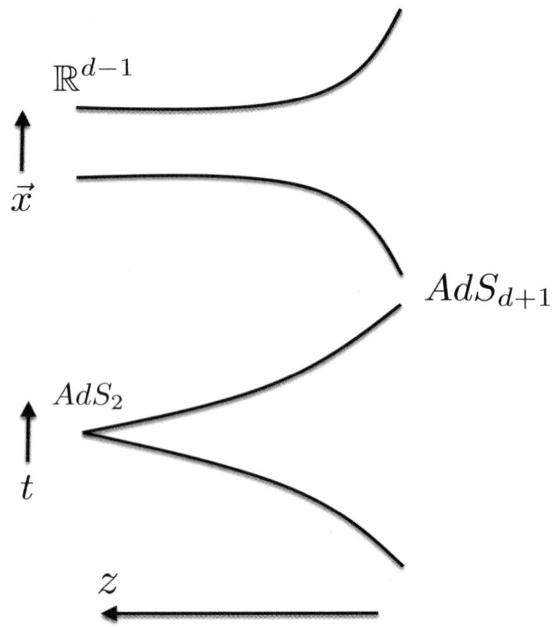

Fig. 13.2. A cartoon picture of the geometry of an extremal charged black hole. The horizon lies at an infinite proper distance away and in the near horizon region the warp for the spatial part approaches a constant, while that for the time direction shrinks to zero at the horizon.

to indicate that there is conformal symmetry only in one dimension (the time direction). However, this hides the fact that the dependence on the spatial directions can have important consequences, and thus we believe that a more descriptive name for this fixed point is a *semi-local quantum liquid* (or SLQL for short),[43] for reasons that we will elaborate on in the next subsection. We will use this term in what follows.

In addition to translational and rotational symmetries of the boundary spatial directions, the SLQL is thus characterized by conformal symmetries of a $(0+1)$-dimensional conformal quantum mechanics including the scaling symmetry in the time direction. It is crucial to note that this conformal invariance has nothing to do with the CFT_d in the ultraviolet, whose conformal invariance was completely broken by the chemical potential. This new conformal symmetry is *emergent*, and has to do with the collective motion of the large number of charged excitations sustaining the background charge density.

Note that SLQL is specified by a single scale μ_*. While μ_* corresponds to the chemical potential for the full theory, near the IR fixed point SLQL, μ_* simply defines a unit for spatial momenta.

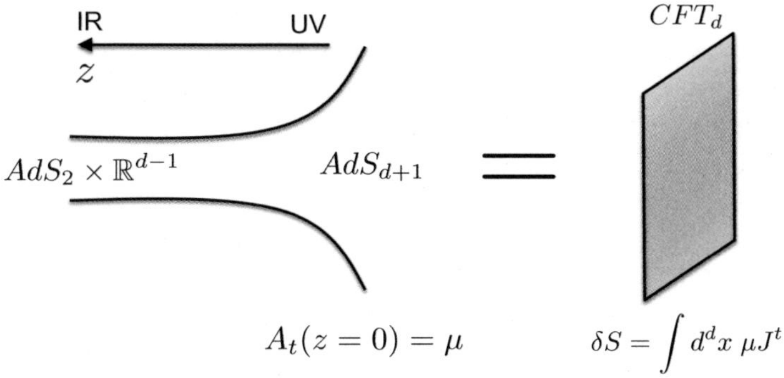

Fig. 13.3. Charged black hole with nonzero gauge field is dual to field theory state with nonzero chemical potential; in the infrared there is an emergent conformal symmetry corresponding to the $AdS_2 \times \mathbb{R}^{d-1}$ part of the geometry.

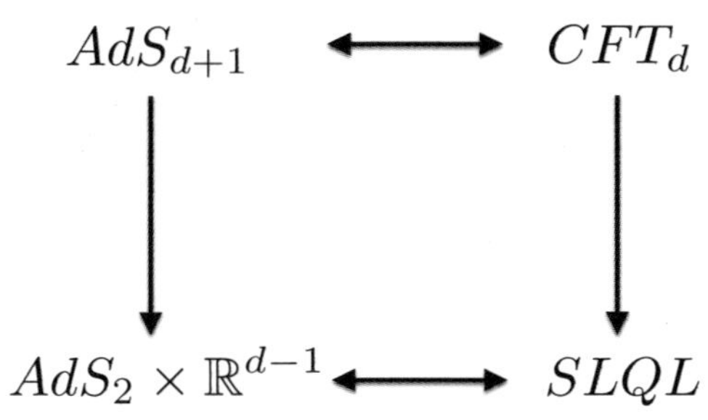

Fig. 13.4. At a finite chemical potential, a CFT_d flows in the IR to SLQL. On the gravity side this is realized geometrically via the flow of the AdS_{d+1} near the boundary to $AdS_2 \times \mathbb{R}^{d-1}$ near the horizon.

13.2.3. *Semi-local quantum liquids*

Consider an operator $\mathcal{O}(t, \vec{x})$ in the boundary theory, which is dual to some bulk field $\phi(t, \vec{x}, z)$ with mass m and charge q. At the UV, i.e. CFT_d, the dimension of $\mathcal{O}$ is given in terms of bulk quantities by (13.27). At the IR fixed point, $\mathcal{O}$ should match to some scaling operator Φ (under scaling symmetry (13.55)) in the SLQL dual to the bulk field ϕ in $AdS_2 \times \mathbb{R}^{d-1}$.

Now in $AdS_2 \times \mathbb{R}^{d-1}$, the Fourier transform[l] $\phi(t, \vec{k}, z)$ of $\phi(t, \vec{x}, z)$ along spatial directions has an effective AdS_2 mass depending on spatial momentum $k = |\vec{k}|$ (as in Kaluza-Klein reduction), which will result in a k-dependent scaling dimension for boundary operators. It is then convenient to label operators by their spatial momentum, with $\phi(t, \vec{k}, z)$ in $AdS_2 \times \mathbb{R}^{d-1}$ dual to some operator $\Phi_{\vec{k}}(t)$ in the SLQL. The Fourier transform $\mathcal{O}_{\vec{k}}(t)$ of $\mathcal{O}(t, \vec{x})$ should then match to $\Phi_{\vec{k}}(t)$ at the IR fixed point. Note that the distinction between $\mathcal{O}_{\vec{k}}$ and $\Phi_{\vec{k}}$ is important; $\mathcal{O}$ is an operator in the full theory, while Φ only exists at the IR fixed point.

The scaling dimension δ_k of $\Phi_{\vec{k}}(t)$ and its correlation functions in SLQL can be obtained from solving the relevant wave equations for ϕ in the $AdS_2 \times \mathbb{R}^{d-1}$ geometry (13.49).[39,43] We present the results here, leaving their derivation to Appendix A.1. One finds that

$$\delta_k = \frac{1}{2} + \nu_k \tag{13.56}$$

with

$$\nu_k = \sqrt{m^2 R_2^2 - q_*^2 + \frac{1}{4} + \frac{k^2 R_2^2}{\mu_*^2 R^2}} = \frac{1}{\sqrt{d(d-1)}} \sqrt{m^2 R^2 - \frac{g_F^2 q^2}{2} + \frac{d(d-1)}{4} + \frac{k^2}{\mu_*^2}} \tag{13.57}$$

and

$$k = |\vec{k}|, \quad q_* = q e_d \ , \tag{13.58}$$

where in the second equality of (13.57) we have used (13.51). The retarded function of $\Phi_{\vec{k}}$ in the SLQL[m] can be written as

$$\mathcal{G}_k(\omega) = c(\nu_k)(-i\omega)^{2\nu_k} \tag{13.59}$$

where

$$c(\nu_k) = 2^{2\nu_k} \frac{\Gamma(-2\nu_k)}{\Gamma(2\nu_k)} \frac{\Gamma(\frac{1}{2} + \nu_k - iq_*)}{\Gamma(\frac{1}{2} - \nu_k - iq_*)} \ . \tag{13.60}$$

Let us now examine what (13.57)–(13.59) tells us regarding the physical properties of a SLQL. Firstly, the dimension (13.57) depends on the momentum momentum k: as a result operators with larger k become less

[l]To simplify notations, we will always use the same symbol to denote the Fourier transform of a field, distinguishing them only by their arguments.

[m]We stress that this is not the full two-point function of the operator $\mathcal{O}$ in the UV theory, but rather just the correlation function at the IR fixed point; the relation between these two is given in (13.75) and is discussed in Sec. 13.2.4.

important in the IR. Also ν_k decreases with q, i.e. an operator with larger q will have more significant IR fluctuations (given the same mass m). From the bulk point of view, this q-dependence in operator dimension is a consequence of the fact that in AdS_2, the electric field following from (13.49) blows up as we approach the AdS_2 boundary.[n]

Secondly, the spectral weight, which is defined by the imaginary part of the retarded function (13.59), scales with ω as a power for any momentum k, which indicates the presence of low energy excitations for all momenta[o] (although with a larger scaling dimension the weight will be more suppressed at larger momenta).

Thirdly, in coordinate space, the system has an infinite correlation time, but a finite correlation length in the spatial directions. This is intuitively clear from the presence of (and lack of) scaling symmetry in the time (spatial) directions. It can be seen more explicitly by examining the Euclidean correlation function $G_E(\tau = it, \vec{x})$ in coordinate space by Fourier transforming (13.59). For this purpose, we note that ν_k can be rewritten as

$$\nu_k = \frac{1}{\sqrt{d(d-1)}\mu_*}\sqrt{k^2 + \frac{1}{\xi^2}}, \tag{13.61}$$

where

$$\xi \equiv \frac{1}{\mu_*}\frac{1}{\sqrt{m^2R^2 - \frac{g_F^2 q^2}{2} + \frac{d(d-1)}{4}}} = \frac{1}{\sqrt{d(d-1)}\nu_{k=0}\mu_*}. \tag{13.62}$$

ν_k has a branch point at $k = i\xi^{-1}$, which under the Fourier transform leads to exponential decay in spatial directions at large distances with a correlation length given by ξ. Explicit calculation yields that $G_E(\tau, \vec{x})$ has two distinct regimes:

(1) For $x \equiv |\vec{x}| \ll \xi$,

$$G_E(\tau, x) \sim \frac{1}{\tau^{2\delta_{k=0}}}. \tag{13.63}$$

(2) For $x \gg \xi$, the correlation function decays at least exponentially as

$$G_E(\tau, x) \sim e^{-\frac{x}{\xi}}. \tag{13.64}$$

[n]In other words, the electric field is non-renormalizable and specifies the corresponding boundary theory.

[o]Exercise for students: compare this result to more "ordinary" gapless systems, like a free massless boson or a Lorentz-invariant CFT

Figure 13.5 provides a heuristic visualization of the behavior: the system roughly separates into independent clusters of size of order ξ, with the dynamics of each cluster controlled by a conformal quantum mechanics, with a power law correlation (i.e. infinite relaxation time) (13.63). Domains separated by distances larger than ξ are uncorrelated with one another. Given that the system has a nonzero entropy density, each cluster has a nonzero entropy that counts the number of degrees of freedom inside the cluster.

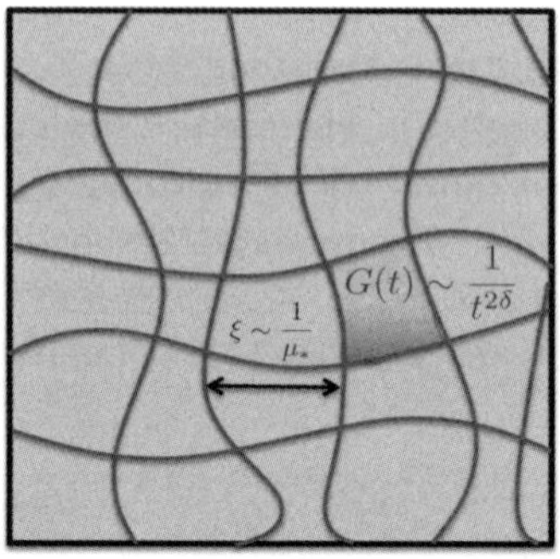

Fig. 13.5. A cartoon picture: the system separates into domains of size $\xi \sim \frac{1}{\mu}$. Within each domain a conformal quantum mechanics governs dynamics in the time direction with a power law correlation (i.e. infinite relaxation time).

While (13.63) and (13.64) can be found by doing Fourier transforms explicitly, they can also be seen geometrically using a geodesic approximation to calculate field-theoretical correlation functions using the Euclidean analytic continuation of (13.49),

$$ds^2 = \frac{R_2^2}{\zeta^2}(d\tau^2 + d\zeta^2) + R^2 \mu_*^2 \, d\vec{x}^2 \, . \tag{13.65}$$

Consider a geodesic that starts at $\zeta \to 0$, moves into AdS_2 at finite ζ, eventually turns around, and returns to the boundary but at a spatial separation x and a Euclidean temporal separation of τ. In the geodesic approximation $G_E \sim e^{-mL(\tau,x)}$ where $L(\tau, x)$ is the proper distance along the geodesic. Since the metric (13.65) is just a direct product, we can simply find the distance moved in each factor and add them using Pythagoras. The distance moved in the $\mathbb{R}^{d-1}$ factor is $R\mu_* x$. A standard calculation shows that the distance moved in the AdS_2 factor is $2R_2 \log\left(\frac{\tau}{\epsilon}\right)$ where ϵ is an

IR cutoff of the AdS_2.[p] Note that since the AdS_2 is embedded in the full spacetime (13.33), the cutoff ϵ should be provided by the scale $\frac{1}{\mu}$. From IR/UV connection, an IR cutoff ϵ of AdS_2 translates into a short-distance cutoff in the SLQL, thus we will restrict to $\tau > \epsilon$. Combining these results we find

$$\begin{aligned} G_E(\tau, x) &\sim \exp\left(-m\sqrt{4R_2^2 \log^2\left(\frac{\tau}{\epsilon}\right) + \mu_*^2 R^2 x^2}\right) \\ &\sim \exp\left(-\sqrt{4\delta^2 \log^2\left(\frac{\tau}{\epsilon}\right) + \frac{x^2}{\xi^2}}\right) \end{aligned} \tag{13.66}$$

with $\delta = mR_2$ and $\xi^{-1} = mR\mu_*$. The geodesic approximation applies to $mR \gg 1$ and in this regime, (13.63) and (13.64) are indeed recovered from (13.66).

The fact that at low energies different points on the $\mathbb{R}^{d-1}$ can be thought of as being in different disconnected domains with size $\xi \sim \frac{1}{\mu}$ can also be seen geometrically as follows. Consider two spacetime points on a hypersurface of given $\zeta \to \infty$. To see whether observers at those locations can communicate with each other we look at time-like geodesics in (13.49) which connect the two points. Simple calculations show that there is a maximal separation in $\mathbb{R}^{d-1}$ directions for two points to communicate with each other, given by

$$\Delta x_{\max} = \pi R_2 \frac{1}{\mu_* R} = \frac{\pi}{\mu_* \sqrt{d(d-1)}} \,. \tag{13.67}$$

The first factor πR_2 in the first equality above is the time for a time geodesic (see Fig. 13.6) to approach the boundary and come back and the second factor $\frac{1}{\mu_* R}$ is the effective velocity in $\mathbb{R}^{d-1}$ (see (13.49)). Equation (13.67) is consistent with (13.61) up to a prefactor.

To summarize, equation (13.59) describes a disordered state in which the space factorizes into independent domains of correlation length ξ. Within each domain one has scale invariance along the time direction. This behavior is reminiscent of various theoretical models based on a large spatial dimension mean field approximation,[32] such as the gapless quantum liquids of Refs. 72–74 and the "local quantum critical point" of Ref. 28.[q] We should emphasize that the scaling behavior within each cluster here describes, however, not the behavior of a single site, but the collective behavior of a large

[p]The easiest way to understand this result is to note that its exponential must reproduce the conformal result, τ^{-2mR_2}.

[q]Some suggestions for a connection have been made in Refs. 75–78.

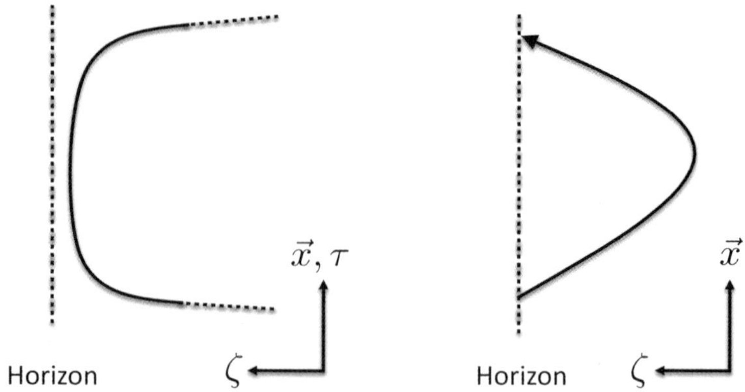

Fig. 13.6. Geodesics in the near horizon geometry. In both cases the horizon is the dotted line at $\zeta \to \infty$, and the boundary is to the right towards $\zeta \to 0$. The left plot gives the geodesic for calculating the Euclidean correlation function (13.66); the right plot gives a timelike geodesic connecting two points on a constant-ζ hypersurface. The maximal boundary time it takes for a timelike geodesic to approach the boundary and come back is πR_2.

number of sites (if one considers our systems as a continuum limit of a lattice) over size of order ξ. This aspect is reflected in that, while there is nontrivial scaling only in the time direction, the scaling dimension (13.57) and correlation functions (13.59) depend nontrivially on k. As discussed earlier around (13.61) it is precisely this dependence that gives the spatial correlation length of the system. Also, while at a generic point in parameter space the dependence of ν_k and $\mathcal{G}_k$ on k is analytic and only through k/μ, as we will see in Sec. 13.4, near certain quantum critical points, the dependence becomes nonanalytic at $k = 0$ and is important for understanding the behavior around the critical point. It is also important to emphasize that, despite the scaling behavior in (13.59) we are describing a *phase*, not a critical point. We thus call it a *semi-local quantum liquid*, or a SLQL for short.

We now consider some generalizations of (13.57) and (13.59). At a finite temperature the background geometry is given by an AdS_2 black hole (13.52) and equation (13.59) generalizes to[39,41]

$$\mathcal{G}_k^{(T)}(\omega) = T^{2\nu_k} g_b \left(\nu_k, \frac{\omega}{2\pi T}\right) \tag{13.68}$$

where g_b is a scaling function given by

$$g_b(\nu_k, y) = (4\pi)^{2\nu_k} \frac{\Gamma(-2\nu_k)}{\Gamma(2\nu_k)} \frac{\Gamma(\frac{1}{2}+\nu_k - iq_*)}{\Gamma(\frac{1}{2}-\nu_k - iq_*)} \frac{\Gamma\left(\frac{1}{2}+\nu_k - iy + iq_*\right)}{\Gamma\left(\frac{1}{2}-\nu_k - iy + iq_*\right)} . \tag{13.69}$$

For a spinor field one finds that, by studying the behavior of the Dirac equation in the AdS_2 region, the dimension δ_k of a spinor operator in SLQL is given by

$$\delta_k = \frac{1}{2} + \nu_k \quad \nu_k = \sqrt{m^2 R_2^2 - q_*^2 + \frac{k^2 R_2^2}{R^2 \mu_*^2}} = \frac{1}{\sqrt{d(d-1)}} \sqrt{m^2 R^2 - \frac{g_F^2 q^2}{2} + \frac{k^2}{\mu_*^2}} . \tag{13.70}$$

Note this is slightly different from the corresponding result for the charged scalar (13.57). Similarly, the retarded Green's function is given by[r]

$$\mathcal{G}_k(\omega, k) = c(\nu_k)(-i\omega)^{2\nu_k} \tag{13.71}$$

with $c(\nu_k)$ now given by

$$c(\nu_k) = 2^{2\nu_k} \frac{\Gamma(-2\nu_k)}{\Gamma(2\nu_k)} \frac{\Gamma(1+\nu_k - iq_*)}{\Gamma(1-\nu_k - iq_*)} \frac{\left(m - \frac{ikR}{\mu_*}\right) R_2 - iq_* - \nu_k}{\left(m - \frac{ikR}{\mu_*}\right) R_2 - iq_* + \nu_k} . \tag{13.72}$$

At finite temperature, equation (13.71) generalizes to

$$\mathcal{G}_k^{(T)}(\omega) = T^{2\nu_k} g_f \left(\nu_k, \frac{\omega}{2\pi T}\right) \tag{13.73}$$

with g_f given by

$$g_f(\nu_k, y) = (4\pi)^{2\nu_k} \frac{\left(m - \frac{ikR}{\mu_*}\right) R_2 - iq_* - \nu_k}{\left(m - \frac{ikR}{\mu_*}\right) R_2 - iq_* + \nu_k} \times \frac{\Gamma(-2\nu_k)}{\Gamma(2\nu_k)} \frac{\Gamma(1+\nu_k - iq_*)}{\Gamma(1-\nu_k - iq_*)} \frac{\Gamma\left(\frac{1}{2}+\nu_k - iy + iq_*\right)}{\Gamma\left(\frac{1}{2}-\nu_k - iy + iq_*\right)} . \tag{13.74}$$

13.2.4. *Low energy behavior of retarded Green functions*

Having understood the behavior of the retarded function at the IR fixed point SLQL, we now turn to the low energy behavior of the retarded function for $\mathcal{O}$ in the full theory, which we will denote as $G_R(\omega, k)$ to distinguish

[r]The following expressions are for half of the components. The expressions for other components are obtained by taking $k \to -k$. See Refs. 39 and 41 for details.

it from $\mathcal{G}_k(\omega)$ at the IR fixed point. We will consider a scalar operator at $T = 0$ and mention the generalization to a finite temperature and a spinor at the end. We are interested in the low frequency regime $\omega \ll \mu$. Note that since we are working in the grand canonical ensemble the frequency ω in our discussion of this paper always refers to energy measured from the chemical potential.[s]

$G_R(\omega, k)$ can be computed explicitly by solving the equation of motion for the corresponding bulk scalar field in the full charged black hole geometry (13.33) and following the procedure outlined in Sec. 13.2.1. Basically our task is to understand how the expansion coefficients A, B in (13.30) behave at low frequencies. We present only the final result here,

$$G_R(\omega, \vec{k}) = \mu_*^{2\nu_U} \frac{b_+(\omega, k) + b_-(\omega, k)\mathcal{G}_k(\omega)\mu_*^{-2\nu_k}}{a_+(\omega, k) + a_-(\omega, k)\mathcal{G}_k(\omega)\mu_*^{-2\nu_k}} + \cdots , \tag{13.75}$$

relegating the details to Appendix A.2. The various objects appearing in this formula deserve explanation. $\mathcal{G}_k(\omega)$ is the retarded Green's function in the SLQL discussed in Sec. 13.2.3 and is proportional to $\omega^{2\nu_k}$ with ν_k defined in (13.57). $a_\pm$ and $b_\pm$ are quantities that arise from the solving the equation of motion in the region of the black hole geometry not too close to the horizon (called the UV region in Appendix A.2); the key fact here is that they are analytic in ω and have a smooth $\omega \to 0$ limit. See Appendix A.2 for further discussion of their properties. Since $a_\pm, b_\pm$ in (13.75) are analytic in ω, we can expand them as follows

$$a_+(\omega, k) = a_+^{(0)}(k) + a_+^{(1)}(k)\omega + a_+^{(2)}(k)\omega^2 + \cdots \tag{13.76}$$

and keep only the leading order term after which (13.75) becomes

$$G_R(\omega, \vec{k}) = \mu_*^{2\nu_U} \frac{b_+^{(0)}(k) + b_-^{(0)}(k)\mathcal{G}_k(\omega)\mu_*^{-2\nu_k}}{a_+^{(0)}(k) + a_-^{(0)}(k)\mathcal{G}_k(\omega)\mu_*^{-2\nu_k}} + \cdots . \tag{13.77}$$

Equation (13.75) contains two sets of data: "universal" data from $\mathcal{G}_k(\omega)$ that depends only on the IR fixed point (or the near-horizon $\mathrm{AdS}_2 \times \mathbb{R}^{d-1}$ region of the black hole geometry), and non-universal data $a_\pm, b_\pm$ which depends on the rest of the black hole geometry and can be thought of as encoding the the effects of the UV degrees of freedom on the low energy physics. This is consistent with general expectations from a Wilsonian understanding of renormalization group flow, and we now show that the

[s]More explicitly, for an operator of charge q, the effective chemical potential is then μq and ω is the energy measured from μq. See also discussion below equation (A.15) in Appendix A.2.

form of (13.75) follows simply from general considerations of a suitable low energy effective action.

We expect that the low energy effective action of the system can be written as

$$S_{eff} = S_{SLQL} + S_{UV} \tag{13.78}$$

where S_{SLQL} is the action for the IR fixed point, for which we do not have an explicit Lagrangian description, but as discussed in Sec. 13.2.3 whose observables can be computable from gravity in $AdS_2 \times \mathbb{R}^{d-1}$. S_{UV} arises from integrating out higher energy degrees of freedom, and can be expanded in terms of scaling operators in S_{SLQL}. The part relevant for $\mathcal{O}$ can be written as[t]

$$S_{UV} = \frac{1}{2}\int \Lambda(k,\omega) J_{\vec{k}}^2 + \int \eta(k,\omega)\Phi_{\vec{k}} J_{-\vec{k}} - \frac{1}{2}\int \xi(k,\omega)\Phi_{\vec{k}}^2 + \cdots \tag{13.79}$$

where we have written the action in momentum space since the dimension of SLQL operator $\Phi_{\vec{k}}(t)$ is momentum-dependent, and the integral signs should be understood as $\int = \int d\omega \int d\vec{k}$. We have introduced a source $J_{\vec{k}}$ for $\mathcal{O}_{\vec{k}}$ and $\cdots$ denotes higher powers of $\Phi_{\vec{k}}$ and J. Note again the distinction between $\mathcal{O}$ and Φ; Φ is the operator at the IR fixed point to which $\mathcal{O}$ matches. Since we are only interested in two-point functions it is enough to keep S_{UV} to quadratic order in Φ and J. For simplicity we have assumed to the quadratic order there is no mixing of $\mathcal{O}$ with other operators.[u] The discussion can be easily generalized to the situation with mixing. The functions $\Lambda(k,\omega), \eta(k,\omega)$ and $\xi(k,\omega)$ parametrize the "UV data" coming from integrating out higher energy modes. In particular they should be *real* and have an analytic expansion in ω with a nonzero $\omega \to 0$ limit, i.e. when written in terms of t, the effective action (13.79) should have a well defined derivative expansion.

The (Euclidean) two-point function for $\mathcal{O}_{\vec{k}}$ can be obtained by differentiating the Euclidean partition function of S_{eff} with respect to $J_{\vec{k}}$. It contains two parts. The first part is simply $\Lambda(\omega,k)$ from the first term in (13.79), which comes purely from UV and is independent of any IR fields. The second part comes from the correlation function $\Phi_{\vec{k}}$ in the SLQL (up to a factor of η). To the order written explicitly in (13.79), the full action (13.78) describes SLQL deformed by a double trace operator

[t]Note the expressions below are written somewhat heuristically to exhibit the main structure. For example, $J_{\vec{k}}^2$ should be interpreted as $J_{\vec{k}} J_{-\vec{k}}$ and similarly with $\Phi_{\vec{k}}^2$.

[u]This corresponds to that the equation for the dual bulk field ϕ does not mix with other fields at linear order.

with coupling ξ. The two-point function of Φ in this deformed theory can be obtained from the general formula (A.41) in Appendix A.3. Combining the two parts and analytically continuing to Lorentizian signature, we find that the retarded function G_R for the full theory can be written as

$$G_R(\omega, k) = \Lambda(k, \omega) + \frac{\eta^2(k, \omega)}{\mathcal{G}_k^{-1}(\omega) + \xi(k, \omega)} \,. \tag{13.80}$$

Equation (13.80) has exactly the same form as the gravity result (13.75) and in fact comparing them we can identify

$$\Lambda = \mu_*^{2\nu_U} \frac{b_+}{a_+}, \quad \xi = \mu_*^{-2\nu_k} \frac{a_-}{a_+}, \quad \eta^2 = \frac{W}{a_+^2} \mu_*^{2\nu_U - 2\nu_k}, \quad W \equiv a_+ b_- - a_- b_+ \,. \tag{13.81}$$

The low energy effective action (13.78), (13.79) with identifications (13.81) can also be derived directly following the formalism of holographic Wilsonian RG introduced recently in Refs. 79 and 80. Note that while our main goal here is to derive the retarded function (13.75), the effective action (13.78)–(13.79) clearly has much wider applications and we will see some of them in later sections.

Since here the spatial directions do not scale, the last term in equation (13.79) is an irrelevant perturbation from SLQL since $\Phi_{\vec{k}}$ has dimension $\frac{1}{2} + \nu_k > \frac{1}{2}$. This is what we should expect; SLQL is the IR fixed point and the last term in (13.79) is simply the lowest irrelevant operator taking it to the UV. Indeed from (13.77) (or equivalently (13.80)) we find that as $\omega \to 0$,

$$\mathrm{Im}\, G_R(\omega, k) = \mu_*^{2\nu_U} \frac{\nu_k}{\nu_U} \frac{1}{(a_+^{(0)}(k))^2} \mathrm{Im}\, \mathcal{G}_k(\omega) + \cdots \propto \omega^{2\nu_k} + \cdots \tag{13.82}$$

where we have used (A.33) and that $a_\pm, b_\pm$ are real. As expected, up to a non-universal prefator (coming from wave renormalization between the UV and IR operators), at low energies the spectral function is given by that of the IR fixed point. In particular, as already emphasized in Sec. 13.2.3 (one paragraph before that of (13.61)), the system has gapless degrees of freedom at any spatial momentum.

It turns out there are circumstances in which (13.82) does not apply, and the physics is in fact much richer. This will be the subject of next subsection.

We conclude this subsection with some generalizations:

(1) At a finite temperature $T \ll \mu$, equation (13.75) can be generalized by replacing $\mathcal{G}_k(\omega)$ there with its finite-temperature generalization

$\mathcal{G}_k^{(T)}(\omega)$ of (13.68). In addition, there will also be analytic finite-T corrections to $a_\pm, b_\pm$, which can be expanded perturbatively in T.

(2) An analogous calculation can be done for a Dirac spinor field in the bulk. Some of the details are different but at the end of the day the formula (13.75) still applies to each of the *eigenvalues* of the Green's function, which is now a matrix in spinor space. For details see Refs. 39 and 81 (see also Sec. 13.3.1 for a discussion).

13.2.5. *Possible low energy physics: Fermi surfaces and scalar instabilities*

As discussed in Sec. 13.1.1 and Sec. 13.1.2, the retarded function for a fermionic operator can be used to probe the existence of a Fermi surface and the nature of low energy excitations around the Fermi surface, while that for a bosonic order parameter provides a convenient diagnostic for possible instabilities, and encodes important dynamical information for quantum phase transitions. If (13.82) were the full story, then that would be somewhat boring, as there will be no Fermi surface or instability (thus no quantum phase transition). However, equation (13.82) breaks down when one of the following occurs and and interesting new physics emerges,

(1) $a_+^{(0)}(k)$ could *vanish* at some momentum;
(2) ν_k which characterizes the scaling dimension of an operator in the SLQL becomes imaginary. Note in this case $a_\pm, b_\pm$ become complex. See (A.35) and (A.36) in Appendix A.2.

From the RG point of view, item 1 implies that the effective action (13.79) breaks down as (13.81) implies that the coefficients of (13.79) become singular in the $\omega \to 0$ limit at the momentum where $a_+^{(0)}(k)$ vanishes. When ν_k is pure imaginary (say equal to $-i\lambda_k$), the dimension of $\Phi_{\vec{k}}$ in the SLQL is now $\frac{1}{2} - i\lambda_k$ and its hermitian conjugate should have dimension $\frac{1}{2} + i\lambda_k$, which makes the last term in (13.79) marginal. Thus in either situation we can no longer conclude that the theory is described by SLQL alone in the IR.

In the rest of this subsection we discuss possible physical interpretations of the above two possibilities assuming they do occur, leaving more in-depth study (including the parameter range they could occur) to Secs. 13.3 and 13.4.

13.2.5.1. *Inhomogeneous black hole hair: Fermi surface v.s. scalar instability*

We now examine the first possibility, i.e. $a_+^{(0)}(k)$ could *vanish* at some momentum. Recall from equation (A.20) in Appendix A.2 that $a_+^{(0)}$ is the leading expansion coefficient near the AdS_{d+1} boundary of the UV region solution $\eta_+^{(0)}$ (A.19). By construction $\eta_+^{(0)}$ is *normalizable* at the horizon. When $a_+^{(0)}$ vanishes $\eta_+^{(0)}$ becomes normalizable also at the AdS_{d+1} boundary and thus is now a genuine normalizable mode in the black hole geometry. One might call it a *black hole hair*. Clearly, equation $a_+^{(0)}(k) = 0$ can only have solutions at most at some discrete values of momentum k. When happening at a nonzero momentum, one finds an inhomogeneous hair. The discussion also applies identically to the spinor case.

Now suppose that $a_+^{(0)}$ indeed vanishes at some value of $k = k_F \neq 0$: i.e. $a_+^{(0)}(k_F) = 0$. Let us consider what happens to (13.75) near k_F. First at $\omega = 0$, we find simply a pole

$$G_R(\omega = 0, k) \approx \mu_*^{2\nu_U} \frac{b_+^{(0)}(k_F)}{\partial_k a_+^{(0)}(k_F)(k - k_F)} + \cdots . \tag{13.83}$$

For a spinor operator, this is a zero-frequency singularity at a shell in k-space, and so is precisely the signature of a Fermi surface! Now further turning on a small frequency ω near $k \sim k_F$ and keeping the first few terms in the relevant expansion of (13.75), we find

$$G_R(\omega, k) \approx \mu_*^{2\nu_U} \frac{b_+^{(0)}(k_F)}{\partial_k a_+^{(0)}(k_F)(k - k_F) + \omega a_+^{(1)}(k_F) + a_-^{(0)}(k_F)\mathcal{G}_{k_F}(\omega)\mu_*^{-2\nu_{k_F}}} + \cdots \tag{13.84}$$

which can be further written as

$$G_R(\omega, k) \approx \mu_*^{2\nu_U} \frac{h_1}{k - k_F - \frac{1}{v_F}\omega - \Sigma(\omega)} + \cdots \tag{13.85}$$

where

$$v_F \equiv -\frac{\partial_k a_+^{(0)}(k_F)}{a_+^{(1)}((k_F)}, \qquad h_1 \equiv \frac{b_+^{(0)}(k_F)}{\partial_k a_+^{(0)}(k_F)} \tag{13.86}$$

and

$$\Sigma(\omega) = h \frac{\mathcal{G}_{k_F}(\omega)}{\mu_*^{2\nu_{k_F}}} = hc(\nu_{k_F}) \left(\frac{\omega}{\mu_*}\right)^{2\nu_{k_F}}, \qquad h \equiv -\frac{a_-^{(0)}(k_F)}{\partial_k a_+^{(0)}(k_F)}. \tag{13.87}$$

Equation (13.85), which has the same form as that for the Fermi liquid (13.7) and Marginal Fermi liquid (13.13), is indeed consistent with the behavior of low energy response near a Fermi surface. $\Sigma(\omega)$ can be interpreted as the self-energy for fermionic excitations near the Fermi surface and is controlled by the retarded function $\mathcal{G}_{k_F}$ of the IR fixed point (i.e SLQL) evaluated at $k = k_F$. Recall now that for a Laudau Fermi liquid (13.8) we have $\Sigma_{FL}(\omega) \sim \omega^2$, following from basic kinematics of interactions about the Fermi surface. Here, however, we have

$$\Sigma \sim \omega^{2\nu_{k_F}} \tag{13.88}$$

with a nontrivial scaling exponent ν_{k_F}. Thus we find a *non-Fermi liquid*; a system with a Fermi surface but whose excitations appear to *not* be Landau quasiparticles. The dispersion and decay width of small excitations near the Fermi surface can be extracted from the poles of (13.85). The nature of such excitations turns out to be rather different depending on the value of ν_{k_F}, which we will discuss in detail in Sec. 13.3, along with when such a k_F could exist.

The above discussion works well for a spinor operator, but how do we interpret (13.83) and (13.85) for a scalar operator? This turns out to be related to an important question for the spinor case, does the pole of (13.85) always lie in the lower complex ω-plane as is required from analyticity? Note at $k = k_F$, the pole lies at the origin of the complex ω-plane. If we move slightly away from $k = k_F$ the pole will move off the origin and into the complex ω plane. What will the motion of the pole be? Based on the following property of the scalar and spinor retarded functions ($\omega > 0$)

$$\text{scalars}: \quad \operatorname{Im} G_R(-\omega) < 0, \quad \operatorname{Im} G_R(\omega) > 0 \tag{13.89}$$

$$\text{spinors}: \quad \operatorname{Im} G_R(-\omega) > 0, \quad \operatorname{Im} G_R(\omega) > 0\ , \tag{13.90}$$

one can show that[39]

(1) For a spinor the pole always lies in the lower-half complex ω-plane.
(2) For a scalar, the pole crosses from the upper half plane to the lower half plane as k is increased from below to above k_F. Thus for a scalar there is a pole for $k < k_F$ in the upper half plane.

We note that (13.89) and (13.90) simply follow from the definition (13.19) that the retarded function for a fermionic (bosonic) operator is defined to be an anticommutator (commutator). Although in the bulk we are dealing with classical equations of scalars and spinors and have not imposed any statistics, the self-consistency of AdS/CFT implies that the classical equations for bulk scalars and spinors should encode the *quantum statistics* of the boundary theory.

A pole for $k < k_F$ in the upper half ω-plane for a scalar gives rise to exponentially growing modes (recall the discussion of Sec. 13.1.2). Thus instead of finding a Fermi surface, a black hole scalar hair at some $k_F \neq 0$ indicates an instability to condensing the scalar operator. In Sec. 13.4 we will see that it is possible to vary parameters of the system so that k_F can be smoothly tuned to zero, after which the instability disappears. The value of the external parameter at which $k_F = 0$ thus corresponds to a quantum critical point, whose critical behavior will be the focus of Sec. 13.4.

13.2.5.2. *Low energy effective theory*

Now let us turn to the low effective theory (13.78)–(13.81) to examine what happens when $a_+^{(0)}$ vanishes. From equation (13.81) and (13.76), when $a_+^{(0)} = 0$, all the coefficients in the S_{UV} diverge as $\omega \to 0$, i.e. the effective action becomes non-local. This means we must have integrated out some gapless modes, which of course should be the small excitations around k_F which we have just identified from the poles of (13.85). In order to have a well-defined local effective action, we need to put such gapless modes back to the low energy effective action. We will introduce a new field Ψ in the low energy theory and "un-integrate" (13.79). Clearly there is no unique way of doing this[v] and the simplest choice is

$$S_{UV} = -\frac{1}{2}\int \kappa(\omega,k)\Phi_{\vec{k}}^2 + \int \lambda(k,\omega)\Phi_{\vec{k}}\,\Psi_{\vec{k}} - \frac{1}{2}\int \Psi_{-\vec{k}}\,K(\omega,k)\,\Psi_{\vec{k}} + \int \Psi J \tag{13.91}$$

where

$$K = \Lambda^{-1} = \frac{a_+}{b_+}, \qquad \lambda = \eta K = \frac{\sqrt{W}}{b_+}, \qquad \kappa = \xi + \frac{\lambda^2}{K} = \frac{b_-}{b_+}\,. \tag{13.92}$$

Now all coefficients in (13.91) have local expansion in ω. This discussion applies to both a spinor or a scalar operator.

Near k_F, K has the expansion

$$K = \frac{1}{h_1}\left(\omega - v_F(k-k_F) + \cdots\right) \tag{13.93}$$

and thus for a spinor operator, the third term in (13.91) can simply be interpreted as the action for a free fermion near its Fermi surface at k_F.

[v] We can for example make a field redefinition in ψ as $\psi \to Z_1\Psi + Z_2\Phi$.

The full low energy effective action (13.78) can now be written as[39,42]

$$S_{eff} = \tilde{S}_{SLQL}[\Phi] + \int \lambda(k,\omega)\Phi_{\vec{k}}\Psi_{\vec{k}} + S_{\text{free fermion}}[\Psi] + \int \Psi J \qquad (13.94)$$

where

$$S_{\text{free fermion}}[\Psi] = -\frac{1}{2h_1}\int dtd\vec{k}\,\Psi_{-\vec{k}}\left(i\partial_t - v_F(k-k_F) + \cdots\right)\Psi_{\vec{k}} \qquad (13.95)$$

is the action for a free fermion near a Fermi surface and

$$\tilde{S}_{SLQL} = S_{SLQL} - \frac{1}{2}\int \kappa(\omega,k)\Phi_{\vec{k}}^2 \qquad (13.96)$$

is the action for Φ which is SLQL deformed by a double-trace term. The low energy action (13.239) describes a free fermion field Ψ *hybridized* with a strongly coupled fermionic operator Φ whose dynamics is in turn controlled by SLQL. (13.239) also offers a simple field theoretical interpretation for the expression (13.85): the self-energy $\Sigma(\omega)$ simply arises from the mixing of Ψ with Φ and is thus given by the retarded function of Φ, see Fig. 13.7. As we will see in Sec. 13.3, the action (13.239) also provides an efficient way to understand various behavior of small excitations near the Fermi surface.

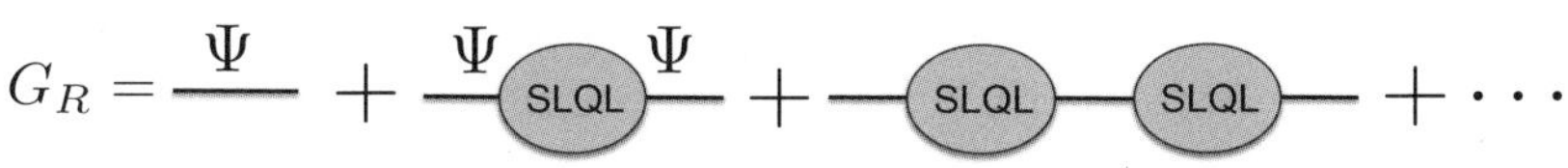

Fig. 13.7. The action (13.239) gives a simple field theoretical interpretation of (13.85) through hybridization between a free and a strongly interacting sectors. The propagator for the free fermion is given by K (13.93).

The above discussion also applies to a scalar case. For $k_F \neq 0$, one is, however, dealing with an unstable state and its physical meaning is not so clear. But for $k_F = 0$, i.e. near a quantum critical point mentioned earlier, then Ψ can be interpreted as the order parameter and the analogue of (13.239) provides an effective action for the order parameter. In particular, the free-fermion part in (13.239) becomes the standard Landau-Ginsburg action. We will discuss this further in Sec. 13.4.

13.2.5.3. *Pair production by black hole: Complex IR dimensions*

Let us now consider what happens when the conformal dimension $\frac{1}{2} + \nu_k$ of an operator in an SLQL becomes complex. Recall that for a scalar the conformal dimension is given by (13.57), which we copy here for convenience,

$$\nu_k = \frac{1}{\sqrt{d(d-1)}}\sqrt{m^2R^2 - \frac{g_F^2 q^2}{2} + \frac{d(d-1)}{4} + \frac{k^2}{\mu_*^2}} \qquad (13.97)$$

with a similar expression (13.70) for a spinor. We thus find that for

$$k_o^2 \equiv \mu_*^2 \left(\frac{g_F^2 q^2}{2} - m^2R^2 - \frac{d(d-1)}{4} \right) > 0 \qquad (13.98)$$

ν_k becomes *imaginary* for $k^2 < k_o^2$ with

$$\nu_k = -i\lambda_k, \qquad \lambda_k = \frac{1}{\sqrt{d(d-1)}\mu_*}\sqrt{k_o^2 - k^2} \, . \qquad (13.99)$$

Note on the gravity side, equation (13.98) corresponds to violating the Breitenlohner-Freedman bound[82] of AdS_2.[83–85,90]

For a spinor (13.98) is replaced by

$$k_o^2 \equiv \mu_*^2 \left(\frac{g_F^2 q^2}{2} - m^2R^2 \right) > 0 \, . \qquad (13.100)$$

Basically we see that the background electric field acts through the charge as an effective "negative mass square", making it possible to have a negative total mass square and resulting in an imaginary conformal dimension.

Using (13.99) we now find from (13.77) the leading small frequency behavior

$$G_R(\omega, k) = \mu_*^{2\nu_U} \frac{b_+^{(0)} + b_-^{(0)} c(\nu_k) \left(\frac{\omega}{\mu_*}\right)^{-2i\lambda_k}}{a_+^{(0)} + a_-^{(0)} c(\nu_k) \left(\frac{\omega}{\mu_*}\right)^{-2i\lambda_k}} \, . \qquad (13.101)$$

As discussed in Appendix A.2, for imaginary ν_k, $a_\pm^{(0)}, b_\pm^{(0)}$ also become complex and in fact

$$a_+^{(0)} = (a_-^{(0)})^* \qquad b_+^{(0)} = (b_-^{(0)})^* \, . \qquad (13.102)$$

As a result, even at $\omega = 0$ we find that the spectral weight is non-vanishing for $k < k_o$

$$\mathrm{Im}\, G_R(\omega = 0, k) = \mu_*^{2\nu_U}\, \mathrm{Im}\, \frac{b_+^{(0)}}{a_+^{(0)}} \neq 0 \, . \qquad (13.103)$$

At finite frequency (13.101) is oscillatory; it is periodic in $\log \omega$ with a period given by $\tau_k = \frac{\pi}{\lambda_k}$. In particular, it is invariant under a *discrete* scale transformation

$$\omega \to e^{n\tau_k}\omega, \qquad n \in \mathbb{Z} \, . \qquad (13.104)$$

We will refer to the region $k < k_o$ as the oscillatory region.

It can be readily checked that (13.101) has an infinite number of exponentially separated poles in the complex ω-half-plane; as (13.104) shows, there is an accumulation point at $\omega = 0$. Again using (13.89) and (13.90) one finds that for a spinor all the poles lie in the *lower* half ω-plane while for a scalar they all lie in the *upper* half ω-plane.[39] Thus we immediately conclude that for a scalar, in the parameter region (13.98), there is an instability to the condensation of the scalar operator. While there appears no obvious pathology for a spinor operator at the current level of discussion, as we will see in Sec. 13.3, there is nevertheless an instability, and at an exponentially small (in terms of N) scale, the system becomes a Fermi liquid.

It is also instructive to understand the above phenomena for both fermions and bosons in terms of bulk language. For this purpose it is convenient to slightly rewrite $\mathrm{AdS}_2 \times \mathbb{R}^{d-1}$ metric (13.49) by introducing

$$\zeta = R_2 e^y \tag{13.105}$$

after which (13.49) becomes

$$ds^2 = -e^{2y}dt^2 + R_2^2 dy^2 + \mu_*^2 R^2 d\vec{x}^2, \qquad A = \frac{e_d}{R_2} e^{-y} dt\,. \tag{13.106}$$

We note that the spatial part of (13.106) is flat and the chemical potential μ_L for a local observer with charge q in the bulk is given by

$$\mu_L = q\sqrt{g^{tt}} A_t = \frac{|q| e_d}{R_2} = \frac{g_F |q|}{\sqrt{2} R} \tag{13.107}$$

where in the last equality we have used (13.51). Note that μ_L does not depend on the radial direction y of the AdS_2.

For a fermion, in the Thomas-Fermi approximation, we fill all the states up to μ_L which is the Fermi energy. This gives the local Fermi momentum

$$k_b^{(F)} = \sqrt{\frac{g_F^2 q^2}{2R^2} - m^2} \tag{13.108}$$

where the local bulk momentum k_b is related to k of the boundary and radial momentum k_y conjugate to y as

$$k_b^2 = \frac{k^2}{\mu_*^2 R^2} + \frac{k_y^2}{R_2^2}\,. \tag{13.109}$$

Note that the bulk Fermi surface defined by (13.108) is one dimensional higher than a momentum shell in the boundary theory. Thus projecting it to the boundary we conclude there should be gapless excitations for all k satisfying

$$k < \mu_* \sqrt{\frac{g_F^2 q^2}{2} - m^2 R^2} \,. \tag{13.110}$$

See Fig. 13.8. Note that the right hand side of the above equation is precisely k_o introduced in (13.100). This explains the nonvanishing of spectral weight in (13.103) as a consequence of the projection of a bulk Fermi surface to the boundary theory. The oscillatory behavior in (13.101) can similarly be understood in terms of the bulk Fermi surface when taking into account of that the local proper energy is red-shifted by a factor e^{-y} compared with the boundary frequency ω (see (13.106)) and the behavior of local wave function.

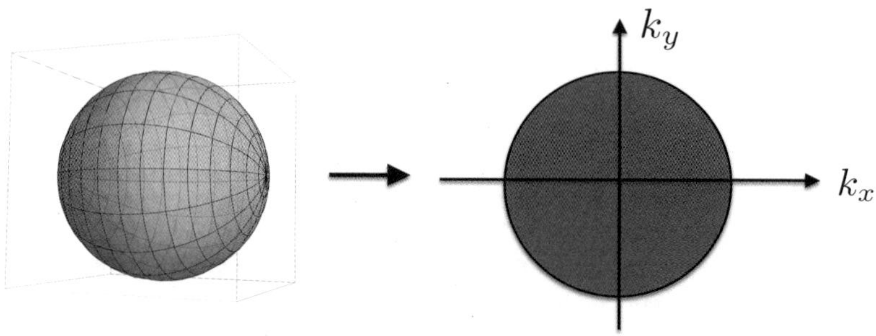

Fig. 13.8. In the $AdS_2 \times \mathbb{R}^2$ (take $d = 3$ for example) region of the extremal black hole geometry, at each point in the bulk there is a local three-dimensional Fermi surface with Fermi momentum k_o, which upon projection to the boundary theory would result in a Fermi disc, in which there are gapless excitations at each point in the *interior* of a disc in the two-dimensional momentum space.

For a scalar, when the local chemical potential (13.107) exceeds the mass of a charged particle

$$\frac{g_F |q|}{\sqrt{2} R} > m \tag{13.111}$$

the system will simply Bose condense. Equation (13.111) agrees with condition (13.98) up to an additive constant.[w] and gives the bulk origin of the

[w]Note that for our local approximation to work well, we need typical local proper wave length to be much smaller than the curvature radius R, which translates into $g_F q$, $mR \gg 1$.

scalar instabilities we found earlier from the poles of boundary retarded functions.

For a charged operator, the above stories can also be understood more heuristically from the black hole geometry as follows. When (13.98) and (13.100) are satisfied one can show that the corresponding quanta for the bulk scalar (or spinor) field can be pair produced by the electric field of the black hole.[86] Suppose that the black hole is positively charged, then the negative-charged particle in a pair will fall into the black hole, while the positive-charged one moves to the boundary of the spacetime. However it cannot escape, as the curvature of AdS pulls all matter towards its center, and thus the particle will eventually fall back towards the black hole. It then has some probability of falling into the black hole or being scattered back toward the boundary. For a scalar it turns out the so-called super-radiace is happening for a negative frequency, i.e. for a wave incident on the black hole, the reflection amplitude becomes greater than the incoming amplitude. Thus this pair production process will be magnified and leads to instability. In contrast, for a spinor, there is no superradiance, this process will eventually reach an equilibrium with a positively charged gas of fermionic quanta hovering outside the horizon. The ground state of this fermionic gas is then described by a bulk Fermi surface described earlier.

13.2.5.4. *Summary*

To summarize, we find two channels for scalar instabilities. The first manifests as the presence of a black hole hair, while the second (when the IR dimension becomes complex) corresponds to the bulk charged scalar condensing near the horizon of a black hole via Bose-Einstein condensation.

For a spinor, a black hole hair gives rise to an isolated boundary Fermi surface, while when there is a bulk Fermi surface (which happens when the IR dimension becomes complex), its projection to the boundary leads to a Fermi fall inside some momentum shell k_o with discrete scaling symmetry in frequency.

The differences between the scalar and spinor do not depend on any specific details and only on their statistics.

13.2.6. *SLQL as a universal intermediate-energy phase*

Since the Einstein-Maxwell system (13.32) is a universal sector common to many theories with a gravity dual, the SLQL phase appears to be universal IR fixed point among a large class of field theories at finite density, independent of their microscopic details. However, from equation (13.47), the extremal charged black hole has a finite zero-temperature entropy density. This implies that the ground states are highly degenerate. In a system without supersymmetry (as it is here) or other apparent symmetries to protect such degeneracies, we expect this nonzero entropy density has to with the large N limit we are working with. In other words, this nonzero zero-temperature entropy density likely reflects the existence of a large number of closely spaced states which are separated from the genuine ground state by spacings which go to zero in the $N \to \infty$ limit. One expects that the system should pick a unique ground state, which may or may not be visible in the large N limit. Given that there are many nearly degenerate low energy states, the precise physical nature of the ground state should be sensitive to the specific dynamics of the individual system.

In a system with supersymmetry, then ground state degeneracy is indeed possible, and the above argument does not apply. There are many such examples in string theory.

There is also another independent general argument,[87] which indicates that any sort of scaling symmetry in the time direction only cannot persist to arbitrarily low energies. On dimensional grounds, the scaling symmetry implies that the density of states of such a theory should have the form

$$\rho(E) = A\delta(E) + \frac{B}{E} \tag{13.112}$$

where A and B are constants.[x] B is nonzero when there are operators which transform nontrivially under the scaling symmetry. With $B \neq 0$, the integral of the density of states is divergent, which means that the scaling symmetry must always break down at sufficiently low energies. Note that this argument is independent of the existence of the ground state degeneracy, which is given by A and would also appear to be indicative of a potential instability (in the absence of supersymmetry), as mentioned above. Even in the presence of supersymmetry, the above argument implies that B must go to zero at a sufficiently low energies, i.e. the system likely has a gap which separates the degenerate ground states with excited states.

Depending on the spectrum of charged or neutral matter fields (their charges and masses etc.), it is known that an extremal charged black hole suffers from various bosonic and fermionic instabilities,[44,88–92] which will lead to a different lower energy state. Some of the instabilities have already

[x] Both A and B should contain a volume factor, but note that spatial directions do not scale. Also note that there is another scale μ which can be combined with the volume to make A and B dimensionless.

been mentioned earlier and we will discuss them in more details in next two sections. The Einstein-Maxwell system (13.32) can also be generalized to an Einstein-Maxwell-dilaton system in which charged black hole is no longer a solution. Instead, one finds a solution with zero entropy density and Lifshitz-type scaling in the interior (at $T = 0$).[93–96] In all of these situations, there exist parameter ranges in which a charged black hole provides a good description for some region of the bulk geometry.

It then appears that among a large class of systems with different microscopics *and* different lower energy states, the charged black hole appears as an intermediate energy state. Thus the SLQL phase may be considered a *universal intermediate energy phase*[43] which connects microscopic interactions with macroscopic, low energy physics, as indicated in Fig. 13.9.[y] It is characterized by a finite spatial correlation length, but an infinite correlation *time* and associated nontrivial scaling behavior (13.59) and (13.68) in the time direction, as well as a nonzero entropy density.

We should emphasize that for a given holographic system there may not always exist an energy range in which SLQL manifests itself as an intermediate state. SLQL behavior is manifest when there exists a hierarchy between the chemical potential μ and the energy scale at which a more stable lower energy phase emerges. On the gravity side there is then an intermediate region of bulk spacetime which resembles that of $AdS_2 \times \mathbb{R}^{d-1}$ (or its finite temperature generalization). In a situation where such a hierarchy does not exist as we will discuss later that SLQL nevertheless provides a useful description for understanding the emergence of the lower energy state.

As emphasized by Anderson some time ago,[98] the existence of a "universal intermediate phase" appears to be a generic phenomenon in nature; the familiar examples include liquid phases of ordinary matter, through which materials settle into different crystal structures at low temperatures. It is thus tempting to speculate that the appearance of a SLQL phase in holographic systems at a finite chemical potential may not be specific to these systems and of wider applications. We will discuss more its possible applications in the conclusion section.

In the next two sections we will see examples of the evolution of the SLQL to various lower energy phases including Fermi liquids, superconductors, and AFM-like states. We find that the emergence of these lower energy phases can be characterized as a consequence of bound states

[y]The idea that an extremal black hole should be interpreted as an intermediate-energy state has been expressed by many people including Refs. 39, 41, 75, 76 and 97.

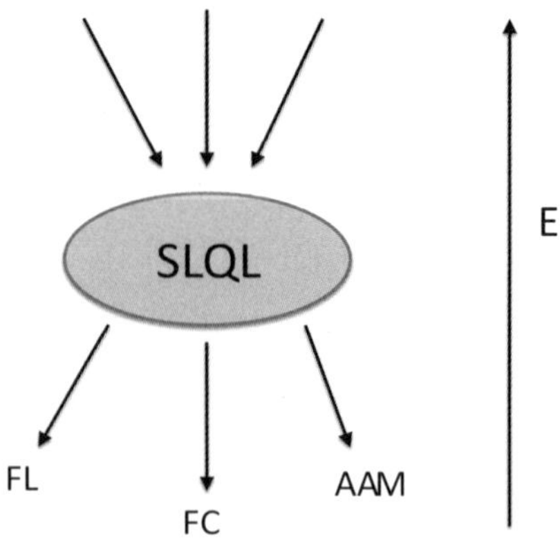

Fig. 13.9. The semi-local quantum liquid phase is a useful description at intermediate scales; many different UV theories (i.e. those with gravity duals described by a Maxwell-Einstein sector) flow to it, and at low energies it settles into one of many different ground states, e.g. a Fermi liquid, a superconducting state, or an antiferromagnetic state, as will be discussed in subsequent sections.

formation in the SLQL. Conversely, SLQL can be considered as a universal deconfined (fractionalized) phase of these lower energy phases.

13.3. Holographic (non-)Fermi Liquids

In this section we consider more explicitly the question when a holographic system has a Fermi surface and properties of small excitation around the Fermi surface. We will see that with a choice of parameters, the IR fixed point SLQL discussed in Sec. 13.2.3 gives rise to non-Fermi liquid behavior which is the same as that of the strange metal phase of cuprate superconductors discussed in Sec. 13.1.1. Furthermore, when N is finite, in a certain parameter range SLQL goes over to a Fermi liquid of heavy fermions at sufficiently low energies through formation of fermionic bound states.

13.3.1. *Setup*

Consider a bulk Dirac spinor field $\psi(z, t, \vec{x})$ dual to a boundary theory spinor operator $\mathcal{O}(t, \vec{x})$. For definiteness we will now specify to $d = 3$. We will assume a standard Dirac action in the bulk

$$S = -\int d^4x\sqrt{-g}\; i(\bar{\psi}\Gamma^M \mathcal{D}_M\psi - m\bar{\psi}\psi), \qquad (13.113)$$

where the covariant derivative $\mathcal{D}_M$ contains couplings to both the background gauge field and the spin connection of the charged black hole geometry (13.33),

$$\mathcal{D}_M = \partial_M + \frac{1}{4}\omega_{abM}\Gamma^{ab} - iqA_M \ . \tag{13.114}$$

Here q is the charge of $\mathcal{O}$ under the $U(1)$ current J^μ. The asymptotic structure of solutions to (13.113) is different from the scalar case discussed above, essentially because it is a first order system with more components. In particular, the number of components of the boundary operator $\mathcal{O}$ is half of that of the bulk spinor ψ,[81] i.e. while ψ has four components, $\mathcal{O}$ has only two components, as expected for a three-dimensional field theory. The scaling dimension Δ of $\mathcal{O}$ is related to the bulk mass m of ψ by

$$\Delta = \frac{3}{2} + mR \ . \tag{13.115}$$

We refer readers to[37,39,81] for detailed analysis of the Dirac equation and only mention the main results here. The boundary theory Green's function can be diagonalized to take the form

$$G_R(\omega, k) = \begin{pmatrix} G_1(\omega, k) & 0 \\ 0 & G_2(\omega, k) \end{pmatrix}, \tag{13.116}$$

where *the momentum k can be taken to be along one of the spatial directions* and both $G_{\alpha=1,2}$ are the ratio of appropriately defined boundary theory expansion coefficients A_α, B_α, as in (13.30)

$$G_\alpha = \frac{B_\alpha}{A_\alpha}. \tag{13.117}$$

The master formula (13.75) applies to both of the eigenvalues $G_{1,2}$, which will have different values for the prefactor in the AdS_2 Green's function and the UV expansion coefficients $a_\pm, b_\pm$. A useful relation is $G_1(\omega, k) = G_2(\omega, -k)$; this means that with no loss of generality we can simply restrict to G_2; when we refer to G_R in the rest of this section this is what we mean. For convenience we reproduce here (13.75),

$$G_R(\omega, \vec{k}) = \mu_*^{\nu_U} \frac{b_+(\omega, k) + b_-(\omega, k)\mathcal{G}_k(\omega)\mu_*^{-2\nu_k}}{a_+(\omega, k) + a_-(\omega, k)\mathcal{G}_k(\omega)\mu_*^{-2\nu_k}}, \tag{13.118}$$

where $\mathcal{G}_k(\omega)$ is given by (13.71)–(13.72) and all of the UV coefficients $a_\pm, b_\pm$ in (13.118) have analytic expansion in ω:

$$a_+(\omega, k) = a_+^{(0)}(k) + a_+^{(1)}(k)\omega + a_+^{(2)}(k)\omega^2 + \cdots \ . \tag{13.119}$$

When $mR \in [0, \frac{1}{2})$ there is also an alternative quantization for the bulk spinor ψ which leads to the conformal dimension of the boundary operator $\mathcal{O}$ to be

$$\Delta = \frac{3}{2} - mR \,. \tag{13.120}$$

Below we will treat both quantizations together by allowing mR to take a negative value in (13.115).

For simplicity below we will also set μ_* to unity to avoid cluttering formulas. It can be easily restored on dimensional ground.

13.3.2. *Existence of Fermi surfaces*

As we discussed in Sec. 13.2.5.1, a boundary system has Fermi surfaces for a spinor operator $\mathcal{O}$, if there exists some k_F, $a_+^{(0)}(k_F) = 0$, i.e. the charged black hole admits an inhomogeneous fermionic hair. To see whether such a k_F indeed exists one needs to solve the static Dirac equation (i.e. at $\omega = 0$) for bulk spinor field ψ in the full charged black hole geometry, which can only be done numerically. Squaring the static Dirac equation and rewriting it as a Schrödinger equation one can then reduce the problem of finding a Fermi surface to finding a bound state in the resulting potential as a function of k. It can then be checked explicitly that such a k_F does exist for certain range of parameters (q, m) and when q is sufficiently large there can also be multiple bound states which corresponds to the system having multiple Fermi surfaces. In Fig. 13.10 we show the values of k_F found numerically for a few choices of mR, taken from Ref. 39. In the figure, the oscillatory region $k < k_o$ (see (13.100)), where ν_k becomes pure imaginary and G_R becomes oscillatory in $\log \omega$ (see (13.101)), is shaded. As discussed in Sec. 13.2.5.3, the oscillatory region can be attributed to a *bulk* Fermi surface in the near horizon AdS_2 region, which gives rise to a "Fermi ball" in the boundary. In contrast, the Fermi surface at k_F from a fermionic hair is isolated.

For standard quantization, i.e. $m > 0$, the parameter region in which a k_F exists always lie inside the parameter region (13.100) for which the oscillatory region exists. But note that k_F is always greater than k_o, i.e. lying outside the oscillatory region in momentum space. For alternative quantization, there exists a parameter range that k_F exists in the absence of an oscillatory region.

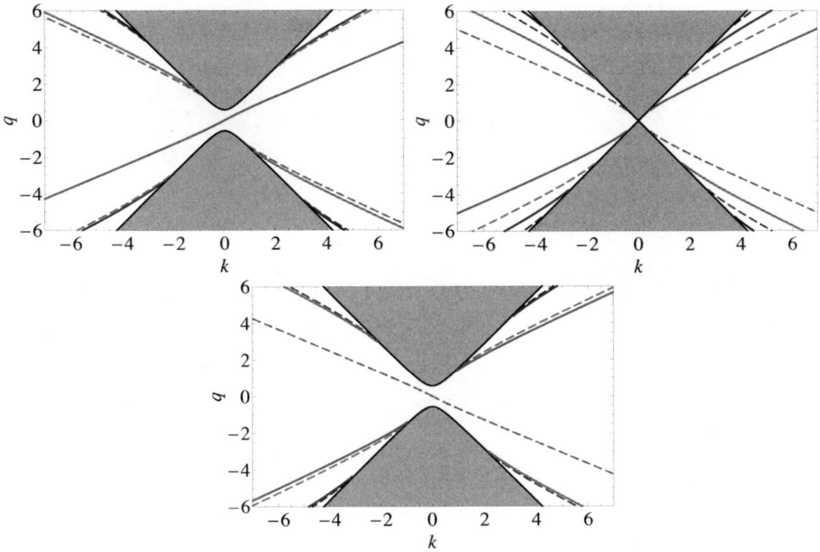

Fig. 13.10. The values of k_F as a function of q for the Green function G_2 are shown by solid lines for $mR = -0.4$, 0, 0.4. In this plot and ones below we use units where $R = 1$, $\mu_* = 1$, $g_F = 1$ and $d = 3$. The oscillatory region, where $\nu_k = \frac{1}{\sqrt{6}}\sqrt{k^2 + m^2 - \frac{q^2}{2}}$ is imaginary, is shaded. Given $G_1(k) = G_2(-k)$, so k_F for G_1 can be read from these plots by reflection through the vertical $k = 0$ axis. The $mR = -0.4$ plot corresponds to alternative quantization for $mR = 0.4$. For convenience we have included in each plot the values of k_F for the alternative quantization using the dotted lines. Thus the first ($mR = -0.4$) and the third plot ($mR = -0.4$) in fact contain identical information; they are related by taking $k \to -k$ and exchanging dotted and solid lines. Also for $m = 0$ the alternative quantization is equivalent to the original one. This is reflected in the middle plot in the fact that the dotted lines and solid lines are completely symmetric. All plots are symmetric with respect to $q, k \to -q, -k$.

13.3.3. *Fermi surfaces with or without long-lived quasiparticles*

Near a k_F and at low frequencies, the retarded function G_R is given by (13.85)–(13.87), which we reproduce here for convenience

$$G_R(\omega, k) \approx \frac{h_1}{k - k_F - \frac{1}{v_F}\omega - \Sigma(\omega)}, \qquad \Sigma(\omega) = h\mathcal{G}_{k_F}(\omega) = hc(\nu_{k_F})\omega^{2\nu_{k_F}} \tag{13.121}$$

with

$$v_F \equiv -\frac{\partial_k a_+^{(0)}(k_F)}{a_+^{(1)}((k_F)}, \qquad h_1 \equiv \frac{b_+^{(0)}(k_F)}{\partial_k a_+^{(0)}(k_F)}, \qquad h \equiv -\frac{a_-^{(0)}(k_F)}{\partial_k a_+^{(0)}(k_F)}. \tag{13.122}$$

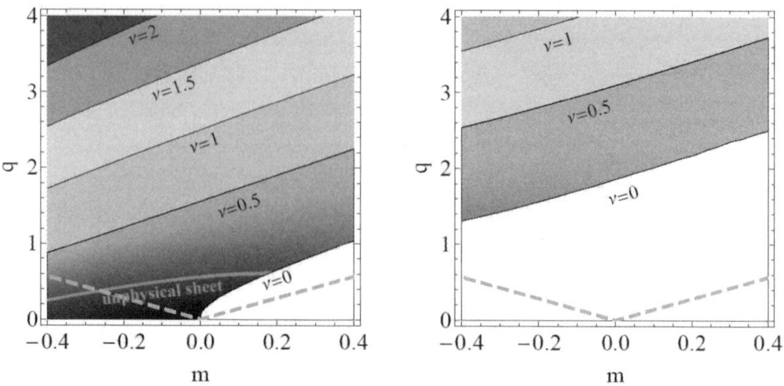

Fig. 13.11. Distribution of ν_{k_F} in the $m-q$ plane. When there are multiple Fermi surfaces, ν_{k_F} is calculated for the one with the largest k_F. The two plots correspond to G_2 and G_1 components. In the white region, there is no Fermi surface. There is no long-lived quasiparticle in the orange region, $\nu_{k_F} < \frac{1}{2}$. In the remainder of the parameter space, there are log-lived quasiparticles.

The self-energy Σ depends on frequency only and is characterized by a nontrivial scaling exponent ν_{k_F}, which can be computed from (13.70) once k_F determined. The result is presented in Fig. 13.11.

Poles of (13.121) in the complex ω-plane, which are determined by

$$k - k_F - \frac{1}{v_F}\omega_c - \Sigma(\omega_c) = 0 \qquad \omega_c(k) = \omega_*(k) - i\Gamma(k) \tag{13.123}$$

give the dispersion of small excitations near the Fermi surface. The real part $\omega_*(k)$ of $\omega_c(k)$ gives the dispersion relation and the imaginary part Γ gives the width of an excitation. Given that in (13.123) only Σ is complex, the width Γ is controlled by the imaginary part of Σ which is in turn is proportional to $\mathrm{Im}\,\mathcal{G}_{k_F}(\omega)$. The physics turns out to be rather different depending on the value of ν_{k_F}, and we now study the different values in turn. For simplicity of notation in the rest of this section we will drop the subscript k_F: $\nu \equiv \nu_{k_F}$ *in this section only.*

13.3.3.1. $\nu > \frac{1}{2}$

If $\nu > \frac{1}{2}$, then the analytic linear-in-ω term in (13.121) dominates over the self-energy, which goes like $\omega^{2\nu}$. Thus the dispersion is essentially controlled by the balancing between the first two terms in (13.123); note however that the sole contribution to the *imaginary* part of the pole comes from the complex self-energy. We find the following expression for the motion of the pole as k is varied near k_F:

$$\omega_c(k) = v_F(k - k_F) - v_F h c(\nu) \left(\frac{v_F(k - k_F)}{\mu_*} \right)^{2\nu} . \tag{13.124}$$

Thus we see that this has a linear dispersion relation $\omega_* = v_F(k - k_F)$, with width $\Gamma \sim \omega^{2\nu}$ that is always much smaller than the energy ω_*: as we approach the pole $\frac{\Gamma(k)}{\omega_*(k)} \sim (k - k_F)^{2\nu - 1} \to 0$ as $k - k_F \to 0$. Furthermore the residue at the pole is always finite and given by

$$Z = -h_1 v_F . \tag{13.125}$$

This is then a long-lived quasiparticle. It is not *qualitatively* different from the excitations of a Fermi liquid, except that its width satisfies $\Gamma \sim \omega^{2\nu}$ rather than $\Gamma_{FL} \sim \omega^2$.

13.3.3.2. $\nu < \frac{1}{2}$

If $\nu < \frac{1}{2}$, the situation is different. Now in (13.121) the non-analytic contribution from $\Sigma(\omega)$ always dominates over the analytic linear-in-ω correction, which we may completely neglect. The solution to (13.123) now takes the form

$$\omega_c = \mu_* \left(\frac{k - k_F}{h c(\nu)} \right)^{\frac{1}{2\nu}} . \tag{13.126}$$

As $c(\nu)$ is complex this pole has both a real and imaginary part; however now they are both coming from the same place, i.e. the AdS_2 Green's function. As a result the dispersion is now nonlinear, $\omega_* \sim (k - k_F)^{\frac{1}{2\nu}}$. Importantly, the width Γ is now always comparable to the frequency ω_*, e.g. for $k - k_F > 0$ we have

$$\frac{\Gamma(k)}{\omega_*(k)} = -\tan\left(\frac{\arg(c(\nu))}{2\nu} \right) = \text{const.} \tag{13.127}$$

A similar expression exists with a different angle for $k - k_F < 0$. Thus as we vary $k - k_F$ through 0, the pole moves along a straight line towards the origin, eventually bouncing off and retreating at another angle (but always in the lower-half plane). The residue at the pole is

$$Z = -\frac{\omega_c h_1}{2\nu(k - k_F)} \propto (k - k_F)^{\frac{1}{2\nu} - 1}, \tag{13.128}$$

and so actually vanishes as we approach the Fermi surface $k - k_F = 0$. The fact that the width of the excitation is always the same order as its energy and that the residue vanishes at the pole indicate that this can *not* considered as a quasiparticle; thus we see that we have an example of a system with a sharp Fermi surface, but with no quasiparticles!

13.3.3.3. $\nu = \frac{1}{2}$: *Marginal Fermi liquid*

We turn now to the threshold case between the two studied above. Here we see that the two terms (analytic and AdS_2) are both naively proportional to ω and are thus becoming degenerate. As is often the case when there is such a degeneracy, a logarithmic term in ω is generated. Basically in (13.121), v_F has a simple zero and $c(\nu)$ in the self-energy Σ (see (13.71)) has a simple pole at $\nu = \frac{1}{2}$; the two divergences cancel and leave behind a finite $\omega \log \omega$ term with a real coefficient,

$$G_R(\omega, k) \approx \frac{h_1}{k - k_F + \tilde{c}_1 \omega \log \omega + c_1 \omega}. \tag{13.129}$$

Here $\tilde{c}_1$ is real and c_1 is complex; for more details on the derivation of this formula see Section VI A of.[39] Remarkably, this is of precisely the form postulated in[12] for the response function of the "Marginal Fermi Liquid" discussed earlier around (13.13). Here this response function emerges from a first-principles holographic calculation (albeit only at the specially tuned value $\nu = \frac{1}{2}$). Note that here the width is suppressed compared to its energy, but only logarithmically $\Gamma \sim \frac{\omega_*}{\log(\omega_*)}$. Similarly, the residue of the pole scales like $Z \sim 1/\log(k - k_F)$ and so also vanishes logarithmically. The potential physical applications of this $\nu = \frac{1}{2}$ point make it very interesting, although from our gravitational treatment so far we do not really have any way to single it out.

13.3.3.4. *Perspective from low energy effective theory*

The qualitative behavior discussed above, in particular, the presence (absence) of long-lived quasiparticles for $\nu > \frac{1}{2}$ ($\nu \leq \frac{1}{2}$) can also be seen from the low energy effective theory (13.239) based on simple dimensional analysis.[42] Since in the SLQL, k does not scale (i.e. has dimension 0), from (13.95) Ψ should have scaling dimension 0 in the IR. Thus the hybrization term $\Psi\Phi$ in (13.239) has dimension $\frac{1}{2} + \nu$ (which is the dimension of Φ). As a result the hybridization coupling is relevant for $\nu < \frac{1}{2}$ and irrelevant for $\nu > \frac{1}{2}$. When the hybridization coupling is relevant, free fermion Ψ is strongly mixed with Φ, resulting a large decay width and a breakdown of the quasiparticle picture. When $\nu = \frac{1}{2}$, the coupling is marginal. Note that while the logarithmic suppression in the decay width was the original motivation for the word "Marginal" in the "Marginal Fermi Liquid" given in;[12] here we find it is due to a marginal hybridization coupling and the word "Marginal" is entirely appropriate.

13.3.4. *Transport of holographic Fermi surfaces*

Having understood the single-particle response functions of our system, we now turn to collective properties, such as transport. A very natural observable to compute is the conductivity; in particular, for Landau Fermi liquids this is essentially always controlled by the single-particle lifetime, and we find for the DC conductivity $\sigma_{FL} \sim T^2$. What is $\sigma(T)$ in our model?

To understand the calculation to follow, it is necessary to understand the answer to the following question: how much of the charge of our system is carried by the Fermi surface that we have just identified? Note from (13.39) that the background charge density of our system scales like $\rho \sim \frac{1}{2\kappa^2}$, where $\frac{1}{\kappa^2}$ is the inverse Newton's constant that multiplies the whole bulk action. In known examples of the duality, this is typically proportional to a positive power of N, i.e. N^2 or $N^{\frac{3}{2}}$; for simplicity let us just assume the former and say that $\rho \sim \mathcal{O}(N^2)$. But k_F identified above has no factors of κ^2 in its definition; thus we conclude that $k_F \sim \mathcal{O}(N^0)$. Since this is the Fermi surface for a singlet operator, it means that the charge density that is associated with our Fermi surface is *also* $\mathcal{O}(N^0)$. Thus we conclude that *our Fermi surface makes up a tiny (i.e. large-N-suppressed) portion of the full charge density.* In our gravitational description, the rest of the charge is stored behind the black hole horizon; in field theory terms, this presumably means that it is stored in some other degrees of freedom about which we know very little except that in some sense there are "N^2 of them".

This means that none of the leading thermodynamic or transport properties will be sensitive to the Fermi surface. We can directly see this from a bulk calculation. The conductivity can be found from the Kubo formula,

$$\sigma(\omega) = \frac{1}{i\omega}\langle J_x(\omega) J_y(-\omega)\rangle_{\text{retarded}}. \tag{13.130}$$

We can find this retarded response function from gravity using the procedures described above; however if we do that it is clear that the leading answer will be of $\mathcal{O}(N^2)$ and will depend only on classical fluctuations of the gauge field, which know only about the black hole and appear to know nothing about the Fermi surface. It is clear that to see the effects of the Fermi surfaces identified above, we have to go to higher order in $1/N^2$; this maps to quantum effects in the bulk. Thus we must perform a one-loop

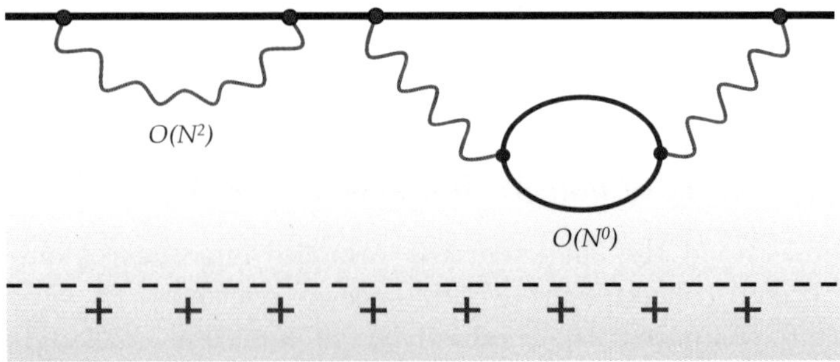

Fig. 13.12. Conductivity from gravity. The current-current correlator in (13.130) can be obtained from the propagator of the gauge field A_x with endpoints on the boundary. Wavy lines correspond to gauge field propagators and the dark line denotes the bulk propagator for the ψ field. The left-diagram is the tree level $\mathcal{O}(N^2)$ contribution while the right diagram includes the contribution from a loop of ψ quanta. The contribution from the Fermi surface associated with boundary fermionic operator $\mathcal{O}$ can be extracted from the diagram on the right.

calculation on the gravity side, including a fermion loop in the gauge boson propagator as shown in Fig. 13.12.

13.3.4.1. *DC conductivity*

In many ways this calculation parallels that of Fermi liquid theory, except for the extra bulk dimension and complications from the curved geometry. A main difference comes from the fact that we are performing the computation in a black hole spacetime, meaning that processes such as Fig. 13.13 where the fermion goes through the horizon can contribute. We do not provide any details of the computation here, only pointing out that the final result can be written in terms of boundary theory quantities as

$$\sigma(\omega) = \frac{C}{i\omega} \int d\vec{k} \int \frac{d\omega_1}{2\pi} \frac{d\omega_2}{2\pi} \frac{f(\omega_1) - f(\omega_2)}{\omega_1 - \omega - \omega_2 - i\epsilon} \times A(\omega_1, k)\Lambda(\omega_1, \omega_2, \omega, \vec{k})\Lambda(\omega_2, \omega_1, \omega, \vec{k})A(\omega_2, \vec{k}) \qquad (13.131)$$

Here $A = \frac{1}{\pi} \mathrm{Im}\, G_R$ is the single-particle spectral function described in the last section, $f(\omega) = \frac{1}{e^{\frac{\omega}{T}}+1}$ is the Fermi distribution function. All radial integrals have been absorbed into an effective vertex Λ, which takes the schematic form $\Lambda \sim \int dr\ \bar{\psi}(r) K_A(r) \psi(r)$, where $\psi(r)$ is a fermion wavefunction and $K_A(r)$ is a bulk-boundary propagator for the gauge field (which also includes a contribution from the graviton, which mixes with

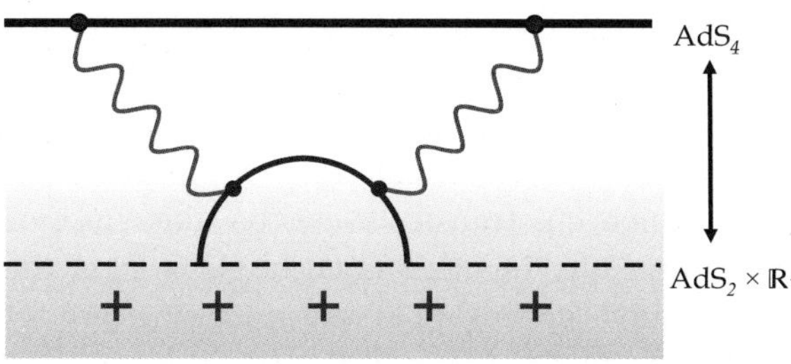

Fig. 13.13. Possible contribution to conductivity in which fermion loop passes through the horizon. This represents the decay of the current due to current-carrying particles decaying by falling into the black hole, and is the dominant contribution to the DC conductivity.

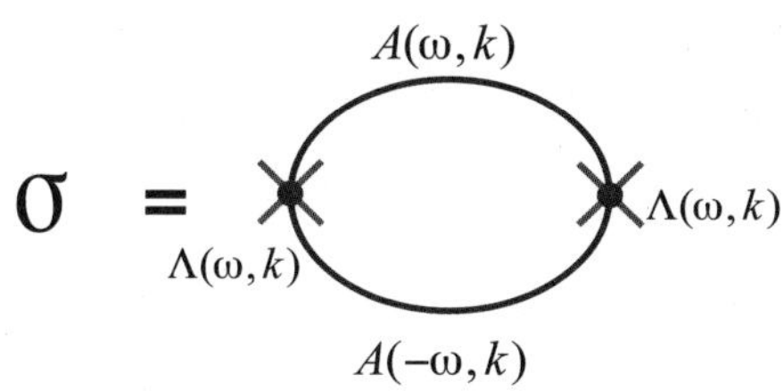

Fig. 13.14. Interpretation of (13.131) in field theory terms; the conductivity can be written as an expression similar to that in Fermi liquid theory, but with exact propagators and renormalized vertexes that can be found from gravity.

it on this background). This expression can be interpreted in boundary theory terms, as shown in Fig. 13.14; the effect of the radial direction is to provide an exact expression for the renormalized vertex Λ.

In general Λ is a complicated function of all its arguments, but in the low temperature limit and near the Fermi surface it becomes a smooth real function of $|\vec{k}|$, independent of ω and T. In this case (13.131) shows that the conductivity is controlled by the single-particle spectral function near the Fermi surface, and we find the expression

$$\sigma_{DC} = \alpha(q, m) T^{-2\nu_{k_F}}. \tag{13.132}$$

Here $\alpha(q, m)$ is a non-universal number that depends on a radial integral over the bulk fermion wavefunctions (and can be found numerically) but the exponent in T is determined by the single-particle decay rate and so by the SLQL. In particular, note that for the Marginal Fermi Liquid case

$\nu_{k_F} = \frac{1}{2}$, we find that the contribution to the resistivity from the Fermi surface is *linear*: $\rho \sim T$, as observed for the cuprates in the strange metal phase and many other non-Fermi liquid materials.[10]

Note that in general the electrical transport scattering rate – which emphasizes large momentum transfer – can be different from the single-particle scattering rate. On the other hand in our system the exponent in (13.132) *is* determined by the single-particle scattering rate. On the gravity side this arises from the fact that the dissipative portion of the correlation function is determined from Fig. 13.13 essentially by how fast the bulk fermions fall into the black hole, which determines the single-particle decay rate and knows nothing of momentum. On the field theory side this is due to the fact that the SLQL sees only the time direction and depends very weakly on momentum.

13.3.4.2. *Optical conductivity*

In the computation above the formula (13.132) for the DC conductivity is rather insensitive to whether or not the Fermi surface has long-lived quasiparticles or not , i.e. the value of ν_{k_F}. The optical conductivity $\sigma(\omega)$ can also be computed along similar lines; this turns out to depend critically on the value of ν_{k_F}.

For $\nu_{k_F} > \frac{1}{2}$ there are two regimes. In the first we hold $\omega T^{-2\nu_{k_F}}$ fixed in the low temperature limit and find then that we can approximate $\sigma(\omega)$ by a Drude form,

$$\sigma(\omega) = \frac{\omega_p^2}{\tau^{-1} - i\omega} \qquad \tau^{-1} = 2v_F \mathrm{Im}\,\Sigma(\omega = 0) \sim T^{2\nu_{k_F}} \tag{13.133}$$

with ω_p^2 an overall constant. This should not be surprising; we are in the regime describable by quasiparticles. In the other regime $\omega \gg T$ (but still $\omega \ll \mu$) we find

$$\sigma(\omega) = i\frac{\omega_p^2}{\omega} + b(i\omega)^{2\nu_{k_F}-2}, \tag{13.134}$$

with b a real constant. The $1/\omega$ term gives us a term proportional to $\delta(\omega)$ with weight consistent with (13.133).

On the other hand, for $\nu_{k_F} < \frac{1}{2}$, when we have no long-lived quasiparticles, we find a scale-invariant form,

$$\sigma(\omega) = T^{-2\nu_{k_F}} F_1\left(\frac{\omega}{T}\right), \tag{13.135}$$

where $F_1(x)$ is a scaling function that is constant as $x \to 0$. For large x it falls off as $x^{-2\nu_{k_F}}$, giving us an optical conductivity $\sigma(\omega) \sim a(i\omega)^{-2\nu_{k_F}}$ with a a real constant. This behavior is consistent with a system with no scales and no quasiparticles.

Finally, for the marginal case $\nu_{k_F} = \frac{1}{2}$, we find behavior similar to (13.135) but with logarithmic violations, i.e.

$$\sigma(\omega) = T^{-1} F_2\left(\frac{\omega}{T}, \log\frac{T}{\mu}\right) \tag{13.136}$$

where again F_2 goes to a constant at small ω. For $\omega \gg T$ in this case we find

$$\sigma(\omega) \sim -C\frac{i}{\omega}\left(\frac{1}{\log\frac{\omega}{\mu}} + \frac{1}{(\log\frac{\omega}{\mu})^2}\frac{1+i\pi}{2}\right) + \cdots . \tag{13.137}$$

13.3.5. *SLQL as a fractionalized phase of Fermi liquids*

In this subsection we further explore the physics of the oscillatory region $k < k_o$ discussed in Sec. 13.2.5.3. Such a region exists when

$$\frac{g_F^2 q^2}{2} > m^2 R^2 \tag{13.138}$$

and it is shown in the shaded regions in Fig. 13.10. In the oscillatory region, the IR scaling dimension of a fermionic operator is complex and the spectral function is nonzero at $\omega = 0$. At a nonzero ω, the correlation function (13.101) is periodic in $\log\omega$ and has an infinite number of poles of in the lower half ω-plane. It can also be readily checked that all the poles have comparable real and imaginary parts and thus cannot be considered as corresponding to quasiparticles. From bulk point of view, this region can be interpreted as corresponding to the projection of a bulk Fermi surface (which is one-dimensional higher) to the boundary. Small excitations at the bulk Fermi surface have a large decay rate since they can fall into the black hole without experiencing any potential barrier.

We will now show that in the parameter range of (13.138) the system in fact goes over to a Fermi liquid of "heavy" fermions at low energies. We will keep the boundary spacetime dimension d general.

From (13.108), the local charge density of fermions is given by

$$\rho_L = D_d (k_b^{(F)})^d, \qquad D_d = qs\frac{\Omega_d}{(2\pi)^d} \tag{13.139}$$

where Ω_d is the volume of a unit sphere in d-dimension and s is the number of possible spin degeneracies. ρ_L in (13.139) is finite and scales as $O(N^0)$

in the large N limit[z] which naively implies that the fermionic backreaction can be ignored at leading order. However, the contribution to the boundary charge density from ρ_L is obtained from integrating it over the radial direction of the black hole, and since in the near horizon region (13.49) the proper radial distance is infinite and the local proper volume of the transverse $\mathbb{R}^{d-1}$ is constant, the total boundary density coming from integrating (13.139) over the radial direction is in fact infinite.[91]

The backreaction of the fermionic gas is thus important *no matter how large N is.* The backreacted geometry can be found by using (13.139) and its corresponding stress tensor as the source for bulk Maxwell and Einstein equations. We refer readers to[91] for details of the calculation and only state the results here. One finds that after taking into account of the backreaction the near-horizon geometry is modified from (13.106) to a Lifshitz geometry

$$ds^2 = -e^{-2y}dt^2 + e^{-\frac{2y}{z}}\mu_*^2 R^2 d\vec{x}^2 + R_2^2 dy^2, \qquad A_t = \frac{e_d}{R_2}e^{-y} \tag{13.140}$$

with

$$z = \frac{d}{\kappa^2 \rho_L}\frac{\sqrt{2}R}{R_2^2} \sim O(N^2) \tag{13.141}$$

where we have only kept leading order terms in $1/z$ expansion. It is important to emphasize that even in the large N limit, the exponential prefector before the second term of (13.140) cannot be set to unity as it becomes important when y is sufficiently large (i.e. $y \sim O(z)$). Also note that as (13.106), (13.140) should be used in the region $y \gtrsim 0$; the geometry matches to the rest of the rest of black hole (UV region) near $y \sim 0$.

In the backreacted geometry (13.140), shown schematically in Fig. 13.15, the local proper volume of the transverse $\mathbb{R}^{d-1}$ goes to zero as $y \to \infty$, resulting in a finite boundary fermionic density of order $O(N^2)$, given by

$$\rho_F = R_2(\mu_* R)^{d-1}\rho_L \int_0^\infty dy\, e^{-\frac{d}{z}y} = \frac{R_2(\mu_* R)^{d-1} z \rho_L}{d}\,. \tag{13.142}$$

Plugging the explicit value (13.141) of z into (13.142) we find that (13.142) becomes identical to the total charge density (13.47), i.e. all the charge density of the system is now carried by the fermioinic gas. The black hole has disappeared! It is now replaced by a naked singularity of zero horizon size and zero charge. The system now also has zero entropy density. Note that such a singularity appears to be of "good" type allowed by string

[z] In contrast the charge density carried by the black hole, given in (13.47) scales as $O(N^2)$.

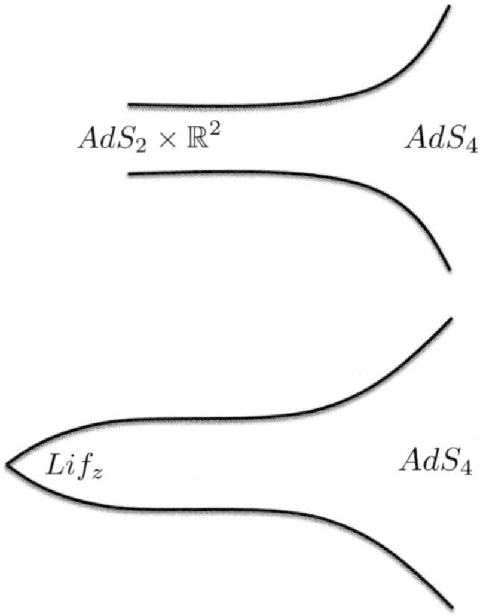

Fig. 13.15. Two different geometries (with $d = 3$): on the top, the $AdS_2 \times \mathbb{R}^2$ describing the SLQL phase; on the bottom, its resolution into a Lifshitz geometry with a finite z given by (13.141). The horizon direction represents the y direction, while the vertical direction represents the transverse $\mathbb{R}^2$. In the plot for the Lifshitz geometry it should be understood that the tip lies at an infinite proper distance away. When z is large as in (13.141), there is a large range of y for which the Lifshitz geometry resembles that of $AdS_2 \times \mathbb{R}^2$. Also note that e^{-y} translates into the boundary theory energy scale. In the Lifshitz geometry, the black hole has disappeared; it is replaced by a naked singularity with zero horizon size and zero charge. All the charged density is now carried by a bulk fermionic gas and the entropy density vanishes at zero temperature.

theory.[99,100] As we will see below, things could still fall into a singularity, which means in the boundary it should be interpreted as some of "bath" which can dissipate things away. While it has zero entropy and charge at order $O(N^2)$, likely such a "bath" still carries some charge and entropy at subleading order.

Recall that in obtaining (13.108) and (13.139) we used the approximation of a local chemical potential (i.e. Thomas-Fermi approximation), which is valid when $qg_F, mR, k_b^{(F)}R$ are taken to be parametrically large. Thus (13.141) and (13.142) are also only valid in this regime.

The Lifshitz geometry (13.140) is well approximated by (13.106) until $y \sim z$. From the boundary theory perspective, this implies that at an energy scale of order $e^{-z} \sim e^{-O(N^2)}$, physics will start deviating from that described by the SLQL. In other words, at sufficiently low energies, the SLQL will order into a new phase which is described by the Lifshitz geometry. What is this new phase? It turns out to be a Fermi liquid of "heavy" fermions[43]! Below we outline the basic arguments and refer readers to Ref. 43 for details (see also Refs. 131 and 132).

The basic idea is again to examine the retarded Green function for the fermionic operator in which we now expect to find new physics at sufficiently low energies. For this purpose one needs to solve the Dirac equation for the bulk fermions in the backreacted geometry (13.140), which for q and mR parametrically large can be reduced to a Schrödinger problem describing a particle of *zero energy* moving in a potential

$$V(y) = \frac{1}{2\mu^2}\left(k^2 e^{\frac{2y}{z}} - k_o^2 - \omega^2 e^{2y} - \sqrt{2}\mu_* q\omega e^y\right) . \tag{13.143}$$

For the $\mathrm{AdS}_2 \times \mathbb{R}^{d-1}$ geometry, the first term is absent.

The final results can be readily understood from the qualitative features of the WKB potential (13.143), as shown in Fig. 13.16 for $k < k_o^2$. First consider $\omega = 0$, the potential starts being an almost negative constant for

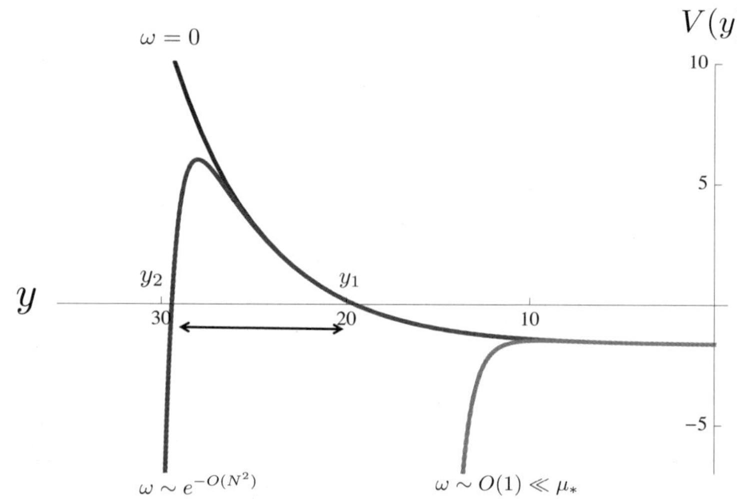

Fig. 13.16. Plots of the WKB potential (13.143) for various values of ω in the Lifshitz region. For $\omega = 0$, the potential is bounded and the Bohr-Sommerfeld quantization gives normalizable modes in the bulk which correspond to different Fermi surfaces in the boundary theory. For an exponentially small ω, although the potential is unbounded from below in the large y region, the excitations have a small imaginary part due to the potential barrier. For small ω but not exponentially small, the potential barrier disappears and the potential becomes the same as that for the $\mathrm{AdS}_2 \times \mathbb{R}^2$.

$y \ll z$, but eventually rises above zero and approaches $+\infty$ as $y \to +\infty$. The Bohr-Sommerfeld quantization condition then leads a series of normalizable bound states at discrete values of k. Each such normalizable mode then gives rise a boundary Fermi surface. Alternatively, from the bulk point of view, the radial modes become quantized due to bounded potential and the continuous bulk Fermi surface indicated in Fig. 13.8 becomes discretized and when projected to the boundary become a series of Fermi surfaces as indicated in Fig. 13.17. Now notice that in (13.143), for any nonzero ω no matter how small, the ω-dependent terms will eventually dominate for sufficiently large y and drive the potential to $-\infty$ as $y \to +\infty$. This implies bulk particles will eventually fall into the singularity of the Lifshitz geometry and the corresponding boundary fermionic excitations should develop a nonzero width. For exponentially small ω, there is still a potential barrier and the tunneling rate is exponentially small. In this range there are long-lived quasiparticles. More explicitly, one finds that near each k_F the retarded function for ω in the range

$$\omega \ll \omega_c(k) \equiv \frac{k_o}{z} \exp\left(-z \log \frac{k_o}{k_F}\right) \tag{13.144}$$

can be written as

$$G_R(\omega, k) = -\frac{Z}{\omega - v_F(k - k_F) + \Sigma} \tag{13.145}$$

with

$$Z, v_F \sim \exp\left(-z \log \frac{k_o}{k_F}\right), \quad \Sigma \sim i \exp\left(-\frac{\sqrt{2} z k_F}{\mu} \left(\frac{k_F}{\omega}\right)^{\frac{1}{z}} - z \log \frac{k_o}{k_F}\right). \tag{13.146}$$

Note that since $z \sim O(N^2)$, these are very heavy fermions[aa] and the decay rate is exponentially small in ω.

For small but not exponentially small ω, the first term in (13.143) is not important for any value of y and the potential barrier disappears. One thus recovers previous results of SLQL. In particular, the decay rate becomes $O(1)$ and there are no long-lived quasiparticles.

This procedure of solving the Dirac equation in the backreacted geometry (13.140) is self-consistent and extracts the leading non-perturbative behavior in $1/N^2$. There are also perturbative loop corrections in the bulk, which give perturbative corrections in $1/N^2$ to the self-energy. We expect

[aa]The exponentially small Fermi velocity v_F implies that the effective mass m_* is exponentially large, i.e. $m_* \sim \exp\left(+z \log \frac{k_o}{k_F}\right)$.

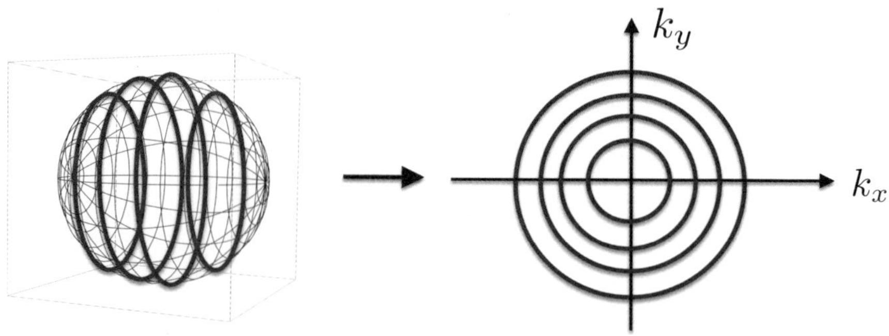

Fig. 13.17. In the Lifshitz geometry, we find a set of discrete states in the bulk in which radial motion is quantized. This results in a family of concentric Fermi surfaces in the boundary theory, which resolves the Fermi disk of Fig. 13.8.

the qualitative features of the above discussion (e.g the family of densely spaced Fermi surfaces etc) to be robust against these corrections, as they have to do with the global structure of the backreacted geometry (13.140). Near the Fermi surface perturbative corrections to the self-energy will give rise to a term $\frac{c}{N^2}\omega^2$ with c some complex $O(1)$ coefficient, which will dominate over the imaginary part of Σ in (13.145) for ω in the range (13.144). Thus we expect that the quasiparticle decay rate should be proportional to ω^2 as in a Landau Fermi liquid.[bb]

We thus see that at low energies the oscillatory region $k < k_o$ splits into a large number of closely spaced Fermi surfaces whose excitations are Landau quasiparticles. In this regime the Luttinger theorem should hold, which can be checked explicitly (see Ref. 43 for details).

Different Fermi surfaces in the family of densely spaced Fermi surfaces can be interpreted as corresponding to *different bound states* generated by the fermionic operator $\mathcal{O}$, as each of them corresponds to a different radial mode in the bulk. Recall that in our set-up, $\mathcal{O}$ is a composite operator of fundamental fields. The number of degrees of freedom for the fundamental fields is $O(N^2)$. Thus the Fermi liquid state can be considered a "confined" state, in which the low energy degrees of freedom are Fermi surfaces from a discrete set of composite fermionic bound states. In contrast, the SLQL is a "deconfined" state in which the composite bound states deconfine and fractionalize into more fundamental degrees of freedom. This also explains

[bb]Nevertheless, one should keep in mind that the much smaller non-perturbative correction does signal some nontrivial underlying physics beyond that of a Landau Fermi liquid.

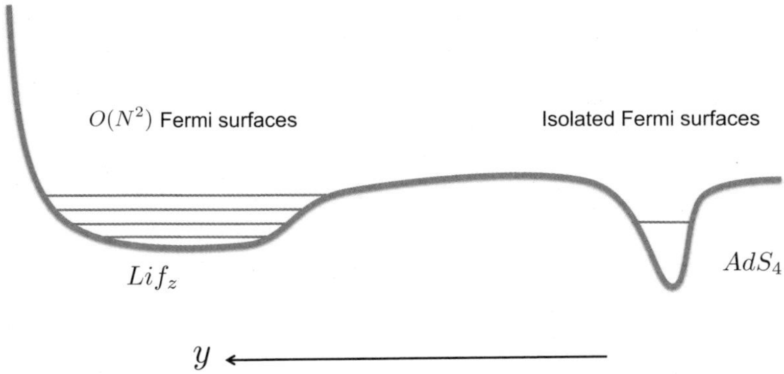

Fig. 13.18. A cartoon of the bound-state structure for the full spacetime, i.e. including the asymptotic AdS4 region. The Fermi surfaces in Sec. 13.3.2 and 13.3.3 appear as bound states in a potential well in the UV region.

the nonzero entropy density of SLQL. In the bulk, the emergence of the fractionalized SLQL phase is reflected in the emergence of a charged black hole description. In the Fermi liquid state, the system is characterized by "heavy" fermions. But such coherent quasiparticles disappear in the SLQL. Instead one finds some kind of quantum soup which is characterized by scaling behavior in the time direction for any bosonic or fermionic operators. In the current context the presence of a large number of Fermi surfaces has to do with a spectrum of densely spaced bound states. The fractionalized picture should be independent of this feature. Thus interpretation of SLQL as a fractionalized phase resonates with that of Ref. 75, although the details are different.

Note that the WKB potential (13.143) only includes the near the horizon region of the backreacted charged black hole geometry, and does not include the Fermi surfaces discussed in Sec.13.3.2. In Fig. 13.18 we show a cartoon of the WKB potential for $\omega = 0$ for the full spacetime, i.e. including the asymptotic AdS_{d+1} region. The isolated Fermi surfaces discussed earlier in Sec. 13.3.2 and 13.3.3 appear as bound states in a potential well outside the near-horizon region. Such isolated Fermi surfaces exist both in the Fermi liquid phase and in the SLQL, i.e. the corresponding fermionic excitations remain confined in the deconfined SLQL phase. At exponentially small energies, the excitations again become Landau quasiparticles, but with an $O(1)$ effective mass.

To summarize, imagine a system with only a single fermionic operator $\mathcal{O}$ satisfying the condition (13.138), and no other instabilities. Then at very low energies, the system is described by a Fermi liquid with $O(N^2)$ densely spaced Fermi surfaces, each of size $O(N^0)$. The quasiparticle excitations have a very large effective mass (proportional to e^{N^2}). There could also be some additional isolated Fermi surfaces with an $O(1)$ effective mass. At small but not exponentially small frequencies, there is a wide energy range over which the system is controlled by the SLQL, with scaling behavior in the time direction and various non-Fermi liquid behavior. In the SLQL phase the clue that the system will eventually settles into a Fermi liquid state is the existence of a region $k < k_o$, where the scaling dimensions for fermions become complex.

13.3.6. *Generalizations*

Before concluding this section, we mention various generalizations which we do not have space to cover.

In the presence of a magnetic field, one can show that fermionic excitations fall into Laudau levels[101–103] and that the free energy displays de-Haas van Alphen oscillations in the inverse magnetic field.[105,106] The latter calculation also requires a one-loop calculation in the bulk. In Ref. 106 it was shown that while the oscillations exist with the expected period, for $\nu_{k_F} < \frac{1}{2}$ the amplitude of the oscillations does not take the textbook Lifshitz-Kosevich form. The fermion response in various holographic superconducting phases has been studied in Refs. 107–113. The superconducting condensate opens a gap in the fermion spectrum around the chemical potential if a certain bulk coupling between the spinor and scalar is included.[108] In particular, in the condensed phase, one finds stable quasiparticles, even when $\nu_{k_F} \leq \frac{1}{2}$. The properties of Fermi surfaces and associated excitations with more general fermionic couplings have also been studied in Refs. 114–116 and in more general geometries in Refs. 117 and 118. Fermi surfaces in 1+1-dimensions were studied in Refs. 119–121. References[122,123] considered the pairing instability of holographic non-Fermi liquids.

By considering spinor fields whose mass and charge grows with N (i.e. $m, q \sim N$) and a gauge coupling of order $1/N$, the Lifshitz geometry (13.140) has been generalized to have a $z \sim O(N^0)$ in Ref. 124 (see also Refs. 125–127). Various aspects of the boundary system dual to such an "electron star" were studied in Refs. 128–132. In particular, it was found[131,132] that the fermionic spectral function exhibits multiple Fermi surfaces with exponentially suppressed (in frequency) decay rates very similar to the behavior discussed earlier around (13.144)–(13.146).

13.4. Scalar Instabilities and Quantum Phase Transitions

In this section we examine possible bosonic instabilities of the finite charge density system. In particular, when there is an instability we describe how to tune external parameters to a quantum critical point and study the critical behavior near a quantum critical point. We will again specialize to $d = 3$. We will mainly follow the discussion of Refs. 44–46 (see also Refs. 133–136).

We will first slightly generalize the discussion of Sec. 13.2.4 by also including double trace deformations in the dual CFT_3

$$S_{CFT_3} \to S_{CFT_3} + \frac{\kappa_+}{2}\int \mathcal{O}^2 . \tag{13.147}$$

Using the general formula (A.42) of Appendix A.3, in the presence of a double trace deformation (13.147), equation (13.75) becomes

$$G_R(\omega,\vec{k}) = \mu_*^{2\nu_U}\frac{b_+(k,\omega)+b_-(k,\omega)\mathcal{G}_k(\omega)\mu_*^{-2\nu_k}}{\tilde{a}_+(k,\omega)+\tilde{a}_-(k,\omega)\mathcal{G}_k(\omega)\mu_*^{-2\nu_k}} \tag{13.148}$$

where

$$\begin{aligned}\tilde{a}_\pm(\omega,k) = a_\pm(\omega,k)+\kappa_+ b_\pm(\omega,k) = \tilde{a}_\pm^{(0)}(k)+O(\omega)\ ,\\ \tilde{a}_\pm^{(0)}(k) = a_\pm^{(0)}(k)+\kappa_+ b_\pm^{(0)}(k)\end{aligned} \tag{13.149}$$

To follow the standard terminology for discussing phase transitions below we will start calling (13.148) susceptibility and use the following notations

$$\chi \equiv G_R(\omega=0,\vec{k}=0), \qquad \chi(\vec{k}) \equiv G_R(\omega=0,\vec{k}), \qquad \chi(\omega,\vec{k}) \equiv G_R(\omega,\vec{k})\ . \tag{13.150}$$

Note that we distinguish the three only by their arguments. From (13.148) we find the momentum-dependent and uniform static susceptibilities are given by

$$\chi(k) = \mu_*^{2\nu_U}\frac{b_+^{(0)}(k)}{\tilde{a}_+^{(0)}(k)}, \qquad \chi = \mu_*^{2\nu_U}\frac{b_+^{(0)}(0)}{\tilde{a}_+^{(0)}(0)} \tag{13.151}$$

and the full dynamical susceptibility takes the form

$$\chi(\omega,\vec{k}) = \mu_*^{2\nu_U}\frac{b_+(k,\omega)+b_-(k,\omega)\mathcal{G}_k(\omega)\mu_*^{-2\nu_k}}{\tilde{a}_+(k,\omega)+\tilde{a}_-(k,\omega)\mathcal{G}_k(\omega)\mu_*^{-2\nu_k}}\ . \tag{13.152}$$

Below we first recall the instabilities discussed in Sec. 13.2.5 for the system at a finite chemical potential and the corresponding quantum critical points. We will then discuss quantum critical behavior near these quantum critical points.

13.4.1. *Finite density instabilities*

In Sec. 13.2.5 we discussed two possible channels for scalar instabilities, which manifest themselves as presence of poles of (13.152) in the *upper* half complex-ω-plane:

(1) The SLQL scaling dimension $\frac{1}{2} + \nu_k$ becomes complex for some k, for which there are an infinite number of poles in the upper half ω-plane. Writing (13.57) as

$$\nu_k = \sqrt{u + \frac{k^2}{6\mu_*^2}}, \qquad u \equiv m^2 R_2^2 + \frac{1}{4} - q_*^2 \tag{13.153}$$

ν_k becomes imaginary for sufficiently small k whenever $u < 0$. For a given m, this always occurs for a sufficiently large q. For a neutral operator $q = 0$, u can be negative for m^2R^2 lying in the window

$$-\frac{9}{4} < m^2R^2 < -\frac{3}{2} \tag{13.154}$$

where the lower limit comes from the stability of vacuum theory (i.e. BF bound of AdS_4) and the upper limit comes from the condition $u < 0$ after using the relation (13.50).

(2) $\tilde{a}_+^{(0)}(k)$ can become zero for some special values of momentum k_F. At $k = k_F$ it is clear from (13.148) that since $\tilde{a}_+^{(0)} = 0$, χ has a singularity at $\omega = 0$. Furthermore since $a_+^{(0)}$ changes sign near $k = k_F$, as discussed in Sec. 13.2.5.1, a pole moves from the upper half ω-plane (for $k < k_F$) to the lower half ω-plane (for $k > k_F$) through $\omega = 0$. In Fig. 13.19 we show some examples of a neutral scalar field for which $\tilde{a}_+^{(0)}$ has a zero at some momentum.

In the parameter range (say for m, q, κ_+) where either (or both) instability appears, the system is unstable to the condensation of operator $\mathcal{O}$ (or in bulk language condensation of ϕ). For a charged scalar the condensed phase corresponds to a holographic superconductor.[50,51] The first instability underlies that of Refs. 88 and 89, while holographic superconductors due to the second type instability was first discussed in Ref. 45. For a neutral scalar, the first type of instability was first pointed out in Ref. 90, and as discussed in Ref. 44 the condensed phase can be used as a model for antiferromagnetism when the scalar operator is embedded as part of a triplet transforming under a global $SU(2)$ symmetry corresponding to spin. For a single real scalar field with a Z_2 symmetry, the condensed phase can be considered as a model for an Ising-nematic phase.

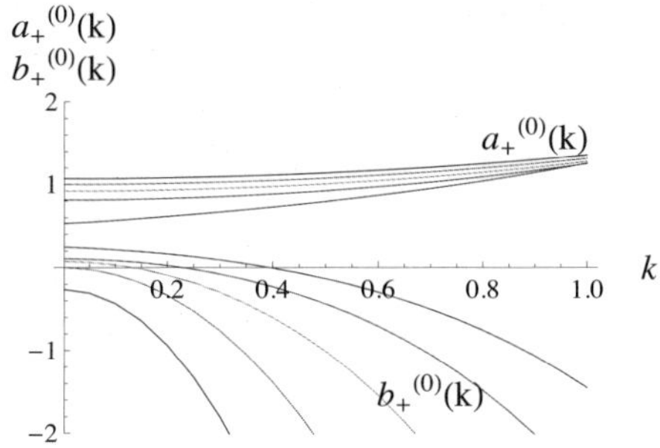

(a) $a_+^{(0)}(k)$ and $b_+^{(0)}(k)$ plotted for different values of m^2R^2; blue is $m^2 = -1.4999$, purple $m^2R^2 = -1$, green is $m^2R^2 = -0.5$, red is $m^2R^2 = 0$ and the black line is $m^2R^2 = 0.5$. There is an analytic argument why $b_+^{(0)}(k=0) = 0$ for $m^2 = 0$.

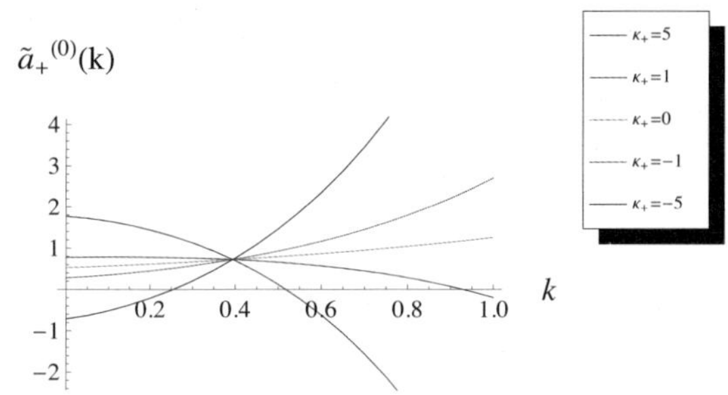

(b) $\tilde{a}_+^{(0)}(k)$ for $m^2R^2 = -1.4999$.

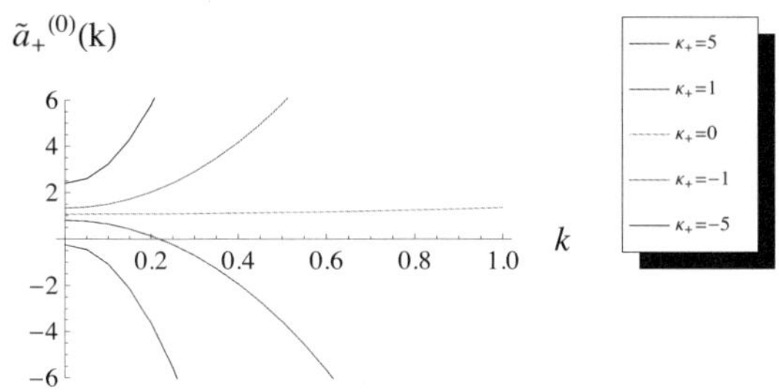

(c) $\tilde{a}_+^{(0)}(k)$ for $m^2R^2 = 0.5$.

Fig. 13.19. Plot of $a_+^{(0)}(k)$, $b_+^{(0)}(k)$ and $\tilde{a}_+^{(0)}(k)$ for various masses and κ_+ for a neutral scalar field as a function of k.

Both types of instabilities can be cured by going to sufficiently high temperature; there exists a critical temperature T_c, beyond which these instabilities no longer exist and at which the system undergoes a continuous superconducting (for a charged scalar) or antiferromagnetic (for a neutral scalar) phase transition. As have been discussed extensively in the literature such finite temperature phase transitions are of mean field type due to that the boundary conditions at a finite-temperature black hole horizon is analytic.

Alternatively one can continuously dial external parameters of the system at zero temperature to get rid of the instabilities.[cc] The critical values of the parameters at which the instabilities disappear then correspond to quantum critical points (QCP) where quantum phase transitions into a superconducting or antiferromagnetic phase occur.

13.4.2. *Quantum critical points*

13.4.2.1. *Bifurcating quantum critical point*

For the first type of instability a quantum critical point occurs when the effective AdS_2 mass becomes zero for $k = 0$,[44,133] i.e. from (13.153), at

$$u = u_c = 0 \ . \tag{13.155}$$

For example for a neutral scalar field (with $q = 0$) this happens at

$$m_c^2 R^2 = -\frac{3}{2} \ . \tag{13.156}$$

Note that while in AdS/CFT models the mass square m^2 for the vacuum theory is typically not an externally tunable parameter, the effective AdS_2 mass square can often be tuned. For example, in the set-up of Ref. 133, the effective AdS_2 mass square can be tuned by dialing an external magnetic field and so is the the example discussed in Ref. 44 when considering a holographic superconductor in a magnetic field. See also Ref. 44 for a phenomenological model. Here we will not worry about the detailed mechanism to realize the $u_c = 0$ critical point and will just treat u as a dialable parameter (or just imagine dialing the mass square for the vacuum theory).

[cc]One might be surprised that in the case of a broken continuous symmetry in $d = (2+1)$ the ordered phase can persist at *any* finite temperature, as one generally expects fluctuations of the Goldstone mode to destroy the order. As there is only one Goldstone mode (as compared to N^2 other fields) this is a $1/N^2$ effect and so is dual to a quantum effect in the bulk; it is shown in Ref. 137 that at one-loop order in the bulk the symmetry is indeed restored over exponentially long length scales.

Our main purpose is to identify and understand the critical behavior around the critical point which is independent of the specific mechanism to realize it. As will be discussed in subsequent sections, as we approach $u_c = 0$ from the uncondensed side ($u > 0$), the static susceptibility remains finite, but develops a cusp at $u = 0$ and if we naively continue it to $u < 0$ the susceptibility becomes complex. Below we will refer to this critical point as a *bifurcating* QCP.

13.4.2.2. *Hybridized quantum critical point*

For the second type of instability, one can readily check numerically that $\tilde{a}_+^{(0)}$ is a monotonically increasing function of k for negative κ_+. See Fig. 13.19. Thus the instability goes away if $\tilde{a}_+^{(0)}(k=0) > 0$, which corresponds to κ_+ satisfying[dd]

$$0 > \kappa_+ \geq \kappa_c \equiv -\frac{a_+^{(0)}(k=0)}{b_+^{(0)}(k=0)} \,. \tag{13.157}$$

At the critical point $\kappa_+ = \kappa_c$

$$\tilde{a}_+^{(0)}(k=0, \kappa_c) = 0 \tag{13.158}$$

and as a result the uniform susceptibility χ in (13.151) diverges. Such a quantum critical point was first discussed in Ref. 45 and elaborated further in Ref. 46. As mentioned in Refs. 45 and 46, the presence of the strongly coupled IR sector described by AdS_2 gives rise to a variety of new phenomena which cannot be captured by the standard Laudau-Ginsburg-Wilson paradigm. For reasons to be clear below, we will refer to such a critical point as a *hybridized* QCP.

13.4.2.3. *A marginal quantum critical point*

We can also tune κ_+ and u together to have a doubly tuned critical point at $u = 0, \kappa_+ = \kappa_c$, where the susceptibility both diverges and bifurcates. As we will see blow, the susceptibility around such a critical point coincide with that of the bosonic fluctuations underlying the "Marginal Fermi Liquid" postulated in Ref. 12 for describing the strange metal region of the high T_c cuprates.

[dd]There is also an instability for $\kappa_+ > 0$, but one can check that it corresponds to a UV instability of the vacuum. Here we are interested in the IR instability due to finite density effect. See Ref. 46 for more discussion.

We note that while understanding the end point of the instability, i.e. the stable phase in which $\mathcal{O}$ is condensed, requires full nonlinear analysis of bulk equations, to diagnose the instability, to identify the onset of instability (i.e. the quantum or finite temperature critical points), and to find the critical behavior near the critical points, linear analysis which gives rise to (13.148) suffices.

We now proceed to describe the critical behavior of various critical points identified in the above. We will use a neutral scalar field for illustration. We should also mention that in all critical points identified above, the uncondensed side is described by an extremal charged black hole. Given the interpretation in Sec. 13.2.6 of an extremal charged black hole as an intermediate state rather than a genuine ground state, the critical behavior discussed below from the uncondensed side should again be interpreted as due to intermediate-energy effects.

13.4.3. *Quantum critical behavior: Bifurcating quantum critical point*

To study the behavior near the critical point $u = 0$ we study the implications of taking $\nu_k \to 0$ in (13.152), i.e. both k^2/μ^2 and u are small.

13.4.3.1. *From uncondensed side*

We first study the critical behavior of the static susceptibility (13.151) by setting $\omega \to 0$ in (13.152) and taking $u \to 0$ from the uncondensed side $u > 0$. From equation (A.31) we find that for small ν_k,

$$\chi(k) = \mu_*^{2\nu_U} \frac{\beta + \nu_k \tilde{\beta}}{\alpha + \nu_k \tilde{\alpha}} + O(\nu_k^2, k^2) \tag{13.159}$$

where $\alpha, \beta, \tilde{\alpha}, \tilde{\beta}$ are numerical constants. Setting $k = 0$ we find the zero momentum susceptibility is given by

$$\chi = \mu_*^{2\nu_U} \frac{\beta + \sqrt{u}\tilde{\beta}}{\alpha + \sqrt{u}\tilde{\alpha}} \tag{13.160}$$

and at the critical point the static susceptibility remains finite, given by

$$\chi|_{u\to 0_+} = \chi_0 \equiv \mu_*^{2\nu_U} \frac{\beta}{\alpha} \,. \tag{13.161}$$

This is in sharp contrast with the critical behavior from the Landau paradigm where one expects that the uniform susceptibility always diverges approaching a critical point. Due to the square root appearing in (13.160),

χ has a branch point at $u = 0$ and bifurcates into the complex plane for $u < 0$. Of course, when $u < 0$, eq. (13.160) can no longer be used, but the fact that it becomes complex can be considered an indication of instability. Furthermore, taking a derivative with respect to u we find that

$$\partial_u \chi = \mu_*^{2\nu_U} \frac{\alpha\tilde{\beta} - \beta\tilde{\alpha}}{2\alpha^2} \frac{1}{\sqrt{u}} = -\mu_*^{2\nu_U} \frac{1}{4\nu_U \alpha^2} \frac{1}{\sqrt{u}} \to \infty, \qquad u \to 0 \qquad (13.162)$$

where we have used (A.34) in the second equality. Thus even though $\chi(u)$ is finite at $u = 0$, it develops a cusp there, as shown in Fig. 13.20. It will turn out convenient to introduce a quantity

$$\chi_* \equiv \mu_*^{2\nu_U} \frac{1}{4\nu_U \alpha^2} = \chi_0 \frac{1}{4\nu_U \alpha\beta} \qquad (13.163)$$

and then (13.162) becomes

$$\partial_u \chi = -\frac{\chi_*}{\sqrt{u}} \ . \qquad (13.164)$$

Similarly, taking derivative over k^2 in (13.159) and then setting $k = 0$, we find that

$$\partial_{k^2} \chi(\vec{k})\big|_{k=0} = -\frac{\chi_*}{6\mu_*^2 \sqrt{u}}, \qquad u \to 0 \ . \qquad (13.165)$$

Note that this divergence is related to the fact for any $u > 0$, $\chi(\vec{k})$ is analytic in k^2, but not at $u = 0$, where $\nu_k \propto k$.

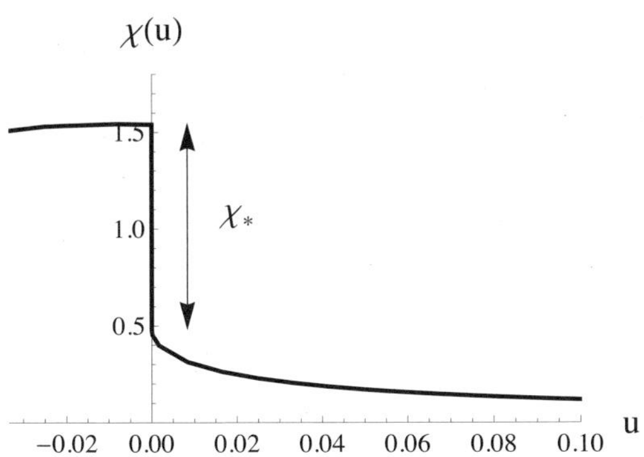

Fig. 13.20. A plot of $\chi(u)$ as a function of u. We also include the behavior on the $u < 0$ side to be discussed a bit later. Note that while there is a cusp in χ approaching the critical point from the uncondensed side ($u > 0$), there is no cusp approaching the critical point from the condensed side ($u < 0$). From both sides the susceptibility is finite at the critical point, but there is a jump in their values.

The above non-analytic behavior at $k = 0$ should have important consequences when we Fourier transform $\chi(\vec{k})$ to coordinate space. Indeed by comparing (13.153) with (13.61), we find that

$$\xi = \frac{1}{\sqrt{6}\mu_* \sqrt{u}} \,. \tag{13.166}$$

Thus as $u \to 0$, the correlation length ξ diverges as $u^{-\frac{1}{2}}$ which is the same as that in a mean field theory.

We now turn to the critical behavior of the susceptibility (13.152) at a nonzero ω near the critical point from uncondensed side $u > 0$. We should be careful with the $\nu_k \to 0$ limit as the factor $\left(\frac{\omega}{\mu_*}\right)^{2\nu_k}$ in the SLQL Green function (13.59) behaves differently depending on the order we take the $\nu_k \to 0$ and $\omega \to 0$ limits. For example, the Taylor expansion of such a term in small ν_k involves terms of the form $\nu_k \log(\omega/\mu_*)$, but in the small ω limit, the resulting large logarithms may invalidate the small ν_k expansion. Expanding the Gamma function and $a_\pm, b_\pm$ to leading order in ν_k, but keeping the full dependence on ω, we find that

$$\chi(\omega, \vec{k}) = \chi_0 \frac{\sinh\left(\nu_k \log\left(\frac{-i\omega}{\omega_b}\right)\right)}{\sinh\left(\nu_k \log\left(\frac{-i\omega}{\omega_a}\right)\right)} + \cdots \tag{13.167}$$

where the energy scales $\omega_{a,b}$ are constants of order μ and χ_0 is uniform susceptibility at the critical point given earlier in (13.161). Considering $\nu_k \to 0$ in (13.167) with a fixed ω, we then find

$$\chi(\omega, \vec{k}) = \chi_0 \frac{\log\left(\frac{\omega}{\omega_b}\right) - i\frac{\pi}{2}}{\log\left(\frac{\omega}{\omega_a}\right) - i\frac{\pi}{2}} + O(u, k^2) \tag{13.168}$$

where the corrections are analytic in both u and k^2. Note that both above expression and (13.167) have a pole at $\omega = i\omega_a$ in the upper half ω-plane. But this should not concern us as our expressions are only valid for $\omega \ll \mu_* \sim \omega_a$. Further taking the $\omega \to 0$ limit in (13.168) then gives

$$\begin{aligned} \chi(\omega, \vec{k}) &= \chi_0 \left(1 + \frac{1}{2\nu_U \alpha\beta} \frac{1}{\log \omega} + \frac{i\pi}{4\nu_U \alpha\beta} \frac{1}{(\log \omega)^2} + \cdots\right) \\ &= \chi_0 + \frac{2\chi_*}{\log \omega} + \frac{i\pi \chi_*}{(\log \omega)^2} + \cdots \end{aligned} \tag{13.169}$$

where we have kept the leading nontrivial ω-dependence in both real and imaginary parts and used (13.163).

Equations (13.168) and (13.169) give the leading order expression at nonzero u (for both signs, as $\chi(\omega, k)$ is analytic at $u = 0$ at a nonzero ω) and k^2 as far as $\nu_k \log \frac{\omega}{\omega_{a,b}}$ remains small. They break down when ω becomes exponentially small in $\frac{1}{\nu}$,

$$\omega \sim \Lambda_{\text{CO}}, \quad \Lambda_{\text{CO}} \sim \mu_* e^{-\frac{1}{\sqrt{u}}} . \tag{13.170}$$

In the regime of (13.170), the susceptibility (13.167) crosses over to

$$\chi(\omega \to 0, \vec{k}) = \chi_0 - 2\nu_k \chi_* - 4\nu_k \chi_* \left(\frac{-i\omega}{2\mu_*} \right)^{2\nu_k} + \cdots \tag{13.171}$$

which is consistent with (13.159). Note that $\cdots$ in the above equation also includes perturbative corrections in ω.

13.4.3.2. *From condensed side*

When $u < 0$, the SLQL scaling dimension of $\mathcal{O}_{\vec{k}}$ becomes complex for sufficiently small k as $\nu_k = \sqrt{u + \frac{k^2}{6\mu_*^2}} = -i\lambda_k$ is now pure imaginary.[ee] For a given nonzero ω and $|u|$ sufficiently small, the corresponding expression for $\chi(k, \omega)$ can be obtained from (13.167) by simply taking $\nu_k = -i\lambda_k$, after which we find

$$\chi(\omega, \vec{k}) = \chi_0 \frac{\sin\left(\lambda_k \log\left(\frac{-i\omega}{\omega_b}\right)\right)}{\sin\left(\lambda_k \log\left(\frac{-i\omega}{\omega_a}\right)\right)} + \cdots . \tag{13.172}$$

While (13.167) is valid to arbitrarily small ω, equation (13.172) has poles in the *upper* half frequency plane (for $k = 0$) at[ff]

$$\omega_n = i\omega_a \exp\left(-\frac{n\pi}{\sqrt{-u}}\right) \equiv i\Lambda_n, \qquad n = 1, 2 \cdots \tag{13.173}$$

with

$$\Lambda_n \sim \mu \exp\left(-\frac{n\pi}{\sqrt{-u}}\right) . \tag{13.174}$$

In particular, we expect (13.172) to break down for $\omega \sim \Lambda_1$, the largest among (13.174), and at which scale the physics of condensate sets in. This

[ee] Note that the choice of branch of the square root does not matter as (13.167) is a function of ν_k^2.

[ff] Note that (13.172) also have poles for non-positive integer n. But at these values ω is either of order or much larger than the chemical potential μ to which our analysis do not apply.

is indeed consistent with an explicit analysis of classical gravity solutions in[44,46,133] where it was found that $\mathcal{O}$ develops an expectation value of order

$$\frac{\langle \mathcal{O} \rangle}{\mu^{\Delta}} \sim \left(\frac{\Lambda_1}{\mu} \right)^{\frac{1}{2}} . \tag{13.175}$$

The exponent $\frac{1}{2}$ in (13.175) is the scaling dimension of $\mathcal{O}$ in the SLQL for $u = 0$, while Δ is its scaling dimension in the vacuum. It was also found in[44,46,133] there are an infinite number of excited condensed states with a dynamically generated scale given by Λ_n and

$$\langle \mathcal{O} \rangle_n \sim \Lambda_n^{\frac{1}{2}} \sim \exp\left(-\frac{n\pi}{2\sqrt{-u}} \right), \qquad n = 1, 2, \cdots \tag{13.176}$$

respectively. Thus the pole series in (13.173) in fact signal a geometric series of condensed states. This tower of condensed states with geometrically spaced expectation values is reminiscent of Efimov states.[138][gg] The largest is in the first state $n = 1$, which is the energetically favored vacuum, as can be checked by calculating the free energy of each state.

One can also compute the response of the system to a static and uniform external source in this tower of "Efimov" states. The result is rather interesting and can be described as follows.[46] One finds that the response in all the "Efimov" states can be read from a pair of continuous spiral curves described parametrically by (for $\sqrt{-u} \ll 1$)

$$\begin{aligned} A &= z_*^{3-\Delta} \frac{\gamma}{\sqrt{-u}} \alpha \sqrt{\frac{z_*}{\zeta_*}} \sin\left(\sqrt{-u} \log \frac{\zeta_*}{z_*} + \sqrt{-u} \frac{\tilde{\alpha}}{\alpha} \right), \\ B &= z_*^{-\Delta} \frac{\gamma}{\sqrt{-u}} \beta \sqrt{\frac{z_*}{\zeta_*}} \sin\left(\sqrt{-u} \log \frac{\zeta_*}{z_*} + \sqrt{-u} \frac{\tilde{\beta}}{\beta} \right) . \end{aligned} \tag{13.177}$$

where A and B denote the source and expectation value for $\mathcal{O}$ respectively, γ is a $O(1)$ constant. ζ_*^{-1} is a dynamical energy scale which parametrizes movement through the solution space; as we vary ζ_*, we trace out a *spiral* in the (A, B) plane. See Fig. 13.21. Since we are considering a system with a Z_2 symmetry $\mathcal{O} \to -\mathcal{O}$, in Fig. 13.21 there is also a mirror spiral obtained from (13.177) by taking $(A, B) \to -(A, B)$. The set of Efimov states are obtained by setting the source $A = 0$, i.e. they can be read from the intersections of the spirals with the B-axis.

[gg]In fact the gravity analysis reduces to a quantum mechanical problem which is exactly the same as that for the Efimov states.

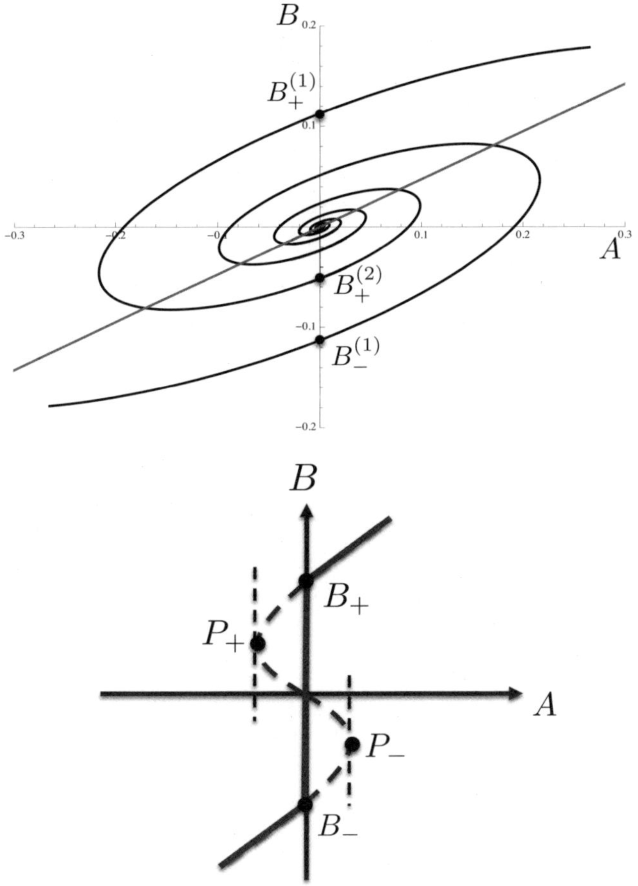

Fig. 13.21. *Top:* the spirals describing responses of the tower of "Efimov" states. The spiral in the $B - A$ plane passing through $B_+^{(1)}$ gives the solution described by (13.177) as ζ_* is varied. The spiral passing through $B_-^{(1)} = -B_+^{(1)}$ gives the mirror curve from $\mathcal{O} \to -\mathcal{O}$ reflection symmetry. The normalizable solutions (i.e. possible condensed states) are given by the intersections of the spiral with respect to the B-axis with $B_\pm^{(1)}$ the ground states and $B_\pm^{(2)}$ the first excited states, etc. The spirals are not drawn to scale and only serve illustration purposes. The red straight line has slope given by (13.178). As $\sqrt{-u} \to 0$ most of the spiral becomes parallel to it. *Bottom:* Response curve in a Landau-Ginsburg mean-field model $\mathcal{L}_{LG} = \frac{1}{2}u\psi^2 + \frac{1}{2}(\partial\psi)^2 + \lambda\psi^4 + \cdots$, where ψ is a Z_2 order parameter and the critical point lies at $u = 0$. The plot is for a small negative u (condensed side). Note that the part to the right of B_+ behaves as $A^{\frac{1}{3}}$ instead of the linear behavior of the left plot since at the critical point the uniform susceptibility diverges. The part between B_+ (B_-) and P_+ (P_-) describes a metastable region while the part between P_+ and P_- describes an unstable region. In the left plot the region between P_+ and P_- is replaced by a pair of infinite spirals which give rise to a tower of infinite excited states.

As $\sqrt{-u} \to 0$, from (13.177), A and B are becoming in phase, and the spiral is being squeezed into a straight line, with limiting slope

$$\left.\frac{B}{A}\right|_{\sqrt{-u}\to 0} = \mu_*^{2\Delta-3}\frac{\beta}{\alpha} = \chi_0 \,. \tag{13.178}$$

This slope agrees with the value found from linear response approaching the critical point from the other side (13.161). This is however *not* the relevant slope for the susceptibility, which should be given by

$$\chi_L = \left.\frac{dB}{dA}\right|_{A=0} \tag{13.179}$$

which in the usual models of spontaneous symmetry breaking, corresponds to the longitudinal susceptibility. From (13.177) we find

$$\chi_L = \mu_*^{2\nu_U}\frac{\beta}{\alpha}\left(1+\frac{\tilde{\alpha}\beta-\alpha\tilde{\beta}}{2\alpha\beta}\right)+\mathcal{O}(u) = \chi_0+\chi_*+\mathcal{O}(u) \tag{13.180}$$

where we have used (A.34) and (13.163). Essentially, even though the spiral is squished into a straight line as we approach the transition, each *intersection* of the spiral with the $A = 0$ axis has a different slope than the limiting slope of the entire spiral. Note that this result is independent of n and in particular applies to $n = 1$, the ground state. Since χ_0 is the value at $u = 0$ from the uncondensed side, we thus find a jump in the value of uniform susceptibility in crossing $u = 0$ and the difference is precisely the same coefficient as the divergent terms in (13.162), which has also appeared in other places such as (13.169).

We see that close to the origin of the spiral in Fig. 13.21, nonlinearities are very important, creating an infinite tower of normalizable states. Let us imagine starting with the ground state[hh] $B_+^{(1)}$ and following the spiral to the right, i.e. we apply an external source A in the *same* direction as the condensate. This will increase B according to (13.179) and (13.180). Note that near the critical point $B_+^{(1)}$ is exponentially small; thus as we increase A further, we will eventually reach a regime where the forced response is much larger than the condensate $B \gg B_+^{(1)}$ but is still much smaller than 1, $B \ll 1$. One thus expects that here the system should not care about the (exponentially small) condensate and the response should simply be given by that at the critical point, i.e. the linear response line given by χ_0. Thus the spiral will approach a straight line parallel to the red straight line in the figure. For more discussion on other parts of the spirals see

[hh] Equivalently we can also start with its Z_2 image $B_-^{(1)}$.

Ref. 46. For comparison we also show the response curve for the standard Landau-Ginsburg story in the right plot of Fig. 13.21.

The existence of a tower of "Efimov" states with geometrically spaced expectation values may be considered as a consequence of spontaneous breaking of the discrete scaling symmetry of the system. With an imaginary scaling exponent, (13.172) exhibits a discrete scaling symmetry with (for $k = 0$)

$$\omega \to e^{\frac{2\pi}{\sqrt{-u}}}\omega \tag{13.181}$$

which is, however, broken by the condensate.[ii] The tower of "Efimov" states may then be considered as the "Goldstone orbit" for this broken discrete symmetry.

13.4.3.3. *Thermal aspects*

One can also look at the critical behavior near the bifurcating critical point at a finite temperature, which is rather similar to with our earlier discussion of finite frequency behavior. We will simply state the results. At $\omega = 0$ but finite T we find

$$\chi^{(T)}(\vec{k}) = \chi_0 \frac{\sinh\left(\nu_k \log\left(\frac{T}{T_b}\right)\right)}{\sinh\left(\nu_k \log\left(\frac{T}{T_a}\right)\right)}, \tag{13.182}$$

where $T_{a,b}$ some constants of order μ. Similarly to (13.172), the expression for $u < 0$ is obtained by analytically continuing (13.182) to obtain

$$\chi^{(T)}(\vec{k}) = \chi_0 \frac{\sin\left(\lambda_k \log\left(\frac{T}{T_b}\right)\right)}{\sin\left(\lambda_k \log\left(\frac{T}{T_a}\right)\right)}. \tag{13.183}$$

And again both (13.182) and (13.183) are analytic at $u = 0$ and reduce to the same function there

$$\chi^{(T)}(\vec{k}) = \chi_0 \frac{\log\frac{T}{T_b}}{\log\frac{T}{T_a}} + O(u, k^2). \tag{13.184}$$

Similar to (13.168), the pole in (13.184) and (13.182) at $T = T_a$ should not concern us as this expression is supposed to be valid only for $T \ll \mu_* \sim T_a$.

[ii]Note that for $n = 1$ state, since the physics of the condensate sets in already at Λ_1, the range of validity for (13.172) is not wide enough for the discrete scaling symmetry to be manifest.

For $u > 0$, at a scale of

$$T \sim \Lambda_{\rm CO} \sim \mu_* e^{-\frac{1}{\sqrt{u}}} \tag{13.185}$$

eq. (13.182) crosses over to an expression almost identical to (13.171) with ω replaced by T. For $u < 0$, at such small temperature scales equation (13.183) has poles at (for $k = 0$)

$$T_n = T_a \exp\left(-\frac{n\pi}{\sqrt{-u}}\right) = \frac{4\mu_*}{\pi} \exp\left(-\frac{n\pi}{\sqrt{-u}} + \frac{\tilde{\alpha}}{\alpha}\right), \qquad n \in \mathbb{Z}^+ . \tag{13.186}$$

The first of these temperatures should be interpreted as the critical temperature

$$T_c = \frac{4\mu_*}{\pi} \exp\left(-\frac{\pi}{\sqrt{-u}} + \frac{\tilde{\alpha}}{\alpha}\right) \tag{13.187}$$

below which the scalar operator condenses. Including frequency dependence, one can check that $\chi^{(T)}(\omega, \vec{k})$ has a pole at

$$\omega_* = -\frac{2i}{\pi}(T - T_n) \tag{13.188}$$

For $T > T_n$ this pole is in the lower half-plane, and it moves through to the upper half-plane if T is decreased through T_n. Thus we see the interpretation of each of these T_n; as the temperature is decreased through each of them, one more pole moves through to the upper half-plane. There exist an infinite number of such temperatures with an accumulation point at $T = 0$; and indeed at strictly zero temperature there is an infinite number of poles in the upper half-plane, as seen earlier in (13.173). Of course in practice once the first pole moves through to the upper half-plane at $T_c = T_1$, the uncondensed phase is unstable and we should study the system in its condensed phase.

One can further study the critical behavior near the finite temperature critical point T_c. Here one finds mean field behavior. In Fig. 13.22 we summarize the finite temperature phase diagram.

13.4.3.4. *Summary: SLQL as a fractionalized phase of Bose condensates*

When the dimension of a scalar operator $\mathcal{O}$ becomes complex in SLQL, SLQL orders at lower energy into a condensed state of the corresponding operator. There are a tower of an infinite number of condensed states. Near the bifurcating critical point, they form a geometric series of Efimov states

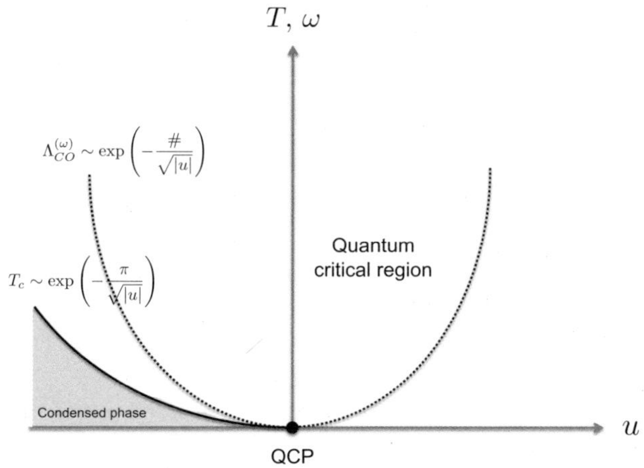

Fig. 13.22. Finite temperature phase diagram with the quantum critical region for bifurcating criticality as a function of u. The dotted line is given by $\Lambda_{\rm CO}$ in (13.170) and (13.185). But note that the discussions here are not enough to determine the $O(1)$ factor in the exponent for $\Lambda_{\rm CO}$.

of exponentially generated scales, whose responses to external sources can be described by a pair of infinite spirals. As in the case of fermionic instability discussed in Sec. 13.3.5, one could interpret different Efimov states as arising from different species of bound states of operator $\mathcal{O}$; at a finite density each species of bound state wants to Bose condense in contrast to the fermionic case where bound particles form Fermi surfaces. The physical picture here is similar to the BEC regime in a strongly interacting ultracold Fermi system where fermions form bound molecules and then Bose condense. In such a situation one expects to see also an intermediate regime in which the system forms bound molecules, but not yet Bose condense. This intermediate stage has not been explicitly identified in the gravity side. It would be interesting to investigate it further.

Approaching from the uncondensed side, a bifurcating point is characterized by a finite uniform susceptibility with a branch point singularity. At the critical point the spectral function for $\mathcal{O}$ is logarithmically suppressed.

13.4.4. *Quantum critical behavior: a hybridized QCP*

In this section we examine the critical behavior around a hybridized QCP (13.158) following Refs. 45 and 46. Expanding $\tilde{a}_+(\omega, k)$ around $\omega = 0$, $k = 0$ and $\kappa_+ = \kappa_c$, we find that the full dynamical susceptibility (13.152)

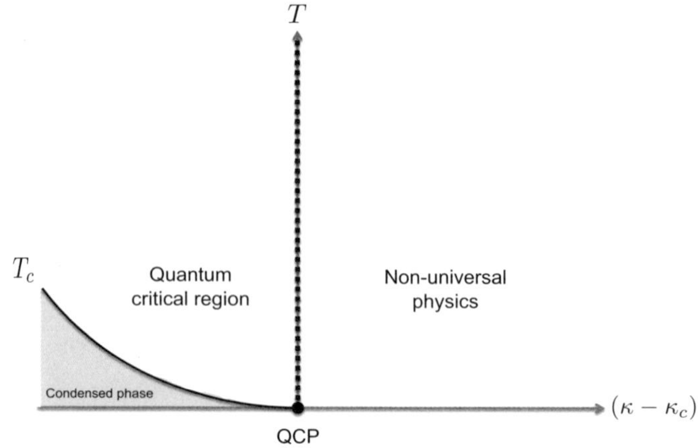

Fig. 13.23. Finite temperature phase diagram with the quantum critical region for marginal criticality at $\nu = 0$ and changing $(\kappa_c - \kappa_+)$. We only observe universal physics for $(\kappa_c - \kappa_+) < 0$, where the double-trace operator is marginally relevant.

can now be written as (for a neutral scalar)

$$\chi(\omega, \vec{k}) \approx \frac{1}{\kappa_+ - \kappa_c + h_k \vec{k}^2 - h_\omega \omega^2 + h\mathcal{G}_k(\omega)} \tag{13.189}$$

with

$$h_k = \left.\frac{\partial_{k^2} \tilde{a}_+^{(0)}(k)}{b_+^{(0)}(k)}\right|_{k=0,\kappa_+=\kappa_c}, \qquad h_\omega \equiv \left.\frac{\tilde{a}_+^{(2)}(k)}{b_+^{(0)}(k)}\right|_{k=0,\kappa_+=\kappa_c},$$
$$h \equiv \left.\frac{\mu_*^{-2\nu_k} \tilde{a}_-^{(0)}(k)}{b_+^{(0)}(k)}\right|_{k=0,\kappa_+=\kappa_c}. \tag{13.190}$$

Recall that the SLQL retarded function $\mathcal{G}_k(\omega) \propto \omega^{2\nu_k}$. Setting $\omega = 0$ in (13.189), the static susceptibility $\chi(k)$ can be written as

$$\chi(\vec{k}) \approx \frac{1}{(\kappa_+ - \kappa_c) + h_k \vec{k}^2} \tag{13.191}$$

which has the standard mean field behavior with the spatial correlation length scaling as

$$\xi \sim (\kappa_+ - \kappa_c)^{-\nu_{crit}}, \qquad \nu_{crit} = \frac{1}{2}\,. \tag{13.192}$$

and critical exponents (see Appendix A.4 for a review of definition of critical exponents)

$$\gamma = 1, \qquad \eta = 0\,. \tag{13.193}$$

At a finite temperature $T \ll \mu$, equation (13.189) generalizes at leading order in T/μ to (see discussion at the end of Sec. 13.2.4),

$$\chi(\omega, \vec{k}; T) \approx \frac{1}{\kappa_+ - \kappa_c + h_k \vec{k}^2 - h_\omega \omega^2 + h_T T + h \mathcal{G}_k^{(T)}(\omega)} \tag{13.194}$$

where $h_T T$ (h_T a constant) comes from (analytic) finite temperature corrections to a_+ and b_+. Finite temperature SLQL retarded function $\mathcal{G}_k^{(T)}(\omega)$ has the form $\mathcal{G}_k^{(T)}(\omega) = T^{2\nu_k} g_b(\frac{\omega}{T}, \nu_k)$ with g_b a universal scaling function (see (13.68) and (13.69)). From explicit gravity calculation one finds various constants in (13.194) have the following behavior: $h_k > 0$, $h < 0$, $h_\omega > 0$ (for $\nu_{k=0} > 1$), and $h_T > 0$ (for $\nu_{k=0} > \frac{1}{2}$).

The discussion in Sec. 13.2.5.2 leading to the low energy effective action (13.239) for the fermionic case can be applied almost identically to a scalar field. In particular, near a hybridized QCP (13.158), K in (13.91) can be expanded near $\omega = 0$ and $k = 0$ as (say for a neutral scalar)

$$K = \frac{\tilde{a}_+(\omega, k)}{b_+(\omega, k)} = \kappa_+ - \kappa_c + h_k \vec{k}^2 - h_\omega \omega^2 + \cdots . \tag{13.195}$$

We thus find that near a hybridized QCP, the low energy effective theory can be written as

$$S_{eff} = \tilde{S}_{SLQL}[\Phi] + \lambda \int dt\, \Phi \psi + S_{LG} + \int \psi J \tag{13.196}$$

where $\tilde{S}_{SLQL}$ is the same as (13.96) (except that Φ is now a scalar operator) and

$$S_{LG} = -\frac{1}{2} \int \psi K(k, \omega) \psi \tag{13.197}$$

can be considered as the quadratic part of the Landau–Ginsburg action for a scalar order parameter ψ near criticality. The full low energy action (13.196) describes the hybridization between a Landau-Ginsburg sector and a strongly coupled sector controlled by SLQL. We can further generalize (13.196) by including the next nonlinear term for ψ in (13.197), i.e.

$$S_{LG} = -\frac{1}{2} \int \psi K(k, \omega) \psi - u \int dt\, \psi^4 + \cdots \tag{13.198}$$

with u some positive constant.

We now proceed to study the scaling behavior of the system using (13.189), (13.194) and (13.196). We first consider the static and finite frequency behavior at zero temperature and then at finite temperature.

13.4.4.1. *Zero temperature: statics*

Let us first look at the the scaling of the expectation value and free energy. This can be done by analyzing the condensed solution on the gravity side.[45] Alternatively, one could use the low energy effective action (13.196) with (13.198), which provides much simplification. We will use the latter approach.

We will consider $k = 0$ and denote $\nu_{k=0}$ simply as ν. In terms of scaling in the IR, the strongly coupled operator Φ has a scaling dimension $\frac{1}{2} + \nu$ and from the second term in (13.196), ψ thus has dimension $\frac{1}{2} - \nu$. Similarly J has the same dimension $\frac{1}{2} + \nu$ as Φ. The spatial momentum k does not scale, thus its IR dimension is zero. Now let us consider ψ develops some nonzero expectation value. From the relative scaling dimensions between Φ and ψ, we can then write Φ as

$$\Phi \sim \psi^{\frac{\frac{1}{2}+\nu}{\frac{1}{2}-\nu}} \tag{13.199}$$

and the free energy density F corresponding to (13.196) can then be written as

$$F \sim C\psi^{\frac{1}{\frac{1}{2}-\nu}} + \frac{1}{2}(\kappa_+ - \kappa_c)\psi^2 + u\psi^4 \tag{13.200}$$

where the first term comes from the $\Phi\psi$ term with C some constant. Equation (13.200) can also be derived from a detailed bulk analysis (see[46]) which also gives that $C > 0$ for $\nu < \frac{1}{2}$. Now notice that for ψ small, the first term dominates over ψ^4 term if $\nu < \frac{1}{4}$, while the Landau-Ginsburg ψ^4 term dominates for $\nu > \frac{1}{4}$. In other words, since the first term is marginal by assignment, ψ^4 term becomes relevant when $\nu > \frac{1}{4}$.[jj]

For $\nu < \frac{1}{4}$ we can ignore the last term in (13.200) and for $\kappa_+ < \kappa_c$ find that

$$\langle \mathcal{O} \rangle \sim \psi \sim (\kappa_c - \kappa_+)^{\frac{\frac{1}{2}-\nu}{2\nu}} \tag{13.201}$$

and as a result

$$F \sim (\kappa_c - \kappa_+)^{\frac{1}{2\nu}} \,. \tag{13.202}$$

[jj]Some readers might worry that higher powers like ψ^6 may also become relevant at some point (for example for $\nu > \frac{1}{3}$). But note that once the last two terms in (13.200) dominate we should reassign the dimension of ψ and the standard Landau-Ginsburg story applies.

Including the source J, which has dimension $\frac{1}{2}+\nu$, the free energy should then be given by a scaling function

$$F = (\kappa_c - \kappa_+)^{\frac{1}{2\nu}} f_1\left(J(\kappa_c - \kappa_+)^{-\frac{\frac{1}{2}+\nu}{2\nu}}\right) = \xi^{-\frac{1}{\nu}} f_2\left(J\xi^{\frac{\frac{1}{2}+\nu}{\nu}}\right) \tag{13.203}$$

where in the second equality we have expressed the free energy in terms of correlation length using (13.192). From (13.203) we can also deduce that at the critical point we should have

$$\langle \mathcal{O} \rangle \sim \psi \sim J^{\frac{\frac{1}{2}-\nu}{\frac{1}{2}+\nu}} \,. \tag{13.204}$$

From (13.201), (13.204) and (13.203) we can collect the values of various scaling exponents (see Appendix A.4 for a review of exponents)

$$\alpha = 2 - \frac{1}{2\nu}, \qquad \beta = \frac{\frac{1}{2}-\nu}{2\nu}, \qquad \delta = \frac{\frac{1}{2}+\nu}{\frac{1}{2}-\nu} \tag{13.205}$$

For $\nu > \frac{1}{4}$, we can ignore the first term in (13.200) and the analysis becomes the standard Landau-Ginsburg one. As a result, the behavior near the critical point becomes those of mean field.[45] We thus find that for $\nu > \frac{1}{4}$,

$$\langle \mathcal{O} \rangle \sim \psi \sim (\kappa_c - \kappa_+)^{\frac{1}{2}} \qquad F \sim -(\kappa_c - \kappa_+)^2 \qquad \langle \mathcal{O} \rangle_{\kappa_+ = \kappa_c} \sim J^{\frac{1}{3}} \tag{13.206}$$

and various exponents become

$$\alpha = 0, \quad \beta = \frac{1}{2}, \quad \delta = 3 \tag{13.207}$$

which agree with the values of (13.205) for $\nu = \frac{1}{4}$.

13.4.4.2. *Finite frequency*

Let us now examine the full dynamical susceptibility (13.189), whose behavior depends on the competition between the analytic contribution $h_\omega \omega^2$ and the non-analytic contribution $\mathcal{G}_k(\omega)$ from eCFT$_1$.

When $\nu \in (0, 1)$, the non-analytic part dominates at low energies and the analytic contribution can be ignored, leading to

$$\chi(\omega, \vec{k}) \approx \frac{1}{(\kappa_+ - \kappa_c) + h_k k^2 + hC(\nu)(-i\omega)^{2\nu}} \tag{13.208}$$

with $C(\nu) < 0$ real for $\nu \in (0, 1)$. We will consider $k = 0$ below. At the critical point $\kappa_+ = \kappa_c$ we find that

$$\chi(\omega, k=0) \sim (-i\omega)^{-2\nu} \,. \tag{13.209}$$

Away from the critical point, the relative magnitude of the two terms (with $k = 0$) in the denominator of (13.208) defines a crossover energy scale

$$\Lambda_{\rm CO}^{(\omega)} \sim |\kappa_c - \kappa_+|^{\frac{1}{2\nu}} \ . \tag{13.210}$$

For $\omega \ll \Lambda_{\rm CO}^{(\omega)}$ we find that

$$\chi(\omega) \sim \frac{1}{\kappa - \kappa_c} + O(\omega^{2\nu}) \tag{13.211}$$

which is the typical behavior in the uncondensed phase, while for $\omega \gg \Lambda_{\rm CO}^{(\omega)}$ we recover the critical behavior (13.209). Thus the crossover scale (13.210) defines the correlation time ξ_τ of the system

$$\xi_\tau \sim \frac{1}{\Lambda_{\rm CO}^{(\omega)}} \sim |\kappa_c - \kappa_+|^{-\frac{1}{2\nu}} \ . \tag{13.212}$$

Comparing the above expression with (13.192) we then find that $\xi_\tau \sim \xi^z$ with the dynamical exponent z given by

$$z = \frac{1}{\nu} \ . \tag{13.213}$$

Of course this exponent can equivalently be seen by balancing the k^2 term and the $\omega^{2\nu}$ term in (13.208). Also note that when $\kappa_+ < \kappa_c$ equation (13.208) has a pole in the upper half plane (since $hC(\nu) > 0$) at

$$\omega_{\rm pole} \sim i\Lambda_{\rm CO}^{(\omega)}. \tag{13.214}$$

Note that on the one hand (13.208) (and its finite temperature generalization) has nontrivial scaling only in the time direction (i.e. only ω/T scaling), which appears to imply $z = \infty$. On the other hand we see the dynamics near the critical point does have a finite dynamical exponent given by (13.213).

When $\nu > 1$, in (13.189), the non-analytic part $\mathcal{G}_k(\omega) \sim \omega^{2\nu}$ from the semi-local sector becomes less important than the analytic corrections $\sim \omega^2$. Now the full dynamical susceptibility is given by

$$\chi(\omega, \vec{k}) \approx \frac{Z}{(\kappa_+ - \kappa_c) + h_k k^2 - h_\omega \omega^2 + hC(\nu)(-i\omega)^{2\nu}} \tag{13.215}$$

which describes a long-lived (nearly gapless) relativistic particle with a small width $\Gamma \sim \omega^{2\nu}$. The dynamical exponent is now $z = 1$. It is interesting that while the free energy already exhibits mean-field behavior for $\nu > \frac{1}{4}$, the dynamical quantity exhibits mean-field behavior only for $\nu > 1$.

This crossover to mean field dynamical behavior at $\nu = 1$ can also be readily seen from the effective action (13.196). For $\nu > 1$, the dimension for

ψ become smaller than $-\frac{1}{2}$, for which the kinetic term $(\partial_t\psi)^2$ (not written explicitly in (13.196)) for ψ becomes relevant and more important than the hybridization term $\Phi\psi$ (which is marginal by definition). Alternatively, we can now assign $-\frac{1}{2}$ as dimension of ψ using $(\partial_t\psi)^2$, under which the hybridization term $\Phi\psi$ will have dimension ν which becomes irrelevant for $\nu > 1$.

13.4.4.3. *Finite temperature*

Let us first look at the static uniform susceptibility at finite T, which is

$$\chi^{(T)} \sim \frac{1}{\kappa_+ - \kappa_c + h_T T + hC(\nu)T^{2\nu}} \,. \tag{13.216}$$

It is interesting that the analytic contribution now dominates for $\nu > \frac{1}{2}$. For $\nu < \frac{1}{2}$ we find that there is a pole at

$$T_c \sim (\kappa_c - \kappa_+)^{\frac{1}{2\nu}} \tag{13.217}$$

for $\kappa_+ < \kappa_c$. It should be interpreted at the critical temperature for a thermal phase transition, above which the instability disappears. From the uncondensed side, such a temperature scale gives the crossover scale to the quantum critical behavior. In this regime we have ω/T scaling with the critical temperature and the finite temperature crossover scale tracking that of zero temperature equation (13.210).

For $\nu > \frac{1}{2}$, we find instead mean field behavior

$$T_c \sim (\kappa_c - \kappa_+) \tag{13.218}$$

and the finite temperature crossover scale becomes

$$\Lambda_{\rm CO}^{(T)} \sim |\kappa_c - \kappa_+| \tag{13.219}$$

which no longer tracks that of zero temperature. In this regime, there is no ω/T scaling and the non-analytic frequency dependence from the SLQL becomes irrelevant compared to leading temperature effects.

This crossover at $\nu = \frac{1}{2}$ can again be readily seen from the effective action (13.196). Finite temperature generate a term $\int dt\, T\psi^2$, which becomes relevant when the dimension of ψ becomes smaller than zero, i.e. for $\nu > \frac{1}{2}$. Alternatively we can now use $T\psi^2$ term to assign dimension 0 to ψ, under which the hybridization term $\Phi\psi$ then becomes irrelevant for $\nu > \frac{1}{2}$ as now the dimension for Φ becomes larger than 1.

We summarize the critical behavior near a hybridized QCP for various values of ν in the following table

Quantity	$\nu \in (0, \frac{1}{4})$	$\nu \in (\frac{1}{4}, \frac{1}{2})$	$\nu \in (\frac{1}{2}, 1)$	$\nu > 1$
ω/T scaling	yes	yes	no	no
$\Lambda_{\rm CO}^{(\omega)}, z$	HYB	HYB	HYB	MFT
T_c	HYB	HYB	MFT	MFT
$\alpha,\ \beta,\ \delta$	HYB	MFT	MFT	MFT
$\gamma,\ \eta, \nu_{crit}$	MFT	MFT	MFT	MFT

In the above "MFT" denotes mean-field behavior and "HYB" (which is for hybridized) refers to the following non-mean-field exponents for zero-T crossover scale $\Lambda_{\rm CO}^{(\omega)}$, the dynamical exponent z and critical temperature T_c,

$$\Lambda_{\rm CO}^{(\omega)} \sim (\kappa_c - \kappa_+)^{\frac{1}{2\nu}}, \quad z = \frac{1}{\nu}, \quad T_c \sim (\kappa_c - \kappa_+)^{\frac{1}{2\nu}} \tag{13.220}$$

and

$$\alpha = 2 - \frac{1}{2\nu}, \quad \beta = \frac{\frac{1}{2} - \nu}{2\nu}, \quad \delta = \frac{\frac{1}{2} + \nu}{\frac{1}{2} - \nu}\,. \tag{13.221}$$

For all values of ν the static susceptibility from the uncondensed side is always given by the mean field behavior

$$\chi(\vec{k}) \approx \frac{1}{(\kappa_+ - \kappa_c) + h_k \vec{k}^2} \tag{13.222}$$

with $h_k > 0$ and the spatial correlation length

$$\xi \sim |\kappa_+ - \kappa_c|^{-\frac{1}{2}}\,. \tag{13.223}$$

Note in particular that for $\nu \in (0, \frac{1}{2})$ one finds an expression

$$\chi(\omega, \vec{k}; T) \approx \frac{1}{\kappa_+ - \kappa_c + h_k \vec{k}^2 + hT^{2\nu} g_b(\frac{\omega}{T})} \tag{13.224}$$

with g_b given by (13.69) (with $q_* = 0$). Equation (13.224) is of the form (13.17) for the dynamical susceptibility of $CeCu_{6-x}Au_x$ with a choice of $2\nu = 0.75$,[45] although the universal function g_b appears to be different from $f(\omega/T)$ there obtained by fitting experimental data.

13.4.5. *Doubly fine-tuning to a marginal critical point*

We now consider tuning $u \to 0$ and $\kappa_+ \to \kappa_c$ at the same time, at some critical value κ_c^* given by

$$\kappa_c^* = -\frac{\alpha}{\beta} \tag{13.225}$$

where we have used the $\nu \to 0$ limit of the expression for κ_c given in (13.157) and α, β are defined in (A.29)–(A.31). We will fix $u = 0$ and vary κ_+.

The dynamical susceptibility can be obtained from the $\kappa_+ = 0$ result at the bifurcating critical point (13.168) and (A.42)

$$\chi^{(\kappa_+)}(\omega, k) = \frac{\log\left(\frac{\omega}{\omega_b}\right) - i\frac{\pi}{2}}{\frac{\alpha}{\beta}\left(\log\left(\frac{\omega}{\omega_a}\right) - i\frac{\pi}{2}\right) + \kappa_+\left(\log\left(\frac{\omega}{\omega_b}\right) - i\frac{\pi}{2}\right)}. \tag{13.226}$$

Now using the definitions of $\omega_{a,b}$, applying the Wronskian relation (A.34), and expanding near $\kappa_+ \sim \kappa_c^* \equiv -\frac{\alpha}{\beta}$, we find:

$$\chi^{(\kappa_+)}(\omega, k) = \frac{\log\left(\frac{\omega}{\omega_b}\right) - i\frac{\pi}{2}}{(\kappa_+ - \kappa_c^*)\left(\log\left(\frac{\omega}{\omega_b}\right) - i\frac{\pi}{2}\right) - \frac{1}{2\nu_U \beta^2}}. \tag{13.227}$$

This contains a good deal of interesting physics. First let us keep $\kappa_+ > \kappa_c^*$ and study the very low-frequency behavior: expanding in powers of the inverse logarithm, we find

$$\chi^{(\kappa_+)}(\omega \to 0, k) \sim \frac{1}{\kappa_+ - \kappa_c^*} + \mathcal{O}\left(\frac{1}{\log \omega}\right). \tag{13.228}$$

Thus the static susceptibility diverges as $\kappa \to \kappa_c^*$, just as for the hybridized critical point discussed above. Away from the critical point the leading contribution to the spectral density appears at order $\log^{-2}(\omega)$,

$$\operatorname{Im} \chi^{(\kappa_+)}(\omega \to 0, k) \sim \frac{\pi}{4(\kappa_+ - \kappa_c^*)\nu_U^2} \log^{-2}\left(\frac{\omega}{\omega_b}\right) + \mathcal{O}\left(\frac{1}{\log^3 \omega}\right). \tag{13.229}$$

This is the same as the corresponding result close to a bifurcating critical point with zero double-trace deformation.

Next, we note that (13.227) contains a pole in the upper half plane at

$$\omega_* = i\omega_b \exp\left(\frac{1}{2\nu_U \beta^2} \frac{1}{\kappa_+ - \kappa_c^*}\right). \tag{13.230}$$

Note however that if $\kappa_+ > \kappa_c^*$, the pole is at an exponentially high energy scale, this low-frequency formula breaks down far before the pole, and the pole is not physical. This is the stable phase. However, if $\kappa_+ < \kappa_c^*$, then close to the critical point the pole is very close to the origin and this formula is valid. The instability pole is now physical, and we see that a new low-energy scale has been exponentially generated. This is reminiscent of the effect of a marginal coupling growing strong in the infrared, as in QCD or the BCS instability. Indeed as discussed in (13.78) and (13.79), we expect in the IR a double trace deformation to SLQL will be generated. At $u = 0 = k$, the double trace deformation

$$\delta S = \frac{\kappa_m^{(IR)}}{2} \int dt\, \Phi^2, \tag{13.231}$$

has dimension 1 and is thus *marginal*. It can be checked explicitly that the value of the effective marginal double-trace parameter $\kappa_m^{(IR)}$ changes sign at $\kappa_+ = \kappa_c^*$ (for details see[46]). Thus what we have found above is that if $\kappa_+ > \kappa_c^*$ then the coupling is marginally irrelevant and has no effect in the IR, whereas if $\kappa_+ < \kappa_c^*$ it is marginally *relevant* and drives the theory to a new fixed point (i.e. the scalar will condense).

Similarly, one can compute the corresponding finite temperature correlator; a precisely analogous calculation gives (see[46] for derivation)

$$\chi^{(\kappa_+)}(\omega, k; T) = \frac{\log\left(\frac{2\pi T}{\omega_b}\right) + \psi\left(\frac{1}{2} - i\frac{\omega}{2\pi T}\right)}{(\kappa_+ - \kappa_c^*)\left(\log\left(\frac{2\pi T}{\omega_b}\right) + \psi\left(\frac{1}{2} - i\frac{\omega}{2\pi T}\right)\right) - \frac{1}{2\nu_U \beta^2}} \tag{13.232}$$

where $\psi(x)$ is the digamma function. From here we can see that for $\kappa_+ < \kappa_c^*$ the static susceptibility diverges at the critical temperature

$$T_c = T_b \exp\left(\frac{1}{2\nu_U \beta^2}\frac{1}{\kappa_+ - \kappa_c^*}\right) \tag{13.233}$$

above which the system is stable. The temperature is set by the same dynamically generated scale.

Now consider the fluctuation spectrum at criticality. Returning to $T = 0$, we now keep $\omega \neq 0$ and sit precisely at the critical point $\kappa = \kappa_c^*$. From (13.227) we now find

$$\chi^{(\kappa_+=\kappa_c^*)}(\omega, k) = -2\nu_U \beta^2 \left(\log\left(\frac{\omega}{\omega_b}\right) - i\frac{\pi}{2}\right). \tag{13.234}$$

The appearance of a pure logarithm in the *numerator* of this propagator at criticality is interesting. Relatedly, at criticality the spectral density

$$\operatorname{Im}\chi^{(\kappa_+=\kappa_c^*)} = \pi\nu_U \beta^2 \operatorname{sgn}(\omega) \tag{13.235}$$

is a pure step function.[kk] This should be contrasted with the situation away from criticality, (13.229), in which there is a logarithmic suppression

[kk]The logarithm jumps by $i\pi$ as we cross through $\omega = 0$, resulting in the step function; note that this was necessary in order to maintain the relation $\omega \operatorname{Im}\chi(\omega) > 0$, true for any bosonic spectral density.

at low frequencies; as expected, by sitting at criticality we find significantly more low frequency modes.

Similarly, setting $\kappa \to \kappa_c^*$ in the finite temperature expression (13.232) we find

$$\chi^{(\kappa_+=\kappa_c^*)}(\omega, k; T) = -2\nu_U \beta^2 \left(\log\left(\frac{2\pi T}{\omega_b}\right) + \psi\left(\frac{1}{2} - i\frac{\omega}{2\pi T}\right)\right). \quad (13.236)$$

Taking the imaginary part and using the identity[11] $\operatorname{Im}\psi\left(\frac{1}{2} + ix\right) = \frac{\pi}{2}\tanh(x)$ we find the expression

$$\operatorname{Im}\chi^{(\kappa_+=\kappa_c^*)}(\omega, k; T) = \pi\nu_U \beta^2 \tanh\left(\frac{\omega}{2\pi T}\right) \quad (13.237)$$

which is simply a smoothed-out version of the step function that we find at zero temperature.

Equation (13.237) implies that

$$\operatorname{Im}\chi^{(\kappa_+=\kappa_c^*)}(\omega, k; T) \sim \begin{cases} \frac{\omega}{T} & \omega \ll T \\ \operatorname{sgn}(\omega) & \omega \gg T \end{cases} \quad (13.238)$$

which is precisely of the form for spin and charge fluctuations in the phenomenological "Marginal Fermi liquid" description of High-T_c cuprates in the strange metal region (as discussed around equation (13.18) in Sec. 13.1.3). Thus the marginal critical point can be viewed as a concrete realization of the bosonic fluctuation spectrum needed to support a Marginal Fermi liquid.

13.5. Discussion and Conclusions

In these lectures, we used gauge/gravity duality to study a class of finite density systems, which are described on the gravity side by a charged black hole. We showed that the duality predicts a universal intermediate-energy phase, called a *semi-local quantum liquid* (SLQL). Such an unstable phase is characterized by a finite spatial correlation length, but an infinite correlation *time* and associated nontrivial scaling behavior in the time direction, as well as a nonzero entropy density. We discussed two sets of phenomena related to SLQL:

(1) Gapless fermionic or bosonic degrees of freedom (described by meant field) hybridized with degrees of freedom in SLQL. In either case, the system can be described by the following low energy

[11]This can be proved using the reflection formula $\psi(1-x) - \psi(x) = \pi\cot(\pi x)$.

effective theory

$$S_{eff} = \tilde{S}_{SLQL}[\Phi] + \int \lambda(k,\omega)\Phi_{-\vec{k}}\Psi_{\vec{k}} + S_{\text{mean field}}[\Psi] \, . \quad (13.239)$$

In the fermionic case, Ψ denotes excitations around a Fermi surface with $S_{\text{mean field}}[\Psi]$ given by that for a free fermion (plus possibly higher order self-interactions). In the bosonic case, Ψ denotes the order parameter around a quantum critical point with $S_{\text{mean field}}[\Psi]$ its Landau-Ginsburg effective theory. In both cases, Φ is some (fermionic or bosonic) operator whose dynamics is controlled by the strongly coupled SLQL. The IR behavior of the theory (i.e. whether one finds a Fermi surface with or without quasiparticles, or a quantum phase transition which obeys locally quantum critical or mean-field behavior) depends crucially on the scaling dimension of Φ in the SLQL. With a suitable choice of the SLQL scaling dimension, IR phenomena which strongly resemble those of high-T_c cuprates in the strange metal phase (including a linear resistivity) and that for $CeCu_{6-x}Au_x$ near a quantum critical point were found.

On the gravity side the gapless degrees of freedom are associated with the appearance of certain scalar or fermionic hair in the black hole geometry, and lie outside the AdS_2 region which describes the SLQL.

(2) Instabilities of the SLQL itself. We discussed examples in which SLQL orders at lower energies to a Fermi liquid, superconductor, or antiferromagnetic-type state. In all cases certain operators develop a complex scaling dimension in the SLQL generating a tower of bound states. In the fermion case, each of these bound states forms a Fermi surface. In the scalar case, they form a tower of Bose-Einstein condensed states. In particular, near the bifurcating critical point where the scalar instability sets in one finds a finite susceptibility and the response is described by an infinite spiral. In all of these examples the lower energy states have no zero-temperature entropy density and SLQL may be considered as a "deconfined" state for them in which the composite bound states deconfine and fractionalize into more fundamental degrees of freedom. This naturally explains the nonzero entropy density for SLQL. As discussed in Sec. 13.2.6, we expect this picture to be generic; while the precise nature of the lower energy

state depends on the specific dynamics of the individual system, the SLQL emerges universally at intermediate energies through *fractionalization*.

Instabilities of SLQL in turn provide novel mechanisms for the emergence of Fermi liquids, superconductors and AFM-type of states at low energies.

On the gravity side, for charged operators the instabilities of the SLQL manifest themselves as pair production of charged particles which then subsequently backreact on the geometry. For a neutral scalar operator, the instability is related to the violation of the BF bound in the AdS_2 region.

It is natural to ask whether some phase similar to the SLQL could underlie some known strongly correlated condensed matter systems, and in particular, whether the observed non-Fermi liquid behavior and novel superconductivities in various systems could be attributed to a similar intermediate-energy phase.

When one encounters scaling behavior in an observable, an important immediate question is whether the behavior is due to intermediate-energy or vacuum effects. Here we provide explicit example in which that Marginal Fermi Liquid behavior – such as that proposed for high-T_c cuprates in the strange metal phase (including linear resistivity) – can arise from intermediate energy effects. Local quantum critical behavior similar to that for $CeCu_{6-x}Au_x$ near a quantum critical point also appears here as due to intermediate energy effects.

In many heavy electron systems, while quantum fluctuations from the quantum critical point corresponding to the onset of magnetism provide a natural starting point for understanding the observed non-Fermi liquid behavior and sometimes novel superconductivity, however, such exotic behavior does not always arise in the proximity of a quantum critical point (see e.g. Ref. 10). There thus appears plenty of room for them to arise from an intermediate-energy phase.

More generally, we expect that candidates for the SLQL to occur include systems which exhibit frustrated or competing interaction terms in their Hamiltonian. Such systems can have a large number of near-degenerate states near the vacuum, similarly to the holographic systems considered here. Also systems which involve strong competition between tendencies towards itinerancy and localization could exhibit the semi-local behavior found here. As mentioned earlier, there are also some tantalizing parallels between properties of the SLQL phase and models from dynamical mean field theory (DMFT),[32] such as scaling in the time direction with spatial directions as spectators, and a finite entropy. The fact that DMFT techniques have been very successful in treating many materials in some intermediate energy region provides some comfort in thinking that an underlying universal intermediate-energy phase like the SLQL could be at work.

Acknowledgments

We thank T. Faulkner, J. McGreevy, D. Vegh, and Q. Si for collaborations. We also want to thank participants of TASI 2010, KITPC workshop for "AdS/CM and other approaches," and IIP school on "Holographic View of Condensed Matter Physics" for many questions and discussions, which helped shape and sharpen these lectures. We also want to thank K. Balasubramanian, H. Ebrahim, T. Grover, G. Kotliar, J. Polchinski, S. Sachdev and T. Senthil for useful discussions. Work supported in part by funds provided by the U.S. Department of Energy (D.O.E.) under cooperative research agreement DE-FG0205ER41360.

A.1. Operator Dimension and Retarded Functions in AdS_2

Here we give a derivation of the scaling dimension (13.56)-(13.57) and retarded Green's function $\mathcal{G}_k(\omega)$ (13.59) for a charged scalar in $AdS_2 \times \mathbb{R}^{d-1}$. The background metric and gauge field is given by (13.49) which we copy here for convenience

$$ds^2 = \frac{R_2^2}{\zeta^2}(-dt^2 + d\zeta^2) + \mu_*^2 R^2 d\vec{x}^2, \qquad A_t = \frac{e_d}{\zeta} . \tag{A.1}$$

We consider the following quadratic scalar action

$$S = -\int d^{d+1}x\sqrt{-g}\,\left[g^{MN}(\partial_M + iqA_M)\phi^*(\partial_N - iqA_N)\phi + m^2\phi^*\phi\right] \tag{A.2}$$

Expanding ϕ in terms of Fourier modes in spatial directions,

$$\phi(t,\vec{x},\zeta) = \int \frac{d\vec{k}}{(2\pi)^d}\, e^{i\vec{k}\cdot\vec{x}}\, \phi_{\vec{k}}(t,\zeta) \tag{A.3}$$

equation (A.2) reduces to an action for $\phi_{\vec{k}}$ in AdS_2

$$S = -\int d^2x\sqrt{-g}\,\left[g^{ab}(\partial_a + iqA_a)\phi_{\vec{k}}^*(\partial_b - iqA_b)\phi_{\vec{k}} + m_{\vec{k}}^2\phi_{\vec{k}}^*\phi_{\vec{k}}\right] \tag{A.4}$$

with a mass square

$$m_k^2 \equiv \frac{k^2}{\mu_*^2 R^2} + m^2, \qquad k^2 = |\vec{k}^2| \,. \tag{A.5}$$

The indices a, b only run over t, ζ. Further going to frequency space one finds that equation of motion from (A.4) can be written as (for notational simplicity we will just denote $\phi_{\vec{k}}(\omega, \zeta)$ as ϕ below)

$$-\partial_\zeta^2 \phi + V(\zeta)\phi = 0 \tag{A.6}$$

with

$$V(\zeta) = \frac{m_k^2 R_2^2}{\zeta^2} - \left(\omega + \frac{q_*}{\zeta}\right)^2 = -\omega^2 - \frac{2\omega q_*}{\zeta} + \frac{\nu_k^2 - \frac{1}{4}}{\zeta^2} \tag{A.7}$$

where

$$\nu_k \equiv \sqrt{m_k^2 R_2^2 - q_*^2 + \frac{1}{4}} = \sqrt{m^2 R_2^2 - q_*^2 + \frac{1}{4} + \frac{k^2 R_2^2}{\mu_*^2 R^2}}, \qquad q_* \equiv q e_d \,. \tag{A.8}$$

Approaching the AdS_2 boundary $\zeta \to 0$, equation (A.6) has general solution

$$\phi_{\vec{k}}(\omega, \zeta \to 0) \sim \alpha \zeta^{\frac{1}{2} - \nu_k} + \beta \zeta^{\frac{1}{2} + \nu_k} \,. \tag{A.9}$$

Compare (A.9) to (13.26); it implies that the conformal dimension of the operator $\Phi_{\vec{k}}$ to $\phi_{\vec{k}}$ has dimension $\delta_k = \frac{1}{2} + \nu_k$, as claimed in (13.56) and (13.57).

Introducing $x = -2i\omega\zeta$, equation (A.6) can be written as

$$\partial_x^2 \phi + \left(-\frac{1}{4} + \frac{iq_*}{x} + \frac{\frac{1}{4} - \nu_k^2}{x^2}\right)\phi = 0 \tag{A.10}$$

which is precisely the Whittaker's equation and the two linearly independent solutions can be written as

$$\phi = c_{\text{in}} \mathrm{W}_{iq_*, \nu_k}(x) + c_{\text{out}} \mathrm{W}_{-iq_*, \nu_k}(-x) \,, \tag{A.11}$$

where $\mathrm{W}_{\lambda,\mu}(z)$ is the Whittaker function. Of these, the function multiplying c_{in},

$$\mathrm{W}_{iq_*, \nu_k}(-2i\omega\zeta) = e^{i\omega\zeta}(-2i\omega\zeta)^{iq_*} + \cdots, \qquad \zeta \to +\infty \tag{A.12}$$

is ingoing at the horizon. Keeping only the ingoing part, the asymptotic behavior of ϕ at $\zeta \to 0$ is given by

$$\phi = \frac{\Gamma(2\nu_k)}{\Gamma(\frac{1}{2} + \nu_k - iq_*)}(-2i\omega\zeta)^{\frac{1}{2} - \nu_k} + \frac{\Gamma(-2\nu_k)}{\Gamma(\frac{1}{2} - \nu_k - iq_*)}(-2i\omega\zeta)^{\frac{1}{2} + \nu_k} \,. \tag{A.13}$$

From (13.30) the retarded function is simply the ratio of two coefficients in the above equation, leading to

$$\mathcal{G}_k(\omega) = (-2i\omega)^{2\nu_k} \frac{\Gamma(-2\nu_k)\Gamma\left(\frac{1}{2}+\nu_k - iq_*\right)}{\Gamma(2\nu_k)\Gamma\left(\frac{1}{2}-\nu_k - iq_*\right)} \tag{A.14}$$

which is (13.59).

A.2. Derivation of the Master Formula

Here we present the derivation of the master formula (13.75), following Ref. 39. The equation of motion (in momentum space) for ϕ with action (A.2) in the charged black hole geometry (13.33) can be written as

$$z^{d+1}\partial_z\left(\frac{f}{z^{d-1}}\partial_z\phi\right) + z^2\left(\frac{(\omega+qA_t)^2}{f} - k^2\right)\phi - m^2R^2\phi = 0\,. \tag{A.15}$$

We will consider $T=0$ with f and A_t given by (13.45). Recall from (13.26) that near the AdS_{d+1} boundary $\phi(z)$ has the standard asymptotic behavior $\phi(z\to 0) \sim Az^{d-\Delta_+} + Bz^{\Delta_+}$ and that the correlator we seek to compute can be written as[mm]

$$G_R(\omega,\vec{k}) = \frac{B}{A} \tag{A.16}$$

provided that ϕ is an in-falling wave at the horizon. Note that since $A_t \to \mu$ as $z\to 0$, the boundary theory energy corresponds to $\omega + q\mu$, i.e. ω should be interpreted as the measured from the effective chemical potential $q\mu$ (which we take to be positive for definiteness).

We are interested in the behavior of G_R in the low frequency limit ($\omega \ll \mu$). However, we cannot directly perform a perturbative expansion in ω in (A.15) since the ω-dependent term becomes singular and dominates at the horizon where $f(z_*) \to 0$, no matter how small ω is. This is not just a technicality; physically this reflects the abundance of critical modes coming from the IR fixed point and is of tremendous importance.

To deal with it, it is useful to isolate the near-horizon $AdS_2 \times \mathbb{R}^{d-1}$ region (13.49) (which will be referred to as the IR region) from the rest of the black hole spacetime (which will be referred to as the UV region). In the IR AdS_2 region the ω dependence will be treated non-perturbatively. The crossover between the two regions happens near the boundary of AdS_2 with $\frac{1}{\mu} \ll \zeta \ll \frac{1}{\omega}$, where $\mu\zeta \sim \frac{z_*}{z_*-z}$ remains big so that we are still in

[mm]We will consider the standard quantization in our discussion below. The generalization to the alternative quantization is immediate.

the near-horizon region, but the ω-dependent term $z_*^2 \frac{\omega^2}{f} \sim \omega^2 \zeta^2$ in (A.15) becomes small. In the UV region we can treat the small ω limit using a standard perturbative expansion with the lowest order equation given by setting $\omega = 0$ in (A.15).

So we start with a solution in the $AdS_2 \times \mathbb{R}^{d-1}$ region and evolve it outwards, matching it to a solution in the UV region to determine the coefficients A and B. The scalar wave equation on the $AdS_2 \times \mathbb{R}^{d-1}$ region (13.49) is given by (A.6). As we are computing a retarded correlator we are looking for an *in-falling* solution at the black hole horizon (i.e. the term proportional to c_{in} in (A.11) even though the explicit form is not important for the current discussion). We now take this infalling solution and evolve it to the boundary of the AdS_2 region, where it can be written as

$$\phi(\zeta) = \zeta^{\frac{1}{2}-\nu_k} + \mathcal{G}_k(\omega)\zeta^{\frac{1}{2}+\nu_k}. \tag{A.17}$$

In (A.17) we have normalized the solution so that the first term has coefficient 1 and then by definition the coefficient of the second term is precisely the retarded Green function in the AdS_2 region we computed earlier in (A.14).

We now need to continue evolution all the way to the asymptotically AdS_{d+1} region to extract the coefficients A and B. We thus need to match the IR solution (A.17) to a solution in the UV region. Now that we are out of the dangerous IR region, to leading order we can set $\omega = 0$ in (A.15) in the UV region,

$$z^{d+1}\partial_z\left(\frac{f}{z^{d-1}}\partial_z\phi\right) + z^2\left(\frac{q^2A_t^2}{f} - k^2\right)\phi - m^2R^2\phi = 0\ . \tag{A.18}$$

The above equation has two independent solutions $\eta_\pm^{(0)}$ which can be specified by their behavior near $z \to z_*$ as

$$\eta_\pm^{(0)}(z) \to \left(\frac{\zeta}{z_*}\right)^{\frac{1}{2}\mp\nu_k}, \qquad z \to z_* \tag{A.19}$$

with the corresponding asymptotic behavior as $z \to 0$ as

$$\eta_\pm^{(0)}(z) \approx a_\pm^{(0)}(k)\left(\frac{z}{z_*}\right)^{d-\Delta_+} + b_\pm^{(0)}(k)\left(\frac{z}{z_*}\right)^{\Delta_+}\ . \tag{A.20}$$

$a_\pm^{(0)}(k)$ and $b_\pm^{(0)}(k)$ thus defined are (dimensionless) functions of k which can be computed numerically.

In the overlapping region both (A.17) and (A.19) apply, which determines the full UV solution to be

$$\phi(z) = \eta_+^{(0)}(z) + \mathcal{G}_k(\omega) z_*^{2\nu_k} \eta_-^{(0)}(z) \,. \tag{A.21}$$

Equation (A.21) can be generalized to higher orders in ω (for details see Ref. 39)

$$\phi(z) = \eta_+(z) + \mathcal{G}_k(\omega) z_*^{2\nu_k} \eta_-(z) \tag{A.22}$$

where

$$\eta_\pm = \eta_\pm^{(0)} + \omega \eta_\pm^{(1)} + O(\omega^2) \tag{A.23}$$

are the two linearly independent perturbative solutions to the full UV region equation. Near $z = 0$, $\eta_\pm$ have the expansion of the form (A.20) with various coefficients $a_\pm^{(0)}, b_\pm^{(0)}$ replaced by $a_\pm, b_\pm$ which have an analytic ω-expansion such as

$$a_+(k,\omega) = a_+^{(0)}(k) + \omega a_+^{(1)}(k) + \cdots \,. \tag{A.24}$$

Note that since both the boundary conditions specifying $\eta_\pm$ (A.19) and the equation (A.15) are real, $a_\pm, b_\pm$ are real. From (A.22) and the expansion of $\eta_\pm$ near $z = 0$ we thus find the boundary theory Green's function to be

$$\boxed{G_R(\omega, \vec{k}) = \mu_*^{2\Delta - d} \frac{b_+(\omega,k) + b_-(\omega,k)\mathcal{G}_k(\omega)\mu_*^{-2\nu_k}}{a_+(\omega,k) + a_-(\omega,k)\mathcal{G}_k(\omega)\mu_*^{-2\nu_k}}} \,. \tag{A.25}$$

We conclude this discussion with some remarks:

(1) At $\omega = 0$, the leading order equation (A.18) in the outer region is in fact the full equation of motion; there is no IR region. $\eta_\pm^{(0)}$ of (A.19) now provide a set of basis for the full equation of motion. Note that by construction $\eta_+^{(0)}$ is *normalizable* at the horizon. When $a_+^{(0)}$ vanishes, from (A.20) $\eta_+^{(0)}$ becomes normalizable also at the boundary and thus is now a genuine normalizable mode in the black hole geometry.

(2) Note that for a neutral scalar with $q = 0$, equation (A.15) only depends on ω^2 and the expansion parameter in (A.23) and (A.24) should be ω^2, i.e.

$$a_+(k,\omega) = a_+^{(0)}(k) + \omega^2 a_+^{(2)}(k) + O(\omega^4) \tag{A.26}$$

and so on.

(3) The functions $a_\pm(\omega, k)$, $b_\pm(\omega, k)$ are obtained by solving equation (A.15) perturbatively in ω in the UV region. Their k-dependence comes from two sources, from dependence on ν_k via the boundary condition (A.19) and from k^2 dependence in the equation (A.15) itself. Since the geometry is smooth throughout the UV region we expect the dependence on both ν_k and k^2 to be analytic. In fact we can think of $b_\pm$ and $a_\pm$ as *functions* of ν_k; i.e. there exists a function $b(\nu_k, k^2, \omega)$, *analytic in all its arguments*, such that $b_\pm = b(\pm\nu_k, k^2, \omega)$. This is clear from the boundary condition (A.19) (and its generalization for higher orders in ω) and from the fact that there is no other dependence on ν_k from the equation of motion itself.

(4) As $\nu_k \to 0$, the two solutions in (A.19) become degenerate, i.e.

$$a_+(k,\omega) \to a_-(k,\omega), \qquad b_+(k,\omega) \to b_-(k,\omega)\ . \tag{A.27}$$

At $\nu_k = 0$, the basis of functions in (A.19) should be replaced by

$$\eta^{(0)}(z) = \left(\frac{\zeta}{z_*}\right)^{\frac{1}{2}}, \qquad \tilde{\eta}^{(0)}(z) = -\left(\frac{\zeta}{z_*}\right)^{\frac{1}{2}} \log\frac{\zeta}{z_*} \tag{A.28}$$

where the asymptotic behavior for them at $z \to 0$ is

$$\eta^{(0)}(z) \approx \alpha\left(\frac{z}{z_*}\right)^{3-\Delta} + \beta\left(\frac{z}{z_*}\right)^{\Delta}, \quad \tilde{\eta}^{(0)}(z) \approx \tilde{\alpha}\left(\frac{z}{z_*}\right)^{3-\Delta} + \tilde{\beta}\left(\frac{z}{z_*}\right)^{\Delta} \tag{A.29}$$

$\alpha, \beta, \tilde{\alpha}, \tilde{\beta}$ are now dimensionless real numbers which can again be found numerically. More explicitly, in the limit of $\nu_k \to 0$ the basis of functions (A.19) can be expanded as

$$\eta_\pm^{(0)} = \eta^{(0)}(z) \pm \nu_k \tilde{\eta}^{(0)}(z) + O(\nu_k^2) \tag{A.30}$$

which leads to

$$b_\pm^{(0)} = \beta \pm \nu_k \tilde{\beta} + \cdots, \qquad a_\pm^{(0)} = \alpha \pm \nu_k \tilde{\alpha} + \cdots\ . \tag{A.31}$$

(5) Coefficients $a_\pm, b_\pm$ are not independent. For example evaluating the Wronskian of (A.15) (for $\omega = 0$)[nn] for $\eta_\pm^{(0)}$ and demanding

[nn] The Wronskian of equation (A.15) is given by

$$W[\phi_1, \phi_2] = \frac{f}{z^2}(\phi_1 \partial_z \phi_2 - \phi_2 \partial_z \phi_1) \tag{A.32}$$

which is independent of z.

that it be equal at infinity and at the horizon, we find the elegant relation:

$$a_+^{(0)}(k)b_-^{(0)}(k) - a_-^{(0)}(k)b_+^{(0)}(k) = \frac{\nu_k}{\nu_U} \,. \tag{A.33}$$

A similar analysis on $\eta, \tilde{\eta}$ results in

$$\alpha\tilde{\beta} - \beta\tilde{\alpha} = -\frac{1}{2\nu_U} \,. \tag{A.34}$$

(6) When $\nu_k = -i\lambda_k$ is pure imaginary (i.e. when (13.98) is satisfied), and the basis of solutions (A.19) now has the form

$$\eta_\pm^{(0)}(z) \to \left(\frac{\zeta}{z_*}\right)^{\frac{1}{2}\pm i\lambda_k}, \qquad z \to z_* \tag{A.35}$$

These boundary conditions are now complex, and thus so are the $\eta_\pm$. As the $\eta_\pm$ actually obey a real wave equation, the full analytic structure is determined by the boundary conditions in the infrared; thus we find that now $\eta_+ = \eta_-^*$. This also implies that $a_\pm, b_\pm$ are complex and

$$a_+^* = a_-, \qquad b_+^* = b_- \,. \tag{A.36}$$

(7) The derivation leading to (A.25) can be immediately generalized to a finite temperature $T \ll \mu$. In the IR region one replaces the equation (A.6) by the corresponding equation for the AdS_2 black hole (13.52). As a result, $\mathcal{G}_k(\omega)$ in (A.17) is now replaced its finite-temperature generalization $\mathcal{G}_k^{(T)}(\omega)$ (13.68). The outer region discussion is exactly the same as before except that one should use finite temperature counterparts of f and A_t in (A.18). For $T \ll \mu$, the outer region equation depends on T analytically and the asymptotic behavior in equations (A.19) and (A.20) is unchanged. Thus for $T \ll \mu$, one simply replaces $\mathcal{G}_k(\omega)$ in (A.25) by $\mathcal{G}_k^{(T)}(\omega)$, but keeping in mind that now $a_\pm, b_\pm$ also have analytic temperature dependence.

A.3. Double Trace Deformation

Consider a large N gauge theory in Euclidean signature deformed by a double-trace operator

$$\delta S = -\frac{1}{2}\kappa \int d^d x\, \mathcal{O}^2 \tag{A.37}$$

where $\mathcal{O}$ is a scalar operator whose Euclidean correlation function (in the absence of deformation) is given by $G(x)$. With this deformation the two point function for $\mathcal{O}$ now becomes:

$$G_\kappa = \frac{1}{Z_\kappa}\left\langle \mathcal{O}(x)\mathcal{O}(0)\, e^{-\frac{1}{2}\kappa\int d^dy\, \mathcal{O}(y)^2}\right\rangle,$$
$$Z_\kappa = \left\langle e^{-\frac{1}{2}\kappa\int d^dy\, \mathcal{O}(y)^2}\right\rangle \tag{A.38}$$

which leads to

$$G_\kappa(x) = \frac{1}{Z_\kappa}\sum_{n=0}^{\infty}\frac{(-\kappa)^n}{2^n n!}\left(\prod_{m=1}^{n}\int d^d y_m\right)\langle \mathcal{O}(x)\mathcal{O}(0)\mathcal{O}(y_1)^2\cdots\mathcal{O}(y_n)^2\rangle. \tag{A.39}$$

The disconnected diagrams cancel between the numerator and denominator, leaving us with only connected diagrams. To leading order in the large N limit, the n-th term in the above equation becomes

$$(-\kappa)^n\int d^dy_1\cdots d^dy_n\, G(x-y_1)\cdots G(y_{n-1}-y_n)G(y_n)\ . \tag{A.40}$$

In momentum space we thus find a simple geometric sum

$$G_\kappa(k) = \sum_{n=0}^{n}(-\kappa)^n G^{n+1}(k) = \frac{G(k)}{1+\kappa\, G(k)} = \frac{1}{G^{-1}(k)+\kappa}\ . \tag{A.41}$$

Analytically continuing (A.41) to the Lorentzian signature and using the standard relation between the Euclidean and retarded functions we conclude that

$$G_R^{(\kappa)}(\omega,\vec{k}) = \frac{1}{G_R^{-1}(\omega,\vec{k})+\kappa} \tag{A.42}$$

where $G_R(\omega,\vec{k})$ and $G_R^{(\kappa)}(\omega,\vec{k})$ are retarded functions for $\mathcal{O}$ before and after the double trace deformation.

For a CFT, the Euclidean correlator $G(k)$ in momentum space for an operator $\mathcal{O}$ can be written as

$$G(k) = C(\Delta)k^{2\Delta-d} \tag{A.43}$$

where where Δ is the scaling dimension of $\mathcal{O}$ and $C(\Delta)$ is some constant factor. Plugging the above equation into (A.41) we find that for $\Delta > \frac{d}{2}$,

$$G_\kappa(k)\to G(k), \qquad k\to 0\ . \tag{A.44}$$

This is expected, since for $\Delta > \frac{d}{2}$, the double trace deformation (A.37) is irrelevant and the theory flows back to the original fixed point in the IR.

When $\Delta < \frac{d}{2}$, for which (A.37) is relevant, we find that in the IR

$$G_\kappa(k) \to \frac{1}{\kappa} - \frac{1}{\kappa^2} G^{-1}(k) + \cdots , \qquad k \to 0 . \tag{A.45}$$

The first term in the above equation is a constant (which gives rise to a contact term in coordinate space) and can be ignored. The second term corresponds to an operator of dimension $d - \Delta$. We thus conclude that under the double trace deformation the system flows to a new CFT in the IR in which $\mathcal{O}$ has dimension $d - \Delta$. Note that two-point functions of other operators (which are orthogonal to $\mathcal{O}$) are unaffected.

In the bullet below that of equation (13.30) in Sec.(13.2.1) we mentioned that for $\nu_U \in (0, 1)$, there are two ways to quantize a scalar field ϕ in AdS. In the standard quantization (Dirichlet boundary condition) the corresponding boundary operator $\mathcal{O}_+$ has dimension $\Delta_+ = \frac{d}{2} + \nu_U$, while in the alternative quantization (Neumann boundary condition), the corresponding boundary operator $\mathcal{O}_-$ has dimension $\Delta_- = \frac{d}{2} - \nu_U$. From our discussion above, upon turing on a double trace operator $\mathcal{O}_-^2$ in the alternative quantization, in the large N limit the system should flow in the IR to a new fixed point in which $\mathcal{O}_-$ has dimension $d - \Delta_- = \Delta_+$ with the dimensions of all other operators unchanged. Clearly this new fixed point should be identified with that corresponding to the standard quantization.

A.4. Review of Critical Exponents

In the vicinity of a critical point we observe scaling behavior of various observable quantities, which is characterized by a set of critical exponents. We list some of the most commonly used exponents in the following. We will denote the external tuning parameter g with which we tune the system to the critical point $g = g_c$. Near the critical point the spatial correlation length diverges as

$$\xi \sim |g - g_c|^{-\nu_{crit}} . \tag{A.46}$$

The energy gap for elementary excitations scales as

$$E_{gap} \sim \xi^{-z} \sim |g - g_c|^{-z\nu_{crit}} , \tag{A.47}$$

where z is called the dynamic critical exponent and determines the characteristic time scale of the approach to equilibrium via $\tau_{eq} \sim 1/E_{gap}$. On the condensed side the order parameter $\mathcal{O}$ also exhibits scaling near the critical point; the corresponding exponent is:

$$\langle \mathcal{O} \rangle \sim |g - g_c|^{\beta} , \tag{A.48}$$

and exactly at the critical point it will depend on the source as

$$\langle \mathcal{O} \rangle \sim J^{1/\delta} \,, \tag{A.49}$$

where the coupling to the external source is $J\mathcal{O}$. The correlation function $\chi = \langle \mathcal{O}\mathcal{O} \rangle$ can also be used to probe the physical properties of the critical point. The next critical exponent we introduce is for χ at zero momentum:

$$\chi(k=0, \omega=0) \sim |g - g_c|^{-\gamma} \,. \tag{A.50}$$

The decay of G_R at the critical point in the free theory would be $1/x^{d-2}$, the deviation from this is characterized by η:

$$\chi(x, \omega=0)|_{g=g_c} \sim \frac{1}{x^{d-2+\eta}} \,. \tag{A.51}$$

To study the scaling of thermodynamic functions we introduce α as:

$$f \sim |g - g_c|^{2-\alpha} \,, \tag{A.52}$$

where f is the free energy density.

References

1. J. M. Maldacena, "The large N limit of superconformal field theories and supergravity," Adv. Theor. Math. Phys. **2**, 231 (1998) [Int. J. Theor. Phys. **38**, 1113 (1999)] [arXiv:hep-th/9711200].
2. S. S. Gubser, I. R. Klebanov and A. M. Polyakov, "Gauge theory correlators from non-critical string theory," Phys. Lett. B **428**, 105 (1998) [arXiv:hep-th/9802109];
3. E. Witten, "Anti-de Sitter space and holography," Adv. Theor. Math. Phys. **2**, 253 (1998) [arXiv:hep-th/9802150]
4. S. S. Gubser, A. Karch, "From gauge-string duality to strong interactions: A Pedestrian's Guide," Ann. Rev. Nucl. Part. Sci. **59**, 145-168 (2009). [arXiv:0901.0935 [hep-th]].
5. J. Casalderrey-Solana, H. Liu, D. Mateos, K. Rajagopal and U. A. Wiedemann, "Gauge/String Duality, Hot QCD and Heavy Ion Collisions," arXiv:1101.0618 [hep-th].
6. P. Coleman, "Introduction to Many-body Physics," http://www.physics.rutgers.edu/~coleman.
7. G. Benfatto and G. Gallavotti, "Renormalization-Group Approach to the Theory of the Fermi Surface," Phys. Rev. B **42** (1990) 9967.
8. J. Polchinski, "Effective Field Theory and the Fermi Surface," arXiv:hep-th/9210046.
9. R. Shankar, "Renormalization group approach to interacting fermions," Rev. Mod. Phys. **66**, 129 (1994).
10. G. R. Stewart, "Non-Fermi-liquid behavior in d- and f-electron metals," Rev. Mod. Phys. **73**, 797 (2001).

11. E. Abrahams and C. M. Varma, "What angle-resolved photoemission experiments tell about the microscopic theory for high-temperature superconductors," PNAS **97**, 5714 (2000).
12. C. M. Varma, P. B. Littlewood, S. Schmitt-Rink, E. Abrahams, and A. E. Ruckenstein, "Phenomenology of the normal state of Cu-O high-temperature superconductors," Phys. Rev. Lett. **63**, 1996 – 1999 (1989).
13. T. Holstein, R. E. Norton and P. Pincus, "de Haas-van Alphen Effect and the Specific Heat of an Electron Gas," Phys. Rev. B **8**, 2649 (1973); P. A. Lee and N. Nagaosa, "Gauge theory of the normal state of high-Tc superconductors," Phys. Rev. B **46**, 5621 (1992); M. Y. Reizer, Phys. Rev. B **40**, 11571 (1989); J. Polchinski, "Low-energy dynamics of the spinon-gauge system," Nucl. Phys. B **422**, 617 (1994) arXiv:cond-mat/9303037; C. Nayak and F. Wilczek, "Non-Fermi liquid fixed point in (2+1)-dimensions," Nucl. Phys. B **417**, 359 (1994) arXiv:cond-mat/9312086, "Renormalization group approach to low temperature properties of a non-Fermi liquid metal," Nucl. Phys. B **430**, 534 (1994) arXiv:cond-mat/9408016; B. I. Halperin, P. A. Lee and N. Read, "Theory of the half filled Landau level," Phys. Rev. B **47**, 7312 (1993); Y. B. Kim, A. Furusaki, X.-G. Wen, and P. A. Lee, "Gauge-invariant response functions of fermions coupled to a gauge field," Phys. Rev. B **50**, 17917–17932 (1994); B. L. Altshuler, L. B. Ioffe and A. J. Millis, "On the low energy properties of fermions with singular interactions," Phys. Rev. B **50** 14048 (1994). [arXiv:cond-mat/9406024].
14. S. S. Lee, "Low energy effective theory of Fermi surface coupled with U(1) gauge field in 2+1 dimensions," Phys. Rev. **B 80**, 165102 (2009) [arXiv:0905.4532 [cond-mat.str-el]].
15. M. A. Metlitski and S. Sachdev, "Quantum phase transitions of metals in two spatial dimensions: I. Ising-nematic order," Phys. Rev. B **82**, 075127 (2010) [arXiv:1001.1153 [cond-mat.str-el]].
16. D. F. Mross, J. McGreevy, H. Liu and T. Senthil, "A controlled expansion for certain non-Fermi liquid metals," Phys. Rev. B **82**, 045121 (2010) [arXiv:1003.0894 [cond-mat.str-el]].
17. J. A. Hertz,"Quantum critical phenomena," Phys. Rev. B **14**, 1165–1184 (1976).
18. S. Sachdev, *Quantum Phase Transitions*, Cambridge University Press, Cambridge (1999).
19. S. L. Sondhi, S. M. Girvin, J.P. Carini, D. Shahar, "Continuous quantum phase transitions,"' Rev. Mod. Phys. **69**, 315-333 (1997) [arXiv:cond-mat/9609279]
20. M. Vojta, "Quantum phase transitions," Rep. Prog. Phys. **66**, 2069 (2003), [arXiv:cond-mat/0309604].
21. Focus issue: Quantum phase transitions, Nature Phys. **4**, 167–204 (2008).
22. P. Gegenwart, Q. Si, and F. Steglich, "Quantum criticality in heavy-fermion metals," Nat. Phys. **4**, 186–197 (2008).
23. H. v. Löhneysen, A. Rosch, M. Vojta, and P. Wölfle, "Fermi-liquid instabilities at magnetic quantum phase transitions," Rev. Mod. Phys. **79**, 1015–1075 (2007).

24. X.-G. Wen, "Vacuum degeneracy of chiral spin states in compactified space,"Phys. Rev. B **40**, 7387 (1989).
25. X.-G. Wen and Q. Niu, "Ground-state degeneracy of the fractional quantum Hall states in the presence of a random potential and on high-genus Riemann surfaces," Phys. Rev. B**41**, 9377 (1990).
26. P. Coleman, "Heavy Fermions: electrons at the edge of magnetism", arXiv:cond-mat/0612006.
27. Q, Si and F. Steglich, "Heavy Fermions and Quantum Phase Transitions," Science **329**, 1161 (2010).
28. Q. Si, S. Rabello, K. Ingersent, and J. L. Smith, "Locally critical quantum phase transitions in strongly correlated metals," Nature **413**, 804–808 (2001).
29. T. Senthil, A. Vishwanath, L. Balents, S. Sachdev, and M. P. A. Fisher, "Deconfined quantum critical points," Science **303**, 1490–1494 (2004).
30. A. Schröder, G. Aeppli, R. Coldea, M. Adams, O. Stock- ert, H. v. Lohneysen, E. Bucher, R. Ramazashvili, and P. Coleman, "Onset of antiferromagnetism in heavy-fermion metals," Nature **407**, 351 (2000).
31. P. Coleman, "Theories of non-Fermi liquid behavior in heavy fermions," Physica **B 259-261**, 353 (1999).
32. A. Georges, G. Kotliar, W. Krauth, M. J. Rozenberg, "Dynamical mean-field theory of strongly correlated fermion systems and the limit of infinite dimensions," Rev. Mod. Phys. **68**, 13-125 (1996);
G. Kotliar, S. Y. Savrasov, K. Haule, V. S. Oudovenko, O. Parcollet, and C. A. Marianetti, "Electronic structure calculations with dynamical mean-field theory," Rev. Mod. Phys. **78**, 865 (2006).
33. J. Bagger and N. Lambert, "Comments On Multiple M2-branes," JHEP **0802**, 105 (2008) [arXiv:0712.3738 [hep-th]]; "Gauge Symmetry and Supersymmetry of Multiple M2-Branes," Phys. Rev. D **77**, 065008 (2008) [arXiv:0711.0955 [hep-th]]; "Modeling multiple M2's," Phys. Rev. D **75**, 045020 (2007) [arXiv:hep-th/0611108].
34. A. Gustavsson, "Algebraic structures on parallel M2-branes," Nucl. Phys. B **811**, 66 (2009) [arXiv:0709.1260 [hep-th]].
35. O. Aharony, O. Bergman, D. L. Jafferis and J. Maldacena, "N=6 superconformal Chern-Simons-matter theories, M2-branes and their gravity duals," JHEP **0810**, 091 (2008) [arXiv:0806.1218 [hep-th]].
36. S.-S. Lee, "A Non-Fermi Liquid from a Charged Black Hole; A Critical Fermi Ball," Phys. Rev. D **79**, 086006 (2009) [arXiv:0809.3402 [hep-th]].
37. H. Liu, J. McGreevy, and D. Vegh, "Non-Fermi liquids from holography," arXiv:0903.2477 [hep-th].
38. M. Cubrovic, J. Zaanen, and K. Schalm, "String Theory, Quantum Phase Transitions and the Emergent Fermi-Liquid," Science **325**, 439–444 (2009) [arXiv:0904.1993 [hep-th]].
39. T. Faulkner, H. Liu, J. McGreevy, and D. Vegh, "Emergent quantum criticality, Fermi surfaces, and AdS2," arXiv:0907.2694 [hep-th].
40. T. Faulkner, N. Iqbal, H. Liu, J. McGreevy and D. Vegh, "Strange metal

transport realized by gauge/gravity duality," Science **329**, 1043-1047 (2010); arXiv:1003.1728 [hep-th].
41. T. Faulkner, N. Iqbal, H. Liu, J. McGreevy and D. Vegh, "Holographic non-Fermi liquid fixed points," arXiv:1101.0597 [hep-th].
42. T. Faulkner and J. Polchinski, "Semi-Holographic Fermi Liquids," arXiv:1001.5049 [hep-th].
43. N. Iqbal, H. Liu and M. Mezei, "Semi-local quantum liquids," arXiv:1105.4621 [hep-th].
44. N. Iqbal, H. Liu, M. Mezei and Q. Si, "Quantum phase transitions in holographic models of magnetism and superconductors," arXiv:1003.0010 [hep-th].
45. T. Faulkner, G. T. Horowitz and M. M. Roberts, "Holographic quantum criticality from multi-trace deformations," arXiv:1008.1581 [hep-th].
46. N. Iqbal, H. Liu and Mark Mezei, to appear.
47. C. P. Herzog, P. Kovtun, S. Sachdev and D. T. Son, "Quantum critical transport, duality, and M-theory," Phys. Rev.Ê D **75**, 085020 (2007) [arXiv:hep-th/0701036].
48. S. Sachdev, M. Mueller, "Quantum criticality and black holes," 0810.2005 [cond-mat.str-el]
49. S. A. Hartnoll, "Lectures on holographic methods for condensed matter physics," Class. Quant. Grav. **26**, 224002 (2009) [arXiv:0903.3246 [hep-th]].
50. C. P. Herzog, "Lectures on Holographic Superfluidity and Superconductivity," J. Phys. A **42**, 343001 (2009) [arXiv:0904.1975 [hep-th]].
51. G. T. Horowitz, "Introduction to Holographic Superconductors," arXiv:1002.1722 [hep-th].
52. S. J. Rey, "String Theory on Thin Semiconductors," Progress of Theoretical Physics Supplement No. 177 (2009) pp. 128-142; arXiv:0911.5295 [hep-th].
53. S. Sachdev, "Condensed matter and AdS/CFT," arXiv:1002.2947 [hep-th].
54. S. A. Hartnoll, "Horizons, holography and condensed matter," arXiv:1106.4324 [hep-th].
55. J. P. Gauntlett, J. Sonner and D. Waldram, "The spectral function of the supersymmetry current (I)," arXiv:1106.4694 [hep-th].
56. R. Belliard, S. S. Gubser and A. Yarom, "Absence of a Fermi surface in classical minimal four-dimensional gauged supergravity," arXiv:1106.6030 [hep-th].
57. I. Bah, A. Faraggi, J. I. Jottar and R. G. Leigh, "Fermions and Type IIB Supergravity On Squashed Sasaki-Einstein Manifolds," JHEP **1101**, 100 (2011) [arXiv:1009.1615 [hep-th]].
58. I. Bah, A. Faraggi, J. I. Jottar, R. G. Leigh and L. A. Pando Zayas, "Fermions and $D = 11$ Supergravity On Squashed Sasaki-Einstein Manifolds," JHEP **1102**, 068 (2011) [arXiv:1008.1423 [hep-th]].
59. J. P. Gauntlett, J. Sonner and T. Wiseman, "Holographic superconductivity in M-Theory," Phys. Rev. Lett. **103**, 151601 (2009) [arXiv:0907.3796 [hep-th]].
60. J. P. Gauntlett, J. Sonner and T. Wiseman, "Quantum Criticality and

Holographic Superconductors in M-theory," JHEP **1002**, 060 (2010) [arXiv:0912.0512 [hep-th]].
61. S. S. Gubser, C. P. Herzog, S. S. Pufu and T. Tesileanu, "Superconductors from Superstrings," Phys. Rev. Lett. **103**, 141601 (2009) [arXiv:0907.3510 [hep-th]].
62. F. Aprile, D. Roest and J. G. Russo, "Holographic Superconductors from Gauged Supergravity," arXiv:1104.4473 [hep-th].
63. A. Donos and J. P. Gauntlett, "Superfluid black branes in $AdS_4 \times S^7$," arXiv:1104.4478 [hep-th].
64. J. Polchinski, "Introduction to Gauge/Gravity Duality," arXiv:1010.6134 [hep-th].
65. J. McGreevy, "Holographic duality with a view toward many-body physics," Adv. High Energy Phys. **2010**, 723105 (2010) [arXiv:0909.0518 [hep-th]].
66. L. Susskind and E. Witten, "The Holographic bound in anti-de Sitter space," arXiv:hep-th/9805114.
67. A. W. Peet and J. Polchinski, "UV / IR relations in AdS dynamics," Phys. Rev. D **59**, 065011 (1999) [arXiv:hep-th/9809022].
68. D. T. Son, A. O. Starinets, "Minkowski space correlators in AdS / CFT correspondence: Recipe and applications," JHEP **0209**, 042 (2002). [hep-th/0205051].
69. See e.g., S. Coleman, "$1/N$", in Aspect of Symmetry, Cambridge Univ. Press, (1985).
70. L. J. Romans, "Supersymmetric, cold and lukewarm black holes in cosmological Einstein-Maxwell theory," Nucl. Phys. B **383**, 395 (1992) arXiv:hep-th/9203018.
71. A. Chamblin, R. Emparan, C. V. Johnson and R. C. Myers, "Charged AdS black holes and catastrophic holography," Phys. Rev. D **60**, 064018 (1999) arXiv:hep-th/9902170.
72. S. Sachdev and J. Ye, "Gapless spin-fluid ground state in a random quantum Heisenberg magnet," Phys. Rev. Lett. **70**, 3339 (1993).
73. L. Zhu and Q. Si, "Critical local-moment fluctuations in the Bose-Fermi Kondo model," Phys. Rev. **B 66**, 024426 (2002); L. Zhu, S. Kirchner, Q. Si, and A. Georges, "Quantum Critical Properties of the Bose-Fermi Kondo Model in a Large-N Limit," Phys. Rev. Lett. **93**, 267201 (2004).
74. S. Burdin, D. R. Grempel, and A. Georges, "Heavy-fermion and spin-liquid behavior in a Kondo lattice with magnetic frustration," Phys. Rev. **B 66**, 045111 (2002).
75. S. Sachdev, "Holographic metals and the fractionalized Fermi liquid," Phys. Rev. Lett. **105**, 151602 (2010) [arXiv:1006.3794 [hep-th]].
76. S. J. Yamamoto, Q. Si, "Global Phase Diagram of the Kondo Lattice: From Heavy Fermion Metals to Kondo Insulators," J. Low Temp. Phys. **161**, 233 (2010) [arXiv:1006.4868].
77. S. Sachdev, "The landscape of the Hubbard model," [arXiv:1012.0299 [hep-th]].
78. L. Huijse, S. Sachdev, "Fermi surfaces and gauge-gravity duality," [arXiv:1104.5022 [hep-th]].

79. T. Faulkner, H. Liu and M. Rangamani, "Integrating out geometry: Holographic Wilsonian RG and the membrane paradigm," arXiv:1010.4036 [hep-th].
80. I. Heemskerk and J. Polchinski, "Holographic and Wilsonian Renormalization Groups," arXiv:1010.1264 [hep-th].
81. N. Iqbal and H. Liu, 'Real-time response in AdS/CFT with application to spinors," Fortsch. Phys. **57**, 367 (2009) [arXiv:0903.2596 [hep-th]].
82. P. Breitenlohner and D. Z. Freedman, "Stability In Gauged Extended Supergravity," Annals Phys. **144**, 249 (1982).
83. S. S. Gubser, "Phase transitions near black hole horizons," Class. Quant. Grav. **22**, 5121 (2005) [arXiv:hep-th/0505189].
84. S. S. Gubser and A. Nellore, "Low-temperature behavior of the Abelian Higgs model in anti-de Sitter space," JHEP **0904**, 008 (2009) [arXiv:0810.4554 [hep-th]].
85. F. Denef and S. A. Hartnoll, "Landscape of superconducting membranes," arXiv:0901.1160 [hep-th].
86. B. Pioline and J. Troost, "Schwinger pair production in AdS(2)," JHEP **0503**, 043 (2005) [arXiv:hep-th/0501169].
87. K. Jensen, S. Kachru, A. Karch, J. Polchinski and E. Silverstein, "Towards a holographic marginal Fermi liquid," arXiv:1105.1772 [hep-th].
88. S. S. Gubser, "Breaking an Abelian gauge symmetry near a black hole horizon," Phys. Rev. D **78**, 065034 (2008) [arXiv:0801.2977 [hep-th]];
89. S. A. Hartnoll, C. P. Herzog and G. T. Horowitz, "Building a Holographic Superconductor," Phys. Rev. Lett. **101**, 031601 (2008) [arXiv:0803.3295 [hep-th]].
90. S. A. Hartnoll, C. P. Herzog and G. T. Horowitz, "Holographic Superconductors," JHEP **0812**, 015 (2008) [arXiv:0810.1563 [hep-th]].
91. S. A. Hartnoll, J. Polchinski, E. Silverstein and D. Tong, arXiv:0912.1061 [hep-th].
92. S. Nakamura, H. Ooguri and C. S. Park, "Gravity Dual of Spatially Modulated Phase," Phys. Rev. D **81**, 044018 (2010) [arXiv:0911.0679 [hep-th]]; H. Ooguri and C. S. Park, "Holographic End-Point of Spatially Modulated Phase Transition," Phys. Rev. D **82**, 126001 (2010) [arXiv:1007.3737 [hep-th]].
93. S. S. Gubser and F. D. Rocha, "Peculiar properties of a charged dilatonic black hole in AdS_5," Phys. Rev. D **81**, 046001 (2010) [arXiv:0911.2898 [hep-th]].
94. K. Goldstein, S. Kachru, S. Prakash and S. P. Trivedi, "Holography of Charged Dilaton Black Holes," JHEP **1008**, 078 (2010) [arXiv:0911.3586 [hep-th]].
95. K. Goldstein, N. Iizuka, S. Kachru, S. Prakash, S. P. Trivedi and A. Westphal, "Holography of Dyonic Dilaton Black Branes," JHEP **1010**, 027 (2010) [arXiv:1007.2490 [hep-th]].
96. C. Charmousis, B. Gouteraux, B. S. Kim, E. Kiritsis and R. Meyer, "Effective Holographic Theories for low-temperature condensed matter systems," JHEP **1011**, 151 (2010) [arXiv:1005.4690 [hep-th]];

R. Meyer, B. Gouteraux and B. S. Kim, "Strange Metallic Behaviour and the Thermodynamics of Charged Dilatonic Black Holes," arXiv:1102.4433 [hep-th];
B. Gouteraux, B. S. Kim and R. Meyer, "Charged Dilatonic Black Holes and their Transport Properties," arXiv:1102.4440 [hep-th].
97. J. McGreevy, "In pursuit of a nameless metal," A Viewpoint on,[75] Physics **3**, 83 (2010).
98. P. W. Anderson, "In Praise of Unstable Fixed Points: The Way Things Actually Work," arXiv:cond-mat/0201431.
99. S. S. Gubser, "Curvature singularities: The Good, the bad, and the naked," Adv. Theor. Math. Phys. **4**, 679 (2000) [arXiv:hep-th/0002160].
100. J. M. Maldacena and C. Nunez, "Supergravity description of field theories on curved manifolds and a no go theorem," Int. J. Mod. Phys. A **16**, 822 (2001) [arXiv:hep-th/0007018].
101. T. Albash and C. V. Johnson, "Holographic Aspects of Fermi Liquids in a Background Magnetic Field," arXiv:0907.5406 [hep-th];
T. Albash and C. V. Johnson, "Landau Levels, Magnetic Fields and Holographic Fermi Liquids," J. Phys. A **43**, 345404 (2010) [arXiv:1001.3700 [hep-th]].
102. P. Basu, J. He, A. Mukherjee and H. H. Shieh, "Holographic Non-Fermi Liquid in a Background Magnetic Field," arXiv:0908.1436 [hep-th].
103. E. Gubankova, J. Brill, M. Cubrovic, K. Schalm, P. Schijven and J. Zaanen, "Holographic fermions in external magnetic fields," arXiv:1011.4051 [hep-th].
104. F. Denef, S. A. Hartnoll and S. Sachdev, "Black hole determinants and quasinormal modes," Class. Quant. Grav. **27**, 125001 (2010) [arXiv:0908.2657 [hep-th]].
105. F. Denef, S. A. Hartnoll and S. Sachdev, "Quantum oscillations and black hole ringing," Phys. Rev. D **80**, 126016 (2009) [arXiv:0908.1788 [hep-th]].
106. S. A. Hartnoll and D. M. Hofman, "Generalized Lifshitz-Kosevich scaling at quantum criticality from the holographic correspondence," Phys. Rev. B **81**, 155125 (2010) [arXiv:0912.0008 [cond-mat.str-el]].
107. J. W. Chen, Y. J. Kao and W. Y. Wen, "Peak-Dip-Hump from Holographic Superconductivity," arXiv:0911.2821 [hep-th].
108. T. Faulkner, G. T. Horowitz, J. McGreevy, M. M. Roberts and D. Vegh, "Photoemission 'Experiments' on Holographic Superconductors," arXiv:0911.3402 [hep-th].
109. S. S. Gubser, F. D. Rocha and P. Talavera, "Normalizable Fermion Modes in a Holographic Superconductor," arXiv:0911.3632 [hep-th].
110. S. S. Gubser, F. D. Rocha and A. Yarom, "Fermion correlators in non-abelian holographic superconductors," arXiv:1002.4416 [hep-th].
111. M. Ammon, J. Erdmenger, M. Kaminski and A. O'Bannon, "Fermionic Operator Mixing in Holographic p-wave Superfluids," JHEP **1005**, 053 (2010) [arXiv:1003.1134 [hep-th]].
112. F. Benini, C. P. Herzog and A. Yarom, "Holographic Fermi arcs and a d-wave gap," arXiv:1006.0731 [hep-th].

113. D. Vegh, "Fermi arcs from holography," arXiv:1007.0246 [hep-th].
114. M. Edalati, R. G. Leigh and P. W. Phillips, "Dynamically Generated Mott Gap from Holography," Phys. Rev. Lett. **106**, 091602 (2011) [arXiv:1010.3238 [hep-th]].
115. M. Edalati, R. G. Leigh, K. W. Lo and P. W. Phillips, "Dynamical Gap and Cuprate-like Physics from Holography," Phys. Rev. D **83**, 046012 (2011) [arXiv:1012.3751 [hep-th]].
116. D. Guarrera and J. McGreevy, "Holographic Fermi surfaces and bulk dipole couplings," arXiv:1102.3908 [hep-th].
117. T. Albash, "Non-Unitary Fermionic Quasinormal Modes at Zero Frequency," arXiv:1002.4431 [hep-th].
118. J. P. Wu, "Holographic fermions in charged Gauss-Bonnet black hole," arXiv:1103.3982 [hep-th].
119. J. R. Wang and G. Z. Liu, "Non-Fermi liquid behavior due to U(1) gauge field in two dimensions," Nucl. Phys. B **832**, 441 (2010) [arXiv:0907.1022 [cond-mat.supr-con]].
120. V. Balasubramanian, I. Garcia-Etxebarria, F. Larsen and J. Simon, "Helical Luttinger Liquids and Three Dimensional Black Holes," arXiv:1012.4363 [hep-th].
121. M. Rangamani and B. Withers, "Fermionic probes of local quantum criticality in one dimension," arXiv:1106.3210 [hep-th].
122. T. Hartman and S. A. Hartnoll, "Cooper pairing near charged black holes," JHEP **1006**, 005 (2010) [arXiv:1003.1918 [hep-th]].
123. E. Gubankova, "Particle-hole instability in the AdS_4 holography," arXiv:1006.4789 [hep-th].
124. S. A. Hartnoll and A. Tavanfar, "Electron stars for holographic metallic criticality," arXiv:1008.2828 [hep-th].
125. J. de Boer, K. Papadodimas and E. Verlinde, "Holographic Neutron Stars," JHEP **1010**, 020 (2010) [arXiv:0907.2695 [hep-th]].
126. X. Arsiwalla, J. de Boer, K. Papadodimas and E. Verlinde, "Degenerate Stars and Gravitational Collapse in AdS/CFT," JHEP **1101**, 144 (2011) [arXiv:1010.5784 [hep-th]].
127. M. Cubrovic, J. Zaanen and K. Schalm, "Constructing the AdS dual of a Fermi liquid: AdS Black holes with Dirac arXiv:1012.5681 [hep-th].
128. S. A. Hartnoll, D. M. Hofman and A. Tavanfar, "Holographically smeared Fermi surface: Quantum oscillations and Luttinger count in electron stars," arXiv:1011.2502 [hep-th].
129. S. A. Hartnoll and P. Petrov, "Electron star birth: A continuous phase transition at nonzero density," arXiv:1011.6469 [hep-th].
130. V. G. M. Puletti, S. Nowling, L. Thorlacius and T. Zingg, "Holographic metals at finite temperature," JHEP **1101**, 117 (2011) [arXiv:1011.6261 [hep-th]].
131. S. A. Hartnoll, D. M. Hofman and D. Vegh, "Stellar spectroscopy: Fermions and holographic Lifshitz criticality," arXiv:1105.3197 [hep-th].
132. M. Cubrovic, Y. Liu, K. Schalm, Y. W. Sun and J. Zaanen, "Spectral probes

of the holographic Fermi groundstate: dialing between the electron star and AdS Dirac hair," arXiv:1106.1798 [hep-th].
133. K. Jensen, A. Karch, D. T. Son and E. G. Thompson, "Holographic Berezinskii-Kosterlitz-Thouless Transitions," arXiv:1002.3159 [hep-th].
134. K. Jensen, "More Holographic Berezinskii-Kosterlitz-Thouless Transitions," Phys. Rev. D **82**, 046005 (2010) [arXiv:1006.3066 [hep-th]].
135. M. Edalati, K. W. Lo and P. W. Phillips, "Neutral Order Parameters in Metallic Criticality in d=2+1 from a Hairy Electron Star," arXiv:1106.3139 [hep-th].
136. K. Jensen, to appear.
137. D. Anninos, S. A. Hartnoll and N. Iqbal, "Holography and the Coleman-Mermin-Wagner theorem," Phys. Rev. D **82**, 066008 (2010) [arXiv:1005.1973 [hep-th]].
138. V. Efimov, "Energy levels arising from resonant two-body forces in a three-body system," Phys. Lett. **B33**, 563 (1970).

Chapter 14

The Fluid/Gravity Correspondence

Shiraz Minwalla*

Tata Institute of Fundamental Research,
School of Natural Sciences,
Homi Bhabha Road, Mumbai 400005, India

Veronika E. Hubeny and Mukund Rangamani

University of Durham, Department of Physics,
Ogen Center for Fundamental Physics,
South Raod, Durham DH1 3LE UK

If supersymmetry turns out to be a symmetry of nature at low energies, the first order of business to measure the soft breaking parameters. But one will also want to understand the symmetry, and its breaking, more microscopically. Two aspects of this problem constitute the focus of these lectures. First, what sorts of dynamics might account for supersymmetry breaking, and its manifestation at low energies. Second, how might these features fit into string theory (or whatever might be the underlying theory in the ultraviolet). The last few years have seen a much improved understanding of the first set of questions, and at least a possible pathway to address the second.

14.1. Introduction

In this chapter we will study a particular long wavelength limit of Einstein's equations with a negative cosmological constant in $d+1$ dimensions. In such a limit we find that Einstein's equations reduce to the equations of fluid dynamics (relativistic generalizations of the famous Navier-Stokes equations) in d dimensions. While the motivation for our study lies within the AdS/CFT correspondence of string theory, the fluid/gravity correspon-

---

*Lectures delivered by S. Minwalla. To be published by Cambridge University Press. Printed with Permission.

dence stands on its own and can be viewed as a map between two classic dynamical systems.

14.1.1. *Prelude: CFT stress tensor dynamics from gravity*

An important consequence of the AdS/CFT correspondence [Ch.AdSCFT] is that the dynamics of the stress(-energy-momentum) tensor in a large class of d-dimensional strongly coupled quantum field theories is governed by the dynamics of Einstein's equations with negative cosmological constant in $d+1$ dimensions. To begin with, we shall try to provide the reader with some intuition for this statement and argue that searching for a tractable corner of this connection leads one naturally to the fluid/gravity correspondence.

In its most familiar example, the AdS/CFT correspondence [Ch.AdSCFT] asserts that $SU(N)$ $\mathcal{N}=4$ Super Yang-Mills (SYM) theory is dual to Type IIB string theory on $\mathrm{AdS}_5 \times \mathbf{S}^5$. It has long been known that in the 't Hooft limit, which involves taking $N \to \infty$ keeping the coupling λ fixed, the gauge theory becomes effectively classical. However, it was widely believed that for any non-trivial gauge theory the resulting classical system would be too complicated to be tractable. The remarkable observation of Maldacena in 1997 was that this field theory intuition is spectacularly wrong. Indeed, not only is the classical system governing $\mathcal{N}=4$ SYM tractable, it is actually a well known theory, viz., classical Type IIB string theory.

Now, even classically, string theory has complicated dynamics; however in the strong gauge coupling ($\lambda \to \infty$) regime, it reduces to the dynamics of Type IIB supergravity (by decoupling the massive string states). More interestingly, Type IIB supergravity on $\mathrm{AdS}_5 \times \mathbf{S}^5$ admits several consistent truncations. The simplest and most universal of these is the truncation to Einstein's equations with negative cosmological constant,

$$E_{\mu\nu} \equiv R_{\mu\nu} - \frac{1}{2}\, R\, g_{\mu\nu} + \Lambda\, g_{\mu\nu} = 0\ , \qquad \Lambda \equiv -\frac{d\,(d-1)}{2\, R_{\mathrm{AdS}}^2}\ . \tag{14.1}$$

(Note that the AdS curvature radius R_{AdS} can be scaled away by a change of units; we therefore set R_{AdS} to unity in the rest of this chapter). Having thus motivated the study of the most beautiful equation of physics, namely Einstein's equations of general relativity, we now confront the question: What does this imply for the field theory?

Recall that according to the AdS/CFT dictionary there is a one-to-one map between single particle states in the classical Hilbert space of string theory and single-trace operators in the gauge theory. For instance the bulk graviton maps to the stress tensor of the boundary theory. Taking the collection of such single trace operators as a whole, one can try to formulate dynamical equations for their quantum expectation values in the field theory. While this can be done in principle, the resulting system is non-local in terms of the intrinsic field theory variables themselves.

However, because we can associate the quantum operators (and their expectation values) of the gauge theory to the classical fields of string theory/supergravity, we know that the set of classical equations we are looking for are just the local equations of Type IIB supergravity on $AdS_5 \times \mathbf{S}^5$ at strong coupling. This reduction, whilst retaining lots of interesting physics, still turns out to be too complicated from the field theory perspective. For one, the space of single trace operators is still infinite dimensional (at infinite N), and relatedly attempting to classify the solution space of Type IIB supergravity is a challenging problem. However, the fact that on the string side we can reduce the system to (14.1), implies that there is a decoupled sector of stress tensor dynamics in $\mathcal{N} = 4$ SYM at large λ.[a]

Actually, there is an infinite number of conformal gauge theories which have a gravitational dual that truncates consistently at the two-derivative level to Einstein's equations with a negative cosmological constant; $\mathcal{N} = 4$ SYM theory is just a particularly simple member of this class. Thus (14.1) describes the *universal decoupled* dynamics of the stress tensor for an infinite number of different gauge theories. In the first part of this chapter we will focus on the study of this universal sector. Later we will generalize to the study of bulk equations with more fields, thereby obtaining richer dynamics at the expense of universality.

Given this association between the dynamics of quantum field theory stress tensors to the dynamics of gravity in negatively curved backgrounds, it is natural to ask – can we do more? Can we for instance classify all possible behaviors of stress tensors? On the gravity side we would have to classify all possible solutions to (14.1); this is a laudable goal and various chapters in this book are aimed at addressing this question using different approaches. We are going to focus on one that naturally follows from the basic organizing principle of physics: separation of scales.

It is well known that in many situations in physics (as well as chemistry, biology, etc.), complicated UV dynamics results in relatively simple

[a]While this is always true in two dimensional field theories, such a decoupling is not generic in higher dimensional field theories (in fact it is not true of $\mathcal{N} = 4$ Yang Mills at weak coupling), and is in itself a surprising and interesting fact about the $\mathcal{N} = 4$ dynamics at strong coupling.

IR dynamics. Perhaps the first systematic exposition of this ubiquitous fact was in the context of finite temperature physics. It has been known for almost 200 years now that the dynamics of nearly equilibrated systems at high enough temperature may be described by an effective theory called hydrodynamics. The key dynamical equation of hydrodynamics is the conservation of the stress tensor

$$\nabla_a T^{ab} = 0 \ , \tag{14.2}$$

where ∇_a is the covariant derivative compatible with the background metric γ_{ab} on which this fluid lives. As this equation is an autonomous dynamical system involving just the stress tensor, it should lie within the sector of universal decoupled stress tensor dynamics.

Given that the AdS/CFT correspondence asserts that this universal sector is governed by (14.1), we are led to conclude that (14.1) must, in an appropriate high temperature and long distance limit which we refer to as the *long wavelength regime*, reduce to the equations of d-dimensional hydrodynamics. Indeed, this expectation has been independently verified in [1] and the resulting map between gravity and fluid dynamics has come to be known as the *fluid/gravity correspondence*. In particular, the specific fluid dynamical equations, dual to long wavelength gravity in the universal sector, have been determined up to the second order in a gradient expansion (cf. §14.4.4). Given any solution to the these fluid dynamical equations, the fluid/gravity map *explicitly* determines a solution to Einstein's equations (14.1) to the appropriate order in the derivative expansion. The solutions in gravity are simply inhomogeneous, time-dependent black holes, with slowly varying but otherwise generic horizon profiles.

The main focus of the present chapter is to explain and present the fluid/gravity map at the full non-linear level following [1] and subsequent work. The connection between these two systems was established and extensively studied much earlier at the linearized level in the AdS/CFT context (following the seminal work [2]). The first hints of the connection between fluid dynamics and gravity at the non-linear level were obtained in attempts to construct non-linear solutions dual to a particular boost invariant flow [3], which provided inspiration for the fluid/gravity map. Such a map was also suggested by the observation that the properties of large rotating black holes in global AdS space are reproduced by the equations of non-linear fluid dynamics [4]. We refer the reader to [5] for a list of developments and references.

14.1.2. *Preview of the fluid/gravity correspondence*

Having provided the reader with a broad, albeit abstract, rationale to associate the dynamics of Einstein's equations to that of a quantum field theoretic stress tensor, we now provide some specifics that set the stage for our discussion.

According to the gauge/gravity dictionary, distinct asymptotically AdS bulk geometries correspond to distinct states in the boundary gauge theory. The pure AdS geometry, i.e., the maximally symmetric negatively curved spacetime, corresponds to the vacuum state of the gauge theory. A large[b] Schwarzschild-AdS black hole corresponds to a thermal density matrix in the gauge theory. This can be easily conceptualized in terms of the late-time configuration a generic state evolves to: in the bulk, the combined effect of gravity and negative curvature tends to make a generic configuration collapse to form a black hole which settles down to the Schwarzschild-AdS geometry, while in the field theory, a generic excitation will eventually thermalize. Note that although the underlying theory is supersymmetric, the correspondence applies robustly to non-supersymmetric states such as the black holes mentioned above. In this sense, supersymmetry is *not* needed for the correspondence.

On the boundary, the essential physical properties of the gauge theory state (such as local energy density, pressure, temperature, entropy current, etc.) are captured by the expectation value of the *boundary stress tensor*, which in the bulk is related to normalizable metric perturbations about a given state. It can be extracted via a well-defined Brown-York type procedure [6] as we review later (see (14.39)).

At the risk of being repetitive we urge the reader to note the distinction between the two separate stress tensors that will enter our analysis. In our framework, the *bulk* stress tensor appearing on the r.h.s. of the bulk Einstein's equation is zero if we are only interested in the universal subsector discussed above. On the other hand, the *boundary* stress tensor T^{ab} is non-zero; it is a measure of the normalizable fall off of the bulk metric at the boundary. Note that the boundary stress tensor does not curve the boundary spacetime à la Einstein's equations since the boundary metric γ_{ab} is non-dynamical and fixed. We will discuss generalizations that allow for

[b]Recall that AdS is a space of constant negative curvature, which introduces a length scale, called the AdS scale R_{AdS}, corresponding to the radius of curvature. The black hole size is then measured in terms of this AdS scale; large black holes have horizon radius $r_+ > R_{\text{AdS}}$. We will be focus on the large black hole limit $r_+ \gg R_{\text{AdS}}$, and therefore consider the *planar* Schwarzschild-AdS black holes.

non-trivial bulk matter in §14.6 when we move outside the universal stress tensor sector.

To describe gravity duals of fluid flows, a useful starting point is the map between the boundary and bulk dynamics in global thermal equilibrium. In the field theory, one characterizes thermal equilibrium by a choice of static frame and a temperature field. On the gravity side, the natural candidates to characterize the equilibrium solution are static (or more generally stationary) black hole spacetimes, as can be seen by demanding regular solutions with periodic Euclidean time circle. The temperature of the fluid is given by the Hawking temperature of the black hole, while the fluid dynamical velocity is captured by the horizon boost velocity of the black hole. For planar Schwarzschild-AdS black holes the temperature grows linearly with horizon size; the AdS asymptotics thus ensures thermodynamic stability as well as providing a natural long wavelength regime.

Now let us try to gently move away from the equilibrium configuration. Starting with the stationary black hole (namely the boosted planar Schwarzschild-AdS_{d+1}) solution, we wish to use it to build solutions where the fluid dynamical temperature and velocity are slowly-varying functions of the boundary directions. Intuitively, this mimics patching together pieces of black holes with slightly different temperatures and boosts in a smooth way so as to get a regular solution of (14.1). In order to obtain a true solution of Einstein's equations, the patching up procedure cannot be done arbitrarily; one is required at the leading order to constrain the velocity and temperature fields to obey the equations of ideal fluid dynamics.[c] Further, the solution itself is corrected order by order in a derivative expansion, a process that likewise corrects the fluid equations. All these steps may be implemented[d] in detail in a systematic boundary gradient expansion. The final output is a map between solutions to negative cosmological constant gravity and the equations of fluid dynamics in one lower dimension, i.e. the fluid/gravity map.

A noteworthy aspect of this construction is that Einstein's equations become tractable due to the long wavelength regime without losing nonlinearity. From the boundary standpoint, one encounters domains of nearly

[c]These constraints are actually the radial momentum constraints for gravity in AdS and imply (14.2). In contrast to the conventional ADM decomposition, we imagine foliating the spacetime with timelike leaves and 'evolve' into the AdS bulk radially.

[d]In the technical implementation of this program, it is important that one respect boundary conditions. We require that the bulk metric asymptote to γ_{ab} (up to a conformal factor) and further be manifestly regular in the part of the spacetime outside of any event horizon.

constant fluid variables; these domains can then be extended radially from the boundary into the bulk and in each such bulk 'tube', illustrated in Figure 14.1, we are guaranteed to have a solution which is close to the equilibrium form. Lest the reader be led astray, we should note that the solutions we construct are perturbative and hence approximate. Nevertheless, they are 'generic' slowly-varying asymptotically-AdS black hole geometries, with no Killing fields.

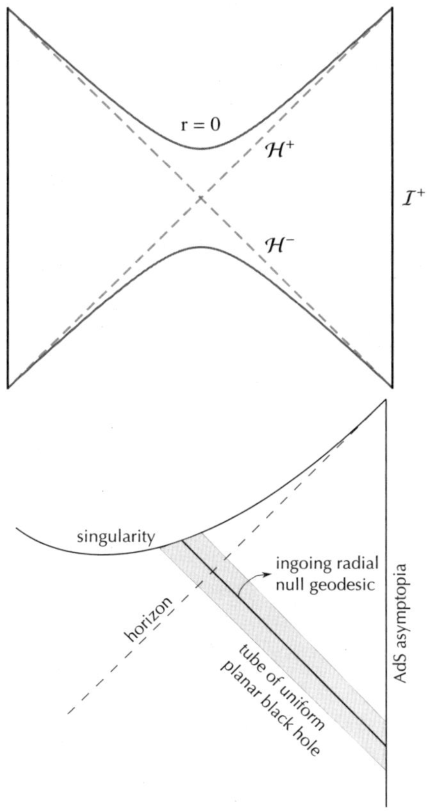

Fig. 14.1. Penrose diagram of the uniform planar black hole (14.17) and the causal structure of the spacetimes dual to fluid dynamics illustrating the tube structure. Dashed line in the second figure denotes the future event horizon, while the shaded tube indicates the region of spacetime over which the solution is well approximated by a corresponding uniform black hole.

A remarkable outcome of the association between generic black holes and fluid flows is that it automatically provides a sensible entropy current with non-negative divergence for hydrodynamics. On the gravitational side, entropy is naturally associated to the area of the event horizon; by pulling back this area form to the boundary, we can equip our fluid with a canonical entropy current.

We now turn to the technical aspects of the fluid/gravity map. Following a review of fluid dynamics and the perturbative construction of gravitational solutions, we finally present the main results (in particular the bulk metric and the boundary stress tensor, to second order in boundary derivatives) in §14.4. The subsequent sections are devoted to describing some implications and extensions of the basic construction.

14.2. Relativistic Fluid Dynamics

To set the stage, let us start by reviewing fluid dynamics, explicating the use of gradient expansion as an organizational principle. At high temperatures every non-trivial quantum field theory (and every experimentally realizable system) equilibrates into a fluid phase, i.e., a translationally invariant phase in which adiabatic displacement of neighboring elements requires no force.

Weakly interacting fluids are composed of a collection of a large number of long lived partonic excitations which continually collide with each other. The time and space intervals between successive collisions of a given parton are called the mean free time, $t_{\rm m}$, and the mean free length, $\ell_{\rm m}$, respectively.[e] Such fluids are characterized by a parton density function in phase space, and the time evolution of this function is governed by the well known Boltzmann transport equations of statistical physics. These equations have an interesting property: arbitrary initial density functions relax to local thermal equilibrium over a time scale of order the mean free time. In other words, for $t \gg t_{\rm m}$, the parton distribution in momentum space approximately reduces, at every point x, to an equilibrium distribution. However the parameters[f] characterizing this equilibrium configuration, the temperature field $T(x)$ and fluid velocity field $u^a(x)$, vary on a length scale that is large compared to $\ell_{\rm m}$. $T(x)$ and $u^a(x)$ are the effective dynamical

[e] In a relativistic system of massless particles, like $\mathcal{N} = 4$ SYM, $t_{\rm m} \sim \ell_{\rm m}$. We will assume this is the case in our discussions below.

[f] For simplicity, in this discussion we assume that the system has no conserved charges other than the stress-energy-momentum tensor, and no other Goldstone-like light degrees of freedom. We discuss generalizations below.

variables of the system at later times; their evolution as a function of time is governed by the equations of fluid dynamics.

Now it turns out that the equations of fluid dynamics may also be derived in a much simpler and more general manner, and so apply even at strong coupling. The main assumption that underlies fluid dynamics is that systems always equilibrate locally over a finite time scale that we continue to refer to as $t_{\rm m}$. While this assumption is true of the Boltzmann transport equations, it is believed to hold more generally also for strongly coupled fluids. It follows immediately that $T(x)$ and $u^a(x)$ are the effective variables for dynamics at length and time scales large compared to $\ell_{\rm m}$ and $t_{\rm m}$. As we will now see, the equations of fluid dynamics follow inevitably out of this conclusion.

14.2.1. *The equations of fluid dynamics and constitutive relations*

The stress tensor in any d-dimensional quantum field theory on a background with metric γ_{ab} obeys the d conservation equations

$$\nabla_a T^{ab} = 0. \tag{14.3}$$

These equations do not constitute a well defined initial value problem for the stress tensor in general as, in $d \geq 2$, we have more variables (the $\frac{1}{2}\,d(d+1)$ independent components of the stress tensor) than equations.[g] In the fluid dynamical limit, however, the stress tensor is determined as a function of d variables, $T(x)$ and $u^a(x)$. Consequently, (14.3), supplemented with a formula for T^{ab} as a function of thermodynamical fields, constitute a complete set of dynamical equations. These are the equations of fluid dynamics.

A constitutive relation that expresses T^{ab} as a function of $T(x)$, $u^a(x)$, and their derivatives turns (14.3) into a concrete set of fluid dynamical equations. In thermal equilibrium the stress tensor T^{ab} is given by

$$T^{ab} = (P + \rho)\, u^a\, u^b + P\,\gamma^{ab} \tag{14.4}$$

where P is the pressure of the fluid and ρ is its energy density. Recall that both ρ and P are known functions of temperature (determined by the

[g]For the special case of conformal field theories the number of variables is reduced by one, as the trace of the stress tensor vanishes. In this case, at $d = 2$, we have as many variables as equations. This observation underlies the special simplicity of CFTs in $d = 2$.

thermodynamic equation of state of the fluid). For a fluid in *local* thermal equilibrium, (14.4) generalizes to

$$T^{ab}(x) = [P(x) + \rho(x)]\, u^a(x)\, u^b(x) + P(x)\, \gamma^{ab} + \Pi^{ab}(x) \tag{14.5}$$

where $P(x) = P(T(x))$, $\rho(x) = \rho(T(x))$ and $\Pi^{ab}(x)$ represents the contributions of derivatives of $T(x)$ and $u^a(x)$ to the stress tensor. This dissipative part Π^{ab} may then be expanded as

$$\Pi^{ab} = \sum_{n=1}^{\infty} \ell_{\rm m}^n\, \Pi^{ab}_{(n)} \tag{14.6}$$

where $\Pi^{ab}_{(n)}$ is defined to be of $n^{\rm th}$ order in derivatives of the fluid dynamical fields. Note that magnitude of $\Pi^{ab}_{(n)}$ relative to the ideal fluid stress tensor is approximately $(\ell_{\rm m}/L)^n$ where L characterizes the length scale of variation of the temperature and velocity fields; consequently terms at higher values of n are increasingly subdominant in the fluid dynamical limit.

The explicit form of the functions $\Pi^{ab}_{(n)}$ can only be derived from a detailed study of the dynamics of the specific system. However the allowed forms for constitutive relations are significantly constrained by symmetry and other general considerations. At first order, for instance, it is possible to assert on very general grounds that

$$\begin{aligned} \Pi^{\langle ab\rangle}_{(1)} \equiv P^a_{\ c}\, P^b_{\ d}\, \Pi^{cd}_{(1)} - \frac{1}{d-1}\, P^{ab}\, P_{cd}\, \Pi^{cd}_{(1)} &= -2\,\eta\,\sigma^{ab} \\ \frac{1}{d-1}\, \Pi^{ab}_{(1)} P_{ab} - \frac{\partial P}{\partial \rho}\, (u_a\, u_b\, \Pi^{ab}_{(1)}) &= -\zeta\,\theta \end{aligned} \tag{14.7}$$

where

$$P^{ab} \equiv u^a u^b + \gamma^{ab} \tag{14.8}$$

is the projector onto space in the local fluid rest frame,

$$\sigma^{ab} = \nabla^{\langle a}\, u^{b\rangle} \equiv P^{ac}\, P^{bd} \left(\nabla_{(c}\, u_{d)} - \frac{1}{d-1}\, P_{cd}\, \theta \right) \tag{14.9}$$

is the fluid shear tensor, $\theta \equiv \nabla_c\, u^c$ is the expansion, and the brackets around the indices $\langle ab\rangle$ denote the symmetric transverse traceless part of the expression. Here η and ζ are arbitrary functions of the temperature, referred to as the shear and bulk viscosity, respectively.

The equation (14.7) is a physically complete specification of the constitutive relations at first order, even though it leaves $P^a_{\ c}\, \Pi^{cd}_{(1)}\, u_d$ and one linear combination of $\Pi^{ab}_{(1)}\, P_{ab}$ and $u_a\, u_b\, \Pi^{ab}_{(1)}$ unspecified. This is because

$T(x)$ and $u^a(x)$ have no intrinsic definition out of equilibrium. All equations of fluid dynamics must be 'field redefinition invariant' (invariant under redefinitions of $T(x)$ and $u^a(x)$ that reduce to identity in equilibrium), and it turns out that the l.h.s. of (14.7) are the only field redefinition invariant data in $\Pi_{(1)}^{ab}$. The other components of $\Pi_{(1)}^{ab}$ can be modified at will by an appropriate field redefinition, and have no physical significance.

The r.h.s. of the two equations in (14.7) represent the most general inequivalent 'tensor' and 'scalar' data that can be constructed from a single derivative of fluid dynamical fields compatible with the conservation equation at first order which imply that $u^a \nabla_a T \propto \theta$.

It is sometimes convenient to fix the field redefinition ambiguity by giving the fields $u^a(x)$ and $T(x)$ unambiguous (but arbitrary) meaning. In the so-called 'Landau Frame' this is achieved by asserting that, at each point,

$$T^a_{\ b}(x)\, u^b(x) = -\rho(x)\, u^a(x)\ . \tag{14.10}$$

This relation defines u^a by identifying it with the unique timelike eigenvector of the stress tensor at any point, and defines the temperature by identifying the corresponding eigenvalue with the energy density.[h] In the Landau frame, which we adopt for most of this chapter, (14.7) simplify to

$$\Pi_{(1)}^{ab} = -2\,\eta\,\sigma^{ab} - \zeta\,\theta\,P^{ab}\ . \tag{14.11}$$

As the equations of fluid dynamics are both local and thermodynamical in nature, they must respect a local form of the second law of thermodynamics. It follows that the equations of fluid dynamics must be accompanied by an entropy current whose divergence is pointwise non-negative in every conceivable fluid flow. At first order, for a charge-free fluid, the constraints imposed by this requirement are a relatively mild set of inequalities on η and ζ: It turns out that the entropy current is constrained to take the form

$$J^a_s = s\,u^a - \frac{1}{T}\,u_b\,\Pi_{(1)}^{ab}\,, \tag{14.12}$$

where s is the entropy density. It is possible to demonstrate that the current in (14.12) is field redefinition invariant to first order. Note that the second term on the r.h.s. of (14.12) vanishes in the Landau frame. Using the Euler relation ($\rho + P = s\,T$) and the Gibbs-Duhem relation ($dP = s\,dT$) of thermodynamics along with the equations of motion, it follows that the

[h]Note that the equations (14.10) are true in equilibrium.

divergence of this entropy current is given by

$$\nabla_a J_s^a = -\nabla_a \left(\frac{u_b}{T}\right) \Pi_{(1)}^{ab} . \tag{14.13}$$

Using (14.7) (or more simply (14.11) in the Landau frame), it is easy to verify that positivity of the entropy current requires that $\eta \geq 0$ and $\zeta \geq 0$. At higher orders in the derivative expansion (for uncharged fluids) and even at first order for more complicated fluids (e.g. charged fluids and superfluids) the requirement of positivity of the entropy current imposes more than a set of inequalities on transport coefficients; it forces linear combinations of otherwise arbitrary transport coefficients to vanish [7].

In the rest of this chapter we will be especially interested in the fluid dynamics of conformal field theories. These theories enjoy three key simplifications. First, as they have no dimensionless parameters, the dependence of all physical quantities (e.g. P, ρ, η) on temperature follows on dimensional grounds. In particular,

$$P = \alpha T^d, \quad \rho = (d-1)\,\alpha T^d, \quad \eta = \eta' \, T^{d-1} \tag{14.14}$$

where α and η' are dimensionless constants. Second, the stress tensor in any CFT is necessarily traceless. It in particular follows from this condition that $\zeta = 0$. Finally, the stress tensor in such theories must transform covariantly under Weyl transformations. This imposes additional restrictions on the stress tensor at higher orders in the derivative expansion. In summary, for a conformal fluid, the stress tensor up to first order has the form

$$T^{ab} = \alpha \, T^d \left(d \, u^a u^b + \gamma^{ab}\right) - \eta' \, T^{d-1} \sigma^{ab} \tag{14.15}$$

where α and η' are pure numbers and $\eta' \geq 0$.

The constraints on allowed forms of the constitutive relations at higher order are more complicated. The most general allowed equations of second order fluid dynamics have largely (but perhaps not completely) been worked out in [8]. In the following we will determine the second order fluid equations for $\mathcal{N} = 4$ Yang Mills at strong coupling using the fluid/gravity duality.

14.2.2. *The Navier-Stokes scaling limit*

An interesting fact about the equations of relativistic (or any other compressible) fluid dynamics is that they reduce to a universal form under a combined low amplitude and long wavelength scaling. Consider a uniform fluid at rest, perturbed so that the amplitude in velocity fluctuations is small (scales like ϵ) and the amplitude in temperature fluctuations is smaller (scales like ϵ^2). We also require that the wavelength of spatial fluctuations is large (scales like $1/\epsilon$) and that their temporal scale even larger (scales like $1/\epsilon^2$). We then take the strict $\epsilon \to 0$ limit. In this limit:

1) The fluid is non-relativistic, as all velocities are parametrically smaller than the speed of light.
2) The fluid is incompressible, as all velocities are parametrically smaller than the speed of sound (recall that a sound wave is a compression wave, and that fluid flows at velocities smaller than the speed of sound are effectively incompressible).
3) The temporal component of the energy conservation equations reduces, at leading order $\mathcal{O}(\epsilon)$ to the continuity equation $\vec{\nabla} \cdot \vec{v} = 0$. We use the symbol $\vec{v}$ for the non-relativistic spatial velocity.
4) The spatial component of the energy conservation equations reduce at leading order, $\mathcal{O}(\epsilon^3)$, to the famous (non-relativistic) Navier-Stokes equations

$$\dot{\vec{v}} + \vec{v} \cdot \vec{\nabla}\, \vec{v} = -\vec{\nabla} P + \nu\, \nabla^2 \vec{v} \tag{14.16}$$

with kinematic viscosity

$$\nu = \frac{\eta}{\rho_0 + P_0},$$

where ρ_0 and P_0 are the background values of the density and pressure of the fluid.

Note that the Navier-Stokes equations are homogeneous neither in amplitude of fluctuations (the convective term is non-linear), nor in derivatives (the viscous term is quadratic in derivatives). All terms retained in (14.16) are equally important in the $\epsilon \to 0$ limit; in particular, the parameter ν may be set to unity by a uniform rescaling of space and time. In contrast, the viscosities η, ζ give a subleading correction to ideal fluid dynamics in the derivative expansion of relativistic fluid dynamics.

Taking the spatial divergence of (14.16) one sees that the pressure may be solved for in terms of the velocity field *on any given time slice*; so the pressure is not an independent degree of freedom. The initial data of the Navier-Stokes equations comprise just the components of a divergence-free velocity field specified on any time slice. This reduction is not surprising since the sound waves are being projected out in this limit.

The incompressible Navier-Stokes equations (14.16) describe a wide variety of sometimes extremely complicated phenomena like turbulence.

Despite the fact that these equations have been studied for almost 200 years, their non-linear phenomenology remains rather poorly understood. One of the hopes of the fluid/gravity correspondence is to shed a new (geometric) light on some of these issues.

14.3. Perturbative Construction of Gravity Solutions

We have seen that fluid dynamics can be treated systematically as the theory of long wavelength fluctuations about thermal equilibrium. We are now going to construct gravitational solutions dual to fluid flows by formalizing this intuition to set up an algorithmic procedure to construct slowly varying dynamical black hole spacetimes as solutions to (14.1).

14.3.1. *Global thermal equilibrium from gravity*

The starting point from the gravitational perspective is a solution that corresponds to global thermal equilibrium. For the moment let us consider a conformal field theory on Minkowski space $\gamma_{ab} = \eta_{ab}$. From the gauge/gravity correspondence we know that the dual geometry in the bulk is the planar Schwarzschild-AdS$_{d+1}$ geometry, whose metric in static coordinates (with $x^a = \{t, y^i\}$) is given by

$$ds^2 = -r^2\, f(r/T)\, dt^2 + \frac{dr^2}{r^2\, f(r/T)} + r^2\, \delta_{ij}\, dy^i\, dy^j\ , \qquad f(r) \equiv 1 - \left(\frac{4\pi}{d\, r}\right)^d . \tag{14.17}$$

This is a one-parameter family of solutions, parameterized in terms of the black hole temperature T, which determines the horizon radius, $r_+ \equiv \frac{4\pi\, T}{d}$, where f vanishes. It is easy to generate a d-parameter family of solutions by boosting (14.17) along the translationally invariant spatial directions y^i, leading to a solution parameterized by a (normalized) timelike velocity field u^a. The parameters which characterize the bulk solution are precisely the basic hydrodynamical degrees of freedom, viz., temperature T and velocity u^a of the black hole. It is easy to see that the solution induces on the Minkowski boundary of the AdS$_{d+1}$ spacetime a stress tensor which precisely takes the ideal fluid form of (14.4) with thermodynamic parameters specified by (14.14). The normalization constant α is fixed by the gravitational theory to be $\alpha = \frac{\pi^d}{16\pi\, G_N^{(d+1)}}$.

Consider this d-parameter family of boosted planar Schwarzschild-AdS$_{d+1}$ geometries, each of which holographically map to an ideal (conformal) fluid living on $\mathbf{R}^{d-1,1}$ (endowed with the Minkowski metric). This fluid is in global thermal equilibrium and one should be able to describe the long wavelength but arbitrary-amplitude fluctuations away from equilibrium via hydrodynamics. This class of fluctuations has a bulk geometric avatar; we now describe an algorithmic procedure that enables us to construct such asymptotically locally AdS_{d+1} black hole geometries which are generically inhomogeneous and dynamical. This procedure crucially relies on the fact that hydrodynamics, as indeed any effective field theory, can be systematically studied in a gradient expansion.

14.3.2. *The perturbation theory*

We start by considering perturbations to the *seed* geometry characterizing equilibrium. Take the boosted planar Schwarzschild-AdS_{d+1} spacetime (14.17), for convenience rewritten in ingoing coordinates so as to remove the coordinate singularity on the horizon, and replace the parameters u_a and T by functions of the boundary coordinates x^a,

$$\begin{aligned} ds^2 = & -2\, u_a(x)\, dx^a\, dr - r^2\, f(r/T(x))\, u_a(x)\, u_b(x)\, dx^a\, dx^b \\ & + r^2\, P_{ab}(x)\, dx^a\, dx^b\,, \end{aligned} \tag{14.18}$$

with $f(r)$ as specified in (14.17) and P_{ab} given by (14.8) with $\gamma_{ab} = \eta_{ab}$. This metric, which we henceforth denote as $g^{(0)}_{\mu\nu}(T(x), u^a(x))$, is not a solution to Einstein's equations. It however has two felicitous features: (i) it is regular for all $r > 0$ and (ii) if the functions $T(x)$ and $u^a(x)$ are chosen so as to have small derivatives, then it can be approximated in local domains by a corresponding boosted black hole solution. These observations lead us to consider an iterative procedure to correct (14.18) order by order in a gradient expansion. We will find that we however cannot specify just any slowly varying $T(x)$ and $u^a(x)$ (recall that x^a included the temporal direction). A true solution to Einstein's equations is obtained only when the functions $T(x)$ and $u^a(x)$ in addition to being slowly varying satisfy a set of equations which happen to be precisely the conservation equations of fluid dynamics. Let us record that we have fixed gauge by setting $g_{rr} = 0$ and $g_{ra} = u_a$.

Since we want to keep track of the derivatives with respect to the boundary coordinates, it is useful to introduce a book-keeping parameter ε and regard the variables of the problem as functions of rescaled boundary coordinates εx^a. At the end of the day ε may be set to unity. With this in mind, let us consider the corrections to the seed metric in a gradient expansion:

$$g_{\mu\nu} = \sum_{k=0}^{\infty} \varepsilon^k g_{\mu\nu}^{(k)}(T(\varepsilon x), u^a(\varepsilon x)), \;\; u^a = \sum_{k=0}^{\infty} \varepsilon^k u^{a\,(k)}(\varepsilon x), \;\; T = \sum_{k=0}^{\infty} \varepsilon^k T^{(k)}(\varepsilon x) \tag{14.19}$$

where the correction pieces $g_{\mu\nu}^{(k)}$, $u^{a\,(k)}$ and $T^{(k)}$ are to be determined by solving Einstein's equations to the kth order in the gradient expansion. The ansatz (14.19) should therefore be inserted into the Einstein's equations (14.1) and the result expanded in powers of ε.

Let us examine the resulting structure in abstraction first. For the sake of argument, assume that we have determined $g_{\mu\nu}^{(m)}$ for $m \leq n-1$ and $T^{(m)}$ and $u^{a\,(m)}$ for $m \leq n-2$. At order ε^n one finds that Einstein's equations reduce to a set of *inhomogeneous linear differential equations* whose structure can be schematically written as

$$\mathbb{H}\left[g^{(0)}(T^{(0)}, u^{a\,(0)})\right] g^{(n)} = s_n \tag{14.20}$$

where we have dropped the spacetime indices for notational clarity (c.f., [1, 5] for the explicit equations). Since each derivative with respect to x^a is accompanied by a power of ε, it follows that the linear operator $\mathbb{H}$ is constructed purely from the data of the equilibrium Schwarzschild-AdS$_{d+1}$ geometry. This means that $\mathbb{H}$ is at most a second-order differential operator with respect to the radial variable r. Moreover, it has to be independent of n. Thus the perturbation theory in ε is ultra-local in the boundary coordinates, implying that we can solve the equations of motion of the bulk spacetime point by point on the boundary!

On the right hand side of (14.20) we collect all order ε^n terms which do not have explicit radial derivatives into a source term s_n, which is then a complicated construct involving contributions from different orders in perturbation theory. It is a local expression of $(n-m)^{\text{th}}$ order in boundary derivatives of $T^{(m)}$ and $u^{a\,(m)}$ for $m \leq n-1$, and ascertaining it is the most substantial part of the computation.

The reader may be puzzled by the following aspect of (14.20): while we have $\frac{d(d+1)}{2}$ equations, we have only $\frac{d(d-1)}{2}$ variables after fixing the gauge redundancy. This implies that a subset of Einstein's equations has a distinguished status as constraint equations, while the remainder are the physical dynamical equations.

To understand this let us examine the differential equations (14.20) by invoking the canonical split of our bulk coordinates $X^\mu = (r, x^a)$. The E_{ra} equations are the momentum constraint equations for 'evolution' in the radial direction. These equations are special in several ways. To start with, they need only be satisfied on a single r slice; the 'dynamical' equations (E_{ab}) then ensure that they will be solved on every r slice. For this reason, it is consistent to study these equations just at the boundary, where they turn out to reduce merely to the equations of conservation of the boundary stress tensor

$$\nabla_a T^{ab}_{(n-1)} = 0 \ . \tag{14.21}$$

Note that at n^{th} order the equations (14.21) depend only on the boundary stress tensor built out of the spacetime metric at order $n-1$. This is because (14.21) has an explicit boundary derivative which carries its own effective power of ε. The net upshot is that the unknown metric $g^{(n)}$ does not enter the equations (14.21) at all (the operator $\mathbb{H}$ in (14.20) vanishes for these solutions). Hence, (14.21) is instead a constraint on the solution already obtained at one lower order in perturbation theory. As we will see below, the solution for $g^{(n)}$ of the dynamical equations at each order in perturbation theory is uniquely obtained in terms of the previous order solution, and so, ultimately, in terms of the velocity and temperature fields that enter the starting ansatz (the zeroth-order term in perturbation theory). Consequently, (14.21) is an equation which constrains the starting velocity and temperature fields, and turns out to be the equation of boundary fluid dynamics.

The remaining equations E_{rr} (the 'Hamiltonian constraint' for radial evolution) and E_{ab} are dynamical equations with the operator $\mathbb{H}$ being a second order differential operator in r. Exploiting the spatial rotational symmetry of the seed solution, these equations can be decoupled and solved by quadratures,

$$g^{(n)} = \text{particular}(s_n) + \text{homogeneous}(\mathbb{H}) \ . \tag{14.22}$$

A unique solution to the dynamical equations is obtained upon specification of boundary conditions: normalizability at infinity and regularity in the interior for all $r > 0$. These turn out to specify the solution completely[i] and one ends up with a regular black hole geometry at each given order in the ε expansion.

In summary, at any order in the perturbative expansion one solves the constraint equations, enforcing fluid dynamical equations on the 'initial' data. One then solves for the corrected metric. This correction feeds into

[i]Modulo the fact that the operator $\mathbb{H}$ has zero modes which are to be accounted for by re-definitions of the background values of T and u^a.

the constraint equations giving corrected equations of fluid dynamics, and so on. The process may be iterated to any desired order, thereby yielding a systematic derivative expansion of the equations of fluid dynamics.

14.4. Results at 2nd Order

Having seen abstractly the iterative procedure which perturbatively corrects the seed metric to obtain a solution to Einstein's equations at arbitrary order in the gradient expansion, we now turn to the results of this construction (for now still considering only energy-momentum transport on the boundary). While our discussion so far has been restricted to the case of a flat boundary metric $\gamma_{ab} = \eta_{ab}$, the observation we made about the ultra-locality of the perturbation theory allows us to immediately generalize to slowly-varying curved boundary metrics. Given a metric γ_{ab} on the boundary, we can exploit the freedom to pass over to a Gaussian normal coordinate chart about the point under consideration, and account for the curvatures which arise starting with the second order in the ε expansion via the computation of appropriate source terms. We will therefore present the results below for this more general setting. Before we do so, we will take the opportunity to review a beautiful technical framework developed by [9] to simplify the results for conformal fluids.

14.4.1. *Weyl-covariant formalism*

The vacuum AdS_{d+1} spacetime is dual to the vacuum state of a conformal field theory. If we are interested in the hydrodynamic description of the latter on a background manifold $\mathcal{B}_d$, then rather than focusing on the metric γ_{ab} of this geometry, we can consider the conformal class of metrics $(\mathcal{B}_d, \mathcal{C})$. On this conformal class there is a natural derivative operator, defined through a Weyl connection, which efficiently keeps track of Weyl transformation properties of various operators. This is all the more natural in the context of fluid dynamics where there is a distinguished vector field, the velocity u^a, defined to be the (normalized) timelike eigenvector of the stress tensor.

Let us first start with local Weyl rescalings of the boundary metric which transforms homogeneously, i.e.,

$$\gamma_{ab} = e^{2\phi}\, \widetilde{\gamma}_{ab} \qquad \Leftrightarrow \qquad \gamma^{ab} = e^{-2\phi}\, \widetilde{\gamma}^{ab}. \tag{14.23}$$

We will call a tensor $\mathcal{Q}$ with components $\mathcal{Q}_{a_1\cdots a_n}^{b_1\cdots b_m}$ conformally covariant and of weight w if it transforms homogeneously under Weyl rescalings of the metric, i.e., $\mathcal{Q} = e^{-w\,\phi}\,\widetilde{\mathcal{Q}}$ under (14.23). The velocity field u^a transforms as a weight 1 tensor while the stress tensor T^{ab} of a conformal fluid has weight $(d+2)$ in d-spacetime dimensions.

One defines a class of torsionless connections, called the Weyl connections, characterized by a connection one form $\mathcal{A}_a$, whose associated covariant derivative $\nabla^{\rm Weyl}$ captures the fact that the metric transforms homogeneously under conformal transformations (with weight -2). In particular, for every metric in the conformal class $\mathcal{C}$,

$$\nabla_a^{\rm Weyl}\gamma_{bc} = 2\,\mathcal{A}_a\,\gamma_{bc}\,. \tag{14.24}$$

Given this derivative structure, we can go ahead and define a Weyl covariant derivative $\mathcal{D}_a = \nabla_a^{\rm Weyl} + w\,\mathcal{A}_a$ which is metric compatible and whose action on tensors transforming homogeneously with weight w (i.e., $\mathcal{Q}^{a\cdots}_{b\cdots} = e^{-w\,\phi}\,\widetilde{\mathcal{Q}}^{a\cdots}_{b\cdots}$) is given by

$$\begin{aligned}\mathcal{D}_c\mathcal{Q}^{a\cdots}_{b\cdots} \equiv\ & \nabla_c\mathcal{Q}^{a\cdots}_{b\cdots} + w\,\mathcal{A}_c\,\mathcal{Q}^{a\cdots}_{b\cdots} \\ & + \left(\gamma_{cd}\,\mathcal{A}^a - \delta^a_c\,\mathcal{A}_d - \delta^a_d\,\mathcal{A}_c\right)\,\mathcal{Q}^{d\cdots}_{b\cdots} + \cdots \\ & - \left(\gamma_{cb}\,\mathcal{A}^d - \delta^d_c\,\mathcal{A}_b - \delta^d_b\,\mathcal{A}_c\right)\,\mathcal{Q}^{a\cdots}_{d\cdots} - \cdots\,. \end{aligned} \tag{14.25}$$

The connection has been defined so that the Weyl covariant derivative of a conformally covariant tensor transforms homogeneously with the same weight as the tensor itself.

In hydrodynamics we will require that the Weyl covariant derivative of the fluid velocity be transverse and traceless,

$$u^a\,\mathcal{D}_a u^b = 0\,, \qquad \mathcal{D}_a u^a = 0\,, \tag{14.26}$$

which enables one to uniquely determine the connection one-form $\mathcal{A}_a$ to be the distinguished vector field

$$\mathcal{A}_a = u^c\,\nabla_c\,u_a - \frac{1}{d-1}\,u_a\,\nabla_c\,u^c = a_a - \frac{1}{d-1}\,\theta\,u_a\,, \tag{14.27}$$

built from the fluid velocity field.

One can rewrite the various quantities appearing in the gradient expansion of the stress tensor in this Weyl covariant notation. For instance, at first order in derivatives, we have the shear and vorticity constructed from the velocity field:

$$\sigma^{ab} = \mathcal{D}^{(a}u^{b)}\,, \qquad \omega^{ab} = -\mathcal{D}^{[a}u^{b]}\,, \tag{14.28}$$

both of which have weight $w = 3$. The fluid dynamical equations, viz., stress tensor conservation, are simply $\mathcal{D}_a\,T^{ab} = 0$ in this Weyl covariant language (which is equivalent to (14.3) since (14.25) with $w = d+2$ gives $\mathcal{D}_a T^{ab} = \nabla_a T^{ab} + T_a{}^a \mathcal{A}^b$ and the conformal fluid stress tensor must be traceless).

14.4.2. *Generic asymptotically AdS black hole metric*

We now have at our disposal all the technical machinery necessary to present the results for the gravity dual of non-linear fluid dynamics. By a suitable choice of gauge (a slight generalization of the Eddington-Finkelstein coordinates), one can express the bulk metric $g_{\mu\nu}$ in the form

$$ds^2 = -2\, u_a(x)\, dx^a \left(dr + \mathfrak{V}_b(r,x)\, dx^b\right) + \mathfrak{G}_{ab}(r,x)\, dx^a\, dx^b\,, \qquad (14.29)$$

where the fields $\mathfrak{V}_a$ and $\mathfrak{G}_{ab}$ are functions of r and x^a which admit an expansion in the boundary derivatives. In the parameterization used in [10] one finds the metric functions are given up to second order in derivatives as:

$$\begin{aligned}
\mathfrak{V}_a &= r\,\mathcal{A}_a - \mathcal{S}_{ac}\, u^c - \mathfrak{v}_1(r/T)\, P_a^{\ b}\, \mathcal{D}_c\, \sigma^c{}_b \\
&\quad + u_a \left[\frac{1}{2}\, r^2\, f(r/T) + \frac{1}{4}\,(1 - f(r/T))\; \omega_{cd}\, \omega^{cd} + \mathfrak{v}_2(r/T)\, \frac{\sigma_{cd}\, \sigma^{cd}}{d-1}\right] \\
\mathfrak{G}_{ab} &= r^2\, P_{ab} - \omega_a{}^c\, \omega_{cb} + 2\,(r/T)^2\, \mathfrak{g}_1(b\,r) \left[\frac{4\pi T}{d}\, \sigma_{ab} + \mathfrak{g}_1(r/T)\, \sigma_a{}^c\, \sigma_{cb}\right] \\
&\quad - \mathfrak{g}_2(b\,r)\, \frac{\sigma_{cd}\, \sigma^{cd}}{d-1}\, P_{ab} - \mathfrak{g}_3(r/T) \left[\mathfrak{T}_{1ab} + \frac{1}{2}\, \mathfrak{T}_{3ab} + 2\, \mathfrak{T}_{2ab}\right] \\
&\quad + \mathfrak{g}_4(r/T)\, \left[\mathfrak{T}_{1ab} + \mathfrak{T}_{4ab}\right].
\end{aligned} \qquad (14.30)$$

Here $\mathcal{S}_{ab} = \frac{1}{d-2}\left(\mathcal{R}_{ab} - \frac{\mathcal{R}}{2(d-1)}\,\gamma_{ab}\right)$ is the Schouten tensor of the boundary metric, where the Weyl covariant curvature tensors are

$$\begin{aligned}
\mathcal{R}_{ab} &= R_{ab} - (d-2)\left(\nabla_a \mathcal{A}_b + \mathcal{A}_a\, \mathcal{A}_b - \mathcal{A}^2\, \gamma_{ab}\right) - g_{ab} \nabla_c\, \mathcal{A}^c - \mathcal{F}_{ab} \\
\mathcal{R} &\equiv \mathcal{R}_a{}^a = R - 2\,(d-1)\, \nabla_c\, \mathcal{A}^c + (d-2)(d-1)\, \mathcal{A}^2
\end{aligned} \qquad (14.31)$$

with $\mathcal{F}_{ab} \equiv \nabla_a\, \mathcal{A}_b - \nabla_b\, \mathcal{A}_a$. Apart from the shear and vorticity tensors (14.28) constructed from the fluid velocity, we also encounter four of the five second order tensors which form a Weyl covariant basis,

$$\begin{aligned}
&\mathfrak{T}_1^{ab} = 2\, u^c\, \mathcal{D}_c \sigma^{ab}\,, \quad \mathfrak{T}_2^{ab} = C^{acbd}\, u_c\, u_d\,, \\
&\mathfrak{T}_3^{ab} = 4\, \sigma^{c\langle a}\, \sigma^{b\rangle}{}_c\,, \quad \mathfrak{T}_4^{ab} = 2\, \sigma^{c\langle a}\, \omega^{b\rangle}{}_c\,, \quad \mathfrak{T}_5^{ab} = \omega^{c\langle a}\, \omega^{b\rangle}{}_c\,.
\end{aligned} \qquad (14.32)$$

Note that the tensor $\mathfrak{G}_{ab}$ is clearly transverse, since it is built out of operators that are orthogonal to the velocity, and it can be inverted via the relation $\left(\mathfrak{G}^{-1}\right)^{ac} \mathfrak{G}_{cb} = P^a_{\ b}$. The induced metric on the boundary in these coordinates takes the form:

$$\gamma_{ab} = \lim_{r\to\infty} \frac{1}{r^2} \left(\mathfrak{G}_{ab} - 2\, u_{(a}\, \mathfrak{V}_{b)}\right), \tag{14.33}$$

which is crucially used to raise and lower the boundary indices.

Finally, the various functions $\mathfrak{g}_i$ and $\mathfrak{v}_i$ appearing in the metric are given in terms of definite integrals once one has inverted the operator $\mathbb{H}$:

$$\begin{aligned}
\mathfrak{g}_1(y) &= \int_y^\infty d\zeta\, \frac{\zeta^{d-1} - 1}{\zeta\, (\zeta^d - 1)} \\
\mathfrak{g}_2(y) &= 2\, y^2 \int_y^\infty \frac{d\xi}{\xi^2} \int_\xi^\infty d\zeta\, \zeta^2\, \mathfrak{g}_1'(\zeta)^2 \\
\mathfrak{g}_3(y) &= y^2 \int_y^\infty d\xi\, \frac{\xi^{d-2} - 1}{\xi\, (\xi^d - 1)} \\
\mathfrak{g}_4(y) &= y^2 \int_y^\infty \frac{d\xi}{\xi\, (\xi^d - 1)} \int_1^\xi d\zeta\, \zeta^{d-3} \Big(1 + (d-1)\, \zeta\, \mathfrak{g}_1(\zeta) + 2\, \zeta^2\, \mathfrak{g}_1'(\zeta)\Big) \\
\mathfrak{v}_1(y) &= \frac{2}{y^{d-2}} \int_y^\infty d\xi\, \xi^{d-1} \int_\xi^\infty d\zeta\, \frac{\zeta - 1}{\zeta^3\, (\zeta^d - 1)} \\
\mathfrak{v}_2(y) &= \frac{1}{2\, y^{d-2}} \int_y^\infty \frac{d\xi}{\xi^2} \Big[1 - \xi\, (\xi - 1)\, \mathfrak{g}_1'(\xi) - 2\, (d-1)\, \xi^{d-1} \\
&\qquad + \big(2\, (d-1)\, \xi^d - (d-2)\big) \int_\xi^\infty d\zeta\, \zeta^2\, \mathfrak{g}_1'(\zeta)^2\Big].
\end{aligned} \tag{14.34}$$

The asymptotic behavior of these functions $\mathfrak{g}_i(r/T)$ and $\mathfrak{v}_i(r/T)$ is important for the stress tensor computation of §14.4.4 and can be found in [10].

14.4.3. *Event horizon and entropy current*

The metric (14.29), (14.30) solves Einstein's equations to second order in the gradient expansion, provided the first order stress tensor (which takes the form (14.15) with the coefficients extracted from the 1st order bulk metric, and given explicitly below in §14.4.4) satisfies the hydrodynamic conservation equations. While this already establishes a firm connection between solutions of Einstein's equations and those of fluid dynamics (in one lower dimension), it is imperative to establish that the bulk geometry we describe is regular everywhere outside the curvature singularity at $r = 0$.

Although one can utilize the behavior of the metric functions and iteratively argue that the sources are regular order by order in perturbation theory, it is convenient to establish once and for all that what one has constructed is a black hole spacetime with a regular event horizon. Doing so involves ascertaining the location of the event horizon. A-priori, this sounds like a tall order, especially given that explicit solution is contingent on having solved the fluid equations. Moreover, as is well known, the event horizon is a teleological concept (it is the boundary of the past of future null infinity) whose determination requires knowing the entire future history of the spacetime. However, with one key assumption of late-time relaxation which is natural from fluid dynamics, it turns out to be possible to determine the location of the event horizon *locally* within our gradient expansion. Apart from showing regularity, this has the additional virtue of enabling us determine a natural entropy current for fluid dynamics [11].

Since generic flows of dissipative fluids tend to approach global equilibrium at late times, it follows that the corresponding event horizon has to approach the radial position determined by the local late-time temperature of the fluid. In particular, we look for a null co-dimension one surface given by the equation $S_{\mathcal{H}}(r,x) = r - r_{\mathcal{H}}(x) = 0$ with the correct asymptotics. The function $r_{\mathcal{H}}(x)$ should be parameterized within the gradient expansion $r_{\mathcal{H}}(x) = \frac{4\pi\, T(x)}{d} + \sum_k \varepsilon^k\, r_{(k)}(x)$. The corrections $r_{(k)}(x)$ are determined by solving the null condition $g^{\mu\nu}\, \partial_\mu S_{\mathcal{H}}\, \partial_\nu S_{\mathcal{H}} = 0$. The resulting equations are algebraic for $r_{(k)}$ and to second order in gradients one finds that, for the solution (14.29)-(14.34),

$$r_{\mathcal{H}}(x) = \frac{4\pi\, T(x)}{d} + \frac{d}{4\pi\, T(x)} \left(\aleph_1\, \sigma_{ab}\, \sigma^{ab} + \aleph_2\, \omega_{ab}\, \omega^{ab} + \aleph_3\, \mathcal{R}\right) \qquad (14.35)$$

with

$$\begin{aligned} \aleph_1 &= \frac{2\,(d^2+d-4)}{d^2\,(d-1)\,(d-2)} - \frac{2\,\mathfrak{v}_2(1)}{d\,(d-1)} \\ \aleph_2 &= -\frac{d+2}{2\,d\,(d-2)}\,, \qquad \aleph_3 = -\frac{1}{d\,(d-1)\,(d-2)}\,. \end{aligned} \qquad (14.36)$$

This indeed establishes that the solutions we have constructed in §14.4.2 to 2nd order qualify to be called inhomogeneous, dynamical black holes.

Note that in general, beyond the leading order, the horizon position and generators are not simply given by the corresponding fluid temperature and velocity (for example, while the horizon generators must be vorticity-free, ω_{ab} need not vanish for the boundary fluid). In some sense, while the black hole horizon is distinguished in the bulk, physics appears simpler when expressed in terms of the fluid data living on the boundary.

Having determined the event horizon of the gravity solution, we immediately have access to an important hydrodynamic quantity, viz., the *entropy current.* For a black hole spacetime it is natural to view the area of the event horizon as an entropy *a la* Bekenstein-Hawking [12, 13]. In fact, by suitably foliating the event horizon with spatial slices (propagated forward by the null generator), we can equivalently talk about an area $(d-1)$-form $a_{\mathcal{H}}$ on these slices. Since we imagine the dual fluid living on the boundary of the spacetime, it is natural to pull-back this area form out to the boundary. A canonical choice is to pull-back along radially ingoing null geodesics [11], which is quite easy to implement for the metric (14.29), where the lines of $x^a =$ constant are precisely such geodesics. We then have a $(d-1)$-form on the boundary which can be dualized to a one-form or equivalently a current J^a_s, which is the entropy current on the boundary. Not only does this definition agree with the equilibrium notion of entropy of the fluid, but also thanks to the area theorem of black hole horizons, we are immediately guaranteed that this current has manifest non-negative divergence as demanded by the second law. The hydrodynamic entropy current takes the general form

$$\begin{aligned} J^a_s &= s\,u^a + \frac{s\,d^2}{(4\pi\,T)^2}\,u^a\,\left(A_1\,\sigma_{cd}\,\sigma^{cd} + A_2\,\omega_{cd}\,\omega^{cd} + A_3\,\mathcal{R}\right) \\ &\quad + \frac{s\,d^2}{(4\pi\,T)^2}\,\left(B_1\,\mathcal{D}_c\,\sigma^{ac} + B_2\,\mathcal{D}_c\,\omega^{ac}\right) + \cdots \end{aligned} \tag{14.37}$$

where s is the entropy density and $A_{1,2,3}$, $B_{1,2}$ are a-priori arbitrary numerical coefficients. While $\mathcal{D}_a J^a_s \geq 0$ only demands that $B_1 + 2\,A_3 = 0$, the gravity solution (14.29) fixes all the coefficients in (14.37) explicitly. In particular, we obtain

$$\begin{aligned} s &= \frac{1}{4\,G_N^{(d+1)}}\left(\frac{4\pi\,T}{d}\right)^{d-1}, \qquad A_1 = \frac{2}{d^2}\,(d+2) - \left(\frac{1}{2}\,\mathfrak{g}_2(1) + \frac{2}{d}\,\mathfrak{v}_2(1)\right), \\ A_2 &= -\frac{1}{2\,d}, \qquad B_1 = -2\,A_3 = \frac{2}{d\,(d-2)}, \qquad B_2 = \frac{1}{d-2}\,. \end{aligned} \tag{14.38}$$

We should note here that there is an ambiguity in pulling back the area-form from the event horizon to the boundary, for one can supplement the pull-back map with a boundary diffeomorphism, which affects the coefficient A_1 above. Since this just relabels boundary points in the gradient expansion, one is tempted to think of this ambiguity as unphysical. However, it is rather curious that if one tries to pull-back the area form from quasi-local horizons one encounters a shifted value of A_1 [14], which suggests that there is perhaps more to this ambiguity than meets the eye.

14.4.4. *Stress tensor of dissipative fluid*

Given an asymptotically locally AdS_{d+1} metric, one can construct a quasi-local boundary tensor which is manifestly conserved and is associated with the stress-energy-momentum tensor of the conformal field theory [6, 15]. To perform the computation one regulates the bulk spacetime by introducing an explicit cut-off at $r = r_\infty$. The boundary stress tensor is given in terms of the extrinsic curvature K_{ab} of this surface, defined in terms of its unit outward pointing normal n^a as $K_{ab} = \gamma_{ac}\, \nabla^c n_b$. In addition to the extrinsic curvature one also has contributions from the counter-terms necessary to obtain a finite boundary stress tensor. Denoting the curvatures of the boundary metric by ${}^\gamma R$ etc., this is given (to 2nd order) as

$$T_{ab} = \lim_{r_\infty \to \infty} \frac{r_\infty^d}{16\pi\, G_N^{(d+1)}} \left[K_{ab} - K\,\gamma_{ab} - (d-1)\,\gamma_{ab} - \frac{1}{d-2}\left({}^\gamma R_{ab} - \frac{1}{2}\, {}^\gamma R\, \gamma_{ab} \right) \right] \tag{14.39}$$

For the gravity duals to fluid dynamics constructed in §14.4.2, one finds that the boundary stress tensor takes the form (14.5) with the dissipative part, at the first and second order, given by this gravitational construction to be

$$\begin{aligned} \Pi^{ab}_{(1)} &= -2\,\eta\,\sigma^{ab} \\ \Pi^{ab}_{(2)} &= \tau_\pi\,\eta\,\mathfrak{T}_1^{ab} + \kappa\,\mathfrak{T}_2^{ab} + \lambda_1\,\mathfrak{T}_3^{ab} + \lambda_2\,\mathfrak{T}_4^{ab} + \lambda_3\,\mathfrak{T}_5^{ab}\,. \end{aligned} \tag{14.40}$$

where the tensors $\mathfrak{T}_i^{ab}$ were defined earlier in (14.32). With the tensor structure determined, one is just left with fixing the six transport coefficients, η, τ_π, κ, and λ_i for $i = \{1, 2, 3\}$, which completely characterizes the flow of a non-linear viscous fluid with a gravitational dual. The transport coefficients for conformal fluids in d-dimensional boundary turn out to be

$$\eta = \frac{1}{16\pi\, G_N^{(d+1)}} \left(\frac{4\pi}{d}\, T\right)^{d-1} ,$$
$$\tau_\pi = \frac{d}{4\pi\, T}\left[1 + \frac{1}{d}\,\text{Harmonic}\left(\frac{2}{d} - 1\right)\right], \qquad \kappa = \frac{d}{2\pi\,(d-2)}\,\frac{\eta}{T} , \tag{14.41}$$
$$\lambda_1 = \frac{d}{8\pi}\,\frac{\eta}{T} , \qquad \lambda_2 = \frac{1}{2\pi}\,\text{Harmonic}\left(\frac{2}{d} - 1\right)\frac{\eta}{T} , \qquad \lambda_3 = 0 .$$

where Harmonic(x) is the harmonic number function. Setting $d = 4$ in the above expressions and using the fact that Harmonic$(-\frac{1}{2}) = -2\log(2)$ together with the replacement $\frac{1}{16\pi\, G_N^{(5)}} = \frac{1}{8\pi^2}\, N^2$, one can obtain the transport coefficients for $SU(N)$ $\mathcal{N} = 4$ Super Yang-Mills theory [1, 16]. This has been used for real data analysis from e.g. RHIC.

One immediate consequence of (14.41) and (14.38) is that our fluid saturates the famous bound on the viscosity to entropy density ratio, $\frac{\eta}{s} \geq \frac{1}{4\pi}$, [17]. This bound is saturated by a large class of two-derivative theories of gravity, and it is indeed experimentally satisfied by all presently-known systems in nature. Intriguingly, cold atoms at unitarity and quark-gluon plasma both come near to saturating this bound [18]. Its status in more general theories is currently under active debate [19].

Moreover, (14.41) reveals further intriguing relations between the coefficients, which hint at the specific nature of any conformal fluid which admits a gravitational dual. For example, the result that $\lambda_3 = 0$ is universal but non-trivial from the fluid standpoint. We also see that $2\,\eta\,\tau_\pi = 4\lambda_1 + \lambda_2$ for all d; this in fact was shown to hold quite generally in a large class of two-derivative theories of gravity (including matter couplings) [20].

14.5. Specific Fluid Flows and Their Gravitational Analog

The construction presented above can be generalized in many interesting ways; however before indicating the most important of these in §14.6, we first pause to discuss some of the special cases of the framework explained in the preceding section. One of the reasons to discuss these special cases is that while we have demonstrated the existence of a map from the equations governing fluid dynamics to those governing the dynamics of gravity, we did not at any stage solve the fluid equations explicitly. The felicitous feature of our construction was the ultra-locality along the boundary directions which allowed us to implement the construction in terms of local solutions to the conservation equations. Construction of novel fluid flows and generic behavior of the relativistic conservation equations are interesting (and perhaps hard) questions. Nevertheless there are some corners where we can gain analytic control which serves not only as a check that the fluid/gravity solution set is non-empty, but also provides a point of contact with previous studies of the hydrodynamic regime in the AdS/CFT literature.

14.5.1. *Linearized setting: quasinormal modes*

Above we have established a map between any solution of the equations of fluid dynamics and long wavelength solutions of Einstein gravity with a negative cosmological constant. In order to find explicit gravitational solutions we need a class of explicit solutions to the equations of fluid dynamics. In this subsection and the next we will study such examples.

It is of course easy to solve the equations of fluid dynamics, derived above, when linearized about static equilibrium. Utilizing translational invariance, we search for solutions of the form

$$u^a = \delta^a_t + \delta^a_j \, \delta v^j \, e^{i(\omega t + k^i y_i)} \,, \qquad T = T_0 + \delta T \, e^{i(\omega t + k^i y_i)} \,,$$

with purely spatial velocity fluctuations δv^j. The resulting linear equations require that the matrix of coefficients $M(\omega, k)$ annihilate the length-d column vector with entries δv^i and δT. So the spectrum $\omega(k)$ is obtained as the roots of the dth order polynomial $\det(M) = 0$. At leading (ideal fluid) order, the d roots to this equation turn out to be $\omega = \pm \frac{k}{\sqrt{d-1}}$ (the sound modes of the fluid) and $\omega = 0$ with degeneracy $d-2$ (the shear modes of the fluid). These modes and their corresponding eigenvectors receive corrections at higher orders in the derivative expansion; in particular the shear modes pick up nontrivial k dependence, $\omega \propto i\,k^2$, at first order. The explicit solutions are easily determined (see [1]). Employing the fluid/gravity map then yields explicit linearized solutions of the Einstein equations (14.1) about the planar black hole background, whose study in fact predates the fluid/gravity map by almost 10 years.

The spectral problem of linearized fluctuations about a black hole solution is of course a well-studied topic (cf., [21]). It is known that due to the presence of a horizon, black holes admit no normal modes. Instead, by imposing regularity at the future event horizon, one finds quasinormal modes, i.e., modes which have complex frequencies characterizing decay of perturbations at late times. Mathematically these are related to the poles of the retarded Green's function computed in the black hole background.

In terms of the gauge/gravity perspective, such quasinormal modes describe the timescale for return to thermal equilibrium in the field theory [22]. Asymptotically AdS black holes host an infinite family of quasinormal modes. All except d of these are 'massive'; their frequency remains finite (and has a finite imaginary part or decay rate $\propto T$) even in the limit $k \to 0$. However, planar black holes admit exactly d special 'massless' quasinormal modes. These so-called hydrodynamic modes can have arbitrarily low frequency at long spatial wavelengths and therefore fall within the long wavelength regime. These are precisely the sound and shear modes described above.

The fact that the dispersion relations of the massless quasinormal modes agrees with hydrodynamic dispersion relations was first demonstrated in the pioneering works [23, 24], who mapped the Schwarzschild-AdS quasinormal modes to sound and shear modes of the dual field theory (perturbed from thermal equilibrium).

14.5.2. *Rotating black holes in global AdS space*

A second class of examples which accord analytic control are explicit solutions corresponding to stationary configurations in hydrodynamics. While on flat space there are no interesting stationary flows other than a uniformly boosted fluid (whose dual is the seed solution we started with), it turns out that by a suitable choice of background geometry one can derive nontrivial flows. We now describe one such flow on the Einstein Static Universe ($\mathbb{R} \times \mathbf{S}^{d-1}$) which allows one to make contact with rotating black holes in asymptotically (globally) AdS spacetimes.

Fluid flows of a conformal relativistic d-dimensional fluid on spatial $\mathbf{S}^{d-1}$ conserve angular momentum in addition to energy. The angular momentum on $\mathbf{S}^{d-1}$ is a rank-d antisymmetric matrix, which can be brought to canonical form by an $SO(d)$ similarity transform (a rotation) and is therefore labeled by its $[d/2]$ inequivalent eigenvalues. For every physically allowed choice of these $[d/2]$ angular momenta together with the energy, an arbitrary fluid flow eventually settles down into an equilibrium stationary configuration.

The stationary configurations of viscous conformal fluids on spheres turn out to be extremely simple. The velocity field is simply that of a rigid rotation. Focusing on the case of $d = 2n$ for concreteness, the metric of a unit $\mathbf{S}^{2n-1}$ may be written in terms of the direction cosines μ_i as

$$ds^2_{\mathbf{S}^{2n-1}} = \sum_{i=1}^{n} \mu_i^2\, d\phi_i^2 + d\mu_i^2\,, \qquad \text{where} \qquad \sum_{i=1}^{n} \mu_i^2 = 1\,. \tag{14.42}$$

In these coordinates the velocity and temperature fields of stationary flows take the form

$$u_s^a\,\partial_a = \gamma\left(\partial_t + \sum_{i=1}^{n} \omega_i\,\partial_i\right), \quad T_s = \gamma\, T_0\,, \quad \gamma = \left(1 - \sum_{i=1}^{m} \omega_i^2 \mu_i^2\right)^{-\frac{1}{2}}. \tag{14.43}$$

This flow is Weyl equivalent to a uniform velocity and temperature configuration on an Einstein Static Universe [4], i.e.,

$$ds^2 = \gamma\left(-dt^2 + ds^2_{\mathbf{S}^{2n-1}}\right), \quad u_s^a\,\partial_a = \partial_t + \sum_{i=1}^{n} \omega_i \partial_i\,, \quad T_s = T_0\,. \tag{14.44}$$

It is not difficult to verify that (14.44) provides a non-dissipative solution to the equations of fluid dynamics at least to second order (the solution is non-dissipative because it is stationary; equivalently the divergence of the entropy current vanishes). The constant parameters ω_i and T_0 of this solution turn out to have thermodynamical significance: they are simply the angular velocity (chemical potential for angular momentum) and temperature of the fluid configuration.

According to the AdS/CFT correspondence, the dual description of a conformal field theory on $\mathbb{R} \times \mathbf{S}^{d-1}$ is simply asymptotically global AdS_{d+1} space. If we pump a large amount of energy and angular momentum into global AdS_{d+1} space and let the system relax, we expect the eventual equilibrium configuration to be that of a large rotating black hole. This reasoning leads to a prediction: the fluid/gravity map applied to (14.44) should produce the (independently known) metric of large rotating AdS_{d+1} black hole, expanded to second order in the derivative expansion. This prediction[j] has been verified in detail in the following manner. It is possible to transform the exactly known metric of the rotating AdS_{d+1} black holes to the fluid/gravity gauge described in §14.3 above. This maneuver in fact

[j] The first observation that the properties of large rotating black holes should be reproduced from fluid dynamics was made in [4], a precursor to the fluid/gravity map. Rotating black holes were further studied in [25] in 4 dimensions, and fully analyzed in d dimensions in [10]. Here for illustration we give the general result of [10], who find the explicit coordinate transformation to rewrite the rotating black hole solution in AdS_{d+1} of [26] in fluid variables.

turns out to greatly simplify the rotating black hole metric which takes the form (14.29) with

$$\begin{aligned} \mathfrak{V}_a(r,x) &= r\,\mathcal{A}_a - \mathcal{S}_{ac}\,u^c - \frac{(4\pi\,T)^d\,r^2\,u_a}{2\,d^2\,\det\left[r\,\delta^b{}_c - \omega^b{}_c\right]}\,, \\ \mathfrak{G}_{ab}(r,x) &= r^2\,P_{ab} - \omega_a{}^c\,\omega_{cb}\,. \end{aligned} \tag{14.45}$$

The above is an exact rewriting of the rotating AdS black holes (in even dimensions) of [26]. The metric (14.45) may be expanded in derivatives simply by expanding the inverse determinant in powers of ω. By truncating this expansion at second order we recover exactly the metric dual to the fluid flow (14.44). It is rather remarkable the full black hole solution can be written economically within the fluid/gravity metric ansatz, perhaps suggesting greater utility for metrics of the form (14.29).

14.5.3. *Non-relativistic fluids*

While relativistic fluids are interesting in astrophysical or high energy plasma physics contexts, most fluids we encounter in everyday situations are non-relativistic. Furthermore, for many practical applications one is usually interested in their dynamics in the incompressible regime, which is attained by projecting out the sound mode. It is natural to ask whether this regime is accessible to fluid/gravity; an affirmative answer is suggested by the discussion in §14.2.2, namely one only needs to implement the Navier-Stokes scaling limit directly in the fluid/gravity solutions. This procedure has been carried out in [27] to obtain the gravitational dual of non-relativistic incompressible fluid flows. In principle this provides a geometric window to explore phenomenologically interesting fluid flows.

14.6. Extensions Beyond Conformal Fluids

The fluid/gravity correspondence was originally derived for the case of conformal fluids which are related to gravitational dynamics in an asymptotically AdS spacetime. Conformal theories are rather special and 'ordinary' fluids one encounters everyday deviate significantly from this behavior. Hence it would be useful to look for extensions of the basic framework which allows for these generalizations. As already indicated earlier, this can be done at the expense of complicating the system of gravitational equations, and accompanying loss of universality. Nevertheless, forays into these areas have revealed very interesting lessons about fluid dynamics in general that transcend the fluid/gravity correspondence. In this section we take stock of some of these developments.

14.6.1. *Non-conformal fluids*

The first generalization we consider is provided by a handy trick to obtain a particular class of non-conformal theories. It turns out that by exploiting the gauge/gravity duality for a special class of theories, viz., theories that naturally arise on the world-volume of Dp-branes, one finds a surprisingly tractable class of non-conformal fluids.

Let us first consider the special case of the D4-brane which is a solution of the equations of 10-dimensional IIA supergravity. IIA supergravity is the dimensional reduction of 11-dimensional supergravity on $\mathbf{S}^1$, and the D4-brane solution is the dimensional reduction of a M5 brane solution that wraps the $\mathbf{S}^1$. The near horizon geometry of the M5 brane solution is $AdS_7 \times \mathbf{S}^4$, and the 11-dimensional equations admit a consistent truncation to 7-dimensional equations of motion (14.1) involving gravitational dynamics with a negative cosmological constant. Compactifying further on an $\mathbf{S}^1$ and restricting to the zero momentum sector on this circle, yields a further consistent truncation of this seven dimensional set of equations. The resulting 6-dimensional equations are simply the Einstein-dilaton equations about the D4-brane background. It follows that the Einstein-dilaton equations constitute a consistent truncation of the equations of IIA supergravity about the D4 near-horizon background (as is easily independently verified).

It turns out that the fluid dynamics dual to the long wavelength fluctuations about the thermal M5 brane solution is simply that computed in §14.3, for the special case $d = 6$. We want to focus on gravitational solutions corresponding to fluid flows independent of one of the boundary directions x^i (the direction of the $\mathbf{S}^1$ in the paragraph above). These lie within the six dimensional consistent truncation described above. But these are simply the gravitational solutions dual to fluid flows on the world volume theory of the D4-brane. In other words, the fluid dynamics of the D4-brane world volume theory is a dimensional reduction of the conformal fluid dynamics on the world volume of the M5 brane. Moreover the gravitational duals to D4-brane fluid flows are very easily obtained from the KK reduction of the results of §14.3. Note that the dimensional reduction of conformal fluid dynamics results in non-conformal fluid dynamics (e.g. the dimensional reduction of a traceless stress tensor generically has non-vanishing trace).

It is an interesting and surprising fact that the discussion of the previous paragraph generalizes to Dp-branes for all p at the purely algebraic level. In every case one can find a consistent truncation of the Einstein-dilaton system which, in a purely formal manner, can be regarded as the reduction of negative cosmological constant Einstein gravity in a higher (sometimes fractional) dimension [28]. This observation immediately yields the fluid descriptions of arbitrary Dp-brane backgrounds as a dimensional reduction of the conformal fluid dynamics derived in §14.3.

14.6.2. *Theories with a deconfinement transition*

So far, we have studied the fluid dynamical description of field theories that are 'deconfined' at every temperature, i.e., the free energy is $\mathcal{O}(N^2)$. Consider, however, a theory like pure Yang Mills at large N which undergoes a first order deconfinement phase transition at finite temperature. Such a system has a dual description in terms of a black hole only above the deconfinement temperature; the low temperature phase is given by a gas of glueballs, and is thermodynamically indistinguishable from the vacuum at leading order in N (free energy is $\mathcal{O}(1)$).

The Scherk-Schwarz reduction of $\mathcal{N} = 4$ Yang Mills on a circle of radius R (with anti-periodic boundary conditions for fermions) is a simple example of such a theory. At strong coupling this theory undergoes a first order deconfinement transition at $T\,R = 2\pi$. The gravity dual of the high temperature phase is simply the $\mathbf{S}^1$ compactification of the AdS_5 planar black hole. The gravity dual of the low temperature phase is a so-called AdS-soliton (a double analytic continuation of the planar Schwarzschild-AdS black hole, in which the role of time and the $\mathbf{S}^1$ direction are interchanged). At temperatures much higher than the phase transition, the effective 3-dimensional low energy theory is simply the dimensional reduction of the 4-dimensional conformal fluid system derived in earlier subsections (just as in §14.6.1).

However, at the phase transition temperature we have a new phenomenon; there exists a new static solution of the equations of motion, a co-dimension one domain wall that interpolates between the AdS-soliton and the $\mathbf{S}^1$ compactification of the planar AdS_5 black hole. Unfortunately this solution has been constructed only numerically [29]. The domain wall is static in these solutions because of a pressure balance on the two sides (recall that the free energy, and hence the pressure, of the two phases are equal at a phase transition temperature). This configuration is that of a fluid with a boundary; the effective low energy fluctuations of this system consist of boundary modes (like waves on the surface of water) in addition to the bulk modes discussed so far in this chapter. At the ideal fluid level the action for boundary degrees of freedom is captured by a single parameter, the surface tension of the boundary (computed from the numerical solution).

Already using the ideal fluid action including boundary terms, it has proved possible to construct many stationary solutions of the fluid equations. These solutions, called plasma-balls and plasma-rings, have dual descriptions as black holes, black rings, and (in higher dimensions) black objects of more exotic topology [30]. The effective action for surface degrees of freedom has not been worked out at higher orders in the derivative expansion, and appears to be an interesting exercise.

One particularly interesting static solution of the equations of ideal fluid dynamics including boundary terms is the plasma-tube; a configuration consisting of a domain wall that interpolates from the vacuum to the high temperature phase at the phase transition temperature, followed by a second parallel domain wall at separation L that interpolates back to the vacuum. Such a fluid configuration is the two dimensional analogue of a 3-dimensional cylindrical tube of fluid, and is well known to undergo a famous fluid dynamical instability (to droplet formation) called the Rayleigh instability. For real fluids such as water, the endpoint of the Rayleigh instability is a series of disconnected droplets. Now the gravitational dual of the plasma tube is an infinitely long black string in 5-dimensional gravity. This solution has the well known Gregory-Laflamme instability [Ch.GL] which, apparently, is dual to the Rayleigh instability in the long wavelength limit. The boundary dual of a series of disconnected droplets, on the other hand, is a series of disconnected black holes. This discussion at least strongly suggests that the end point of the Gregory-Laflamme instability consists of localized black holes [31]. Note that the fluid description breaks down near the 'pinch off' point; the actual description of topology change in this process requires the use of the full field theory (e.g. details of interactions between water molecules in the case of water).

14.6.3. *Charged fluids and anomalies*

Under the AdS/CFT correspondence a global symmetry in the boundary maps to a gauge symmetry in the bulk. This suggests that there should be a duality between the long wavelength asymptotically AdS planar black hole solutions of the Einstein-Maxwell theory (with negative cosmological constant) and the equations of charged fluid dynamics. This is a useful extension as fluids of interest in experimental situations conserve one or more $U(1)$ charges in addition to energy and momentum. For instance, the flow of air in the atmosphere conserves air molecule number.

It is conceptually straightforward to generalize the set-up of section §14.3 to the study of locally thermalized charged planar black holes. The starting point is the construction of spacetimes that tubewise approximate Reissner-Nordström AdS black hole solutions with locally varying temperature, chemical potential and velocity. A perturbation expansion entirely analogous to the one outlined in §14.3 then constructs the gravitational solutions to the Einstein-Maxwell-Chern-Simons theory (which forms a consistent truncation of IIB supergravity on $AdS_5 \times \mathbf{S}^5$) dual to charged fluid flows order by order in the derivative expansion [32, 33]. This procedure also determines the equations of charged fluid dynamics to order by order in the derivative expansion. The results of this analysis turned out to throw up a surprise purely from the viewpoint of charged first order fluid dynamics, as we now explain.

The equations of charged fluid dynamics are the conservation of the charge current

$$\nabla_a J^a = 0 \tag{14.46}$$

together with the conservation of the stress tensor (14.3). Concrete fluid dynamical equations require constitutive relations that express the field redefinition invariant parts of the stress tensor and charge current in terms of expressions of first order in the derivative of fluid fields. The charge current for charge density q takes the form

$$J^a = q\, u^a + J^a_{diss} \,, \tag{14.47}$$

where J^a_{diss} represents the contribution of terms with one or more derivatives of the fluid fields to the charge current, and is to be viewed as the charge current analogue of Π^{ab}; similarly to (14.6) we expand

$$J^a_{diss} = \sum_{n=1}^{\infty} \ell_{\rm m}^n \, j^a_{(n)} \,. \tag{14.48}$$

Standard textbook analyses assert that the most general allowed form at first order for the constitutive relations of a relativistic charged fluid are the first equation of (14.7) along with

$$P^a_{\ c}\left(j^c_{(1)} + \frac{q}{\rho+P}(u_b\,\Pi^{bc}_{(1)})\right) = \kappa\, V^a_1 \,, \quad V^a_1 \equiv -P^{ab}\nabla_b \frac{\mu}{T} + \frac{F^{ab}u_b}{T} \tag{14.49}$$

while the second equation of (14.7) gets modified to

$$\frac{1}{d-1}\,\Pi^{ab}_{(1)} P_{ab} - \frac{\partial P}{\partial \rho}\,(u_a\, u_b\, \Pi^{ab}_{(1)}) + \frac{\partial P}{\partial q}(u_a j^a_{(1)}) = -\beta\, \nabla_c\, u^c\ . \tag{14.50}$$

In (14.49) F^{ab} is the non-dynamical background electromagnetic field that couples to the $U(1)$ current J^a in (14.46) and μ is the chemical potential of the fluid. Provided that

$$\eta \geq 0, \quad \kappa \geq 0, \quad \beta \geq 0,$$

these expressions are consistent with the positivity of divergence of the fluid entropy current

$$J^a_{can} = s\, u^a - \frac{1}{T}\, u_b\, \Pi^{ab}_{(1)} - \frac{\mu}{T}\, j^a_{(1)}\,, \tag{14.51}$$

using the alleged relation

$$\nabla_a J^a_{can} = -\nabla_a \left(\frac{u_b}{T}\right) \Pi^{ab}_{(1)} - \left(\nabla_a \left(\frac{\mu}{T}\right) - \frac{F_{ab} u^b}{T}\right) j^a_{(1)}. \tag{14.52}$$

However it was found by explicit computation that the fluid dual to the asymptotically AdS-Einstein-Maxwell-Chern-Simons system has constitutive relations that differ from those of (14.49) in the following fashion: the r.h.s. of (14.49) includes new terms proportional to fluid vorticity ω^a and rest frame magnetic field B^a where

$$\omega^a = \frac{1}{2}\epsilon^{abcd}\, u_b\, \partial_c u_d\ , \qquad B^a = \frac{1}{2}\epsilon^{abcd}\, u_b\, F_{cd}\ .$$

In a beautiful paper [34] pointed out the reason for the appearance of these new terms. When the $U(1)$ current has a global $U(1)^3$ triangle anomaly (as is true of the field theory dual to a bulk system with a 5-dimensional Chern-Simons term), (14.52) has an additional term on its r.h.s. proportional to this anomaly. This term spoils the positivity of the divergence of the canonical entropy current in the presence of such a field. It is however consistent with the positivity of the divergence of a modified entropy current provided that modifications are also made to the r.h.s. of (14.49). More concretely, positivity of the entropy current in every conceivable circumstance requires that, in addition to the first equation of (14.7) and (14.50),

$$\begin{aligned} J^a_S &= J^a_{can} + \sigma_\omega\, \omega^a + \sigma_B\, B^a \\ P^a_{\ c}\left(j^c_{(1)} + \frac{q}{\rho+P}(u_b\, \Pi^{bc}_{(1)})\right) &= \kappa\, V^a_1 + \tilde{\kappa}_\omega\, \omega^a + \tilde{\kappa}_B\, B^a \end{aligned} \tag{14.53}$$

where

$$\begin{aligned}
\sigma_\omega &= c\,\frac{\mu^3}{3T} + T\,\mu\,k_2 + T^2\,k_1 \\
\sigma_B &= c\,\frac{\mu^2}{2T} + \frac{T}{2}\,k_2 \\
\tilde{\kappa}_\omega &= c\left(\mu^2 - \frac{2}{3}\,\frac{q}{\rho+P}\,\mu^3\right) + T^2\left(1 - \frac{2q}{\rho+P}\,\mu\right)k_2 - \frac{2q}{\rho+P}\,k_1 \\
\tilde{\kappa}_B &= c\left(\mu - \frac{1}{2}\,\frac{q}{\rho_n+P}\,\mu^2\right) - \frac{T^2}{2}\,\frac{q}{\rho+P}\,k_2
\end{aligned} \tag{14.54}$$

and k_1 and k_2 are integration constants. Further imposition of CPT invariance forces k_2 to vanish.

This explanation accounts for the additional transport coefficients in the AdS/CFT duality, but applies more generally to every fluid flow with a $U(1)^3$ anomaly. The effect of these new transport coefficients may well turn out to have experimentally measurable effects in the relativistic heavy ion collisions or in neutron or quark stars.

14.6.4. *Holographic superfluid hydrodynamics*

It was pointed out by [35] that charged asymptotically AdS_5 planar black holes are sometimes unstable in the presence of charged scalar fields. The endpoint of this instability is a hairy black hole: a black hole immersed in a charged scalar condensate. The AdS/CFT correspondence maps the hairy black hole to a phase in which a global $U(1)$ symmetry is spontaneously broken by the vacuum expectation value of a charged scalar operator (see [Ch.CM] for further discussion). In condensed matter physics a phase with a spontaneously broken global $U(1)$ symmetry is referred to as a superfluid.

The variables of relativistic superfluid dynamics consist of two velocity fields, the normal fluid velocity u^a and a superfluid velocity field u_s^a, together with a temperature and chemical potential field. The superfluid velocity is the unit vector in the direction of $-\xi_a$ where ξ_a is the gradient of the phase of the scalar condensate. Conservation of the stress tensor and charge current together with the assertion that ξ_a is curl free constitute the equations of superfluid dynamics. These equations form a closed dynamical system once they are supplemented with constitutive relations that express the stress tensor, charge current and the component of ξ_a along the normal velocity, as functions of the fluid dynamical variables.

It has proved possible to apply the fluid/gravity map to hairy black holes to derive the constitutive relations for holographic superfluids, with interesting results. The theory of perfect superfluids was worked out by Landau and Tisza in the 1940s. In a beautiful recent work [36] have used the equations of Einstein gravity to re-derive Landau-Tisza equations for superfluids that admit a holographic description. The theory of first order dissipative corrections to the equations of Landau-Tisza superfluidity was most completely spelled out in [37]. Calculations done within the fluid/gravity framework have led to the realization that the 13 parameter Clark-Putterman equations derived therein miss one parameter (under the assumption of parity invariance for the superfluids) and 6 more parameters (if the superfluids are not assumed to preserve parity). A completely satisfactory framework for superfluid hydrodynamics has been developed only very recently [7], and the fluid/gravity map has played a major role in this development.

14.7. Relation to Other Developments

Having surveyed the fluid/gravity correspondence and its various applications, we finally describe connections with other approaches.

14.7.1. *Implications for Israel-Stewart formalism*

One useful application of the fluid/gravity correspondence is an improvement on the 'causal relativistic hydrodynamics', also known as the Israel-Muller-Stewart formalism [38, 39]. To appreciate the context, recall that a conventional theory of relativistic dissipative (i.e., irreversible) hydrodynamics, which is first order in time derivatives, is described in terms of a parabolic system of differential equations, leading to instantaneous propagation of signals. While these apparently a-causal modes lie outside of the long wavelength regime of validity of the hydrodynamical formulation as discussed in [40], they nevertheless lead to conceptual and computational problems. To capture the dissipative physics, [38] observed that second order terms are needed in the entropy current. These render the system hyperbolic, thereby providing a good initial value formulation. However, the particular terms added are not all the possible ones consistent with the symmetries, so as such, the construction is somewhat ad-hoc. Indeed, [16] observed in the context of conformal fluid, that the terms added do not maintain conformal invariance of the system, manifesting the incompleteness of the approach. The fluid/gravity construction in effect prescribes the correct completion to render the full system causal, as well as manifestly consistent with the symmetries. We expect that due to the gravitational dual, causality will be guaranteed at all orders in the derivative expansion.

14.7.2. *The black hole membrane paradigm*

Perhaps the most salient feature of the fluid/gravity correspondence is the fact that the horizon dynamics (which in this case prescribes the dynamics of the entire spacetime) is governed by hydrodynamics. At the face of it, such type of relation is not new; in fact for several decades relativists have explored the idea that spacetime, or important aspects thereof like black hole horizons, might resemble a fluid. Early indications include black hole thermodynamics [12, 13] developed in the 70's, analog models of black holes [41] initiated in the early 80's, and most strikingly the black hole Membrane Paradigm [42, 43] formulated in the late-70's. The latter realizes the idea that for external observers, black holes behave much like a fluid membrane, endowed with physical properties such as viscosity, conductivity, and so forth. In particular, the dynamics of this membrane is governed by the familiar laws of fluid dynamics, namely the compressible Navier-Stokes equations.

Motivated by the superficial similarity between the membrane paradigm and the fluid/gravity correspondence, recently [44, 45] have attempted to formulate a precise derivation of the former. In [44] Einstein's equations in the bulk are projected onto a null hypersurface and then expanded in gradients along the hypersurface. On the other hand [45] show that one can systematically find a solution to vacuum Einstein's equations which describes the near-horizon geometry of a generic non-degenerate black hole in the long wavelength regime.

Within the fluid/gravity correspondence, the entire spacetime evolution is mapped to the dynamics of a conformal fluid, which, albeit reminiscent of the membrane paradigm, has one important twist: the membrane lives on the *boundary* of the spacetime (which is unambiguously defined and admits a fluid description with well-defined dynamics), and gives a perfect mirror of the full bulk physics. This "membrane at the end of the universe" picture is a natural consequence of the holographic nature of the fluid/gravity correspondence.

14.7.3. *Blackfolds*

As in the fluid/gravity correspondence and the membrane paradigm type ideas, the blackfold approach to constructing higher-dimensional black holes (discussed in [ch.BF] of this book) likewise asserts that the effective theory describing the long wavelength dynamics of black hole horizons can be expressed in terms of fluid dynamics. However, there are several important differences between these descriptions. Since the blackfold 'fluid' pertains to the effective world-volume dynamics of an extended black object as seen from far away, the intrinsic dynamics typically has to be supplemented by extrinsic dynamics, describing how the blackfold embeds in the ambient spacetime. On the other hand, in the fluid/gravity correspondence the fluid resides on the boundary of asymptotically (locally) AdS spacetime, so there is no issue with extrinsic dynamics.

Moreover, although the blackfold formalism is most naturally formulated in asymptotically flat spacetime, by a suitable separation of scales, one can in principle consider blackfolds with any asymptotics. In contrast, the fluid/gravity correspondence concerns asymptotically AdS black holes. On the other hand, unlike all the above-mentioned approaches, fluid/gravity is the only one where there is a known physical microscopic origin to the fluid: it is the effective behavior of the dual field theory residing on the AdS boundary.

14.8. Summary

The fluid/gravity correspondence provides a natural way to map solutions of fluid dynamics into those of gravity, enabling one to construct time-dependent, inhomogeneous black hole solutions to Einstein's equations, retaining full non-linearity. An interesting aspect of the construction is the manner in which classical gravity can be moulded to fit naturally with effective field theory intuition to extract approximate solutions. While the construction itself arose from the gauge/gravity correspondence, it is clear that it can be implemented in greater generality.

Apart from providing interesting insights into the dynamics of gravity, the map has played an important role in clarifying various issues in fluid dynamics. The role of quantum anomalies in hydrodynamical transport, and generalizations of fluid dynamics to systems with spontaneously broken symmetries, are two examples where the fluid/gravity map has served to elucidate the underlying physics cleanly. The physical points seem much simpler to understand from the gravitational perspective; aided by this intuition one can re-evaluate the hypotheses of traditional descriptions of fluids.

The fluid/gravity map suggests several extremely interesting technical, as well as conceptual, questions for the future. Some of these are: Does the gravitational viewpoint shed any light on turbulent fluid flows, or questions about singularities that develop in finite time from smooth initial data in fluid dynamics? Is there a path integral formulation of fluid dynamics at finite N, and how does it map to the path integral of bulk gravity? Are the corrections to the classical equations of gravity constrained by the requirement of positivity of divergence of an 'entropy current' on an event horizon (analogously to, and perhaps even dual to, fluid dynamics)? It seems likely that many interesting results remain to be discovered in this general area.

14.9. Epilogue: Einstein and Boltzmann

As we have emphasized throughout, the equations of fluid dynamics, for which we have an independent field theory intuition, are dual to a long wavelength limit of Einstein's equations (14.1). It is then natural to ask what is the field theoretical interpretation of the full dynamical system of equations (14.1)? We believe that these equations may be conceptually thought of as the strong coupling analogue of (a decoupled sector of) the Boltzmann transport equations.

It is well known that the linearization of the Boltzmann transport equations about equilibrium yields an infinite set of 'quasinormal modes', i.e., solutions to the equations of motion that all decay to zero (returning the system back to equilibrium) at late times. Exactly d of these quasinormal modes are massless (in the sense that they are static in the infinite wavelength limit). In textbooks on statistical mechanics, fluid dynamics is sometimes derived as the non-linear theory of this finite set of Boltzmann 'quasinormal modes'. The remaining quasinormal modes are 'fast modes' that decay away on a time scale set by the mean free path of kinetic theory.

Similarly, the linearization of the equations of gravity about the planar black hole has d massless quasinormal modes and an infinite set of massive quasinormal modes. In direct analogy with the work on the Boltzmann equations, the fluid/gravity correspondence constructs the equations of fluid dynamics as the non-linear theory of these massless modes (effectively by integrating out the massive modes, order by order). For this reason it is natural to think of the full set of Einstein's equations in the presence of an event horizon (including all quasinormal mode degrees of freedom) as the strong coupling analogue of the Boltzmann transport equations.

An important property of the Boltzmann transport equations is that they are irreversible; they obey the Boltzmann H-theorem (which asserts that a certain functional of kinetic variables called H always increases in time and is maximum in equilibrium). In direct analogy, Einstein's equations, together with the assumption of regularity of future event horizons (and physical energy conditions), always obey the classic area increase theorem of general relativity. This suggests that the better analogy is between the Boltzmann transport equations and Einstein's equations *plus the condition of regularity of the future event horizon.* The last condition breaks the time reversal invariance of Einstein's equations. In fact, the requirement that the future event horizon stay regular was a crucial element in our implementation of the fluid/gravity map. The Boltzmann theorem has a local analogue in fluid dynamics; it maps to the statement that the equations of fluid dynamics are accompanied by a local entropy current that whose divergence is everywhere non-negative. The area increase theorem of general relativity can be used to construct such an entropy current for the fluid dynamics generated from the fluid/gravity map.

Just like the Boltzmann equations, the system of gravitational equations (14.1) can be used to study the approach to equilibrium from a highly non equilibrated starting point. In simple studies of equilibration using Einstein's equations [46, 47], the equilibration time, measured by the time taken for fluid dynamics to take over as the effective description, turns out to be extremely rapid. In more complicated situations the equilibration process displays sharp phase transitions associated with Choptuik phenomena. Indeed the equations (14.1) undoubtedly contain a host of dynamical delights for the intrepid gravitational and statistical physicist; it seems clear that the fluid/gravity map is merely the tip of an iceberg's worth of connections between gravity and statistical physics.

References

[1] S. Bhattacharyya, V. E. Hubeny, S. Minwalla, and M. Rangamani, "Nonlinear Fluid Dynamics from Gravity," *JHEP* **0802** (2008) 045, `arXiv:0712.2456 [hep-th]`.

[2] G. Policastro, D. T. Son, and A. O. Starinets, "The shear viscosity of strongly coupled N = 4 supersymmetric Yang-Mills plasma," *Phys. Rev. Lett.* **87** (2001) 081601, `arXiv:hep-th/0104066`.

[3] R. A. Janik and R. B. Peschanski, "Asymptotic perfect fluid dynamics as a consequence of AdS/CFT," *Phys. Rev.* **D73** (2006) 045013, `arXiv:hep-th/0512162`.

[4] S. Bhattacharyya, S. Lahiri, R. Loganayagam, and S. Minwalla, "Large rotating AdS black holes from fluid mechanics," *JHEP* **09** (2008) 054, `arXiv:0708.1770 [hep-th]`.

[5] M. Rangamani, "Gravity and Hydrodynamics: Lectures on the fluid-gravity correspondence," *Class.Quant.Grav.* **26** (2009) 224003, `arXiv:0905.4352 [hep-th]`.

[6] V. Balasubramanian and P. Kraus, "A stress tensor for anti-de Sitter gravity," *Commun. Math. Phys.* **208** (1999) 413–428, `arXiv:hep-th/9902121`.

[7] J. Bhattacharya, S. Bhattacharyya, S. Minwalla, and A. Yarom, "A theory of first order dissipative superfluid dynamics," `arXiv:1105.3733 [hep-th]`.

[8] P. Romatschke, "Relativistic Viscous Fluid Dynamics and Non-Equilibrium Entropy," *Class.Quant.Grav.* **27** (2010) 025006, `arXiv:0906.4787 [hep-th]`.

[9] R. Loganayagam, "Entropy Current in Conformal Hydrodynamics," *JHEP* **0805** (2008) 087, `arXiv:0801.3701 [hep-th]`.

[10] S. Bhattacharyya, R. Loganayagam, I. Mandal, S. Minwalla, and A. Sharma, "Conformal Nonlinear Fluid Dynamics from Gravity in Arbitrary Dimensions," *JHEP* **0812** (2008) 116, `arXiv:0809.4272 [hep-th]`.

[11] S. Bhattacharyya, V. E. Hubeny, R. Loganayagam, G. Mandal, S. Minwalla, *et al.*, "Local Fluid Dynamical Entropy from Gravity," *JHEP* **0806** (2008) 055, `arXiv:0803.2526 [hep-th]`.

[12] J. D. Bekenstein, "Black holes and entropy," *Phys. Rev.* **D7** (1973) 2333–2346.

[13] S. W. Hawking, "Particle Creation by Black Holes," *Commun. Math. Phys.* **43** (1975) 199–220.

[14] I. Booth, M. P. Heller, G. Plewa, and M. Spalinski, "On the apparent horizon in fluid-gravity duality," `arXiv:1102.2885 [hep-th]`.

[15] M. Henningson and K. Skenderis, "The holographic Weyl anomaly," *JHEP* **07** (1998) 023, `arXiv:hep-th/9806087`.

[16] R. Baier, P. Romatschke, D. T. Son, A. O. Starinets, and M. A. Stephanov, "Relativistic viscous hydrodynamics, conformal invariance, and holography," *JHEP* **0804** (2008) 100, `arXiv:0712.2451 [hep-th]`.

[17] P. Kovtun, D. T. Son, and A. O. Starinets, "Viscosity in strongly interacting quantum field theories from black hole physics," *Phys. Rev. Lett.* **94** (2005) 111601, `arXiv:hep-th/0405231`.

[18] T. Schafer and D. Teaney, "Nearly Perfect Fluidity: From Cold Atomic Gases to Hot Quark Gluon Plasmas," *Rept.Prog.Phys.* **72** (2009) 126001, `arXiv:0904.3107 [hep-ph]`.

[19] A. Buchel, R. C. Myers, and A. Sinha, "Beyond eta/s = 1/4 pi," *JHEP* **0903** (2009) 084, `arXiv:0812.2521 [hep-th]`.

[20] M. Haack and A. Yarom, "Nonlinear viscous hydrodynamics in various dimensions using AdS/CFT," *JHEP* **0810** (2008) 063, `arXiv:0806.4602 [hep-th]`.

[21] E. Berti, V. Cardoso, and A. O. Starinets, "Quasinormal modes of black holes and black branes," *Class.Quant.Grav.* **26** (2009) 163001, arXiv:0905.2975 [gr-qc].

[22] G. T. Horowitz and V. E. Hubeny, "Quasinormal modes of AdS black holes and the approach to thermal equilibrium," *Phys. Rev.* **D62** (2000) 024027, arXiv:hep-th/9909056.

[23] G. Policastro, D. T. Son, and A. O. Starinets, "From AdS/CFT correspondence to hydrodynamics," *JHEP* **09** (2002) 043, arXiv:hep-th/0205052.

[24] G. Policastro, D. T. Son, and A. O. Starinets, "From AdS/CFT correspondence to hydrodynamics. II: Sound waves," *JHEP* **12** (2002) 054, arXiv:hep-th/0210220.

[25] S. Bhattacharyya, R. Loganayagam, S. Minwalla, S. Nampuri, S. P. Trivedi, *et al.*, "Forced Fluid Dynamics from Gravity," *JHEP* **0902** (2009) 018, arXiv:0806.0006 [hep-th].

[26] G. Gibbons, H. Lu, D. N. Page, and C. Pope, "The General Kerr-de Sitter metrics in all dimensions," *J.Geom.Phys.* **53** (2005) 49–73, arXiv:hep-th/0404008 [hep-th].

[27] S. Bhattacharyya, S. Minwalla, and S. R. Wadia, "The Incompressible Non-Relativistic Navier-Stokes Equation from Gravity," *JHEP* **0908** (2009) 059, arXiv:0810.1545 [hep-th].

[28] I. Kanitscheider and K. Skenderis, "Universal hydrodynamics of non-conformal branes," *JHEP* **0904** (2009) 062, arXiv:0901.1487 [hep-th].

[29] O. Aharony, S. Minwalla, and T. Wiseman, "Plasma-balls in large N gauge theories and localized black holes," *Class. Quant. Grav.* **23** (2006) 2171–2210, arXiv:hep-th/0507219.

[30] S. Lahiri and S. Minwalla, "Plasmarings as dual black rings," *JHEP* **05** (2008) 001, arXiv:0705.3404 [hep-th].

[31] V. Cardoso and O. J. Dias, "Rayleigh-Plateau and Gregory-Laflamme instabilities of black strings," *Phys.Rev.Lett.* **96** (2006) 181601, arXiv:hep-th/0602017 [hep-th].

[32] J. Erdmenger, M. Haack, M. Kaminski, and A. Yarom, "Fluid dynamics of R-charged black holes," *JHEP* **0901** (2009) 055, arXiv:0809.2488 [hep-th].

[33] N. Banerjee, J. Bhattacharya, S. Bhattacharyya, S. Dutta, R. Loganayagam, *et al.*, "Hydrodynamics from charged black branes," *JHEP* **1101** (2011) 094, arXiv:0809.2596 [hep-th].

[34] D. T. Son and P. Surowka, "Hydrodynamics with Triangle Anomalies," *Phys.Rev.Lett.* **103** (2009) 191601, arXiv:0906.5044 [hep-th].

[35] S. S. Gubser, "Breaking an Abelian gauge symmetry near a black hole horizon," *Phys.Rev.* **D78** (2008) 065034, arXiv:0801.2977 [hep-th].

[36] J. Sonner and B. Withers, "A gravity derivation of the Tisza-Landau Model in AdS/CFT," *Phys.Rev.* **D82** (2010) 026001, arXiv:1004.2707 [hep-th].

[37] S. J. Putterman, *Superfluid hydrodynamics*, vol. 3. North-Holland Publishing Co., 1974.

[38] W. Israel, "Nonstationary irreversible thermodynamics: A Causal relativistic theory," *Ann. Phys.* **100** (1976) 310–331.

[39] W. Israel and J. M. Stewart, "Transient relativistic thermodynamics and kinetic theory," *Ann. Phys.* **118** (1979) 341–372.
[40] R. P. Geroch, "On Hyperbolic "Theories" of Relativistic Dissipative Fluids," `gr-qc/0103112`.
[41] W. G. Unruh, "Experimental black hole evaporation," *Phys. Rev. Lett.* **46** (1981) 1351–1353.
[42] K. S. Thorne, R. H. Price, and D. A. MacDonald, *Black Holes: The Membrane Paradigm.* Yale University Press, New Haven, 1986.
[43] T. Damour, "Black Hole Eddy Currents," *Phys. Rev.* **D18** (1978) 3598–3604.
[44] C. Eling and Y. Oz, "Relativistic CFT Hydrodynamics from the Membrane Paradigm," *JHEP* **1002** (2010) 069, `arXiv:0906.4999 [hep-th]`.
[45] I. Bredberg, C. Keeler, V. Lysov, and A. Strominger, "From Navier-Stokes To Einstein," `arXiv:1101.2451 [hep-th]`.
[46] P. M. Chesler and L. G. Yaffe, "Horizon formation and far-from-equilibrium isotropization in supersymmetric Yang-Mills plasma," *Phys.Rev.Lett.* **102** (2009) 211601, `arXiv:0812.2053 [hep-th]`.
[47] S. Bhattacharyya and S. Minwalla, "Weak Field Black Hole Formation in Asymptotically AdS Spacetimes," *JHEP* **0909** (2009) 034, `arXiv:0904.0464 [hep-th]`.

Student Seminars at TASI 2010

6/3/10

Michael Gary — Quantum Gravity: Kicking Ass and Taking Names
Paul McGuirk — Why AdS/CFT is Wrong in Every Way
Dusan Simic — Supersymmetry Breaking in Systems with Emergent Gravity
Gokce Basar — Great Scott!: Time Travel with Doc Brown

6/7/10

Gang Xu — How I Learned to Stop Worrying and Love String Cosmology
Richard Eager — Generalized Gauge-Gravity Duality
Idse Heemskerk — Holography from Conformal Field Theory

6/8/10

Mark Mezei — Holographic Quantum Phase Transitions
Nabil Iqbal — Bulk Quantum Effects in Applied Holography
Sho Yaida — Holographic Lattices, Dimers, and Glasses

6/14/10

David Marsh — Sequestering or Not
Sohang Gandhi — Holographic Systematics of Compactification Effects
Nima Lashkari — Decay of AdS in Topologically Massive Gravity
Yi Yin — Dimensional Deconstruction and Gauge/gravity Duality

6/15/10

Jae Hyuk Oh — Dilaton Cosmologies and Their Field Theory Duals
Bart Horn — Axion Monodromy Inflation with Trapping and Fluxes

Vladmir Rosenhaus — Eternal Inflation and Holography
Daniel Harlow — Metastability in Anti-de Sitter Space

6/16/10

Abhijit Gadde — S-duality and TFT
George Bruhn — 3-Algebras in N = 5,6,8 CS Theories
Anindya Dey — Mirror Symmetry, D-branes, and Quivers
Steve Avery — D0-D6 Case Study: An Exactly Soluble System

6/17/10

Anastasios Taliotis — Heavy Ion Collisions with Transverse Dynamics in Evolving AdS Geometries
Andreas Stergian — Gauge Mediation of SUSY
Curtis Asplund — Quantum Dynamics of Moduli: A Toy Model

6/21/10

Dan Wohns — Resonant Tunneling in QFT
Marko Djuric — Deep Inelastic Scattering after AdS/CFT
Yang Zhang — Dual Identities in Scattering Amplitude via Heterotic String Theory

6/22/10

Eric Perlmutter — Scaling Solution with Domain Wall Holography
Koushik Balasubramanian — The Particle Number in Galilean Holography
Stefan Leighenauer — Saturating the Holographic Entropy Bound
Xi Dong — dS Holography